Global Positioning System: Theory and Applications
Volume II

Global Positioning System: Theory and Applications
Volume II

Edited by

Bradford W. Parkinson
Stanford University, Stanford, California
James J. Spilker Jr.
Stanford Telecom, Sunnyvale, California

Associated Editors:
Penina Axelrad
University of Colorado, Boulder, Colorado
Per Enge
Stanford University, Stanford, California

Volume 164
PROGRESS IN
ASTRONAUTICS AND AERONAUTICS

Paul Zarchan, Editor-in-Chief
Charles Stark Draper Laboratory, Inc.
Cambridge, Massachusetts

Published by the
American Institute of Aeronautics and Astronautics, Inc.
370 L'Enfant Promenade, SW, Washington, DC 20024-2518

Third Printing

Copyright © 1996 by the American Institute of Aeronautics and Astronautics, Inc. Printed in the United States of America. All rights reserved. Reproduction or translation of any part of this work beyond that permitted by Sections 107 and 108 of the U.S. Copyright Law without the permission of the copyright owner is unlawful. The code following this statement indicates the copyright owner's consent that copies of articles in the volume may be made for personal or internal use, on condition that the copier pay the per-copy fee ($2.00) plus the per-page fee ($0.50) through the Copyright Clearance Center, Inc., 222 Rosewood Drive, Danvers, Massachusetts 01923. This consent does not extend to other kinds of copying, for which permission requests should be addressed to the publisher. Users should employ the following code when reporting copying from this volume to the Copyright Clearance Center:

1-56347-107-8/96 $2.00 + .50

Data and information appearing in this book are for informational purposes only. AIAA is not responsible for any injury or damage resulting from use or reliance, nor does AIAA warrant that use or reliance will be free from privately owned rights.

ISBN 1-56347-107-8

Progress in Astronautics and Aeronautics

Editor-in-Chief
Paul Zarchan
Charles Stark Draper Laboratory, Inc.

Editorial Board

John J. Bertin
U.S. Air Force Academy

Leroy S. Fletcher
Texas A&M University

Richard G. Bradley
Lockheed Martin Fort Worth Company

Allen E. Fuhs
Carmel, California

William Brandon
MITRE Corporation

Ira D. Jacobsen
Embry-Riddle Aeronautical University

Clarence B. Cohen
Redondo Beach, California

John L. Junkins
Texas A&M University

Luigi De Luca
Politecnico di Milano, Italy

Pradip M. Sagdeo
University of Michigan

Martin Summerfield
Lawrenceville, New Jersey

Dedication

To Anna Marie, Elaine, Virginia, and Tim

Table of Contents

Preface xxxi

Part III. Differential GPS and Integrity Monitoring

Chapter 1. Differential GPS 3
Bradford W. Parkinson and Per K. Enge, *Stanford University, Stanford, California*

Introduction 3
  Standard Positioning Service Users 3
  Precise Positioning Service Users 4
  Major Categories of Differential GPS 4
Code-Phase Differential GPS 7
  User Errors Without Differential GPS 7
  Reference Station Calculation of Corrections 10
  Application of Reference Correction 11
Analysis of Differential GPS Errors 11
  Receiver Noise, Interference, and Multipath Errors for Differential GPS 12
  Satellite Clock Errors for Differential GPS 16
  Satellite Ephemeris Errors for Differential GPS 17
  Ionospheric Errors for Differential GPS 20
  Troposhere Errors for Differential GPS 23
  Local Area Differential GPS Error Summary 24
Carrier-Phase Differential GPS 27
  Attitude Determination 27
  Static and Kinematic Survey 28
  Near Instantaneous Determination of Integers 30
Radio Technical Commission for Maritime Services Data Format for Differential
    GPS Data 31
  Radio Technical Commission for Maritime Services Message Types 1, 2, and 9 . 32
  Types 18, 19, 20, and 21 Messages 34
Datalinks 34
  Groundwave Systems 34
  VHF and UHF Networks 36
  Mobile Satellite Communications 39
Differential GPS Field Results 41
  Short-Range Differential Code-Phase Results 41
  Long-Range Differential Code-Phase Results 42
  Dynamic Differential Carrier-Phase Results 43
Conclusions 47

Appendix: Differential GPS Ephemeris Correction Errors Caused by Geographic
  Separation 47
References 49

Chapter 2. Pseudolites 51
Bryant D. Elrod, *Stanford Telecom, Inc., Reston, Virginia*
and A. J. Van Dierendonck, *AJ Systems, Los Altos, California*

Introduction 51
Pseudolite Signal Design Considerations 52
   Previous Pseudolite Designs 52
   New Pseudolite Signal Design 53
Integrated Differential GPS/Pseudolite Considerations 57
   Pseudolite Siting 57
   Pseudolite Time Synchronization 58
   User Aircraft Antenna Location 62
   Pseudolite Signal Data Message 63
   GPS/Pseudolite Navigation Filter Considerations 64
Pseudolite Testing 65
   Pseudolite Interference Testing 65
   Pseudolite Data Link Testing 67
   Navigation Performance Testing 68
Appendix A: Interference Caused by Cross Correlation Between C/A Codes 70
Appendix B: Interference Caused by Pseudolite Signal Level 74
Appendix C: Navigation Filter Modeling with Pseudolite Measurements 76
References 78

Chapter 3. Wide Area Differential GPS 81
Changdon Kee, *Stanford University, Stanford, California*

Introduction 81
Wide Area Differential GPS Architecture and Categories 82
   Wide Area Differential GPS Architecture 82
   Wide Area Differential GPS Categories 85
   User Message Content and Format 87
   Error Budget 88
Master Station Error Modeling 88
   Ionospheric Time Delay Model for Algorithms A or B 89
   Ephemeris and Satellite Clock Errors for Algorithms A, B, or C 92
Simulation of Algorithm B 95
   Simulation Modules 95
   Ionospheric Error Estimation Results 100
   Navigation Performance 101
   Summary of Results 104
Test Using Field Data to Evaluate Algorithm C 104
   Locations of the Receiver Sites 105
   Test Results 105
   Latency and Age Concern 111

Conclusion 112
References 114

Chapter 4. Wide Area Augmentation System 117
Per K. Enge, *Stanford University, Stanford, California* and
A.J. Van Dierendonck, *AJ Systems, Los Altos, California*

Introduction 117
Signal Design 120
    Link Budget and Noninterference with GPS 120
    Data Capacity 123
    Loop Threshold 124
Ranging Function 124
Nonprecision Approach and Error Estimates 126
Precision Approach and Vector Corrections 128
    Vector Corrections 129
    Precision Approach Integrity 130
Wide Area Augmentation System Message Format 131
    Parity Algorithm 134
    Message Type 2 Fast Corrections and User Differential Range Errors 135
    Type 25: Long-Term Satellite Error Corrections Message 135
    Type 26: Ionospheric Delay Error Corrections Message 136
    Type 9: WAAS Satellite Navigation Message 137
    Applied Range Accuracy Evaluation 137
Summary 138
Appendix: Geostationary Satellite Ephemeris Estimation and Code-Phase Control .. 139
References 142

Chapter 5. Receiver Autonomous Integrity Monitoring 143
R. Grover Brown, *Iowa State University, Ames, Iowa*

History, Overview, and Definitions 143
Basic Snapshot Receiver Autonomous Integrity Monitoring Schemes and
    Equivalences 145
    Range Comparison Method 146
    Least-Squares-Residuals Method 147
    Parity Method 148
    Maximum Separation of Solutions 150
    Constant-Detection-Rate/Variable-Protection-Level Method 151
Screening Out Poor Detection Geometries 152
Receiver-Autonomous Integrity Monitoring Availability for Airborne Supplemental
    Navigation 155
Introduction to Aided Receiver-Autonomous Integrity Monitoring 156
Failure Isolation and the Combined Problem of Failure Detection and Isolation 158
    Introductiory Remarks 158
    Parity Method and Failure Detection and Isolation 158
    Calculation of the P Matrix 161
    Failure Detection and Exclusion Algorithm 163
References 164

Part IV. Integrated Navigation Systems

Chapter 6.   Integration of GPS and Loran-C 169
Per K. Enge, *Stanford University, Stanford, California* and F. van Graas,
Ohio University, Athens, Ohio

Introduction 169
   Calibration of Loran Propagation Errors by GPS 171
   Cross-Chain Synchronization of Loran-C Using GPS 171
   Combining Pseudoranges from GPS and Loran-C for Air Navigation .. 171
Loran Overview 172
Calibration of Loran Propagation Errors by GPS 174
Cross-Rate Synchronization of Loran 176
Combining GPS Pseudoranges with Loran Time Differences 179
   Navigation Equations 179
   Probability of Outage Results 182
Summary 184
References 185

Chapter 7.   GPS and Inertial Integration 187
R. L. Greenspan, *Charles Stark Draper Laboratories, Cambridge, Massachusetts*

Benefits of GPS/Inertial Integration 187
   Operation During Outages 189
   Providing All Required Navigation Outputs 190
   Reduced Noise in GPS Navigation Solutions 190
   Increased Tolerance to Dynamics and Interference 191
GPS Integration Architectures and Algorithms 191
   Integration Architectures 191
   Integration Algorithms 194
   Embedded Systems 197
Integration Case Studies 199
   GPS/Inertial Navigation Systems Navigation Performance in a Low-Dynamics
   Aircraft 199
   Using GPS for In-flight Alignment 206
   Integrated Navigation Solutions During a GPS Outage 213
Summary 217
References 218

**Chapter 8.   Receiver Autonomous Integrity Monitoring Availability
for GPS Augmented with Barometric Altimeter Aiding and
Clock Coasting** 221
Young C. Lee, *MITRE Corporation, McLean, Virginia*

Introduction 221
Methods of Augmentations 222
   Augmented Geometry for Barometric Altimeter Aiding 222
   Barometric Altimeter Aiding with GPS-Calibrates Pressure Altitude Data 223

Barometric Altimeter Aiding with Local Pressure Input 227
Augmented Geometry for Clock Coasting 228
Simultaneous Use of Barometric Altimeter Aiding and Clock 229
Definitions of Function Availability 229
Navigation Function 229
Receiver Autonomous Integrity Monitoring Detection Function 229
Receiver Autonomous Integrity Monitoring Function 230
Results 230
Parameters of Interest 230
Discussion of Results 231
Summary and Conclusions 235
Appendix: Statistical Distribution of the Height Gradients 239
References 241

Chapter 9. GPS and Global Navigation Satellite System (GLONASS) 243
Peter Daly, *University of Leeds, Leeds, England, United Kingdom* and Pratap N. Misra, *Massachusetts Institute of Technology, Lexington, Massachusetts*

Introduction to the Global Navigation Satellite System 243
History of Satellite Navigation Systems 243
Orbits 244
History of Launches 247
Signal Design 248
Message Content and Format 252
Satellite Ephemerides 253
Satellite Almanacs 254
GPS/GLONASS Onboard Clocks 255
Performance of GLONASS and GPS + GLONASS 258
Introduction 258
Requirements of Civil Aviation 259
Integrated Use of GPS and GLONASS 260
Performance of GLONASS and GPS and GLONASS 261
Summary 271
Acknowledgments 271
References 271

Part V. GPS Navigation Applications

Chapter 10. Land Vehicle Navigation and Tracking 275
Robert L. French, *R. L. French & Associates, Fort Worth, Texas*

Application Characteristics and Markets 275
Commercial Vehicle Tracking 275
Automobile Navigation and Route Guidance 277
Intelligent Vehicle Highway Systems 279

Historical Background 281
    Early Mechanical Systems 281
    Early Electronic Systems 282
Enabling/Supporting Technologies 283
    Dead Reckoning 284
    Digital Road Maps 286
    Map Matching 288
    Integration with GPS 291
    Mobile Data Communications 292
Examples of Integrated Systems 294
    Etax Navigator™/Bosch Travelpilot™ 294
    Toyota Electro-Multivision 296
    TravTek Driver Information System 297
    NavTrax™ Fleet Management System 298
References 299

Chapter 11. Marine Applications 303
Jim Sennott and In-Soo Ahn, *Bradley University, Peoria, Illinois* and
Dave Pietraszewski, *United States Coast Guard Research and
Development Center, Groton, Connecticut*

Marine Navigation Phases and Requirements 303
Marine DGPS Background 304
Global Positioning Systems-Assisted Steering, Risk Assessment, and Hazard Warning
    System 305
Vessel and Sensor Modeling 308
    Vessel Dynamics Model 308
    Standardized Sensor Model 310
    Combined Ship and Sensor Model 311
Waypoint Steering Functions 312
    Filter and Controller Design 313
    Sensor/Ship Bandwidth Ratio and Straight-Course Steering 314
    Comparative Footprint Channel Clearance Width Distributions 315
Hazard Warning and Risk Assessment Functions 321
    Risk Assessment 321
    Hazard Warning 323
Summary 323
References 324

Chapter 12. Applications of the GPS to Air Traffic Control 327
Ronald Braff, *MITRE Corporation, McLean, Virginia,* J. David Powell,
Stanford University, Stanford, California, and Joseph Dorfler,
Federal Aviation Administration, Washington, DC

Introduction 327
Air Traffic Control System 327
General Considerations 329
    Operational Requirements 331
    Government Activities 333

Air Navigation Applications 334
  En Route, Terminal, and Nonprecision Approach Operational Considerations and
  Augmentations 335
  Precision Approach Operational Considerations and Augmentations 344
  Other Navigation Operational Considerations 354
  Area and Four-Dimensional Navigation 361
Surveillance 362
  Current Surveillance Methods 362
  Surveillance via GPS 366
Summary of Key Benefits 370
References 370

Chapter 13. GPS Applications in General Aviation 375
Ralph Eschenbach, *Trimble Navigation, Sunnyvale, California*

Market Demographics 375
  Airplanes 375
  Pilots 375
  Airports 377
Existing Navigation and Landing Aids (Non-GPS) 377
  Nondirectional Beacons (NDB) 377
  Very High Frequency Omnidirectional Radio 378
  Distance-Measuring Equipment 379
  Long-Range Radio Navigation 379
  Omega 380
  Approaches 380
Requirements for GPS in General Aviation 384
  Dynamics 385
  Functionality 385
  Accuracy 385
  Availability, Reliability, and Integrity 386
Pilot Interface 387
  Input 387
  Output 388
GPS Hardware and Integration 388
  Installation Considerations 388
  Number of Channels 389
  Cockpit Equipment 389
  Hand-held 389
  Panel Mounts 389
  Dzus Mount 390
Differential GPS 390
  Operational Characteristics 391
  Ground Stations 391
  Airborne Equipment Features 391
Integrated Systems 392
  GPS LORAN 392
  GPS/Omega 392

Future Implementations 393
   Attitude and Heading Reference System 393
   Approach Certification 393
   Collision Avoidance 393
   Autonomous Flight 394
Summary 394
References 395

Chapter 14. Aircraft Automatic Approach and Landing Using GPS 397
Bradford W. Parkinson and Michael L. O'Connor, *Stanford University, Stanford, California*, and Kevin T. Fitzgibbon, *Technical University, São Jose Dos Compos, Brazil*

Introduction 397
   Autolanding Conventionally and with GPS 397
   Simulations Results Presented 398
Landing Approach Procedures 399
   Instrument and Microwave Landing Systems 399
   GPS Approach 401
Aircraft Dynamics and Linear Model 401
   State Vector 401
   Control Vector 402
   Disturbance Vector 402
   Measurement Vector 403
   Equations of Motion 403
   Wind Model 403
   Throttle Control Lag 404
   Glide-Slope Deviation 404
Autopilot Controller 405
   Linear Quadratic Gaussian and Integral Control Law Controllers . 405
   Regulator Synthesis 405
GPS Measurements 407
Results 409
   Cases Simulated 409
   Landing with GPS Alone 410
   Landing with GPS Plus Altimeter 410
   Landing with Differential GPS 411
   Landing with Carrier-Phase 411
   Linear Quadratic Gaussian vs Integral Control Law 413
Conclusions and Comments 414
Appendix A: Discrete Controllers 415
Appendix B: Discrete Time Optimal Estimator 421
Appendix C: Numerical Values for Continuous System 422
Bibliography 425
References 425

Chapter 15. Precision Landing of Aircraft Using Integrity Beacons 427
Clark E. Cohen, Boris S. Pervan, H. Stewart Cobb, David G. Lawrence, J. David Powell, and Bradford W. Parkinson, *Stanford University, Stanford, California*

Overview of the Integrity Beacon Landing System 427
    Centimeter-Level Positioning 428
    History of the Integrity Beacon Landing System 429
    Doppler Shift and Geometry Change 429
Required Navigation Performance 429
    Accuracy 429
    Integrity 430
    Availability 431
    Continuity 431
Integrity Beacon Architecture 432
    Droppler Marker 432
    Omni Marker 432
Mathematics of Cycle Resolution 434
    Observability Analysis 434
    Matrix Formulation 435
Experimental Flight Testing 438
    Quantification of Centimeter-Level Accuracy 438
    Piper Dakota Experimental Flight Trials 439
    Federal Aviation Administration Beech King Air Autocoupled Approaches 442
    Automatic Landings of a United Boeing 737 444
    Flight Test Summary and Observations 447
Operations Using Integrity Beacons 447
    Integrity Beacon Landing System Landing Sequence 448
Integrity Beacon Landing System Navigation Integrity 450
    Receiver Autonomous Integrity Monitoring 450
    System Failure Modes 453
    Quantifying Integrity 454
    Signal Interference 456
Conclusion 458
References 458

Chapter 16. Spacecraft Attitude Control Using GPS Carrier Phase 461
E. Glenn Lightsey, *NASA Goddard Space Flight Center, Greenbelt, Maryland*

Introduction 461
Design Case Study 462
Sensor Characteristics 464
    Antenna Placement 466
    Sensor Calibration 466
    Multipath 467
    Sensor Accuracy 467

Dynamic Filtering 468
Vehicle Dynamics 468
   Gravity Gradient Moment 470
   Aerodynamic Moment 471
   System Natural Response 472
Control Design 474
   Control Loop Description 474
   Simulation Results 475
Conclusion 477
Acknowledgments 479
References 479

Part VI. Special Applications

Chapter 17. GPS for Precise Time and Time Interval Measurement 483
William J. Klepczynski, *United States Naval Observatory, Washington, DC*

Introduction 483
Universal Coordinated Time 484
Role of Time in the GPS 485
Translation of GPS Time to Universal Coordinated Time 487
GPS as a Clock in the One-Way Mode 489
Common-View Mode of GPS 490
Melting-Pot Method 498
Problem of Selective Availability 498
Future Developments 499
References 500

Chapter 18. Surveying with the Global Positioning System 501
Clyde Goad, *Ohio State University, Columbus, Ohio*

Measurement Modeling 502
Dilution of Precision 508
Ambiguity Search 509
Use of Pseudoranges and Phase 510
   Review 511
   Three-Measurement Combinations 514
Antispoofing? 516
A Look Ahead 516
References 517

Chapter 19. Attitude Determination 519
Clark E. Cohen, *Stanford University, Stanford, California*

Overview 519
Fundamental Conventions for Attitude Determination 521
Solution Processing 523

Cycle Ambiguity Resolution 524
   Baseline Length Constraint 525
   Integer Searches 525
   Motion-Based Methods 525
   Alternative Means for Cycle Ambiguity Resolution 531
Performance 531
   Geometrical Dilution of Precision for Attitude 532
   Multipath 532
   Structural Distortion 533
   Troposphere 533
   Signal-to-Noise Ratio 534
   Receiver-Specific Errors 534
   Total Error 534
Applications 535
   Aviation 535
   Spacecraft 537
   Marine 537
References 537

Chapter 20. Geodesy 539
Kristine M. Larson, *University of Colorado, Boulder, Colorado*

Introduction 539
Modeling of Observables 540
Reference Frames 542
Precision and Accuracy 547
Results 550
   Crustal Deformation 550
   Earth Orientation 553
Conclusions 554
Acknowledgments 554
References 554

Chapter 21. Orbit Determination 559
Thomas P. Yunck, *Jet Propulsion Laboratory, California Institute of Technology, Pasadena, California*

Introduction 559
Principles of Orbit Determination 560
   Dynamic Orbit Determination 560
   Batch Least Squares Solution 562
   Kalman Filter Formulation 563
   Dynamic Orbit Error 565
   Kalman Filter with Process Noise 565
Orbit Estimation with GPS 567
   Carrier-Pseudorange Bias Estimation 567
   Kinematic Orbit Determination 569
   Reduced Dynamic Orbit Determination 571
   Orbit Improvement by Physical Model Adjustment 572
Direct Orbit Determination with GPS 573

Precise Orbit Determination with Global Positioning Systems 574
    Global Differential Tracking 574
    Fine Points of the Global Solution 576
    Precise Orbit Determination 577
    Single-Frequency Precise Orbit Determination 580
    Extension to Higher Altitude Satellites 584
    Highly Elliptical Orbiters 585
Dealing with Selective Availability and Antispoofing 586
    Antispoofing 586
    Selective Availability 586
Summary 588
Acknowledgments 589
References 589

Chapter 22. Test Range Instrumentation 593
Darwin G. Abby, *Intermetrics, Inc. Holloman Air Force Base, New Mexico*

Background 593
Requirements 595
    Test Requirements 595
    Training Requirements 596
Range Instrumentation Components 598
    Global Positioning Systems Reference Station 598
    Data Links 599
    Test Vehicle Instrumentation 599
    Translator Systems 599
    Digital Translators 601
Differential Global Positioning Systems Implementations 604
Existing Systems 604
    Department of Defense Systems 604
    Commercial Systems 614
    Data links 619
Accuracy Performance 619
    Position Accuracy 619
    Velocity Accuracy 620
Future Developments 621
    National Range 621
Kinematic Techniques 622
References 622

Author Index 625

Subject Index 627

Table of Contents for Companion Volume I

Preface xxxi

Part I. GPS Fundamentals

Chapter 1. Introduction and Heritage of NAVSTAR, the Global Positioning System 3
Bradford W. Parkinson, *Stanford University, Stanford, California*

Background and History	3
Predecessors	4
Joint Program Office Formed, 1973	6
Introductory GPS System Description and Technical Design	10
Principals of System Operation	10
GPS Ranging Signal	11
Satellite Orbital Configuration	13
Satellite Design	14
Satellite Autonomy: Atomic Clocks	14
Ionospheric Errors and Corrections	16
Expected Navigation Performance	16
High Accuracy/Carrier Tracking	18
History of Satellites	19
Navigation Technology Satellites	19
Navigation Development Satellites—Block I	19
Operational Satellites—Block II and IIA	19
Replacement Operational Satellites—Block IIR	20
Launches	20
Launch Vehicles	20
Initial Testing	22
Test Results	22
Conclusions	24
Applications	24
Military	24
Dual Use: The Civil Problem	24
Pioneers of the GPS	26
Defense Development, Research, and Engineering—Malcolm Currie and David Packard	26
Commander of SAMSO, General Ken Schultz	26
Contractors	26
Joint Program Office Development Team	27
Predecessors	27
Future	28
References	28

Chapter 2. Overview of GPS Operation and Design 29
J. J. Spilker Jr., *Stanford Telecom, Sunnyvale, California* and Bradford W. Parkinson, *Stanford University, Stanford, California*

Introduction to GPS	29
Performance Objectives and Quantitative Requirements on the GPS Signal	29

xxi

Satellite Navigation Concepts, Position Accuracy, and Requirement Signal Time Estimate Accuracy	31
GPS Space Segment	36
GPS Orbit Configuration and Multiple Access	36
GPS Satellite Payload	38
Augmentation of GPS	40
GPS Control Segment	40
Monitor Stations and Ground Antennas	41
Operational Control Center	42
GPS User Segment	43
GPS User Receiver Architecture	43
Use of GPS	45
GPS Signal Perturbations—Atmospheric/Ionospheric/Tropospheric Multipath Effects	49
Ionospheric Effects	49
Tropospheric Effects	52
Multipath Effects	52
Other Perturbing Effects	54
References	54

Chapter 3. GPS Signal Structure and Theoretical Performance 57
J. J. Spilker Jr., *Stanford Telecom, Sunnyvale, California*

Introduction	57
Summary of Desired GPS Navigation Signal Properties	57
Fundamentals of Spread Spectrum Signaling	59
GPS Signal Structure	67
Multiplexing Two GPS Spread Spectrum Signals on a Single Carrier and Multiple Access of Multiple Satellite Signals	68
GPS Radio Frequency Selection and Signal Characteristics	69
Detailed Signal Structure	73
GPS Radio Frequency Receive GPS Power Levels and Signal-to-Noise Ratios	82
GPS Radio Frequency Signal Levels and Power Spectra	82
Satellite Antenna Pattern	84
Signal Specifications	87
User-Receiver Signal-to-Noise Levels	88
Recommendations for Future Enhancements to the GPS System	93
Detailed Signal Characteristics and Bounds on Pseudorange Tracking Accuracy	94
Cross-Correlation Properties—Worst Case	94
Coarse/Acquisition-Code Properties	97
Bounds on GPS Signal Tracking Performance in Presence of White Thermal Noise	106
Appendix: Fundamental Properties of Maximal Length Shift Registers and Gold Codes	114
References	119

Chapter 4. GPS Navigation Data 121
J. J. Spilker Jr., *Stanford Telecom, Sunnyvale, California*

Introduction	121
Overall Message Content of the Navigation Data	121
Navigation Data Subframe, Frame, and Superframe	123
Detailed Description of the Navigation Data Subframe	132
Subframe 1—GPS Clock Correction and Space Vehicle Accuracy	132
GPS Ephemeris Parameters—Subframes 2 and 3	136
Subframes 4 and 5—Almanac, Space Vehicle Health, and Ionosphere Models	139
Time, Satellite Clocks, and Clock Errors	149
Mean Solar, Universal Mean Sidercal, and GPS Time	149
Clock Accuracy and Clock Measurement Statistics	152
Satellite Orbit and Position	159
Coordinate Systems and Classical Orbital Elements	159
Classical Keplerian Orbits	162

Perturbation of Satellite Orbit	164
Ionospheric Correction Using Measured Data	169
Dual-Frequency Ionospheric Correction	169
Appendix	173
References	175

Chapter 5. Satellite Constellation and Geometric Dilution of Precision 177
J. J. Spilker Jr., *Stanford Telecom, Sunnyvale, California*

Introduction	177
GPS Orbit Configuration, GPS-24	178
GPS Orbit-Semi-Major Axis	178
GPS Orbit—Satellite Phasing	180
GPS Satellite Visibility and Doppler Shift	181
Bound on Level of Coverage for 24 Satellites	182
GPS Satellite Visibility Angle and Droopler Shift	183
GPS-24 Satellite Visibility	184
Augmentation of the GPS-24 Constellation	187
Constellation of 30 GPS Satellites	187
Coverage Swath for an Equatorial Plane of Satellites	187
Satellite Ground Traces	189
Geometric Dilution of Precision Performance Bounds and GPS-24 Performance	190
Bounds on Geometric Dilution of Precision—Two Dimensions	192
Bounds on Geometric Dilution of Precision—Three Dimensions	197
Position Dilution of Precision with an Accurate Clock	204
Position Dilution of Precision for the GPS-24 Constellation	205
References	207
Bibliography	207

Chapter 6. GPS Satellite and Payload 209
M. Aparicio, P. Brodie, L. Doyle, J. Rajan, and P. Torrione, *ITT, Nutley, New Jersey*

Spacecraft and Navigation Payload Heritage	209
Concept	209
Relation to Earlier Non-GPS Satellites	209
Overview of Payload Evolution	209
On-Orbit Performance History	210
Navigation Payload Requirements	211
GPS System	211
GPS Performance	211
GPS Signal Structure	213
Payload Requirements	214
Block IIR Space Vehicle Configuration	215
Navigation Payload Architecture	216
Block IIR Payload Design	216
Payload Subsystems	216
Mission Data Unit	223
L-Band Subsystem	229
Characteristics of the GPS L-Band Satellite Antenna	234
Coverage Area	234
Antenna Pattern	234
Antenna Evolution	234
Crosslinks	236
Primary and Secondary Functions	237
Autonomous Navigation	238
Future Performance Improvements	242
Additional Capabilities	242
References	243

Chapter 7. Fundamentals of Signal Tracking Theory 245
J. J. Spilker Jr., *Stanford Telecom, Sunnyvale, California*

Introduction	245
GPS User Equipment	245
GPS User Equipment-System Architecture	246
Alternate Forms of Generalized Position Estimators	249
Maximum Likelihood Estimates of Delay and Position	251
Overall Perspective on GPS Receiver Noise Performance	252
Interaction of Signal Tracking and Navigation Data Demodulation	255
Delay Lock Loop Receivers for GPS Signal Tracking	256
Coherent Delay Lock Tracking of Bandlimited Pseudonoise Sequences	256
Noncoherent Delay Lock Loop Tracking of Pseudonoise Signals	272
Quasicoherent Delay Lock Loop	280
Coherent Code/Carrier Delay Lock Loop	284
Carrier-Aided Pseudorange Tracking	287
Vector Delay Lock Loop Processing of GPS Signals	290
Independent Delay Lock Loop and Kalman Filter	291
Vector Delay Lock Loop (VDLL)	293
Quasioptimal Noncoherent Vector Delay Lock Loop	298
Channel Capacity and the Vector Delay Lock Loop	305
Appendix A: Maximum Likelihood Estimate of Delay and Position	305
Appendix B: Least-Squares Estimation and Quasioptimal Vector Delay Lock Loops	310
Appendix C: Noncoherent Delay Lock Loop Noise Performance with Arbitrary Early-Late Reference Spacing	312
Appendix D: Probability of Losing Lock for the Noncoherent DLL	321
Appendix E: Colored Measurement Noise in the Vector Delay Lock Loop	323
References	325

Chapter 8. GPS Receivers 329
A. J. Van Dierendonck, *AJ Systems, Los Altos, California*

Generic Receiver Description	329
Generic Receiver System Level Functions	329
Design Requirements Summary	331
Technology Evolution	335
Historical Evolution of Design Implementation	335
Current Day Design Implementation	335
System Design Details	337
Signal and Noise Representation	338
Front-End Hardware	340
Digital Signal Processing	348
Receiver Software Signal Processing	365
A Signal-Processing Model and Noise Bandwidth Concepts	365
Signal Acquisition	367
Automatic Gain Control	368
Generic Tracking Loops	369
Delay Lock Loops	372
Carrier Tracking	378
Lock Detectors	390
Bit Synchronization	395
Delta Demodulation, Frame Synchronization, and Parity Decoding	396
Appendix A: Determination of Signal-to-Noise Density	399
Appendix B: Acquisition Threshold and Performance Determination	402
References	405

Chapter 9. GPS Navigation Algorithms 409
P. Axelrad, *University of Colorado, Boulder, Colorado*, and R. G. Brown,
Iowa State University, Ames, Iowa

Introduction	409

Measurement Models 410
    Pseudorange 410
    Doppler 411
    Accumulated Delta Range 412
    Navigation Delta Inputs 412
Single-Point Solution 412
    Solution Accuracy and Dilution of Precision 413
    Point Solution Example 415
Users Process Models 417
    Clock Model 417
    Stationary User or Vehicle 418
    Low Dynamics 419
    High Dynamics 420
Kalman Filter and Alternatives 420
    Discrete Extended Kalman Filter Formulation 421
    Steady-State Filter Performance 422
    Alternate Forms of the Kalman Filter 423
    Dual-Rate Filter 423
    Correlated Measurement Noise 424
GPS Filtering Examples 424
    Buoy Example 425
    Low Dynamics 427
    Unmodeled Dynamics 427
    Correlated Measurement Errors 430
Summary 430
References 433

Chapter 10. GPS Operational Control Segment 435
Sherman G. Francisco, *IBM Federal Systems Company, Bethesda, Maryland*

Monitor Stations 439
Master Control Station 445
Ground Antenna 447
Navigation Data Processing 449
System State Estimation 457
Navigation Message Generation 463
Time Coordination 464
Navigation Product Validation 465
References 465

Part II. GPS Performance and Error Effects

Chapter 11. GPS Error Analysis 469
Bradford W. Parkinson, *Stanford University, Stanford, California*

Introduction 469
Fundamental Error Equation 469
    Overview of Development 469
    Derivation of the Fundamental Error Equation 470
Geometric Dilution of Precision 474
    Derivation of the Geometric Dilution of Precision Equation 474
    Power of the GDOP Concept 474
    Example Calculations 475
    Impact of Elevation Angle on GDOP 477
Ranging Errors 478
    Six Classes of Errors 478
    Ephemeris Errors 478
    Satellite Clock Errors 478

Ionosphere Errors	479
Troposhere Errors	479
Multipath Errors	480
Receiver Errors	480
Standard Error Tables	480
Error Table Without S/A: Normal Operation for C/A Code	481
Error Table with S/A	481
Error Table for Precise Positioning Service (PPS Dual-Frequency P/Y Code)	482
Summary	482
References	483

Chapter 12. Ionospheric Effects on GPS 485
J. A. Klobuchar, *Hanscom Air Force Base, Massachusetts*

Introduction	485
Characteristics of the Ionosphere	485
Refractive index of the ionosphere	488
Major Effects on Global Positioning Systems Caused by the Ionosphere	489
Ionospheric Group Delay—Absolute Range Error	489
Ionospheric Carrier Phase Advance	490
Higher-Order Ionospheric Effects	491
Obtaining Absolute Total Electron Content from Dual-Frequency GPS Measurements	493
Ionospheric Doppler Shift/Range-Rate Errors	495
Faraday Rotation	496
Angular Refraction	497
Distortion of Pulse Waveforms	498
Amplitude Scintillation	499
Ionospheric Phase Scintillation Effects	502
Total Electron Content	503
Dependence of Total Electron Content on Solar Flux	504
Ionospheric Models	506
Single-Frequency GPS Ionospheric Corrections	509
Magnetic Storms Effects on Global Positioning Systems	510
Differential GPS Positioning	511
Appendix: Ionospheric Correction Algorithm for the Single-Frequency GPS Users	513
References	514

Chapter 13. Troposheric Effects on GPS 517
J. J. Spilker Jr., *Stanford Telecom, Sunnyvale, California*

Troposheric Effects	517
Introduction	517
Atmospheric Attenuation	520
Rainfall Attenuation	521
Tropospheric Scintillation	522
Tropospheric Delay	523
Path Length and Delay	524
Tropospheric Refraction Versus Pressure and Temperature	528
Empirical Models of the Troposphere	534
Saastamoinen Total Delay Model	534
Hopfield Two Quartic Model	534
Black and Eisner (B&E) Model	536
Water Vapor Zenith Delay Model—Berman	538
Davis, Chao, and Marini Mapping Functions	538
Altshuler and Kalaghan Delay Model	539
Ray Tracing and Simplified Models	541
Lanyi Mapping Function and GPS Control Segment Estimate	541
Model Comparisons	544
Tropospheric Delay Errors and GPS Positioning	544
References	545

Chapter 14. Multipath Effects 547
Michael S. Braasch, *Ohio University, Athens, Ohio*

Introduction	547
Signal and Multipath Error Models	548
Pseudorandom Noise Modulated Signal Description	549
Coherent Pseudorandom Noise Receiver	549
Noncoherent Pseudorandom Noise Receiver	553
Simulation Results	554
Aggravation and Mitigation	558
Antenna Considerations	558
Receiver Design	560
Multipath Data Collection	560
Acknowledgments	566
References	566

Chapter 15. Foliage Attenuation for Land Mobile Users 569
J. J. Spilker Jr., *Stanford Telecom, Sunnyvale, California*

Introduction	569
Attenuation of an Individual Tree or Forest of Trees—Stationary User	571
Foliage Attenuation—Mobile User	575
Probability Distribution Models for Foliage Attenuation—Mobile User	576
Measured Models—Satellite Attenuation Data	580
Measured Fading for Tree-Lined Roads—Mobile Users	581
References	582

Chapter 16. Ephemeris and Clock Navigation Message Accuracy 585
J. F. Zumberge and W. I. Bertiger, *Jet Propulsion Laboratory, California Institute of Technology, Pasadena, California*

Control Segment Generation of Predicted Ephemerides and Clock Corrections	585
Accuracy of the Navigation Message	586
Global Network GPS Analysis at the Jet Propulsion Laboratory	587
Accuracy of the Precise Solution	588
Comparison of Precise Orbits with Broadcast Ephemerides	590
Comparison of the Precise Clocks with Broadcast Clocks	593
Summary and Discussion	595
Appendix: User Equivalent Range Error	597
References	598

Chapter 17. Selective Availability 601
Frank van Graas and Michael S. Braasch, *Ohio University, Athens, Ohio*

Goals and History	601
Implementation	601
Characterization of Selective Availability	602
Second-Order Gauss-Markov Model	605
Autoregressive Model	608
Analytic Model	614
Recursive Autoregressive Model (Lattice Filter)	615
Selective Availability Model Summary	619
References	620

Chapter 18. Introduction to Relativistic Effects on the Global Positioning System 623
N. Ashby, *University of Colorado, Boulder, Colorado,* and
J. J. Spilker Jr., *Stanford Telecom, Sunnyvale, California*

Introduction	623
Objectives	623
Statement of the GPS Problem	625
Introduction to the Elementary Principles of Relativity	627
Euclidean Geometry and Newtonian Physics	627
Space-Time Coordinates and the Lorentz Transformation	629
Relativistic Effects of Rotation in the Absence of a Gravitational Field	640
Principle of Equivalence	649
Relativistic Effects in GPS	657
Relativistic Effects on Earth-Based Clocks	659
Relativistics Effects for Users of the GPS	676
Secondary Relativistic Effects	683
References	695

Chapter 19. Joint Program Office Test Results 699
Leonard Kruczynski, *Ashtech, Sunnyvale, California*

Introduction	699
U.S. Army Yuma Proving Ground (YPG)	700
Reasons for Selection of Yuma Proving Ground	701
Lasers	701
Range Space	702
Joint Program Office Operating Location	702
Satellite Constellation for Test Support	703
Control Segment Responsiveness to Testing Needs	703
Trajectory Determination YPG	704
Real-Time Estimate	704
Best Estimate of Trajectory	705
Validation of Truth Trajectory Accuracy	705
Ground Truth	705
Phase I Test (1972–1979)	707
Ground Transmitters	707
Navy Testing for Phase I	710
Tests Between Phase I and Phase II (1979–1982)	711
Weapons Delivery	711
Differential Tests	711
Phase II: Full-Scale Engineering Development Tests (1982–1985)	713
Summary	714
Bibliography	715

Chapter 20. Interference Effects and Mitigation Techniques 717
J. J. Spilker Jr. and F. D. Natali, *Stanford Telecom, Sunnyvale, California*

Introduction	717
Possible Sources of Interference	719
Frequency Allocation in Adjacent and Subharmonic Bands	719
Receiver Design for Tolerance to Interference	720
Receiver Systems	720
Quantizer Effects in the Presence of Interference	724
Effects of Interference on the GPS C/A Receiver	745
Effects of the C/A-Code Line Components on Narrow-Band Interference Performance	745
Narrow-Band Interference Effects—Spectra of Correlator Output	748
Interference Effects—Effects on Receiver-Tracking Loops	752

Detection of Interference, Adaptive Delay Lock Loop, Adaptive Frequency Notch Filtering, and Adaptive
    Null Steering Antennas 756
  Adaptation of the Delay Lock Loop and Vector Delay Lock Loop 757
  Rejection of Narrow-Band Interference by Adaptive Frequency Nulling Filters 757
  Adaptive Antennas for Point Source Interference 759
  Augmentation of the GPS Signals and Constellation 767
Appendix: Mean and Variance of the Correlator Output for an M-Bit Quantizer 768
References 771

Author Index 773

Subject Index 775

Preface

Overview and Purpose of These Volumes

Of all the *military* developments fostered by the recent cold war, the Global Positioning System (GPS) may prove to have the greatest positive impact on everyday life. One can imagine a 21st century world covered by an augmented GPS and laced with mobile digital communications in which aircraft and other vehicles travel through "virtual tunnels," imaginary tracks through space which are continuously optimized for weather, traffic, and other conditions. Robotic vehicles perform all sorts of construction, transportation, mining, and earth moving functions working day and night with no need for rest. Low-cost personal navigators are as commonplace as hand calculators, and every cellular telephone and personnel communicator includes a GPS navigator. These are some of the potential positive impacts of GPS for the future. Our purpose in creating this book is to increase that positive impact. That is, *to accelerate the understanding of the GPS system and encourage new and innovative applications.*

The intended readers and users of the volumes include all those who seek knowledge of GPS techniques, capabilities, and limitations:

- Students attending formal or informal courses
- Practicing GPS engineers
- Applications engineers
- Managers who wish to improve their understanding of the system

Our somewhat immodest hope is that this book will become a standard reference for the understanding of the GPS system.

Each chapter is authored by an individual or group of individuals who are recognized as world-class authorities in their area of GPS. Use of many authors has led to some overlap in the subject matter which we believe is positive. This variety of viewpoints can promote understanding and contributes to our overall purpose. Books written by several authors also must contend with variations in notation. The editors of the volume have developed common notations for the important subjects of GPS theory and analysis, and attempted to extend this, where possible, to other chapters. Where there are minor inconsistencies we ask for your understanding.

Organization of the Volumes

The two volumes are intended to be complementary. Volume I concentrates on fundamentals and Volume II on applications. Volume I is divided into two parts: the first deals with the operation and theory of basic GPS, the second section with GPS performance and errors. In Part I (GPS Fundamentals), a summary of GPS history leads to later chapters which promote an initial understanding of the three GPS segments: User, Satellite, and Control. Even the best of systems has its limitations, and GPS is no exception. Part II, GPS Performance and Error Effects, is introduced with an overview of the errors, followed by chapters devoted to each of the individual error sources.

Volume II concentrates on two aspects: augmentations to GPS and detailed descriptions of applications. It consists of Parts III to VI:

- III. Differential GPS and Integrity Monitoring
- IV. Integrated Navigation Systems
- V. GPS Navigation Applications
- VI. Special Applications

Parts III and IV expand on GPS with explanations of supplements and augmentations to the system. The supplements enhance accuracy, availability, or integrity. Of special interest is differential GPS which has proven it can provide sub-meter (even centimeter) level accuracies in a dynamic environment. The last two sections (V and VI) are detailed descriptions of the major applications in current use. In the rapidly expanding world of GPS, new uses are being found all of the time. We sincerely hope that these volumes will accelerate such new discoveries.

Acknowledgments

Obviously this book is a group undertaking with many, many individuals deserving of our sincere thanks. In addition to the individual authors, we would especially like to thank Ms. Lee Gamma, Mr. Sam Pullen, and Ms. Denise Nunes. In addition, we would like to thank Mr. Gaylord Green, Dr. Nick Talbot, Dr. Gary Lennon, Ms. Penny Sorensen, Mr. Konstantin Gromov, Dr. Todd Walter, and Mr. Y. C. Chao.

Special Acknowledgment

We would like to give special acknowledgment to the members of the original GPS Joint Program Office, their supporting contractors and the original set of engineers and scientists at the Aerospace Corporation and at the Naval Research Laboratory. Without their tenacity, energy, and foresight GPS would not be.

B. W. Parkinson
J. J. Spilker Jr.
P. Axelrad
P. Enge

Part III. Differential GPS and Integrity Monitoring

Chapter 1

Differential GPS

Bradford W. Parkinson* and Per K. Enge†
Stanford University, Stanford, California 94305

I. Introduction

DIFFERENTIAL GPS (DGPS) is a technique that significantly improves both the accuracy and the integrity of the Global Positioning System. The most common version of DGPS is diagrammed in Fig. 1. As shown, DGPS requires high-quality GPS "reference receivers" at known, surveyed locations. The reference station estimates the slowly varying error components of each satellite range measurement and forms a correction for each GPS satellite in view. This correction is broadcast to all DGPS users on a convenient communications link. Typical ranges for a local area differential GPS (LADGPS) station are up to 150 km. Within this operating range, the differential correction greatly improves accuracy for all users, regardless of whether selective availability (SA) is activated or is not (see Chapter 11, the companion volume, on error analysis). This improvement arises because the largest GPS errors vary slowly with time and are strongly correlated over distance. Differential DGPS also significantly improves the "integrity," or truthfulness, of GPS for all classes of users, because it reduces the probability that a GPS user would suffer from an unacceptable position error attributable to an undetected system fault. (Integrity is the probability that the displayed position is within the specified or expected error boundaries.)

A. Standard Positioning Service Users

The most dramatic DGPS improvement is found for the Standard Positioning Service (SPS) when SA is activated. Although an SPS receiver itself is capable of range measurement precision of approximately 0.5 m, the normal ranging errors include slowly varying biases attributable to all six of the error classes

---

Copyright © 1995 by the authors. Published by the American Institute of Aeronautics and Astronautics, Inc., with permission. Released to AIAA to publish in all forms.
*Professor of Aeronautics and Astronautics and of the Hansen Experimental Physics Laboratory. Director of the GPS Program.
†Professor of Aeronautics and Astronautics.

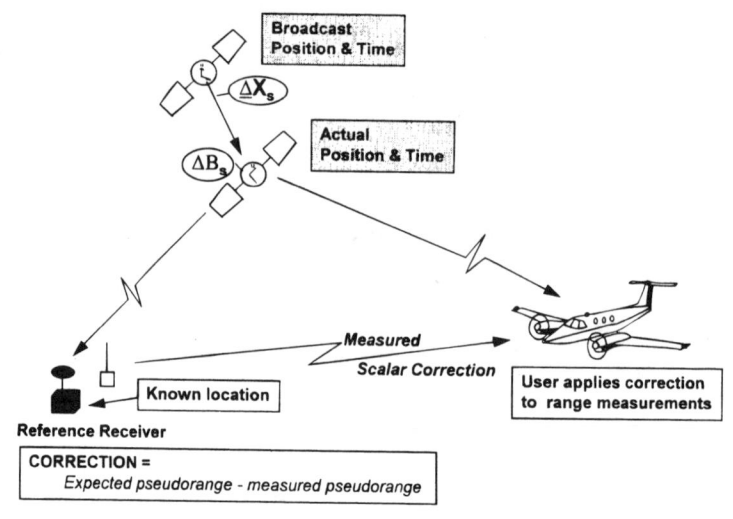

Fig. 1 Local area differential GPS. The reference receiver calibrates the correlated errors in ranging to the satellites. These filtered errors are then transmitted to the user as range corrections.

described in Chapter 11 of the companion volume. These are dominated by SA, with one sigma *ranging* errors typically measured to be about 21 m. Without differential corrections, these SA-dominated biases limit the *horizontal positioning* accuracy of the SPS to 100 m (approximate 95 percentile level). With differential corrections, the SPS navigation accuracy can be improved to better than 1 m (1 σ), provided that the correction age is less than 10 s, and the user is within 50 km of the reference station. As the corrections age, or the geographic separation from the reference station increases, the accuracy of DGPS degrades. This degradation with range is graceful; thus LADGPS provides adequate accuracy for some applications at ranges of up to 1000 km.

B. Precise Positioning Service Users

As mentioned, DGPS also improves the performance of the Precise Positioning Service (PPS). Without differential corrections, the PPS is significantly more accurate than the SPS, because PPS users do not suffer from SA. In addition, PPS receivers can use measurements at *both* GPS frequencies to reduce the effect of ionospheric delays. Nonetheless, differential corrections can still provide significant improvements to the PPS accuracy, which is nominally 15 meters SEP (spherical error probable, which is the radius of a sphere that is expected to contain 50% of the errors). Expected accuracies with DGPS are about the same as SPS: they range from 1 to 5 m, depending upon the system design.

C. Major Categories of Differential GPS

There are many DGPS techniques and applications.[1,2] The major techniques are broadly characterized in the following subsections.

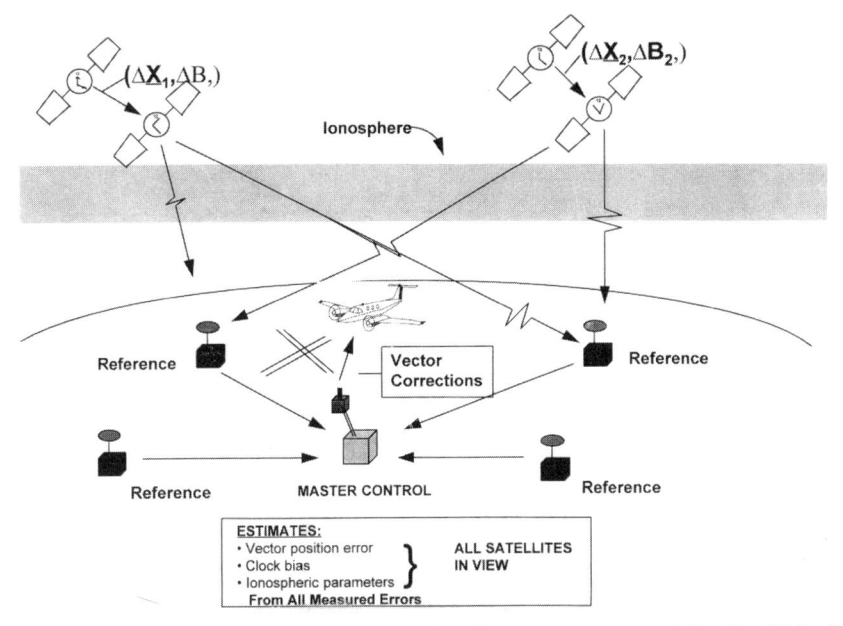

Fig. 2 Wide area differential GPS concept. One type, sponsored by the FAA, is known as the wide-area augmentation system (WAAS).

1. Local Area Differential GPS

Most DGPS systems use a single reference station to develop a *scalar* correction to the code-phase* measurement for each satellite. This approach is shown in Fig. 1. If the correction is delivered within 10 s, and the user is within 1000 km, the user accuracy should be between 1 and 10 m. This capability (shown in Fig. 3) is detailed further in Sec. II of this chapter. An additional technique uses inexpensive, ground-based transmitters that broadcast a GPS signal at the L_1 or L_2 frequency. These are called pseudosatellites or *pseudolites* (PL) and act as an additional ranging source as well as a datalink. Pseudolites provide significant improvements in geometry[3] and accuracy; one technique is described under test results and discussed in a later chapter on precision landing.

2. Wide Area Differential GPS

As shown in Fig. 2, networks of reference stations can be used to form a *vector* correction for each satellite. This vector consists of individual corrections for the satellite clock, three components of satellite positioning error (or ephemeris), and parameters of an ionospheric delay model. The validity of this correction

---

*The modern technique for receiver code-phase measurements is to use "carrier aiding," which filters the noisy code-phase measurements with the smoother carrier measurements. This is not to be confused with pure carrier-tracking techniques described further later.

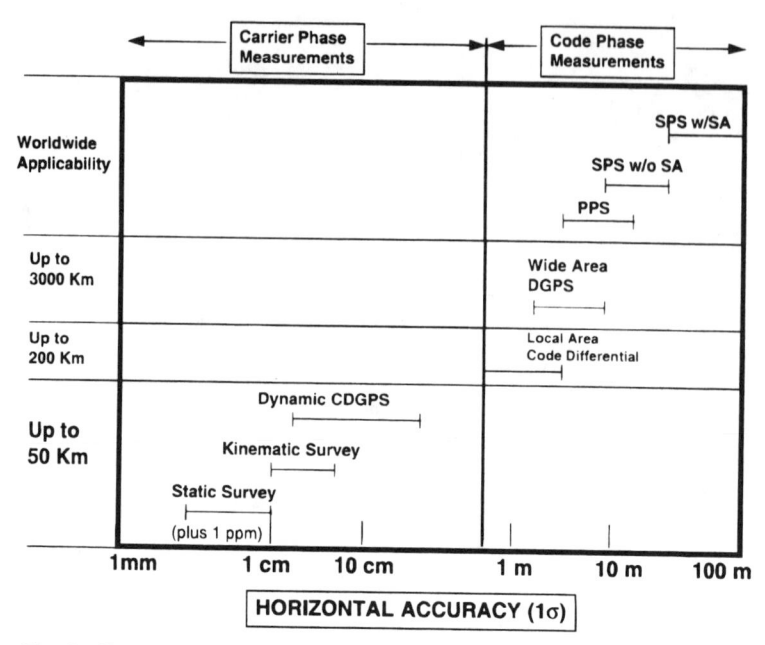

Fig. 3 Summary of expected differential GPS concepts and accuracies.

still decreases with increased *latency** or age of the correction. However, compared to a scalar correction, a vector correction is valid over much greater geographical areas. This concept is called wide area DGPS, or WADGPS.[4] Such networks will be used for continental or even world-hemisphere coverage, because they require many fewer reference stations than a collection of independent systems with one reference station each. Moreover, they require less communication capacity than the equivalent network of LADGPS systems. Wide area GPS is a subject unto itself, and it is described in detail in Chapters 3 and 4 of this volume.

3. Carrier-Phase Differential GPS

Users with very stringent accuracy requirements may be able to use a technique called carrier-phase DGPS or CDGPS. These users measure the phase of the GPS carrier relative to the carrier phase at a reference site; thus achieving range measurement precisions that are a few percent of the carrier wavelength (typically about one centimeter). These GPS phase comparisons are used for vehicle attitude determination and also in survey applications, where the antennas are separated by tens of kilometers. If the antennas are fixed, then the survey is called *static*, and millimeter accuracies are possible, because long averaging times can be used to combat random noise. If the antennas are moving, then the survey is *kinematic*,

---

*Latency is the total time from the reference station measurement of error to the actual application in the user receiver. It includes the calculation time and any communications delay.

and shorter time constants must be used with some degradation of accuracy. These static and kinematic capabilities are included in Fig. 3. Several carrier-phase techniques for aircraft precision landing have also been demonstrated. Carrier-phase DGPS is introduced in Sec. IV of this chapter and is further described in Chapters 4, 15, 18, and 19 of this volume.

4. Organization of the Chapter

This chapter introduces DGPS, and many of the remaining chapters apply or further develop this important technique. Section I of this chapter describes the measurements of a code-phase differential system. In Sec. III the error analysis for a LADGPS is developed. Accuracy degradation for "aged" corrections and for user displacements from the reference station are quantified. Section IV introduces CDGPS by describing GPS phase interferometry for attitude determination as well as static and kinematic survey. It also introduces techniques for resolving the λ, or wavelength ambiguity, which must be determined to realize centimeter-level accuracies. Section V describes standardized data formats for the transmission of local area differential corrections, and Sec. VI provides an overview of DGPS broadcast systems. Section VII provides a small sample of the huge number of DGPS field results reported in the literature.

II. Code-Phase Differential GPS

It is useful to summarize the expected user errors in a form that allows analysis of differential system accuracy. Errors can be categorized as either correlated between receivers or uncorrelated. Only the correlated errors can be corrected with DGPS. Even the nominally correlated errors lose that correlation if they are significantly delayed in application (*temporally decorrelated*) or are applied to a receiver significantly separated from the reference station (*geographically decorrelated*). This section provides estimates of these decorrelation factors.

A. User Errors Without Differential GPS

This section draws heavily on the development of Chapter 11, the companion volume, which should be used as a reference. A code-tracking receiver actually measures the raw difference between the user's biased clock and the transmitted time of the start of the satellite code phase (which is part of the satellite message). This quantity is called raw *pseudorange* ρ. With the speed of light c used to convert time to distance, this is expressed as follows:

$$\rho = c \cdot (t_{Au} - t_{Ts}) \qquad (1)$$

where t_{Au} = Arrival time measured by the *u*ser, s; t_{Ts} = uncorrected value of satellite *T*ransmission time, s; and u,s represents the user and the *s*th satellite. This is measured (or raw) pseudorange, which equals the true range D from the user u to satellite s plus an unknown offset between the user clock b_u and the satellite clock B_s. Additional time delays are caused by the ionosphere I and the troposphere T, as well as noise, multipath, and/or interchannel errors in the user's receiver v:

$$\rho = D + c \cdot (b_u - B) + c \cdot (T + I + v) \qquad (2)$$

The true geometric range (in xyz Cartesian coordinates) is given by the following:

$$D = \sqrt{(x_s - x_u)^2 + (y_s - y_u)^2 + (z_s - z_u)^2}$$

This can also be written in a more convenient way for calculations:

$$D = |\bar{r}_s - \bar{r}_u| = \bar{1}_s \cdot [\bar{r}_s - \bar{r}_u] \qquad (3)$$

where $\bar{r}_s$ = true satellite position (included in user message); $\bar{r}_u$ = true user position; and $\bar{1}_s$ = true unit vector from users to satellite.

As an aside, Eq. (2) is modified as follows if carrier phase is the basic measurement. The carrier cycles are counted and converted to range so that the quantity $\phi_{u,s}$, formed by counting zero crossings of the reconstructed radio frequency (rf) carrier, is in meters.

$$\phi_{u,s} = D + c \cdot (b_u - B_s) + c \cdot (T - I + v_{u,s}^{(\phi)}) + N_{u,s} \cdot \lambda \qquad (4)$$

Note that the *sign* of the ionospheric *group* delay is changed for this *phase* delay. Also note the addition of $N_{u,s} \cdot \lambda$, where N is an unknown integer that counts carrier wavelengths, and λ is the carrier wavelength of 19.2 cm. Tropospheric errors affect both types of measurements equally. Also note the superscript on the receiver noise. Carrier measurement noise is not the same as that measured for the code; in fact, it is usually orders of magnitude smaller.

Returning to code-phase measurements, the measurement represented by Eq. (2) is adjusted to form *corrected pseudorange*. This is formed by correcting the measurement for estimates of some of the raw errors:

$$\rho_{c_{u,s}} = \rho_{u,s} + c \cdot \hat{B}_s - c \cdot (\hat{T}_s + \hat{I}_s) \qquad (5)$$

where the ($\wedge$) is used to denote estimates of satellite timing error or estimates of ionospheric or tropospheric delays.

A user without DGPS, then, forms four or more of these measurements into a matrix equation as developed in Chapter 11 of the companion volume:

$$G\hat{\bar{x}} = (\mathbf{A} \cdot \bar{R} - \bar{\rho}_c) \qquad (6)$$

and solves* for estimated position as follows:

$$\hat{\bar{x}} = (G^T G)^{-1} G^T (\mathbf{A} \cdot \bar{R} - \bar{\rho}_c) \qquad (7)$$

---

*Note that this pseudoinverse collapses to $\hat{\bar{x}} = G^{-1}(\mathbf{A} \cdot \bar{R} - \bar{\rho}_c)$ if the number of satellites equals four.

where

$$\hat{\bar{x}}_{4\times1} \equiv \begin{bmatrix} \hat{\bar{r}}_u \\ -c\cdot\hat{b}_u \end{bmatrix}; \qquad G_{n\times4} \equiv \begin{bmatrix} \hat{\bar{1}}_{s1}^T & 1 \\ \hat{\bar{1}}_{s2}^T & 1 \\ \vdots & \vdots \\ \hat{\bar{1}}_{sn}^T & 1 \end{bmatrix};$$

$$A_{n\times 3n} \equiv \begin{bmatrix} \hat{\bar{1}}_{s1}^T & & 0 \\ & \hat{\bar{1}}_{s2}^T & \\ & & \ddots & \\ 0 & & & \hat{\bar{1}}_{sn}^T \end{bmatrix}; \qquad \bar{R}_{3n\times1} \equiv \begin{bmatrix} \hat{\bar{r}}_{s1} \\ \hat{\bar{r}}_{s2} \\ \vdots \\ \hat{\bar{r}}_{sn} \end{bmatrix}$$

Note that G, the geometry matrix, is determined by the estimated directions to each of the visible satellites. A is a matrix of the satellite locations which have been received as part of the satellite broadcast, and $\hat{\bar{\rho}}_c$ is the corrected pseudorange to each satellite, arranged as a vector. We can derive the following vector to be the pseudorange error ($\Delta\bar{\rho}$):

$$\Delta\bar{\rho} \triangleq c\cdot(-\Delta\bar{B} + \Delta\bar{I} + \Delta\bar{T} + \bar{v}) + \epsilon\cdot(\bar{R} - \bar{P}) + A\cdot\Delta\bar{R} \qquad (8)$$

where the first four Δ's are the residual errors caused by satellite clock (including SA), ionosphere, and troposphere (after any corrections). The vector v includes receiver errors and multipath, and we also define

$$\epsilon_{n\times 3n} \equiv \begin{bmatrix} \Delta\hat{\bar{1}}_{s1}^T & & 0 \\ & \Delta\hat{\bar{1}}_{s2}^T & \\ & & \ddots & \\ 0 & & & \Delta\hat{\bar{1}}_{sn}^T \end{bmatrix}; \qquad \bar{P}_{3n\times1} \equiv \begin{bmatrix} \bar{r}_u \\ \bar{r}_u \\ \vdots \\ \bar{r}_u \end{bmatrix}$$

Note that in Eq. (8), the satellite position error is the last term, and the error in calculation caused by an erroneous unit vector is the next-to-last term. This is the error in the range to the satellite. The error in the user position calculation is then given as follows:

$$\Delta\bar{x} = (G^TG)^{-1}G^T\Delta\bar{\rho} \qquad (9)$$

The major purpose of all DGPS systems is to estimate the user's stand-alone ranging error $\Delta\bar{\rho}_c$ so that a more accurate pseudorange can be used to estimate position.*

---

*There have been differential systems that corrected position rather than pseudorange. This is not a good design approach, because it assumes that both reference and user employ the same set of satellites, or else that all combinations of satellites are provided as part of the correction message.

Table 1 GPS errors with selective availability for the Standard Positioning Service

Error Source	1 sigma position error (m)		
	BIAS	RANDOM	TOTAL
Ephemeris Data	2.1	0.0	2.1
Satellite Clock	20.0	0.7	20.0
Ionosphere	4.0	0.5	4.0
Troposphere	0.5	0.5	0.7
Multipath	1.0	1.0	1.4
Receiver Measurement	0.5	0.2	0.5
Reference Station Errors	0.0	0.0	0.0
Pseudo-Range Error (RMS)	20.5	1.4	20.6
Filtered PRE (RMS)	20.5	0.4	20.5
Total Vertical Error	VDOP = 2.5		51.4 m
Total Horizontal Error	HDOP = 2.0		41.1 m

Table 1 summarizes the errors that contribute to both stand-alone GPS and DGPS before the DGPS correction. As shown, the total pseudorange error can be approximated by taking the square root of the sum of the individual errors squared. The total pseudorange error (1σ) for a GPS user without differential corrections is approximately 21 m. Clearly, SA is the dominant error source for the user without differential corrections. This table should be compared to Table 8 of this chapter, which shows the residual errors after applying DGPS corrections.

B. Reference Station Calculation of Corrections

The reference receiver turns the problem around. Its receiver antenna is located in a known position* relative to the desired reference frame. It then solves Eq. (6) for the unknown corrections to the raw pseudorange vector $\bar{\rho}$ using†: $G\hat{\bar{x}}_T = (A \cdot \bar{R} - \bar{\rho} - \Delta\hat{\bar{\rho}}_R)$, thus we have the following:

$$\Delta\hat{\bar{\rho}}_R = A \cdot \bar{R} - \bar{\rho} - G\hat{\bar{x}}_T$$

This is the fundamental reference station calculation. This $\Delta\hat{\bar{\rho}}_R$ (estimated ranging error) is then transmitted to the user, who applies it to his measured psuedoranges.

---

*The choice of reference frames is up to the system designer. Any convenient frame is acceptable, but all reference points should be located to the same desired accuracy. For example, location of a runway relative to the reference antenna should have a consistent level of accuracy.

†It is best for both the reference and the user to use raw pseudoranges. Any corrections applied must be consistent at both places.

The "known" reference station position is *four-dimensional*. That is, it includes a local time correction. Any consistent timing error for *all* pseudorange corrections will only affect the user clock (see Chapter 11, the companion volume, for the reason). Assuming that the user is interested only in the user's geographical position, this clock correction is, therefore, arbitrary. This time bias is usually slowly "steered" so that the magnitude of the largest correction is minimized.

C. Application of Reference Correction

Several important points must be made about applying this correction:

1) Any pseudorange corrections that are in addition to the reference correction must be agreed upon by both user and reference, and they must be applied in exactly the same way. For example, in the formation of the "raw" pseudorange measurements $\bar{\rho}$, the broadcast ionospheric delay correction must be the same for both the user and the reference station. The safest course of action is for neither to apply corrections. Assuming this is the case, the user applies the correction that has been received ($\Delta\hat{\bar{\rho}}_R$) to the measured raw psuedoranges. The user's fundamental equation becomes $G\hat{\bar{x}}_u = [A \cdot \bar{R} - (\bar{\rho} + \Delta\hat{\bar{\rho}}_r)]$. The solution for the $n = 4$ case is $\hat{\bar{x}}_u(t) = G^{-1}[A \cdot \bar{R} - (\bar{\rho}(t) + \Delta\hat{\bar{\rho}}_R(t - \tau)]$. The τ has been introduced to highlight the delay between measurement and application.

2) The reference and the user *must be using the same satellite ephemeris*. Because these are periodically revised, a well-designed DGPS system will continue to broadcast corrections based on both old and new ephemerides during transition periods. The user can then use either correction while he completes gathering the new ephemeris data.

3) The reference station must take great care to not introduce additional user errors by including effects that are not measured by the user. An example would be multipath error induced by reflections from buildings near the reference station antenna.

4) The time of the reference station correction should be passed to the user as part of the correction message. This will allow the user to both evaluate integrity and properly apply any time-rate-of-change information.

III. Analysis of Differential GPS Errors

In general, the reference station gathers corrections that are geographically separated from the user and are delayed in application. A first-order relationship between the reference correction and the user's best correction is as follows:

$$\Delta\bar{\rho}_u \cong \Delta\hat{\bar{\rho}}_R + \frac{\partial(\Delta\bar{\rho})}{\partial\bar{x}} \cdot \delta\bar{x} + \Delta\dot{\bar{\rho}}_R \cdot \Delta t + \delta\bar{\rho}$$

The first term $\Delta\hat{\bar{\rho}}_R$ is the correction estimated by the reference station. The next three terms represent deviations from a perfect correction. These errors are referred to as types 1, 2, and 3, and they are defined as follows:

- *Type 1: Decorrelation with Distance.* The term $[\partial(\Delta\bar{\rho})]/(\partial\bar{x}) \cdot \delta\bar{x}$ is the error attributable to the vector gradient of corrections (the decorrelation with distance from the reference site).

- **Type 2: Decorrelation with Time.** The term $\Delta\dot{\rho}_R \cdot \Delta t$ is the error attributable to the time rate of change of the corrections (decorrelation with time). This effect is frequently called *latency*. To cope with time decorrelations, most DGPS systems broadcast the measured time rate of change of corrections as part of the communicated message. This first-order correction usually achieves an accuracy of about 0.5 m for a 10 s delay. Higher-order derivatives can be transmitted, but their estimates are noisy, and the prediction process is deliberately made difficult by the high-frequency changes induced by SA.

- **Type 3: Uncorrelated Errors (not correctable with DGPS).** The last term represents errors at the user that are not correlated with those measured at the reference. This term can be viewed as the error for user and reference if they were next to each other and there were no delay in application of the corrections. Note that type 3 errors at *both* the user and reference station contribute to the total DGPS user error.

These error types are not necessarily mutually exclusive. For example, ionospheric errors are both type 1 and type 2, because they decorrelate with both time and distance.

The subsequent sections analyze each of these error sources and present estimates of residual errors after differential corrections. For each, the distance and time decorrelation factors are also estimated. The preceding definitions of decorrelation types are used in these discussions.

A. Receiver Noise, Interference, and Multipath Errors for Differential GPS

Receiver noise, interference, and multipath are grouped together because *they constitute the noise floor* for DGPS. These errors are almost totally of type 3. They have very short decorrelation distances; thus noise, interference, and multipath at the reference station are not usually correlated with those effects at the mobile receiver.

Special care must be taken with type 3 errors in the reference station. Any effects in the reference correction *will be directly added to the user error, because they will be uncorrelated errors that are included in the broadcast correction.* Therefore, the elimination of these effects in the reference receiver is a primary design goal. Fortunately, two techniques—carrier aiding and narrow correlator spacing—can minimize these effects. Their use has significantly reduced this DGPS noise floor. Both techniques can be used with mobile receivers as well. These are discussed in the following two sections.

Multipath arises when GPS signals travel over multiple paths from the satellite to the receiver. Some of the signals are delayed relative to the "direct" signal, because they have traveled paths that include a reflection. The reflecting object might be a building, ship, aircraft, or truck, or it might be the surface of the sea or of a runway. In general, the strongest reflections occur close to the receiver. If these reflected signals are delayed by more than 1.5 μs (about 500 m of increased path length), they will be suppressed in the decorrelation process, because the autocorrelation of the C/A-code is nearly zero for delays greater than $1^1/_2$ chips. However, if they are delayed by less than 1.5 µs, their impact depends upon their amplitude, the amount of delay, and the persistence of the reflection. This persistence can be quantified as the *correlation time*.

Multipath errors, particularly in the reference station, should be the major issue in selecting and siting reference antennas. Certain modern antennas have substantial improvements in sidelobe suppression, which helps further eliminate multipath before it can enter the receiver. Avoiding antenna sites close to reflective materials can also help greatly. These considerations should be regarded as the primary defense against multipath.

1. Random Errors and Carrier Aiding

Code and carrier measurements both suffer from random observation noise, which is denoted v for the code phase and ϕ for the carrier. These random variables model the impact of thermal noise, multiple access interference, and multipath. In the absence of multipath, the standard deviation of the carrier noise is 1 cm or less compared to over 1 m in the unaided code. Therefore, the carrier-phase measurement is much more precise than the code measurement, but the carrier measurement does suffer from the mentioned integer ambiguity $N \cdot \lambda$ caused by the unknown number of carrier phase cycles between the user and the satellite.

Carrier aiding is a technique that uses the precision of the carrier observations to smooth the observed code-phase measurements. The following (fading memory) recursion is an example of a filter that is used:

$$\hat{\rho}_{u,s}(t_k) = \frac{1}{L} \rho_{u,s}(t_k) + \frac{L-1}{L} (\hat{\rho}_{u,s}(t_{k-1}) + \phi_{u,s}(t_k) - \phi_{u,s}(t_{k-1}))$$

where

$$\hat{\rho}_{u,s}(t_0) = \rho_{u,s}(t_0)$$

The first term of the recursion is the current code-phase measurement weighted by $1/L$, where L is a large number, perhaps 100 or 200. The current code-phase measurement receives a relatively low weighting because the carrier-phase difference in the second term predicts the future value of the pseudorange with very high accuracy. The forward prediction does not suffer from any integer ambiguity because the *carrier difference* is used. Moreover, under most conditions, well-designed GPS phase–lock-loops (PLL) rarely suffer from cycle slips that would degrade accuracy.

This carrier-aiding technique should not be confused with rate-aiding techniques, which use integrated Doppler measurements. Indeed, the carrier-phase measurements maintain phase coherency and do not suffer from accumulated error growth caused by accumulated measurement noise. Nonetheless, the forward prediction will eventually degrade because of code–carrier divergence (attributable to the ionosphere). In fact, the weighting constant L must be carefully chosen to balance the very low noise of the carrier measurements with the accumulation of code–carrier divergence. In the absence of significant multipath, carrier aiding bounds the standard deviation of the pseudorange error to a few tenths of a meter.

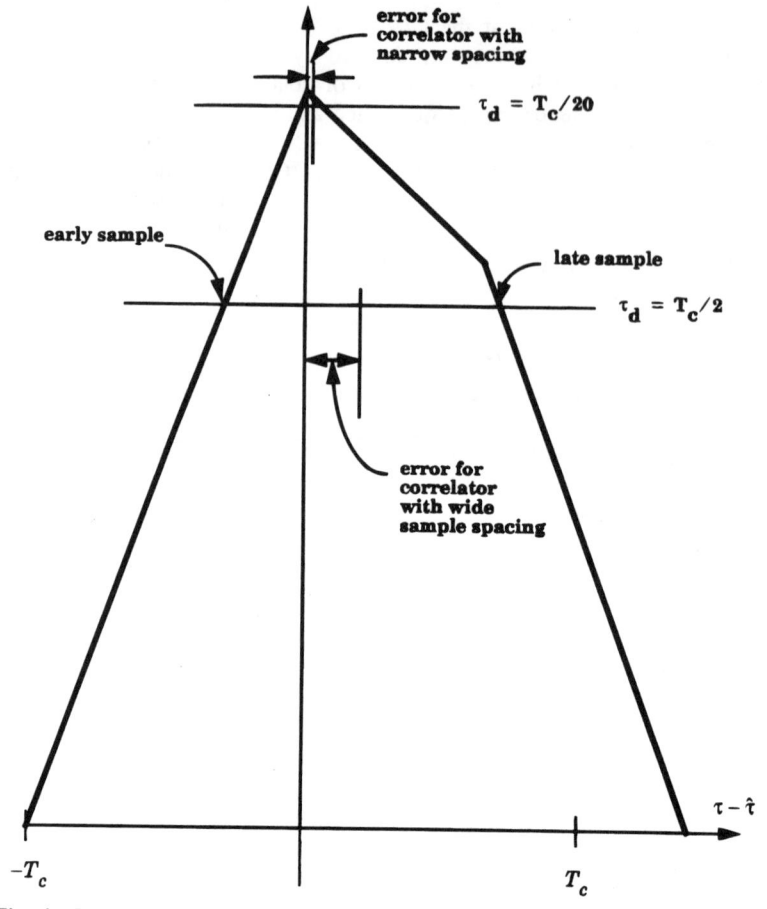

Fig. 4 Use of narrow correlator spacing to mitigate the effects of multipath.

If L_2 frequency measurements are used to track ionospheric delays, the time constant of the filter can be much greater, thus better precision can be achieved.

On a moving vehicle, the multipath correlation time may be very small (because the differential path length is changing rapidly), and carrier aiding may be quite effective in averaging any disturbances. In fact, antenna designs that intentionally randomize the phase difference between the direct and delayed signals are being considered for moving platforms. At fixed sites, the correlation time tends to be significantly longer; thus carrier aiding is not as effective. Of course, the antenna can be more carefully located and designed to have very low gain at low or negative elevation angles in order to combat multipath at a fixed location.

2. *Multipath and Narrow Correlator Spacing*

In addition to antenna selection and siting, there is a receiver-processing technique that can be used to mitigate (somewhat) the effects of multipath.

As discussed by Refs. 5–7, multipath interference can be reduced further by minimizing the time between early and late correlator samples. This is known as *narrow correlator spacing*. A sample of the correlation function in the presence of multipath is shown in Fig. 4. As shown, the multipath interference distorts the shape of the correlation function, which is symmetric in the absence of multipath. The advantage of narrow correlator spacing can be seen in the figure. If the correlators are separated by 1.0 T_c, then the early and late samples will settle at the location indicated, and the error caused by multipath can be quite large. In contrast, if the correlator spacing is 0.1T_c, then the correlator samples will settle near the peak, and the error will generally be smaller than 1 m.

3. Summary of Receiver Noise and Multipath Errors

Table 2 summarizes this class of ranging errors under the following assumptions:
1) The user has a state-of-the-art, multichannel receiver with a modern digital signal processor.
2) The reference station has taken great care to reduce multipath susceptibility, as described in the preceding subsections.
3) The magnitude of the random error in the reference station is also found in the user's receiver (but is uncorrelated) and multiplies this statistic by the square root of 2.

Table 2 Errors in DGPS caused by receiver noise and multipath for a well-designed user equipment receiving corrections from a well-designed reference station[a]

	Without DGPS Correction with or without SA Clock Dither		Zero Baseline Zero Latency DGPS (Type 3)		Decorrelation with Latency (Type 2)		Geographic Decorrelation (m/100 km) (Type 1)
	Bias (m)	Random (m)	Bias (m)	Random (m)	Vel. (m/s)	Accel. (m/s^2)	
Receiver noise	0.5	0.2	0.5	0.3	0.0	0.0	0.0
Multipath	0.3 to 3.0	0.2 to 1.0	0.3 to 3.0	0.2 to 1.0	0.0	0.0	0.0

[a]All effects are one-sigma errors. Bias is a steady value with persistence of more than 5 s. Random is the measurement-to-measurement variation in the user or reference receiver, sometimes called "white noise." Using carrier smoothing or other averaging techniques, the random errors can be significantly reduced in modern receivers.

Because these errors are all type 3, there is no decorrelation because of latency or separation, as indicated in the three right-hand columns in the table. (Later tables follow this same format.)

B. Satellite Clock Errors for Differential GPS

Satellite clock errors are differences in the true signal transmission time and the transmission time implied by the navigation message. In the absence of SA, these errors are small and change slowly. During periods when SA was not activated, clock errors of about 1–2 m and correlation times of about 5 min have been measured.

In the presence of SA, clock errors of 20–30 m are not unusual. *Differential corrections can be very effective against clock errors, because their validity decreases only with time and not with distance.* In other words, this error is exclusively type 2. Because SA has relatively large, fairly random velocity and acceleration magnitudes, it totally dominates the latency-induced error growth. The DGPS positioning error, therefore, grows as the DGPS correction ages.

Most DGPS implementations are relatively simple and predict future values of the pseudorange correction from the current values of the pseudorange and its rate. In this case, the residual pseudorange error growth attributable to SA is approximated as $1/2\ at^2$, where a is the rms acceleration (a random variable) of SA, and t is the age of the correction in seconds. Typically, SA has exhibited an acceleration a error in range (1σ) of about 4 mm/(s)^2. Consequently, if the latency is 10 s, then the pseudorange error (1σ) attributable to SA is expected to grow to approximately 0.2 m.

Somewhat more accurate DGPS systems have used system identification techniques in real time to build more sophisticated models of SA. Three good examples are the following: 1) a second-order Gauss–Markov model[8]; 2) an autoregressive moving average (ARMA) model[9]; and 3) a technique using autoregressive (AR) models and lattice filters.[10]

All of these models are still limited by the deliberate uncertainties in the *true*, presumably nonlinear SA model. However some improvement can be realized by transmitting the particular *estimator* model's parameters to the users as well as the measured current state of the SA offset. The user's receiver can then reconstruct the current approximation to the SA model to make more sophisticated predictions.

Figure 5 shows the standard deviation of the DGPS range error as a function of the age of the correction. The error for a user without differential corrections is the horizontal line at about 34 m.* The curve marked "two state" is for a differential user who employs the simple prediction based on the current value of SA and its rate. Finally, the curve marked "optimal prediction" is for a differential user who uses a more complete estimator model (such as referenced above) for predicting future values of SA. It assumes that the estimator model parameters have been "optimally" estimated by including the known statistics of SA.

As shown in Fig. 5, DGPS can reduce the pseudorange error provided that the correction is delivered promptly. Note that the initial growth of the error

---

*This value is larger than typical SA errors which are closer to 23 m (1σ).

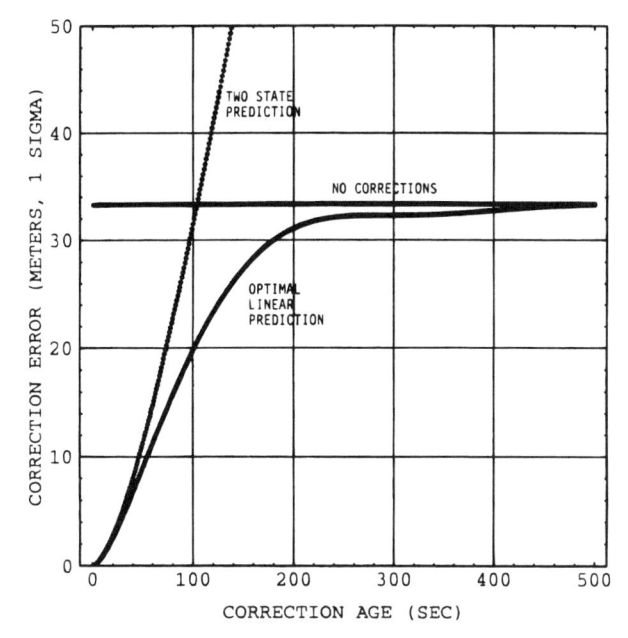

Fig. 5 Growth of horizontal differential GPS range error caused by selective availability as a function of the age of the correction.

(first 30 s) for both cases is parabolic: it grows as time squared. In fact, a delay of 20 s will lead to an error standard deviation of about 3 m in pseudorange, which corresponds to a (2 drms) positioning error of approximately 10 m.

The error for the optimal prediction never exceeds that for a user without corrections. In contrast, two-state prediction will give larger errors than nondifferential processing if the rate term is used to predict too far into the future. However, the error for the two-state prediction is very close to that for optimal prediction for smaller correction ages. For example, the differential error grows to over 10 m if the correction age exceeds 50 s, but for ages less than 50 s, the two-state prediction is almost as good as the optimal prediction.

Table 3 summarizes the statistics for DGPS satellite clock errors.

C. Satellite Ephemeris Errors for Differential GPS

As mentioned earlier, the navigation message contains errors. We have asserted that errors in the satellite clock data can be corrected by DGPS. Furthermore, these *clock corrections* are valid regardless of the distance between the monitor and the user. In other words, there is no decorrelation with displacement between reference and user. On the other hand, if the errors are in the satellite ephemeris data, then the validity of the corrections will decrease as the distance between the user and reference station increases.

In the appendix to this chapter, there is a detailed development of the *scalar* errors in DGPS ranging corrections as a vector function of the *vector* errors in

Table 3 Clock errors before and after differential GPS corrections[a]

	Without DGPS Correction with or without SA Clock Dither		Zero Baseline Zero Latency DGPS (Type 3)		Decorrelation with Latency (Type 2)		Geographic Decorrelation (m/100 km) (Type 1)
	Bias (m)	Random (m)	Bias (m)	Random (m)	Vel. (m/s)	Accel. (m/s^2)	
Satellite Clock Errors	21.0	0.1	0.0	0.14	0.21	0.004	0.0

[a]The random error is increased, because the noise in both the user and reference receivers are added after the corrections.

satellite position and the vector displacement (sometimes called the *baseline*) between reference station and user. The correct first-order expansion for these errors is the following:

$$\Delta E \triangleq \frac{-\Delta \bar{R}^T \cdot \bar{d}}{R} \qquad (10)$$

where $\Delta E \triangleq$ scalar error in DGPS correction; $\bar{d} \triangleq [\Delta \bar{r}_{r,u} - (\bar{1}_s \cdot \Delta \bar{r}_{r,u})\bar{1}_s]$; $\Delta \bar{R} \triangleq$ satellite position error vector; $R \triangleq$ range from reference station to satellite; $\bar{1}_s \triangleq$ unit vector from reference station to satellite; and $\Delta \bar{r}_{r,u} \triangleq$ displacement from reference station to user.

Equation (10) shows the decorrelation of errors as the user moves away from the reference station. If we consider the magnitudes of these quantities, the error in the DGPS correction is bounded by the following:

$$\max |\Delta \hat{\bar{\rho}}_R - \Delta \hat{\bar{\rho}}_u| = \Delta E \leq \frac{|\Delta \bar{r}_{r,u}| \cdot |\Delta \bar{R}|}{R}$$

This equation is valid for separations of less than 1500 km (the usual case for scalar corrections); larger separations are treated below. This error may be quite conservative depending upon the exact orientation of the vectors in Eq. (10). Further feeling for this "worst-case" relationship is given by Fig. 6, which plots worst-case user position error (after DGPS correction) vs satellite position error. For example, a 100-m satellite positioning error at 100 km separation between user and reference produces user errors *in the worst case* of less than 1 m.

1. Selective Availability Effects on Ephemeris

Typical satellite ephemeris errors are usually less than 10 m. Although SA could be applied by creating errors in the ephemeris message, this technique

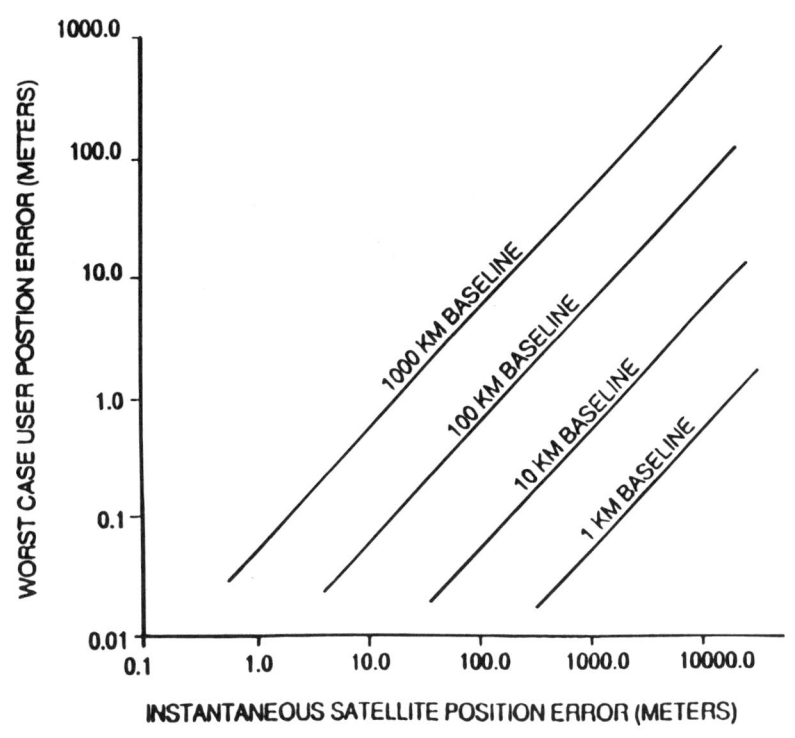

Fig. 6 Worst-case differential GPS errors for various distances from reference to user.

apparently has not been used. This is because any errors in the ephemeris would be slowly changing, and hence, strongly correlated over many minutes. Therefore, corrections for these errors would be valid for extended periods, which defeats the purpose of SA. It is assumed that the worst case, if SA were used, would limit the ephemeris message error to 100 m.

2. Maximum Errors

The maximum separation between a user and a reference station that can still have common view of all possible satellites is determined by their minimum elevation angles, or *mask* angles. Figure 7 shows a reference station and a user with a central angle separation of 142 deg (2.48 radians). This is the maximum common view separation, assuming the user and reference station both have elevation mask angles of 5 deg. The maximum errors caused by this extreme separation have three components that correspond to the three components of satellite ephemeris error before differential corrections. Even at this extreme separation, only the component parallel to the baseline (the vector between reference and user) is not completely canceled by scalar DGPS. This is emphasized in Table 4.

Table 5 summarizes the residual errors attributable to satellite ephemeris after DGPS corrections are applied.

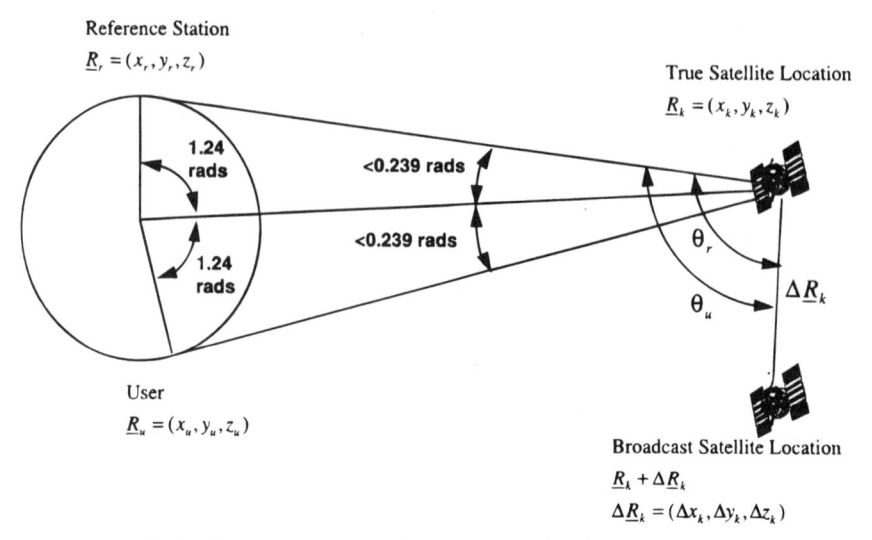

Fig. 7 Worst-case separation of user and reference station.

The small type 2 error only occurs when there are large ephemeris errors. The expected velocity and accelerations shown in Table 5 are limited by the ephemeris message, which acts as a low-pass filter on the error, effectively limiting the magnitude of these effects.

In summary, SA manipulation of the ephemeris data in the navigation message could cause larger spatial decorrelation of the DGPS correction, but such manipulation is unlikely to cause meaningful temporal decorrelation. If SA is not applied to the ephemeris message, this is a negligible source of error for DGPS, provided the user is within 500 km of the reference station.

D. Ionospheric Errors for Differential GPS

Free electrons in the ionosphere produce a group delay in the GPS signal, which is a significant error source. The ionosphere is usually modeled as a

Table 4 Impact of ephemeris errors at maximum separation for scalar corrections

Satellite Ephemeris Component	DGPS error at maximal baseline separation (12,040 km) as a percentage of raw error
Radial (away from earth center)	0%
Parallel to baseline ("in plane")	47%
Perpendicular to above ("out of plane")	0%

Table 5 Residual differential GPS errors for satellite ephemeris

	Without DGPS Correction with or without SA Clock Dither		Zero Baseline Zero Latency DGPS (Type 3)		Decorrelation with Latency (Type 2)		Geographic Decorrelation (m/100 km) (Type 1)
	Bias (m)	Random (m)	Bias (m)	Random (m)	Vel. (m/s)	Accel. (m/s^2)	
Ephemeris Errors -SA not applied	10.0 extreme case	0.0	0.0	0.0	negl.	negl.	< 0.05
Ephemeris errors -SA applied to Ephemeris	100.0 extreme case	0.0	0.0	0.0	< 0.01	< 0.001	< 0.5

relatively thin blanket located at about 350 km above the Earth. Its effective vertical delay varies from a few meters in the early morning hours to 10–20 m at the maximum, which occurs about 2 h past local solar noon. This vertical delay must be multiplied by an "obliquity factor," which accounts for the angle with which the signal penetrates the blanket.

Under extreme conditions, the ionosphere can delay the satellite signal by many tens of meters because of the following: 1) solar storms during periods of solar maximum; 2) low elevation angles (high obliquity factor); or 3) peak delay conditions in the early afternoon.

More typically, vertical delays throughout a 24-h period are in the 4–10 m range. Without differential corrections, about 50–75% of this error can be removed by using a standard model and coefficients available in the navigation message (see Refs. 11 and 12, and Chapter 12 in the companion volume). A dual-frequency, P-code receiver can directly measure the delay and make a correction that should be accurate to about 1 m. As long as *both* or *neither* the user and reference station make a dual-frequency correction, the impact on DGPS should be errors of less than 1 m for separations of less than 100 km. It should be noted that long-range users have been successful in using differential ionosphere and troposphere models. These have reduced the geographic correlation. An example is discussed at the end of this chapter along with other test results.

Differential corrections for ionospheric delays will be in error because of the following.

1) The GPS signals received by the reference and user pass through ("pierce") the ionosphere blanket at different locations.

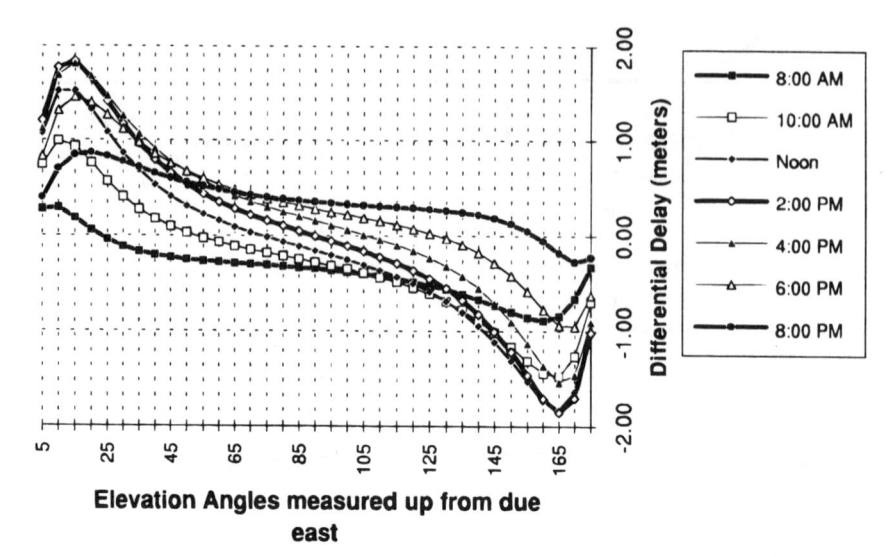

Fig. 8 The expected difference in ranging, m, attributable to the ionosphere for a 100-km separation due east. Each curve is for a different time of day at the reference station.

2) The incidence angle of the signal through the blanket is different (this is quantified by the obliquity factor).*

3) Latency provides outdated corrections (this is usually a smaller effect).

The impact of these effects is strongly a function of the time of day and has a small, relatively constant magnitude in the early morning hours.

1. Simulation of Ionospheric Decorrelations

With differential corrections, the size of the residual pseudorange error for the ionosphere depends most strongly upon the separation of the user and the reference station and the elevation angles of the satellites. Figure 8 predicts the size of this residual as a function of elevation angle and time of day. Larger separation distances will scale approximately linearly.

In this figure, the standard ionospheric model[12] is used to predict the signal delay at the reference station and at the user. As shown, as the elevation angle of the satellite decreases, the nominal ionospheric delay increases. If the user is assumed to be due east of the reference station, then the difference between the reference delay and the user delay also increases. Perhaps much of this residual delay can be modeled and removed (see test results in Sec. VII.B of this chapter); however, a residual error of $0.5 \times 10^{-6}|\Delta \bar{r}_{r,u}|$ to $5 \times 10^{-6}|\Delta \bar{r}_{r,u}|$ is expected (one sigma), where $|\Delta \bar{r}_{r,u}|$ is the reference station-to-user separation.

---

*The obliquity factor is the ratio of delays at any elevation angle to the vertical delay. It varies from 1.0 at 90 deg to about 3.0 at 5 deg. It is weakly a function of mean iononspheric height.

2. Measured Ionospheric Decorrelations

Ionospheric spatial decorrelation has been measured by Ref. 13, and these early measurements are summarized here. At ranges of 500 km, the residual errors were less than 1.8 m 95% of the time and less than 4.0 m 99% of the time. This effort to characterize differential residuals caused by the ionosphere is ongoing, and many years of data collection will be required for a complete characterization. However, these preliminary results suggest that the residual pseudorange error 1σ will be approximately $2 \times 10^{-6} |\Delta \bar{r}_{r,u}|$. As such, this spatial decorrelation is approximately equal to the decorrelation that would be introduced if the satellite ephemeris error were around 50 m.

3. Summary of Ionospheric Errors for Differential GPS

Under 50-km separation, the ionosphere is not a significant problem for dynamic DGPS systems. For the static surveyor, care should be taken beyond about 10 km, although the error (at two parts per million of range) is considerably better than a first-order survey. This is summarized in Table 6.

Table 6 Residual ionospheric errors for differential GPS[a]

	Without DGPS Correction with or without SA Clock Dither		Zero Baseline Zero Latency DGPS (Type 3)		Decorrelation with Latency (Type 2)		Geographic Decorrelation (m/100 km) (Type 1)
	Bias (m)	Random (m)	Bias (m)	Random (m)	Vel. (m/s)	Accel. (m/s^2)	
Ionospheric Errors (raw Ionosphere)	2 to 10 (times obliquity)	< 0.1 (times obliquity)	0.0	< 0.14	< 0.02	neglig.	< 0.2

[a]The values are typical 1σ estimates. The decorrelation with separation (between user and reference), although small, is the largest of the geographic decorrelation tersm.

E. Troposphere Errors for Differential GPS

The index of refraction of the lower atmosphere under standard conditions is not quite unity (it is typically 1.0003), and it depends upon temperature, pressure, and humidity. At low satellite elevation angles (below about 10 deg), tropospheric refraction can result in significant delays. Fortunately, most of this delay can be removed using a simple model that depends only upon satellite elevation angle and not on the local pressure, temperature, or humidity.

Without differential corrections, this model typically removes 90% of the delay, but the unmodeled error can reach 2–3 m at about 5 deg elevation. With differential corrections, the residual error is almost always very small. However, if the signal ray paths to the user and reference station traverse volumes with significantly different meteorological parameters, the error could be troublesome for demanding applications. For example, if the reference station and user are at significantly different altitudes (several thousand feet), then variations in the index of refraction could be significant. In these cases, the DGPS user should apply a differential tropospheric model that accounts for the altitude difference. These sensitivities are summarized in Table 7.

F. Local Area Differential GPS Error Summary

For convenience, the effects of various error sources on DGPS are summarized in Table 8. Two comments should be made. First, a poorly designed DGPS system will consistently be worse than these estimates of performance. Design deficiencies can occur in many elements, but the most common problems tend to be associated with the DGPS communications link. These are treated in more detail in Sec. VI of this chapter. Second, the table summarizes expected (one-sigma) values of pseudorange error. Because many of the underlying error sources are random, there will be times when they are better and times when they are worse.

Figure 9 shows the *ranging* error growth with latency *and* distance (it conservatively assumes that SA is applied to both clock and ephemeris). The *position* error suffered by a GPS (or DGPS) user is proportional to these pseudorange measurement errors, but it also depends upon the geometry of the user and the satellites. As discussed in Chapters 2, 5, and 11 of the companion volume, the measures that describe the degradation caused by geometry are known as dilution of precision (DOP) values. In fact, the position error is approximately equal to the relevant DOP value times the pseudorange error.

Many users are not comfortable with one-sigma values of pseudorange error and prefer the two-sigma values of positioning error that approximate the 95th

Table 7 Residual differential GPS user pseudorange errors caused by tropospheric delays

	Without DGPS Correction with or without SA Clock Dither		Zero Baseline Zero Latency DGPS (Type 3)		Decorrelation with Latency (Type 2)		Geographic Decorrelation (m/100 km)
	Bias (m)	Random (m)	Bias (m)	Random (m)	Vel. (m/s)	Accel. (m/s^2)	(Type 1)
Tropospheric errors	2 multiplied by obliquity	<0.1 multiplied by obliquity	0.0	<0.14	neglig.	neglig.	<0.1

DIFFERENTIAL GPS

Table 8 Summary of residual differential GPS pseudorange errors

	Without DGPS Correction		Zero Baseline Zero Latency DGPS (Type 3)		Decorrelation with Latency (Type 2)		Geographic Decorrelation (meters/100 km) (Type 1)
	Bias (meters)	Random (meters)	Bias (meters)	Random (meters)	Velocity (m/s)	Acceler. (m/s^2)	
Receiver Noise	0.5	0.2	0.5	0.3	0.0	0.0	0.0
Multipath	0.3 to 3.0	0.2 to 1.0	0.4 to 3.0	0.2 to 1.0	0.0	0.0	0.0
Satellite Clock Errors	21.0	0.1	0.0	0.14	0.21	0.004	0.0
Satellite Ephemeris Errors S/A not applied	10.0 extreme case	0.0	0.0	0.0	negl.	negl.	<0.05
Satellite Ephemeris errors -S/A applied to Ephemeris	100.0 extreme case	0.0	0.0	0.0	<0.01	<0.001	<0.5
Ionospheric Errors (raw Ionosphere)	2 to 10 meters (times obliquity)	<0.1 (times obliquity)	0.0	<0.14	<0.02	neglig.	<0.2
Tropospheric errors	2 meters (times obliquity)	<0.1 (times obliquity)	0.0	<0.14	neglig.	neglig.	<0.2

percentile; that is, the position error that is expected to be exceeded no more than 5% of the time. This is called the 2drms value. An example for horizontal positioning is as follows:

$$\text{Horizontal 2drms} = 2\sqrt{\sigma_x^2 + \sigma_y^2} = 2 \text{ HDOP} \cdot \sigma_p$$

If the application requires three-dimensional positioning, then we use the following:

$$\text{Spherical 2drms} = 2\sqrt{\sigma_x^2 + \sigma_y^2 + \sigma_z^2} = 2 \text{ PDOP} \cdot \sigma_p$$

Because a typical value for horizontal *DOP* is 2.0, the 2drms horizontal position error for a user without differential corrections is about 85 m.* In contrast, the 2drms horizontal accuracy for a differential user within 50 km of the reference station varies from 1 to 5 m, depending upon the age of the correction and the quality of the system. A conservative LADGPS error budget is shown in Table 9.

---

*While this is driven by the SA-induced satellite clock errors that have been experimentally observed, there is no guarantee that it could not be larger (or smaller). The agreement with the U.S. Department of Defense (DOD) is that it will not exceed a 2drms horizontal error value of 100 m.

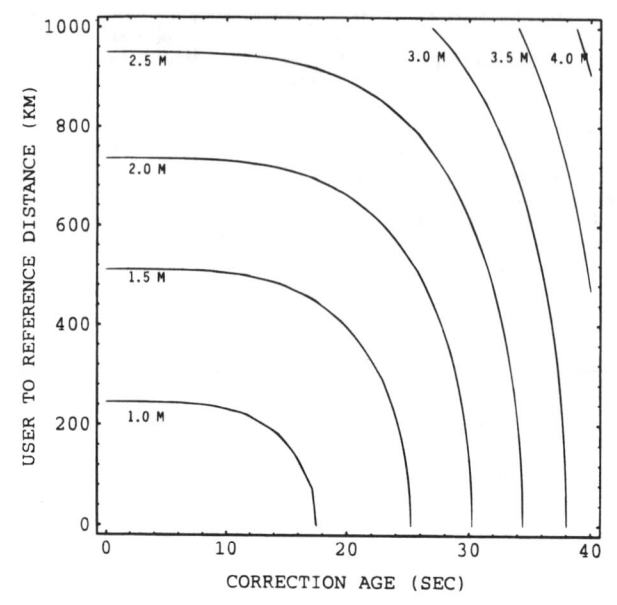

Fig. 9 Growth in pseudorange errors from age (latency) and distance.

Table 9 Differential GPS error budget for users within 50 km of the reference station

Error Source	1 sigma Error (meters)		
	Bias	Random	Total
Ephemeris Data	0.0	0.0	0.0
Satellite Clock	0.0	0.7	0.7
Ionosphere	0.0	0.5	0.5
Troposphere	0.0	0.5	0.5
Multipath	1.0	1.0	1.4
Receiver Measurement	0.0	0.2	0.2
Reference Station Errors	0.3	0.2	0.4
UERE (RMS)	1.0	1.4	1.8
Filtered UERE (RMS)	1.0	0.4	1.1
1 sigma errors--vertical	VDOP = 2.5		2.8
--horizontal	HDOP = 2.0		2.2

DIFFERENTIAL GPS

IV. Carrier-Phase Differential GPS

As mentioned in the introduction to this chapter, users with very stringent accuracy requirements may be able to use carrier-phase DGPS. These users measure the phase of the GPS carrier and compare that to the carrier phase measured at a reference site. This process can achieve range measurement precisions that are a few percentage points of the carrier wavelength (less than 0.5 cm). However, the antenna separation usually exceeds one wavelength (19 cm). The estimated position (or attitude) is, therefore, ambiguous, because the number of integer wavelengths contained in the phase difference is unknown. To be useful, carrier-phase DGPS requires resolution of this integer ("$n\lambda$" or $N_{u,s} \cdot \lambda$) ambiguity.

This section introduces carrier-phase DGPS by discussing: 1) attitude determination based on GPS phase interferometry; 2) static and kinematic survey based on single- and double-phase differences; and finally 3) techniques for resolving the integer ambiguity. All of these techniques and applications are introductory, with deeper coverage in subsequent chapters.

A. Attitude Determination

The difference in the GPS carrier phase received at two or more nearby antennas can be used to determine the attitude of a vessel or aircraft. This type of phase interferometer is shown in Fig. 10, which depicts two antennas, r and u. These antennas are connected to a rigid body, and they are constrained by the body-fixed vector $\Delta \bar{r}_{u,r}$.

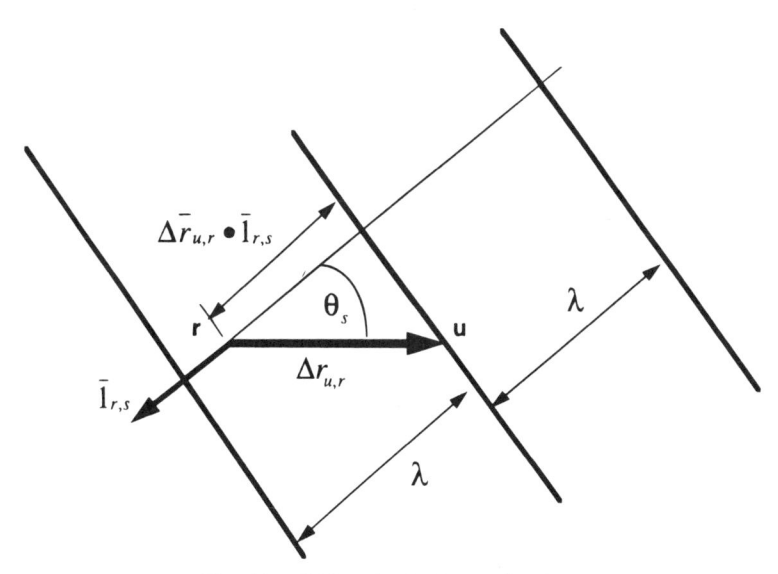

Fig. 10 GPS attitude determination.

A single-phase difference is measured as follows:

$$\Delta\phi_s = \phi_{u,s} - \phi_{r,s}$$
$$= \Delta\bar{r}_{u,r} \cdot \bar{1}_{r,s} + b_{u,r} + (N_{u,s} - N_{r,s}) \cdot \lambda + v_{u,s}^{(\phi)} - v_{r,s}^{(\phi)}$$
$$= |\Delta\bar{r}_{u,r}| \cdot \cos\theta_s + b_{u,r} + (N_{u,s} - N_{r,s}) \cdot \lambda + v_{u,s}^{(\phi)} - v_{r,s}^{(\phi)} \quad (11)$$

The quantity $b_{u,r}$ is the constant, but unknown, "line-bias" between the two antennas. In addition, the quantity $(N_{u,s} - N_{r,s})$ is an integer ambiguity in the number of whole wavelengths that may exist in the measured phase difference. This integer ambiguity exists whenever $|\Delta\bar{r}_{u,r}| \geq \lambda$. If the problem is three-dimensional, and only a single pair of antennas are used, then rotations about two perpendicular axes of rotation are required to solve for these unknowns. Fortunately, if three antennas are placed on two perpendicular baselines, then a single rotation of the vehicle (for example, an aircraft) can be used to resolve all these nuisance parameters.[14]

After the integer ambiguities are resolved and the biases are calibrated the only remaining measurement errors are receiver noise, interference, and local multipath. The other error sources are effectively eliminated, because the proximity of the antennas ensures that none of the error mechanisms that result from spatial separation is significant. Furthermore, in the preferred implementation, both antennas feed a single co-located receiver. Therefore, latency is negligible, and temporally decorrelated errors are not significant.

In a sense, then, attitude determination is the ultimate in DGPS, because the errors encountered are the ultimate noise floor of any DGPS system. As discussed in Ref. 14, the standard deviation of the angle errors introduced by noise is as follows:

$$\sigma_\phi = \frac{\lambda}{2\pi \cdot |\Delta\bar{r}_{u,r}|} \cdot \sqrt{\frac{BW_N}{C/N_0}} \quad (12)$$

where BW_N is the noise equivalent bandwidth, and C/N_0 is the carrier power-to-noise power density ratio. Additionally, the standard deviation of the angle errors introduced by multipath is as follows:

$$\sigma_\phi \leq \sigma_R/|\Delta\bar{r}_{u,r}| \quad (13)$$

where σ_R is the differential range error caused by multipath. At signal-to-noise ratios (SNR) and noise bandwidths that are reasonable for GPS, *these two error sources result in predicted errors less than a tenth of a degree* for baselines of 10 m. Such results have been obtained in practice.[14]

B. Static and Kinematic Survey

For carrier-phase DGPS, the mobile user receives the carrier-phase measurements (raw or reduced) from the reference station and uses them to form single and then double differences.[15] The single difference is between user and reference; the second difference is among locally received satellites. Basically, single differences are used to eliminate strongly correlated error sources, and double differences are used to eliminate the difference between reference and user clocks.

Further processing is required to resolve the "integer ambiguity." From Eq. (11), the single differences are given by the following:

$$\Delta\phi_{s1} = \phi_{u,s1} - \phi_{r,s1} = [(|\bar{r}_{u,s_1}| + b_u - B_{s_1} - I_{u,s_1} + T_{u,s_1} + N_{u,s_1} \cdot \lambda + v_{u,s_1}]$$
$$- [(|\bar{r}_{r,s_1}| + b_r - B_{s_1} - I_{r,s_1} + T_{r,s_1} + N_{r,s_1} \cdot \lambda + v_{r,s_1}]$$
$$\cong |\bar{r}_{u,s_1}| - |\bar{r}_{r,s_1}| + (b_u - b_r) + (N_{u,s_1} - N_{r,s_1}) \cdot \lambda + (v^{(\phi)}_{u,s_1} - v^{(\phi)}_{r,s_1})$$
$$\cong \Delta\bar{r}_{r,u} \cdot \bar{1}_{r,s_1} + (b_u - b_r) + (N_{u,s_1} - N_{r,s_1}) \cdot \lambda + (v^{(\phi)}_{u,s_1} - v^{(\phi)}_{r,s_1})$$
$$= |\Delta\bar{r}_{r,u}| \cdot \cos\theta_{r,s_1} + (b_u - b_r) + (N_{u,s_1} - N_{r,s_1}) \cdot \lambda + (v^{(\phi)}_{u,s_1} - v^{(\phi)}_{r,s_1}) \quad (14)$$

The second equation assumes that the errors attributable to the ionosphere, troposphere, and satellite clock cancel completely, because the distance between user and reference station is small (tens of kilometers) and the corrections are prompt (a few seconds at most). The third equation assumes that the vectors from the user to the satellite are parallel to the vector from the reference station to the satellite. This small angle assumption is valid for small separations.

The user wants to estimate the position of the user relative to the reference station. In other words, the estimates of interest are the three components of $\Delta\bar{r}_{u,r}$. The angle between the reference station and the satellite θ_{r,s_1} is known with adequate accuracy from the satellite navigation message. The clock difference $(b_u - b_r)$ and the integer ambiguity $(N_{u,s_1} - N_{r,s_1})$ are unknown nuisance parameters. Finally, the observation noise $(v^{(\phi)}_{u,s_1} - v^{(\phi)}_{r,s_1})$ has zero mean and very small standard deviation (a few centimeters).

The double differences for two satellites are given by the following:

$$\nabla\Delta\phi_{s_1,s_2} = \Delta\phi_{s_1} - \Delta\phi_{s_2} = |\Delta\bar{r}_{r,u}| \cdot (\cos\theta_{r,s_1} - \cos\theta_{r,s_2}) + N_{s_1,s_2} + v^{(\phi)}_{s_1,s_2}$$

where

$$N_{s_1,s_2} = N_{u,s_1} - N_{r,s_1} - N_{u,s_2} + N_{r,s_2}$$
$$v^{(\phi)}_{s_1,s_2} = v^{(\phi)}_{u,s_1} - v^{(\phi)}_{r,s_1} - v^{(\phi)}_{u,s_2} + v^{(\phi)}_{r,s_2} \quad (15)$$

As shown, the double differencing eliminates the difference between the user and reference clock. As such, one of our two nuisance parameters has been removed. However, it still suffers the integer ambiguity; thus double differences from different times (over about 30 min) are typically used by surveyors to eliminate this final nuisance parameter.*

Double differences between multiple pairs of satellites over two adequately long periods provide enough information to solve for the three components of $\Delta\bar{r}_{u,r}$. The number of satellites and double differences required depends upon whether or not the user and reference are moving. In general, N satellites provide $N-1$ independent double differences for each observation time. If the observations are adequately spaced in time, then N satellites provide $2(N-1)$ independent

---

*As we might expect, "triple differences" are the differences between double differences measured at different times. Such differences do not suffer from integer ambiguity provided that all carrier cycles have been counted; that is, no cycle slips have occurred. This is explained further in Chapter 18 of this volume.

equations. In the static case, there are 3 + (N-1) unknowns, so four satellites are required.

C. Near-Instantaneous Determination of Integers

The commercial market for carrier-phase DGPS has concentrated on survey applications, because they do not require position fixes in real time. However, many other applications would like to use the accuracy of carrier-phase DGPS, but they require that the accuracy be achieved almost instantaneously. Specifically, Category II and III precision landing of aircraft may well benefit from carrier-phase DGPS, but they require that the integer ambiguities be reliably identified during approach so that the accuracy is at the submeter level in the final landing phase.

Two likely concepts for identifying the integers in real time are summarized here. The first concept uses pseudosatellites or "pseudolites." These are ground-based transmitters that generate a GPS signal, typically at the primary frequency (L_1). The pseudolite may be placed under the approach path so that the aircraft passes over it perhaps 1000–5000 m from the touchdown point. With such a placement, the line-of-sight vector from the aircraft to the pseudolite sweeps out a large angle as the aircraft passes over it. This large change of direction can serve the same function as the satellite motion used by surveyors to resolve integer ambiguities. Although the surveyor must wait at least 30 min to experience such a sweep, the angular change in the landing scenario is created in less than a minute by the rapid motion of the aircraft over a pseudolite below. This technique has achieved close to a 100% demonstrated success factor in solving for the integers. Resulting demonstrations in aircraft are providing centimeter-level, three-dimensional positioning accuracy. This technique is the basis of the "Integrity Beacon Landing System," which is briefly discussed later in this chapter and detailed in Chapter 15 of this volume.

The second concept is called "wide-laning." It is a method to simplify the *multiple-satellite search* for feasible integer ambiguities. This technique has also demonstrated the ability to greatly reduce the time required for integer determination. Wide-laning is done by multiplying and filtering the L_1 and L_2 signals to form a beat frequency signal. This beat frequency is equal to 347.82 MHz and has a wavelength of 86 cm, which is significantly longer than the wavelength of either the L_1 (19 cm) or L_2 (24 cm) carriers. Consequently, resolution of the integers can be accomplished by using code observations to determine the integer ambiguity of the wider "lanes" formed by the beat frequency signal. These, in turn, greatly reduce the volume that must be searched for the L_1 integers. A difficulty with this method is that L_2 is usually broadcast with encrypted modulation. As a result, sophisticated techniques of cross correlation, squaring, or partially resolving the encryption must be used. All of these tend to be noisy, especially because L_2 has less power than the primary L_1 signal. The wide-laning resolution of integers may, therefore, not be as reliable as mandated by some of the more stringent integrity requirements. Wide-laning is further discussed in Chapter 18 of this volume.

V. Radio Technical Commission for Maritime Services Data Format for Differential GPS Data

Several data formats have been established for the transmission of DGPS corrections. For example, a proposed format for data transmitted by the WAAS is described in Chapter 4, this volume. However, the standards established by Special Committee 104 of the Radio Technical Commission for Maritime Services (RTCM) have been the most widely used,[16,17] and this section summarizes those standards. The committee report recommends which data to broadcast, a message format, and a set of rules for applying the corrections.

Some existing, private DGPS use variants of the RTCM message or entirely different formats. Examples include applying a standard of 32- rather than 30-bit words and reducing the overhead associated with the headers.

The RTCM messages contain pseudorange corrections and range-rate corrections for each satellite in view of the reference station. They also include satellite health, estimated accuracy, and the "age" of the data being used by the reference station. The format also allows for messages that contain auxiliary data.

All RTCM messages consist of a string of 30-bit words, and a typical correction message for six satellites is about 16 words or 480 bits in length. Each 30-bit word consists of 24 databits and 6 parity bits. The parity scheme is the same ($n = 32$, $k = 24$, $d = 4$) extended Hamming code used in the GPS navigation message, and it is described in Ref. 18. Transmission errors are usually detected by the parity algorithm, and erroneous messages are discarded by the DGPS receiver. This action protects the DGPS user from incorrect data, but it increases message delay, which in turn decreases the accuracy of the corrections.

Also common to all RTCM messages is the two-word header shown in Fig. 11. As shown, the header includes: an 8-bit preamble; message type, which identifies which of 64 message types is being sent; and the station identification of the DGPS reference station. It also includes modified Z-count, which is the reference time for the message parameters; sequence number for the message,

FIRST WORD OF EACH MESSAGE

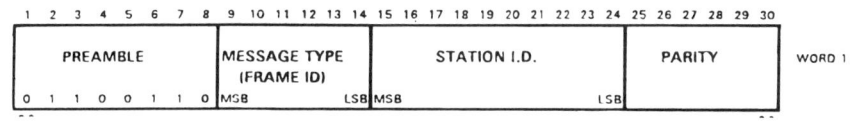

1 2 3 4 5 6 7 8	9 10 11 12 13 14	15 16 17 18 19 20 21 22 23 24	25 26 27 28 29 30	
PREAMBLE	MESSAGE TYPE (FRAME ID)	STATION I.D.	PARITY	WORD 1
0 1 1 0 0 1 1 0	MSB LSB	MSB LSB		

First Bit Transmitted   Last Bit Transmitted

SECOND WORD OF EACH MESSAGE

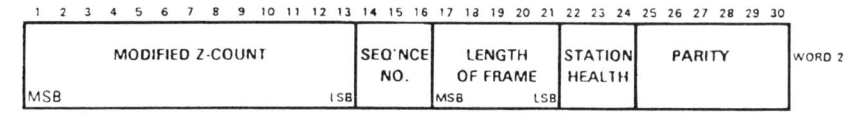

1 2 3 4 5 6 7 8 9 10 11 12 13	14 15 16	17 18 19 20 21	22 23 24	25 26 27 28 29 30	
MODIFIED Z-COUNT	SEQ'NCE NO.	LENGTH OF FRAME	STATION HEALTH	PARITY	WORD 2
MSB LSB		MSB LSB			

Fig. 11 Format of the standard Radio Technical Commission for Maritime Services header for differential GPS messages.

which aids in frame synchronization; length of message, which is the message length in words; and the reference station health.

The RTCM message types are listed in Table 10, and they are detailed in Ref. 16. Of these, the Type 1, 2, 9, 18, 19, 21, and 22 messages are the most important and are summarized below.

A. Radio Technical Commission for Maritime Services Message Types 1, 2, and 9

Message Types 1, 2, and 9 all use the format shown in Fig. 12 and all provide pseudorange corrections (PRC) and range-rate corrections (RRC).

As shown, five words are required to send the corrections for three satellites. As such, the length of a Type 1, 2, or 9 message in bits is given by the following:

$$M_{1,2,9} = 30(2 + \lceil 5N/3 \rceil) \qquad (16)$$

where $\lceil x \rceil$ denotes the least integer, which is greater than x, and N is the number of satellites in view of the reference station. These messages assume that the receiver will add the following correction to the measured pseudorange for the appropriate satellite:

$$\text{PRC}(t) = \text{PRC}(t_0) + (t - t_0) \cdot \text{RRC}(t_0) \qquad (17)$$

where t is the time at which the correction is applied, and t_0 is the Z-count contained in the second word of the header.

The scale factor assigns a value to the least significant bit (LSB) used by the corrections. If the scale factor is set to 0, then the LSB of the pseudorange

Table 10 DGPS message types defined by RTCM special committee 104

Message type number	Current status	Title
1	Fixed	Differential GPS Corrections
2	Fixed	Delta Differential Corrections
3	Fixed	Reference Station Parameters
4	Tentative	Surveying
5	Tentative	Constellation Health
6	Fixed	Null Frame
7	Tentative	Beacon Almanacs
8	Tentative	Pseudolite Almanacs
9	Fixed	High-Rate Differential Corrections
10	Reserved	P-Code Differential Corrections
11	Reserved	C/A Code L1, L2 Delta Corrections
12	Reserved	Pseudolite Station Parameters
13	Tentative	Ground Transmitter Parameters
14	Tentative	Surveying Auxiliary Message
15	Tentative	Ionosphere (Troposphere) Message
16	Fixed	Special Message
17	Tentative	Ephemeris Almanac
18 to 59	—	Undefined
60 to 63	Reserved	Differential Loran-C Messages

DIFFERENTIAL GPS

SCALE FACTOR ↓					
UD RE	SATELLITE ID	PSEUDORANGE CORRECTION		PARITY	WORDS 3, 8, 13 OR 18

		SCALE FACTOR ↓				
RANGE-RATE CORRECTION	ISSUE OF DATA (IOD)	UD RE	SATELLITE ID	PARITY	WORDS 4, 9, 14 OR 19

PSEUDORANGE CORRECTION	RANGE-RATE CORRECTION	PARITY	WORDS 5, 10, 15 OR 20

SCALE FACTOR ↓					
ISSUE OF DATA (IOD)	UD RE	SATELLITE ID	PSEUDORANGE CORR. (UPPER BYTE)	PARITY	WORDS 6, 11, 16 OR 21

PSEUDORANGE CORR. (LOWER BYTE)	RANGE-RATE CORRECTION	ISSUE OF DATA (IOD)	PARITY	WORDS 7, 12, 17 OR 22

o o o

RANGE-RATE CORRECTION	ISSUE OF DATA (IOD)	FILL	PARITY	WORDS N+2 IF N1 = 1, 4, 7 OR 10

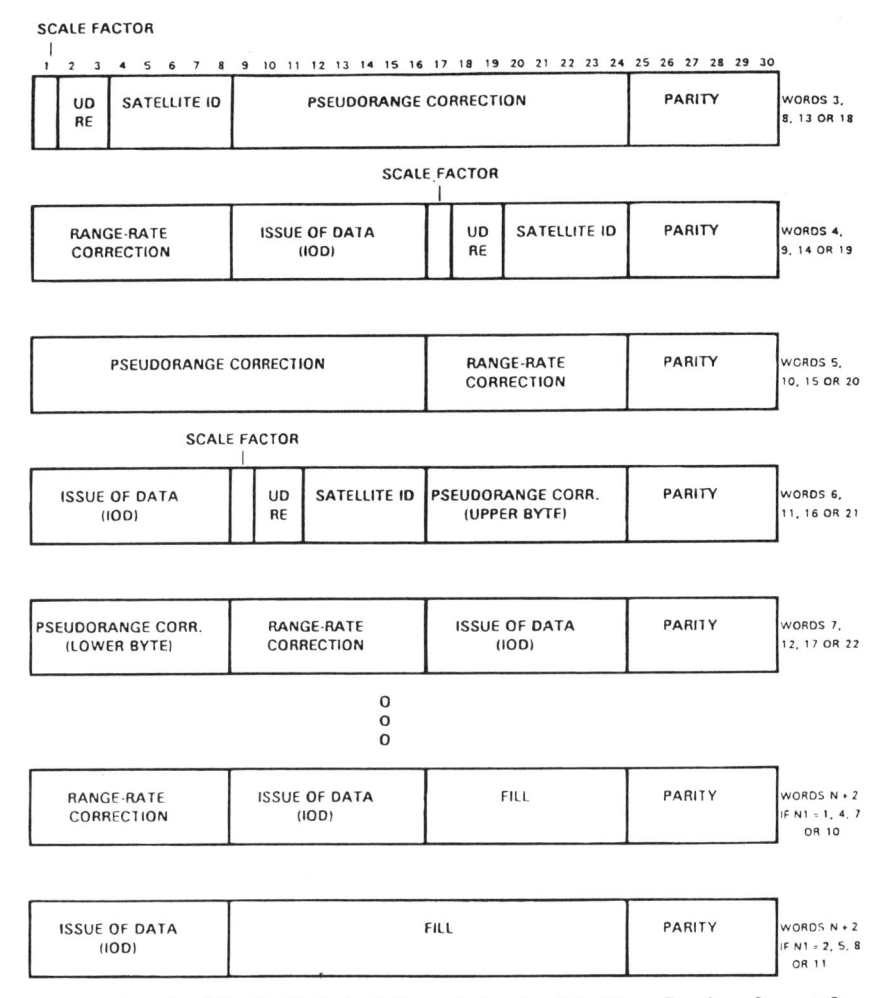

ISSUE OF DATA (IOD)	FILL	PARITY	WORDS N+2 IF N1 = 2, 5, 8 OR 11

Fig. 12 Standard Radio Technical Commission for Maritime Services format for message Types 1, 2, and 9.

correction is 0.02 m, and the LSB of the range rate correction is 0.002 m/s. In this case, the ranges of the pseudorange and range-rate corrections are ± 655.36 m and ± 0.256 m/s. If the scale factor is set to 1, then the LSB values change to 0.32 m and 0.032 m/s, respectively, and the ranges change to ± 10485.76 m and ± 4.096 m/s.

The issue of data (IOD) in the type 1, 2, and 9 messages identifies the timing of the data in the GPS navigation message used by the reference station. It is included in the message so that the user equipment can ensure that it is using clock and ephemeris data from the same navigation message that the reference station is using. The type 1 message uses the most recent navigation data available to the reference station. After a change in the IOD of the navigation message, type 2 messages are interleaved with type 1 messages, and they contain corrections based on the old clock and ephemeris data. In this way, they provide a bridge for users who have yet to acquire the new navigation data.

Type 9 messages differ from type 1 and 2 messages because they can be used to send corrections for a subset of the satellites in view of the reference station; whereas type 1 and 2 messages contain corrections for all satellites in view. Type 9 messages can be used to send corrections for any satellites suffering from high correction rates relative to the rest of the satellites. This helps maintain DGPS system accuracy. Type 9 can also be used to reduce the impact of transmission errors on noisy DGPS broadcast channels.[19]

B. Type 18, 19, 20, and 21 Messages

The RTCM message types 18, 19, 20, and 21 are for the real-time broadcast of differential carrier phase corrections, and they are sometimes known as the "real-time kinematic messages." These messages have three-word headers instead of two. The first two words of the header are the same as described above, and the third word contains a GPS time-of-measurement field that increases the resolution of the Z-count in the second word of the header. After the three-word header, each message contains two words for each satellite. Thus, the total message length in bits is given by the following:

$$M_{18,19,20,21} = 30(3 + 2N) \qquad (18)$$

where N is the number of satellites in view of the reference station.

For type 18 and 19 messages, the two words contain the raw carrier and pseudorange measurements made by the reference receiver. For type 20 and 21 messages, the two words contain measurements that have been corrected by the satellite ephemerides contained in the navigation message.

The two words also carry flags that indicate whether the corrections are for L_1 or L_2: C/A- or P-code; half or full wave L_2 carrier-phase measurements; or ionospheric free pseudorange. The flags also describe the carrier-smoothing interval for the pseudoranges and pseudorange corrections.

VI. Datalinks

This brief summary of DGPS broadcast techniques is an updated version of the more complete survey provided in Ref. 20. In general, most broadcast alternatives may be categorized as follows: 1) low- and medium-frequency (LF and MF) groundwave systems; 2) VHF and UHF networks; and 3) mobile satellite communications.

These alternatives are discussed in the following subsections. Two specific broadcast techniques of importance are not treated here because they are covered in separate chapters. These are pseudolites, which are the subject of Chapter 2 of this volume, and the WAAS, which is the subject of Chapter 4 of this volume.[21]

A. Groundwave Systems

Low-frequency, medium frequency, and high-frequency (HF) designate the bands from 30–300 kHz, 300–3000 kHz, and 3–30 MHz, respectively. All three bands are characterized by groundwave and skywave propagation, where the groundwave follows the surface of the Earth and the skywave reflects off the ionosphere.

Low-frequency and MF groundwave propagation have been very successfully used to broadcast DGPS data. At these frequencies, the groundwave affords reliable and predictable coverage well beyond the radio horizon, where the DGPS corrections themselves are still valid. As such, groundwave systems are well matched to the DGPS application. Two specific groundwave systems are described in the following subsections.

High-frequency broadcast of DGPS corrections is not commonplace for a number of reasons. First, the HF groundwave is rapidly attenuated by propagation over land or over any significant land segments. Second, skywave systems suffer from a "skip zone," so the transmitter must be located outside of the coverage area. Third, multiple skywaves can destructively interfere and cause signal "fades." Indeed, a reliable skywave broadcast would require frequency diversity, and multiple frequencies are difficult to license in the active HF band.

1. Marine Radiobeacons

Existing marine radiobeacons are being used to broadcast DGPS corrections to marine users. The correction data use minimum shift keying (MSK) to modulate the signal digitally from marine radiobeacons (which operate in the 283.5–325 kHz band) creating "DGPS/radiobeacons."

A DGPS/radiobeacon broadcast network is attractive for many reasons:

1) Because marine radiobeacons have been used by mariners for direction finding for many years, they are widespread and well located for critical navigation functions. Fortunately, the DGPS modulation can be added to the broadcast without interfering with the original direction-finding function of the signal.[22] Additionally, the DGPS function falls within the primary radionavigation allocation of the marine radiobeacon band.

2) The DGPS/radiobeacon *service* will be inexpensive. Broadcast of DGPS data from an existing radiobeacon requires only the addition of a GPS reference station, an MSK modulator, and an integrity monitor. Even in those cases where a new radiobeacon is required, the transmitter and antenna are relatively inexpensive.

3) The DGPS/radiobeacon *receiver* is inexpensive to manufacture. In fact, the first commercial receiver was available in the summer of 1991 for around $7000. By the summer of 1992, a competing product was available for $3500. The price had fallen to $1900 by the spring of 1993, and a low-capability receiver was available for $500 in the spring of 1994.

4) The DGPS/radiobeacon signal propagates in the groundwave mode and can be received reliably at ranges well beyond the visual horizon. Atmospheric noise eventually limits the range of most DGPS/radiobeacons, but the seawater range of experimental installations is usually over 300 km.

For these reasons, DGPS/radiobeacons are being installed worldwide. Indeed, the International Association of Lighthouse Authorities (IALA) has worldwide responsibility for marine radiobeacons, and they have established an international standard for the DGPS/radiobeacon signal. Finland and Sweden established the first operational DGPS/radiobeacon system, which helps guide car ferries across the Baltic Sea from Stockholm to Helenski.

By 1996, the U.S. Coast Guard will have deployed over 50 DGPS/radiobeacons to cover the coastal areas of the conterminous U.S. (CONUS), Hawaii, Puerto Rico, the Great Lakes, and much of Alaska. The Coast Guard system will include coverage of Prince William Sound, where it will be part of a Vessel Traffic System designed to prevent recurrences of the Exxon Valdez tragedy. Differential GPS/radiobeacons will also be placed to cover many inland waterways such as the Mississippi, Ohio, and Missouri Rivers, where they will help the U.S. Army Corps of Engineers maintain these routes. Other nations deploying DGPS/radiobeacon systems include Australia, Norway, Germany, the Netherlands, England, Poland, and Egypt.

2. *2 MHz Groundwave Systems*

A number of commercial DGPS broadcast systems employ groundwave propagation using frequencies around 2 MHz. The band from 1.8–2.0 MHz has been designated for commercial radio location purposes. These systems use frequency diversity, forward error correction, and other forms of temporal diversity to combat the atmospheric noise and fading that characterize the upper MF band. With such precautions, these broadcasts can have groundwave ranges of up to 700 km over seawater, and they have become very popular in the survey industry.

B. VHF and UHF Networks

1. Operating Ranges

Radios that operate in the very high-frequency (VHF, 30–300 MHz) and ultrahigh-frequency (UHF, 300–3000 MHz) ranges can reliably communicate data over short distances. Very roughly, a VHF or UHF radio can communicate to the radio horizon, which is equal to $D(km) = 4.12 (\sqrt{h_1(m)} + \sqrt{h_2(m)})$, where $h_1(m)$ and $h_2(m)$ are the heights of the receiving and transmitting antennas over a spherical Earth.

Over water, "ducting" phenomena make possible reliable signal reception over longer ranges than predicted by this formula. For example, a 4-W transmitter on the coast at water level can usually be received by a ship 40 miles from the coast. However, overland propagation is limited by line of sight, and range prediction over land requires a detailed path profile. Propagation along a coast also requires an analysis of the path profile to determine whether any obstructions exist.

Very-high-frequency and UHF radio channels are afflicted by a number of generic problems. First, the VHF or UHF signal can be shadowed by valleys, hills, buildings, or even trees. Second, multipath fading can impair quality, particularly if the mobile receiver is at relatively long range. Such fading can be mitigated using antenna, frequency, or time diversity.

At shorter ranges, spread-spectrum equipment that conforms to Part XV of the Federal Communications Commission (FCC) regulations can be used. Such an option is attractive because Part XV allows unlicensed use of the 902–928 MHz band, provided that the total transmitted power is below 1 W. Additionally, spread-spectrum modulation mitigates multipath, resists interference, and can coexist with other services using the same band. However, the range of Part XV systems is limited to around 5 km, because the output power is limited to 1 W.

In general, if UHF or VHF area coverage beyond the line of sight is required, then networks of transmitters or repeaters are required. Such networks would be expensive to establish from scratch, but some suitable in-place networks may exist. A viable DGPS network must offer widespread coverage, data latencies below 10 s, inexpensive user equipment, proven mobile performance, and relatively low overall cost.

2. Special Mobile Radio Systems

Special mobile radio systems (SMRS) are terrestrial radio networks that operate in the 800-MHz band for special commercial applications. Two example SMRS are the Advanced Radio Data Information Service (ARDIS) and Motorola Data Plus. The ARDIS has significant coverage in 400+ metropolitan areas (including Hawaii and the Virgin Islands). Motorola Data Plus covers 90% of the interstate highway system and approximately 60–70% of the land mass of CONUS. Unfortunately, both ARDIS and Motorola Data Plus are packet switch networks, and the data delay can be 10 s or higher. For DGPS, with SA activated, the delay is probably unacceptable. With SA off, the effect of latency is much less, and up to 30 s of delay may be tolerable for many users.

3. Television Vertical Blanking Interval

The Public Broadcasting System uses the vertical blanking interval (VBI) of the television picture to provide the national data broadcast system, National Datacast. This service can broadcast 9600 bps of data per vertical line, and there are 6–10 lines per station. This service is well suited for the broadcast of DGPS data in a number of respects:

1) The PBS can deliver data to all their affiliates via a satellite broadcast system.

2) PBS covers 97% of TV households over the air and has federal and state mandates to cover areas without service. Consequently, their nationwide coverage is quite good.

3) The National Datacast system was designed for small data latency: 2 s from bit arrival at PBS to user decoder output.

4) The remote equipment is small and costs only about $300.

5) National Datacast can be used to broadcast from individual PBS TV stations, and the cost is significantly lower than for a national broadcast.

However, National Datacast messages sent to mobile users have only been partially investigated and are not used regularly. An important limitation is that PBS TV stations only broadcast 18 h a day.

4. Cellular Radio

Cellular radio is spreading across the United States to provide telephone and data service to mobile users. It divides a given coverage area into small cells (hence, its name), and each cell is served by a single transmitter. Each cell is given a fixed set of frequency pairs (one for reception and one for transmission), and adjacent cells are given different frequency sets. However, cells not immediately adjacent can reuse the original set of frequencies; thus the overall system achieves fairly efficient use of the radio spectrum. A mobile receiver that crosses a cell boundary is "handed over" to the transmitter in the middle of the next cell. This handover is automatically handled and coordinated by a network of fixed receivers, but handovers can disrupt or terminate the DGPS transmission.

Cellular radio suffers the same drawbacks as all VHF and UHF systems: signal blockage in cities and rough terrain and multipath fading, particularly at longer distances. However, the main limitation to the use of cellular telephones is the very limited available coverage. Indeed, coverage is strong in urban areas and the Gulf of Mexico. However, areas without appreciable cellular markets are completely without coverage. Certainly, cellular coverage will increase, but only when market conditions are suitable.

5. Frequency Modulation Subcarrier

The FM broadcast stations can broadcast data by placing a subcarrier 66–96 kHz above their main carrier. This service is called Special Commercial Authorization (SCA) and is used to broadcast "muzak", financial data (for Dow Jones), and news data (for Reuters). Special Committee 104 of the RTCM proposed FM subcarrier broadcast of DGPS information to automobiles. Two commercial services based on this promising concept are described in Refs. 23 and 24.

6. Mode-S

Mode-S is a transponder datalink that operates at the secondary radar frequencies of 1030 MHz for the uplink and 1090 MHz for the downlink. Originally, the Mode-S data uplink was designed for brief transmissions from air traffic controllers to individual aircraft. For example, an uplink message could direct changes in the assigned altitude. The downlink may carry acknowledgments from the aircraft as well as requests for new altitudes. Broadcast of DGPS data using Mode-S seems sensible for larger air carriers, because Mode-S equipment is required for air carriers with 30 or more seats, and most commuter air carriers plan to have Mode-S transponders by 1995.[25]

Normally, Mode-S communication with a given aircraft is only possible when the ground station antenna beam is pointed toward that aircraft. This occurs only once every 4 s for airport surveillance radars and once every 10–12 s for enroute radars. This limitation is troublesome for the broadcast of DGPS data, but flight trials have demonstrated that the DGPS data can be broadcast from an omnidirectional antenna added to the ground station.[25]

7. Very-High-Frequency Broadcast for Precision Landing of Aircraft

Very-high-frequency radio can be used to broadcast differential corrections to aircraft that are conducting precision approaches and landings. As discussed below, this signal can certainly provide enough capacity to send code-phase corrections to serve the modest requirements of Category I landing. It also has enough capacity to send carrier-phase corrections at very high update rates. As such, it may be able to serve the very demanding requirements of Category II and III landings as well.

DIFFERENTIAL GPS

In the United States, VHF broadcast of DGPS data will use unused channels in the band reserved for the VHF omnidirectional ranging (VOR) system. The VOR band extends from 112 to 117.95 MHz, and the original channel spacing was 0.1 MHz. However, the channel spacing was reduced to 0.05 MHz when improvements in filtering and clock stability made possible greater rejection of adjacent channel interference. Differential GPS broadcasts are allowed to use some of these "new" channels.

The VHF signal will use eight-level differential phase-shift keyed modulation (D8PSK). This signal is of constant amplitude, but the modulation can cause the phase to take any of eight evenly spaced values on the unit circle. These are given by $\{n\pi/4\}_{n=0}^{7}$, and the information is contained in the change in phase. The signaling rate is 10.5 KHz, but there are 3 bits contained in each phase change; thus, the bit rate is 31.5 Kbps. This signal is designed to operate with a 25-KHz spacing between channels, so potentially three DGPS services could exist in the spectral space between two VOR signals with 100-KHz spacing.

Some of this 31.5 Kbps must be given over to forward error correction, overhead, and parity. Even so, this signal can carry carrier-phase as well as code-phase corrections at very high update rates. Consequently, it may be able to serve the most demanding of DGPS applications.

A data format for Special Category I (SCAT-I) landings using the so-called Differential Instrument Approach System (DIAS) has been designed by Special Committee 159 of the RTCA. This format is well described in the "Minimum Aviation System Performance Standards" (MASPS) published by the RTCA.[26] It is not described here, because it strongly resembles the RTCM format described in a previous section of this chapter. Its most notable departures from the RTCM format are: the use of 48 bits per satellite rather than 40, the specification of a particular forward error correcting code, and the use of longer parity fields.

C. Mobile Satellite Communications

1. Inmarsat

Mobile satellite services are currently being used to broadcast DGPS corrections over wide areas, and satellite broadcast of DGPS data will probably grow in importance. Indeed, the Inmarsat satellites are geosynchronous satellites with wide-area coverage out to high latitudes. By 1993, survey companies were already using these Inmarsat satellites to provide the DGPS coverage shown in Fig. 13. In time, the WAAS, described in Chapter 4 of this volume, may provide DGPS data over a virtually worldwide area.

The cost of satellite delivery of DGPS data depends strongly upon the required satellite power, which in turn depends upon the data rate, the size of the mobile antenna, and the coverage footprint. This is because bigger antennas have more gain, and thus require less satellite power to deliver the same quality of service. The key cost parameter is the G/T ratio of the mobile terminal, which is the antenna gain G divided by the noise temperature of the entire terminal T.

For example, the Inmarsat system includes three types of user terminals: Inmarsat-A, Inmarsat-M, and Inmarsat-C. The Inmarsat-A terminal is the largest terminal and has the largest antenna gain; thus it is the cheapest to use. A "lightweight," compact version of this terminal weighs 75 pounds and includes

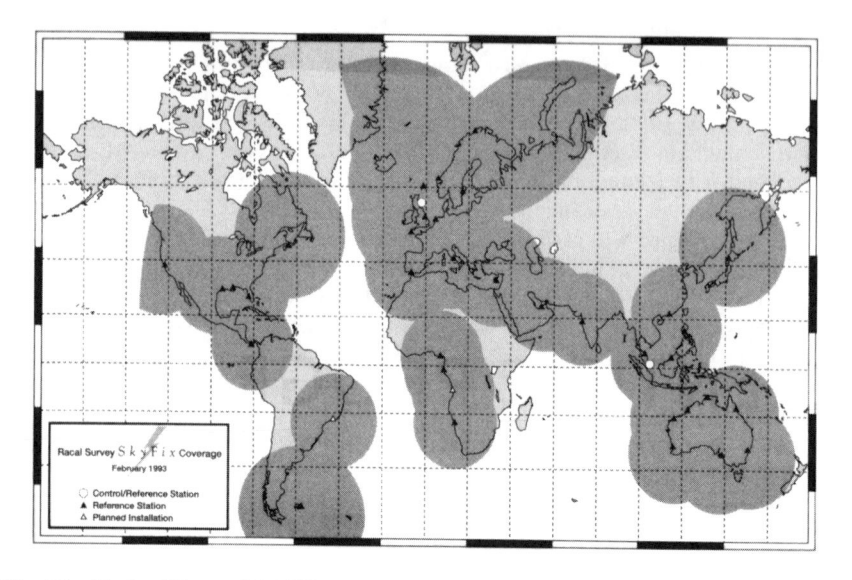

Fig. 13 Early differential GPS coverage offered by commercial survey companies using Inmarsat.

a 1.2-m dish. Clearly, it would not be suitable for operation on a small boat or aircraft. However, an Inmarsat-A terminal has $G/T = -4$ dB. For a 2400-bps link, the corresponding satellite power requirement is about 9 dBw, which is quite low. If the required bit rate is only 240 bps, then the satellite power requirement drops to -1 dBw.

The Inmarsat-M terminal is omnidirectional in elevation, but it scans in azimuth. Consequently, it has $G/T = -12$ dB. It would require approximately 15 dBw of satellite power for 2400 bps. However, it is suitable for many mobile applications.

Finally, the Inmarsat-C terminal is omnidirectional in elevation and azimuth, which means that it is suitable for use on almost any platform. However, it has $G/T = -25$ dB, which means that approximately 27.5 dBw of satellite power would be required to achieve 2400 bps.

By 1993, several survey companies were using Inmarsat satellites to deliver DGPS data to mobile users. In general, these services required the mobile user to carry an Inmarsat-A terminal (or near equivalent), because the satellite power was rather low and the satellite was in geosynchronous orbit, which means that the path loss was high. This limits the DGPS application to those vessels capable of carrying this large, expensive terminal.

2. *Other Mobile Satellite Services*

Many new mobile satellite services are being developed that offer the promise of smaller mobile terminals at reasonable costs. For example, the American Mobile Satellite Corporation (AMSC) is planning to offer spot coverage to North

America using geostationary satellites. In fact, AMSC will use four spot beams to cover CONUS, one spot beam to cover Alaska, one beam to cover Mexico, and four additional beams to cover Canada. In this case, the increased gain of the satellite transmit antenna decreases the G/T requirement for the mobile terminal, and consequently smaller terminals can be used.

Further in the future, a network of low-Earth orbiting satellites (LEO) will provide coverage from an altitude of a few thousand kilometers. These satellites are not geosynchronous; so a few dozen will be required to provide complete continuous coverage. However, their proximity also means that the path loss is not nearly as great as for geosynchronous satellites, which fly at 40,000 km. This reduced path loss greatly decreases the G/T requirement for the mobile antenna, thus reducing its size and cost.

VII. Differential GPS Field Results

This section samples the large number of DGPS field results obtained in the last few years. The first subsection describes some differential code-phase results for DGPS at short ranges.[19] The second subsection shows the impact of sending corrections to a user over 1500 km away from a single reference station.[27] The third subsection describes results obtained while designing a flight reference system based on differential carrier phase.[28] The final subsection describes an aircraft landing system that uses psuedolites to resolve the integer ambiguities.[29]

A. Short-Range Differential Code-Phase Results

The results in this subsection are for code-phase DGPS at shorter ranges. As such, they show the impacts of SA, receiver noise, and multipath, but not the effects of ionospheric refraction or errors in the satellite ephemeris. They were obtained by Ref. 19 while assessing the performance of DGPS when the corrections were broadcast by DGPS/radiobeacons, with real-world correction message failures being fairly commonplace. In fact, these results compare DGPS accuracy when RTCM type 1 messages are used in this noisy environment to the case when RTCM type 9 messages are used.

Figure 14 shows the North position error when both links have a marginal signal-to-noise ratio (8 dB) and when RTCM message failures are frequent. Both the type 1 and type 9 links suffer from message failures, but the impact is significantly greater on the type 1 link. From 8.21 to 8.23, the position error for the type 1 link exhibits an error sawtooth, where each "tooth" corresponds to one or two type 1 message failures followed by a successful message. This error growth is caused by SA. In this case, the *standard deviation of the North error for the type 1 and type 9 links is 0.46 and 0.17 m respectively.*

The type 1 link is less robust because each type 1 message is carrying corrections for all the satellites in view of the reference station. Consequently, a single noise burst on the radiobeacon channel destroys all of the corrections, and two noise bursts can destroy all the corrections for two consecutive message intervals. In contrast, the type 9 messages are only carrying three corrections at most. A single noise burst only destroys three corrections, and two noise bursts are unlikely to destroy the same set of corrections twice in a row. As a result of these

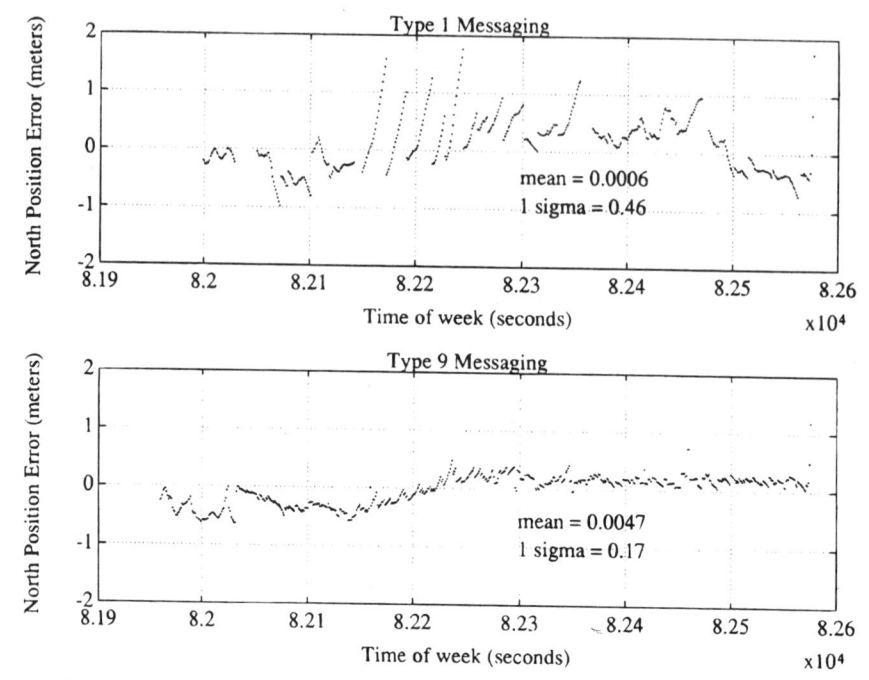

Fig. 14 Test results for radiobeacon DGPS under adverse communications.

considerations, the U.S. Coast Guard has standardized on the type 9 message format for their DGPS broadcasts.

B. Long-Range Differential Code-Phase Results

The results in this subsection are for code-phase DGPS at ranges greater than 1000 km. They were provided by Dr. Lee Ott of the John Chance Corporation, a part of the Fugro McClellan group. The John Chance Corporation specializes in providing support services to survey operations in North America. Accordingly, they have a network of DGPS reference stations that ring the Gulf of Mexico and the United States.

The DGPS data in Figs. 15 and 16 are for differential corrections developed at a reference station in Duluth, MN and applied at a receiver in Houston. Duluth is at the west end of Lake Superior and is more than 900 miles from Houston. Figure 15 is for July 25, 1993, and Fig. 16 is for August 9, 1993. Both figures show four individual time series over 24 h. The top time series is the HDOP, and the bottom trace is the number of satellites visible. The second trace is the error in East–West (in meters), and the third trace is the error in North–South (in meters).

Figure 15 shows the ΔE and ΔN performance when differential corrections are applied without accounting for the differences in the ionosphere or troposphere at these two rather distant sites. As shown, the standard deviations of the ΔE and ΔN errors are 1.8 and 2.7 m respectively. However, the ΔE and ΔN *biases* are -0.8 and -5.6 m respectively. Note that these biases represent the ionosphere and troposphere errors, which change relatively slowly with time.

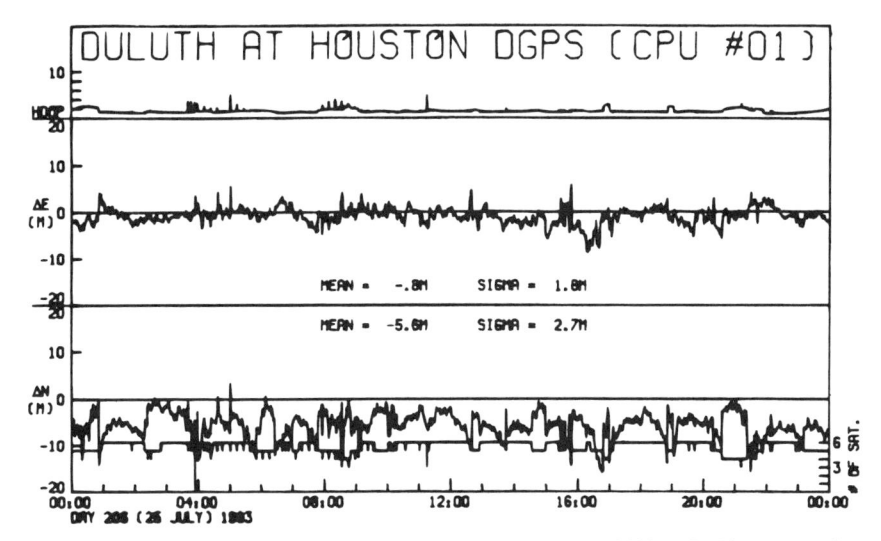

Fig. 15 Long-range differential GPS test results without additional adjustments for ionosphere and troposphere.

Figure 16 shows the ΔE and ΔN performance when the differential corrections are applied with an additional correction for the differential ionosphere and troposphere conditions. In this case, the standard deviations of the ΔE and ΔN errors are 1.7 and 2.3 m, which is about the same as the errors without any additional adjustment. However, the ΔE and ΔN biases decrease dramatically to -0.1 and 0.2 m, respectively. This performance over a 900-mile baseline is remarkable.

C. Dynamic Differential Carrier-Phase Results

The results in this subsection are for carrier-phase DGPS. Two sets of data are presented. The first is for a flight reference system application. The second is for a carrier-phase system that uses a simple pseudolite (called an *integrity beacon*) to resolve integer ambiguities. These carrier-phase DGPS systems provide the accuracy required for category II and III autolanding systems are feasible.

1. Flight Reference System

These results were obtained by Ref. 28 while developing a flight reference system. A flight reference system can be used to evaluate high-performance aircraft positioning systems. Examples of systems to be evaluated include aircraft approach and landing systems, such as the instrument landing system (ILS), the microwave landing system (MLS), and other DGPS systems. It can also be used to calibrate test ranges that may employ laser, infrared, or optical trackers.

This DGPS system made carrier-phase measurements at L_1, L_2, and the difference frequency $L_1 - L_2$. These measurements could be used to resolve the integer ambiguities rapidly because of the difference in the wavelengths at these three

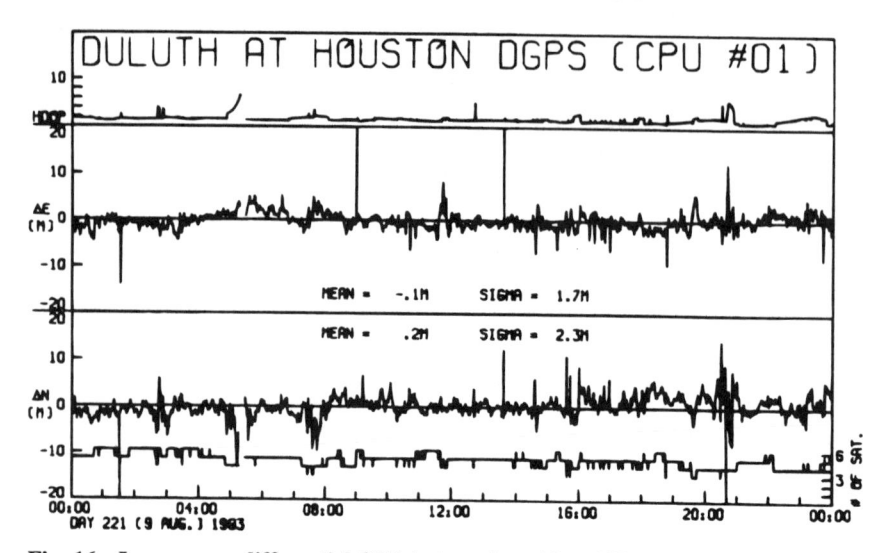

Fig. 16 Long-range differential GPS test results with a differential model for ionosphere and troposphere applied.

frequencies. Specifically, the L_1 and L_2 observations have wavelengths of 19 and 24 cm respectively, and the $L_1 - L_2$ (or wide lane) observation has a wavelength of 84 cm. The receiver searches for the set of integers that gives the same position fix for all three sets of observations. In addition, it uses code-phase DGPS data to initialize the search and constrain the search area.

Figure 17 shows the vertical difference between the DGPS fix and the laser fix at the Wallops Flight Facility. It shows the difference for two different approaches. This difference is shown as a function of distance to runway threshold. The large differences far from the threshold exist because the laser has not yet acquired the laser reflector on the aircraft. As shown, the vertical difference where GPS errors are usually largest is seldom greater than 2 ft.

2. Integrity Beacon Landing System

Perhaps the ultimate dynamic positioning system has been developed at Stanford University and is known as the Integrity Beacon Landing System, or IBLS.[29] This differential system consistently produces *high-dynamic accuracies of better than 10 cm* in three dimensions. Unfortunately, the system is about an order of magnitude better than the best laser trackers, so full dynamic verification is not possible but is inferred from the carrier-tracking loop errors.

These accuracies are achieved by positioning a pair of simple, low-power transmitters below the final landing pattern. This is shown in Fig. 18. As the aircraft flies through the reception pattern, it resolves the integer ambiguity with a very high success rate (well above 99%). After reception, the integers are held through aircraft landing. Figure 19 shows the results of tests of the system using the NASA Ames laser-tracking system as a reference. The solid lines are the specified accuracies of the tracker. In essence, the IBLS system is calibrating

DIFFERENTIAL GPS 45

VERTICAL DIFFERENCE GPS/LASER @ WFF - 6 May 1993 - R689 - Run 23.2 & 23.3

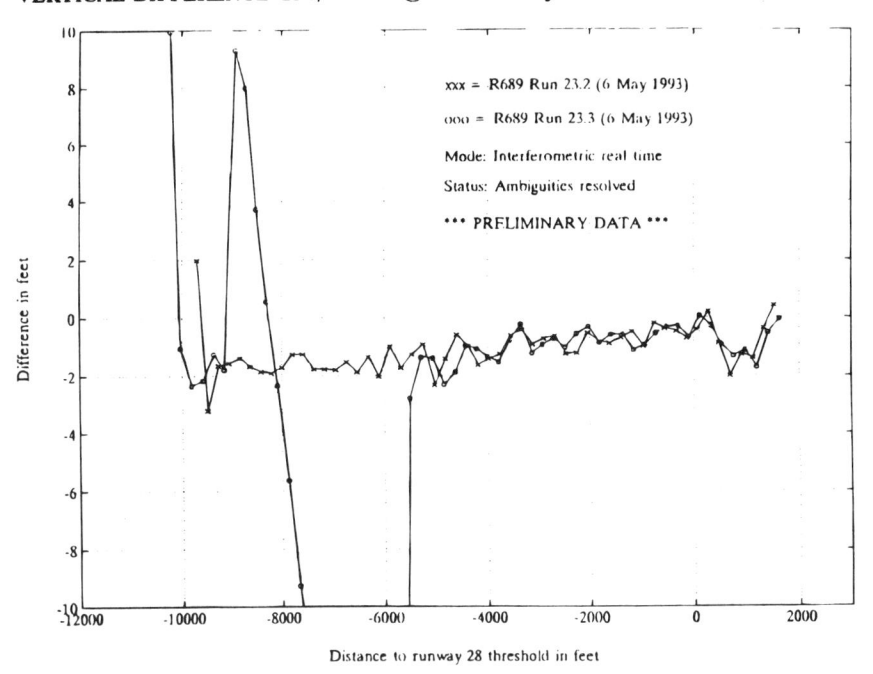

xxx = R689 Run 23.2 (6 May 1993)
ooo = R689 Run 23.3 (6 May 1993)
Mode: Interferometric real time
Status: Ambiguities resolved

*** PRELIMINARY DATA ***

Distance to runway 28 threshold in feet

Fig. 17 Results of carrier-phase differential GPS for aircraft landing.

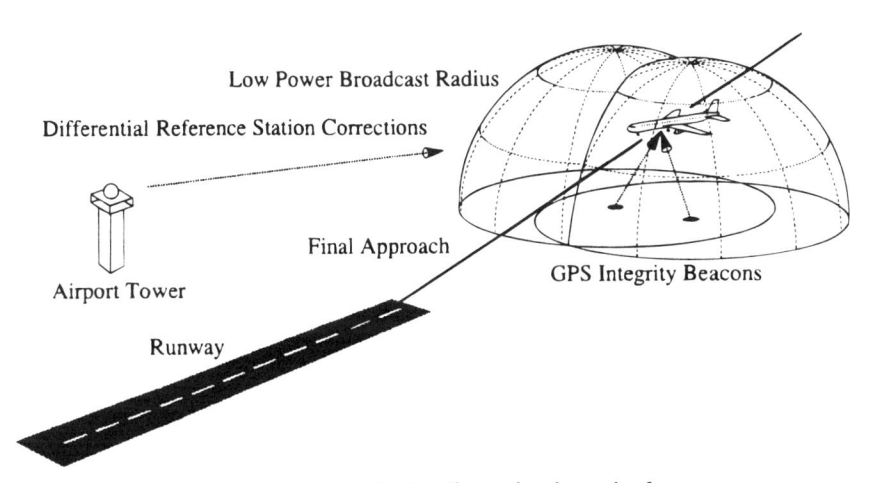

Fig. 18 Configuration for landing using integrity beacons.

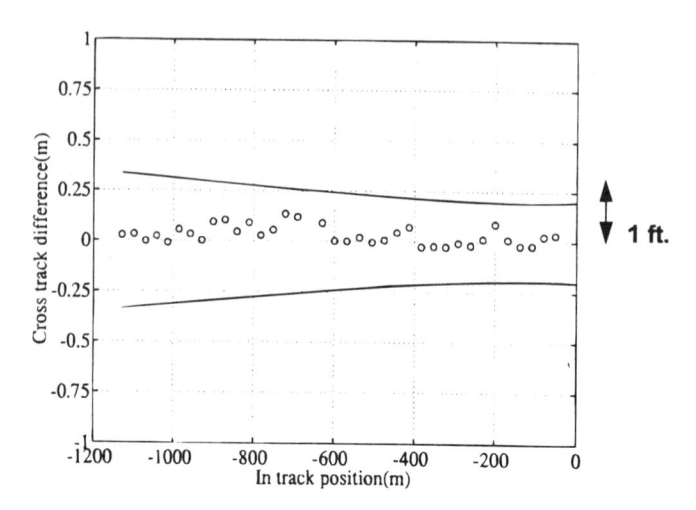

a) Critical vertical dimension; note that the laser tracker apparently has a small "tilt" error (no GPS error we know of would exhibit this trend)

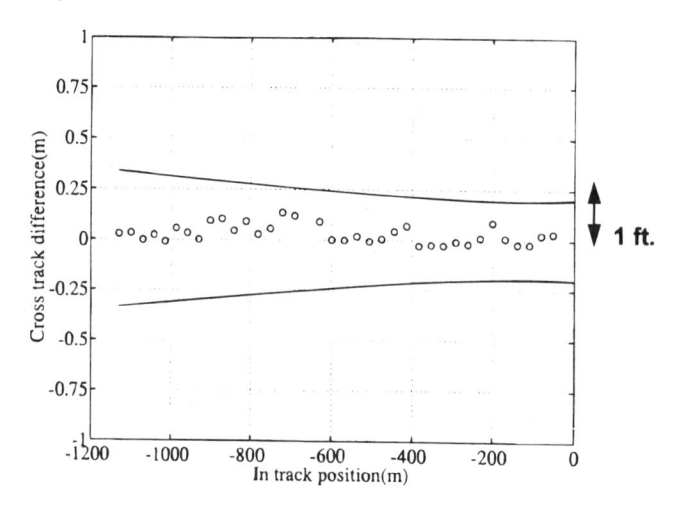

b) Tracking results against a laser tracker (Ames) while the aircraft is landing (*lines are the specified accuracy* of the tracker); the GPS output was probably +/- 0.02m or better in the approach and during the go-arounds (we make 12 landings)

Fig. 19 Test results using the NASA Ames laser-tracking system as a reference.

DIFFERENTIAL GPS

the laser-tracking system and demonstrating accuracies better than 10 cm. This system was also incorporated into a United Airlines' Boeing 737 that had a full autoland capability. Of 111 landings attempted, 110 were carried out to touchdown. The only exception was caused by a GPS satellite switching "off." In this case, the system correctly detected and flagged the integrity violation, and the aircraft aborted the landing. Further discussion of the IBLS is featured in Chapter 15 of this volume.

VIII. Conclusions

Differential GPS is the most accurate dynamic positioning system in use today. Its accuracy seems to be better than the best laser trackers if carrier-tracking techniques are used. Increasingly, users who require these accuracies will provide a powerful market force to simplify them and increase their availability. Perhaps most important for GPS users is the improved integrity that is a consequence of a well-designed DGPS system.

Appendix: Differential GPS Ephemeris Correction Errors Caused by Geographic Separation

This appendix derives the expression for the change in scalar ranging error as the user is displaced from the reference station. It assumes knowledge of elementary vector calculus; however, the result is very intuitive and is used to set some simple bounds on the expected error.

The vector relationships are depicted in Fig. A1. The reference station is located at r and the satellite is at s. The satellite error $\Delta \bar{R}$ is shown as a displacement of the satellite.

The scalar ranging error E caused by a vector satellite ephemeris error ($\Delta \bar{R}$) is the dot product of a unit vector from the user (or reference station) to the satellite $\bar{1}_s$ with the vector ephemeris error. Using matrix notation

$$E = \Delta \bar{R}^T \cdot \bar{1}_S = \Delta \bar{R}^T \cdot \frac{\bar{R}}{R} \tag{A1}$$

where

$$\bar{R} = \bar{R}_S - \bar{r} \tag{A2}$$

and

$$\bar{1}_S = \frac{\bar{R}}{R} \tag{A3}$$

To find the change in the error (ΔE) caused by the difference between reference station and user, form the partial derivative of ΔE [Eq. (A1)] with respect to the reference location $\bar{r}$:

$$\frac{\partial E}{\partial \bar{r}} = \Delta \bar{R}^T \cdot \left[\frac{\left(\frac{\partial \bar{R}}{\partial \bar{r}}\right)}{R} - \left(\frac{\partial R}{\partial \bar{r}}\right) \cdot \frac{\bar{R}}{R^2} \right] \tag{A4}$$

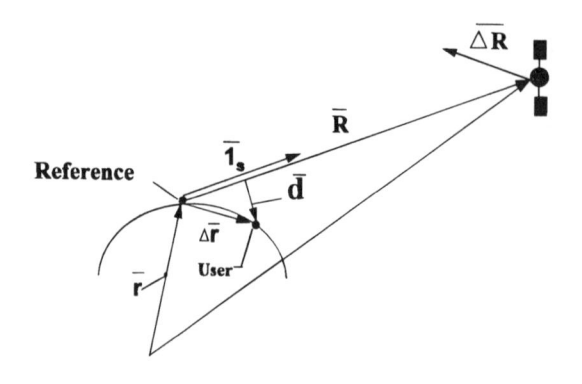

Fig. A1 Vector relationships for ephemeris errors.

Then

$$\Delta E \cong \frac{\partial E}{\partial \bar{r}} \cdot \Delta \bar{r} \tag{A5}$$

The first partial derivative is found from Eq. (A2). With I as the 3×3 identity matrix, we get the following:

$$\frac{\partial \bar{R}}{\partial \bar{r}} = -I \tag{A6}$$

The second partial derivative in Eq. (A4) is found by expanding Eq. (A2) into the following scalar magnitude:

$$R = \sqrt{\bar{R}^T \cdot \bar{R}} = \sqrt{\bar{R}_S^T \cdot \bar{R}_S - 2\bar{R}_S^T \cdot \bar{r} + \bar{r}^T \cdot \bar{r}} \tag{A7}$$

Then the following relationship can be verified:

$$\frac{\partial R}{\partial \bar{r}} = -\frac{1}{R}(\bar{R}^T \cdot I) \tag{A8}$$

Using results (A6) and (A8) into (A4) and then into (A5), we find the following:

$$\Delta E = \Delta \bar{R}^T \cdot \left[\frac{-\Delta \bar{r}}{R} + \frac{1}{R}(\bar{R}^T \cdot \Delta \bar{r}) \cdot \frac{\bar{R}}{R^2} \right] \tag{A9}$$

Using the definition (A3), we find:

$$\Delta E = \frac{-\Delta \bar{R}^T}{R} [\Delta \bar{r} - (\bar{1}_S^T \cdot \Delta \bar{r}) \bar{1}_S]$$

Let

$$\bar{d} \triangleq [\Delta \bar{r} - (\bar{1}_S^T \cdot \Delta \bar{r}) \bar{1}_S]$$

Then

$$\Delta E = \frac{-\Delta \overline{R}^T \cdot \overline{d}}{R} \qquad (A10)$$

The error is, therefore, the dot product of the ephemeris error vector with the vector $\overline{d}$, divided by the distance to the satellite. This vector $\overline{d}$ has a simple explanation. It is the component of $\Delta \overline{r}$ perpendicular to the satellite. Figure A1 clarifies this concept

The usual technique used to predict this error is to bound the quantity ΔE by observing that the vector $\overline{d}$ is less than $\Delta \overline{r}$ and the dot product magnitude is less than the magnitude of the two vectors involved, so Eq. (A10) shows us that

$$\Delta E \leq \frac{|\Delta \overline{r}| \cdot |\Delta \overline{R}|}{R} \qquad (A11)$$

This is a convenient way to characterize the worst case *range decorrelation*. Of course, the geometric effects (PDOP) must be included to calculate the expected position errors.

References

[1] Kalafus, R. M., Vilcans, J., and Knable, N., "Differential Operation of NAVSTAR GPS," *Navigation,* 1984.

[2] Beser, J., and Parkinson, B. W., "The Application of NAVSTAR/GPS Differential GPS in the Civilian Community," *Navigation,* Vol. 29, No. 2, 1982.

[3] Klein, D., and Parkinson, B. W., "The Use of Pseudo-Satellites for Improving GPS Performance," *Proceedings of the 40th Annual Meeting of the Institute of Navigation,* (Cambridge, MA), Institute of Navigation, Washington, DC, June 1984.

[4] Kee, C., Parkinson, B., and Axelrad, P., "Wide Area Differential GPS," *Proceedings of ION-GPS 90,* (Colorado Springs, CO), Institute of Navigation, Washington, DC, Sept. 19–21, 1990.

[5] Van Dierendonck, A. J., Fenton, P., and Ford, T., "Theory and Performance of Narrow Correlator Spacing in a GPS Receiver," *Navigation,* Vol. 39, No. 3, 1993.

[6] Van Nee, R. D. J., "GPS Multipath and Satellite Interference," *Proceedings of the 48th Annual Meeting of the Institute of Navigation,* Institute of Navigation, Washington, DC, Aug. 1992.

[7] Braasch, M., "Characterization of GPS Multipath Errors in the Final Approach Environment," *Proceedings of ION GPS-92* (Albuquerque, NM), Institute of Navigation, Washington, DC, Sept. 16–18, 1992.

[8] Matchett, G., "Stochastic Simulation of GPS Selective Availability Errors," *TM,* TASC Contract DTRS-57-83-C-00077, June 1985.

[9] Braasch, M., "A Signal Model for GPS," *Navigation,* Vol. 37, No. 4, Winter 1990–1991.

[10] Chou, H-T., "An Adaptive Correction Technique for Differential Global Positioning System," Ph.D. Dissertation, Stanford Univ., Stanford, CA, SUDAAR 613, June 1991.

[11] Feess, W. A., and Stephens, S. G., "Evaluation of GPS Ionospheric Time Delay Algorithm for Single Frequency Users," *Proceedings, IEEE PLANS '86, IEEE Position, Location, and Navigation Symposium* (Las Vegas, NV), IEEE, New York, Nov. 1986.

[12] Klobuchar, J., "Design and Characteristics of the GPS Ionospheric Time Delay Algorithm for Single Frequency Users," *Proceedings, IEEE PLANS '86, IEEE Position, Location, and Navigation Symposium* (Las Vegas, NV), IEEE, New York, Nov. 1986.

[13] El-Arini, M. B., O'Donnell, P., Kellam, P., Klobuchar, J. A., Wisser, T. C., and Doherty, P. H., "The FAA Wide Area Differential GPS (WADGPS) Static Ionospheric Experiment," *Proceedings of the 1993 National Technical Meeting of the Institute of Navigation*, (San Francisco, CA), Institute of Navigation, Washington, DC, Jan. 1993.

[14] Cohen, C. E., "Attitude Determination Using GPS," Ph.D. Dissertation, Stanford Univ., Stanford, CA, Dec. 1992.

[15] Allison, T., Eschenbach, R., Hyatt, R., and Westfall, B., "C/A Code Dual Frequency Surveying Receiver—Architecture and Performance," *Proceedings, IEEE PLANS '88, IEEE Position, Location, and Navigation Symposium*, IEEE, New York, 1988.

[16] Anon., "Data Standard for Differential GPS Corrections," Radio Technical Commission Maritime, Special Committee 104, 1992.

[17] Kalafus, R. M., Van Dierendonck, A. J., and Pealer, N. A., "Special Committee 104 Recommendations for Differential GPS Service," *Navigation*, Vol. 33, No. 1, 1986.

[18] GPS ICD.

[19] Enge, P. K., Young, D., Sheynblatt, L., and Westfall, B., "DGPS Field Trials, Which Compare Type 1 and Type 9 Messaging," *Navigation*, Vol. 40, No. 4, Winter 1993-94.

[20] Lanigan, C. A., Pflieger, K., and Enge, P. K., "Real-Time Differential Global Positioning System (DGPS) Datalink Alternatives," *Proceedings of ION GPS-90*, (Colorado Springs, CO), Institute of Navigation, Washington, DC, Sept. 19–21, 1990.

[21] Enge, P., Van Dierendonck, A. J., and Kinal, G., "A Signal Design for the GIC Which Includes Capacity for WADGPS Data," *Proceedings of ION GPS-92* (Albuquerque, NM), Navigation, Washington, DC, Sept. 16–18, 1992.

[22] Enge, P. K., Ruane, M. F., and Sheynblatt, L., "Marine Radiobeacons for the Broadcast of Differential GPS Data," *Proceedings, IEEE PLANS '86, IEEE Position, Location, and Navigation Symposium* (Las Vegas, NV), IEEE, New York, 1986, pp. 368–376.

[23] Weber, L., and Tiwari, A., "Performance of a FM Sub-carrier (RDS) Based DGPS System," *Proceedings of ION GPS-93*, (Salt Lake City, UT), Institute of Navigation, Washington, DC, Sept. 22–24, 1993.

[24] Galyean, P., "The Acc-Q Point DGPS System," *Proceedings of ION-GPS-93* (Salt Lake City, UT), Institute of Navigation, Washington, DC, Sept. 22–24, 1993.

[25] Bayliss, E., "Incorporation of GNSS Technology in C/N/S-Demonstration of DGPS Precision Approach," MIT Lincoln Laboratory, 1993.

[26] Anon., "DGNSS Instrument Approach System: Special Category I (SCAT-1)," RTCA Incorporated, 1993.

[27] Lee Ott, personal communication, 1992.

[28] Van Graas, F., Diggle, D., and Hueschen, R., "Interferometric GPS Flight Reference/Autoland System: Flight Test Results," *Navigation*, Vol. 41, No. 1, 1994.

[29] Cohen, C. E., Pervan, B., Cobb, H. S., Lawrence, D., Powell, J. D., and Parkinson, B. W., "Real-Time Cycle Ambiguity Resolution Using a Pseudolite for Precision Landing of Aircraft with GPS," *Proceedings of the Second International Symposium on Differential Satellite Navigation System, DSNS '93*, Amsterdam, The Netherlands, March 1993.

Chapter 2

Pseudolites

Bryant D. Elrod*
Stanford Telecom, Inc., Reston, Virginia 22090
and
A. J. Van Dierendonck[†]
AJ Systems, Los Altos, California 94024

I. Introduction

THE Global Positioning System (GPS) has reached operational status and is now being used internationally as a means for accurate, world-wide, all-weather positioning and navigation based on pseudonoise signals transmitted by a constellation of 24 satellites. Pseudolites (PLs) are ground-based transmitters that can be configured to emit GPS-like signals for enhancing the GPS by providing increased accuracy, integrity, and availability.[11,21]

Accuracy improvement can occur because of better local geometry, as measured by a lower vertical dilution of precision (VDOP), which is important in aircraft precision approach and landing applications. Accuracy and integrity enhancement can also be achieved by employing a PL's integral data link to support differential (DGPS) modes of operation and timely transmittal of integrity warning information. Availability is increased because a PL provides an additional ranging source to augment the GPS constellation.

Although the use of PLs offers many potentially significant benefits, a number of technical issues must also be addressed. One is the PL signal power level and the associated "near–far" problem that a user receiver may experience, depending upon the dynamic range of signal strength encountered as the distance to a PL changes. Other issues include deployment requirements, signal data rate, signal integrity monitoring, and user antenna location and sensitivity. This chapter addresses PLs from the perspectives of signal design considerations, integrated DGPS/PL considerations, and testing activities for assessing the reality and mitigation of various technical issues associated with using PLs.

Copyright © 1995 by Stanford Telecommunications, Inc. Published by the American Institute of Aeronautics and Astronautics, Inc. with permission.
*Vice President, GPS Navigation Systems.
[†]Systems Consultant.

II. Pseudolite Signal Design Considerations

Pseudolites operating within the GPS frequency bands (L_1: 1565–1585 MHz or L_2: 1217–1237 MHz) can be configured to serve a limited area with a power level low enough to preclude appreciable interference to standard GPS signals. One application is the PL bubble concept proposed for operating near the ends of airport runways to augment real-time kinematic positioning based on GPS carrier-phase measurements[3] (also see Chapter 15, this volume).

For PLs designed to cover a larger area, such as an entire airport or terminal area, potential interference to GPS signals is a key technical issue. Although operation outside the GPS bands is a possibility, this option ultimately adds rf front end complexity and cost to user equipment.

In this chapter, a PL signal structure is described that can operate within the L_1 band and mitigate or virtually eliminate the near–far issue. First, however, a brief review is given of previous implementations used on DOD test ranges and other approaches proposed for civil applications.

A. Previous Pseudolite Designs

Table 1 compares two DOD implementations and three other proposed approaches in terms of the diversity of techniques employed for interference mitigation: signal (code, frequency, time) and spatial separation. Pseudolites, called ground transmitters (GTs) during the Phase I GPS program, were used to augment GPS for testing user equipment at Yuma, AZ before there were enough satellites for navigation.[4] The PL signal structure was the same as used on the GPS satellites, except for the data content. In those days, interference to the satellite signals was avoided by using different gold codes (Code Division Multiple Access, or CDMA) and keeping the user equipment under test at an adequate distance from the four GTs to prevent dynamic range sensitivity problems. The same was true in more recent Space Defense Initiative (SDI) testing with GTs built by Stanford Telecom (STel) to augment GPS for DOD's Range Applications Program. However, a Time Division Multiple Access (TDMA) signal (on–off scheme) was also implemented to prevent interfering with a co-located GPS receiver that provided real-time synchronization to the GPS satellites. At transmit times, this receiver simply blanked out the signal.

The use of PLs for civil aviation applications was first proposed in 1984.[5] The "near–far" problem was recognized and addressed at that time, along with suggested solutions, but not carried any further. Three signal diversity techniques suggested as a solution were as follows:

1) Pulsed signals with random or fixed cycle rates, a TDMA variation

2) Signals transmitted at a frequency offset from L_1 (1575.42 MHz), but within the same frequency band as GPS, a variation of frequency division multiple access (FDMA)

3) Alternative codes that have a longer sequence than the existing GPS codes, a variation of CDMA

The first two alternatives were favored, with a preference toward the first. Although technically viable, the FDMA approach, including the possible use of L_2 (1227.6 MHz) or frequencies outside the GPS band, was considered more costly given the state of GPS receiver technology at the time. The CDMA

Table 1 Pseudolite interference mitigation techniques—previous designs

Pseudolite signal implementation (I) or recommendation (R)	Interference mitigation techniques			
	Code division multiple access	Frequency division multiple access	Time division multiple access	Pseudolite spatial separation
(I) Yuma ground transmitters for the Phase I GPS tests (1977)	Different gold codes	None	None	Unspecified
(I) Stanford Telecom ground transmitters for Space Defense Initiative (1989)	Different gold codes	None	Pulsed (simple on–off)	Unspecified
(R) Reference 5	Longer sequence codes	L_2 or $L_1 \pm 15$ MHz	Pulsed at random or fixed rate	Unspecified
(R) Reference 6	Different gold codes	None	Pulsed with: 10% duty cycle, random pattern	≥ 30 km
(R) Reference 7	Different gold codes	$L_1 \pm \Delta f$ ($\Delta f \geq 30$ kHz)	Pulsed with: 1:11 duty cycle, fixed pattern, fixed offset between PLs	No constraint

approach with different pseudorandom noise (PRN) codes could be part of the diversity solution, but longer sequence codes would not add significant margin against cross-correlation interference.

As part of the Radio Technical Commission for Maritime (RTCM) user activities, a more definitive pseudolite signal structure was proposed in 1986.[6] All three of the above multiple access techniques were considered, but pulsed TDMA was the only approach recommended, because it made the least impact on the design of GPS receivers based on the state of technology at the time. Subsequently, flaws in that TDMA scheme were observed with respect to a class of "nonparticipating" receivers, some of which are still in use today. This led to a modified TDMA scheme, which was proposed in 1990.[7] Despite these proposals, fear of the near–far problem remained, and rightly so, because a limited interference margin can still exist with only code (CDMA) and pulsing (TDMA) employed.

B. New Pseudolite Signal Design

Fortunately, GPS receiver technology has advanced to a point that the FDMA approach is now viable, which improves the solution to the near–far interference issue significantly. As a result, a more effective signal structure has been proposed that combines: good C/A codes a frequency offset that takes advantage of the code cross-correlation properties, and a good pulsing scheme.[8]

1. C/A Codes

A recent study identified 19 C/A codes that were considered best (in terms of cross-correlation level) of the 1023 possible codes in the GPS C/A-code family.[9] These codes were selected for use with the FAA's wide area augmention system (WAAS), which will augment GPS with a ranging signal transmitted at L_1 from geostationary orbit. Although the 19 codes selected were the best, there remain 712 balanced codes, almost as good, that could be used for PLs. (See also Chapter 3 in the companion volume for additional discussion of Gold codes and cross-correlation properties.)

2. Frequency Offset

The proposed offset is 1.023 MHz on either side of L_1 at 1575.42 MHz, which places the PL carrier in the first null of the GPS satellite C/A-code spectrum. This offset is much more effective and one that at least some current GPS receivers can accommodate, because it resembles a large Doppler frequency offset. Furthermore, it virtually eliminates any code cross-correlation with the GPS satellite signals. This spectrum is shown in Fig. 1, which is a typical spectrum of a GPS C/A code [PRN 2]. It was reported years ago that the cross-correlation between C/A codes at different frequencies was simply the magnitude of another C/A-code spectral line at the offset frequency.[10] This property has since been verified. Because all the codes have similar spectral characteristics, the spectrum of any C/A code tells the story. As indicated in Fig. 1 and discussed further in Appendix A, the spectral lines near the null are well below −80 dB and are 60 dB lower than those at a zero frequency offset. This property has been used to

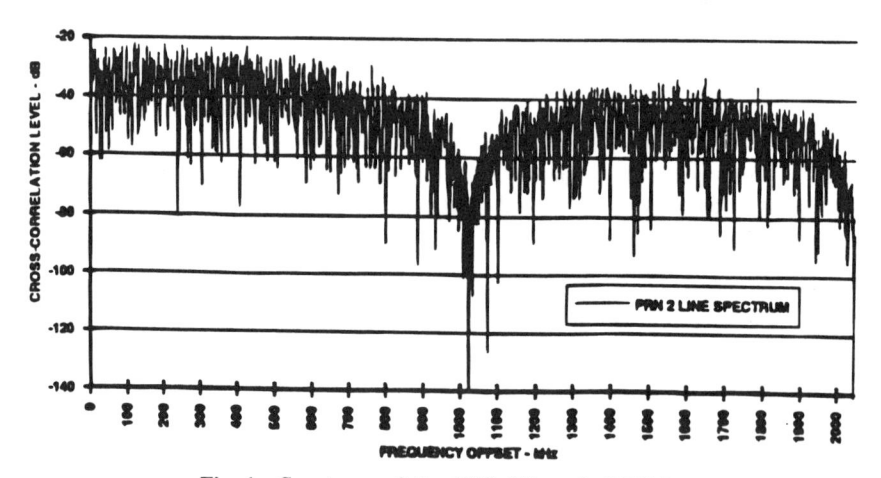

Fig. 1 Spectrum of the GPS C/A-code PRN 2.

advantage in the Global Orbiting Navigation Satellite System (GLONASS) signal structure.[11] All the GLONASS signals have the same maximum length codes with the first nulls at 511 kHz offsets from the carrier. The GLONASS FDMA signals are spaced at 562.5 kHz intervals, 51.5 kHz from the nulls of the adjacent signal.

Recently, it has been shown that cross-correlation levels can be somewhat higher than indicated above when the code chip boundaries are not aligned.[12] However, when the carrier frequency is offset by 1.023 MHz, and the carrier/code frequency ratio of 1540 is maintained, the code of the PL is shifting with respect to the satellite codes at a rate of over 664 chips per second. Thus, any cross-correlation is noise-like and averages to a lower rms level. As pointed out, this is still interference.[12] An in-band frequency offset of 1.023 MHz lowers the rms interference by about 8 dB, but does not eliminate it. However, it does eliminate cross-correlation problems.

3. Pulsing Scheme

The FDMA approach described above is a variation of an earlier proposal, which suggested a continuous signal with a longer code on the fringe of the allocated L_1 band at 1560 or 1590 MHz. However, a *continuous* in-band PL signal could still cause interference problems, depending upon its power level compared to GPS satellite signals. For example, a PL signal whose power is set to be received at 37 km (20 n.mi.) will be 60 dB stronger at 37 m (0.02 n.mi.), which will be approximately 30–40 dB above the noise. Thus, although offsetting the frequency is a good idea, TDMA pulsing may still be required to avoid this situation. Furthermore, if TDMA is used, there is no reason to operate on the fringe of the L_1 allocated band, because that large a frequency offset and a longer code are both undesirable due to the additional burden on the GPS receiver, even with modern technology.

With a good pulsing scheme, the impact on the reception of GPS signals can be made essentially transparent. The receiver treats it as a continuous signal, provided that it is designed to suppress pulse interference, as most modern receivers are, even if by accident. As noted in the RTCM work,[6] any hard-limiting receiver or "soft-limiting" receiver will clip the pulses and limit their effect, but still pass more than enough pulse power to track the PL signal itself. A "soft-limiting" receiver clips the incoming signal-plus-noise at two to three times the rms noise level, resulting in more, but still negligible degradation in signal-to-noise performance.

All modern digital receivers are either hard-limiting or possess the soft-limiting property through precorrelation quantization. Although this is not true for the older, analog military receivers, their wideband automatic gain control (AGC) suppresses the pulses for the same effect. Any cross-correlation problems in those receivers caused by PL transmissions are eliminated with the proposed frequency offset.

The pulsing scheme presented here differs from previous proposals.[6,7] It accounts for the fact that the PL message symbol rate could be 50 N sps (where $N = 1, 2, 4, 5, 10,$ or 20). The recommended pulse pattern is illustrated by the Xs shown in Fig. 2. These repeat every 11 ms, and thus, would never be synchro-

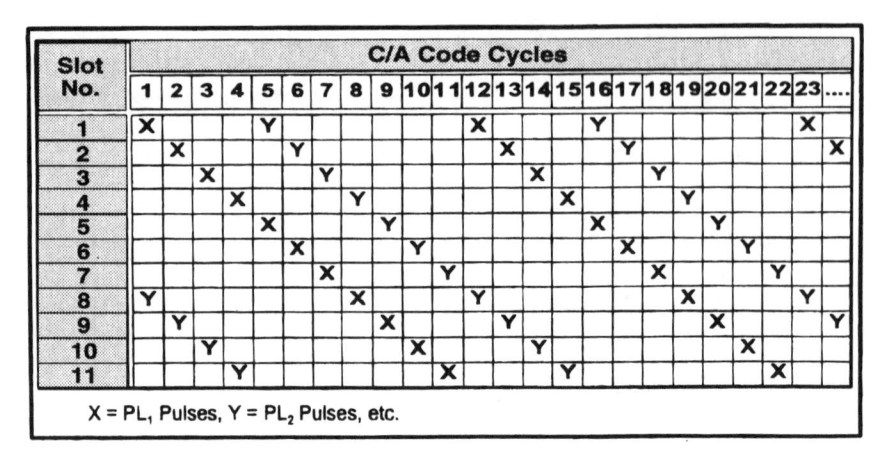

Slot No.	C/A Code Cycles																								
	1	2	3	4	5	6	7	8	9	10	11	12	13	14	15	16	17	18	19	20	21	22	23	....
1	X			Y					X			Y				X					Y			
2		X			Y					X			Y				X					Y		
3			X			Y					X			Y				X					Y	
4				X			Y					X			Y				X					
5					X			Y					X			Y				X				
6						X			Y					X			Y				X			
7							X			Y					X			Y				X		
8	Y							X			Y					X			Y				X	
9		Y							X			Y					X			Y				X
10			Y							X			Y					X			Y			
11				Y							X			Y					X			Y		

X = PL$_1$ Pulses, Y = PL$_2$ Pulses, etc.

Fig. 2 **Illustration of pseudolite pulse pattern.**

nous with a received GPS bit pattern. Each code cycle (1 ms) is divided into 11 slots, each with a width of 1/11,000 s (90.90909 μs). The pulses would never be synchronous with a received GPS bit pattern, as would occur with the RTCM design.[6]

Only one of these slots contains a pulse, so the duty cycle is 1/11. There will be 20/N pulses per symbol, but only one is required because the receiver would integrate the energy over the entire symbol period. Because every slot is filled once every 11 ms, the entire C/A code would be received every 11 ms. The clocking rate for the slots is 1/93 of the C/A-code chipping rate of 1.023 MHz, or 11 kHz. It is noted in Appendix B that with a duty cycle of 1/11, the loss of GPS satellite signal-to-noise ratio, in either an analog receiver with pulse suppression, or a digital receiver with natural *soft-limiting*, is less than 1.5 dB when close to the PL.

Potential mutual interference when multiple PLs are installed in an area was reported in the RTCM work.[6] With the RTCM pulse pattern, a minimum distance between PLs would be required, because the transmission of pulses simultaneously from each PL could result in the simultaneous reception of multiple pulses. But, that need not be, because the pulse timing of multiple PLs can be offset. Unfortunately, the RTCM pattern had irregular times between pulses, so that there would still be some simultaneous receptions.

However, with the pulse pattern shown in Fig. 2, a suitable pulse-timing offset would prevent simultaneous reception from ever occurring, except at large distances where the PL signal would be of little consequence. For example, consider two PLs where the pulse timing was offset by 4 ms as indicated by the Xs and Ys in Fig. 2. Because the minimum transmit time separation between X and Y pulses is 4/11 ms, only receivers with a differential range more than 110 km from the PLs would encounter simultaneous reception. At that distance, the received powers of either PL would be negligible.

4. Pseudolite Carrier Tracking

A potential misconception about pulsing a PL signal is that it will prevent the tracking of PL carrier phase. This is not true if the pulses occur at a high enough rate, which is the case in the proposed scheme. One or more pulses will always be integrated along with noise over each symbol. The result is transparent to the tracking loops, because the phase change due to Doppler uncertainty over a symbol is negligible. To the tracking loops, it looks like a continuous signal.

5. Pseudolite Transmit Power

Given an average received power P_r through a receiving antenna (with gain G_a) at a distance d (in n.mi.) from a PL, the average transmitted power (P_T) at $L_1 \pm \Delta f$ is as follows:

$$P_T \approx P_r + 20 \log_{10}\left(\frac{4\pi d}{\lambda_1}\right) - G_a$$

$$= P_r + 20 \log_{10} d + 101.75 - G_a$$

where $\lambda_1 = 0.00010275$ n.mi. is the signal wavelength corresponding to L_1. As an example, consider an average received power of -130 dBm at 20 n.mi. (37 km) through an antenna gain of -10 dB (assumed for a small negative elevation angle). The average transmitted power is 7.77 dBm, or about 6 mW. The peak power for a duty cycle of 1/11 is then 18.18 dBm, or about 66 mW.

III. Integrated Differential GPS/Pseudolite Considerations

The introduction of PLs has two key objectives: signal augmentation and data link enhancement. The first is to increase the number of available signals and, thereby, improve or maintain the geometrical quality for user position determination. The second is to provide an integrated capacity for supplying key data to users for GPS (and PL) integrity warning and differential corrections to improve positioning accuracy via code-based local differential GPS (LDGPS) and potentially carrier phase-based kinematic differential GPS (KDGPS) techniques.

Although the definition of a compatible signal format is one of the most critical requirements for meeting these objectives, there are other implementation-related aspects that need consideration as well. This section addresses five of these: PL siting, PL time synchronization, antenna location on a user aircraft, the PL data message, and the filter algorithm for integrated GPS/PL measurement processing.

A. Pseudolite Siting

It is well known that the GPS geometry (quantified in terms of HDOP and VDOP factors) will vary over time and user location, even with a full 24-satellite constellation[13] (also see Chapter 5, the companion volume). It is also known that at times significant degradation (VDOP >> 6) can occur if fewer satellites are active. The utility of PLs for geometric enhancement lies in the fact that lower DOP values with less temporal variation can be achieved.

Figures 3 and 4 illustrate representative HDOP and VDOP profiles determined for a full constellation[13] and with one satellite inactive for cases where the user employs the best four GPS satellites or "all-in-view" with augmentation by one or two PLs.[2] In this illustration, the user was assumed to be at an altitude of 200 ft (the decision height for a Cat. I approach) with the PLs located 1 n.mi. ahead and/or 2 n.mi. behind. It is evident that in this situation, a two-PL augmentation would enable HDOP ≤ 1.0 and VDOP ≤ 1.5 *continuously*. At those levels, code-based LDGPS with the capability to correct uncorrelated pseudorange errors to $\leq 1m$ (2σ) would meet the current Cat. III (horizontal) and Cat. II (vertical) sensor accuracy requirements of 4.1m and 1.4m (2σ), respectively for aircraft precision approaches.[14] With the addition of ranging-capable satellites in geostationary orbit (e.g., via the FAA's WAAS implementation) to offset one or two inactive GPS satellites an even lower VDOP is possible with favorable implications for meeting the most stringent Cat. III (vertical) accuracy requirement of 0.4m (2σ).[14]

An alternate approach to meeting the Cat. III vertical accuracy requirement for precision approaches is based on KDGPS techniques with PL augmentation. Recent flight tests have demonstrated that a varying user/PL geometry will permit rapid resolution of carrier-range ambiguities to provide KDGPS-based measurements with centimeter-level precision[3] (also see Chapter 15, this volume). The configuration adopted for this technique utilizes a *bubble* concept in which two PLs located ahead of the runway threshold on each side of the glideslope transmit a standard, unpulsed L_1-C/A signal at a low power level sufficient to be received within a hemispherical bubble centered at each PL. A descending aircraft would enter the common bubble zone, acquire the PL signals, resolve GPS carrier-phase ambiguities, and emerge to continue the approach with KDGPS-based corrections supplied via a separate (non-PL) datalink. As proposed, four active PLs per runway would be needed to support landings in either direction.

On the other hand, wider coverage *large bubble* PLs with interference mitigation inherent in the signal design could provide service to multiple runways over an entire airport region. Location of the PL antenna off runway and at suitable elevation would also be desirable from the standpoint of user antenna requirements (discussed in Sec. II.C). Pseudolite-siting requirements (number and location) to accommodate multirunway situations, user antenna considerations, and local constraints are key issues needing further analysis and testing.

B. Pseudolite Time Synchronization

Synchronizing a PL's clock to GPS time can be achieved in two ways—one in which a PL is collocated with an LDGPS reference receiver (RR), and one in which it is remote from the RR that is tracking its transmitted signal. In the latter case, the RR sends corrections to the remote "slave" PL to correct itself and also sends message data (e.g., code- and/or carrier-based DGPS corrections) to be transmitted. These two different PL configurations are illustrated in Fig. 5. In the co-located configuration, the RR shares the transmit/receive antenna with the PL, which also allows for self-calibration.

The type of configuration used would depend upon whether or not there is more than one PL at a given local region. If there is only one, the collocated

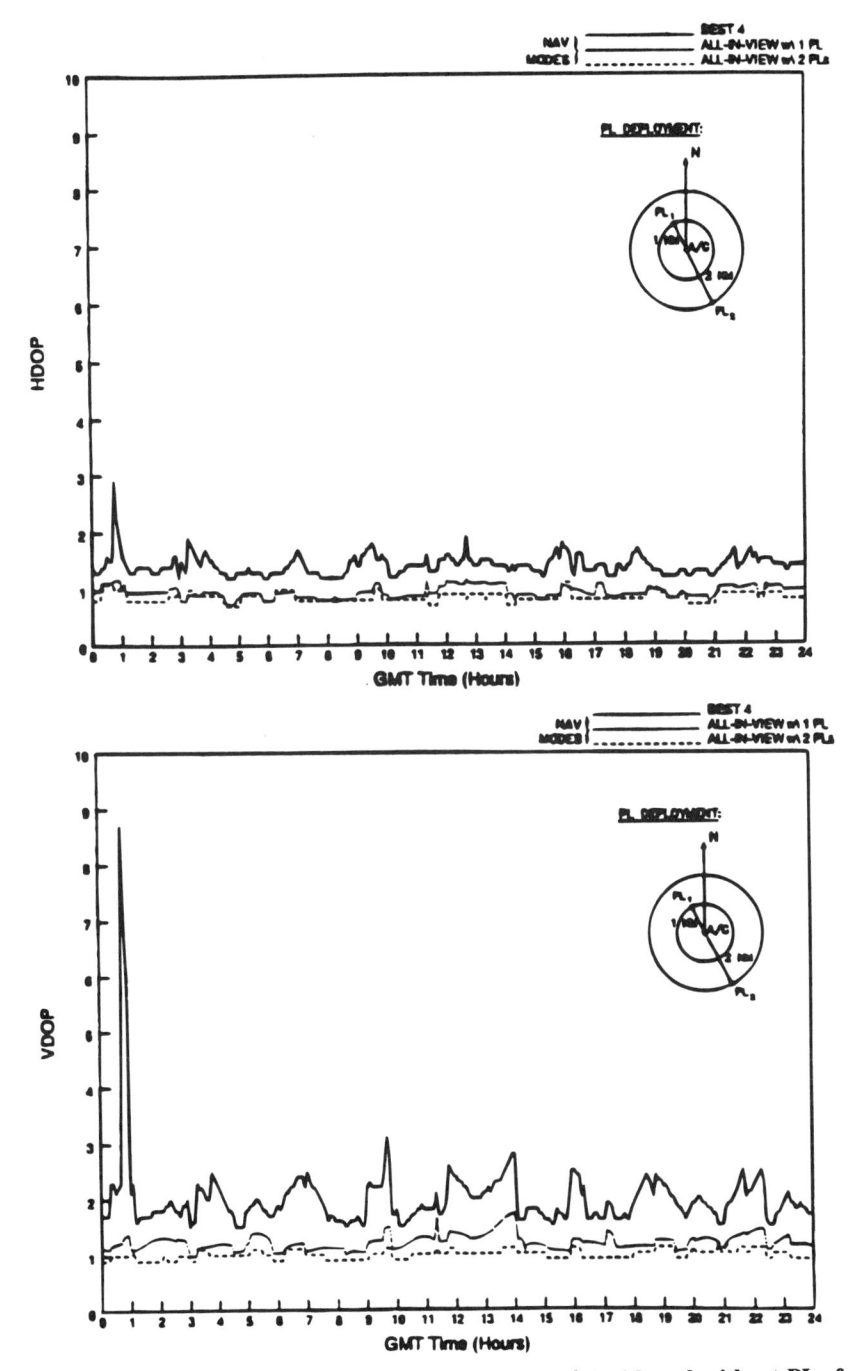

Fig. 3 Worst-case dilution-of-precision profiles at 35°N with and without PLs for a full (24) GPS constellation.

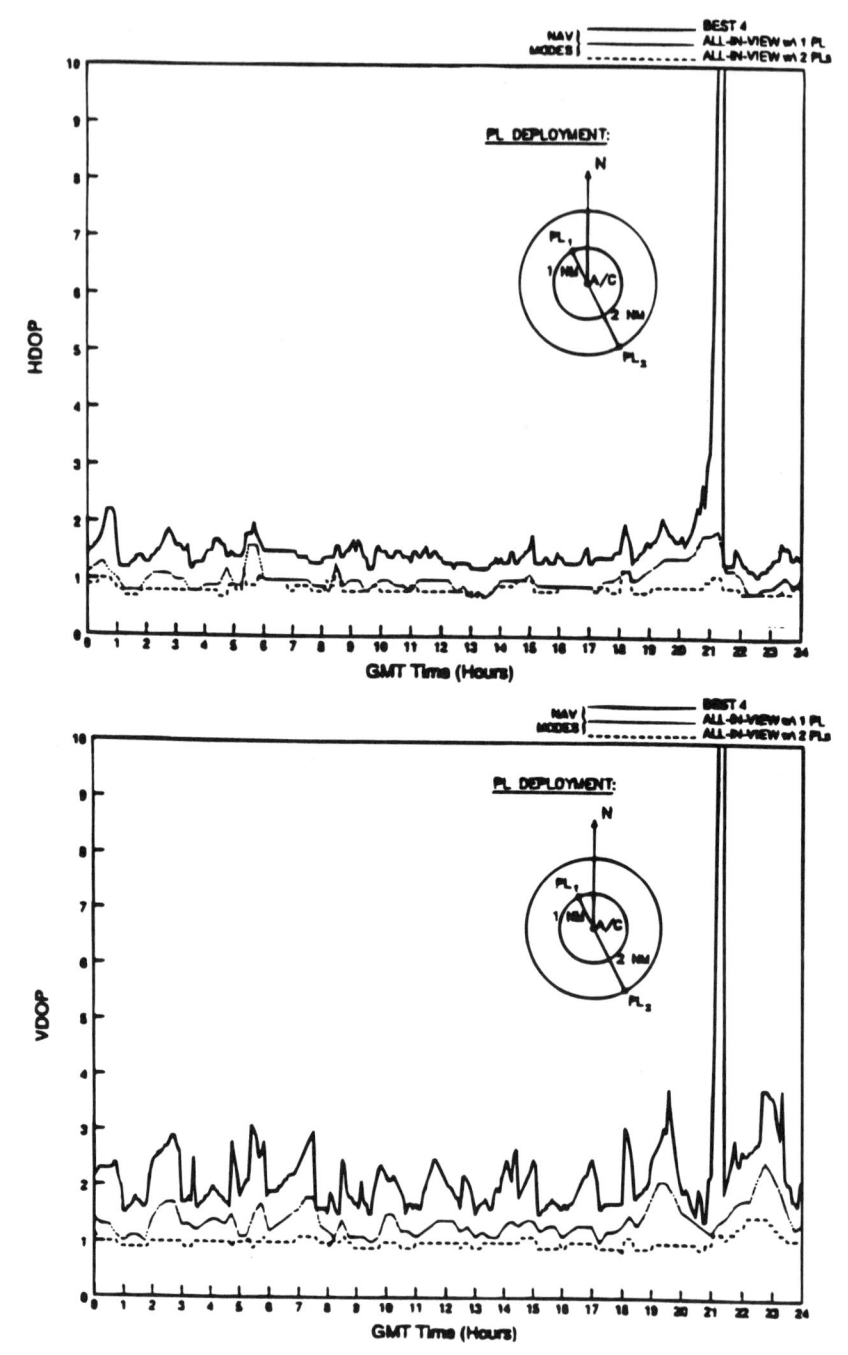

Fig. 4 Worst-case dilution-of-precision profiles at 35°N with and without PLs for a GPS constellation with one inactive satellite.

PSEUDOLITES 61

a) Master PL with Reference Receiver

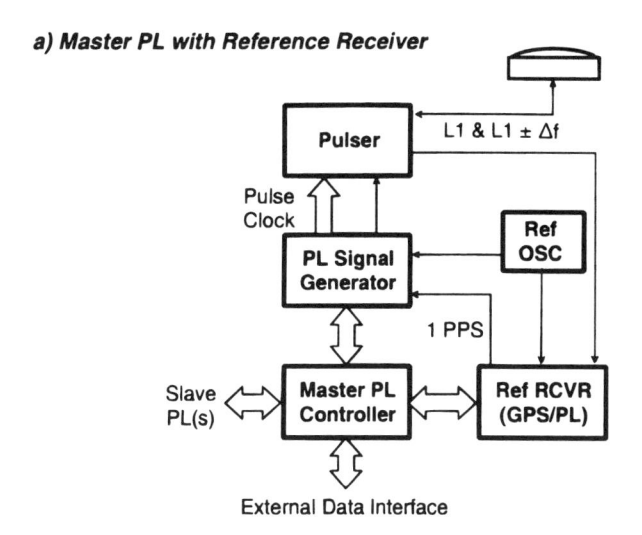

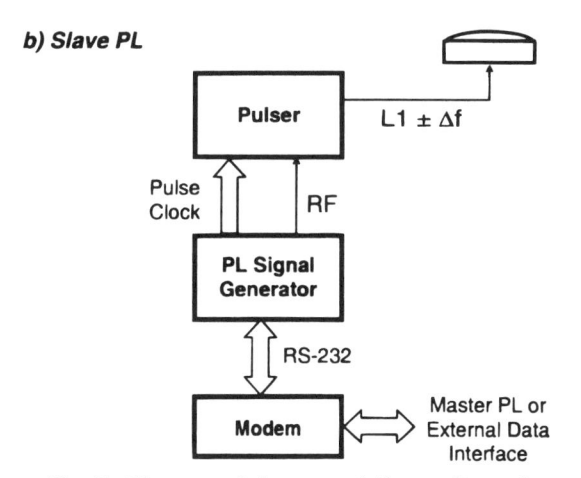

Fig. 5 Master and slave pseudolite configurations.

approach is more desirable, especially if line-of-sight (LOS) visibility to the RR might be a problem. If there is more than one PL, the remote approach with a common RR for synchronization is more desirable. However, if LOS visibility problems exist, having a RR collocated with each PL would enable time synchronization via GPS common-view, time-transfer techniques. However, only one RR can be used for deriving satellite LDGPS corrections. In this case, master and slave PLs would be designated with the master clock and LDGPS corrections derived centrally and distributed to each remote PL, which would slave its time to the master RR clock. This approach is similar to that used by various DOD test and training ranges coordinated by the Range Applications Joint Program Office (RAJPO), for which P-code PLs were developed by Stanford Telecom.

1. Master Pseudolite Configuration

Both the RR and the PL signal generator derive their timing coherently from the same stable frequency standard. The signal generator/pulser electronics module pulses the transmission of the PL signal in order to minimize interference to both participating and nonparticipating GPS receivers, as discussed in Sec. IB3. This multiplexing also allows the RR to receive the satellite signals via the same antenna. In fact, by providing a suitable calibration path, the RR can also track the output of the signal generator. In this way, the collocated PL is self-calibrating, and the transmitted PL signal will be synchronized to the same clock that is used to derive the differential GPS corrections. This is true even in the multiple-PL scenario, where the slave PLs receive differential corrections from the master PL. In this case, the time solutions of the slave PL receivers will be referenced to the master PL's clock, because the differential corrections are computed with respect to that clock.

2. Slave Pseudolite Configurations

The slave PLs need not have receivers if they can be tracked by the RR. The RR simply supplies corrections to the PL for correcting its clock via a number-controlled oscillator (NCO) and provides the satellite differential corrections for modulation on the PL signal. Because the RR can update the PL continuously, a good quality crystal oscillator will suffice as its frequency reference. Otherwise, its configuration is identical to the co-located configuration but without the RR and the self-calibration path.

C. User Aircraft Antenna Location

Reception of the PL signal by a user will be affected by the aircraft antenna location and the PL position relative to its approach path. Ideally, the PL signals would be received by a top-mounted antenna, the same one used for receiving the satellite signals. This could be accommodated by locating the PL antenna at an appropriate elevation and offset from the glideslope. Even if the line of sight to the PL is below the aircraft antenna horizon, the increased signal level in the vicinity of the PL will tend to cancel the loss in antenna gain.

If the aircraft were to pass directly over a PL, a larger angular gradient would be available to support the positioning process. However, a bottom-mounted antenna and a separate front end to interface the antenna(s) to the receiver may be needed. A configuration with a bottom-mounted antenna would likely have more sensitivity to ground-reflected multipath and interference. A top-mounted antenna is less sensitive, because reflective surfaces on the aircraft are typically small relative to the C/A-code chip width (293 m). In addition, the use of a bottom-mounted antenna would introduce additional sensitivity to lever arm uncertainty and knowledge of aircraft attitude information. On the other hand, it could be used directly, if a PL were to be employed only as a datalink. A bottom antenna *is* required for the small bubble concept.[3] Obviously, further

PSEUDOLITES

analysis and testing are needed before the aircraft antenna location issue is resolved, one way or the other.

D. Pseudolite Signal Data Message

A PL offers the possibility for an order-of-magnitude increase in the data rate (up to 1000 bps vs 50 bps for GPS) that can be supported via a GPS-compatible signal with essentially a firmware change in the user receiver. Validation of data link performance at this rate is a key test objective. A closely related issue is the PL message capacity required to support GPS and PL integrity updates and DGPS corrections (code and/or carrier).

Some have proposed that for the latter, a 2400 bps data link would be needed to supply raw carrier-phase and other GPS information from a reference receiver at a 0.5–1.0 Hz update rate. Efforts by RTCM SC-104 (Special Committee 104) and others contend that KDGPS could be supported with much lower data transfer requirements ($\leq$ 1 kbps) based on the use of carrier phase *corrections,* not raw data.[15] A related issue for testing is whether PL aiding of carrier-phase ambiguity resolution and cycle slip maintenance procedures could reduce the data requirements still further.

1. General Format

For efficiency, the general format is patterned after the WAAS format[16] with three differences—it may or may not include forward error correction (FEC), a 7-bit distributed time word is added and the data rate can be higher than 250 bps, up to 1 kbps without FEC. This general PL format is shown in Fig. 6. The time word is added as a convenience to the user receiver, because PL time is not the same as GPS time. This is because the code-chipping rate is offset in frequency by approximately $\pm$ 664 chips per second, the feature that eliminates cross-correlation with the GPS signals. The actual PL frequency can be chosen (at an offset from the spectral null of 710.4166667 Hz) so that the PL week is exactly 393 s shorter or longer than the GPS week. In fact, PL time is, at any GPS time, as follows:

$$t_{PL} = \left(1 \pm \frac{393}{604,800}\right) t_{GPS}$$

depending upon which null is selected. The PL time is distributed over three 250-bit subframes making up a 21-bit word. The 21-bit time word represents the

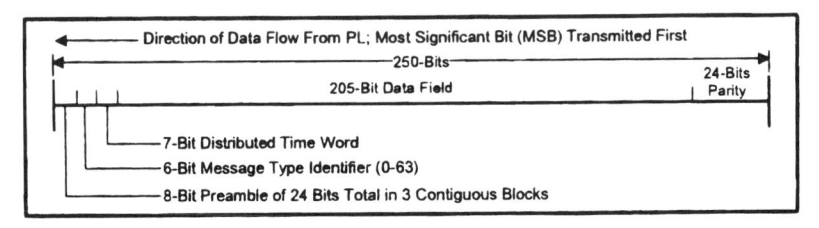

Fig. 6 Pseudolite message format.

PL time of week at the start of the currently transmitted 24-bit preamble, starting with 0 at the beginning of the week. This PL time word also serves as the reference time for the data in the messages.

The 24-bit parity is the same as the cyclic redundancy check (CRC) parity specified for the WAAS.[16] The data field consists of 205 information-bearing bits.

2. Message Types

To avoid confusing the message types with those of the WAAS, the message type numbers start at 40. Table 2 lists a tentative set of PL message types. Every message would include a certain number of GPS/PL signal integrity flags to provide a short time to alarm capability.

3. Message Content

The contents of the messages are somewhat different from the message contents defined for the RTCA SCAT-I (Special Category I) DGPS.[17] This is to accommodate the 205-bit data fields and to provide data that are more consistent with Category II and III precision approach requirements. The SCAT-I messages are quite inefficient in requiring too much signal bandwidth. Preliminary message contents show more than adequate bandwidth in using the maximum 1 kbps capability of the proposed PL signal. Every message broadcast contains integrity flags for several PRN numbers, with a minimum of 11, including the transmitting PL. This allows for a positive integrity indication at least once per 0.5 s. PRN numbers that accompany the flags do not have to be broadcast in order, so an alarm can always be inserted in any 0.25 s message providing a maximum time to alarm of 0.5 s.

Message Type 42 consists of a slight modification to the content of the corresponding RTCM carrier-phase corrections message.[15] Each can accommodate four satellites/pseudolites. Therefore, both pseudorange and carrier-phase corrections can be broadcast.

E. GPS/Pseudolite Navigation Filter Considerations

When pseudorange (or carrier-range) measurements are processed by the user navigation filter, modeling to accommodate the nonlinearity in the measurement

Table 2 Pseudolite message types

Type	Contents
40	Don't use this PL for anything (PL testing)
41	Integrity flags/Pseudorange corrections
42	Integrity flags/Carrier-phase corrections (if required)
43	Integrity flags/PL location and PRN assignment
44	Integrity flags/PL almanacs
45	Integrity flags/Precision approach path definition
46	Integrity flags/Special message
63	Null message—Alternating 1s and 0s

model can be ignored in the case of satellites, but not necessarily for PLs. As the user range to the PL diminishes during an approach, the impact of the nonlinearity is to introduce an apparent bias into the measurements. If this error is comparable to the measurement error, then a standard extended Kalman filter (EKF) will yield inferior performance. Filter divergence may occur, as the combination of measurement and nonlinearity error exceeds the filter's own computation of rms error, and it rejects new measurement data.

One approach to preventing filter divergence of this sort is to choose sufficiently large a priori measurement variances that include worst-case nonlinearity effects. Unfortunately, this requires identification of what *is* worst case. More importantly, high constant measurement variances will cause sluggish performance at other times when there is a negligible nonlinear effect present.

Another approach, as outlined in Appendix C, is to increase the filter sophistication by using a Gaussian second-order (GSO) filter that is similar to a (linearized) EKF but includes a quadratic component of the measurement nonlinearity. The key benefit is that it offers improved performance when the nonlinear effect is present but reverts naturally to a standard EKF when it is not. A possible downside is that more software and processing time are required, although some approximations are possible depending upon the scenario. Pseudolite siting relative to the approach trajectory will also be a factor in this.

IV. Pseudolite Testing

The PL concept for augmenting GPS, as described above, is intended to provide users with the following potential benefits: 1) enhanced local area navigation performance via integrated GPS/PL positioning using single or multiple PL signals with airportwide or even terminal area coverage depending upon the power level and antenna configuration (s) employed; 2) more data link capacity to support DGPS operations directly (code and/or carrier-based) at a multiple N of 50 bps (where $N = 2, 4, 5, 10$, or 20); and 3) mitigation of potential interference to standard GPS signals through the code, frequency, and time diversity techniques employed in the PL signal design.

Initial testing to verify these features has been conducted under a research and development project sponsored by the FAA,[18] and further comprehensive testing is planned. The following subsections describe the test results.

A. Pseudolite Interference Testing

Initial laboratory tests to evaluate PL interference impacts on GPS receivers was conducted by applying simulated GPS and PL signals at various power levels to several commercial receivers. After initial signal acquisition and steady-state operation by the GPS receiver was reached, a PL signal was injected at gradually increasing power levels. Signal quality reported in terms of GPS carrier power-to-noise power density ratio (C/N_0) was recorded as a function of average PL signal power. The plots in Fig. 7 show the test results obtained for two receivers under the following PL signal conditions: 1) no pulsing, no frequency offset; 2) no pulsing, frequency offset applied; 3) pulsing applied, no frequency offset; and 4) pulsing and frequency offset applied.

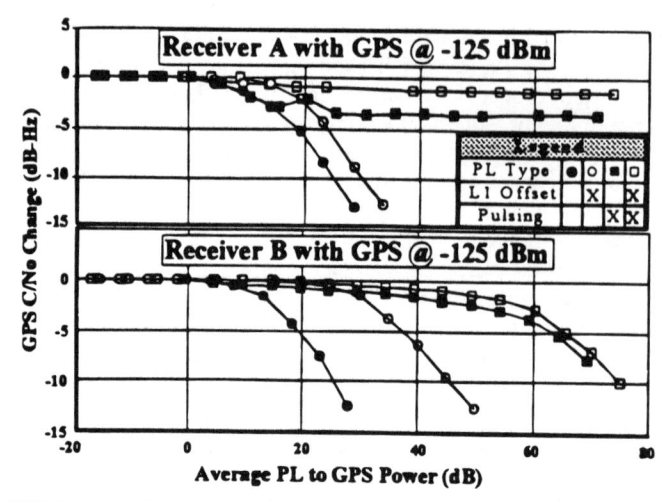

Fig. 7 GPS degradation vs pseudolite signal power with and without pulsing and frequency offset.

With no pulsing or frequency offset applied, the degradation is significant at just a 20 dB advantage in PL over GPS signal power. With either pulsing or frequency offset applied, the degradation is less, although pulsing is the more effective feature. With both applied, the degradation in reported C/N_0 from the no PL signal case is small (≤ 2 dB) over a range of 60 dB or more in the ratio of PL to GPS signal power. The results for this case from testing four receivers (two L_1-C/A only and two L_1/L_2 cross-correlating types) are shown in expanded scale in Fig. 8. When compared to a user receiving a PL signal at a range of 20 n.mi. (37 km) at a level comparable to GPS, this means that a degradation of only 2 dB would be experienced by a user only 0.02 n.mi. (37 m) away but receiving the PL signal at 60 dB higher power!

Note that receiver A's degradation with a continuous signal, or with either a frequency offset or pulsing, but not both, is more than that of receiver B. This

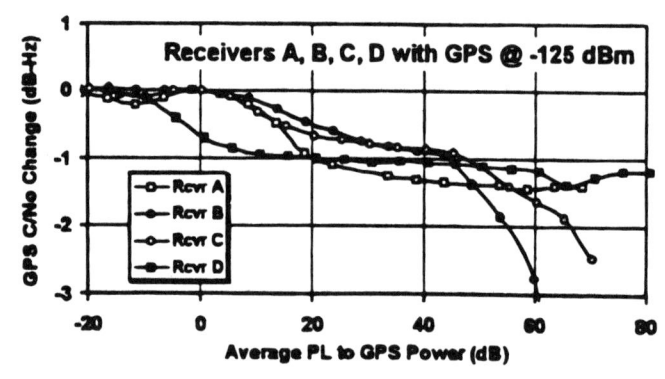

Fig. 8 GPS degradation vs pseudolite signal: 4 receivers with pulsing and frequency offset.

illustrates the difference between *soft-limiting* and *hard-limiting*. Receiver A uses multibit sampling. However, with both features turned on, receiver A's performance is quite good. Note also that receiver B's performance tails off as the PL signal gets quite strong (> 50 dB above GPS). Receiver B is a discontinued model, and receiver C is its replacement. Receiver C did not exhibit the same behavior, which was probably because of slow saturation recovery in the receiver B's front end. For receivers with fast enough saturation recovery, the results agree well with Eq. (B12) and Eq. (B13) of Appendix B.

The PL interference test results discussed above were obtained with *high-end* receivers capable of reporting signal quality data. Testing of other relatively low-cost commercial receivers was also done on a qualitative basis. These receivers were placed at varying distances from an antenna radiating a PL signal with pulsing and frequency offset features applied or not. With the PL signal off, each receiver was set up for normal GPS operations. With the PL signal on at peak power, each receiver was moved toward the PL antenna, and the separation distance was recorded when anything anomalous appeared on its display. The results for the four units tested in this manner showed no effect until within 1–5 m of the PL antenna at a peak radiated power of +15 dBm.

B. Pseudolite Data Link Testing

Tests were performed to verify the capability to transmit, receive, and demodulate the message data encoded on a PL signal with the pulsing, frequency offset, and higher data rate features described above. During laboratory tests at Stanford Telecom, a signal generator/pulser unit was interfaced directly to a GPS/PL-capable receiver. Pseudolite signals with a fixed data message were transmitted at different power levels to simulate operations at various PL/user ranges. Results of these initial tests indicated virtually error-free data reception, with the received signal quality (C/N_0) at a level of 35 dB-Hz or more.

To support planning for more comprehensive PL testing and evaluation, preliminary flight trials of a DGPS data link provided by one PL were arranged. A GPS/PL reference receiver and the signal generator/pulser unit used for the laboratory tests were installed at the FAA Technical Center (FAATC) and configured to transmit the PL signal from an antenna on the hanger roof. The test aircraft was an FAA Aerocommander (N50) with a top-mounted, low-profile GPS antenna, a GPS/PL-capable receiver, and data processing/storage equipment. For test purposes, the PL message data (LDGPS) corrections) were encoded in a 250-bit WAAS message format and transmitted without forward error correction at 250 bps.

Flight profiles flown with the test aircraft included straight-in approaches to Runway 6 at FAATC and various holding patterns ranging up to 29 km (15 n.mi.) from the PL. Received PL message data and signal quality data (C/N_0) were recorded onboard during flight segments, such as A, B, and C in Fig. 9.

Corresponding C/N_0 profiles and occurrences of individual data message errors are shown in Fig. 10. The PL data error occurrences appear to be predominantly associated with PL antenna gain reduction/obstruction phenomena during turns. Future flight trials will help address the impact of aircraft antenna type on PL signal-tracking performance. The test aircraft will be equipped to receive PL

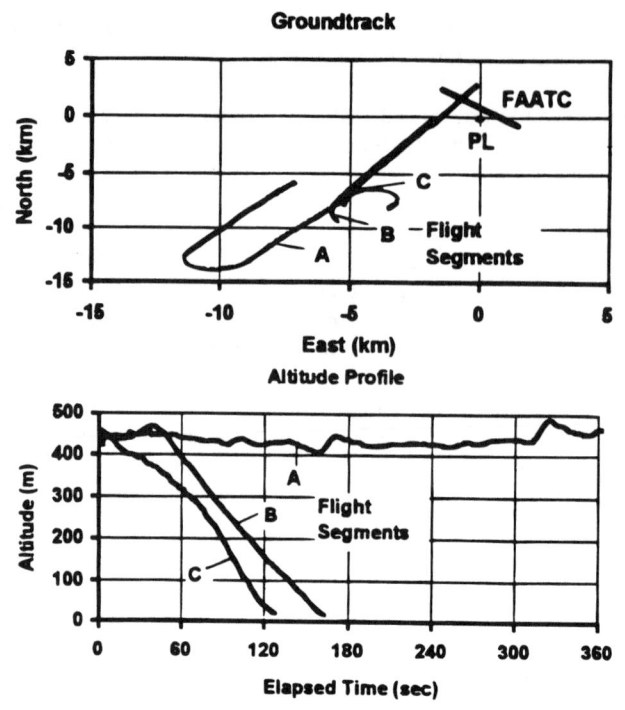

Fig. 9 Examples of flight test segments at the FAA Technical Center during pseudolite data link tests.

signals from top- and bottom-mounted, low-profile antennas, and a high-profile (blade) antenna. These trials will also assess PL data link reliability at other data rates (500 and 1000 bps).

C. Navigation Performance Testing

The initial laboratory tests and preliminary flight testing have focused on demonstrating the feasibility of PL tracking and interference mitigation to GPS signals with the pulsing and L_1 offset features applied. Additional aspects that remain to be assessed include the following:
1) PL code and carrier measurement quality
2) accuracy performance enhancement in LDGPS/PL and KDGPS/PL modes
3) sensitivities to PL siting (number/location) and user antenna (type/location)

Future flight trials are planned to assess navigation performance in LDGPS/PL and KDGPS/PL modes, and the sensitivity to PL siting and type/location of user antenna employed.

Independent KDGPS-based tracking of the test aircraft will be used as a truth source for assessing the consistency of PL ranging data (code and carrier) throughout a terminal area flight envelope.

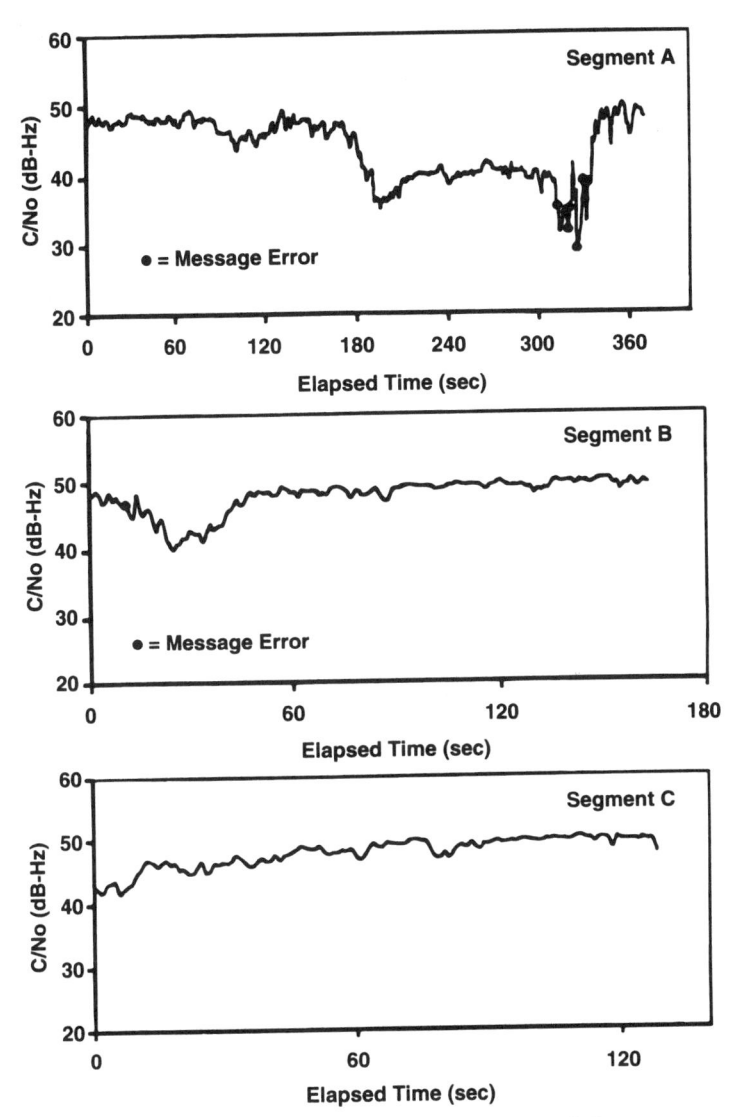

Fig. 10 Pseudolite signal profiles during data link tests.

Appendix A: Interference Caused by Cross Correlation Between C/A Codes

The effect that one C/A code has on another with respect to cross-correlation properties was described in Ref. 10 (see also Chapter 3, the companion volume). A more rigorous derivation is provided here based on the receiver signal-processing model shown in Fig. A1.

In this signal-processing model, the input signals at ①, in terms of in-phase and quadraphase components, are the following desired signal:

$$I_{i1}(t) = A_i C_i(t)\cos(2\pi\Delta f_i t + \phi_i)$$
$$Q_{i1}(t) = A_i C_i(t)\sin(2\pi\Delta f_i t + \phi_i) \qquad (A1)$$

and the undesired signal

$$I_{j1}(t) = A_j C_j(t + \tau_j(t))\cos[2\pi(\delta f_j + \Delta f_j)t + \phi_j]$$
$$Q_{j1}(t) = A_j C_j(t + \tau_j(t))\sin[2\pi(\delta f_j + \Delta f_j)t + \phi_j] \qquad (A2)$$

where

$$A_i, A_j = \text{signal amplitudes}$$
$$C_i(t), C_j(t) = \text{signal C/A codes}$$
$$\Delta f_i, \Delta f_j = \text{signal Dopplers}$$
$$\phi_i, \phi_j = \text{signal phases}$$
$$\delta f_j = \text{frequency offset of undesired signal}$$
$$\tau_j(t) = \text{time offset between signals}$$

The Doppler removal process eliminates the desired signal's Doppler and phase so that the signal components at ② in Fig. A1 are as follows:

$$I_{i2}(t) = A_i C_i(t)$$
$$Q_{i2}(t) = 0 \qquad (A3)$$

$$I_{j2}(t) = A_j C_j(t + \tau_j(t))\cos[2\pi(\delta f_j + \Delta f_j - \Delta f_i)t + \phi_j - \phi_i]$$
$$Q_{j2}(t) = A_j C_j(t + \tau_j(t))\sin[2\pi(\delta f_j + \Delta f_j - \Delta f_i)t + \phi_j - \phi_i] \qquad (A4)$$

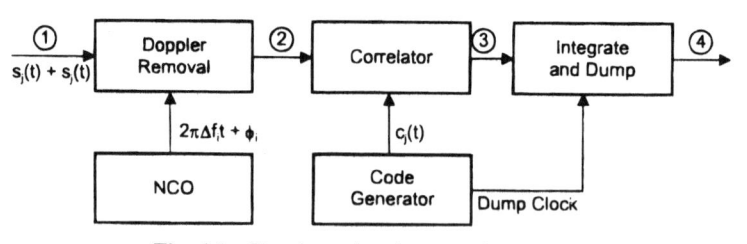

Fig. A1 Receiver signal-processing model.

PSEUDOLITES

where the Doppler difference is as follows:

$$\Delta f_{ij} = \delta f_j + \Delta f_j - \Delta f_i \tag{A5}$$

For simplicity, assume that the code transitions line up. Thus, for the moment, assume the following:

$$\tau_j(t) = NT_c \tag{A6}$$

where N is an integer, and T_c is a chip width (1/1,023,000 s). Also, assume full correlation for signal i. Then, at ③ we have the following:

$$I_{i3}(t) = A_i \tag{A7}$$
$$Q_{i3}(t) = 0$$

$$I_{j3}(t) = A_j C_j(t + NT_c) C_i(t) \cos(2\pi \Delta f_{ij} t + \phi_j - \phi_i) \tag{A8}$$
$$Q_{j3}(t) = A_j C_j(t + NT_c) C_i(t) \sin(2\pi \Delta f_{ij} t + \phi_j - \phi_i)$$

Now, because of the cycle-and-add property of the C/A-codes, Eq. (A8) becomes the following:

$$I_{j3}(t) = A_j C_k(t) \cos(2\pi \Delta f_{ij} t + \phi_j - \phi_i) \tag{A9}$$
$$Q_{j3}(t) = A_j C_k(t) \sin(2\pi \Delta f_{ij} t + \phi_j - \phi_i)$$

where $C_k(t)$ is another code in the same family, which is different for each value of N. The signal components at ④ are then as follows:

$$I_{i4}(t) = A_i T \tag{A10}$$
$$Q_{i4}(t) = 0$$

$$I_{j4}(t) = A_j \int_0^T C_k(t) \cos(2\pi \Delta f_{ij} t + \phi_j - \phi_i) \, dt \tag{A11}$$
$$Q_{j4}(t) = A_j \int_0^T C_k(t) \sin(2\pi \Delta f_{ij} t + \phi_j - \phi_i) \, dt$$

where T is a multiple M of 1 ms C/A-code repetition periods.

Power in the two correlated signals is given as follows:

$$2P_{i4} = I_{i4}^2 + Q_{i4}^2 = A_i^2 T^2 \tag{A12}$$

$$2P_j(\Delta_{ij}) = I_{j4}^2 + Q_{j4}^2$$
$$= A_j^2 \left\{ \left[\int_0^T C_k(t) \cos(2\pi f_{ij} t) \, dt \right]^2 + \left[\int_0^T C_k(t) \sin(2\pi f_{ij} t) \, dt \right]^2 \right\} \tag{A13}$$

Note that, through expansion using trigonometric identities, the dependence upon the phase difference has been removed in Eq. (A13), which resembles the expression for a Fourier power spectrum component at the Doppler difference. The code repeats at a 1-kHz rate, and the integration over each 1 ms code period is identical for Doppler differences of multiples of 1 kHz. Thus, the ratio of Eq. (A13) to Eq. (A12) the cross-correlation power ratio, can be reduced to the following:

$$P_{ij}(n) = \frac{A_j^2}{A_i^2} \left\{ \left[\frac{1}{2L} \int_0^{2L} C_k \cos\left(\frac{n\pi}{L} t\right) dt \right]^2 + \left[\frac{1}{2L} \int_0^{2L} C_k \sin\left(\frac{n\pi}{L} t\right) dt \right]^2 \right\} \quad (A14)$$

where

$$\Delta f_{ij} = n/2L = 1000\, n \text{ Hz}$$

$$2L = 0.001 \text{ s} = T/M$$

The term in the brackets of Eq. (A14) can be recognized as the nth power spectrum component. Obviously, Eq. (A13) can take on values for Doppler differences other than multiples of 1 kHz, depending upon the value of T, but peaks at the 1 kHz lines. That is, there is cross-correlation at other Doppler differences, but less than at the 1 kHz lines. Because of the navigation message databits, M is limited to 20. If T were allowed to be infinite (very large), the cross correlation at these other values would approach zero. For example, if T was a large multiple of Doppler difference cycle periods, integration over each cycle would pick up a different portion of the code, or multiple codes plus a different portion of the code. If we take a Doppler difference of 50 Hz and integrate over 20 ms, then integration over the last 10 ms would cancel that over the first 10 ms, because the codes would simply be flipped. This is true for any Doppler difference that is a submultiple of 1 kHz and an integration time that is an integer number of Doppler difference cycle periods. For other Doppler differences, other than the multiples of 1 kHz, a longer integration period is required for eventual cancellation.

Equation (A14) was evaluated for equal amplitude signals for all relative code phases as defined in Eq. (A6) for two specific GPS C/A-code pairs. The first pair (PRN6/PRN28) is considered to be the *worst* pair of GPS codes; whereas the second pair (PRN7/PRN201) is made up of the *best* GPS code and the *best* selected for Inmarsat-3.[9] For each pair, the maximum spectrum components over all possible integer code phases ($0 \leq N \leq 1022$), are plotted in Fig. A2 for Doppler differences up to 20 kHz. The maximum spectrum components are also plotted in Fig. A3, for Doppler differences in the range 1017–1029 kHz, which corresponds to one of the signals transmitting in the first null of the other. This ± 6 kHz Doppler range represents the maximum expected from satellite and user motion. The average over all N computed for the PRN6/PRN28 pair is also plotted in Fig. A3 for comparison. The average is approximately equal to the following spectral line envelope:

$$S(f) = \frac{1}{1023} \frac{\sin^2(n\pi/1023)}{(n\pi/1023)^2} \quad (A15)$$

which is about 8–9 dB below the maximum values. Note that in both figures, it doesn't seem to matter which code pair is used when determining the worst-case

PSEUDOLITES 73

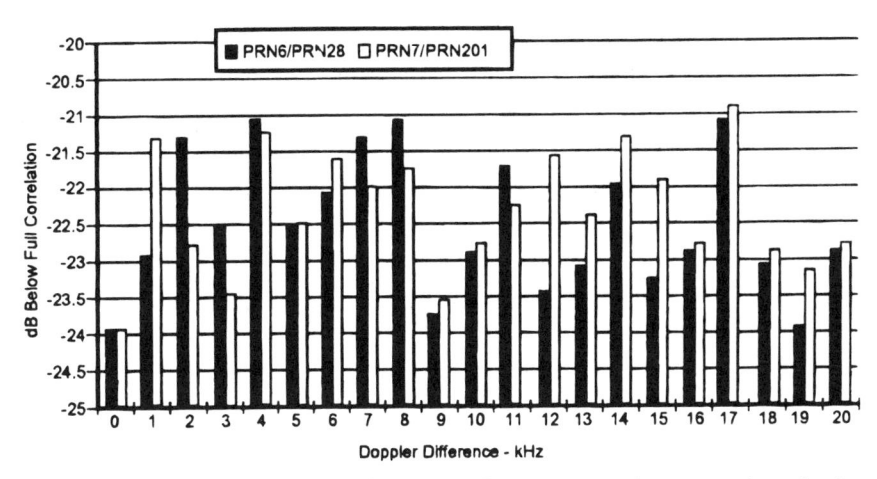

Fig. A2 Maximum cross-correlation spectral components for two code pairs in Doppler difference range of 0–20 kHz.

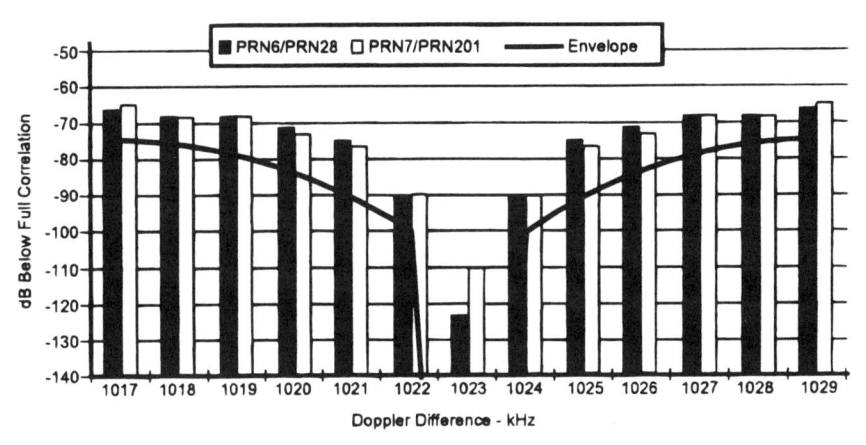

Fig. A3 Maximum cross-correlation spectral components for two code pairs in Doppler difference range of 1017–1029 kHz.

magnitudes. This is because each of the 1023 code phases ($0 \leq N \leq 1022$) results in a different code. Consequently, codes with bad (i.e., large) spectral components are generated in each case, and some of these codes are unbalanced, as well.

Although some of the *worst*-case components shown in Fig. A3 slightly exceed the predicted −80 dB level stated in Sec. IIB1, they are still typically below −70 dB. Within ±2 kHz of the null, they are, in fact, below −80 dB. More significantly, all components are a good 50–60 dB lower than the level near center frequency. This is in addition to the margin realized from pulsing with a 1/11 duty cycle. (See Appendix B)

The derivation presented above applies when the cross-correlation code transitions are lined-up, which, of course, they rarely will be. Spilker[10] points out that, at the n kHz carrier frequency differences, there also exists a code frequency shift, which is less by a factor of 1540. Thus, for example, at the 1 kHz carrier frequency difference, there is a code frequency difference of $1000/1540 \approx 0.65$ chips per second. Thus, the code transitions will not stay lined-up. This is especially true for a large frequency difference of 1.023 MHz, in which case the code frequency difference is approximately 664 chips per second. McGraw[12] pointed out that cross-correlation levels can be even higher when the code transitions are not lined up. However, because of the rapidly changing time relationships between the codes, the resulting cross-correlation becomes noise-like, and simply becomes a noise interference. The effect of this interference is addressed in Appendix B.

Appendix B: Interference Caused by Pseudolite Signal Level

One C/A-code signal can interfere with another if it is strong enough, independent of the cross-correlation. This can certainly happen in the case of a PL signal, which may interfere with satellite signals as well as with other PL signals. In other words, a PL signal is a noise source that may jam the other signals unless measures are taken to mitigate this effect. The following provides an assessment of the impact of that jamming and a method for minimizing it; namely, by pulsing the PL signal.

Background

In general, the spreading process in a receiver's correlator is defined as a signal (noise or otherwise) being passed through a filter described with a frequency response equal to the spectral density of the PRN code. That is, the interference noise density at the output of the correlator is as follows:

$$N_{0I} = \int_{-\infty}^{\infty} S_c(f) S_I(f) \, df \qquad (B1)$$

where $S_c(f)$ is the spectral density of the reference PRN code and $S_I(f)$ is the density of the interference or noise. The reference C/A-code has a discrete spectral density that can be described as follows:

$$S_c(f) = \sum_{n=-\infty}^{\infty} c_n \delta(f - 1000\, n) \qquad (B2)$$

where the c_n are spectral line coefficients: $\delta(f)$ is the dirac delta function; and the c_n vary about the envelope of Eq. (A15) in Appendix A. The reference C/A-code spectral density has the property that

$$\int_{-\infty}^{\infty} S_c(f) \, df = 1 \qquad (B3)$$

First consider bandlimited thermal noise with density N_0 with a two-sided intermediate frequency (IF) bandwidth of B_I (brick wall filter).

$$N_{0T} = N_0 \int_{-B_I/2}^{B_I/2} S_c(f) \, df \le N_0 \qquad (B4)$$

The variation of the C/A-code spectral lines averages out over the wide bandwidth.

PSEUDOLITES

A similar equation applies for wide- and narrow-band interference with a spectral density as follows:

$$S_I(f) = \frac{P_I}{f_u - f_l} \tag{B5}$$

for upper and lower frequency limits f_u and f_l (converted to IF) and total interference power P_I (relative to the signal power S).

$$N_{0I} = \frac{P_I}{f_u - f_l} \int_{\max(-\alpha_1,\alpha_3)}^{\min(\alpha_1,\alpha_2)} S_c(f)\, df \tag{B6}$$

where

$$\alpha_1 = B_I/2,\ \alpha_2 = f_u - f_{IF}\ \text{and}\ \alpha_3 = f_l - f_{IF}.$$

For narrow bandwidth noise interference centered at the GPS frequencies (i.e., somewhat less than the code chipping rate $1/T_c$), this equation becomes as follows:

$$N_{0I} \approx P_I T_c \tag{B7}$$

This is only true for the C/A code where interference bandwidths are on the order of 100 kHz or greater because of the variations in the line spectrum of the codes.

Pseudolite Interference

Now assume that the interference is another C/A-code signal with the following spectral density:

$$S_{PL}(f) = P_{PL} \sum_{n=-N_B}^{N_B} c'_n \delta(f - 1000\, n \pm 1.023 \times 10^6) \tag{B8}$$

where P_{PL} is the received PL signal power, and N_B indicates the band-limiting effect. Because the summations in Eqs. (B2) and (B8) are over a large number of spectral components, it suffices to use the average envelope of Eq. (A15), divided by 1000 to spread the components into a continuous spectral density, for evaluation. Then, Eq. (B1) can be approximated with the following integral:

$$N_{0PL} \approx P_{PL} T_c^2 \int_{-B_I/2}^{B_I/2} \frac{\sin^2(\pi f T_c)\sin^2[\pi(f \pm 1/T_c)T_c]}{(\pi f T_c)^2[\pi(f \pm 1/T_c)T_c]^2}\, df \tag{B9}$$

This can be compared to the interference noise density from one GPS satellite j to another satellite i at the normal frequency given by the following:

$$N_{0S_{ij}} \approx P_{S_j} T_c^2 \int_{-B_I/2}^{B_I/2} \frac{\sin^4(\pi f T_c)}{(\pi f T_c)^4}\, df \tag{B10}$$

In general, Eqs. (B9) and (B10) must be evaluated numerically. This was done over a wide bandwidth of $\pm 10/T_c$, resulting in the following relationship:

$$10 \log_{10}\!\left(\frac{N_{0PL}}{N_{0S_{ij}}}\right) = 10 \log_{10}\!\left(\frac{P_{PL}}{P_{S_j}}\right) - 8.19\ \text{dB} \tag{B11}$$

which indicates that the PL interferes by 8.19 dB less by transmitting in the null, than it would if it were transmitting at the same frequency as the satellites. However, this is not enough!

The first term of Eq. (B11) can become significant as the user receiver comes closer to a PL. This is a key reason for PL pulse modulation, which has the effect of only interfering a percentage of the time equal to the pulse duty cycle, provided that the receiver clips the pulses. This results in a loss in received satellite C/N_0 of either:

$$L = -10 \log_{10}\left[1 - DC + \frac{2R_{\Delta f}}{3}\left(\frac{L_{max}}{L_N}\right)^2 \frac{DC}{(1-DC)^2} B_I T_c\right] \quad \text{(B12)}$$

or

$$L = 10 \log_{10}\left[1 + \frac{2R_{\Delta f}}{3}\frac{P_{PL}}{N_0} T_c \cdot DC\right] \quad \text{(B13)}$$

depending upon whether or not the PL signal is saturating the receiver's analog-to-digital (A/D) sampler (*soft* or *hard-limiting*), where $R_{\Delta f}$ is the reduction in interference realized using a frequency offset (0.1517, if ± 1.023 MHz, and 1, if on frequency), L_{max} is the maximum sampler threshold level, L_N is the one sigma noise level in terms of the threshold level and P_{PL}/N_0 is the average received PL signal-to-noise density. Eq. (B12) represents the loss when the PL pulses are saturating the A/D and Eq. (B13) represents the loss when they are not. Note that in the former case, the loss is proportional to the ratio of IF bandwidth to the code bandwith.

Cross-correlation can still occur, even if the pulses are clipped. However, as was shown in Appendix A, the cross-correlation is reduced substantially by PL transmission in the first null of the satellite C/A-code/spectrum ($L_1 \pm 1.023$ MHz). Pseudolite transmission at higher nulls is also possible, but this begins to add more complexity to a receiver designed to process both GPS and PL signals. Another key reason for pulsing is to prevent capturing the front end of the user's receiver. This is especially important for hard-limiting receivers (1-bit samplers) that normally do not employ front-end AGC circuits.

Appendix C: Navigation Filter Modeling with Pseudolite Measurements

A GPS pseudorange measurement (and carrier range) is generally a nonlinear function of the satellite position vector for (r_s) and the user position vector (r_u). This measurement is modeled by the following:

$$z = g + b_u + v \quad \text{(C1)}$$

where

$$g = |r_s - r_u| \quad \text{(C2)}$$

is the geometric range; b_u is the user clock offset from GPS time, and v represents the composite of uncorrected measurement errors caused by atmospherics, satellite timing offsets, multipath, and noise. Given the satellite ephemeris and an a priori estimate of the user location (r_u), then g can be represented by the Taylor series expansion:

$$g = g_0 + h\Delta r_u + \Delta r_u^T J \Delta r_u/2 + \cdots \tag{C3}$$

where

$$\Delta r_u = r_u - r_u^- \tag{C4}$$

$$g_0 = |r_s - r_u^-| \tag{C5}$$

$$h = \partial g/\partial r_u|_{r_u = r_u^-} \tag{C6}$$

$$J = \partial^2 g/\partial r_u \partial r_u|_{r_u = r_u^-} \tag{C7}$$

Linear Measurement Model

Given the large user/satellite separation and slowly changing geometry, common practice is to employ an extended (linearized) Kalman filter that encompasses only the first two terms for g in Eq. (C3). The standard EKF equations for updating the a priori estimate of user position (r_u) and clock bias (b_u) are as follows:

$$x_u^+ = x_u^- + k(z - g_0)$$
$$k = P^- h^T/(hP^- h^T + \sigma_v^2)$$
$$P^+ = (I - kh)P^-(I - kh)^T + k(\sigma_v^2)k^T \tag{C8}$$

where

$$x_u \equiv \begin{pmatrix} r_u \\ b_u \end{pmatrix} \tag{C9}$$

and

$$h \equiv [(r_s - r_u)^T/g \quad 1] \tag{C10}$$

Also, P^- is the covariance of the a priori state estimate (x_u), and σ_v^2 is the variance of the measurement error v, assumed to be Gaussian white noise.

Nonlinear Measurement Model

When PL measurements are introduced, the user/PL range is much less and more dynamic. Consequently, a Gaussian second-order filter[19] can be employed that accounts for the measurement nonlinearity with the quadratic component included

$$z - g_0 = h\Delta x_u + \Delta r_u^T J \Delta r_u/2 + v \tag{C11}$$

The corresponding GSO filter equations are expressed by the following:[20]

$$x_0^+ = x_u^- + k(z - g_0 - \eta)$$

$$k = P^- h^T/(hP^- h^T + \sigma_v^2 + \sigma_n^2)$$

$$P^+ = (I - kh)P^-(I - kh)^T + k(\sigma_v^2 + \sigma_\eta^2)k^T$$

$$\eta = \text{Tr}[JP^-]/2$$

$$\sigma_\eta^2 = \text{Tr}[JP^- JP^-]/2 \qquad \text{(C12)}$$

where x_u, h, and J are as defined in this appendix. The new components account for the bias and measurement variance introduced by the quadratic nonlinearity.

References

[1]Van Dierendonck, A. J., Elrod, B. D., and Melton, W. C., "Improving the Integrity, Availability and Accuracy of GPS Using Pseudolites," *Proceedings of NAV '89* (London, UK), Royal Institute of Navigation, London, UK, Oct. 17–19, 1989, (RION Paper 32).

[2]Schuchman, L., Elrod, B. D., and Van Dierendonck, A. J., "Applicability of an Augmented GPS for Navigation in the National Airspace System," *Proceedings of IEEE*, Vol. 77, No. 11, Nov. 1989, pp. 1709–1727.

[3]Cohen, C. A., et al., "Real-Time Cycle Ambiguity Resolution using a Pseudolite for Precision Landing of Aircraft Using GPS," *Proceedings of DSNS '93*, Amsterdam, The Netherlands, March 31–April 3, 1993.

[4]Harrington, R. L., and Dolloff, J. T., "The Inverted Range: GPS User Test Facility," *Proceedings, IEEE Position, Location, and Navigation Symposium*, (PLANS '76, San Diego, CA), IEEE, New York, Nov. 1976, pp. 204–211.

[5]Klein, D., and Parkinson, B. W., "The Use of Pseudo-Satellites for Improving GPS Performance," *Global Positioning System*, Vol. II, Institute of Navigation, 1986, pp. 135–146.

[6]Stansel, T. A., Jr., "RTCM SC-104 Recommended Pseudolite Signal Specification," *Global Positioning System*, Vol. III, Institute of Navigation, 1986, pp. 117–134.

[7]Van Dierendonck, A. J., "The Role of Pseudolites in the Implementation of Differential GPS," *Proceedings, IEEE Position, Location, and Navigation Symposium*, (PLANS '90, Las Vegas, NV), IEEE, New York, March 1990.

[8]Elrod, B. D., and Van Dierendonck, A. J., "Testing and Evaluation of GPS Augmented with Pseudolites for Precision Landing Applications," *Proceedings of DSNS'93*, Amsterdam, The Netherlands, March 31, 1993.

[9]Nagle, J., Van Dierendonck, A. J., and Hua, Q. D., "Inmarsat-3 Navigation Signal C/A Code Selection and Interference Analysis," *Navigation*, Winter 1992–93, pp. 445–461.

[10]Spilker, J. J., Jr., "GPS Signal Structure and Performance Characteristics," *Global Positioning System*, Vol. I, Institute of Navigation, 1980, pp. 29–54.

[11]"Global Orbiting Navigation Satellite System (GLONASS) Interface Control Document," RTCA Paper 518-91/SC159-317 (Handout—Jan. 1992).

[12]McGraw, G. A., "Analysis of Pseudolite Code Interference Effects for Aircraft Precision Approaches," *Proceedings of ION 50th Annual Meeting* (Colorado Springs, CO), Institute of Navigation, Washington, DC, June 6–8 1994.

[13]Green, G. B., et al., "The GPS 21 Primary Satellite Constellation," *Navigation*, Spring 1989, pp. 9–24.

[14]Anon, "1992 Federal Radionavigation Plan." U.S. Dept. of Transportation, Dec. 1992, pp. 2–15.

[15]Anon., "Recommendations of the Carrier Phase Working Group to RTCM SC-104," RTCM Paper 170-921/SC104-92, Aug. 10, 1992.

[16]Anon., "Wide Area Augmentation System Signal Specification," App. 2 in U.S. DOT/FAA Specification for the Wide Area Augmentation System (WAAS), FAA-E-2892, May 9, 1994.

[17]Anon., "Minimum Aviation System Performance Standards—DGNSS Instrument Approach System: Special Category I (SCAT-I)," RTCA/DO-217, RTCA, Inc., Washington, DC, Aug. 27, 1993.

[18]Elrod, B. D., Barltrop, K. J., and Van Dierendonck, A. J., "Testing of GPS Augmented with Pseudolites for Precision Approach Applications," Proceedings of ION GPS '94 (Salt Lake City, UT), Institute of Navigation, Washington, DC, Sept. 20–23, 1994.

[19]Jazwinski, A. H., *Stochastic Processes and Filtering Theory*, Academic Press, New York, 1970, pp. 340–346.

[20]Widnall, W. S., "Enlarging the Region of Convergence of Kalman Filters Employing Range Measurements," *AIAA Journal*, Vol. 2, No. 3, 1973, pp. 283–287.

Chapter 3

Wide Area Differential GPS

Changdon Kee*
Stanford University, Stanford, California 94305

I. Introduction

IN addition to reducing cost and complexity, the GPS is expected to improve the accuracy of navigation greatly for land, marine, and aircraft users. Under normal operating conditions, it can provide positioning accuracies in the range of 15–25 m. However, with selective availability (SA) the errors incurred by typical civilian users have been found to be 100 m or more. In some situations, especially for precision landing of an aircraft or for harbor navigation, these accuracies are insufficient.

Differential GPS (DGPS) is a means for improving navigation accuracy in a local area. A single DGPS monitor station at a known location can compute a range error correction for each GPS satellite in view. These error corrections are then broadcast to users in the vicinity, as depicted in Fig. 1. By applying the corrections to the signals received, a user can typically improve the accuracy down to the 2–5 m level (see Chapter 1 of this volume and Refs. 1 and 2). However, as the distance between the user and the monitor station increases, range decorrelation occurs, and accuracy degrades. This increased error is caused by ephemeris error, ionospheric time delay error, and tropospheric error. As the user and reference station separate, the projection of the ephemeris error onto the user–satellite line of sight is no longer the same as that projected onto the monitor station–satellite line of sight, as illustrated in Fig. 2. (The accuracy of the range correction broadcast by a DGPS monitor station degrades with distance. The figure shows an ephemeris error δR that produces a small range error δR_s at the monitor station but a larger range error δR_u at the user location. If the user were to employ the range correction broadcast by the monitor station, a residual range error of $\delta R_s - \delta R_u$ would remain.)

Copyright © 1994 by the author. Published by the American Institute of Aeronautics and Astronautics, Inc., with permission. Released to AIAA to publish in all forms.
*WADGPS Algorithm Development Group Leader, WADGPS Laboratory, Department of Aeronautics and Astronautics, HEPL (GP-B).

The maximum range error difference δR_{error} between the monitor and user is given by the following:

$$\max(\delta R_{error}) \approx \frac{d}{D} \delta R \qquad (1)$$

where δR is the magnitude of the satellite ephemeris error; d is the separation between the user and the monitor station; and D is the distance from the user to the GPS satellite.

If the two receivers are widely separated, the lines of sight through the ionosphere are also different, resulting in differences in the ionospheric delay observed. A similar, but smaller effect occurs for the tropospheric delay.

Beyond a separation distance of 100 km, a scalar range error correction is not sufficiently accurate to realize the full potential of DGPS. In fact, hundreds of monitor stations would be required to provide standard single-station DGPS aiding across the entire United States. Wide area differential GPS (WADGPS)[3] provides a powerful means for bridging the gap between unaided performance and high-accuracy navigation in the vicinity of a correction station. Now the Federal Aviation Administration (FAA) is planning to implement WADGPS in the National Airspace System by 1997. Various WADGPS techniques have been suggested by Refs. 4–8.

The following sections describe the WADGPS architecture, and master station algorithms. Then user message content and format, and WADGPS error budget are discussed. The next sections describe WADGPS simulations and evaluations using actual field data. The next chapter in this volume discusses implementation of WADGPS for the National Airspace System.

II. Wide Area Differential GPS Architecture and Categories

Instead of calculating a scalar range error correction for each satellite, as is done in DGPS, WADGPS provides a vector of error corrections composed of a three-dimensional ephemeris error and clock offset for each GPS satellite, plus ionospheric time delay parameters. The accuracy of the WADGPS correction is nearly constant within the monitored region, and degrades gracefully on the perimeter.

A. Wide Area Differential GPS Architecture

The WADGPS network includes at least one master station, a number of monitor stations, and communication links. Each monitor station is equipped with a high-quality clock and a high-quality GPS receiver capable of tracking all satellites within the field of view. The GPS measurements are taken at each monitor station and sent to the master station. The master station computes GPS error components, based on the known monitor station locations and the information collected. The computed error corrections are transmitted to the users via any convenient communication link, such as satellite, telephone, or radio.

Figure 3 provides an overview of the WADGPS, and Fig. 4 shows the flow of information between the system components. The process can be summarized as follows:

1) Monitor stations at known locations collect GPS pseudoranges from all satellites in view.

WIDE AREA DIFFERENTIAL GPS

Fig. 1 Overview of differential GPS.

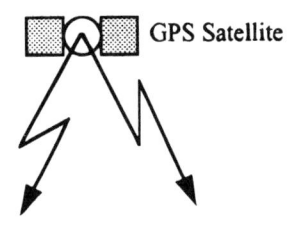

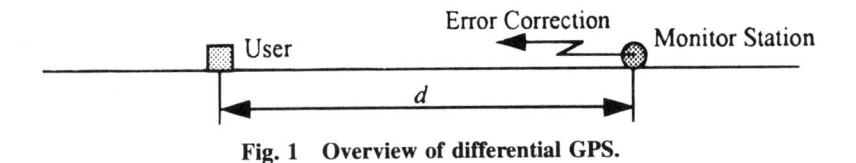

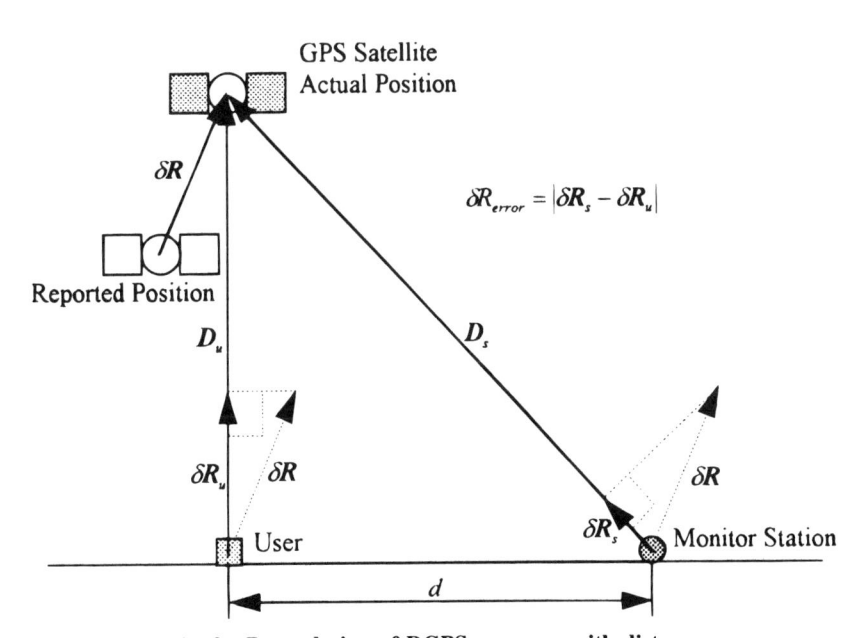

$$\delta R_{error} = |\delta R_s - \delta R_u|$$

Fig. 2 Degradation of DGPS accuracy with distance.

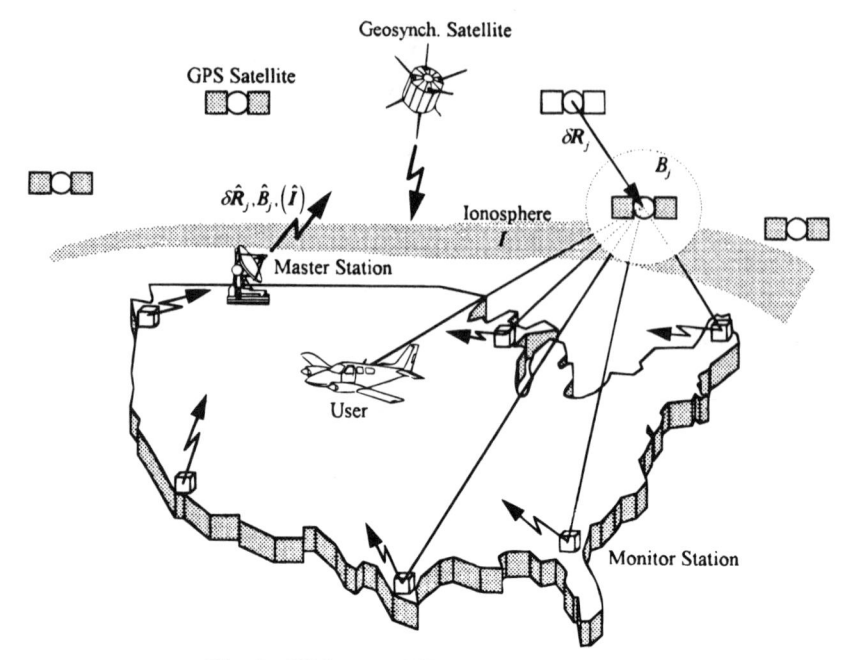

Fig. 3 Wide area differential GPS concept.

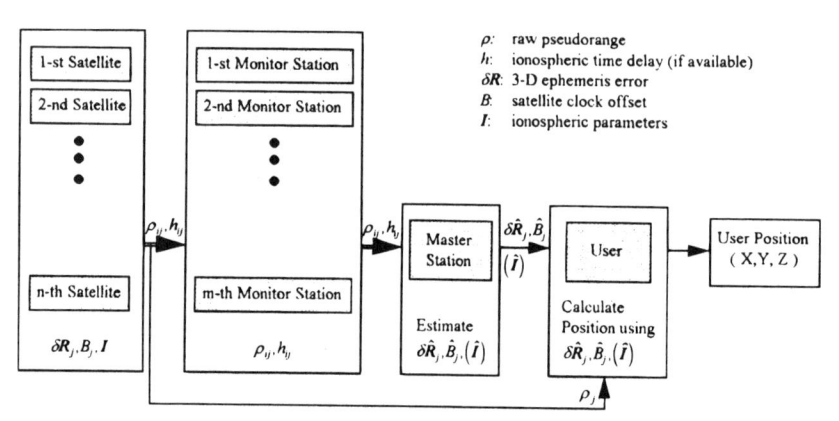

Fig. 4 Block diagram of WADGPS components.

WIDE AREA DIFFERENTIAL GPS

2) Pseudoranges and dual-frequency ionospheric delay measurements (if available) are sent to the master station.
3) Master station computes an error correction vector.
4) Error correction vector is transmitted to users.
5) Users apply error corrections to their measured pseudoranges and collected ephemeris data to improve navigation accuracy.

B. Wide Area Differential GPS Categories

The WADGPS system can be categorized by the estimator located at the master station. In the design of WADGPS, we must address such issues as the receiver required for monitor station and user, the estimation speed, which corresponds to the update rate of error corrections, and the navigation accuracy. Because the most important application of WADGPS is aviation, the navigation accuracy should be the major concern.

The master station estimates the three-dimensional ephemeris errors, the satellite clock errors, the monitor station receiver clock errors, and, optionally, the ionospheric time delay parameters. It does not transmit the monitor station receiver clock errors.

A single-frequency receiver normally provides L_1 pseudorange and continuous carrier phase as outputs. A dual-frequency receiver provides not only L_1 pseudorange and continuous carrier phase but also L_2 pseudorange and continuous carrier phase from which the ionospheric time delay can be calculated as if it were an extra output, but dual-frequency receivers are far more expensive than single-frequency receivers.

A single-frequency receiver can, in principle, estimate the ionospheric time delay by measuring the dispersive effect of the ionospheric time delay on the received code and carrier. Using the fact that the ionospheric time delays in the pseudorange and continuous carrier phase are equal in size and have opposite signs, we may be able to estimate ionospheric time delay with a single-frequency receiver[9,10] at the expense of a loss in navigation accuracy. However, the single-frequency technique needs further study. Because dual-frequency receivers (provided they are at both the monitor station and the user) directly give the ionospheric time delays, they can save time that would have to be spent on estimating the ionospheric time delay parameters in the master station and can, therefore, improve the navigation accuracy.

Three WADGPS algorithms (A, B, and C) and their performances are summarized in Table 1 and Table 2, respectively, and each algorithm is discussed in the following subsections.

1. Algorithm A

This algorithm allows both monitor station and user to use single-frequency receivers that do not provide ionospheric time delays as extra measurements. In this algorithm, the master station estimates the three-dimensional ephemeris errors, the satellite clock errors, and the ionospheric time delay parameters in one large filter using pseudoranges as the only measurement vector. Processing

Table 1 Wide area differential GPS algorithms

	Algorithms		
	A	B	C
Variables to be estimated			
Three-dimensional ephemeris errors	Yes	Yes	Yes
Satellite clock error + SA	Yes	Yes	Yes
Ionospheric parameters	Yes	Yes	No
Variables to be transmitted (error corrections)			
Three-dimensional ephemeris errors	Yes	Yes	Yes
Satellite clock error + SA	Yes	Yes	Yes
Ionospheric parameters	Yes	Yes	No
Number of master station estimators	One	Two	One
Size of master station estimator	Large	Small	Small
Required receiver			
Monitor station	Single-frequency	Dual-frequency[a]	Dual-frequency[a]
User	Single-frequency	Single-frequency	Single-frequency[b] Dual-frequency[a]

[a]There are several receivers, such as the Trimble 4000SSE, and Allan Osborne, Rogue, that can measure the ionospheric time delay even when the P-code is encrypted.
[b]Ionospheric time delay estimation with a single-frequency receiver has been demonstrated by Refs. 9 and 10.

time is the longest among three algorithms because the observation matrix is large. The transmission message consists of the ephemeris errors, the satellite clock errors, and the ionospheric parameters. There are some advantages in terms of lower cost to using this algorithm, the penalty being higher computational load and worse accuracy.

2. Algorithm B

By using the extra measurement of ionospheric time delay from a dual-frequency receiver in the monitor station we can separate the one large estimator used in Algorithm A into two small estimators. The estimation of ionospheric time delay parameters is one process, and the estimation of the three-dimensional ephemeris errors, the satellite clock errors, and the monitor station receiver clock errors is another, separate, process.[3] For the resulting algorithm (algorithm B), a dual-frequency receiver is required in the monitor station, but the user needs only a single-frequency receiver. Because the ionospheric time delays are separate measurements in a dual-frequency receiver, they are used to estimate the ionospheric parameters directly, and the pseudoranges, corrected for ionospheric time delay and tropospheric error, are fed into the other filter, which estimates ephemeris errors and clock errors.

WIDE AREA DIFFERENTIAL GPS 87

Table 2 Performances of wide area differential GPS algorithms

Performance	Algorithms		
	A	B	C
Computational load	Heavy	Light	Lightest
Navigation accuracy	Good	Better	Best

Relative to algorithm A, algorithm B has better accuracy and reduced computational load. These advantages result from the extra ionospheric measurements and the estimator is divided into two small ones, resulting in reduced matrix sizes. The transmission message to users consists of three-dimensional ephemeris errors and satellite clock errors, as well as ionospheric parameters. In principle, single- or dual-frequency users can take advantage of these error corrections to improve positioning accuracy.

3. Algorithm C

If the mobile users can measure the ionospheric time delay, the ionospheric parameters do not need to be estimated in the master station, and as a result, only three-dimensional ephemeris errors and satellite clock errors need to be estimated there. The resulting algorithm (algorithm C) requires a dual-frequency receiver in the monitor station. Because there is no ionospheric parameter estimation, the transmission message does not contain ionospheric parameters, and therefore, the computational load is smaller than that of algorithm B. A user may be equipped with a dual-frequency receiver, but such a receiver may not be required. We may be able to estimate ionospheric time delay with a single-frequency receiver at the expense of a loss in navigation accuracy.[9] The single-frequency technique needs further study, however.

If the user estimate of ionospheric time delay is accurate, then algorithm C is the most accurate. It is most accurate because it does not fit the ionosphere to the model, as is done in algorithm B. This is especially important in the equatorial and polar regions. Also T_{gd} may be an error source for algorithm B. T_{gd} is a time delay between L_1 and L_2 frequencies in the GPS satellite and is included in the ionospheric time delay measurement from dual-frequency receiver unless it is carefully calibrated and taken off the raw ionospheric time delay measurement. Thus, of the three algorithms discussed, algorithm C provides the best accuracy for users with dual-frequency receivers, but users may opt for a single-frequency receiver depending upon how much accuracy they desire.

C. User Message Content and Format

Transmission of the WADGPS correction could be accomplished by any of the following: geosynchronous satellite broadcast (see next chapter), FM subcarrier,[11] or any other suitable broadcast system. The correction could be converted to the standard differential message format developed by the Radio Technical Commission for Maritime Service Special Committee 104 (RTCM 104).[12] This

Table 3 Wide area differential GPS correction message content

Message		Update rate
SV ID	PRN	Every 5–10 s
Time tag (GPS time)	Time of transmission	
SV clock offset	Offset	Every 5–10 s
	Offset rate	
SV position error	X component	Every 1–5 min
(in WGS-84 frame)	Y component	
	Z component	
Ionospheric parameters	Eight parameters	Every 2–5 min

allows use of receivers designed to meet the current DGPS industry standard without significant modifications.

Clock offsets including SA[13] have been observed to have variation with time constants on the order of three minutes. Thus, an update rate of 0.1–0.2 Hz is sufficient to eliminate the clock error, assuming that users compute WADGPS error correction rate based on prior correction message and apply the correction rate to calculate the present error correction.

Ephemeris errors have been observed to have variations with time constants on the order of 0.5–6 h. Thus, an update rate of 1–5 min is sufficient to eliminate the ephemeris errors.

Usually the total electronic content at zenith varies very slowly (on the order of 6–12 h), but the scintillation of the ionosphere, an abrupt change of the ionosphere in a small region, can make it difficult to estimate. Space vehicle identification and a time tag in GPS time are attached to the beginning of each message. Suggested message content for transmission of the WADGPS correction is shown in Table 3. A more detailed message format is given in the next chapter.

D. Error Budget

The navigation accuracy that a user can achieve using WADGPS is summarized in Table 4. Selective availability is included in the satellite clock offset because part of it is generated by satellite clock dithering. Receiver noise can be decreased by averaging 10 measurements in time. The multipath effect can be reduced by smoothing the code with continuous carrier phase information. This can be achieved with a Hatch/Eshenbach or Kalman filter.

III. Master Station Error Modeling

The key to WADGPS is the formulation and computation of the error correction vector by the master station. This correction consists of a three-dimensional ephemeris error and clock bias for each GPS satellite in view of one or more of the monitor stations, plus eight ionospheric time delay parameters. These parameters are estimated based upon the information gathered by the monitor stations. In addition to the error correction vector, the master station must also estimate the offset of each monitor station clock from a single reference.

The following subsections describe the sources of error, the models used by the master station, and the techniques for estimating the model parameters. For algorithms A and B, the master station computes the correction vector, which is ephemeris and clock errors, and ionospheric time delay parameters, in a 2-step process. In the first step, the parameters in the ionospheric model are identified by a nonlinear static estimation (NSE) algorithm or a recursive filter. Also, there is an alternative ionospheric time delay estimation algorithm, which uses modified interpolation technique.[14] The estimated ionospheric delays are then used to adjust the raw measurements from each of the stations. The second stage solves for the ephemeris and clock errors for each of the GPS satellites observed by the network, using a batch least squares (BLS) solution or recursive filter.

A. Ionospheric Time Delay Model for Algorithms A or B

As GPS satellite signals traverse the ionosphere, they are delayed by an amount proportional to the number of free ions encountered (total electron content). The ion density is a function of local time, magnetic latitude, sunspot cycle, and other factors. Its peak occurs at 2:00 p.m. local time.

Klobuchar developed a simple analytical model for ionospheric time delay, which we have used as the basis for the WADGPS ionospheric correction model.[15] His model yields an ionospheric time delay prediction that reduces the rms error by at least 60% for the entire northern hemisphere.[16] We can improve this accuracy by performing a parameter fit optimized for the region of interest.

In Klobuchar's model, the vertical ionospheric time delay is expressed by the positive portion of a cosine wave plus a constant night-time bias, as follows[15]:

$$T_{ij} = A_1 + A_2 \cos[2\Pi(\tau - A_3)/A_4] \qquad (2)$$

where T_{ij} = ionospheric time delay in vertical direction at the intersection of the ionosphere with the line from the ith station to the jth satellite; $A_1 = 5 \times 10^{-9}$ seconds (night-time value); $A_2 = \alpha_1 + \alpha_2\phi_M + \alpha_3\phi_M^2 + \alpha_4\phi_M^3$ (amplitude); $A_3 =$ 14:00 local time (phase); $A_4 = \beta_1 + \beta_2\phi_M + \beta_3\phi_M^2 + \beta_4\phi_M^3$ (period); $\phi_M =$ geomagnetic latitude of ionosphere subpoint; α_i, β_i = ionospheric parameters (I) transmitted by the GPS satellites or by the master station; and, τ = local time.

Table 4 Wide area differential GPS error budget

Source	Error budget, m
Ephemeris errors	0.4
Satellite clock offset/selective availability	0.2
Ionospheric time delay	0.5
Tropospheric error	0.3
Receiver noise[a]	0.2
Multipath effect[b]	0.1
UERE, rms	0.77
Navigation accuracy, rms (HDOP = 1.5)	1.2

[a]Receiver noise is based on averaging 10 measurements.
[b]Multipath effect can be reduced by smoothing the code with continuous carrier phase information.

A typical vertical time delay profile generated by this model is shown in Fig. 5. The delay shown corresponds to an L_1 signal coming from a satellite directly above the observer. To estimate the actual ionospheric time delay h_{ij}, for a given satellite elevation angle, we must scale T_{ij} by the appropriate obliquity factor Q_{ij}, which is defined as the secant of the zenith angle at the mean ionospheric height, as follows:

$$h_{ij} = T_{ij}(I) \cdot Q_{ij}(\theta) \tag{3}$$

where h_{ij} = ionospheric time delay from ith station to jth satellite; $Q_{ij} = 1/\sin[\sin^{-1}\{r_e/(r_e + h_{iono}) \cos \theta\}]$ = obliquity factor from ith station to jth satellite; r_e = radius of the Earth; h_{iono} = height of the average ionosphere; θ = elevation angle; and, $I = [\alpha_1, \ldots, \alpha_4, \beta_1, \ldots, \beta_4]^T$ = ionospheric parameter.

The τ and ϕ_M of Eq. (2) are constant at each time-step.

The task of the master station is to generate the eight parameters, $[\alpha_1, \ldots, \alpha_4, \beta_1, \ldots, \beta_4]$, which, when substituted in the Klobuchar model, will yield the best ionospheric delay estimate for the region covered by the WADGPS network.

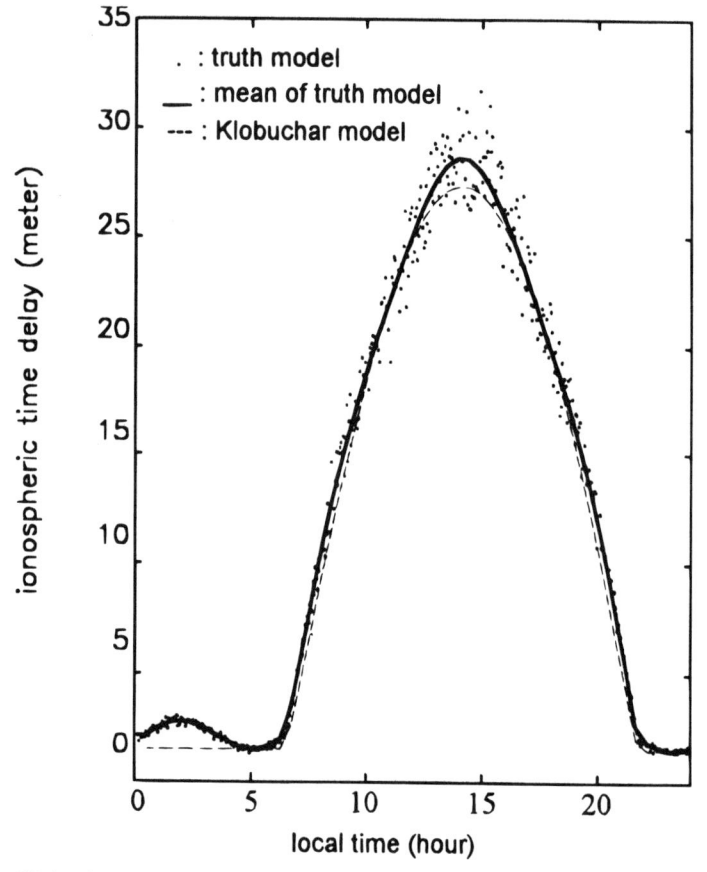

Fig. 5 Klobuchar model (cosine curve) and truth model of ionospheric time delay (at Stanford, California, elevation angle = 90 deg).

1. Ionospheric Time Delay Measurement Equation

By collecting h_{ij} for satellite $i = 1, \ldots, m$, and station $j = 1, \ldots, n$, the following ionospheric time delay measurement equation can be obtained:

$$d = h(I) + v \tag{4}$$

where $d = [d_{11}, \ldots, d_{ln} \vdots, \ldots, \vdots d_{m1}, \ldots, d_{mn}]^T$; d_{ij} = ionospheric time delay from ith station to jth satellite measured using dual-frequency technique; $h(I) = [h_{11}(I), \ldots, h_{ln}(I) \vdots, \ldots, \vdots h_{m1}(I), \ldots, h_{mn}(I)]^T$; and v = measurement noise.
The linearized form of Eq. (4) is as follows:

$$\delta d = H \cdot \delta I + v$$

where

$$I = I_0 + \delta I$$

$I_0 = [\alpha_{1_0}, \ldots, \alpha_{4_0}, \beta_{1_0}, \ldots, \beta_{4_0}]^T =$ nominal ionospheric parameters

$\delta I = [\delta\alpha_1, \ldots, \delta\alpha_4, \delta\alpha_4, \delta\beta_1, \ldots, \delta\beta_4]^T =$ increment of ionospheric parameters

$$\delta d = d - h(I_0)$$

$$H = \left.\frac{\partial h}{\partial I}\right|_{I=I_0} = [T_{11}^T \cdot Q_{11}, \ldots, T_{ln}^T \cdot Q_{ln} \vdots, \ldots, \vdots T_{m1}^T \cdot Q_{m1}, \ldots, T_{mn}^T \cdot Q_{mn}]^T$$

$$T_{ij} = [T_{ij\alpha_1}, \ldots, T_{ij\alpha_4}, T_{ij\beta_1}, \ldots, T_{ij\beta_4}]^T$$

$$T_{ij\alpha_k} = \left.\frac{\partial T_{ij}}{\partial \alpha_k}\right|_{I=I_0}$$

$$T_{ij\beta_k} = \left.\frac{\partial T_{ij}}{\partial \beta_k}\right|_{I=I_0}$$

$v =$ measurement noise

2. Nonlinear Static Estimation of Ionospheric Parameters

A nonlinear static estimation technique can be applied to the problem of fitting the ionospheric parameters to the data collected by the monitor stations.[17] We define the state x and measurement z as follows:

$$x = I = [\alpha_1, \ldots, \alpha_4, \beta_1, \ldots, \beta_4]^T \tag{6}$$

$$z = d = [d_{11}, \ldots, d_{ln} \vdots, \ldots, \vdots d_{m1}, \ldots, d_{mn}]^T \tag{7}$$

The algorithm to find the solution may be formulated as follows:
1) Guess x.
2) Evaluate $h(x)$ and H.
3) $P = (M^{-1} + H^T V^{-1} H)^{-1}$.
4) $\partial J/\partial x = M^{-1}(x - \bar{x}) - H^T V^{-1}[z - h(x)] \equiv GR.\_$
5) If $|GR| \leq \epsilon$, then set $\hat{x} = x$ and stop. Otherwise $\bar{x} = x$.
6) Replace x by $(x - P \cdot GR)$.
7) Go to (2).

B. Ephemeris and Satellite Clock Errors for Algorithms A, B, or C

The GPS navigation message broadcast by the satellites provides a means for computing the satellite positions in the WGS-84 Coordinate frame.[18] These reported positions are in error because of the limitations of the GPS control segment's ability to predict the satellite ephemeris, and potentially also because of intentional degradation of the reported parameters under SA. The GPS satellite ephemeris errors can be estimated through a network of monitor stations, by essentially using GPS upside-down. Just as a user can determine its position and clock bias based on the ranges to the known locations of four or more GPS satellites, four or more monitor stations viewing the same satellite from known locations, can be used to estimate the satellite position, clock offset, and monitor station clock offsets.

The measured pseudorange ρ_{ij}, from ith monitor station to jth GPS satellite, after being adjusted for atmospheric error and multipath error, is modeled by the following:

$$\rho_{ij} = D_{ij} \cdot e_{ij} - B_j + b_i + n_{ij}$$
$$= [(R_j + \delta R_j) - S_i] \cdot e_{ij} - B_j + b_i + n_{ij} \quad (8)$$

where ρ_{ij} = measured pseudorange; D_{ij} = range vector from ith monitor station to jth satellite; e_{ij} = range unit vector from ith monitor station to jth satellite; R_j = jth satellite location calculated from the GPS message; δR_j = ephemeris error vector of jth satellite; S_i = known ith monitor station location; B_i = satellite clock offset; b_j = monitor station clock offset; and, n_{ij} = measurement noise. This is illustrated in Fig. 6.

Define x for all the monitor stations ($i = 1, \ldots, n$) and the GPS satellites ($j = 1, \ldots, m$) as follows:

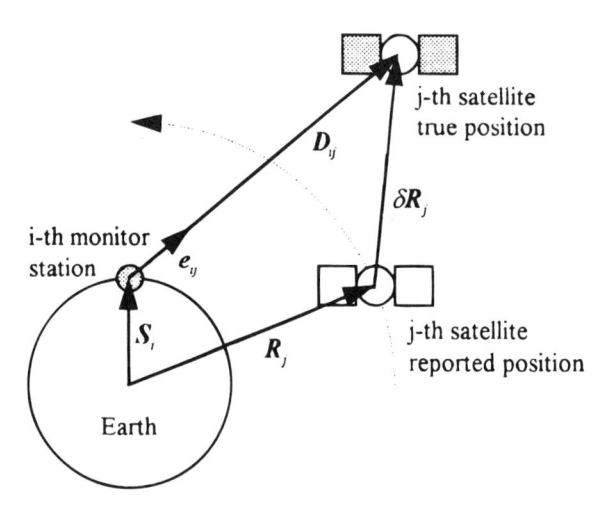

$$x = [\delta R^T \quad B^T \quad b^T]^T \quad (9)$$

Fig. 6 GPS ephemeris errors.

where

$$\delta R = [\delta R_1^T \quad \delta R_2^T \quad \cdots \quad \delta R_m^T]^T$$

$$B = [B_1 \quad B_2 \cdots B_{n-1}]^T \cdot$$

$$b = [b_1 \quad b_2 \cdots b_m]^T$$

If we gather all the measurement Eqs. (8) for all the monitor stations ($i = 1, \ldots, n$) and the GPS satellites ($j = 1, \ldots, m$) and rearrange them, we will get a matrix equation as follows:

$$\begin{bmatrix} E_1 & -I & I_1 \\ E_2 & -I & I_2 \\ \vdots & \vdots & \vdots \\ E_n & -I & I_n \end{bmatrix} x = D - \begin{bmatrix} E_1 & 0 & 0 & 0 \\ 0 & E_2 & 0 & 0 \\ 0 & 0 & \ddots & 0 \\ 0 & 0 & 0 & E_n \end{bmatrix} P \qquad (10)$$

where

$$E_i = \begin{bmatrix} e_{i1}^T & 0 & 0 & 0 \\ 0 & e_{i2}^T & 0 & 0 \\ 0 & 0 & \ddots & 0 \\ 0 & 0 & 0 & e_{im}^T \end{bmatrix} \quad (m \times 3m)$$

$$I = \begin{bmatrix} 1 & 0 & 0 & 0 \\ 0 & 1 & 0 & 0 \\ 0 & 0 & \ddots & 0 \\ 0 & 0 & 0 & 1 \end{bmatrix} \quad (m \times m)$$

$$I_i = \begin{bmatrix} 0 & \cdots & 1 & \cdots & 0 \\ 0 & \cdots & 1 & \cdots & 0 \\ \vdots & \cdots & \vdots & \cdots & \vdots \\ 0 & \cdots & 1 & \cdots & 0 \end{bmatrix} \overset{(i\text{th column})}{} \quad [m \times (n-1)] \quad (\text{for } i = 1, \ldots, n-1)$$

$$I_n = 0 \quad [(m \times (n-1)] \quad (\text{for } i = n)$$
$$D = [D_1^T \quad D_2^T \cdots D_n^T]^T$$
$$D_i = [\rho_{i1} \quad \rho_{i2} \cdots \rho_{im}]^T$$
$$P = [P_1^T \quad P_2^T \cdots P_n^T]^T$$
$$P_i = [(R_1 - S_i)^T \quad (R_2 - S_i)^T \cdots (R_m - S_i)^T]^T$$

In the above equations the matrix I_n is set to be 0 matrix because all the clock errors are relative and are estimated on the basis of the nth monitor station clock.

If we define the system matrix H and measurement z as follows

$$H = \begin{bmatrix} E_1 & -I & I_1 \\ E_2 & -I & I_2 \\ \vdots & \vdots & \vdots \\ E_n & -I & I_n \end{bmatrix} \tag{11}$$

$$z = D - \begin{bmatrix} E_1 & 0 & 0 & 0 \\ 0 & E_2 & 0 & 0 \\ 0 & 0 & \ddots & 0 \\ 0 & 0 & 0 & E_n \end{bmatrix} P \tag{12}$$

then Eq. (10) becomes

$$z = Hx \tag{13}$$

If the ith monitor station cannot see the jth satellite, the corresponding row element of the vector z and row vector of the matrix H in the Eq. (13) must be eliminated.

The master station uses a BLS technique to estimate the three-dimensional ephemeris error vector and clock bias for each GPS satellite within view of the network. If there are more measurements than the unknowns (three-dimensional errors, satellite clock offset, and monitor station clock offset) in the WADGPS network, the observation equation for that satellite is overdetermined, and the solution is picked to minimize the measurement residual sum of squares.

$$x = (H^T H)^{-1} H^T z \tag{14}$$

If there are fewer measurements than the unknowns (Fig. 7), the solution is underdetermined, and the optimal estimate minimizes the two-norm of the error solution.

$$x = H^T (H H^T)^{-1} z \tag{15}$$

In the underdetermined case, the corrections for ephemeris errors and clock offsets are not accurate, but the user positioning is still accurate with these corrections because for the user, only the projection of the error correction vector on the line of sight to the satellite is important.

If the monitor stations are confined to the continental United States, users near the coastal monitor stations will be using satellites that are underdetermined, and therefore, accuracy will degrade. Consequently, we recommend locating monitor stations over a wider area than the system designed for the users.

WIDE AREA DIFFERENTIAL GPS 95

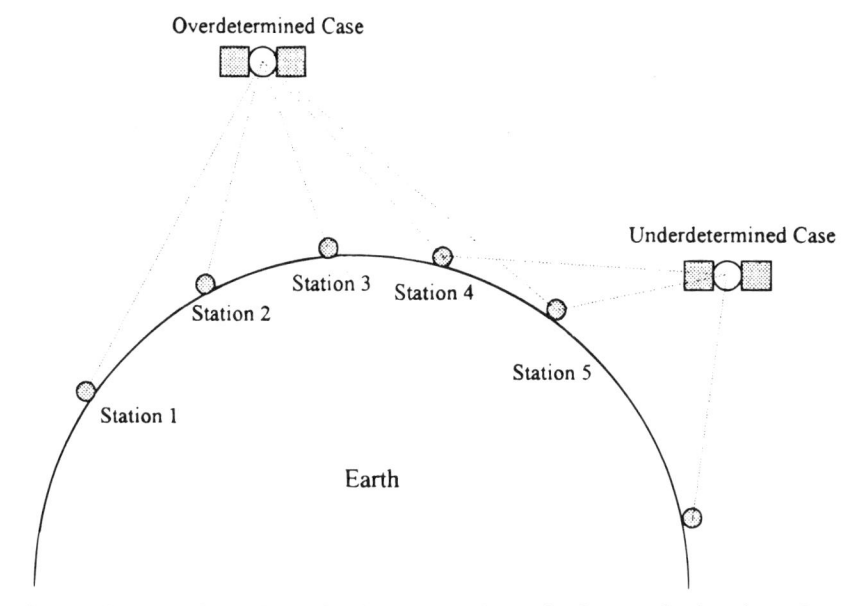

Fig. 7 Example of overdetermined and underdetermined cases of estimating ephemeris errors.

IV. Simulation of Algorithm B

The performance of the WADGPS network employing algorithm B was evaluated using a computer simulation. The simulation was run for 12 h starting at 6:00 a.m. Pacific standard time (PST).

A. Simulation Modules

The simulation is composed of four modules describing the GPS satellites, the monitor stations, the master station, and the users. A block diagram is shown in Fig. 8. The truth model error specifications are listed in Table 5.

1. GPS Satellite Module

The GPS 21 primary satellite constellation is modeled in the simulation.[19] The ephemeris reported by the GPS module to the monitor stations and the users is equal to the true ephemeris corrupted by an error vector. Each ephemeris error vector component is produced by passing white noise through a first-order shaping filter with time constant of 1800 s and standard deviation of 20 m.[20]

Each satellite clock offset is also modeled by white noise input to a first-order shaping filter, this time with $t = 200$ s,[13] and standard deviation of 30 m. These values account for possible effects caused by SA.

The ionospheric delay is modeled according to the Klobuchar model. An average ionospheric height of 350 km is assumed. In addition to the delay predicted by this model, two terms are included in our truth model to account for higher frequency

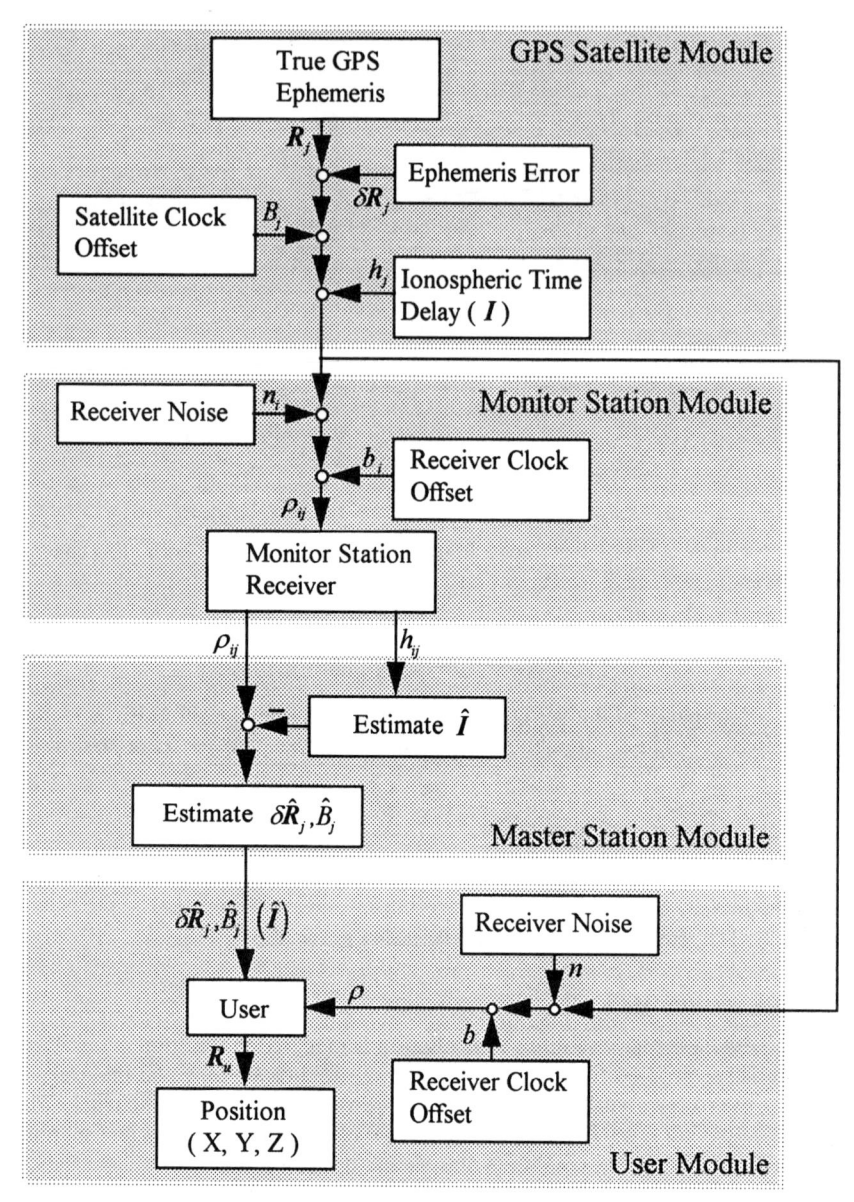

Fig. 8 Block diagram of WADGPS computer simulation.

WIDE AREA DIFFERENTIAL GPS

Table 5 True model error specifications

Error source	Error model	Time constant, s	Min., m	Max., m	rms, m
Three-dimensional satellite ephemeris errors	1st-order Markov process	1800			20
Satellite clock offset	1st-order Markov process	200			30
Ionospheric time delay[a]	Klobuchar's model plus spatial sinusoidal bias and white noise	6 h	1.5 at zenith	30 at zenith	
Tropospheric error	Modeled as receiver noise[b]				
Monitor station receiver clock offset	2nd-order Markov process		$h_0 = 2.0 \times 10^{-22}$ $h_{-1} = 4.0 \times 10^{-26}$ $h_{-2} = 1.5 \times 10^{-33}$		
User receiver clock offset	2nd-order Markov process		$h_0 = 9.4 \times 10^{-20}$ $h_{-1} = 1.8 \times 10^{-19}$ $h_{-2} = 3.8 \times 10^{-21}$		
Receiver noise	White noise				0.2
Multipath	Modeled as receiver noise				

[a]Ionosphere is varying with time constant of 6 h, and 5% of one-fifth period spatial sinusoidal bias $(0.05 \{A_1 + A_2 \cos[2\Pi(\tau - A_3)/(0.2 \, A_4)]\})$ and 5% of white noise (b0.05$T_{ij} \times N(0,1)$, $N(0,1)$ is Gaussian noise that has zero mean and one standard deviation) were added.
[b]A more sophisticated model is under development

ionospheric variations that have been observed in experimental data. The first is a sinusoidal error with amplitude of 5% of the cosine peak, and period of one-fifth of the cosine period of Eq. (2). The second is a random error with zero mean and standard deviation equal to 5% of the sum of the cosine terms. Ionospheric parameters I are varying from the nominal values with time constant of 6 h and result in maximum vertical ionospheric time delay 30 m and minimum, 1.5 m. A typical vertical time delay profile generated by this model is shown in Fig. 5.

2. Monitor Station Module

The monitor station module generates the pseudorange measurements and ionospheric delays observed by the monitor station receivers. The monitor station receiver clock offset relative to GPS time is modeled by white noise input to a second-order Markov process based on Ref. 21, as follows:

$$\begin{bmatrix} x_1 \\ x_2 \end{bmatrix}_{k+1} = \begin{bmatrix} 1 & \tau \\ 0 & 1 \end{bmatrix} \begin{bmatrix} x_1 \\ x_2 \end{bmatrix}_k + \begin{bmatrix} w_1 \\ w_2 \end{bmatrix}_k$$

$$Q = E\{w \cdot w^T\} = \begin{bmatrix} Q_{11} & Q_{12} \\ Q_{12} & Q_{22} \end{bmatrix}$$

$$Q_{11} = \frac{h_0}{2\tau} + 2h_{-1}\tau^2 + \frac{2}{3}\pi^2 h_{-2}\tau^3$$

$$Q_{12} = 2h_{-1}\tau + \pi^2 h_{-2}\tau^2 \qquad (16)$$

$$Q_{22} = \frac{h_0}{2\tau} + 2h_{-1} + \frac{8}{3}\pi^2 h_{-2}\tau$$

$$h_0 = 2.0 \times 10^{-22}$$

$$h_{-1} = 4.0 \times 10^{-26}$$

$$h_{-2} = 1.5 \times 10^{-33}$$

where x_1 = clock offset, s; x_2 = average frequency; and, τ = sampling time. The receiver clock offset parameters h_0, h_{-1}, h_{-2} are based on a typical rubidium standard. The receiver noise is assumed to be white with zero mean and standard deviation of 0.2 m. This is based on averaging over 10 measurements at 1-s intervals.

For this simulation, 15 monitor stations are assumed, located at LORAN or VOR stations across the United States including Alaska and Hawaii. Figure 9 illustrates the location, and Table 6 lists the latitude and longitude of each station.

3. Master Station Module

The master station module collects inputs from the monitor station module, and implements the ionospheric and ephemeris error estimation algorithms described in Secs. III and IV. Ionospheric delay parameters, and estimated ephemeris and clock errors are provided to the user module.

4. User Module

The user module simulates the operation of the user receiver. The user clock error is assumed to be white noise input to a second-order Markov process based on Ref. 21, with Eq. (16) $h_0 = 9.4 \times 10^{-20}$; $h_{-1} = 1.8 \times 10^{-19}$; and $h_{-2} = 3.8 \times 10^{-21}$, where these receiver clock offset values are based on a typical quartz standard.

The receiver noise is assumed to be white with zero mean and standard deviation of 0.2 m. This is based on averaging over 10 measurements at 1-s intervals. The user applies the eight parameters in the Klobuchar model to the raw pseudorange and adjusts the ephemeris parameters received from the GPS satellite module by the correction vector sent by the master station. Then the user forms a least-squares position solution using measurements to all the satellites within his field of view. The performance of the WADGPS is evaluated by

WIDE AREA DIFFERENTIAL GPS 99

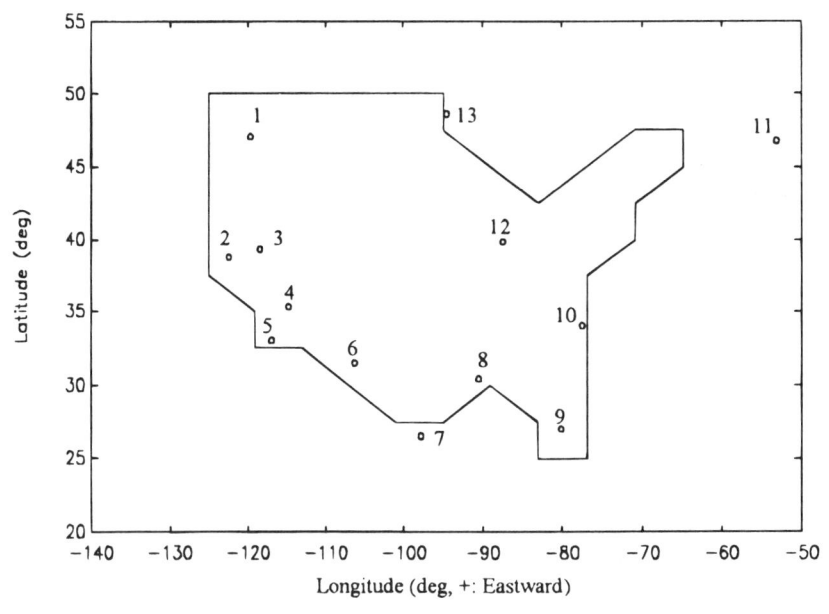

Fig. 9 Locations of monitor stations in United States (Narrow Cape and Upolo Point are not shown on the map).

Table 6 Locations of monitor stations

Location	Latitude	Longitude	LORAN site	VOR site
1) George, WA	47:04 N	119:45 W	√	
2) Middletown, CA	38:47 N	122:30 W	√	
3) Fallon, NV	39:33 N	115:50 W	√	
4) Searchlight, NV	35:19 N	114:48 W	√	
5) San Diego, CA	33:00 N	117:00 W		√
6) El Paso, TX	31:30 N	106:20 W		√
7) Raymondville, TX	26:32 N	97:50 W	√	
8) Grangeville, LA	30:43 N	90:50 W	√	
9) Jupiter, FL	27:02 N	80:07 W	√	
10) Carolina Beach, NC	34:04 N	77:55 W	√	
11) Cape Race, Newfoundland, Canada	46:47 N	53:10 W	√	
12) Dana, IN	39:51 N	87:29 W	√	
13) Baudette, MN	48:37 N	94:33 W	√	
14) Narrow Cape, Kodiak Is., AK (not shown on the map)	57:26 N	152:22 W	√	
15) Upolo Pt., HI (not shown on the map)	20:15 N	155:23 W	√	

comparing the error in this solution to the error that would have been obtained by using the raw measurements directly.

A typical user who would benefit from the WADGPS, has a single-frequency, C/A-code GPS receiver and a quartz oscillator. Eighty-one stationary users are modeled at locations distributed uniformly across the United States (Fig. 10). All users are assumed to have an elevation mask angle of 6.5 deg, which is typical antenna visibility for an aircraft.

B. Ionospheric Error Estimation Results

The first step in evaluating the performance of the WADGPS is to see how well it determines the ionospheric errors. Figures 11 and 12 show contour plots of the actual and estimated vertical ionospheric delays superimposed on a map of the United States. These represent snapshots from the 12-h simulation at 5 p.m. PST and 2 p.m. PST, respectively. The contour lines are labeled by the ionospheric delay in 3-m increments. Note that the actual values of the ionospheric delay increase from east to west as we get closer to local noon.

In the 5 p.m. plot, we can see that the nonlinear static estimator algorithm does well at estimating the delay because the estimated contours are within 1.5 m of the true error contours. The performance results are summarized in Table 7.

Figure 11 also shows the improvement in ionospheric delay estimates toward the center of the country. This can be attributed to the larger number of monitor stations that can observe satellite signals passing through the central region as compared to satellites in the far eastern or western parts of the sky.

Figure 12 shows similar results for 2 p.m. PST. In this case, however, we notice local contours of varying heights that are not estimated. These small areas of variation in the ionospheric delay are generated by the random noise introduced in the truth model. Because the standard deviation of this variation is set at 5% of the nominal value for the local time of day, the maximum random error is as

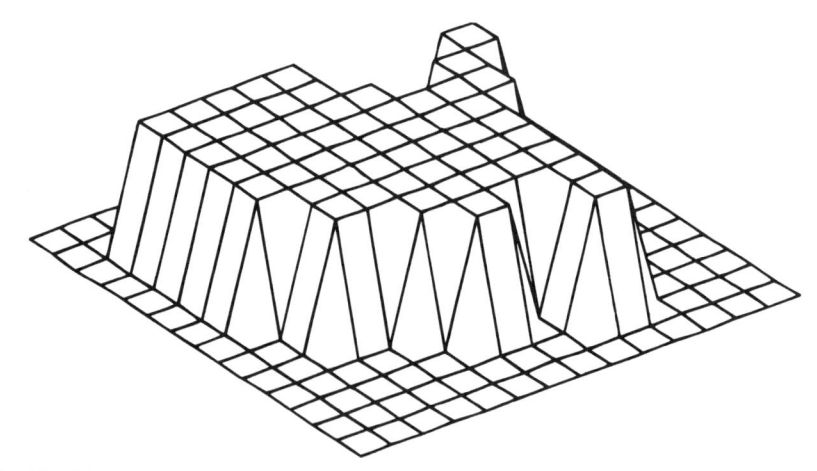

Fig. 10   Mesh plot of the continental United States. (Each point on the grid represents one of the 81 simulated user's positions.)

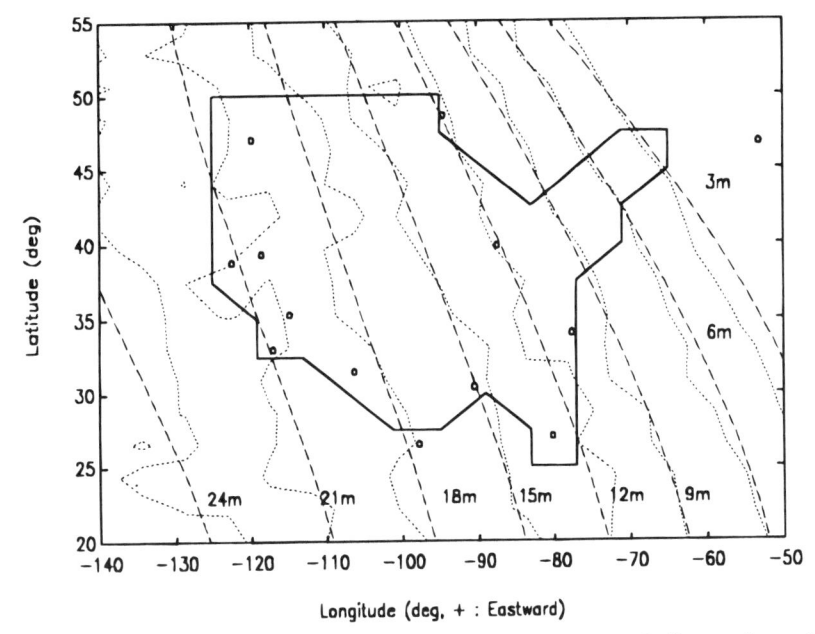

Fig. 11 Ionospheric time delay estimates (5:00 p.m. PST). (This figure shows the vertical ionospheric delay in meters as generated by the truth model and the NSE. A map of the United States and the monitor station locations is also shown. The dotted line is the true delay contour; the dashed line is the NSE estimate.)

much as 4.5 m at 2 p.m. There is no way for our rather sparse WADGPS network model to estimate these simulated random, high-frequency, localized components of the ionospheric error.

C. Navigation Performance

The objective of the WADGPS system is to improve navigation accuracy for users. The simulation results indicate that this goal can be achieved using the proposed system.

Figures 13–18 provide a very compact summary of the simulation results. In these mesh plots, each grid point within the outline of the United States represents 1 of the 81 user locations we considered. The height of the grid point above the surface corresponds to the magnitude of the error at the grid location. These heights reflect rms or maximum error for that user over the entire 12-h simulation period.

Figures 13 and 16 show the uncorrected navigation performance of typical across the United States. The rms of the positioning errors for all user locations, over the entire 12-h period, is 82 m in vertical direction, and 42 m in the horizontal direction. As we might expect, the error magnitudes are fairly uniform over the entire area. As is common in GPS navigation, the vertical error is approximately twice as large as the horizontal error because of the larger geometrical dilution of precision in the vertical direction (VDOP).

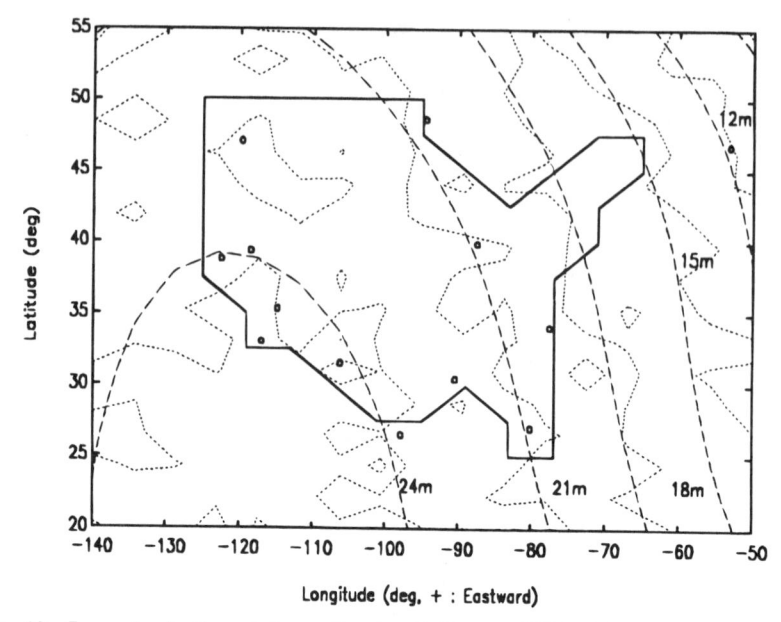

Fig. 12 Ionospheric time delay estimates (2:00 p.m. PST). (This figure shows the vertical ionospheric delay in meters as generated by the truth model and the NSE. A map of the United States and the monitor station locations is also shown. The dotted line is the true delay contour; the dashed line is the NSE estimate.)

Table 7 Root-mean-square and maximum errors in ionospheric estimates

PST local time		7:00–8:00 a.m.	10:00–11:00 a.m.	1:00–2:00 p.m.	4:00–5:00 p.m.
Nonlinear static	max,[a] m	1.0	1.4	1.3	1.3
Estimation	rms,[b] m	0.3	0.5	0.5	0.4

[a]max[abs($z_i - \hat{z}_i$)] where z_i is the true vertical ionosphere measurement and $\hat{z}_i$ the estimated value.
[b]rms(z_i)

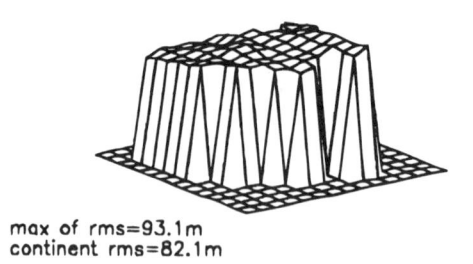

max of rms=93.1m
continent rms=82.1m

Fig. 13 Root-mean-square value of stand-alone GPS vertical positioning errors.

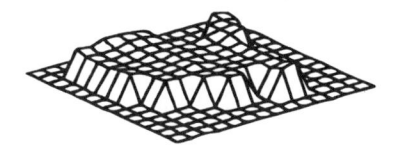

WIDE AREA DIFFERENTIAL GPS 103

max of rms=1.9m
continent rms=1.5m

Fig. 14 Root-mean-square value of WADGPS vertical positioning errors.

max of max=9.1m
continent rms of max=6.3m

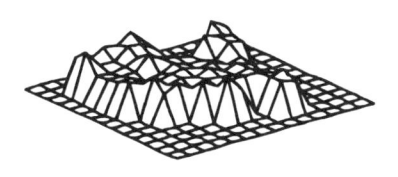

Fig. 15 Maximum WADGPS vertical positioning errors.

max of rms=45.4m
continent rms=42.3m

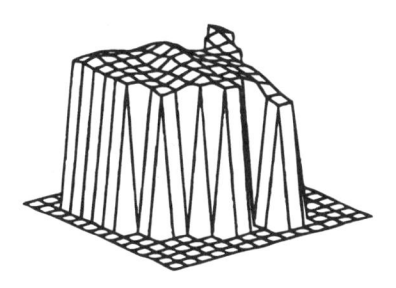

Fig. 16 Root-mean-square value of stand-alone GPS horizontal positioning errors.

max of rms=1.5m
continent rms=1.1m

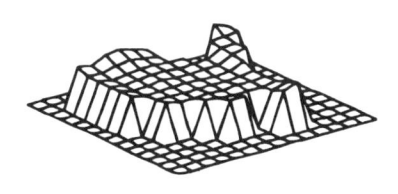

Fig. 17 Root-mean-square value of WADGPS horizontal positioning errors.

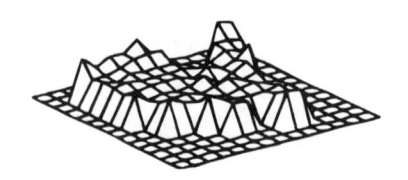

max of max=6.0m
continent rms of max=3.6m

Fig. 18 Maximum WADGPS horizontal positioning errors.

Figures 14–15 and 17–18 show the significantly improved navigation accuracy achieved using WADGPS with the NSE ionospheric estimation algorithm and ephemeris and clock bias BLS algorithm. Continental rms averages of the vertical and horizontal position errors have been reduced from 45.4 m and 42.3 m to 1.5 m and 1.1 m, respectively. Figures 15 and 18 show the maximum values of the errors. The largest vertical and horizontal errors anywhere in the United States over the entire 12-h period are 9.1 m and 6.0 m, respectively.

One of the most striking features of the plots of WADGPS corrected errors is the concave shape of the error mesh. In general, the navigation performance in the center of the United States is better than along the coasts. This is because the satellites viewed by users in this region are also visible from a larger number of monitor stations, and with better geometry [lower geometric dilution of precision (GDOP)] than their coastal counterparts.

One exception to this observation occurs in the southwestern United States. This region exhibits better positioning accuracies than other edges of the country because of the high density of monitor stations (Figs. 14–15 and 17–18). Likewise, the north central part of the country is noticeably worse than average because of the relative sparsity of monitor stations.

D. Summary of Results

A 12-h simulation was run starting at 6:00 a.m. PST and ending at 6:00 p.m. PST. The NSE technique was used to determine the ionospheric delay parameters. Table 8 summarizes the simulation results.

Simulation results over a 12-h period indicate that stand-alone GPS positioning errors can be reduced by over 95% using WADGPS without degradation caused by separation between the monitor station and the users. These results indicate that WADGPS can provide accurate ionospheric delay estimates as well as positioning errors comparable to local area differential GPS operations.

V. Test Using Field Data to Evaluate Algorithm C

Previously collected field test data (from Dec. 10, 1992 to Feb. 12, 1993 using the GPS Global Tracking Network) were processed to evaluate WADGPS performance. The GPS Global Tracking Network has more than 30 sites distributed worldwide, which are equipped with P-code receivers and of which locations are known to within a few centimeters, and the Jet Propulsion Laboratory (JPL)

WIDE AREA DIFFERENTIAL GPS

Table 8 Summary of positioning errors

	Nonlinear static estimation			
	Vertical		Horizontal	
	Maximum[a]	rms[b]	Maximum[a]	rms[b]
Root-mean-square of stand-alone positioning error, m	93.1	82.1	45.4	42.3
Root-mean-square value of WADGPS positioning error, m	1.9	1.5	1.5	1.1
WADGPS error/stand-alone error, %	2.6	1.9	3.0	2.1
Max. value of WADGPS positionng error, m	9.1	6.3	6.0	3.6

[a]Maximum for the continental United States.
[b]Root-mean-square for the continental United States.

collects data from all the network sites. Sampling time for the most sites was 30 s, and the available measurements were L_1 and L_2 P-code pseudoranges and L_1 and L_2 continuous carrier phases. C/A-code pseudoranges and Doppler measurements were not available and SA was on during the field test. Because meteorological data were not available, the temperature, pressure, and humidity of each site were inferred by location and time of day.

Algorithm C was used for this test. It provides the best accuracy because the user uses a dual-frequency receiver to measure ionospheric time delays very accurately. As such, the results presented here provide a estimate of the best possible performance of WADGPS.

A. Locations of the Receiver Sites

Among over 30 Rogue receiver sites, 7 were picked in North America and Hawaii for the field test. Six sites were chosen as monitor stations for WADGPS because their sites are evenly distributed and their geometry constitutes a rough circle. One site, ALBH (Albert Head, Canada), was picked as user because it is near the center of the circle, and therefore, could demonstrate the potential of WADGPS. The minimum baseline between the user (ALBH) and a monitor station (JPLM) was 1632 km. The locations of the receiver sites are listed in Table 9 and the corresponding map is in Fig. 19.

All the coordinates of the receiver locations were given in the International Terrestrial Reference Frame (ITRF) 91 coordinate frame instead of the WGS-84 frame in which the GPS ephemeris is computed. The ephemeris errors in the ITRF91 frame are different from those calculated in the WGS-84 frame, but using the same coordinate frame for the monitor stations and the user avoids unexpected positioning errors.

B. Test Results

As soon as a set of data from monitor stations became available, the master station estimated the ephemeris errors and satellite clock errors and transmitted

Table 9 Locations of the P-code Receiver Sites

Site type	Station ID	City	Nation	Latitude deg	Longtitude deg	Baseline from, ALBH, km
Monitor	ALGO	Algonquin	Canada	46.0 N	78.0 W	3363
Monitor	FAIR	Fairbanks, AK	USA	65.0 N	147.5 W	2318
Monitor	JPLM	Pasadena, CA	USA	34.1	118.1 W	1632
Monitor	KOKB	Kokee, HI	USA	22.1 N	159.7 W	4245
Monitor	RCM2	Richmond, FL	USA	25.6 N	80.4 W	4414
Monitor	YELL	Yellowknife	Canada	62.5 N	114.5 W	1661
User	ALBH	Albert Head	Canada	48.4 N	123.5 W	0

the error corrections to users. All the data were postprocessed as if it were in real time. Normally it took about 3–4 s to compute the error corrections for each epoch using PC-486/25 computer. However, the data were applied without delay.

A total of six satellites were in view from ALBH during the field test. Typically six to nine satellites will be seen from a receiver when GPS is in full operation 1994. Figure 20 shows azimuth vs. elevation plot during the test period.

Only six monitor stations were used in this test, which is considerably fewer than 15 stations used in the simulations (Sec. IV). Not all the satellites were in view from more than 4 monitor stations, which is the underdetermined case. In that case, the estimates of the ephemeris errors and clock offsets were not accurate, but the user positioning was still accurate with the WADGPS corrections.

We tested a total of 12 days, and the results are summarized in Table 10. We show the worst case (1/23/93) in Figs. 21–23, and one of the best cases (1/13/93) in Figs. 24–26.

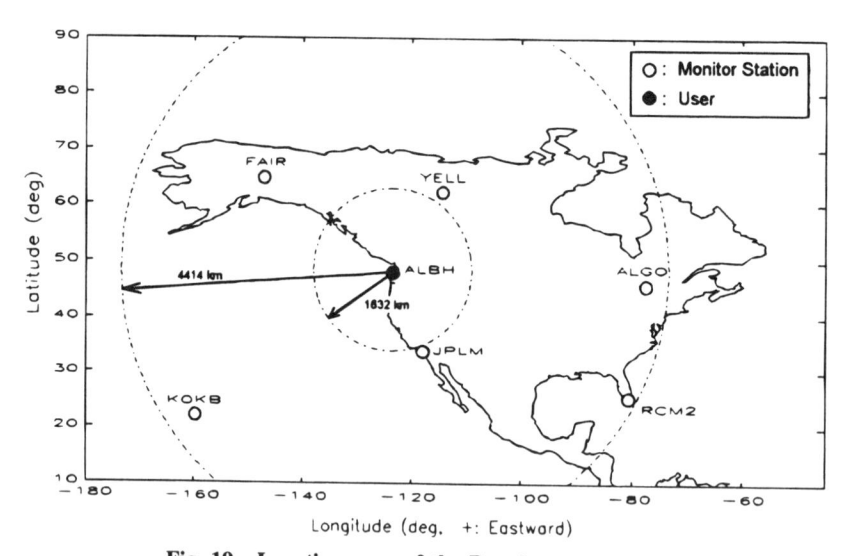

Fig. 19 Location map of the P-code receiver sites.

WIDE AREA DIFFERENTIAL GPS 107

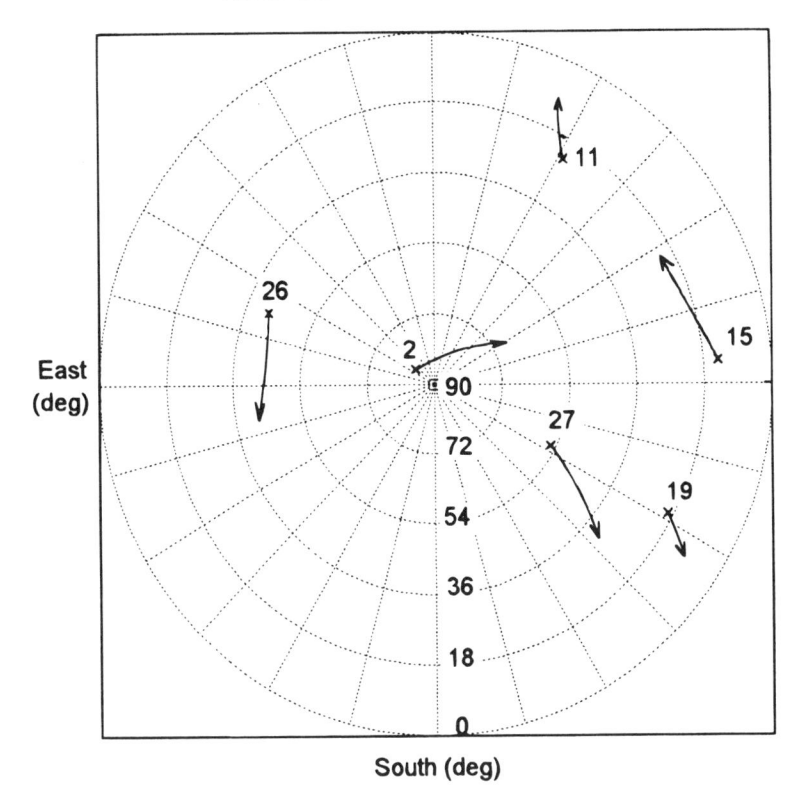

Fig. 20 Azimuth vs elevation plot (ALBH, 1/23/93).

Table 10 Summary of navigation errors at ALBH using Algorithm C and zero latency (1632 km baseline, 12/10/92–2/12/93)

Date	GPS time	GPS, three-dimensional rms, m	WADGPS, three-dimensional	WADGPS/ GPS, %
12/10/92	4:10 a.m.–4:36 a.m.	57.0	0.71	1.2
1/6/93	4:34 a.m.–5:00 a.m.	55.4	0.84	1.5
1/11/93	3:12 a.m.–3:42 a.m.	52.9	1.11	2.1
1/12/93	3:15 a.m.–3:38 a.m.	50.1	1.32	2.6
1/13/93	3:38 a.m.–4:10 a.m.	62.0	0.86	1.4
1/14/93	3:34 a.m.–4:05 a.m.	48.5	1.51	3.1
1/23/93	3:00 a.m.–3:55 a.m.	65.7	2.18	3.3
1/29/93	3:01 a.m.–3:47 a.m.	81.3	1.45	1.8
1/30/93	3:07 a.m.–3:55 a.m.	51.3	1.99	3.9
1/31/93	3:03 a.m.–3:47 a.m.	68.5	1.58	2.3
2/11/93	3:22 a.m.–3:59 a.m.	73.9	0.94	1.3
2/12/93	3:20 a.m.–3:55 a.m.	67.5	0.94	1.4
Total average		61.2	1.29	2.2

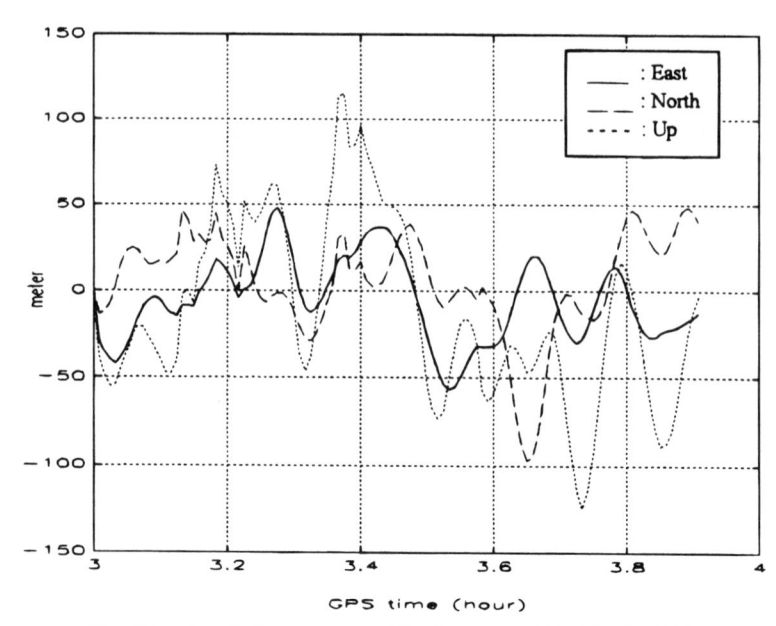

Fig. 21 Stand-alone user positioning error (ALBH, 1/23/93).

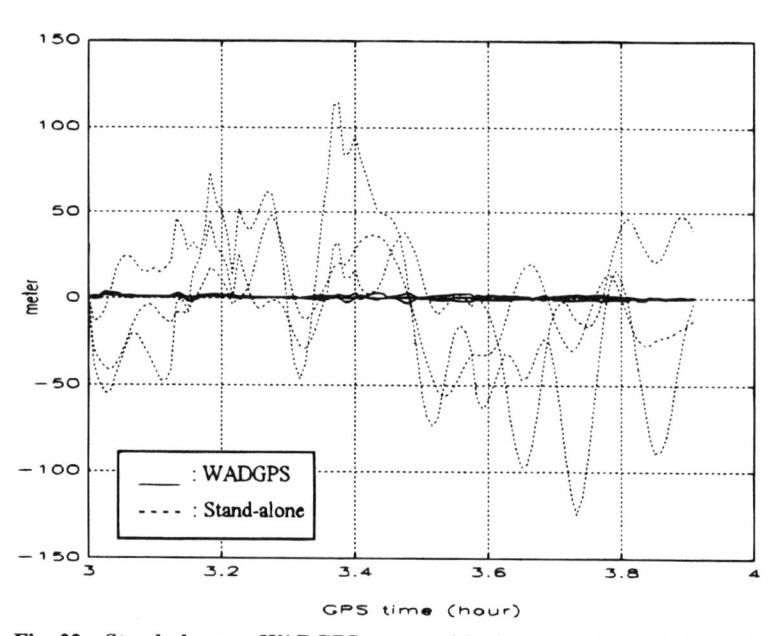

Fig. 22 Stand-alone vs WADGPS user positioning error (ALBH, 1/23/93).

WIDE AREA DIFFERENTIAL GPS 109

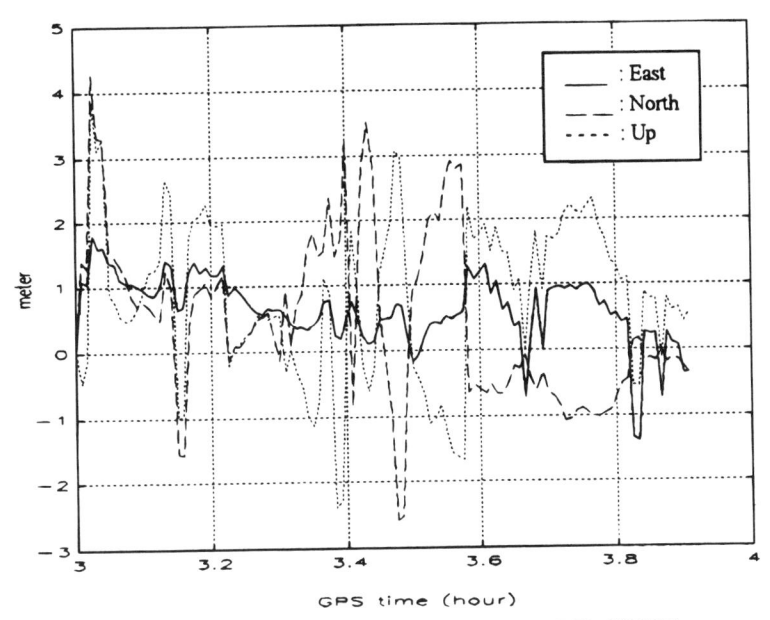

Fig. 23 WADGPS user positioning error (ALBH, 1/23/93).

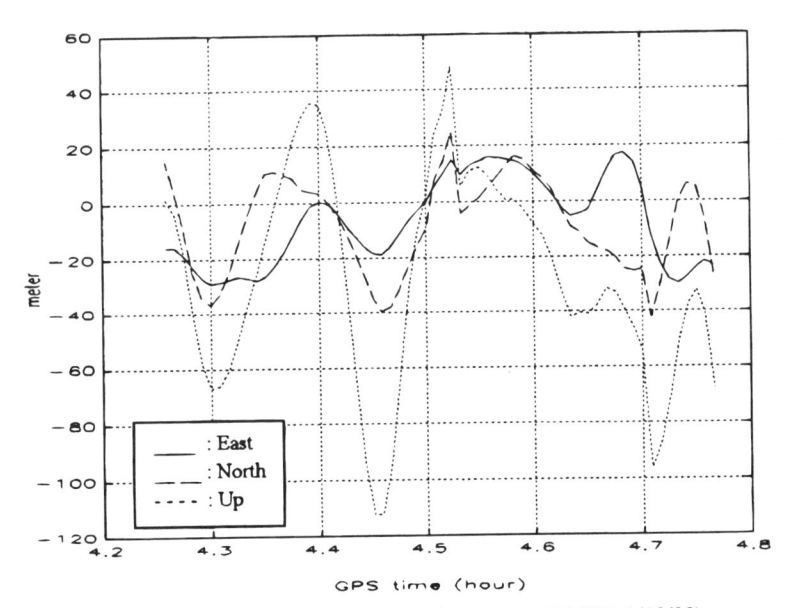

Fig. 24 Stand-alone user positioning error (ALBH, 1/13/93).

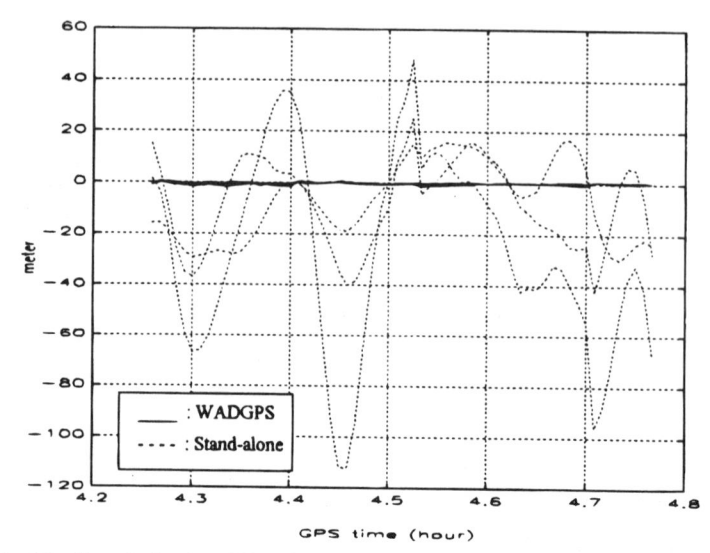

Fig. 25 Stand-alone vs WADGPS user positioning error (ALBH, 1/13/93).

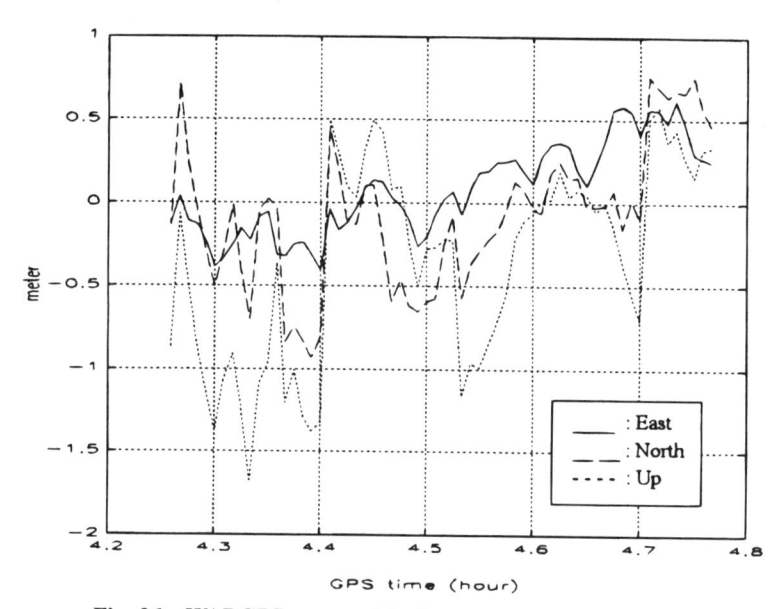

Fig. 26 WADGPS user positioning error (ALBH, 1/13/93).

Figures 21 and 24 show the stand-alone user positioning errors at night at ALBH on January 23, 1993 and January 13, 1993, respectively. Their rms three-dimensional positioning errors for stand-alone user are 65.7 m and 62 m, which is typical under SA. The navigation errors contain large oscillations with time constants of approximately 2 min, indicating the presence of SA.

Figures 22 and 25 show the significantly improved WADGPS vs. stand-alone GPS navigation errors, and Figs. 23 and 26 show the only WADGPS positioning errors with much smaller scale. The three-dimensional rms positioning errors of WADGPS user for the days, January 23 and January 13, are 2.18 m and 0.86 m, respectively. The shortest baseline from the monitor station to ALBH is 1632 km, which is very long, but WADGPS provides nonspatially degrading error corrections to user.

Overall, a normal GPS three-dimensional positioning error, 61.2 m, is reduced to 1.29 m using WADGPS. In other words about 98% of the normal GPS three-dimensional positioning error can be eliminated using WADGPS.

Two important error sources for WADGPS positioning errors are multipath and tropospheric refraction. Multipath can be reduced by using a choke ring antenna and the Hatch/Eshenbach filter but cannot be totally eliminated. Better tropospheric models can reduce the tropospheric error, but a residual error still exists.

C. Latency and Age Concern

The results of the test showed that WADGPS can achieve navigation accuracy on the order of one meter even for a 1632-km baseline with zero latency, which is not achievable for local area DGPS. However, in practice, it is impossible for users to apply the error corrections at the same epoch at which they are estimated in the master station. We define latency as the time taken to estimate the correction parameters plus the time spent for the WADGPS error corrections to arrive at the users via geosynchronous satellite. Actually, users have to use the old correction message until the new correction message arrives. So we define age, total time delay, as latency plus the time interval from when the old correction message arrived till when users apply a new correction message. Figure 27 shows the definition of latency and age.

Usually the maximum age for 5–10 s of latency is 10–20 s. The major source of the error caused by the time delay is SA, which has a 2–3 min time constant and is the fastest changing error source.

The GPS Global Tracking Network data taken on December 10, 1992 was used to investigate the latency effects on WADGPS navigation accuracy. Because

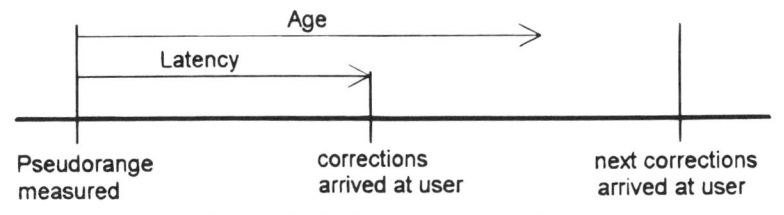

Fig. 27 Definition of latency and age.

the sampling time of the data was 30 s, the tests were repeated with latencies of 30–180 s with 30-s increments. We used two different error correction techniques. The first technique keeps the error corrections constant and after a certain time delay applies these fixed corrections to the users. The second technique estimates the error–rate corrections as well as error corrections and uses both terms to predict an error correction for the user.

Fig. 28 gives the effect of the different latency values on the three-dimensional positioning accuracies, and Fig. 29 shows the effect of the latency on an enlarged scale. Table 11 summarizes the effect of latency on WADGPS navigation accuracy.

Figure 29 indicates that a 1.5–3 m three-dimensional rms positioning accuracy can be achieved with 5–10 s of latency, which corresponds to 10–20 s of age, if WADGPS is used with error–rate corrections (rather than with constant error corrections). On the worst case day (1/23/93), 3–5 m three-dimensional rms positioning accuracy can be achieved with 5–10 s of latency using WADGPS.

VI. Conclusion

This chapter has introduced the WADGPS concept, which includes one or more master stations, monitor stations, and a broadcast system. Each functional component of WADGPS was described. Then the user message content and its format were proposed, and a listing of the error budget for WADGPS was given. The procedures used at the master station for estimating ionospheric time delay parameters and ephemeris and satellite clock errors were discussed in detail.

Three different real-time WADGPS algorithms have been proposed, which depend on receiver type in the monitor stations and users and broadcasting

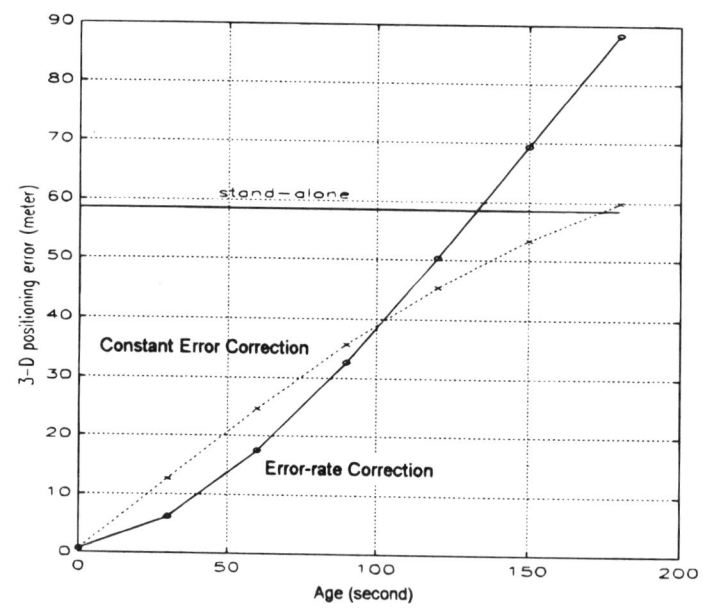

Fig. 28 Age effect on WADGPS (ALBH, 12/10/92).

WIDE AREA DIFFERENTIAL GPS 113

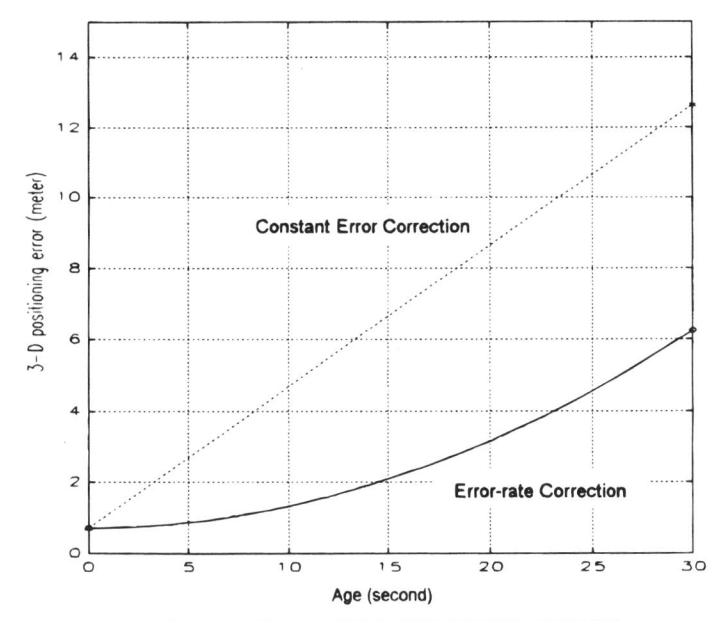

Fig. 29 Age effect on WADGPS (ALBH, 12/10/92).

parameters from the master station to the users. Algorithm B was chosen to simulate WADGPS because it includes the effect of the ionospheric time delay in WADGPS navigation accuracy. algorithm C, which estimates only three-dimensional ephemeris errors and satellite clock offset including SA assuming that users can measure the ionospheric time delays by themselves, was evaluated using field data because it provides a best possible bound for WADGPS performance.

The GPS Global Tracking Network data collected by P-code receivers were used to evaluate algorithm C. Six monitor stations that provide good geometry and one user site near the center of the network were picked in North America and Hawaii to demonstrate WADGPS performance. A batch least squares and minimum norm solution were used in the estimation.

The test results indicate that stand-alone GPS positioning error, which is 61.2 m, can be reduced to on the order of a meter (1632-km baseline) using algorithm C with zero latency. However, latency will degrade the WADGPS navigation

Table 11 Three-dimensional WADGPS rms positioning errors in the existence of latency (age) for two different error correction techniques (ALBH, 12/10/92)

Latency, s	0	30	60	90	120	150	180
Constant error correction	0.71	12.63	24.64	35.68	45.38	53.64	59.85
Error–rate correction	0.71	6.25	17.49	32.67	50.33	69.33	88.33

accuracy, but a study of the latency showed that 2–4 m of three-dimensional rms positioning accuracy can be achieved with 5–10 s of latency, which corresponds to 10–20 s of age, if WADGPS is used with error–rate corrections. Of course, algorithm C assumes that users have dual-frequency receivers or can estimate the ionospheric time delays accurately. If this assumption is not true, the positioning accuracy will be further degraded, but still should be better than 10 m.

The prediction of WADGPS above has been verified in flight trials by Stanford Telecommunication and the FAA[22,33] so that WADGPS will be used as the basis of the Wide Area Augmentation System (WAAS) to be operational by 1997, which is detailed in the next chapter.

References

[1]Chou, H., "An Adaptive Correction Technique for Differential Global Positioning System," Ph.D. Dissertation, Stanford University, Stanford, CA, June, 1991.

[2]Kremer, G. T., Kalafus, R. M., W. Loomis, P. V., and Reynolds, J. C., "The Effect of Selective Availability on Differential GPS Corrections," *Navigation*, Vol. 37, No. 1, 1990, pp. 39–52.

[3]Kee, C., Parkinson, B. W., and Axelrad, P., "Wide Area Differential GPS," *Navigation*, Vol. 38, No. 2, 1991.

[4]Kee, C., and Parkinson, B. W., "Algorithms and Implementation of Wide Area Differential GPS," *Proceedings of ION GPS-92* (Albuquerque, NM), ION, Washington, DC, Sept. 16–18, 1992, pp. 565–572.

[5]Kee, C., and Parkinson, B. W., "High Accuracy GPS Positioning in the Continent: Wide Area Differential GPS," Differential Satellite Navigation Systems 93 (DSNS-93) Conference, Amsterdam, The Netherlands, April 1993.

[6]Kee, C., and Parkinson, B. W., "Static Test Results of Wide Area Differential GPS," *Proceedings of ION GPS-93* (Salt Lake City, UT), ION, Washington, DC, Sept. 22–24, 1993.

[7]Brown, A., "Extended Differential GPS," *Navigation*, Vol. 36, No. 3, 1989.

[8]Loomis, P. V. W., Denaro, R. P., and Saunders, P., "Worldwide Differential GPS for Space Shuttle Landing Operations," *IEEE PLANS '90, IEEE Position, Location, and Navigation Symposium* (Las Vegas, NV, IEEE, New York, March, 1990.

[9]Cohen, C., Pervan, B., and Parkinson, B. W., "Estimation of Absolute Ionospheric Delay Exclusively though Single Frequency GPS Measurements," *Proceedings of ION GPS-92* (Albuquerque, NM), ION, Washington, DC, Sept. 16–18, 1992, pp. 325–330.

[10]Xia, R. "Determination of Absolute Ionospheric Error using a Single Frequency GPS Receiver," *Proceedings of ION GPS-92*, (Albuquerque, NM), ION, Washington, DC, Sept. 16–18, 1992, pp. 483–490.

[11]Weber, L., and Tiwari, A., "Performance of a FM Sub-Carrier (RDS) Based DGPS System," *Proceedings of ION GPS-93*, (Salt Lake, UT), ION, Washington, DC, Sept. 22–24, 1993, pp. 1285–1292.

[12]Kalafus, R. M., "Special Committee 104 Recommendations for Differential GPS Service," *Navigation*, Vol. 3, 1986, pp. 101–116.

[13]Chou, H., "An Anti-SA Filter for Non-differential GPS Users," *Proceedings of ION GPS-90* (Colorado Springs, CO), ION, Washington, DC, Sept. 19–21, 1990.

[14]El-Arini, M., O'Donnell, P. A., Kellam, P. M., Klobuchar, J. A., Wisser, T. C., and Doherty, P. H., "The FAA Wide Area Differential GPS (WADGPS) Static Ionospheric

Experiment," *National Technical Meeting of the Institute of Navigation,* (San Francisco, CA), ION, Washington, DC, Jan. 1993, pp. 485–496.

[15]Klobuchar, J., Air Force Geophysical Laboratory, "Design and Characteristics of the GPS Ionospheric Time Delay Algorithm for Single Frequency Users," *IEEE PLANS '86, Position, Location, and Navigation Symposium* (Las Vegas, NV), IEEE, New York, Nov. 4–7, 1986.

[16]Stephens, S. G., and Feess, W. A., "An Evaluation of the GPS Single Frequency User Ionospheric Time Delay Model," *IEEE PLANS '86, Position, Location, and Navigation Symposium,* (Las Vegas, NV), IEEE, New York, Nov. 4–7, 1986.

[17]Bryson, A. E., "Optimal Estimation and Control Logic in the Presence of Noise" (Lecture Notes, Stanford Course AA278B), Spring 1989.

[18]"Navstar GPS Space Segment/Navigation User Interfaces," Rockwell International Corporation, ICD-GPS-200, Nov. 30, 1987.

[19]Green, G. B., Massatt, P. D., and Rhodus, N. W., "The GPS 21 Primary Satellite Constellation," *Navigation,* Vol. 36, No. 1, Spring, 1989.

[20]Wells, D., "Guide to GPS Positioning," Canada GPS Associates, 1987.

[21]Van Dierendonck, A. J., McGraw, J. B., and Brown, R. G., "Relationship between Allan Variances and Kalman Filter Parameters," *Proceedings of the 16th annual PTTI Applications & Planning Meeting,* Maryland, Nov. 1984.

[22]Lage, M. E., and Elrod, B. D., "The FAA's WIB/WDGPS Testbed and Recent Test Results," *Proceedings of ION GPS-93,* (Salt Lake, UT), ION, Washington, DC, Sept. 16–18, 1993, pp. 487–493.

[23]Loh, R., Persello, F., and Wollschleger, V., "FAA's Wide Area Augmentation System (WAAS) Summary of Ground & Flight Test Data," *Proceedings of ION GPS-94* (Salt Lake, UT), ION, Washington, DC, Sept. 16–18, 1994, pp. 99–106.

Chapter 4

Wide Area Augmentation System

Per K. Enge*
Stanford University, Stanford, California 94305
and
A. J. Van Dierendonck†
AJ Systems, Los Altos, California 94024

I. Introduction

IN time, the Global Positioning System (GPS) will be used for a wide variety of aircraft operations. However, aircraft use of any satellite-based navigation system raises significant concern with respect to integrity (hazardous but undetected faults), reliability (continuity of service), time availability, and accuracy. After all, a single satellite malfunction would affect users over a huge geographic area. A navigation system with integrity warns its users if position errors may be greater than a prespecified "alarm limit." Clearly, radionavigation systems used by aviators must have integrity, and the integrity requirement depends on whether the system is the primary navigation aid or supplements another system.

The wide area augmentation system (WAAS) is a safety-critical system consisting of a signal-in-space and a ground network to support enroute through precision approach air navigation. It is designed to augment GPS so that it can be used as the primary navigation sensor. The WAAS augments GPS with the following three services: a ranging function, which improves availability and reliability; differential GPS corrections, which improve accuracy; and integrity monitoring, which improves safety.

The WAAS is shown in Fig. 1. As shown, the WAAS broadcasts GPS integrity and correction data to GPS users and also provides a ranging signal that augments GPS. In the near future, the WAAS signal will be broadcast to users from geostationary satellites. Inmarsat-3 satellites will be the first to carry the WAAS navigation payload, and they are scheduled to be launched in 1996.

Copyright © 1995 by the authors. Published by the American Institute of Aeronautics and Astronautics, Inc. with permission.
*Professor (Research), Department of Aeronautics and Astronautics.
†Systems Consultant.

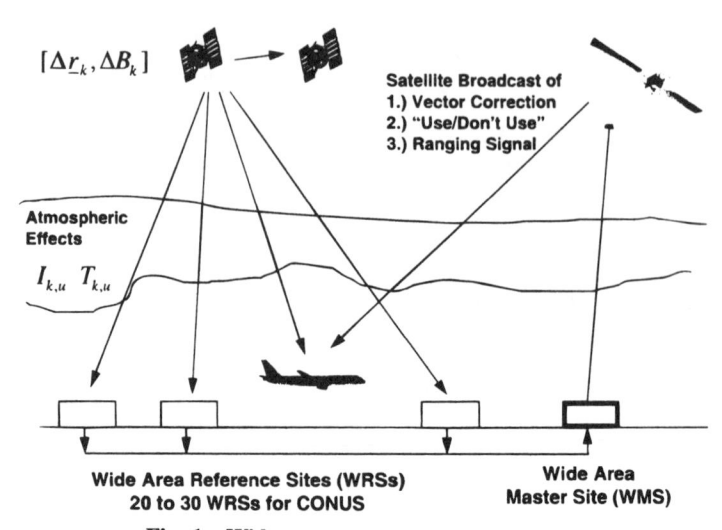

Fig. 1 Wide area augmentation system.

The WAAS ranging signal is GPS-like and will be received by slightly modified GPS receivers. More specifically, it will be at the GPS L_1 frequency and will be modulated with a spread spectrum code from the same family as the GPS C/A-codes. The code phase and carrier frequency of the signal will be controlled so that the WAAS satellites will provide additional range measurements to the GPS user. The WAAS signal will also carry data that contain differential corrections and integrity information for all GPS satellites as well as the geostationary satellite(s).

The ground network shown in Fig. 1 develops the differential corrections and integrity data broadcast to the users. Wide area reference stations (WRS) are widely dispersed data collection sites that receive and process signals received from the GPS and geostationary satellites. An example network of WRS for the United States is shown in Fig. 2. The WRS forward their data to data-processing sites referred to as wide area master stations (WMS or central processing facilities).

The WMS process the raw data to determine integrity, differential corrections, residual errors, and ionospheric delay information for each monitored satellite. They also develop ephemeris and clock information for the geostationary satellites. All these data are packed into the WAAS message, which is sent to navigation Earth stations (NES). The NES uplink this message to the geostationary satellites, which broadcast the "GPS-like" signal described earlier.

Taken together, the differential corrections and the improved geometry provided by the geostationary satellites will improve user accuracy to better than 10 m (2drms) in the vertical, which is adequate for aircraft Category I precision approach. The integrity data will improve user safety by flagging GPS satellites that are behaving incorrectly and cannot be corrected. In fact, the WAAS can deliver health warnings to the pilot within 6 s of a GPS satellite malfunction.

WIDE AREA AUGMENTATION SYSTEM

○ WIDE-AREA REFERENCE STATION (WRS)
⊗ WIDE-AREA MASTER STATION (WMS)
□ NAVIGATION EARTH STATION (NES)

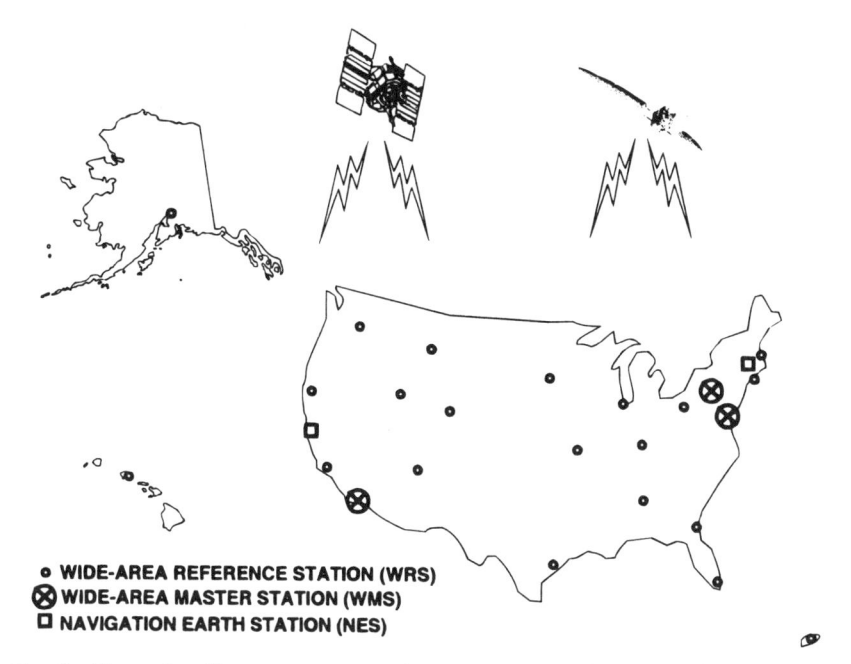

Fig. 2 Example wide area augmentation system ground network for the United States.

Section II of this chapter details the WAAS signal design, including the link budget, design of the spectrum-spreading codes, and forward error correction. Section III describes the ranging function of the WAAS signal and quantizes the improvement in position-fixing availability made possible by the WAAS ranging sources. It also includes some interesting results that specify the required accuracy of the differential corrections as a function of phase of flight.

Section IV discusses nonprecision approach, where stand-alone GPS provides sufficient accuracy, and the WAAS data only need provide integrity. Section V discusses Category I precision approach, where the WAAS must provide vector corrections to achieve the accuracy requirements. Section VI describes the WAAS data format, which was approved by Working Group 2 of the Radio Technical Committee for Aeronautics (RTCA) in the Spring of 1994, and Sec. VII contains a brief summary. Finally, an appendix briefly describes the estimation of the geostationary satellite ephemeris and clock offsets.

This chapter draws on the work of the RTCA Special Committee 159, which is charged with studying the integrity of GPS and writing performance standards for the airborne equipment.[1,2] In particular, this chapter is based on the efforts of Working Group 2, which is responsible for the WAAS.

Importantly, this chapter does not describe the algorithms that generate the vector corrections, nor does it analyze the accuracy of a positioning system that uses vector corrections. These important topics are covered in Refs. 3 and 4 and the previous chapter.

II. Signal Design

The WAAS signal must achieve the following:
1) Not interfere with the reception of GPS signals by existing GPS receivers or by other GPS receivers that are not designed to receive the WAAS signal
2) Provide an additional ranging measurement
3) Provide the highest possible capacity for integrity data and differential corrections
4) Be received by a modified GPS receiver with minimum complexity

To achieve these goals, the biphase shift keyed (BPSK) signal given in Eq. (1) seems to be the strongest candidate among the many discussed by Working Group 2.

$$s(t) = \sqrt{2C} X(t) D(t) \cos(\omega_L t + \theta) \tag{1}$$

In this equation, C is the power of the single carrier, and $D(t)$ is the data waveform that modulates the carrier. The carrier phase is θ, and $X(t)$ is a Gold sequence with the GPS chipping rate of 1.023×10^6 chips/s. The carrier frequency is $\omega_L \approx 2\pi\ 1575.42 \times 10^6$ rad/s, which is the same as the GPS L_1 frequency.

A. Link Budget and Noninterference with GPS

As shown in Table 1 and Fig. 3 (see Ref. 5), the received WAAS power through a typical GPS antenna will vary from -161 dBW to -157 dBW depending on the elevation angle. The signal is assumed to be received at a low elevation angle of 5 deg. The table also shows the probability of bit error $[Pr(\epsilon)]$ as a function

Table 1 Link power budget for the wide area augmentation system

Received WAAS carrier power	-161 dBW
Receiver antenna gain at 5 deg	-4 dBic
Cable/filter losses	-1 dB
Other implementation losses	-1 dB
Net effective carrier power (C)	-167 dBW
Boltzmann's constant	-228.6 dBW/K Hz
Equivalent noise temperature T_{eq}	$+28$ dB K
Noise power spectral density N_0	-200.6 dBW/Hz
C/N_0	33.6 dB Hz
Uncommitted margin	4.0 dB
Final C/N_0	29.6 dB Hz
100 b/s	
$E_b/N_0 = C/N_0 - 10 \log_{10}(100)$	9.6 dB
$Pr(\epsilon)$ without forward error correction (FEC)	$\approx 7 \times 10^{-6}$
$Pr(\epsilon)$ with forward error correction (FEC)[6]	$< 10^{-12}$
250 b/s	
$E_b/N_0 = C/N_0 - 10 \log_{10}(250)$	5.6 dB
$Pr(\epsilon)$ without forward error correction (FEC)	$\approx 2.0 \times 10^{-3}$
$Pr(\epsilon)$ with forward error correction (FEC)[6]	$\approx 1.5 \times 10^{-8}$

WIDE AREA AUGMENTATION SYSTEM

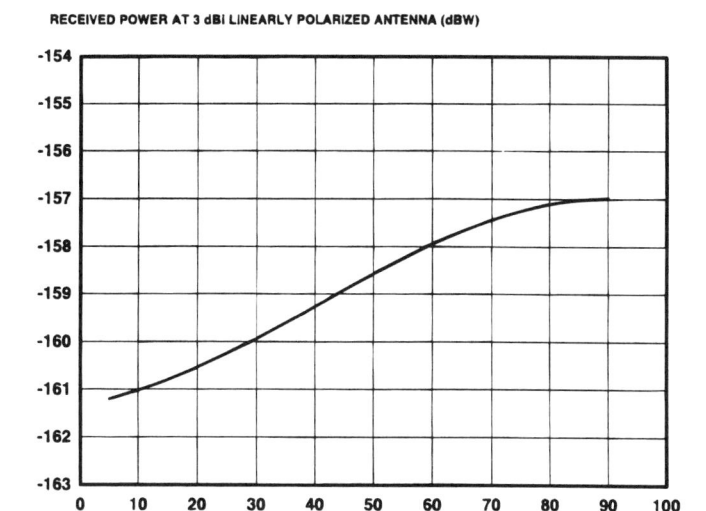

RECEIVED POWER AT 3 dBi LINEARLY POLARIZED ANTENNA (dBW)

Fig. 3 Expected typical received power levels from the wide area augmentation system.

of data rate. The forward error correction scheme is a $R = 1/2$ convolutional code with a constraint length of 7 and 3 bit soft decisions. This yields C/N_0 ratios of 33.6 dB Hz to 37.6 dB Hz, which in turn corresponds to E_b/N_0 ratios of 5.6–9.6 dB at 250 b/s.

The WAAS signal will not interfere with existing GPS receivers or future GPS receivers that are not designed to receive the WAAS signal. The WAAS power levels are generally weaker than the current and expected power levels for GPS. Even the highest possible WAAS power level is no more than 7 dB above the minimum specified GPS C/A power level of -164 dBW. Additionally, the strongest WAAS power level (-157 dBW) is approximately 20 dB weaker than the ambient noise power in the 2 MHz C/A-code bandwidth (and 30 dB weaker than the ambient noise in the 20 MHz P-code bandwidth).

Because the GPS and WAAS power levels are nearly equal, code division multiple-access (CDMA) can be used to separate the signals and prevent interference. As described in Chapter 3 of the companion volume, the GPS satellites share their L_1 band by using CDMA rather than time division multiple access (TDMA) or frequency division multiple access (FDMA). The GPS satellites use nearly orthogonal codes from a family of 1025 Gold sequences, and the WAAS satellites will broadcast unused Gold sequences from this same family.

Spread-spectrum multiple access is based on the lack of correlation between the different codes that belong to a given set of "signature" sequences. This property is depicted in Fig. 4, which assumes that K signals are sharing a channel, and so the received signal is given by the following:

$$r(t) = s_1(t) + \sum_{k=2}^{K} s_k(t - \tau_k) + n(t)$$

$$r(t) = s_{k=1}(t) + \sum_{k=2}^{K} s_k(t - \tau_k) + n(t)$$

$s_{k=1}(t)$

Z decision statistic

$$SNR = \frac{(E\{Z\})^2}{Var\{Z\}}$$

$$= \frac{E_b}{N_0}\left[1 + \frac{2(K-1)}{3N}\frac{E_b}{N_0}\right]^{-1}$$

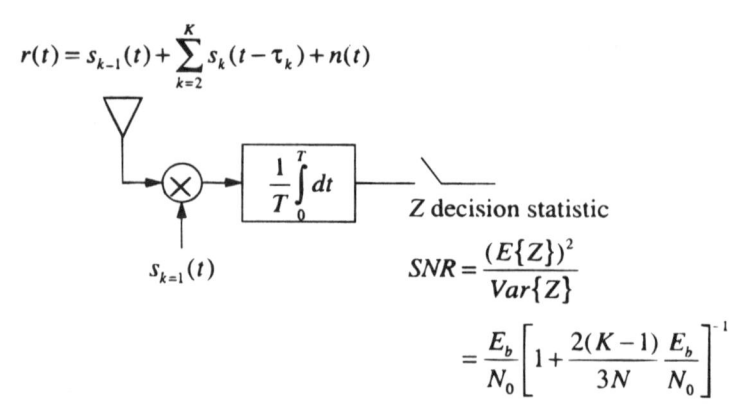

Fig. 4 The WAAS signal uses code division multiple access to share the L_1 band with GPS.

In this equation, $s_1(t)$ is the signal from the "desired" satellite, and $\{s_k(t - \tau_k)\}_{k=2}^{K}$ are the signals from the competing satellites. The channel also adds additive white Gaussian noise $n(t)$, which has a power spectral density of $N_0/2$. As shown in Fig. 4, the receiver is matched to the signal from the $k = 1$ user and integrates over T seconds.

The signal-to-interference ratio for the receiver shown in Fig. 4 can be computed based on the following assumptions. First, the desired and competing signals are received with equal power. (The equal power assumption can be removed without any real difficulties, but we retain it for simplicity.) Second, the energy in all K signals is given by the following:

$$E_s = \frac{1}{T}\int_0^T s_k^2(t)dt$$

Third, all signature sequences are comprised of N random binary bits, where these bits are independent of other bits in their own sequence and any other satellite's sequence. These binary bits modulate the carrier using biphase shift keying. Finally, the decision statistic Z is the sampled output of the integrator shown in Fig. 4.

The signal-to-noise (SNR) ratio of that decision statistic will be

$$SNR = \frac{(E\{Z\})^2}{Var\{Z\}}$$

$$= \frac{E_s}{N_0}\left[1 + \frac{2(K-1)}{3N}\frac{E_s}{N_0}\right]^{-1}$$

where the full analysis can be found in Ref 7. The signal-to-noise ratio in the absence of multiple-access interference is E_s/N_0; so the $K - 1$ competing users contribute the term inside the square brackets. As shown, the multiple-access interference is manageable provided that N is large compared to K. For GPS, $N = 1023$, and K is typically 8 or 9; thus multiple-access interference is small.

Importantly, if two or three WAAS satellites are added, then K is typically equal to 10 or 12, and the multiple-access interference will remain small.

Of course, GPS does not use random sequences. Rather, the GPS satellites draw their signature sequences from a set of Gold codes, and the WAAS signatures will be unused sequences from the same set. Indeed, any sequence in a Gold set can only take three cross-correlation values with any other sequence in the set, and these three values are shown in Fig. 5.

If the codes achieve their worst-case alignment in time, the magnitude of the cross-correlation is upper bounded by 65/1023, and the processing gain is $20 \log_{10}(1023/65) \approx 24$ dB. This means that the correlation level of the weakest GPS C/A signal will always be at least 17 dB over the strongest WAAS signal. The margin of the P code will be significantly greater, because of the tenfold increase in the chipping rate and the absence of cross-correlation.

A preferred set of codes from the GPS Gold code family has been identified for the WAAS.[8] These codes have the minimum number of shifts where the maximum autocorrelation occurs, and they have the minimum number of shifts where the maximum cross-correlation with the GPS codes occur.

In summary, the WAAS will not interfer with GPS, because the differences in the received power levels are small compared to the spread spectrum processing gains. In fact, WAAS received power levels are generally below that of GPS.

B. Data Capacity

The WAAS signal has enough capacity to carry both the differential corrections and the integrity data required to augment GPS. Recall that the differential corrections will improve user accuracy to better than 10 m (2 drms) in the vertical,

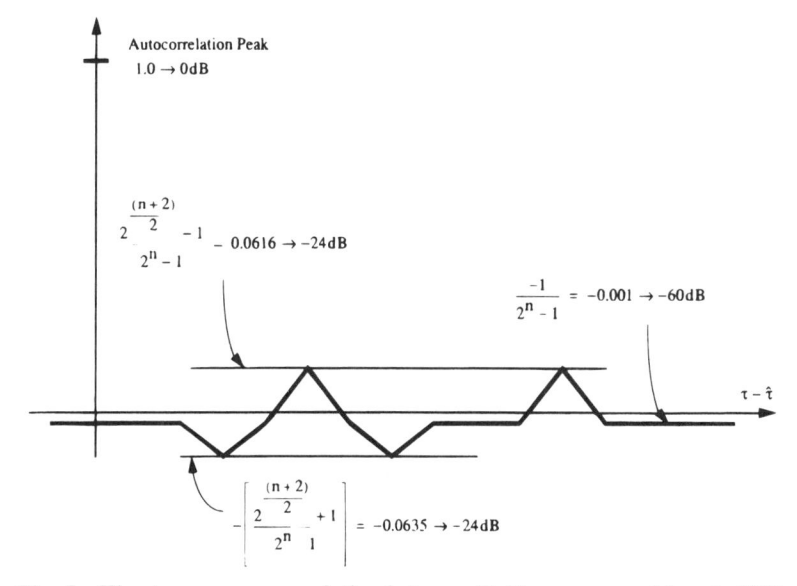

Fig. 5 Worst case cross-correlation between Gold sequences of length 1023.

which is adequate for aircraft Category I precision approach. The integrity data will improve user safety by flagging GPS satellites that are behaving incorrectly and cannot be corrected. For the reasons discussed in Sec. V and VI, these data require approximately 250 b/s of throughput.

Table 1 includes the probability of bit error for data carried by the WAAS as a function of the data rate. As shown, the probability of bit error for 100 b/s data would be sufficiently low even without the use of forward error correction. However, the corrections discussed in Secs. V and VI require 250 b/s; so forward error correction must be used to reach acceptably low bit error probabilities.

The error correction code used in Table 1 is the $R = 1/2$, convolutional code with constraint length 7 because the corresponding transmission rate is 500 symbols per second (sps), which is a multiple of 50 b/s and a submultiple of the C/A-code repetition rate of 1000 Hz. This code introduces a decoding delay of five constraint lengths or 35 databits, which is 0.14 s at 250 b/s. Incidentally, the $Pr(\epsilon)$ values in Table 1 assume three-bit soft decisions, and hard decisions would cost 2 dB.

This convolutional code is attractive, because it is well known and is a standard for satellite communications. Accordingly, decoders are available as integrated circuits for a few tens of dollars. As an alternative, this code can easily be realized in software at the 250 b/s data rate required for the WAAS.

C. Loop Threshold

Although forward error correction lowers the signal-to-noise ratio at which bit errors are made, it does not lower the signal-to-noise ratio at which the phase–lock-loops (PLL) lose lock. The PLL threshold can be estimated from the following equation for the variance of the phase estimate from a Costas Loop:

$$\sigma_\phi^2 = \frac{B_L N_0}{C} \left[1 + \frac{N_0}{2CT} \right] \text{rad}^2$$

where B_L is the loop noise bandwidth in Hertz and $1/T$ is the data or symbol rate. If this equation is set to a threshold phase variance of $(0.25)^2$ rad^2, then the SNR threshold is as follows:

$$C/N_0 = 8B_L[1 + \sqrt{1 + 1/(8B_L T)}]$$

or

$$[C/N_0]_{dB} = 10 \log_{10}(C/N_0) \text{ dB Hz}$$

This loop threshold is plotted in Fig. 6. Note that for nominal loop bandwidths (10–15 Hz), there is about a 2 dB difference between the 50 and 500 Hz predetection bandwidth performances.

III. Ranging Function

As described in Sec. II, the WAAS signal will be at the GPS L_1 frequency and will be modulated with a length 1023 Gold sequence with the same chipping rate as the GPS signals. The phase of the Gold code will be synchronized to GPS, and the carrier frequency will be controlled to facilitate carrier aiding. This

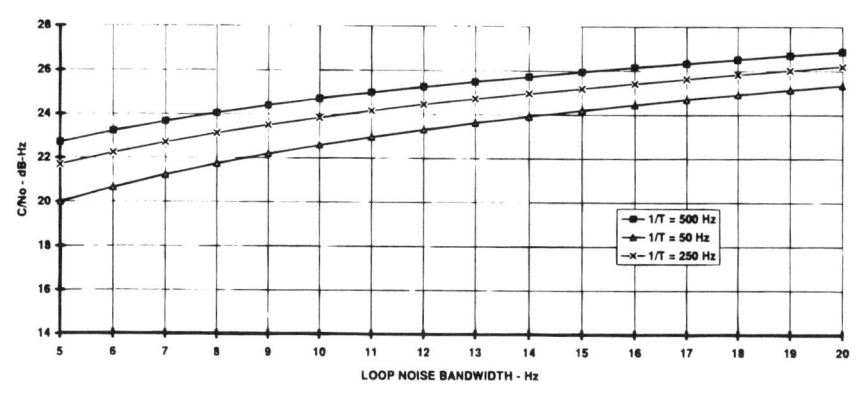

Fig. 6 Approximate tracking threshold for a Phase–lock-loop tracking the WAAS Signal (no dynamics).

code phase control is described in the Appendix to this chapter which is based on the more detailed discussion in Ref. 9.

With such synchronization, the WAAS satellites will provide new range measurements, and improve the *time availability* of position fixing. This improvement is shown in Fig. 7, which is from Ref. 10. Figure 7 shows the availability of a given level of horizontal dilution of precision (HDOP) in the conterminous United States (CONUS) when averaged over 24 hours. It incorporates the effect of possible satellite failures, where GPS satellite failures are based on the model described in Ref. 11. Wide area augmentation system satellite failures are based on the model described by Ref. 12.

Figure 7 shows availability versus HDOP when 0, 1, 2, or 3 geostationary satellites are used with 24 GPS satellites (21 primary plus 3 active spares). The geostationary satellites are located in optimal or nearly optimal locations for covering CONUS. As shown, if 2 geostationary satellites are parked in optimal locations, then an HDOP of 3.2 or better is achieved 99.999% of the time. If no geostationary satellites are used, then an HDOP of 3.2 is only achieved approximately 99.9% of the time. As such, the geostationary satellites reduce the probability of outage by two orders of magnitude.

The results shown in Fig. 7 for two WAAS satellites are probably achievable at reasonable cost. The assumed location of 60 deg W is very close to the planned location for the Inmarsat-3 satellite for the Atlantic Ocean West (55°W). The location at 100°W is near the center of the North American domestic satellite arc, where many system operators have geostationary satellites.

If a barometric altimeter is used, then lower HDOPs can be achieved with an availability of 99.999%.[10] Clearly, if the Glonass satellites, described in Chapter 9 of this volume, are added to the constellation, then lower HDOPs or higher availabilities are also possible.

The horizontal or vertical position error of a user can be bounded based on the corresponding DOP distribution and the pseudorange error standard deviation.[10]

$$HPE_d \leq 2HDOP_d\sigma_{PR}$$

$$VPE_d \leq 2VDOP_d\sigma_{PR}$$

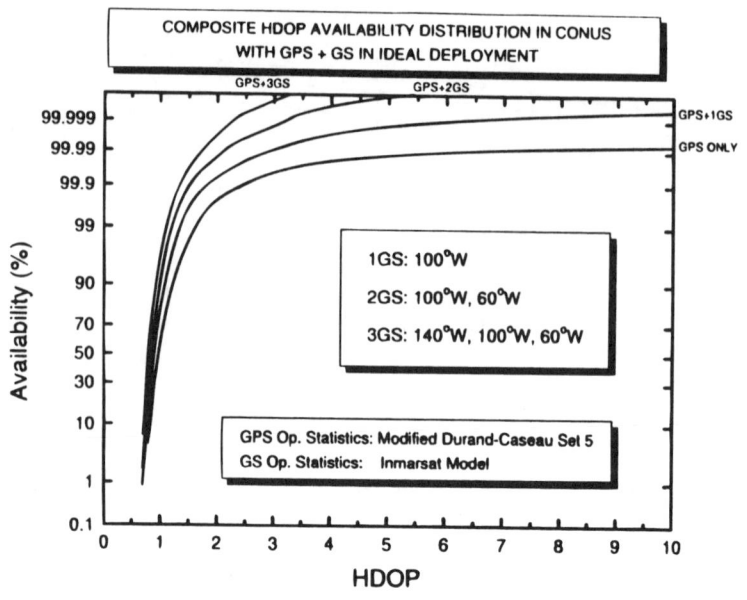

Fig. 7 Availability vs horizontal dilution of precision.[10] This analysis is for ranging WAAS satellites located as shown.

where HPE_d and $HDOP_d$ are the horizontal position error and horizontal dilution of precision not exceeded with probability d. Conversely, these values are exceeded with probability $1 - d$. VPE_d and $VDOP_d$ are similar variables for the vertical dimension. These relationships have been used to plot HPE and VPE vs σ_{PR} in Figs. 8 and 9.

As discussed in the next two sections, the HPE and VPE data shown in Figs. 8 and 9 can be used to derive requirements on the accuracy of the corrections provided by the WAAS. Section IV discusses nonprecision approach, where stand-alone GPS provides enough accuracy, and the WAAS data only need to provide integrity. Section V discusses Category I precision approach, where the WAAS must provide vector corrections to meet the accuracy requirements.

IV. Nonprecision Approach and Error Estimates

Nonprecision approach traditionally uses a baroaltimeter for vertical position information, and a ground-based radio navigation aid for horizontal information. In fact, the horizontal information is typically derived from any of the following (or combinations thereof): a VHF omnidirectional radio (VOR) navigation aid; distance-measuring equipment (DME); Loran-C; or automatic direction finding (ADF) to a nondirectional beacon (NDB). Global Positioning System receivers can be certified for nonprecision approach if they conform to the performance requirements in Technical Standard Order C-129, which was released in 1993. These receivers still derive vertical information from a baroaltimeter, and GPS provides the horizontal information. However, these TSO-C129 receivers are for

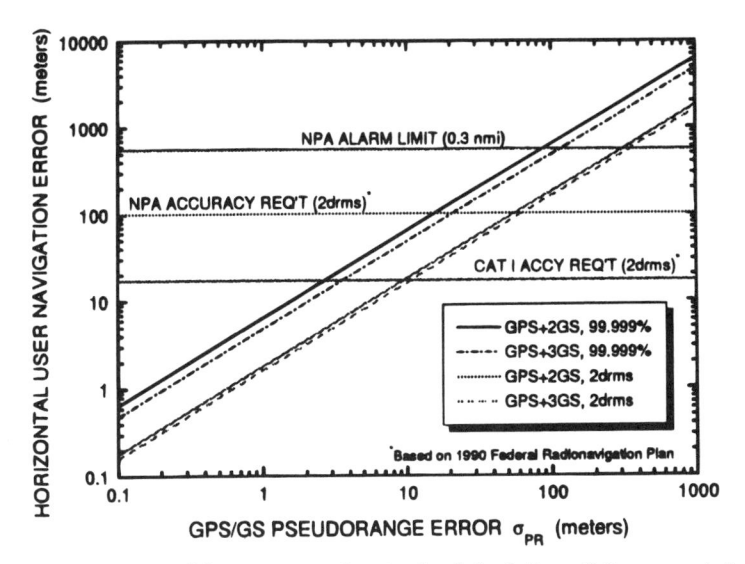

Fig. 8 Horizontal position error vs the standard deviation of the corrected pseudorange.[10]

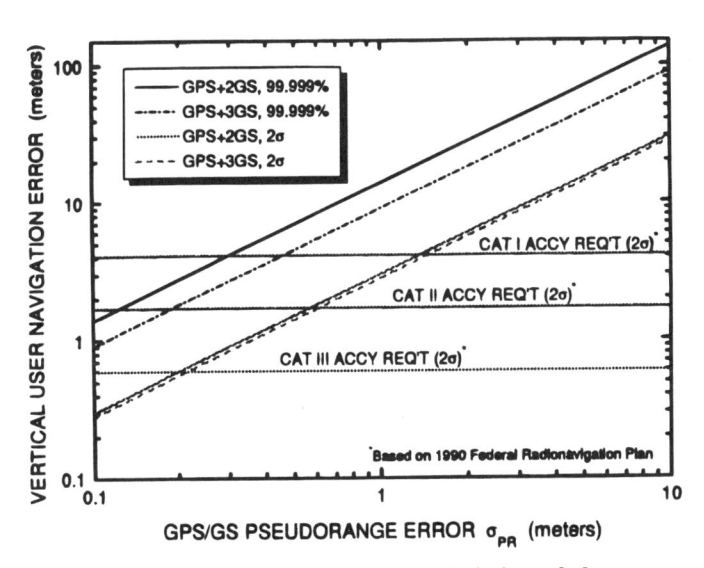

Fig. 9 Vertical position error vs the standard deviation of the corrected pseudorange.[10]

supplemental use only, and some augmentation of GPS, such as the WAAS, is required for GPS to be the primary navigation system.

Strawman specifications for nonprecision approach are superposed onto Fig. 8. As shown, these tentative specifications call for the horizontal error to be less than 100 m 95% of the time, and less than 560 m 99.999% of the time. If GPS is augmented with two WAAS satellites, then the 95 and 99.999% horizontal specifications require that $\sigma_{PR} \leq 60$ ms and $\sigma_{PR} \leq 90$ m, respectively. These derived requirements can be met without the use of differential corrections.

For nonprecision approach, the WAAS does not need to send correction data, but it must warn the user of any GPS malfunctions. Such integrity data could take some very simple forms. For example, it could simply warn the user not to use GPS for navigation, or it could warn the user not to use a specific GPS satellite. These warnings could apply to all phases of flight or they could specify the affected phase. Although these strategies are simple, they will result in conservative alarm thresholds. After all, the ground segment would need to send "don't use" messages for satellites that were unsuitable for use anywhere in the coverage area.

More sophisticated integrity data leave the final "use/don't use" decision to the aircraft receiver, which can account for its own geometry and satellite selection. In this case, the WAAS would send pseudorange error estimates for each satellite. The user would compute its uncorrected position; then it would apply the error estimates as corrections, and compute its corrected position. If the two position fixes differed by more than a certain threshold, then the receiver could discard one satellite at a time until the position fixes were close enough.[13] The receiver would declare the system unavailable only if it could not produce agreement. Consequently, the transmission of error estimates for each satellite would yield a much lower false alarm rate than the transmission of the simpler "don't use" messages described in the last paragraph.

A dense network of reference stations is not required to generate the error estimates for nonprecision approach, because the error estimate need only be moderately accurate. In fact, 10–12 reference station may well provide adequate worldwide coverage.[14] However, denser networks, such as the one shown in Fig. 2, are required to provide sufficient accuracy for precision approach.

V. Precision Approach and Vector Corrections

Unlike nonprecision approach, precision approach is based on a smooth glide path with a constant rate of descent. This glide path, typically 3 deg, passes through a "decision height" at which the pilot must decide whether or not to complete the landing. The pilot must execute a missed approach unless visual references have become available. Precision approach is divided into three categories depending on the decision height and the "runway visual range" (RVR). Category I conditions exist when the decision height is at 200 feet or above, and the RVR is 2400 feet or greater. Category II conditions exist when the decision height is between 100 and 200 feet, and the RVR is 1200 feet or greater. Category III conditions exist when the visibility is poorer, and include conditions with zero visibility. The requirements for Category I, II, and III precision approach are superposed on Fig. 9, which shows vertical position error versus σ_{PR}.

WIDE AREA AUGMENTATION SYSTEM 129

As shown in Fig. 9, the 2 drms (95%) requirements for Category I precision approach require that $\sigma_{PR} \leq 3.0$ m. These requirements can be met using either *scalar* corrections over a local area or *vector* corrections over larger areas. As shown in Fig. 10, local broadcast of scalar corrections can provide Category I capability, but the validity of the corrections decreases with distance. This spatial decorrelation is detailed in Chapter 1 of this volume. Briefly, the error sources that cause the correction to decorrelate spatially are: errors in the GPS ephemeris data, ionospheric refraction, and tropospheric refraction. If a broadcast system is to provide Category I service to a large area (hundreds of miles across), it must deliver *vector* corrections. These corrections are described later in this section and in Sec. VI.

Also shown in Fig. 9, the requirements for Category II or III precision approach demand an extremely accurate pseudorange measurement. These applications require some variety of local area system (very close to the runway) and are beyond the capability of the WAAS.

A. Vector Corrections

Vector corrections carry the following components for each satellite:

Satellite clock offset: This term does not decorrelate spatially, but decorrelates temporally because of selective availability (SA). In fact, the satellite clock corrections must be sent much more frequently than any other component of the vector correction.

Satellite ephemeris: Estimates of all three components of Δr_k are sent, where this vector connects the true location of the kth satellite to the location given by the satellite's navigation message. In component form, these data do not decorrelate spatially and decorrelate very slowly in time. In fact, the update rate for the

$[\Delta \underline{r}_k, \Delta B_k]$

Atmospheric Effects $I_{k,u}$ and $T_{k,u}$

Local Broadcast of
1.) Scalar Correction
2.) Use/Don't Use

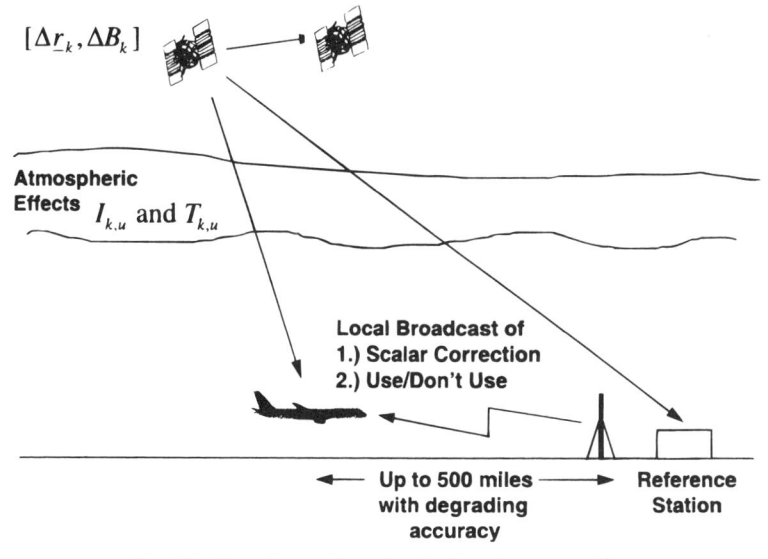

◄──── Up to 500 miles ────► Reference Station
with degrading
accuracy

Fig. 10 Local area broadcast of scalar corrections.

ephemeris data is not set by concerns over accuracy. Rather, it is set to ensure that the time to first position fix is not too large.

The vector corrections also carry separate data that describe the ionosphere. These data may be a grid of ionospheric samples or a set of coefficients for an orthonormal function description of the ionospheric delay. The data do not decorrelate rapidly in time, and an update interval of 2–5 minutes is probably adequate.

As described in Chapter 1 of this volume, ionospheric corrections lose their validity as the user moves away from the reference station. This concern will dictate the spacing between the reference stations and the amount of ionospheric data sent. In fact, a likely set of reference stations could be located at the Federal Aviation Administration's Air Route Traffic Control Centers (ARTCC). This network is shown in Fig. 2 and inlcudes approximately 24 reference stations.

Vector corrections are designed to serve a large geographical area, and as such they are sometimes called wide area differential GPS (WADGPS) corrections. Wide area differential GPS is well described in the previous chapter, which details this system concept.

B. Precision Approach Integrity

Vector corrections will certainly provide enough accuracy for Category I approach. However, their use raises an interesting integrity issue because the aircraft must now fly on the corrected position fix. In other words, the WAAS has seemingly combined the positioning function with the integrity function, whereas these two functions are traditionally independent. A full analysis of Category I integrity when using the WAAS is beyond the scope of this chapter. However, adequate integrity can be realized with a combination of ground monitoring and receiver autonomous monitoring, as briefly described in the following paragraphs.

1. Ground Monitoring

The set of ground monitors shown in Figs. 1 and 2 help realize the integrity function by setting and broadcasting a "don't use" flag for any satellite. This flag is set only if the ground network cannot develop differential corrections with confidence. For example, if the error acceleration will render the correction invalid before the next update, the master station will set the "don't use" flag for that satellite.

Additionally, the ground monitors could be partitioned into two groups: reference stations and integrity monitors. The latter group would be dedicated to integrity checking and would be independent of the reference stations. The observations made by these integrity monitors are not used to develop the corrections, and if there are any difficulties with the corrected fixes, they simply cause a "don't use" message to be sent.

Alternatively, the ground monitors could all be reference stations, and data from all of the ground stations would be used to form the WAAS corrections. In this case, integrity would be determined by computing the measurement residuals from each reference station. In other words, if a transmitted correction was very different from the correction suggested by a single reference station, a "don't use" message would be sent.

In either case, the density of the WAAS ground stations will be eventually determined by the *integrity* requirements of Category I precision approach. Specifically, they must be dense enough to guarantee that no large error has been introduced into a local area that cannot be detected by the nearest one (or two) monitors.

2. Receiver Autonomous Integrity Monitoring

Receivers using the WAAS can confirm the integrity of their position fix by using receiver-autonomous integrity monitoring (RAIM), which is described in the next chapter. For Category I approach, the aircraft applies the differential corrections sent by the WAAS and then determines whether or not the corrected data supplied by the different satellites are *consistent*. This autonomous action protects the aircraft against certain error mechanisms that may not be noticed by a sparse set of ground monitors.

As described in the next chapter, such a consistency check requires that at least five satellites are in view of the aircraft. It also requires that the vertical dilution of precision (VDOP) of all five subsets be acceptably small, where the subsets are formed by deleting one of the satellites in view. Consequently, the time availability of this self-contained integrity monitoring is smaller than the time availability of position fixing. In fact, the low time availability of autonomous fault detection is a main reason that the GPS constellation must be augmented.

As shown in Figs. 11 and 12, the additional pseudoranges provided by the geostationary satellites dramatically improve the availability of autonomous failure detection. Figure 11 is for the 24-satellite GPS constellation with no satellite failures and no additional geostationary ranging sources. Figure 12 is for the same set of GPS satellites but includes four Inmarsat-3 satellites as well as geostationary satellites at the following locations: 100°W and 135°W.

Both figures show the the maximum vertical dilution of precision (VDOP) among the five subsets formed by deleting one satellite at a time. This number is called the "maximum subset VDOP", and the figures show the daily value of maximum subset VDOP, which is not exceeded 99.9% of the time.

As shown in these figures, geostationary satellites significantly increase availability or equivalently decrease the outage probability. Furthermore, if some auxiliary onboard sensor allows the aircraft to coast through short outages, then the probability of outage is further decreased. However, neither figure accounts for satellite failures; and this possibility must be accounted for in a full analysis.[11]

VI. Wide Area Augmentation System Message Format

For each GPS satellite, the WAAS message contains separate corrections for the quickly varying component of the pseudorange error (mostly clock) and the slowly varying component of the pseudorange error (mostly ephemeris). The WAAS message also carries estimates of the ionospheric delay for a grid of locations.

The WAAS message format serves an extremely ambitious goal: provide Category I precision approach accuracy over a continental area using a data rate of

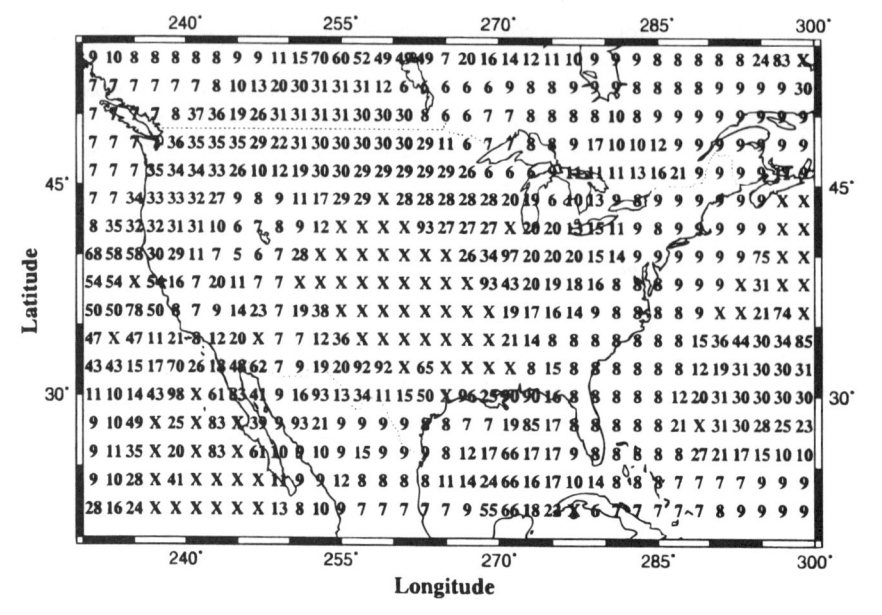

Fig. 11 Subset VDOP for 24 GPS Satellites. The subset VDOP is the maximum VDOP obtained when one satellite is deleted from the set of satellites in view. The value displayed in the figure is the daily value of subset VDOP, which is not exceeded 99.9% of the time. All 24 GPS satellites are assumed to be healthy.

only 250 b/s. To do this, the message stream must carry corrections for all 24 GPS satellites and any GNSS (Global Navigation Satellite System) satellites in orbit. In contrast, local area DGPS datalinks typically use 100 b/s or more to provide a similar capability to ranges of 200–300 km. Moreover, local links need only carry corrections for the 6–12 satellites in view of the single reference station.

The WAAS message format is extremely efficient for the following reasons:

Fast corrections: These messages carry the quickly varying component of the pseudorange errors for each satellite (mostly clock error). They must be sent much more frequently than any other message (every 6–10 s), but they do not decorrelate spatially. Consequently, a single, very short message suffices for the entire footprint of the geostationary WAAS satellite. The GPS error components, which do decorrelate spatially, are corrected by separate messages that need not be sent frequently.

No rate corrections: Unlike most local area DGPS data formats, the WAAS carries no rate corrections for the quickly varying component of the satellite error. As shown in Ref. 16, it is more efficient for the user receiver to estimate the rate by differencing the most recent fast corrections.

Masks: A "mask" is used to designate which satellite belongs to which slot in the fast correction messages. A mask is used to assign slots, because the WAAS data format must be able to send corrections efficiently for all GNSS satellites in orbit. At times, all 24 satellites in the GPS primary 21 (+ 3) satellite constellation can be seen by a combination of users in the footprint of a geostation-

WIDE AREA AUGMENTATION SYSTEM   133

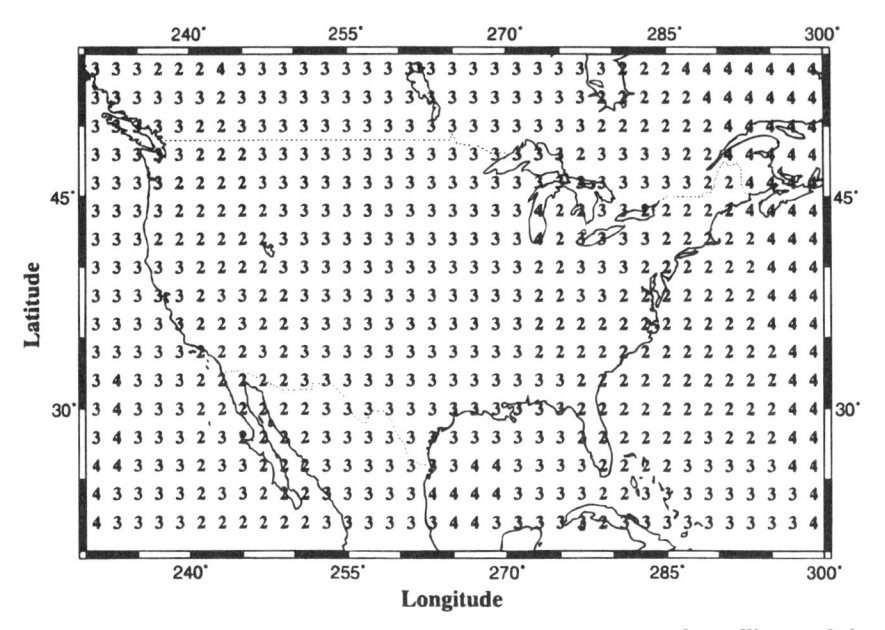

Fig. 12 Subset VDOP for 24 GPS satellites plus 4 Inmarsat-3 satellites and 2 additional geostationary satellites. The subset VDOP is the maximum VDOP obtained when one satellite is deleted from the set of satellites in view. The value displayed in the figure is the daily value of subset VDOP, which is not exceeded 99.9% of the time. All 24 GPS satellites are assumed to be healthy.

ary satellite.[17] A similar ionospheric mask is used to associate each slot in the ionospheric correction message with a geographic location.

Geostationary navigation message: In contrast to the GPS satellites, the WAAS satellites are geostationary. Hence, they have much lower accelerations than the GPS satellites, and their ephemeris need not be updated as frequently.

Parity: As described shortly, the WAAS message uses a much more powerful algorithm for detecting transmission errors than GPS or most local area DGPS datalinks. WAAS error detection only adds 24 parity bits to 226 databits, whereas most local area DGPS data formats add 6 bits to 24 databits. As such, the overhead for parity is reduced from $6/30 = 0.20$ to $24/250 \approx 0.1$.

The basic WAAS message is shown in Fig. 13. As shown, all WAAS messages are 250 bits in length. At the data rate of 250 b/s, the duration of a WAAS

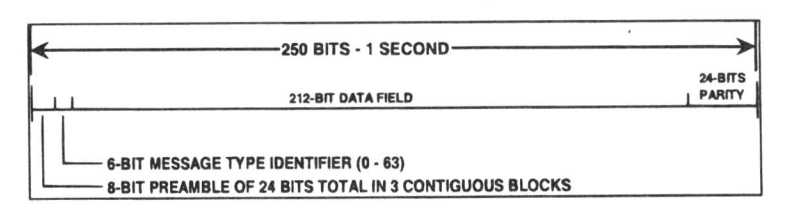

Fig. 13 Basic WAAS datablock format.

message is 1 s, and the start of the message block is synchronous with the 6 s GNSS time epoch.

Each block consists of the following:
1) 8-bit (distributed) preamble
2) 6-bit message type
3) 212-bit datafield
4) 24-bit CRC (cyclic redundancy check) parity

The preamble is a 24-bit unique word, distributed over three successive blocks. An 8-bit preamble is adequate because the WAAS message is 1 s in duration and remains in synchronism with the GPS time epoch of 6 s. The message type field is 6-bits long, which allows for 64 different messages. The currently defined message types are summarized in Table 2.

The following subsections describe the parity field, and then they briefly discuss the type 1, 2, 3, 9, 24, 25, and 26 messages. All message types are detailed in Ref. 5.

A. Parity Algorithm

The WAAS uses a much stronger parity algorithm than the extended Hamming code used in the GPS navigation message (and typically used for the transmission of differential corrections). The GPS parity algorithm uses the extended Hamming code to add 6 parity bits to a field of 26 databits. It achieves a minimum distance of 4, which means that any combination of 3 or fewer bit errors can be detected. However, if an interference or noise burst coincides with one of the 30-bit words, and the received data are completely garbled, then the probability that

Table 2 Wide area augmentation system

Type	Contents
0	Don't use GEO for anything (for testing)
1	Pseudorandom noise (PRN) mask assignments, set up to 50–212 bits
2	Fast pseudorange error estimates
3–8	Reserved for future messages
9	GEO navigation message (x,y,z, time, etc.)
10–11	Reserved for future messages
12	WAAS network universal coordinated time (UTC) offset parameters
13–16	Reserved for future messages
17	GEO satellite almanacs
18	Ionospheric pierce point mask 1
19	Ionospheric pierce point mask 2
20	Ionospheric pierce point mask 3
21	Ionospheric pierce point mask 4
22	Ionospheric pierce point mask 5
23	User differential range errors (UDRE) zone radii and weights
24	Mixed fast/long-term satellite error estimates
25	Long-term satellite error estimate
26	Ionospheric delay error estimate
27–63	Reserved for future messages

the algorithm fails to detect the error burst is $p_{fd} = 2^{-6} = 1.56 \times 10^{-2}$ (see Ref. 6). This probability is too large for a system designed to make GPS fail-safe.

To reduce the probability of failing to detect a bit error, the parity scheme in Fig. 13 uses 24 bits and thus reduces the probability of failing to detect an interference burst to $2^{-20} = 9.54 \times 10^{-7}$. As described earlier, the overhead for parity is reduced from $6/30 = 0.20$ to $24/250 \approx 0.096$. The WAAS error detection algorithm does increase the data latency of the integrity data, because 250 bits must be collected by the receiver for a parity check instead of only 30. However, this increases transmission time by only 0.88 s at the WAAS datarate of 250 b/s.

B. Message Type 2 Fast Corrections and User Differential Range Errors

The format for type 2 messages is shown in Fig. 14. These messages carry the quickly varying component of the pseudorange errors for each satellite (mostly clock errors). Each message delivers 16-bit datablocks for 13 satellites. The 16-bit datablock consists of a 12-bit correction and 4 bits of user differential error indication (UDREI). Each correction shall have 0.25-m/resolution and a range of ± 255.5 m. A value of +255.5 (011111111111) indicates that the satellite is not currently observed by the ground network. A −256 (100000000000) is a "don't use" message for that satellite.

As shown, message type 2 also contains a two-bit block identification that indicates those satellites to which the corrections apply. Block 0 contains the corrections for the first 13 satellites designated by the mask, Block 1 contains the fast corrections for satellites 14–26, and so forth. If six or fewer corrections remain at any time, they should be contained in a type 24 mixed corrections message. The last half of message type 24 is reserved for the long-term corrections described below.

In the absence of an emergency, the fast corrections could be sent as infrequently as every 10 s. However, if a satellite fails suddenly, then a fast correction message that carries the emergency "don't use" flag can be completed within 6 s.

C. Type 25: Long-Term Satellite Error Corrections Message

This message carries the slowly varying clock and ephemeris components of the pseudorange errors for each satellite. Its format is shown in Fig. 15. As shown, the datafield begins with a velocity code. If this bit is set to 0, then the message will contain the slowly varying clock and ephemeris errors for four satellites. If it is set to 1, then the message contains error *and* error rate estimates, but only for two satellites.

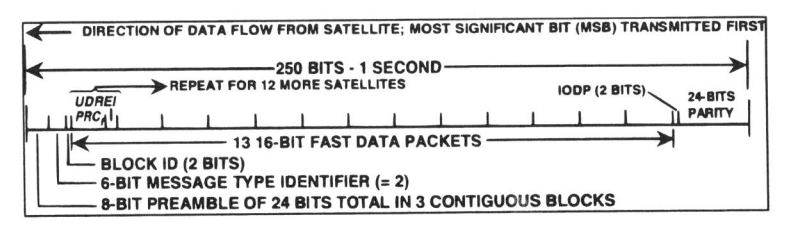

Fig. 14 Message type 2—fast corrections and UDRE.

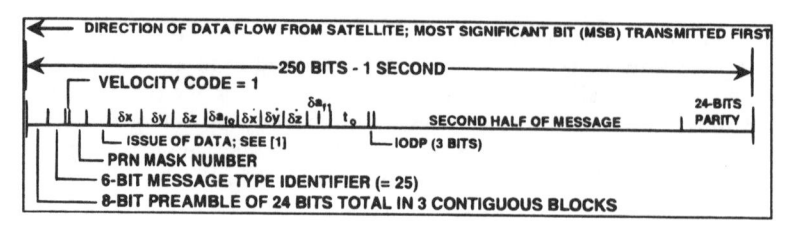

```
|<--- DIRECTION OF DATA FLOW FROM SATELLITE; MOST SIGNIFICANT BIT (MSB) TRANSMITTED FIRST
|<-------------------250 BITS - 1 SECOND----------------------->|
   VELOCITY CODE = 1
                            δa_f1                                          24-BITS
   |  | δx | δy | δz |δa_f0|δẋ|δẏ|δż| | t_0 ||    SECOND HALF OF MESSAGE   | PARITY
            ISSUE OF DATA; SEE [1]   IODP (3 BITS)
            PRN MASK NUMBER
         6-BIT MESSAGE TYPE IDENTIFIER (= 25)
         8-BIT PREAMBLE OF 24 BITS TOTAL IN 3 CONTIGUOUS BLOCKS
```

Fig. 15 Message type 25—long-term corrections with velocity code set to 1.

The details of the long-term error correction message are given in Table 3. The ephemeris and ephemeris rate corrections are given in Earth-centered, Earth-fixed (ECEF) coordinates as (Δx_k, Δy_k, Δz_k) and ($\Delta \dot{x}_k$, $\Delta \dot{y}_k$, $\Delta \dot{z}_k$). In Table 3, the clock and clock rate corrections are denoted a_{f0} and a_{f1} respectively. The velocity corrections should be used to predict future ephemeris and clock values using t_0 as the reference time.

D. Type 26: Ionospheric Delay Error Corrections Message

These messages give the WAAS estimates of vertical delay caused by the ionosphere and the 99.5% accuracy of this estimate. The vertical delays are for ionospheric pierce points (IPP), which are specified by the ionospheric mask messages (type 18, 19, 20, 21, and 22). The mask messages specify which of 929 possible IPP locations are being used by the type 26 messages.[5] For example, approximately 80 IPP are required to provide Category I accuracy throughout the National Airspace System (NAS).

Table 3 Long-term error correction message (half message when velocity code is set to 1)

Parameter	Number of bits	Scale factor LSB	Effective range	Units
For each of 1 or 2 satellites	106			
Velocity code	1			discrete
PRN mask number	6	1	50	discrete
Issue of data	8	1	255	discrete
Δx_k (ECEF)	9	0.5	±128	m
Δy_k (ECEF)	9	0.5	±128	m
Δz_k (ECEF)	9	0.5	±128	m
a_{f0}	10	2^{-30}	$\pm 2^{-21}$	s
$\Delta \dot{x}_k$ (ECEF)	7	2^{-10}	±0.0625	m/s
$\Delta \dot{y}_k$ (ECEF)	7	2^{-10}	±0.0625	m/s
$\Delta \dot{z}_k$ (ECEF)	7	2^{-10}	±0.0625	m/s
a_{f1}	7	2^{-38}	$\pm 2^{-32}$	s/s
Time of applicability t_0	16	16	604,784	s
IODP	3	1	0–7	discrete

LSB denotes least significant bit; ECEF denotes Earth-centered Earth-fixed; and IODP denotes issue of data PRN where PRN denotes pseudorange noise code.

WIDE AREA AUGMENTATION SYSTEM

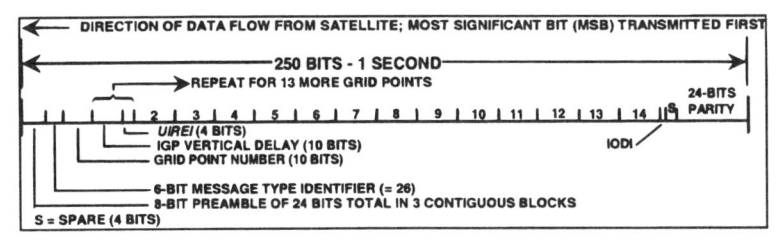

```
┌─── DIRECTION OF DATA FLOW FROM SATELLITE; MOST SIGNIFICANT BIT (MSB) TRANSMITTED FIRST
│◄──────────────── 250 BITS - 1 SECOND ────────────────►
│    ┌──► REPEAT FOR 13 MORE GRID POINTS
│    │                                                              24-BITS
│ |  | | 2 | 3 | 4 | 5 | 6 | 7 | 8 | 9 | 10 | 11 | 12 | 13 | 14 |   PARITY
│    └── UIREI (4 BITS)
│        └── IGP VERTICAL DELAY (10 BITS)                        IODI
│            └── GRID POINT NUMBER (10 BITS)
│        └── 6-BIT MESSAGE TYPE IDENTIFIER (= 26)
│            8-BIT PREAMBLE OF 24 BITS TOTAL IN 3 CONTIGUOUS BLOCKS
S = SPARE (4 BITS)
```

Fig. 16   Message type 26—ionospheric corrections.

A type 26 message is shown in Fig. 16 and described in Table 4. It carries 13 vertical delay estimates, where each of these is 10 bits and covers a range of 64 m with a resolution of 0.125 m. It also carries the corresponding vertical delay accuracy indicators, where each of these is 4-bits long. The type 26 message also includes 10 bits to identify where the pierce point falls within the current ionospheric mask.

E. Type 9: WAAS Satellite Navigation Message

The navigation message for the geostationary WAAS satellite is shown in Fig. 17, and its contents are detailed in Table 5. The location and velocity of the satellite is described using ECEF coordinates, because transmission of ECEF coordinates is more efficient for a geostationary satellite than transmission of Keplerian elements.

The WAAS navigation message also includes the time of week (TOW) and week number, which represent the time of transmission at the start of the block carrying the message. The TOW can be used to initialize receiver timing during signal acquisition. The TOW is also the reference time for applying the velocity terms to predict position. This message also includes an accuracy exponent to estimate the accuracy of the position information.

F. Applied Range Accuracy Evaluation

The data in the various WAAS messages can be combined to estimate the 99.5% accuracy of the corrected pseudorange. The applied range accuracy (ARA)

Table 4   Type 26 ionospheric delay model parameters

Parameter	Number of bits	Scale factor LSB	Effective range	Units
Pierce point number	10	—	—	discrete
IODI	2	1	0–3	discrete
For each of 13 more grid points	14	—	—	—
Vertical delay estimate	10	1	929	discrete
User ionospheric vertical Accuracy Indicator (UIREI)	4	1	0–15	discrete

IODI denotes issue of data ionosphere; UIREI denotes user ionospheric range error indicator.

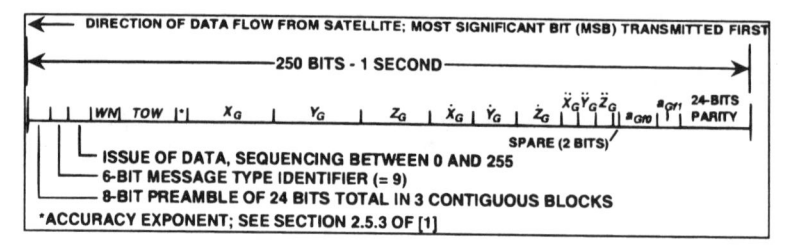

```
┌──────────────────────────────────────────────────────────────────────────┐
│ ←─── DIRECTION OF DATA FLOW FROM SATELLITE; MOST SIGNIFICANT BIT (MSB) TRANSMITTED FIRST
│ ←─────────────── 250 BITS - 1 SECOND ──────────────→                     │
│ | |WM| TOW |*| X_G | Y_G | Z_G | Ẋ_G | Ẏ_G | Ż_G |Ẍ_GŸ_GZ̈_G| a_Gf₁ | a_Gf₀ | 24-BITS PARITY |
│                                                        SPARE (2 BITS)    │
│         ISSUE OF DATA, SEQUENCING BETWEEN 0 AND 255                      │
│         6-BIT MESSAGE TYPE IDENTIFIER (= 9)                              │
│         8-BIT PREAMBLE OF 24 BITS TOTAL IN 3 CONTIGUOUS BLOCKS           │
│        *ACCURACY EXPONENT; SEE SECTION 2.5.3 OF [1]                      │
└──────────────────────────────────────────────────────────────────────────┘
```

Fig. 17 Message type 9—Geostationary satellite navigation message.

for satellite i is as follows:

$$\text{ARA}_i = \sqrt{\text{UDRE}_i^2 + F_i^2 \, \text{UIREI}_i^2} \text{ m}$$

The UDRE_i estimate the accuracy of the combined fast and slow corrections and are broadcast in message type 2. They are for the worst location in the coverage region. The UIREI_i estimate the 99.5% accuracy of the ionospheric delay estimates. This estimate must be multiplied by the satellite's obliquity factor F_i, because the UIREI_i value is the error for the vertical ionospheric delay at the pierce point.

VII. Summary

This chapter describes a signal design for a ranging WAAS. In particular, it demonstrate that the proposed WAAS signal can be received by slightly modified GPS receivers without interfering with nonparticipating receivers. The WAAS signal does not interfere, because the received WAAS signal has approximately the same power as GPS signals, and CDMA is used to share the L_1 channel.

The WAAS signal can be controlled to add a valuable range measurement to the navigation solution of airborne receivers. This augmentation significantly improves the availability of position fixing, as well as autonomous fault detection.

Table 5 Wide area augmentation system satellite navigation message parameters

Parameter	Number of bits	Scale factor LSB	Effective range	Units
X (ECEF)	32	0.02	±42,949673	m
Y (ECEF)	32	0.02	±42,949673	m
Z (ECEF)	32	0.02	±42,949673	m
$\dot{X}$ (ECEF)	18	0.0002	±26.2144	m/s
$\dot{Y}$ (ECEF)	18	0.0002	±26.2144	m/s
$\dot{Z}$ (ECEF)	18	0.0002	±26.2144	m/s
TOW	20	1	604,799	s
Week number	10	1	1023	weeks
Accuracy	4			
Issue of data	8	1	255	discrete
Spare	20			

If forward error correction is used, then the WAAS signal has a data capacity of 250 b/s, which is enough to carry the vector corrections required to make possible category I precision approach. In fact, the WAAS uses an extremely efficient data format to carry the required information over the available capacity. This format separates the fast corrections (mostly clock) for each satellite from the slow corrections (mostly ephemeris). It also has separate messages for the ionospheric corrections.

Appendix: Geostationary Satellite Ephemeris Estimation and Code-Phase Control

This appendix briefly describes the WAAS satellite ephemeris estimation process and how the WAAS code phase is controlled. It is based on Ref. 9, where a more detailed explanation is provided. In this appendix, the ephemeris Δr_{geo} and clock offset (ΔB_{geo}) of the WAAS satellite are the estimanda of interest. These are estimated using WAAS pseudorange residuals measured at the remote reference stations and forwarded to the master station. The reference station clock offsets Δb_m are "nuisance parameters" and must be estimated using GPS common-view time transfer.

This appendix begins by discussing the use of GPS common-view time transfer to estimate the reference station clock offsets. Then it discusses the determination of the WAAS ephemeris and clock offset when the WAAS is a processing transponder with the clock onboard the satellite. Finally, it discusses WAAS ephemeris estimation and code phase control for a bent-pipe transponder where the clock is on the ground.

For simplicity, this appendix assumes that: the WMS is colocated with the Mth reference station; the WAAS master clock is also located at the WMS; and the uplink to the geostationary satellite is at this same site.

Reference Station Clock Synchronization

The WMS uses common-view time transfer to estimate the difference between the reference station clocks and the master clock located at the WMS. More specifically, it uses the following GPS pseudorange residuals from the M reference stations to estimate the clock offsets:

$$\{(\Delta_{k,m} = \Delta \underline{r}_k \cdot \underline{1}_{k,m} + \Delta b_m - \Delta B_k + v_{k,m})_{k=1}^{K_m}\}_{m=1}^{M} \quad (A1)$$

In Eq. (A1) $\Delta \underline{r}_k$ is the vector that connects the true location of the kth satellite to its location according to the navigation message. In Eq. (A1), $\underline{1}_{k,m}$ denotes the unit vector from the kth satellite to the mth reference station. Additionally, Δb_m and ΔB_k are the offsets in the reference station and satellite clocks. The measurement noise is given by $v_{k,m}$; K_m is the number of satellites in view of the mth reference station; and M is the total number of reference stations.

These GPS pseudoranges are called "iono free", because dual-frequency measurements have been used to estimate the ionospheric delay accurately, and this delay is removed from the measured pseudoranges. In addition, they are called "tropo free", because local meteorological measurements have been used to estimate the tropospheric delay accurately, and this delay is also removed.

The master station uses common-view time transfer to estimate the difference between the reference station clocks and the master clock located at the WMS. An estimate of this difference for the mth reference station is given by the following:

$$\Delta \hat{b}_{m,M} = \frac{1}{K_{M,m}} \sum_{k=1}^{K_{M,m}} (\Delta_{k,M} - \Delta_{k,m}) \quad \text{(A2)}$$
$$\approx \Delta b_M - \Delta b_m$$

where $K_{M,m}$ is the number of satellites in common view of the WMS and the mth reference station. This family of estimated clock offsets $(\Delta \hat{b}_{m,M})_{m=1}^{M-1}$ is very important. It is used to eliminate the clock differences in the observations from the reference stations.

It is hoped that the master clock is kept in close synchronism with GPS time. This can be done by averaging GPS observations at the master station to steer an atomic clock or ensemble of atomic clocks. Alternatively, it could be done if there were a more direct connection from the master clock to GPS time as kept by the U.S. Naval Observatory. For example, if a reference station is located at USNO, then it could serve as the timing reference for the entire system.

Processing Transponder

The master station will also estimate the ephemeris of the geostationary WAAS satellite from the pseudorange residuals provided by the reference stations. If the WAAS satellite is a processing transponder with the clock onboard the satellite, then the WAAS residuals from the reference stations are the following:

$$(\Delta_{geo,m} = \Delta \underline{r}_{geo} \cdot \underline{1}_{geo,m} + \Delta b_M - \Delta B_{geo} + v_{geo,m})_{m=1}^M \quad \text{(A3)}$$

where ΔB_{geo} is the WAAS satellite clock offset relative to the data in the WAAS navigation message, and $\underline{r}_{geo}$ is the error in the broadcast WAAS ephemeris. This equation assumes that the reference station clock offsets have been removed. These measurements are "tropo-free," because local meteorological measurements have been used to estimate the tropospheric delay accurately, and these delay estimates have been subtracted from the measured residuals.

The ionospheric delay is removed from the WAAS measurements in one of two ways. First, the WAAS downlink will include a signal at C-band. As such, dual-frequency measurements can be used to estimate the ionospheric delay. However, the C-band signal will be considerably weaker than the L_1-band signal; thus a dish antenna will be required. Alternatively, the GPS measurements at the reference station can be used to build an accurate model of the local ionosphere, and that model can be used to estimate the WAAS ionosphere.

These WAAS residuals from all M reference stations form the following linear system of equations:

$$\begin{bmatrix} \Delta_{geo,1} \\ \Delta_{geo,2} \\ \vdots \\ \Delta_{geo,M} \end{bmatrix} = \begin{bmatrix} \underline{1}_{geo,1} & 1 \\ \underline{1}_{geo,2} & 1 \\ \vdots & \vdots \\ \underline{1}_{geo,M} & 1 \end{bmatrix} \begin{bmatrix} \Delta \underline{r}_{geo} \\ \Delta b_M - \Delta B_{geo} \end{bmatrix} + \underline{v}$$

$$= G_{geo} \Delta \underline{r}_{geo} + \underline{v} \quad \text{(A4)}$$

WIDE AREA AUGMENTATION SYSTEM

This system of equations is solved by inverting or "pseudo-inverting" the design matrix G_{geo}, which depends solely on the geometry of the WAAS satellite relative to the reference stations. The resulting estimates of $\Delta \underline{r}_{geo}$ and $\Delta b_M - \Delta B_{geo}$ will have minimum mean square error. The operational master site will use a Kalman filter based on these equations to estimate these states as well as their rates as a function of time. The WMS then includes these data in the type 9 message described earlier.

Bent-Pipe Transponder

If the WAAS satellite is a bent-pipe transponder with the clock on the ground, then the WAAS residuals from the reference stations are as follows:

$$\{\Delta_{geo,m} = \Delta \underline{r}_{geo} \cdot 1_{geo,m} + \Delta b_M + \nu_{geo,m}$$
$$+ (|r_{M,geo}| + I_{M,geo} + T_{M,geo} - B_{geo} - B_{uplink} - B_{down} + \Delta \tau)\}_{m=1}^{M} \quad (A5)$$

The first part of this Eq. (A5) is the normal one-way GPS pseudorange residual. It is the residual after the following terms have been removed: the nominal one-way range from the WAAS satellite to the mth reference station; the clock offset between the mth reference station clock and the master clock; the downlink ionosphere; and the downlink troposphere.

The second line in Eq. (A5) above contains the terms unique to a bent-pipe transponder and which require some special attention. For Inmarsat-3, the WAAS signal is uplinked from the ground at C-band, translated from C-band to L-band in the satellite transponder, and broadcast to the users at L-band. Consequently, the uplink range ($|\underline{r}_{M,geo}|$) from the control site to the WAAS satellite is included in the pseudorange. The uplink ionosphere and troposphere $I_{M,geo}$ and $T_{M,geo}$ are also included, as is the transponder delay B_{geo}. The uplink and downlink hardware delays appear as B_{uplink} and B_{down} respectively. Finally, $\Delta \tau$ is the control, which the WMS exerts over the WAAS code phase.

Most importantly, all of the terms unique to the bent-pipe transponder are common-mode. In other words, they are the same for all M reference stations. Consequently, the system of equations used above for ephemeris estimation can also be used with a bent-pipe transponder. Then the WMS will actually control the transmitted code phase $\Delta \tau$ so that the WAAS "clock" state is nulled. This control policy yields a residual at the user that is compatible with the GPS code phase residuals after correction.

If the WAAS satellite is a bent-pipe transponder, then the master station must also control the frequency of the WAAS uplink signal. This control must yield a signal that can be used by a GPS receiver to "carrier-aid" their code tracking loops. Such control can be achieved by using the dual-frequency WAAS observables to remove the ionospheric delay and by maintaining coherency between the carrier and the code at the master site.

References

[1] RTCA, "Minimum Aviation System Performance Standards—GPS Integrity Channel Working Group," RTCA/DO-202, 1988.

[2] RTCA, "Minimum Operational Performance Standards (MOPS) for Airborne Supplemental Navigation Equipment Using GPS," RTCA 204-91/SC159-293. (These MOPS are modified by Technical Standard Order TSO-C129, released Dec. 10, 1992.)

[3] Kee, C., Parkinson, B. W., and Axelrad, P., "Wide Area Differential GPS," *Navigation*, Vol. 38, No. 2, 1991, pp. 123–143.

[4] Kee, C., and Parkinson, B. W., "High Accuracy GPS Positioning in the Continent: Wide Area Differential GPS," *Proceedings of the Second International Symposium on Differential Satellite Navigation Systems (DNSN-93)*, (Amsterdam), March 1993.

[5] RTCA Special Committee 159, Working Group 2, "Wide Area Augmentation System Signal Specification," March 1994.

[6] Peterson, W. W., and Weldon, E. J., "Error Correcting Codes," MIT Press, 1972.

[7] Sarwate, D. V., and Pursley, M. B., "Crosscorrelation Properties of Pseudorandom and Related Sequences," *Proceedings of the IEEE*, Vol. 68, 1980, pp. 593–619.

[8] Nagle, J., Van Dierendonck, A. J., and Hua, Q. D., "Inmarsat-3 Navigation Signal C/A Code Selection and Interference Analysis," *Proceedings of the Annual Technical Meeting of the Institute of Navigation*, (San Diego, CA), Jan. 1992, to appear in *Navigation*.

[9] Van Dierendonck, A. J., and Elrod, B. D., "Ranging Signal Control and Ephemeris/Time Determination for Geostationary Navigation Payloads," *Proceedings of the 1994 Institute of Navigation Technical Meeting*, (San Diego, CA), Institute of Navigation, Washington, DC, Jan. 1994, pp. 393–402.

[10] Phlong, W. S., and Elrod, B. D., "Availability Characteristics of GPS and Augmentation Alternatives," *Proceedings of the 1993 National Technical Meeting of the Institute of Navigation*, (San Francisco), Institute of Navigation, Washington, DC, Jan. 1993.

[11] Durand, J. M., Michal, T., and Bouchard, J., "GPS Availability, Part I," *Navigation*, Vol. 37, No. 2, 1990, pp. 123–140.

[12] Kinal, G. V., "A Note on Satellite Reliability as Applied to the Inmarsat-3 Navigation Payload," Paper presented to Working Group 2 of RTCA SC-159, Jan. 1992.

[13] Virball, V. G., Enge, P. K., Michalson, W., and Levin, P., "A Fault Detection and Exclusion Algorithm to be Used with the GPS Integrity Channel," Record of the IEEE Position Location and Navigation Symposium, Las Vegas, April 11–15, 1994.

[14] Enge, P., Levin, P., Kalafus, R., McBurney, P., Daly, P., and Nagle, J., "Architecture for a Civil Integrity Network Using Inmarsat," *Proceedings of the Third International Technical Meeting of the Satellite Division of the Institute of Navigation*, (Colorado Springs, CO), Institute of Navigation, Washington, DC, Sept. 1990, pp. 287–296.

[15] Ananda, M., Munjal, P., Siegal, B., Sung, R., and Woo, K. T., "Proposed GPS Integrity and Navigation Payload on DSCS," Record of the 1993 IEEE Military Communications Conference, Boston, Oct. 11–14, 1993.

[16] Hegarty, C., "Optimal Differential GPS for a Data Rate Constrained Broadcast Channel," *Proceedings of the ION GPS-93, Sixth International Technical Meeting of the Satellite Division of the Institute of Navigation*, (Salt Lake City, UT), Institute of Navigation, Washington, DC, Sept. 22–24, 1993, pp. 1527–1535.

[17] Van Dyke, K., "Satellite Visibility Analysis," Paper presented to RTCA SC 159 Working Group 2, RTCA Paper 596-92/SC 159-389, Aug. 18, 1992.

Chapter 5

Receiver Autonomous Integrity Monitoring

R. Grover Brown*
Iowa State University, Ames, Iowa 50011

I. History, Overview, and Definitions

NAVIGATION system integrity refers to the ability of the system to provide timely warnings to users when the system should not be used for navigation.[1] The basic Global Positioning System (GPS) as described in Part I provides integrity information to the user via the navigation message, but this may not be timely enough for some applications, especially in civil aviation. Therefore, additional means of providing integrity are being planned, and two different approaches are being considered. One of these is the receiver autonomous method, now referred to simply as RAIM (receiver autonomous integrity monitoring). A variety of RAIM schemes have been proposed, and all are based on some kind of self-consistency check among the available measurements. Of course, there must be some redundancy of information in order for RAIM to be effective. The other approach to providing an independent assurance of integrity is to have a network of ground monitoring stations whose primary purpose is to monitor the health of the GPS satellites. Appropriate integrity information is then transmitted to users via a radio link of some sort. This is referred to as the GPS integrity channel (GIC). We will be primarily concerned with RAIM in this section. A discussion of GIC is presented in Chapter 24, this volume.

Work on RAIM began in earnest in the latter half of the 1980s, and a wealth of papers have appeared in the navigation literature since then. We do not attempt to give a complete bibliography here. One has only to browse through the proceedings of the meetings of the Institute of Navigation (ION) beginning in 1986 to get a reasonably complete history of the evolution of RAIM. We should note that early papers did not use the term RAIM. The term "self-contained" was more common then. The acronym "RAIM" was first suggested in 1987 by R.M. Kalafus, and it has been used almost universally ever since.[2] Two early papers presented at an ION meeting in 1986 are of special interest, because they

---

Copyright © 1993 by the American Institute of Aeronautics and Astronautics, Inc. All rights reserved.
*Distinguished Professor Emeritus, Electrical and Computer Engineering Department.

illustrate two different approaches to RAIM.[3,4] The Lee paper[3] is a good example of what is now referred to as a snapshot scheme. With this method, only current redundant measurements are used in the self-consistency check. On the other hand, the Brown and Hwang paper[4] presents a scheme where both past and present measurements, along with a priori assumptions with regard to vehicle motion, are used in the RAIM decision. Such schemes are loosely referred to as averaging or filtering schemes. The snapshot approach has gained more acceptance than the other in recent times, so we concentrate on it here. It has the advantage of not having to make any questionable assumptions about how the system got to its present state. It matters only that the system is in a particular state "now," and the RAIM decision as to failure or no-failure is based on current observations only.

The theoretical structure for RAIM is statistical detection theory. Two hypothesis-testing questions are posed: 1) Does a failure exist? and 2) If so, which is the failed satellite? (It is usually assumed that there is only one failure at a time.) The answer to question 1 is sufficient for supplemental navigation because, presumably, there is an alternative navigation system to fall back on if a failure is detected. However, in the case of sole-means navigation, both questions 1 and 2 must be answered. Here, the errant satellite must be identified and eliminated from the navigation solution, so that the aircraft can proceed safely with an uncontaminated GPS solution. As might be expected, determining which satellite has failed is more difficult than simple failure detection, and it requires more measurement redundancy. Most of this section is devoted to the detection function as it pertains to the supplemental navigation application. Then there is a short discussion at the end of the section on failure isolation and its application in sole-means navigation.

The Radio Technical Commission for Aeronautics (RTCA) published its Minimum Operational Performance Standards (MOPS) for GPS as a supplemental navigation system in July 1991.[5] Table 1 shows its specifications for GPS integrity.

As indicated in Table 1, three distinct phases of flight are considered, and each has its own set of specifications. Failure here is defined to mean that the solution horizontal radial error is outside a specified limit, which is called "alarm limit."

Table 1 Radio Technical Commission for Aeronautics global positioning system integrity specifications for supplemental navigation

Phase of flight	Performance item			
	Alarm limit	Maximum allowable alarm rate	Time to alarm	Minimum detection probability
En route (oceanic, domestic, random, & J/V routes)	2.0 n.mi.	0.002/h	30 s	0.999
Terminal	1.0 n.mi.	0.002/h	10 s	0.999
RNAV approach, nonprecision	0.3 n.mi.	0.002/h	10 s	0.999

RECEIVER AUTONOMOUS INTEGRITY MONITORING 145

(Note that this is not exactly the same as saying that a particular satellite pseudorange error is outside a specified bound. In other words, failure is defined in the solution domain rather than the measurement domain.) The maximum allowable alarm rate in the table refers to the usual false alarm rate with no satellite malfunction. The 0.002 per hour figure is interpreted as meaning the false alarm probability associated with any independent sample will be no greater than 1/15,000. The detection probability is a conditional probability and is defined as (1 − miss probability). The specifications require that both the detection and alarm-rate specifications must be met globally at all times. Indirectly, this says that if, at any time or location, the satellite geometry is such that it cannot support both of these specifications simultaneously, then that geometry must be declared as inadmissible for integrity purposes. When combined, the two requirements are quite severe, and this causes the RAIM availability to be less than desirable for the nonprecision approach phase of flight.[6] Note that there is no firm availability requirement specified in Table 1.

Next, we look briefly at some of the major snapshot RAIM schemes that have been proposed over the past few years. The discussion in all cases is qualitative to the extent possible. Key references are cited, and the reader can refer to them for mathematical details.

II. Basic Snapshot Receiver Autonomous Integrity Monitoring Schemes and Equivalences

Three RAIM methods have received special attention in recent papers on GPS integrity. These are the following: 1) the range comparison method first introduced by Lee,[3] 2) the least-squares-residuals method suggested by Parkinson and Axelrad,[7] and 3) the parity method as described in the context of GPS in papers by Sturza and Brown.[8,9] All three methods are snapshot detection schemes because they assume that noisy redundant range-type measurements are available at a given sample point in time. Also, in all cases, the algebraic problem is linearized about some nominal value of vehicle position and clock bias. The basic measurement relationships are then described by an over-determined system of linear equations of the form

$$y = G x_{true} + \epsilon \qquad (1)$$

where n = the number of redundant measurements; y = the difference between the actual measured range (or pseudorange) and the predicted range based on the nominal user position and the clock bias (y is an $n \times 1$ vector); x_{true} = three components of true position deviation from the nominal position plus the user clock bias deviation (x_{true} is a 4×1 vector); ϵ = the measurement error vector caused by the usual receiver noise, vagaries in propagation, imprecise knowledge of satellite position and satellite clock error, selective availability, and, possibly, unexpected errors caused by a satellite malfunction (ϵ is an $n \times 1$ vector); and G = the usual linear connection matrix arrived at by linearizing about the nominal user position and clock bias. (G is an $n \times 4$ matrix.)

This basic measurement equation is common to all the RAIM methods described presently. In doing so, it is convenient to use the six-in-view-case for tutorial purposes. The generalization to $n = 5$ or $n > 6$ is fairly obvious, so this is not discussed in detail.

A. Range Comparison Method

Imagine having six satellites in view. We would then have six equations in four unknowns. Now, suppose that we solve the first four equations (as if there were no noise) and obtain a solution that satisfies the first four equations. (The ordering of the six equations is immaterial.) The resulting solution could then be used to predict the remaining two measurements, and the predicted values could then be compared with the actual measured values. If the two differences (residuals) are small, we have near-consistency in the measurements, and the detection algorithm declares "no failure." On the other hand, if either or both of the residuals are large, it declares "failure." This is the essence of the range comparison method. It only remains to quantify what we mean by "small" and "large" and assess the decision rule's performance.

Proceeding further with the six-in-view example, we see that the two residuals that we would obtain with the range comparison method can be thought of as a two-tuple, and they represent a point in a test-statistic plane as shown in Fig. 1.

We now look for a decision rule that divides the plane into two distinct regions, one corresponding to the "no failure" hypothesis, and the other corresponding to the "failure" hypothesis. A common way to choose the decision boundary is to let it be an equal probability density contour, conditioned on the assumption of no satellite malfunction. If the statistics of the noise are Gaussian, the contour will be elliptical, as shown in Fig. 1, and the particular contour chosen is the one that sets the alarm rate at the desired value. For example, the alarm rate could be set at 1/15,000, as specified in the RTCA MOPS.[5] Setting the decision boundary quantitatively then firms up the decision rule, and all that remains is to assess the algorithm's performance under the alternative hypothesis. This is somewhat complicated, so we do not pursue it further here. The elliptical contour (or closed surface in hyperspace) makes the decision rule a bit awkward computationally. However, conceptually, the rule is simple: decide "no failure" if the test statistic lies inside the hypersurface; decide "failure" if it lies outside the hypersurface.

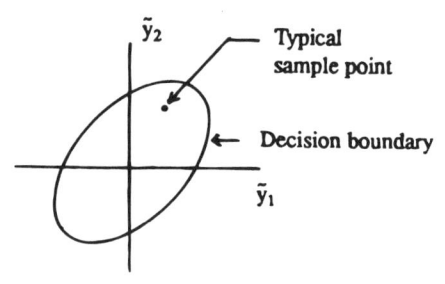

Fig. 1 Test statistic plane for the six-in-view case ($\tilde{y}_1$ and $\tilde{y}_2$ are the observed residuals).

B. Least-Squares-Residuals Method

Just as with the range comparison example, imagine a six-in-view situation where we have six equations in four unknowns. Now, instead of solving the first four equations as if there were no noise, say we look at the least squares "solution." This is well known and is given in Refs. 7 and 10.

$$\hat{x}_{LS} = (G^T G)^{-1} G^T y \qquad (2)$$

The least-squares solution can now be used to predict the six measurements in accordance with

$$(\text{predicted } y) = G\hat{x}_{LS} \qquad (2a)$$

Six residuals are then formed in the measurement domain in much the same manner as was done in the range comparison method. The residuals can then be grouped together as a 6 × 1 vector, which we will call w. Substituting $\hat{x}_{LS}$ from Eq. (2) into Eq. (2a) then leads to the equation for w.

$$w = y - (\text{predicted } y) = [I - G(G^T G)^{-1} G^T] y \qquad (3)$$

This is the linear transformation that takes the range measurement y into the resulting residual vector. (It can be easily verified that a similar equation also takes the measurement error ϵ into w.) In our example w is a six-tuple. However, Parkinson and Axelrad[7] show that there are constraints among the elements of w. For example, in the case at hand, if the elements of ϵ are independent zero-mean Gaussian random variables with the same variance, the sum of the squares of the elements of w has an unnormalized chi-square distribution with only two-DOF, rather than the six degrees that might be expected at first glance.

The sum of the squares of the residuals plays the role of the basic observable in the RAIM method under discussion, and Parkinson and Axelrad called it SSE (for sum of squared errors). We do also; i.e.,

$$\text{SSE} = w^T w \qquad (4)$$

This basic observable has three very special properties that are important in the least-squares-residuals decision rule.

1) SSE is a nonnegative scalar quantity. This makes for a simple decision rule. All we must do is partition the positive semi-infinite real line into two parts, one for "no failure," and the other for "failure." The dividing point is called the threshold.

2) If all elements of ϵ have the same independent zero-mean Gaussian distributions, then the statistical distribution of SSE is completely independent of the satellite geometry for any n. This makes it especially simple to implement a constant alarm-rate algorithm. All we must do is precalculate the thresholds (partitions) that yield the desired alarm rate for the various anticipated values of n. Then the real-time algorithm sets the threshold appropriately for the number of satellites in view at the moment.

3) For the zero-mean Gaussian assumption mentioned in 2, SSE has an unnormalized chi-square distribution with $(n - 4)$ degrees of freedom.

It should also be apparent that any other scalar variable monotonically related to SSE could also be used as the test statistic. Parkinson and Axelrad[7] suggested using $\sqrt{SSE/(n-4)}$ as the test statistic.[7] Later in Sec. III we also use $\sqrt{SSE/(n-4)}$ as our test statistic because this yields a linear relationship (rather than quadratic) between a satellite bias error and the associated induced test statistic. This is a convenience because the connection between the satellite bias error and the resultant radial position error is also linear.

To determine the threshold, we first use chi-square statistics and find the threshold for the value of n of interest, using SSE as the test statistic. This value can then be converted to the corresponding threshold for the $\sqrt{SSE/(n-4)}$ test statistic simply by dividing the SSE value by $(n-4)$ and square rooting the result. Such thresholds (rounded to integer values) are given in Table 2.

In summary, the least-squares-residuals RAIM method is especially simple in its implementation, because its test statistic is a scalar, regardless of the number of satellites in view. Calculating the test statistic involves some matrix manipulations, but these are no worse than calculating GDOP, PDOP, etc., which is done routinely as a background computation in most current GPS receivers.

C. Parity Method

The parity RAIM method as described by Sturza[9] is more formal and less heuristic than the other two methods. It is somewhat similar to Lee's range comparison method[3], except that the way in which the test statistic is formed is different. In the parity scheme, we first perform (conceptually) a linear transformation on the measurement vector y as follows:

$$\begin{bmatrix} \hat{x}_{LS} \\ --- \\ p \end{bmatrix} = \begin{bmatrix} (G^T G)^{-1} G^T \\ --------- \\ P \end{bmatrix} [y] \qquad (5)$$

Clearly, the upper partitioned part of the transformation yields the usual least-squares solution $\hat{x}_{LS}$. The lower partitioned part, which yields p, is the result of operating on y with a special $(n-4) \times n$ matrix P whose rows are mutually orthogonal, unity in magnitude, and also mutually orthogonal to the columns of G. We ignore how P is found for the moment and concentrate on the $(n-4) \times 1$ vector p, which is called the parity vector. (In our six-in-view example, p would be a two-tuple.) The very special way in which it is formed (i.e., as Py) gives p some special properties. Under the usual assumption that the elements

Table 2 Approximate thresholds (test statistic = $\sqrt{SSE/(n-4)}$, noise σ = 33m, alarm rate = 1/15,000)

Number of satellites in view, n	Chi-square degrees of freedom	Threshold, m
5	1	132
6	2	102
7	3	90
8	4	82
9	5	77

of the random forcing function ϵ are independent similar zero-mean Gaussian random variables, the following statements can be made:

$$E[p] = 0 \tag{6}$$

$$E[pp^T] = \text{cov } p = \sigma^2 I \tag{7}$$

where σ^2 is the variance associated with any particular element of ϵ. Conceptually, in the parity method we use p as the test statistic. But, because of the special properties stated in Eqs. (6) and (7), we need not look at the individual elements of p; they are all decoupled and have the same variance σ^2. For simple detection, we obtain all the information we need about p merely by looking at its magnitude, or its magnitude squared. Thus, in the parity method, the test statistic for detection reduces to a simple scalar, just as was the case with the least-squares-residuals method. Furthermore, Sturza shows that the sum of the squares of the elements of p and Parkinson and Axelrad's SSE[7] are identical,[9] i.e.,

$$p^T p = w^T w = \text{SSE} \tag{8}$$

This is to say that although the dimensionalities of p and w are different, their magnitudes are the same. The significance of Eq. (8) is simply this: if all we are interested in is the test statistic $p^T p$, then we do not actually have to go to the trouble of finding the orthogonal transformation P that leads to p. Instead, we can just use SSE (ignoring p and P) and get the same result as if we were to go to the work of finding P, then forming p as Py, and, finally, forming $p^T p$. Clearly, forming SSE directly in the measurement-residual space is easier.

In summary, in the detection application, the least-squares-residuals and parity methods lead to identical observables. Thus, with similar threshold settings, the two methods must yield identical results. Now, all that is needed to tie the range comparison method to the other two is to show that a linear transformation can be found that will always take the $(n - 4)$ vector range comparison test statistic into the parity vector p. This is illustrated in Fig. 2.

It suffices here to say that such a transformation exists (see Ref. 11 for the derivation). Then, assuming that the decision boundaries in the two spaces are chosen to yield the same false alarm rate, it should be apparent that the range comparison decisions will be identical with those obtained from the parity and least-squares-residuals methods. The differences among the three detection meth-

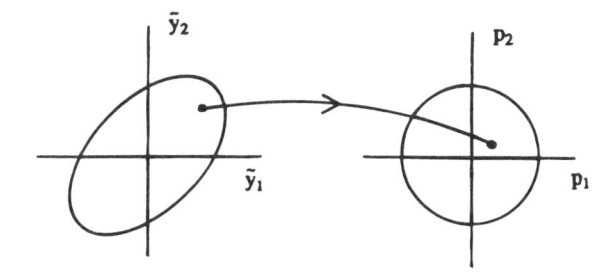

Fig. 2 Mapping from range-comparison space to parity space.

ods are, thus, primarily conceptual and a matter of computational convenience. For the sake of brevity, we simply refer to any of the three methods as the *parity detection method* henceforward.

D. Maximum Separation of Solutions

A RAIM method distinctly different from the parity method was suggested by Brown and McBurney in 1987.[12] It is probably the most heuristic of all the failure detection schemes. Assume that no more than one satellite has failed. Then, if there are n satellites in view, consider the n subset solutions obtained by omitting one satellite at a time from the full set. If a failure exists, the failed satellite will be omitted from one of the subsets, and the solution thus obtained will be a "good" solution. All other subset solutions will contain the failed satellite, and they will be in error to various degrees. Now, imagine the pseudorange error in the failed satellite gradually increasing with time. We would then expect the subset solutions to begin to spread apart with time, and a measure of this would be the maximum separation observed among the n solutions. (Note that one solution would remain near truth, because it does not contain the failed satellite.) On the other hand, if there is no failure present, the solutions should remain bunched around the true position. Thus, the maximum observed solution separation in the horizontal plane can be used as a test statistic. It is scalar and nonnegative, and it only remains now to set the threshold that separates the "no-failure" decision from the "failure" decision.

A. Brown[13] presented an interesting heuristic method of setting the threshold and assessing the radial error protected for the maximum separation scheme. It is illustrated for a five-in-view situation in Fig. 3. Suppose that under normal conditions we are assured that any four-satellite solution with a reasonable horizontal dilution of precision (HDOP) yields a solution within the 100-m 2-drms accuracy specification [with selective availability (SA)]. Then the maximum possible solution separation among the five solutions is 200 m, as shown in Fig. 3. Thus, let us set the test statistic threshold at 200 m. Then, any observed maximum separation greater than 200 m will be declared "failure." Furthermore, we are also assured that any horizontal radial error greater than 300 m will be detected because one of the solutions will always be within 100 m of truth, and

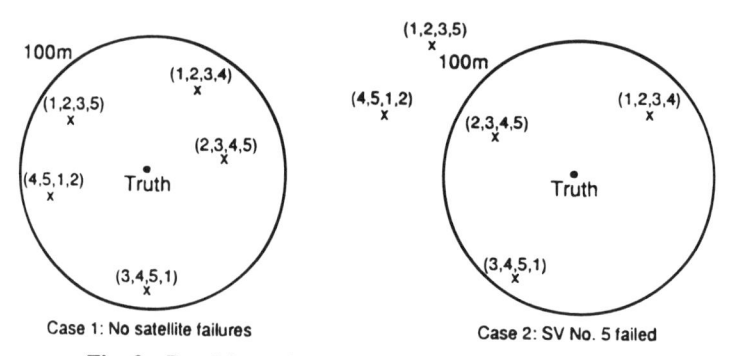

Fig. 3 **Possible navigation solutions with five satellites.**

the threshold on maximum separation has been set at 200 m. This establishes 300 m (and above) as the radial error being protected by the maximum separation scheme.

The preceding argument is approximate because 100-m 2-drms does not mean 100%. Also, the matter of false alarms has not even been considered. Nevertheless, the preceding reasoning led to radial-error-protection levels that were remarkably close to those obtained independently by Brown and McBurney using Monte Carlo simulation.[12,13] The results in both cases were somewhat optimistic, but this is understandable because the integrity specifications that were finally adopted by the RTCA group in 1991 were much more stringent than those that were being considered in 1987.

The maximum-solution-separation RAIM method is more difficult to analyze mathematically than is the parity method. This may account for the dearth of papers on the subject in the past few years. Much is now known about RAIM availability under conditions of the RTCA specifications and using the constant-false-alarm parity approach.[6] The same cannot be said for the maximum-solution-separation method.

E. Constant-Detection-Rate/Variable-Protection-Level Method

In September 1990, Brenner[14] presented a snapshot RAIM scheme that differs significantly from the four methods just presented. Brenner's approach starts out with the parity vector as the basic test statistic, and a threshold is set to yield the desired constant alarm rate, just as discussed previously. It is at this point that Brenner's method diverges from the others. In effect, he poses the question, "If we were to also keep the detection probability constant (at 0.999, for example) as the satellite geometry varies, what is the smallest radial error that we could protect and still stay within the desired specifications?" Brenner called this smallest-radial-error-protected the *detection level*, and it varies with the satellite geometry. (Detection level is also referred to as *protection radius* in some papers.) Furthermore, it can be calculated on-line (approximately, at least) on an essentially continuous basis in much the same manner as GDOP, PDOP, etc. This additional information is not required in the RTCA integrity specifications, but it is certainly permissible to display it in the cockpit as background information.

A typical plot of the calculated detection level as a function of time is shown in Fig. 4. The figure shows how the variable-detection-level method would be used within the context of the RTCA specifications where fixed, discrete alarm

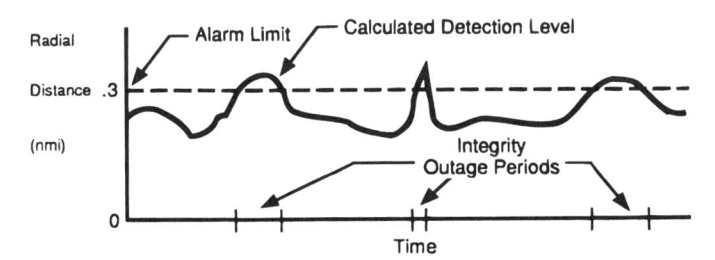

Fig. 4 Variable detection level (alarm limit for nonprecision approach is illustrated).

limits are set for each phase of flight. The calculated detection level is shown as the solid curve in Fig. 4. Whenever it exceeds the specified alarm limit (dashed line), the alarm would be annunciated. Assuming that the detection level is also displayed, it would obvious to the flight crew that the alarm was caused by inadequate satellite geometry and not a bona fide satellite failure. Of course, an alarm would also be annunciated if the test statistic exceeded its threshold, and this would be indicative of a satellite malfunction. The extent to which the variable-detection-level idea will be adopted by the aviation community remains to be seen.

III. Screening Out Poor Detection Geometries

The integrity requirements in the RTCA MOPS for GPS supplementary navigation are demanding.[5] They state that both the required detection probability and false alarm rate have to be met for *any* location and at *all* times. If the satellite geometry is such that both specifications cannot be met, then the alarm must be annunciated, indicating that integrity cannot be assured. Such geometries are referred to here as being *inadmissible*. It should be noted that these poor detection geometries might yield perfectly good navigation solutions; however, they simply do not have the appropriate redundancy to provide a good integrity check.

Inadmissible geometries detract from RAIM availability. Thus, it is important that they be screened out carefully. We do not want to throw out the good with the bad, so to speak. In early papers on GPS integrity, $HDOP_{max}$ (or $PDOP_{max}$) was frequently used as the screening criterion. The intuitive argument for this criterion is that with n satellites in view, consider the HDOPs associated with the n subset solutions obtained by omitting one satellite at a time. Let the largest of these HDOPs be denoted as $HDOP_{max}$. Now, if $HDOP_{max}$ is abnormally large, this means that the associated solution's projection onto the range of the missing satellite will be unreliable (i.e., noisy). This, in turn, makes it difficult to detect a failure (say, a modest bias error) on the missing satellite. A ceiling value is set on $HDOP_{max}$, and all geometries whose $HDOP_{max}$ exceed the ceiling value are declared inadmissible. One difficulty with this criterion is that it is not clear which of the DOPs should be used in the criterion. Perhaps GDOP would be more appropriate than HDOP, because the whole four-variable solution error reflects into projected range error on the missing satellite. Also, there is the matter of how a range error on a particular satellite projects into the horizontal solution, which is of primary interest in the avionics application. The net result of all this is that $HDOP_{max}$ is only a coarse measure of the quality of the detection geometry.

Two new (and better) methods of screening out poor detection geometries were introduced in the 1990–1991 period. One of these is called the δH_{max} method, and it comes from Sturza and A. Brown.[8,15] The other is referred to as the ARP method, and it was introduced independently by R. G. Brown et al.[16] It was shown later that the two methods are exactly equivalent for constant false-alarm-rate algorithms with similar thresholds.[11] These new methods of screening out poor detection geometries work out to be more exact than the older $HDOP_{max}$ criterion in terms of a sorting out the good geometries from the poor ones. This is demonstrated in Ref. 17.

RECEIVER AUTONOMOUS INTEGRITY MONITORING 153

The δH_{max} method is justified algebraically in Ref. 15, and it proceeds as follows.
1) Compute the HDOPs associated with the n subset solutions. Call these HDOP$_i$, $i = 1, 2, \ldots, n$.
2) Compute the HDOP associated with the full n-satellite least-squares solution. Call it HDOP.
3) Then

$$\delta H_{max} = \text{Max}_i \, [\text{HDOP}_i^2 - \text{HDOP}^2]^{1/2} \qquad (9)$$

The parameter δH_{max} then becomes an inverse measure of the quality of the satellite geometry for failure detection purposes (small values are best). A ceiling value for δH_{max} is then set in accordance with the integrity specifications, and if the calculated δH_{max} for the geometry at hand exceeds the ceiling value, the geometry is declared inadmissible; otherwise, it is admissible.

The ARP method is more intuitive than δH_{max} and it derives from geometric considerations when we look at a plot of position radial error vs test statistic. In the ARP method, it is convenient to work with a test statistic that is derived from SSE (or the magnitude of the parity vector squared) rather than SSE itself. Therefore, for screening purposes consider the test statistic to be

$$\sqrt{\text{SSE}/(n-4)} \qquad (10)$$

With this definition, the radial position error and the test statistic have the same dimensions, and their ratio is dimensionless. It can be easily shown that a bias error on any particular satellite projects linearly into both the position-error and test-statistic domains. The slope, which relates the induced position error to the test statistic, can be readily calculated from the satellite geometry,[16] and it will be different for each satellite, as illustrated in Fig. 5a. For failure detection purposes, it should be clear that the satellite whose bias error causes the largest slope is the one that is the most difficult to detect. It is the one that produces the largest position error (which we want to protect) for a given test statistic (which is what we can observe). We call the slope associated with the most-difficult-to-detect satellite SLOPE$_{max}$.

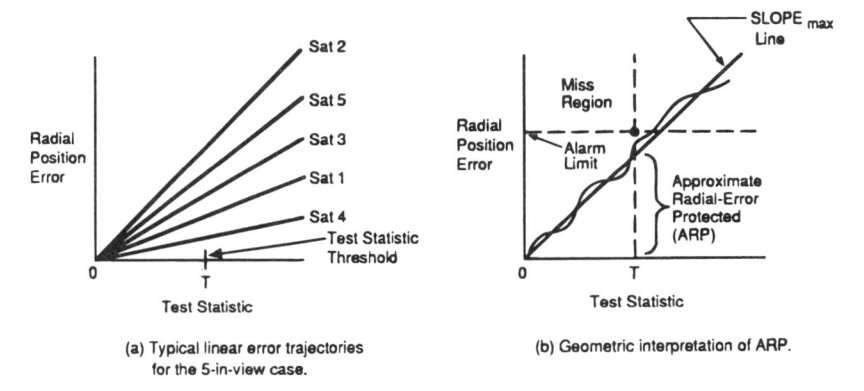

Fig. 5 Radial error vs test statistic plots illustrating SLOPE$_{max}$ and ARP.

Now suppose the threshold has been set to yield the allowed false alarm rate, and we then apply a ramp-type bias error to the SLOPE$_{max}$ satellite. If we had no noise to contend with, the radial error trajectory would move up along the linear SLOPE$_{max}$ line as shown in Fig. 5b. It is desirable, of course, that the trajectory not intrude into the miss region, or an undetected failure would occur. For the noiseless case just described, it can be seen that the alarm limit line could be moved down to the intersection of the SLOPE$_{max}$ line with the vertical threshold line. The ordinate of the intersection is the smallest radial error that we could protect under these ideal conditions (assuming that the threshold is fixed). We call this the approximate radial-error protected, or simply ARP. Clearly, from Fig. 5b it can be seen that ARP is given by

$$\text{ARP} = \text{SLOPE}_{max} \times \text{threshold} \tag{11}$$

Now, to be more realistic, the noises on the other satellites cause the actual trajectory to follow a wavy line, which is also shown in Fig. 5b. The random deviation from the ideal linear trajectory will be large if the noises are large (e.g., with SA turned "on"), and vice versa. Therefore, as a practical matter, the alarm limit line must be kept comfortably above the ARP value to keep the wavy trajectory out of the miss region. For the current RTCA specifications, and with SA present, it has been found empirically that the actual radial error that can be protected is related to ARP by the following approximate equation.[11]

$$\text{Actual radial error protected} \approx 1.7 \times \text{ARP} \tag{12}$$

This approximation is only within about 5%, but it is a useful rule of thumb to get a quick estimate of the position error that can be protected, given the calculated ARP figure for the satellite geometry at hand. (Note that the 1.7 multiplication factor would be considerably less with SA not present.)

Ceiling values for the ARP criterion have been determined by extensive simulation,[17] and they are repeated in Table 3. Note that these values are a function of the number in view as well as the phase of flight. Also, an exact relationship between ARP and δH_{max} is derived in Ref. 11, and it is

$$\text{ARP} = \sqrt{n - 4} \times \text{threshold} \times \delta H_{max} \tag{13}$$

This equation may be used to convert the ARP ceiling values in Table 3 to corresponding δH_{max} ceiling values for those who prefer to work with δH_{max} rather than ARP. Note, however, that "threshold" in Eq. (13) is computed using

Table 3 ARP ceiling values

Phase of flight	Number of satellites in view		
	5	6	7 (or more)
Nonprecision approach	328	339	352
Terminal	1077	1135	1135[a]
Enroute	2159	2262	2262[a]

[a]These numbers are conservative estimated values. Very few seven-in-view geometries have ARP values this large.

$\sqrt{SSE/(n-4)}$ as the test statistic rather than just SSE. Results given in Ref. 17 demonstrate that ARP (or equivalently, δH_{max}) is more discriminating in screening out bad detection geometries than is $HDOP_{max}$. This is to say that the older $HDOP_{max}$ criterion unnecessarily eliminates some geometries that should, by rights, be admissible.

IV. Receiver-Autonomous Integrity Monitoring Availability for Airborne Supplemental Navigation

The results of an extensive study of RAIM availability for airborne supplemental navigation were presented by Van Dyke in 1992.[6] The ARP screening criterion was used in the study, and the ceiling values given in Table 3 (Sec. III) were used to distinguish between admissible and inadmissible geometries. The study involved five geographic areas: the continental United States (CONUS), the North Atlantic, Europe, the Central East Pacific, and the North Pacific, and a mask angle of 7.5 deg was used in the study. An abridged version of Van Dyke's results for unaided RAIM is presented in Table 4. The percentages in this table were obtained by averaging over the five geographic areas, and they have been rounded to the nearest 0.1%. Note that unaided RAIM availability for nonprecision approach is only about 94%, even for the best satellite configurations. This is obviously less than desirable.

The Van Dyke paper[6] also gives availability percentages for baro-aided RAIM over CONUS. The results are repeated here in Table 5. Fundamentally, any RAIM scheme is merely a consistency check among a group of redundant measurements. A baro-altitude measurement can be thought of as a range measurement to the center of the Earth, so it can also be brought into the group of measurements being used in the consistency check. Details about exactly how this is done are discussed in Sec V. For now, it suffices to compare the availability results for nonprecision approach with and without baro-aiding. Note the dramatic improvement obtained by adding baro-altitude information to the suite of measurements. For both 24-satellite constellations (i.e., 21 primary and optimized 21 + 3) the percent availability improves from 94% to 99%.

Table 4 Average unaided global positioning system receiver-autonomous integrity monitoring availability in percentages[a]

	Phase of flight		
Constellation	En route	Terminal	Nonprecision approach
21 Primary	99.7	99.4	94.4
21 Primary–1 failure	99.1	98.2	87.5
21 Primary–2 failures	97.7	95.7	82.0
21 Primary–3 failures	96.3	93.2	73.9
Optimal 21	97.4	94.9	78.1
Optimized 21 + 3	99.8	99.5	93.8

[a]Ref. 6.

Table 5 Receiver-autonomous integrity monitoring availability over the continental United States with baro-aiding in percentages[a]

Constellation	Phase of flight		
	En route	Terminal	Nonprecision approach
21 Primary	99.99	99.45	99.02
21 Primary–1 failure	99.87	98.48	96.82
21 Primary–2 failures	99.79	96.42	93.90
21 Primary–3 failures	99.39	94.32	90.43
Optimal 21	99.96	95.34	96.13
Optimized 21 + 3	99.99	99.69	99.20

[a]Ref. 6.

V. Introduction to Aided Receiver-Autonomous Integrity Monitoring

It was mentioned in Sec. IV that non-GPS measurements can also be added to the suite of measurements being used for the RAIM consistency check. We concentrate here on snapshot range-type measurements such as might be obtained form GLONASS, baro-altitude, or Loran (with master stations synchronized to GPS time). Then other types of measurements such as inertial and time (clock coasting) are discussed briefly. We use GLONASS as an example of the use of non-GPS measurements in the RAIM consistency check. The methodology for incorporating baro-altitude or Loran measurements into the suite of measurements follows.

The fundamental measurement in GLONASS is pseudorange, just as in GPS. The main difference (as it affects RAIM) is the variance of the measurement error. Assume for the moment that our RAIM scheme uses one GLONASS measurement in addition to the usual suite of GPS measurements. This simply adds one linearized equation of the form

$$y_g = -C_{xg} \Delta x - C_{yg} \Delta y - C_{zg} \Delta z + \Delta T + \epsilon_g \quad (14)$$

where Δx, Δy, Δz, and ΔT are elements of the x vector; C_{xg}, C_{yg}, and C_{zg} are direction cosines between the user east, north, and up axis and the line of sight to the satellite; and ϵ_g is the measurement error of the non-GPS measurement. We assume here that the offset between GPS time and GLONASS time is known, so ΔT is only the receiver clock bias as before (in units of range). The y_g quantity of the left side of Eq. (14) is the usual linearized measurement: i.e., the difference between the actual measurement and the predicted value based on the nominal x about which the linearization takes place [see Eq. (1)]. For the sake of simplicity, let us assume that ϵ_g has exactly half the standard deviation of that associated with the GPS measurements. However, the previous unified RAIM theory discussed in Sec. II is based on the assumption of independent measurement errors, all having the same variance. Therefore, something has to be done to the GLONASS measurement equation to bring it into line with the other equations. It should be apparent that this can be accomplished by multiplying both sides of Eq. (14) by 2. The result is

RECEIVER AUTONOMOUS INTEGRITY MONITORING

$$2y_g = (-2C_{xg}) \Delta x + (-2C_{yg}) \Delta y + (-2C_{zg}) \Delta z + (2) \Delta T + 2\epsilon_g \quad (15)$$

where the coefficients that make up the extra row of the new G matrix are shown in parentheses for emphasis. The left side of Eq. (15) says that we must now consider the GLONASS linearized measurement scaled up by a factor of 2 as the "measurement" in our set of modified linear equations. In effect, this gives extra weight to the measurement residual on the GLONASS satellite. This is as it should be. According to our assumption, it is a more accurate measurement than the corresponding GPS measurements (with SA turned on).

This same measurement rescaling procedure can be followed for any other snapshot-type range or pseudorange measurement. It is worth mentioning that baro-altitude is a range measurement (in contrast to pseudorange), so the clock bias term does not appear in the measurement equation in this case. It should also be mentioned that this procedure does not apply (directly, at least) to nonsnapshot-type measurements; i.e., those that involve past as well as present measurement information.

Inertial navigation systems (INS) and GPS receiver clocks both have unstable error characteristics. This gives rise to some special problems when these sources of information are coupled into RAIM schemes. This can be illustrated for a receiver clock with the following scenario. Suppose that the aircraft has enjoyed a period of good satellite geometry during the first portion of the flight, and the clock bias and drift have been accurately calibrated during this period by the receiver Kalman filter. The RAIM integrity checks have indicated that everything is normal. Then comes a short period when the satellite geometry is not admissible for RAIM purposes. The question arises: "Why not merely treat the predicted clock bias (based on calibrations just prior to entering the bad geometry period) as a noisy measurement and add it to the measurement suite, thus making RAIM viable during the otherwise bad geometry period?" This would seem to make sense intuitively, provided, of course, that the clock is reasonably stable. The weakness in this reasoning is that the GPS measurements were used to calibrate the clock, and they could have drifted off considerably before triggering the alarm. Even for nonprecision approach, which is the most demanding phase of flight, the alarm limit is 0.3 n.mi. This is roughly an order of magnitude greater than normal GPS position error. The situation is even more exaggerated for the terminal and enroute phases of flight where the alarm limits are 1 and 2 n.mi, respectively. Thus, the clock calibration that we can actually rely on may not be as good as we might think initially. We are not assured that an errant satellite has not pulled the position and clock bias solutions considerably out of tolerance just prior to the bad geometry period. A similar problem would exist with INS-aided RAIM if the INS is continually updated with GPS which is normally the case. The basic problem is that the aiding sources are not independent of the satellite measurements that they are trying to verify. The beauty of aiding RAIM with baro-altitude, GLONASS, or Loran is that the errors in these sources are genuinely independent of the GPS satellite errors.

We will not try to predict how effective either INS or clock coasting might prove to be in the overall RAIM problem. The cautions just mentioned should be kept in mind, however. There have been a number of imaginative papers on the subject recently, and it is likely that there will be more as we get into the period when GPS will be used for sole-means navigation.[18,19]

VI. Failure Isolation and the Combined Problem of Failure Detection and Isolation

A. Introductory Remarks

All of the preceding discussion about RAIM has been directed toward the failure *detection* problem. This is sufficient for supplemental navigation where, presumably, there is an alternative navigation system to fall back on in case a failure is detected. Simple detection is not enough for sole-means navigation, however. There, the integrity system must also be able to isolate (i.e., identify) the errant satellite so that it can be removed from the navigation solution. This combined problem of failure detection and isolation (FDI) is often referred to simply as the FDI problem. The detection half of the problem has been studied in great detail over the past few years, and much is now known about RAIM performance for supplemental navigation.[6] Availability is the key measure of performance for supplemental navigation, because it is assumed that the detection and false alarm rate specifications will be met by properly screening out poor detection geometries.

The performance of RAIM for sole-means navigation has not been assessed as thoroughly as it has for supplemental navigation. There are two reasons for this. First, specific requirements for FDI have not been recommended by RTCA Special Committee 159 as yet. When these recommendations do arrive, it is likely that they will not be identical with those for supplemental navigation. For example, a false alarm is just a nuisance matter in supplemental navigation; on the other hand, in the sole-means case an unnecessary alarm that leads to a RAIM outage or an inferior navigation solution could be a serious safety matter.[20] Thus, the whole RAIM specifications matter must be reconsidered for sole-means navigation.

A second reason for lack of good performance data for FDI is that the methodology for solving the isolation half of the problem is still evolving. There is still plenty of opportunity for innovation on this problem, and new papers on FDI appear regularly at current navigation meetings. We do not attempt here to predict exactly where FDI research will lead ultimately. However, there are some basics of parity theory that will no doubt play a role in the final FDI solution, so we now continue the parity discussion that began in Sec. II.

B. Parity Method and Failure Detection and Isolation

Parity theory provides an especially useful geometric perspective in the FDI problem. Recall that we only use the magnitude of the parity vector (or a related quantity) as the test statistic for detection. We see presently that the direction of the parity vector provides useful information about the identity of the failed satellite.

Recall from Eq. (5) that the parity vector p is given by

$$p = Py \qquad (16)$$

where y is the measurement vector and P is the special transformation that takes

us from the n dimensional measurement space to the $(n - 4)$ dimensional parity space. (We discuss the computation of P later in Sec. VI.C.) By definition, P has some very special properties.
1) The rows of P are orthogonal to the columns of G.
2) The rows of P are mutually orthogonal (i.e., with each other).
3) The rows of P are normalized so that each of their magnitudes (i.e., Euclidean or 2-norm) is unity.
Note that if y in Eq. (16) is replaced with its equivalent $(Gx_{\text{true}} + \epsilon)$, the orthogonality property dictates that $PGx_{\text{true}} = 0$; and thus

$$p = P\epsilon \qquad (17)$$

In effect, the definition of P makes it such that the true value of x is blocked in projecting y into p, and all we are left with is the measurement *error* projected into the parity space. Note that the dimension of the parity space is four less than that of the measurement space in this application. Thus, the projection of the measurement error into a useful test statistic is much easier to visualize in parity space than in the measurement-residual space (and with no loss of information).

Next, omit the measurement noise for the moment and consider the effect of a bias error b on a single satellite. For purposes of illustration, let us say that we have six-in-view, just as in the example in Sec. II. In this case, P is a 2 × 6 matrix. Suppose we put the bias on satellite four (i.e., the fourth element of the column vector ϵ). The resultant projection into parity space is then

$$p(\text{for bias on sat. 4}) = \begin{bmatrix} p_{14} \\ p_{24} \end{bmatrix} b \qquad (18)$$

(lower case p with *two* subscripts denotes elements of P).

It can be seen from Eq. (18) that the parity vector induced by a bias on satellite four must lie along a line whose slope is p_{24}/p_{14}. This is shown in Fig. 6.

A similar argument applies to the parity vector that would be induced by placing a bias on any of the other five satellites. Each satellite has its own characteristic bias line, with a slope determined by the elements of the respective column vector of P, i.e.,

$$\text{Slope of char. line for sat. } i = p_{2i}/p_{1i}, i = 1,2, \ldots, 6 \qquad (19)$$

A decision rule for identifying the failed satellite is now obvious.

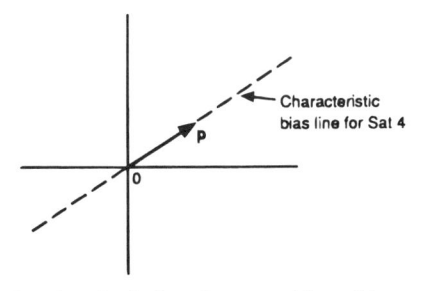

Fig. 6 Parity space showing the induced p caused by a bias on satellite 4 (no noise).

Decision rule: The failed satellite is the one whose characteristic bias line lies along the observed parity vector p.

Let us now be more realistic and include noise in our FDI example. To illustrate the difficulty of meeting reasonably stringent (but hypothetical) specifications, suppose we say the alarm rate must be 1/15,000 (or less) and the desired detection probability is 0.999, just as in the supplemental navigation case. Furthermore, assume that we use a parity detection rule, and then follow it immediately with a "most-likely" isolation rule to identify the failed satellite. Most-likely here will mean the satellite whose characteristic bias line is closest to the observed parity vector. The probability of successful isolation of the failed satellite is to be at least as high as the detection probability of 0.999. For convenience, we will use $|p|$ (or $\sqrt{SSE}$) as our test statistic.

Figure 7 shows the six characteristic bias lines for one of the six-in-view geometries used in the Van Dyke availability study.[6] The ARP value for this geometry is 1130, so it is admissible for detection purposes for both terminal and enroute flight (see Table 3). Now, suppose we consider a bias on satellite 4 that is just sufficient to trigger the alarm without noise. This induces a component vector in the parity domain that emanates from the origin and terminates on the alarm circle, as shown in Fig. 7. We then add to this a random Rayleigh noise vector (mainly caused by SA) whose mode is 33 m. The sum then terminates on the smaller dashed circle shown in Fig. 7. We know nothing about the direction of the random noise vector except that is uniformly distributed in angle. Now imagine performing a Monte Carlo experiment, keeping the bias and satellite geometry fixed and choosing the noise vectors at random. This would produce a cloud of data points centered around the center of the dashed (i.e., the noise)

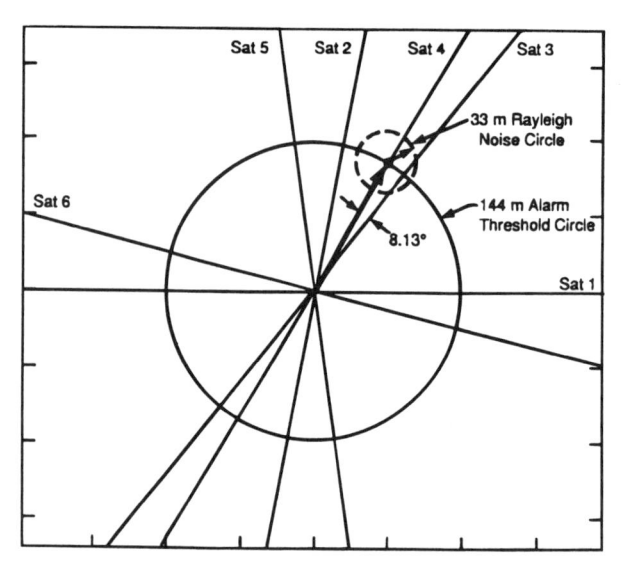

Fig. 7 Characteristic bias lines in parity space—six-in-view example (test stastic = $|p|$, measurement noise $\sigma = 33$ m).

circle. Even without doing any calculations, we can see visually that a sizable fraction of the data would lie closer to the satellite 3 line than the satellite 4 line, and thus, these data would result in incorrect isolations. The fraction of misidentifications in this example would be about two orders-of-magnitude greater than desired! Thus, an FDI algorithm that simply tacks a single isolation appendage onto a detection stage does not even come close to satisfying reasonable specifications for this geometry.

As might be expected, there are certain equivalencies in the isolation part of FDI as well as in detection. Brown and Sturza[15] point out that the geometric "closest to the observed parity vector" decision rule can be replaced with an equivalent algebraic rule in the measurement-residual space. Which rule is to be used is purely a matter of computational convenience. The beauty of the parity approach lies in its vivid geometric interpretation. For example, it is obvious from Fig. 7 what must be done to improve the isolation performance. We must either 1) make the alarm threshold circle larger; or 2) make the noise circle smaller, or 3) add more satellites to the measurement suite to obtain a higher-order parity space (e.g., seven characteristic bias lines will diverge faster in three-space than six lines do in two-space). Also, the effect of the satellite bias error is clear from Fig. 7. Small biases are not a problem, because they (plus the noise) do not trigger the alarm. On the other hand, a very large bias will push the observed parity vector far from the origin, and a near-perfect identification will take place. Thus, very small and very large bias errors are not problems; the intermediate ones are those that are difficult to detect and isolate with certainty. All of these conclusions as to how various parameters interact in the FDI problem are obvious from one simple picture in parity space.

We do not pursue various avenues for improving the basic parity FDI scheme any further here. The main lesson to be learned from the preceding example is this. The direct snapshot use of the parity vector for both detection and isolation looks simple initially. Only after we put realistic numbers into the problem does the real difficulty becomes apparent. The tight specifications make this a truly difficult problem. It is likely that more sophisticated algorithms, such as the two-stage approach suggested in a recent paper by Lee,[20] will have to be developed before RAIM will become viable for sole-means navigation.

C. Calculation of the P Matrix

We now present two methods of computing the P matrix. One of these is formal and involves linear algebra theory; the other is less formal, but equally valid.

1. Formal Method

QR factorization decomposes a matrix into orthogonal and triangular factors. This is discussed in such works as Golub and Van Loan,[21] so we omit the details here. We begin with G, the linear connection matrix that connects the state space to the measurement space [see Eq. (1)]. It is first factored in QR form; i.e.,

$$G = QR \tag{20}$$

where Q is the orthogonal factor. Consider next the transpose of Q, which we denote as Q^T.

Then

$$P = \text{Bottom}(n - 4) \text{ rows of } Q^T \qquad (21)$$

This completes the formal method of computing P. (We note that QR factorization is one of the "built-in" functions in MATLAB®, which makes the formal approach especially easy for those who have access to MATLAB.[22])

2. Heuristic Method

A suitable P matrix that satisfies all the requirements stated in Sec. VI.B can be found by simply using these requirements as constraints in the choice of the elements of P. (It is seen presently that one of these elements is arbitrary in our six-in-view example.) In words, an algorithm for finding P is as follows.

A. Begin with the bottom row of P (six elements):
  1) Let element one be zero and element two be unity.
  2) Use the orthogonal relationship between the rows of P and the columns of G, and write four linear equations relating the remaining unknown elements of the last row of P to the respective elements of G.
  3) Solve the equations and write an unnormalized bottom row of P as $[0 \ 1 \ p'_{23} \ p'_{24} \ p'_{25} \ p'_{26}]$.
  4) Now normalize the unnormalized row by multiplying by a scale factor that makes its magnitude unity. This yields the final bottom row of P, which we denote with the usual notation $[0 \ p_{22} \ p_{23} \ p_{24} \ p_{25} \ p_{26}]$.

B. Now go the next to the bottom row of P.
  1) Let element one be unity.
  2) Use the orthogonality relationships between this row and G and the previously determined bottom row of P, and write five linear equations in the remaining unknown elements of this row.
  3) Solve the equations and write an unnormalized equation for the next to bottom row as $[1 \ p'_{12} \ p'_{13} \ p'_{14} \ p'_{14} \ p'_{15} \ p'_{16}]$.
  4) Now normalize the unnormalized row. This yields $[p_{11} \ p_{12} \ p_{13} \ p_{14} \ p_{15} \ p_{16}]$.

The determination of P is now complete. Note that there are 12 degrees of freedom in P and only 11 constraining equations. Thus, one element is arbitrary. By letting the first element in the bottom row be zero, we automatically place the characteristic bias line for satellite 1 on the zero-angle reference. Obviously, we could have put the zero in any of the other five positions, thus causing those respective bias lines to be on the reference axis. In doing so, we would not change the relative spacing among the bias lines. The effect would simply be a rotation of the whole characteristic bias line picture by an appropriate amount to bring the desired bias line to the zero-angle axis.

The extension of this algorithm to the seven-in-view and higher-order cases is straight-forward. For example, in the seven-in-view case, P is a 3×7 matrix with 21 elements. There are 15 orthogonal and 3 normalizing constraints that must be satisfied. Thus, 3 elements may set equal to zero (within some restrictions). A reasonable way to choose the zeros is to put them in left portion of the unnormalized P as follows:

$$P(\text{unnormalized}) = \begin{bmatrix} 1 & p'_{12} & p_{13} & \cdots & p'_{17} \\ 0 & 1 & p'_{23} & \cdots & p'_{27} \\ 0 & 0 & 1 & \cdots & p'_{37} \end{bmatrix} \quad (22)$$

This choice will force the first column vector of P to be aligned with the reference axis in 3-space; the second column vector will be normal to the third axis in 3-space; and the remaining columns will generally be vectors with three nontrivial elements. The extension to the higher-order cases is fairly obvious, so this is not pursued further.

One final comment is in order. The P matrix is not unique. However, the relative spacing among the characteristic bias lines is unique, and this does not change when a rotation or mirror-image reflection-type transformation is made that takes one P matrix into another equally valid P matrix.

D. Failure Detection and Exclusion Algorithm

Before leaving the subject of FDI, it should be mentioned that a variation on the basic FDI scheme just described has recently appeared in the literature.[23] This variation is usually referred to as FDE for failure detection and exclusion (in contrast to isolation). This is something of a play on words, but there are subtle differences that warrant treating FDE separately from FDI.

The basic idea of FDE can be illustrated with an eight-in-view example. Suppose that the normal detection suite operates with only six satellites (as suggested in Ref. 23), and an alarm occurs. Instead of immediately trying to isolate the bad satellite, the algorithm simply searches among all the other six-satellite subsets for one that will satisfy its self-consistency test, thus assuring that the bad satellite has been excluded. Two satellites are excluded, but the algorithm does not attempt to determine which of the two is the bad one. In this sense, the algorithm does not isolate the individual offender, it only separates it into a group of possible offenders. Furthermore, if the bad satellite just happened to be one of the pair that was excluded in the original suite of six prior to the failure, then there would be no alarm and no search would be necessary, an obvious fringe benefit. Contrast this with the usual FDI algorithm that would be using all eight satellites in its detection suite. It would be forced to detect and isolate the bad satellite and run the attendant risk of making a mistake.

Another attractive feature of the FDE philosophy is that it will accomodate multiple satellite failures, to a limited extent at least. For example, with eight-in-view and six satellites normally in the detection suite, two concurrent failures could be detected and excluded. The algorithm would have to search through all combinations of eight things taken six at a time (28) to do so, but this is routine. With nine-in-view, three failures could be detected, and so forth. All of this comes with a price, however. The computational effort escalates rapidly with the number in view, and it is still not clear how this extra capability affects the RAIM availability.

As mentioned earlier, the state of the art relative to both FDI and FDE is still evolving, and it has not been decided at this time exactly what RAIM scheme will be recommended by the RTCA committee studying the matter. However, parity space plays an important role in both FDI and FDE, so it is essential that those working with RAIM understand the fundamentals of parity-space methods.

References

[1]"1990 Federal Radio Navigation Plan," Document DOT-VNTSC-RSPA-90-3/DOD-4650.4, U.S. Depts. of Transportation and Defense.

[2]Kalafus, R. M., "Receiver Autonomous Integrity Monitoring of GPS," Project Memorandum DOT-TSC-FAA-FA-736-1, U.S. DOT Transportation Systems Center, Cambridge, MA, 1987.

[3]Lee, Y. C., "Analysis of Range and Position Comparison Methods as a Means to Provide GPS Integrity in the User Receiver," *Proceedings of the Annual Meeting of the Institute of Navigation* (Seattle, WA), June 24–26, 1986, pp. 1–4.

[4]Brown, R. G., and Hwang, P. Y. C., "GPS Failure Detection by Autonomous Means Within the Cockpit," *Proceedings of the Annual Meeting of The Institute of Navigation* (Seattle, WA), June 24–26, 1986, pp. 5–12.

[5]"Minimum Operational Performance Standards for Airborne Supplemental Navigation Equipment Using Global Positioning System (GPS)," Document RTCA/DO-208, Radio Technical Commission for Aeronautics, Washington, DC, July 1991.

[6]Van Dyke, K. L., "RAIM Availability for Supplemental GPS Navigation," *Proceedings of the 48th Annual Meeting of the Institute of Navigation*, June 29–July 1, 1992, Washington, DC.

[7]Parkinson, B. W., and Axelrad, P., "Autonomous GPS Integrity Monitoring Using the Pseudorange Residual," Navigation (Washington), Vol. 35, No. 2, 1988, pp. 255–274.

[8]Sturza, M. A., and Brown, A. K., "Comparison of Fixed and Variable Threshold RAIM Algorithms," *Proceedings of the Third International Technical Meeting of the Institute of Navigation*, Satellite Division, ION GPS-90 (Colorado Springs, CO), Sept. 19–21, 1990, pp. 437–443.

[9]Sturza, M. A., "Navigation System Integrity Monitoring Using Redundant Measurements," Navigation (Washington). Vol. 35, No. 4, 1988–1989, pp. 483–501.

[10]Brown, R. G., and Hwang, P. Y. C. "Introduction to Random Signals and Applied Kalman Filtering," 2nd ed., John Wiley & Sons New York, 1992.

[11]Brown, R. G., "A Baseline RAIM Scheme and a Note on the Equivalence of Three RAIM Methods," *Proceedings of the National Technical Meeting of the Institute of Navigation* (San Diego, CA), Jan. 27–29, 1992, pp. 127–137.

[12]Brown, R. G., and McBurney, P. W., "Self-Contained GPS Integrity Check Using Maximum Solution Separation as the Test Statistic," *Proceedings of the Satellite Division First Technical Meeting, The Institute of Navigation* (Colorado Springs, CO), 1987, pp. 263–268.

[13]Brown, A. K., "Receiver Autonomous Integrity Monitoring Using a 24-Satellite GPS Constellation," *Proceedings of the Satellite Division, First Technical Meeting, The Institute of Navigation* (Colorado Springs, CO), 1987, pp. 256–262.

[14]Brenner, M., "Implementation of a RAIM Monitor in a GPS Receiver and an Integrated GPS-IRS," *Proceedings of the Third International Technical Meeting of the Satellite*

Division of the Institute of Navigation, ION GPS-90 (Colorado Springs, CO), Sept. 19–21, 1990, pp. 397–406.

[15]Brown, A. K., and Sturza, M., "The Effect of Geometry on Integrity Monitoring Performance," *Proceedings of the Institute of Navigation Annual Meeting*, June 1990.

[16]Brown, R. G., Chin, G. Y., and Kraemer, J. H., "Update on GPS Integrity Requirements of the RTCA MOPS," *Proceedings of the 4th International Technical Meeting of the Satellite Division of the Institute of Navigation*, ION GPS-91 (Albuquerque, NM), Sept. 11–13, 1991.

[17]Chin, G. Y., Kraemer, J. H., and Brown, R. G., "GPS RAIM: Screening Out Bad Geometries Under Worst-Case Bias Conditions," *Proceedings of the 48th Annual Meeting of the Institute of Navigation*, June 29–July 1, 1992.

[18]Diesel, J. W., "A Synergistic Solution to the GPS Integrity Problem," *Proceedings of the National Technical Meeting of the Institute of Navigation* (Phoenix, AZ), Jan. 22–24, 1991, pp. 229–236.

[19]McBurney, P. W., and Brown, R. G., "Receiver Clock Stability: An Important Aid in the GPS Integrity Problem," *Proceedings of the National Technical Meeting of the Institute of Navigation* (Santa Barbara, CA), Jan. 26–29, 1988, pp. 237–244.

[20]Lee, Y. C., "Receiver Autonomous Integrity Monitoring (RAIM) Capability for Sole-Means GPS Navigation in Oceanic Phase of Flight," *IEEE 1992 Position Location and Navigation Symposium (PLANS) Record*, Monterey, CA, 1992, pp. 464–472.

[21]Golub, G. H., and Van Loan, C. F., *Matrix Computations*, 2nd ed., The Johns Hopkins University Press, Baltimore, 1989.

[22]*386-MATLAB Users Guide*, The Math Works, South Natick, MA, Oct. 15, 1990.

[23]Van Graas, F., and Farrell, J. L., "Baseline Fault Detection and Exclusion Algorithm," *Proceedings of the 49th Annual Meeting of the Institute of Navigation* (Cambridge, MA), June 21–23, 1993.

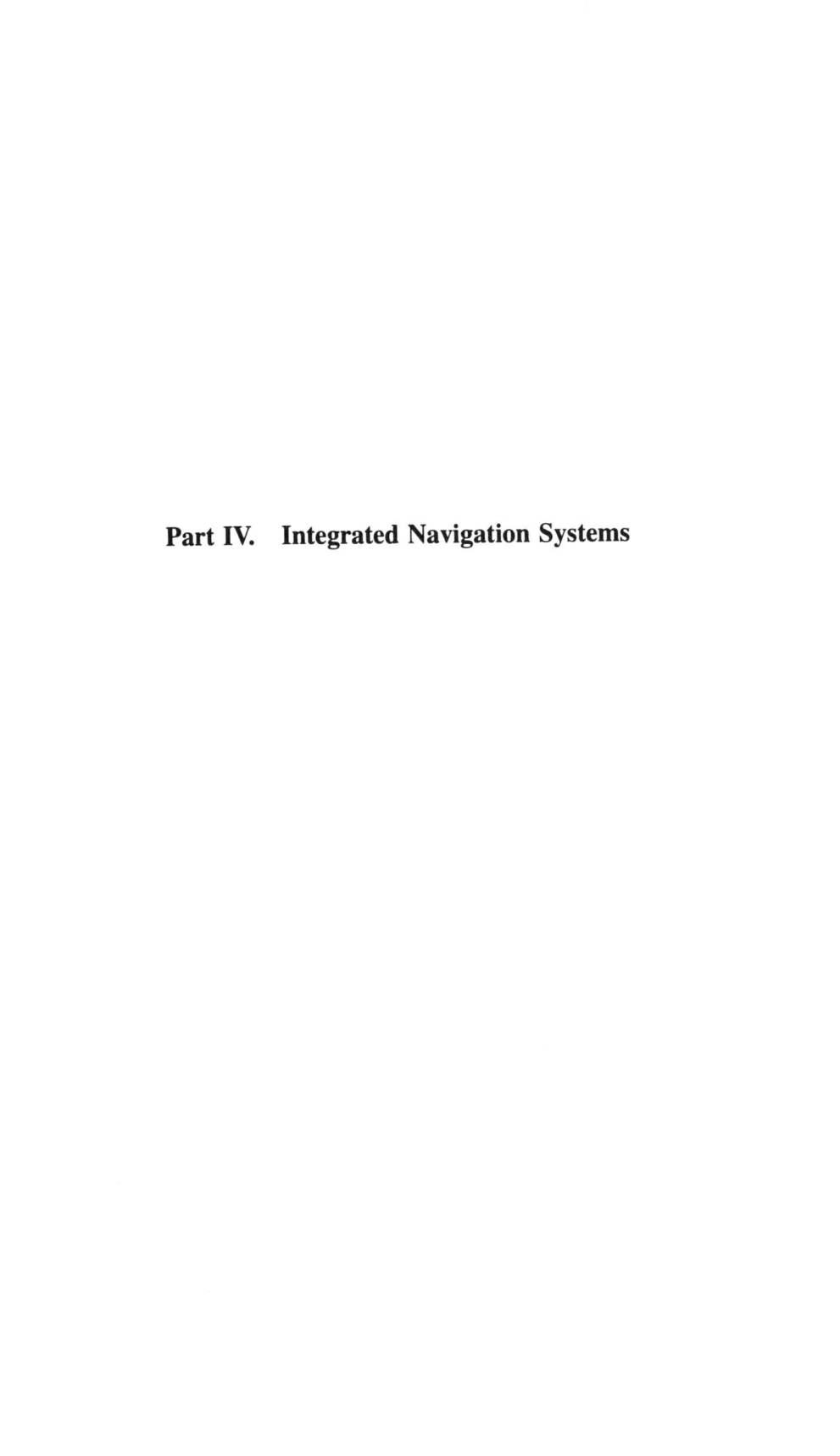

Part IV. Integrated Navigation Systems

Chapter 6

Integration of GPS and Loran-C

Per K. Enge*
Stanford University, Stanford, California 94305
and
F. van Graas†
Ohio University, Athens, Ohio 45701

I. Introduction

FROM 1945 through the 1970s, the Long Range Navigation System (Loran-C) was developed by the United States and the Soviet Union (the Soviet system is called Chayka) primarily for military use. In the 1970s, Loran-C was declared an official national system by the United States and Canada. These decisions spurred a wealth of civilian applications and tremendous commercial development. The integrated circuit and microprocessor improved the performance of Loran-C receivers, while greatly reducing their size and price. Solid-state Loran transmitters were introduced, and these had much greater reliability and efficiency than their vacuum tube predecessors.

Today, Loran provides service to nearly a million maritime, airborne, and terrestrial users throughout most of the northern hemisphere. The coverage of Loran is shown in Fig. 1, where the coverage contours are for the extremely reliable groundwave coverage. Skywave propagation does provide extended coverage, but it is not as reliable.

Recently, the United States has added four new Loran transmitters to cover the middle of the United States (to fill the "midcontinent gap").[1] With these new transmitters, five or more Loran signals are available 95% of the time over more than 95% of the conterminous United States (CONUS).

Furthermore, the Far East Loran Technical Group, which includes Japan, the Peoples Republic of China, Korea, and the Commonwealth of Independent States, has agreed to build and operate six Loran chains under the auspices of the International Association of Lighthouse Authorities. As part of this effort, Japan and Korea have already contracted to build three solid-state Loran transmitters to replace older tube-type transmitters.

Copyright © 1994 by the authors. Published by the American Institute of Aeronautics and Astronautics, Inc. with permission.
*Professor (Research), Department of Aeronautics and Astronautics.
†Associate Professor, Department of Electrical and Computer Engineering.

Fig. 1 Worldwide coverage of Loran-C.

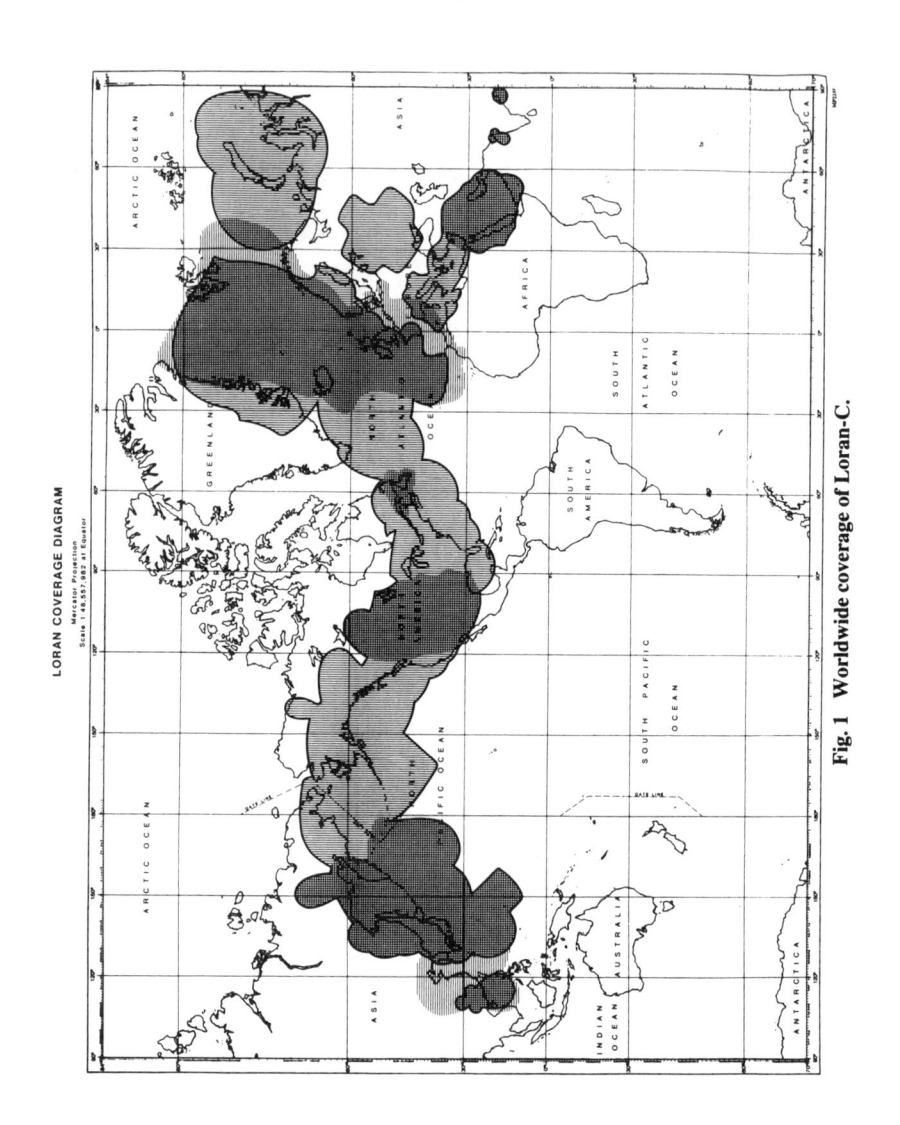

INTEGRATION OF GPS AND LORAN-C   171

In February of 1992, the Council of the European Communities adopted a decision on radionavigation systems for Europe, which stated "support efforts to set up a worldwide radionavigation system including European regional Loran-C chains with the purpose of enlarging worldwide Loran-C coverage in order to improve the safety of navigation and protection of the marine environment." As a consequence, nations in northwestern Europe are expanding their Loran coverage with modern transmitters.

In the summer of 1992, India installed two new Loran chains, which serve the waters near Bombay and Calcutta.

As described throughout this book, GPS is currently being installed and will be fully operational in the mid-1990s. These two vital radionavigation systems have much to offer each other.

A. Calibration of Loran Propagation Errors by GPS

Loran-C receivers measure the arrival time of groundwave radio signals emitted from a network of synchronized terrestrial transmitters. With repeatable accuracy measured in tens of meters, Loran provides excellent service to users navigating relative to landmarks that have been previously marked using Loran-C. However, its absolute (or geodetic) accuracy is currently limited to approximately one-quarter nautical mile (460 m) by uncertainties in the groundwave propagation speed. Fortunately, the Global Positioning System (GPS) can be used as a very accurate and convenient truth system to calibrate these uncertainties. Section III of this paper describes the results of such a calibration in the Gulf of Maine.

B. Cross-Chain Synchronization of Loran-C Using GPS

Loran-C transmitters are grouped into different chains, and each chain transmits groups of radio frequency pulses. Currently, Loran-C transmissions from different chains are not tightly synchronized. Consequently, Loran receivers can only compute time differences between signals from the same chain. If the receiver can receive many stations within a single chain, then this limitation is not troublesome. However, if the receiver is near the edge of coverage or if a transmitter has failed, then it can be helpful to remove this limitation and allow the receiver to compute "cross-chain" (or cross-rate) time differences. The time transfer capability of the GPS is one way of accomplishing the required cross-chain synchronization. In Sec. IV of this chapter, we examine (by example) the impact of cross-chain synchronization on Loran-C coverage.

C. Combining Pseudoranges from GPS and Loran-C for Air Navigation

Loran and GPS are currently widely used by aviators. Moreover, Loran-C is approved as a primary aid to navigation for certain airport approaches. However, Loran by itself does not qualify as a primary navigation system for terminal or enroute air navigation because one or more Loran signals may become unavailable. Signal loss may be attributable to "cycle slip," transmitter outage, or high noise. Furthermore, GPS by itself does not qualify as a primary civilian system because the time availability and integrity of stand-alone GPS are inadequate.

Section V of this chapter describes how information from the two systems can be combined in a hybrid receiver to give a system with improved coverage, availability and reliability.

Section II of this chapter provides an overview of Loran-C. Then Secs. III, IV, and V develop the above described techniques for using GPS and Loran together. Finally, Sec. VI is a brief summary.

II. Loran Overview

The Loran signal is shown in Fig. 2. As shown in Fig. 2a, Loran transmitters (also known as a Loran station or LORSTA) broadcast radio frequency pulses. Each pulse has a duration of approximately 200 μs and a center frequency of 100 KHz. The receiver identifies and tracks the arrival time of the sixth zero crossing of the pulse. This is the essential and fundamental measurement of any Loran receiver.

As shown in Fig. 2b, each transmitter periodically emits a group of eight (or nine) pulses. Loran transmitters are grouped into chains, and every transmitter in a given chain sends its group of pulses every group repetition interval (GRI), where the GRI varies from 50 to 100 ms. In fact, chains are distinguished by their unique GRIs. Each chain contains one master and two to five secondary stations, where the secondary transmissions are synchronized to the master transmission. A typical chain contains a master station M and three secondaries X, Y and Z. Secondary X emits its pulse group NED_X μs after the master transmits, where NED stands for nominal emission delay. Secondaries Y and Z are also synchronized and transmit NED_Y and NED_Z μs after the master.

As shown in Fig. 2c, phase codes control the polarity of the transmitted pulses. These phase codes distinguish the master signal from the secondary signals, and they repeat every two groups. The design of these sequences is described in an excellent article.[2]

A Loran-C user receiver measures the time difference (TD) between the arrival of the pulse groups from the master station and the secondary stations. The transmitter locations and nominal emission delays are well known, and the propagation speed of the Loran pulse can be estimated accurately. Therefore, each measured time difference defines a hyperbolic line of position (LOP) for the user. The intersection of two such LOPs defines the user's position in two dimensions. Loran is incapable of providing accurate estimates of altitude, because of the geometry of land-based transmitters relative to any user close to the surface of the Earth.

The transmitted Loran-C signal has a groundwave component that travels along the surface of the Earth, and a skywave component that is reflected off the ionosphere. The skywave component is not suitable for accurate position fixing because variations in the height and density of the ionosphere make the travel time of the skywave difficult to predict. Fortunately, the travel time of the groundwave is stable and predictable, and the design of the Loran pulse allows the receiver to separate the groundwave from the skywave. The groundwave attenuates as it propagates over Earth with finite ground conductivity. This attenuation limits each chain's range to approximately 500 nautical miles of the master station over land and 800 nautical miles over sea.

INTEGRATION OF GPS AND LORAN-C 173

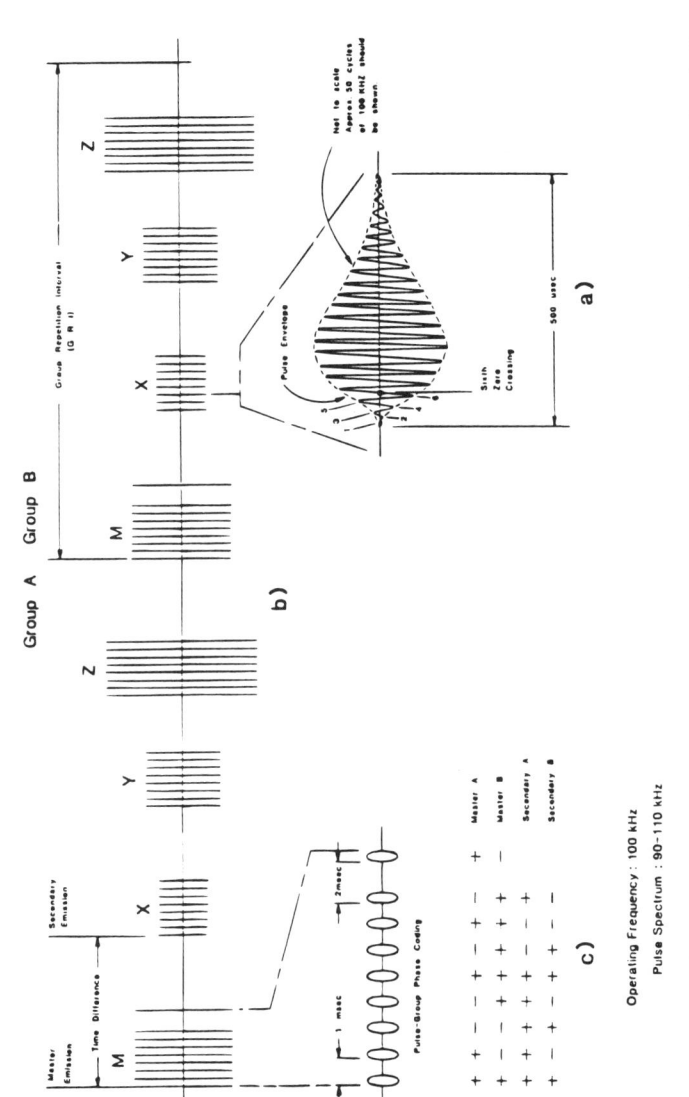

Fig. 2 Loran signal design: a) an individual Loran pulse, where the Loran reciever tracks on the sixth zero crossing of the 100 KHz carrier; b) illustration of how the individual pulses are grouped together with nine pulses for the master stations M and eight pulses for the secondary stations X, Y and Z; and c) illustration of the phase code that shifts the polarity of the pulses within a group. As shown, the master and secondary phase codes are unique. Additionally, the phase codes repeat every two groups.

Loran accuracy is limited by random errors and bias errors. Random errors are caused by noise and interference. They cause the time of the sixth zero crossing to vary rapidly relative to the time constant of the tracking loops in the receiver. Bias errors can also be caused by interference, but in the main they are caused by propagation effects. They cause the time of the sixth zero crossing to be offset and are difficult to remove by averaging. In other words, the decorrelation time of the propagation errors are long compared to the user platform dynamics.

Atmospheric noise is caused by lightning and is the most powerful natural noise source in the Loran band. The errors attributable to atmospheric noise are minimized by placing nonlinear signal processing elements in the receiver. These nonlinearities clip or limit the impulsive bursts characteristic of lightning noise.

Man-made signals near the Loran band are also important, particularly in Europe. The impact of this interference is minimized by careful design of the filters in the front end of the receiver. Interference at known frequencies can also be mitigated by careful selection of the group repetition intervals. Additionally, man-made interference and atmospheric noise are both reduced by making the time constants of the tracking loops as large as possible commensurate with the dynamics of the receiver platform. All told, atmospheric and man-made noise limit the short-term repeatable accuracy of Loran to tens of meters.

Propagation effects introduce errors that vary slowly or are bias errors. Indeed, the total travel time of the Loran-C signal (neglecting the difference between the transmitter and receiver clocks) is modeled as follows:

$$t_{tot} = t_{PF} + t_{SF} + t_{ASF} = t_{SALT} + t_{ASF} \qquad (1)$$

The first term t_{PF} is known as the primary factor and is the travel time of a Loran-C signal moving at the speed of light in air with no boundary effects. The secondary factor t_{SF} is the additional time needed to travel over an all-seawater path. The "salt water model" $t_{SALT} = t_{PF} + t_{SF}$ is the total travel time over an all-seawater path. This term can be accurately modeled, because seawater variations have little effect on the travel time of the Loran pulse.[3]

The additional secondary factor t_{ASF} is the additional time needed to traverse any land segments. Moreover, t_{ASF} depends on the conductivity of the underlying ground and the overland distance traveled by the Loran pulse.[4-6] It grows to three or four μs after the Loran signal has traveled 200–300 km over typical land. The U.S. Defense Mapping Agency (DMA) publishes lookup tables for t_{ASF} for all U.S. Loran-C chains.[7] These tables give oversea estimates of t_{ASF} and are based on a model of ground conductivity in the US. With these data, the absolute accuracy (the accuracy of a position estimate with respect to the Earth's coordinates) of Loran is approximately 400 m.[8]

The additional secondary factor also varies with time as weather and climate change the effective conductivity of the land. Fortunately, these variations are small compared to the overall uncertainty in t_{ASF}. Consequently, Loran-C's repeatable accuracy (the accuracy with which a user can return to a position whose coordinates have been previously measured with Loran-C) is 18–90 m.[8]

III. Calibration of Loran Propagation Errors by GPS

Global Positioning System position fixes can be used to estimate t_{ASF} accurately, and thus to improve the absolute accuracy of Loran-C. In fact, Figs. 3 and 4

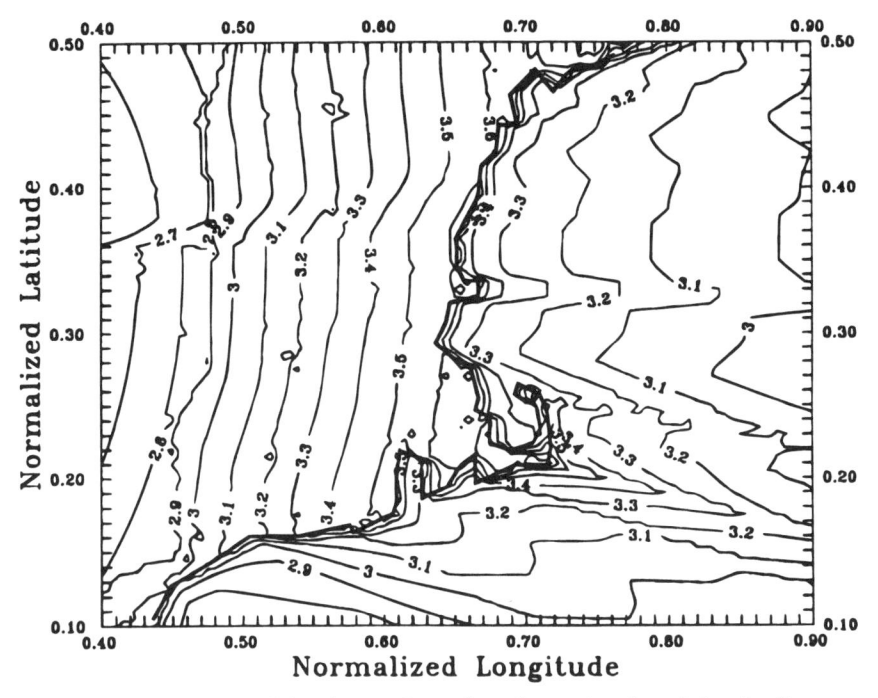

Fig. 3 Prediction of additional secondary phase factor (t_{ASF} in μs) for the Seneca–Nantucket time difference based on simultaneous GPS/Loran observations in the Gulf of Maine.

show t_{ASF} estimates in μs for New England based on GPS/Loran position fixes in the Gulf of Mexico. These figures are based on data collected by the U.S. Geological Survey (USGS) and processed as described by Refs. 9 and 10.

Briefly, the USGS collected Loran and GPS data in the Gulf of Maine over a 30-day period in October of 1985.[11] The USGS data include two-dimensional GPS position estimates and estimates of the GPS positional dilution of precision (PDOP). It also includes the Loran-C chain "9960" M–X and M–W time differences. The 9960 chain is identified by its group repetition interval—99.60 ms. The 9960 master station M is located in Seneca, New York, and secondaries X and W are located in Nantucket, Massachusetts and Caribou, Maine, respectively. The GPS measurements are used to compute t_{SALT} for each time difference in real time. Moreover, t_{SALT} is subtracted from the observed Loran travel time to produce an accurate estimate of $t_{ASF} + \Delta NED$, where ΔNED is the error in the published nominal emission delay.

The algorithm developed by Refs. 9 and 10 uses the GPS-based observations of $t_{ASF} + \Delta NED$ to calibrate a model that predicts t_{ASF} as a function of the conductivities for the N ground segments over which the signal propagates. The N conductivities and ΔNED can be adjusted to minimize the squared error between the model's output and the USGS observations.

Without calibration, the maximum and rms absolute errors of Loran in the Gulf of Maine are around 700 and 500 m, respectively, depending on the choice

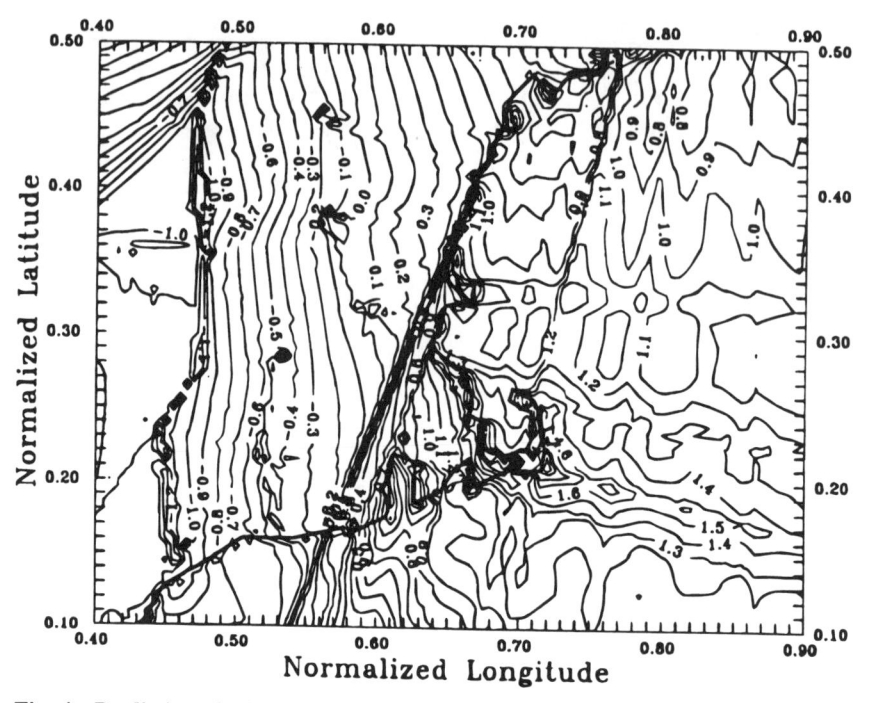

Fig. 4 Prediction of additional secondary phase factor (t_{ASF} in μs) for the Seneca–Caribou time difference based on simultaneous GPS/Loran observations in the Gulf of Maine.

of land conductivity. Significant improvements in the absolute accuracy of Loran can be achieved even with very simple calibrations. If the land conductivities are fixed a-priori and ΔNED (a single parameter) is optimized the maximum, and rms absolute errors fall to around 250 and 60 m, respectively. Alternatively, land can be treated as a single conductivity and this conductivity can be adjusted to reduce offshore additional secondary phase factor (ASF) errors. The performance of this practice results in maximum and rms errors of around 300–100 ms, respectively. More complicated approaches, which adjust multiple conductivities and ΔNED are also discussed in Refs. 9 and 10.

The GPS calibration can be used to form fixed correction tables (databases), which can be used to significantly improve the absolute accuracy of Loran for Loran-only missions. Indeed, a similar approach has been used to estimate t_{ASF} for the northern coast of Scotland.[12] Alternatively, a hybrid GPS/Loran receiver could perform such calibrations in real time. When GPS coverage is strong and integrity guaranteed, GPS could continuously estimate t_{ASF}. The absolute accuracy of Loran calibrated in this way would be nearly equal to GPS and could be used to help an aircraft coast through a GPS outage.

IV. Cross-Rate Synchronization of Loran

Currently, the transmission times of the Loran masters are allowed to drift as long as they stay within ±2.5 μs of universal coordinated time (UTC). The

INTEGRATION OF GPS AND LORAN-C   177

transmission times of the secondary stations are more tightly controlled. They are controlled so that constant master–secondary time differences are maintained at certain system area monitors (SAM). For example, the Nantucket secondary station is controlled so that the Seneca–Nantucket time difference at Sandy Hook, New York is 26999.78 µs. This SAM control results in excellent repeatable accuracy for Loran users near the SAM.

In this section, we investigate the possible value of synchronizing the transmission from master stations in different chains. This analysis of cross-chain synchronization requires consideration of the following modes of receiver operation:

Single rate: The receiver can only use stations from a single GRI.

Two pair fixing: The receiver is also capable of using four stations, where two are in one GRI, and the other two are in a different GRI (a two pair fix).

Chain independent: The receiver can use any three or four stations regardless of which GRI they are in.

Single-rate receivers and receivers capable of cross-pair fixing exist today, and no change in chain timing is required to support these receivers. Chain-independent operation alone requires cross-chain synchronization.

Clearly, coverage for chain independent operation will always be at least as large as coverage for the other two receivers. However, is the chain-independent coverage ever significantly larger than coverage provided by the less flexible receiver? In other words, does a receiver that makes use of cross-chain synchronization provide more coverage than a receiver available today? To answer this, we plotted Loran coverage for an area that includes the Northwest United States and Southwest Canada (from 35 to 55°N and from 135 to 110°W). These coverage plots assume that the new North Central Chain and South Central Chain, which fill the midcontinent gap are operational. Finally, they assume that the station at Boise City is also dual rated with the Great Lakes Chain.

Figures 5 and 6 show coverage for the three modes of operation: single rate, two pair fixing, and chain independent. Both are for those noise levels exceeded 10% of the time annually. However all transmitters are healthy in Fig. 5, whereas the transmitter at George (47°N, 118°W) has failed in Fig. 6.

Both figures are for master independent receivers. Some Loran receivers are master dependent, whereas others are master independent. The master-dependent receivers must receive the signals from a Loran master station. They may require the master signal to identify which secondary is which, because the phase codes from Loran secondaries are identical and carry no identification information.

The figures assume that a signal is not useable if the received signal-to-noise ratio is less than −10 dB. They also assume that the Loran groundwaves suffer greater attenuation than predicted by an inverse distance law, because they are traveling over poorly conducting terrain. In fact, the figures assume that the conductivity of the underlying ground is 0.003 S/m. Figures 5 and 6 assume the atmospheric noise field strength is 49 dBµ (decibels above 1 microvolt per meter) when measured in a 20 kHz bandwidth. These noise levels are exceeded about 10% of the time annually in this geographical area as predicted by Ref. 13.

Figure 5 assumes that all of the stations are healthy. As shown in Fig. 5, the chain-independent receiver increases coverage slightly relative to two pair fixing in the western portion of the coverage area. The improvement is not dramatic, but under normal circumstances we should not expect it to be. After all, the

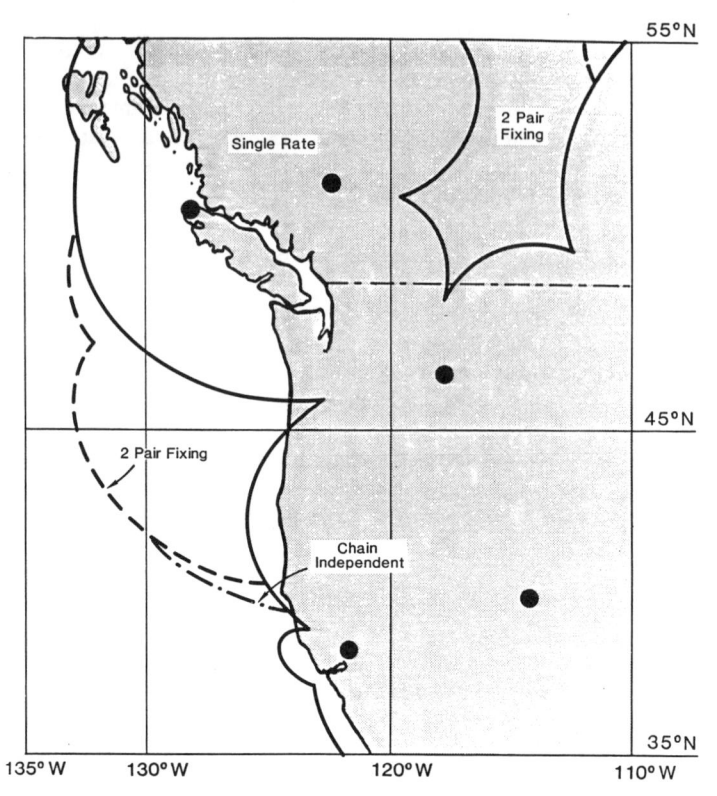

Fig. 5 Loran-C coverage in the U.S. Northwest for single-rate, two-pair fixing and chain-independent receivers. The noise level is exceeded 10% of the time annually, and all stations are healthy. The Loran transmitters are shown as dots.

Pacific, North Central and South Central chains have been designed to give excellent Loran coverage to all users in the western United States.

Figure 6 shows the coverage if the Loran station at George (Washington state) fails, and the noise field is 49 dBμ. Now, the chain-independent Loran receiver improves coverage dramatically. It seems that cross-rate synchronization would be very valuable in this case.

If a master station fails and the receiver is master dependent, then the advantage of cross-rate synchronization might not be as large. (The station at George is not a master). This follows, because the failure of a master would make all the stations in that rate useable. However, master-independent, single-rate receivers exist, and it seems likely that master-independent, chain-independent receivers can be realized.

As described in this section, cross-chain synchronization improves the availability of position fixing for a Loran-only receiver. However, combining GPS and Loran measurements in a combined receiver results in much more dramatic improvements. Such a receiver is described in the next section.

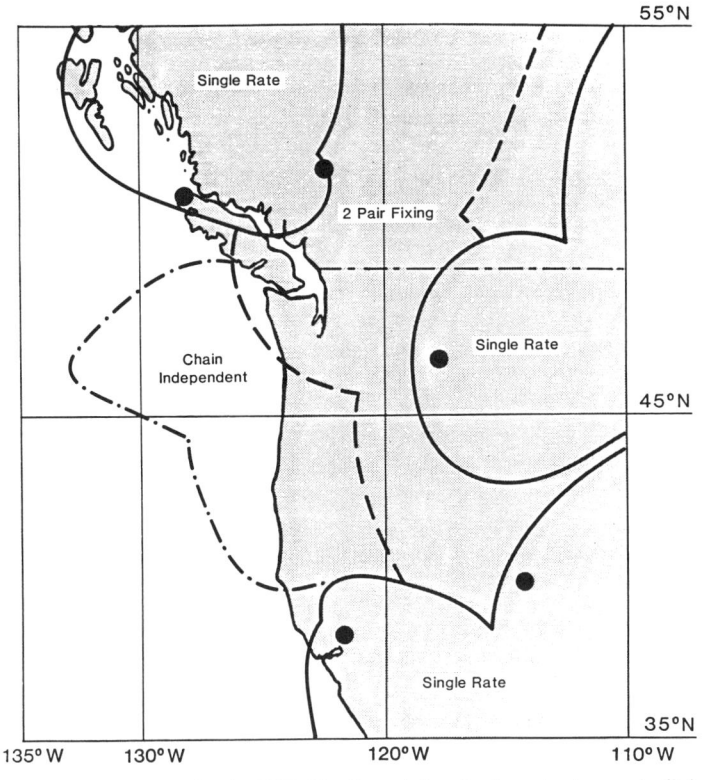

Fig. 6 Loran-C coverage in the U.S. Northwest for single-rate, two-pair fixing and chain-independent receivers. The noise level is exceeded 10% of the time annually. The Loran transmitters are shown as dots, but the station at George (47°N, 118°W) has failed.

V. Combining GPS Pseudoranges with Loran Time Differences

The planned availability of 98% with the GPS constellation is not sufficient for a primary navigation system. Furthermore, the availability of receiver autonomous integrity monitoring (RAIM) is significantly less than 98%. However, combining GPS with Loran-C in the user equipment significantly improves both the availability and the integrity of the position solution. Also, this type of integration provides dissimilar redundancy.

A. Navigation Equations

Global Positioning System and Loran-C measurement data can be combined in two ways: 1) GPS pseudoranges and Loran-C pseudoranges; and 2) GPS pseudoranges and Loran-C time differences (TDs). The first option potentially allows for the highest possible accuracy of the integrated solution.[14,15] To achieve this accuracy, it is necessary to synchronize the time of transmission of all Loran-C transmitters. In the United States, the time of transmission (TOT) of the

Loran-C master stations is held to within 100 ns with respect to UTC, but the transmissions of the secondary stations are under control of system area monitors. A SAM is located in the primary (marine) user area, and it adjusts the time of transmission of a secondary station so that the measured TD at the SAM is held to within ±50 ns of the controlling standard time difference (CSTD).[16] This provides a stable and accurate TD for users close to the line-of-position on which the SAM is located, but it causes the TDs at other locations to change as a function of varying propagation delays.[17] Furthermore, the Loran-C ASF corrections for TDs are available, whereas ASF corrections for pseudorange measurements do not exist at the present time. For these reasons, TDs should be used for Loran-C position calculations, unless the Loran chain uses TOT control.

A radionavigation range measurement is given by the following:

$$r_i = \sqrt{(x - x_i)^2 + (y - y_i)^2 + (z - z_i)^2} \qquad (2)$$

where x, y, z is the three-dimensional user position; x_i, y_i, z_i is the position of the transmitting station; and r_i is the geometric range between the user and the station. A hyperbolic line of position is obtained by measuring the TD between the times of arrival of signals from two different transmitting stations:

$$\text{td}_{i,j} = \left(\frac{b_{i,j} - r_i + r_j}{c} \right) + \text{CD} \qquad (3)$$

where $\text{td}_{i,j}$ is the time difference observation for stations i and j; c is the speed of propagation of the radiowaves; $b_{i,j}$ is the geometric distance between the two stations; r_i and r_j are given by Eq. (2); and CD is the coding delay. The CD is constant, and is inserted by the secondary station to ensure that the transmissions of Loran stations do not overlap within the service area of the chain.

Note that for users at sea level, the Loran-C signals travel great-circle paths. To compensate for this, the transmitter locations are projected onto a locally level plane with respect to the user position estimate at distances equal to the great-circle distances to the transmitters.[18]

Next, the measurement equations are linearized to arrive at the position solution. An a priori estimate of the user position is used to form a Taylor series expansion, of which only the first-order terms are kept.

$$r_i = r_i + \left. \frac{\partial r_i}{\partial x} \right|_{x,y,z} \delta x + \left. \frac{\partial r_i}{\partial y} \right|_{x,y,z} \delta y + \left. \frac{\partial r_i}{\partial z} \right|_{x,y,z} \delta z \qquad (4)$$

The a priori position estimate is used to calculate the estimate of the distance to the station r_i.

Equation (1) can now be linearized as follows:

$$\delta r_i = \left[\frac{x - x_i}{r_i} \quad \frac{y - y_i}{r_i} \quad \frac{z - z_i}{r_i} \right] \begin{bmatrix} \delta x \\ \delta y \\ \delta z \end{bmatrix} \qquad (5)$$

INTEGRATION OF GPS AND LORAN-C

If an unknown clock offset exists in the range measurement, the measurement is called a pseudorange:

$$pr_i = r_i + cb \quad (6)$$

where b is the unknown clock offset. Linearizing equation Eq. (5) results in a slightly different measurement equation:

$$\delta pr_i = \begin{bmatrix} \dfrac{x - x_i}{r_i} & \dfrac{y - y_i}{r_i} & \dfrac{z - z_i}{r_i} & 1 \end{bmatrix} \begin{bmatrix} \delta x \\ \delta y \\ \delta z \\ c\delta b \end{bmatrix} \quad (7)$$

A similar procedure is used to linearize the time difference equation:

$$\delta td_{i,j} = \begin{bmatrix} \dfrac{x - x_j}{r_j} - \dfrac{x - x_i}{r_i} \\ \dfrac{y - y_j}{r_j} - \dfrac{y - y_i}{r_i} \\ \dfrac{z - z_j}{r_j} - \dfrac{z - z_i}{r_i} \\ 0 \end{bmatrix}^T \begin{bmatrix} \delta x \\ \delta y \\ \delta z \\ c\delta b \end{bmatrix} \quad (8)$$

Equations (6) and (7) relate a change in the user state to changes in the range and time difference measurements. In general, Eqs. (6) and (7) can be written as follows:

$$\delta y_i = g_i \begin{bmatrix} \delta x \\ \delta y \\ \delta z \\ c\delta b \end{bmatrix} \quad (9)$$

where y_i is a measurement, and g_i is a row vector corresponding to that measurement. If all the measurements are included, Eq. (8) becomes the following:

$$\delta \underline{y} = G \delta \underline{x} \quad (10)$$

where $\underline{y}$ is a vector containing the measurements and $\underline{x}$ is the user state vector. G is a matrix containing data related to the geometry of the transmitting stations with respect to the user, as given by the row vectors g_i.

Equation (9) can be used to solve for the user state vector iteratively based on the following steps:

1) Start with the user state estimate $\hat{\underline{x}}$ and the measurement vector $\underline{y}$.
2) Convert the Loran-C transmitter coordinates to a locally level plane with $\hat{\underline{x}}$ as the origin.
3) Convert the GPS satellite coordinates to the same locally level plane.
4) Calculate the estimated measurement vector $\hat{\underline{y}}$ using $\hat{\underline{x}}$, GPS satellite positions, and Loran-C transmitter positions.
5) Calculate the partial derivative matrix G; the rows of G are given in Eqs. (7) and (8).

6) Calculate the user state update from the following:

$$\Delta \underline{x} = (G^T G)^{-1} G^T (\underline{y} - \underline{\hat{y}}) \tag{11}$$

7) Update the user state as follows;

$$\underline{\hat{x}} = \underline{\hat{x}} + \Delta \underline{x} \tag{12}$$

8) If the magnitude of the update in step 7 is too large, go to step 4.

9) Use the new user state estimate in the locally level plane to update the user position in latitude, longitude, and height.

10) If the magnitude of the update in step 9 is too large, go to step 2.

11) Repeat steps 1–10 for the next set of measurements. (See Ref. 19 for a detailed description of this algorithm.)

To accomodate different measurement variances, Eq. (10) is left multiplied by a weighting matrix W^{19}:

$$W \delta \underline{y} = W G \delta \underline{x} \tag{13}$$

In general, W could be derived from the measurement noise covariance matrix, but in most applications, it is sufficient simply to use a diagonal matrix, where the diagonal elements are the inverses of the measurement noise standard deviations.

B. Probability of Outage Results

This subsection quantifies the probability of outage (unity minus availability) for a hybrid GPS/Loran receiver. It presents a pair of figures that show the probability of outage for position fixing and autonomous fault detection.

For additional information, the reader is referred to Refs. 15 and 20. The first paper originally published the figures discussed in this subsection and gives a more detailed description of the underlying assumptions. It also includes results on the probability of autonomous fault isolation, which are not summarized here. The second paper also contains an excellent discussion of autonomous fault detection using Loran and GPS.

Probability of outage is the fraction of time/area for which the specified level of service is not available, and it is equal to 1 minus availability. For example, an availability of 0.999999 corresponds to a probability of outage of 10^{-6} or 0.0001%. This probability of outage means that service outages at an average location will last for approximately 30 s per year. An outage probability of 10^{-6} seems very low, but recall that hybrid GPS/Loran could serve as a primary air navigation system, and the requirements for such a system are very severe.

Figure 7 shows the probability of outage for position fixing vs the 2 drms accuracy of the all-in-view fix. If only GPS is used, then this probability is defined as follows.

$$Pr[2(\sigma_p)\text{HDOP} > \text{Accuracy}] \tag{14}$$

where HDOP is the horizontal dilution of precision of the GPS solution. If both systems are used, then the probability of outage is as follows

$$Pr[2(\sigma_p)\text{WHDOP} > \text{Accuracy}]$$

where WHDOP is the weighted HDOP for the combined solution.

INTEGRATION OF GPS AND LORAN-C 183

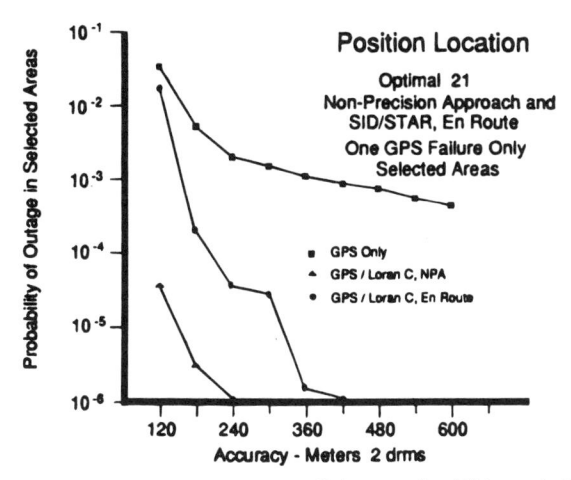

[Graph: Probability of Outage in Selected Areas vs Accuracy - Meters 2 drms, 120 to 600. Position Location, Optimal 21, Non-Precision Approach and SID/STAR, En Route, One GPS Failure Only, Selected Areas. Curves: GPS Only, GPS/Loran C, NPA, GPS/Loran C, En Route]

Fig. 7 Probability of outage for the position fixing service. This probability averages over the following underlying random variables: GPS satellite failures, LORSTA failures, atmospheric noise level, latitude, longitude, and time of day.

Both expression were evaluated with a GPS pseudorange error $1\sigma_p$ of 30 m, and this is a reasonable estimate for the (GPS) standard positioning service (SPS) with selective availability (SA). In both equations, probability is with respect to the following underlying random variables: GPS satellite failures, Loran Station (LORSTA) failures, atmospheric noise level, latitude, longitude, and time of day.

Figure 7 gives the probability of outage averaged over all of the selected geographical areas and time windows described in Ref. 15. It assumes that the receiver has full knowledge of all time offsets except its own clock bias. Additionally, Fig. 7 allows a single satellite shutdown from the optimal 21 constellation described in Ref. 21.

The top (worst) curve gives the outage probability for GPS position fixing, and the other two describe the performance of the GPS/Loran hybrid system. The best hybrid curve is for nonprecision approach and assumes that additional secondary phase factor (ASF) is well known, because of a Loran monitor at the destination airport. The middle curve is for the enroute case and assumes that ASF-related errors increase by 1ns/km (1σ) of range from the transmitters. These larger ASF error prevents the Loran aiding from providing as much improvement at high-accuracy levels.

Figure 8 gives probability of outage for the autonomous fault detection capability described in Chapter 5 of this volume. In contrast to position fixing, autonomous fault detection requires that all the signal subsets created by deleting a single signal have good geometry. Consequently, if GPS alone is used to detect GPS malfunctions (or faults), then we define the outage probability as follows:

$$Pr\{\max_{k \in S} \text{HDOP}(k) > X\} \quad (15)$$

where HDOP (k) is the HDOP when satellite k is deleted, and S is the set of satellites in view. Hence, $\max_{k \in S}$ HDOP (k) is the maximum HDOP of the satellites that remain when each satellite in view is deleted one at a time.

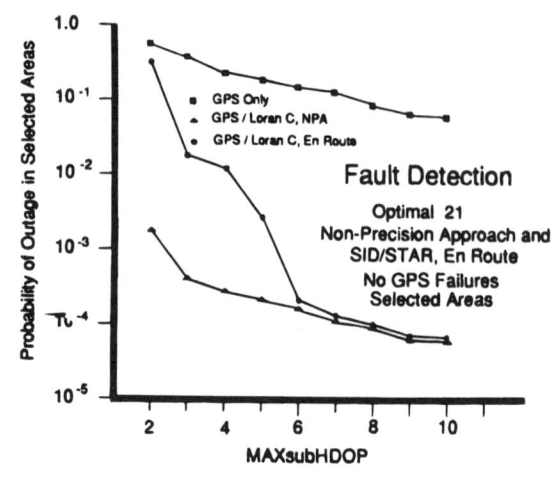

Fig. 8   Probability of outage for the autonomous fault detection fixing service. This probability averages over the following underlying random variables: GPS satellite failures, LORSTA failures, atmospheric noise level, latitude, longitude, and time of day.

If GPS and Loran are used to detect GPS malfunctions, then the probability that soft fault detection is unavailable is given by the following:

$$Pr\{\max_{k \in S} \text{WHDOP}(k) > X\} \qquad (16)$$

where WHDOP (k) is the WHDOP when satellite k is deleted. In this case, the deleted transmitter only comes from the set of GPS satellites because Loran stations broadcast "aviation blink," which warns the user if any Loran signals are outside of specification.

Figure 8 shows the independent variable X from equations YY and ZZ on the horizontal axis. If the probability of false alarm and probability of missed detection are specified, then X can be related to the "protection limit" provided by the fault detection algorithm.[22-25] In general, Brown and Schmid[26] suggest that X is roughly equal to the desired protection limit in meters divided by 100. Figure 8 shows that Loran aiding does greatly reduce the probability of outage for fault detection. The improvement is greatest for a system that enjoys ASF calibration at the destination airport.

Although Figs. 7 and 8 assume TOT control for Loran-C, the results for TD operation should be very similar. The reason for this is that in the United States, almost all Loran-C transmitters are dual-rated, which means that one transmitter participates in two different chains.

VI.   Summary

In this chapter, we have discussed and analyzed the substantive benefits of integrating GPS and Loran. First, GPS may be used to calibrate the propagation uncertainties that traditionally have limited the absolute accuracy of Loran-C.

Second, GPS time transfer can be used to synchronize transmissions from Loran transmitters in different chains. This cross-chain synchronization enables the measurement of cross-chain time differences by the Loran receiver, and these measurements could be very valuable if the Loran system is stressed by high noise or station failures.

Finally, GPS pseudoranges can be combined with Loran time differences in the user equipment. This combination greatly improves the availability of high-accuracy position fixing and autonomous fault detection and isolation. Of course, this improvement is especially dramatic in areas and times where and when GPS coverage is weak. Moreover, it is not very sensitive to whether or not there is an unknown time offset between GPS and Loran. The improvement in availability is greatest when errors attributable to Loran propagation uncertainties have been controlled through some means.

References

[1]Sedlock, A. J., "Mid-continent Loran-C Expansion," *Proceedings of the IEEE 1986 PLANS—Position, Location, and Navigation Symposium* (Las Vegas, NV), Institute of Electrical and Electronics Engineers, NY, Nov. 1986.

[2]Frank, R. L., "Current Developments in Loran-C," *Proceedings of the IEEE*, Vol. 71, Oct. 1983, pp. 1127–1139.

[3]McCullough, J., Irwin, B., and Bowles, R., "Loran-C Latitude–Longitude Conversion at Sea: Programming Considerations," *Proceedings of the Eleventh Annual Technical Symposium of the Wild Goose Association*, Washington, DC, 1982.

[4]Van der Pol, B., and Bremmer, H., "Diffraction of Electromagnetic Waves from an Electrical Point Source Round a Finitely Conducting Sphere." *Philosophical Magazine*, Vol. 7, No. 24, 1937, p. 825; No. 25, 1939, p. 817.

[5]Samaddar, S. N., "The Theory of Loran-C Ground Wave Propagation—A Review." *Navigation*, Vol. 26, No. 3, 1979, p. 173.

[6]Brunavs, P., "Phase Lags of 100 kHz Radiofrequency Ground Wave and Approximate Formulas for Computation," written communication, 1977.

[7]"Loran-C Correction Table for the Northeast U.S.A. (9960)," DMA Stock LCPUB2211200-C, 1988.

[8]"Federal Radionavigation Plan," DOT-TSC-RSPA-88-4, Dec. 1988.

[9]Pisano, J. J., "Using GPS to Calibrate Loran-C," Master's Thesis, Worcester Polytechnic Institute, Worcester, MA, Aug. 1990.

[10]Pisano, J. J., Enge, P. K., and Levin, P. L., "Using GPS to Calibrate Loran-C," *IEEE Transactions on Aerospace and Electronics Systems*, Vol. 27, No. 4, 1991.

[11]Irwin, B., and McCullough, J. "Gulf of Maine GPS/Loran-C Measurement Set," private communication, 1989.

[12]Last, D., and Ward, N., "The Use of DGPS for Mapping Loran-C Additional Secondary Phase Factors," First International Symposium on Real-Time Differential Applications of the Global Positioning System," Braunschweig, Germany, Sept. 1991.

[13]Spaulding, A. D., and Washburn, J. S., "Atmospheric Radio Noise: Worldwide Levels and Other Characteristics," National Telecommunications and Information Administration, NTIA Rept. 85-173.

[14]Van Graas, F., "Sole Means Navigation through Hybrid Loran-C and GPS," *Navigation*, Vol. 35, No. 2, 1988.

[15]Enge, P. K., Vicksell, F. B., Goddard, R. B., and Van Graas, F., "Combining Pseudoranges from GPS and Loran-C for Air Navigation," *Navigation,* Vol. 37, No 1, 1990.

[16]"Loran-C User Handbook," U.S. Department of Transportation, United States Coast Guard, Commandant Publication P16562.6, Washington, DC, 1992.

[17]Vicksell, F. B., and Goddard, R. B., "Implementation and Performance of the TOT Controlled French Loran Chain," *Proceedings of the Fifteenth Annual Technical Symposium of the Wild Goose Association,* New Orleans, LA, Oct. 21–24, 1986.

[18]Van Graas, F., "Hybrid GPS/Loran-C: A Next Generation of Sole Means Air Navigation," Ph.D. Dissertation, Nov. 1988. Athens, OH, Ohio University.

[19]Lawson, C. L., and Hanson, R. J., *Solving Least Squares Problems,* Prentice-Hall, Englewood Cliffs, NJ, 1974.

[20]Brown, R. G., and McBurney, P. W., "Loran-Aided GPS Integrity," *Proceedings of the Satellite Division First International Technical Meeting, Institute of Navigation* (Colorado Springs, CO), Institute of Navigation, Washington, DC, Sept. 1988.

[21]Green, G. B., Massatt, P. D., and Rhodus, N. W., "The GPS 21 Primary Satellite Constellation," *Navigation,* Vol. 35, No. 1, 1988.

[22]Brown, A. K., "Receiver Autonomous Integrity Monitoring Using a 24-Satellite GPS Constellation," *Proceedings of the Satellite Division First Technical Meeting, Institute of Navigation* (Colorado Springs, CO), Institute of Navigation, Washington, DC, Sept. 1987.

[23]Brown, R. G., and McBurney, P. W., "Self-Contained GPS Integrity Check Using Maximum Solution Separation as the Test Statistic," *Navigation,* Vol. 35, No. 1, 1988.

[24]Kalafus, R., and Chin, G. Y., "Performance Measures of Receiver Autonomous GPS Integrity Monitoring," *Proceedings of the 1988 National Technical Meeting of the Institute of Navigation* (Santa Barbara, CA), Institute of Navigation, Washington, DC, 1988.

[25]Lee, Y. C., "Performance Analysis of Self-Contained Methods for GPS Integrity Function," MITRE Corporation TR MTR-88W89, Nov. 1988.

[26]Brown, A. K., and Schmid, T., "Integrity Monitoring of GPS Using a Barometric Altimeter," *Proceedings of the National Technical Meeting of the Institute of Navigation* (Santa Barbara, CA), Institute of Navigation, Washington, DC, Jan. 1988.

Chapter 7

GPS and Inertial Integration

R. L. Greenspan*
Charles Stark Draper Laboratories, Cambridge, Massachusetts 02139

THE Global Positioning System (GPS) and inertial navigation systems (INS) have complementary operational characteristics. Even a modest attempt to combine their functionality in an integrated navigation system can produce a system performance superior to either one acting alone. However, because of the costs of such benefits, it is fitting to inquire about trade-offs that would justify the investment. Trade studies typically address the following questions:

1) What benefits of GPS/inertial integration are important in the application being considered?

2) What configuration of data paths (integration architecture) is appropriate for the application?

3) How complex are the integration algorithms required to provide the desired level of performance, with options for growth to meet future requirements?

This chapter devotes one section to address each one of these questions. Because of space limitations, the presentation is qualitative, with only limited recourse to the underlying mathematical structures required to understand integration filtering and the performance evaluation of an integrated navigation system. Wherever possible, the reader is directed to other chapters in this text for those details, or to the literature. Furthermore, the properties of GPS user equipment (UE) and inertial navigation systems cited here are generic rather than specific, and they are representative of technology circa 1993.

I. Benefits of GPS/Inertial Integration

The design of any complex navigation system for civilian or military markets reflects the designer's judgment of the best trade-off among the following factors:
1) Cost
   a) Development (nonrecurring)
   b) Life-cycle (recurring)
2) Installation constraints
   a) Volume, weight, power consumption

Copyright © 1994 by the American Institute of Aeronautics and Astronautics, Inc. All rights reserved.
*Director, Electrical Design and Sensor Development Directorate.

b) Interfaces
3) Performance
   a) Mission requirements/mission environment
   b) Reliability/graceful degradation
   c) Options for improvement

The following remarks emphasize the performance considerations because that is the area where the benefits of GPS/Inertial integration are most evident. However, cost and installation factors are often decisive. These are raised throughout the chapter wherever they are a significant differentiator between alternative integration techniques. Ultimately, the system designer must justify his or her design as being the best way to satisfy the design problem. It is of critical importance that the authorities who are managing the design team surface all requirements and constraints, both present and anticipated, so that informed and timely choices can be made among the alternatives.

The GPS system can provide a suitably equipped user with a position, velocity, and time (PVT) solution whose errors are generally smaller than those of any alternative navigation system. This performance is achieved in all weather, at any time of the day, and under specified conditions of radio-frequency interference, signal availability, and vehicle dynamics. Why then would we undertake the cost and complexity of integrating GPS UE with any other navigation sensor, and, in particular, with an inertial navigator?

The goal of integration is to provide more robust, and possibly more accurate, navigation service than is possible with stand-alone sensors. In particular, integration may be the only way to achieve the following[1-6]:

1) Maintain a specified level of navigation performance during outages of GPS satellite reception.

2) Provide a complete six-degree-of-freedom navigation solution (translational and rotational motion) at a higher output rate than is conventionally available from GPS alone.

3) Reduce the random component of errors in the GPS navigation solution.

4) Maintain the availability of a GPS solution in the presence of severe vehicle dynamics and interference.

Most civilian and non-Department of Defense Government GPS users have access to the Standard Positioning Service (SPS) only, which is subject to intentional degradation of accuracy (but not precision) of pseudorange and delta-range measurements. GPS users who are not authorized to use the GPS Precise Positioning Service (PPS) will not be able to benefit fully from performance benefits attributable to GPS/INS integration. Moreover, the two following constraints exist. First, the bandwidth of GPS code and carrier tracking loops cannot be reduced below the minimum required to track dithered GPS signals. This prevents the most aggressive use of INS aiding to reduce the dynamics tracked by the loops so that loop bandwidths can be reduced for purposes of increased radio-frequency interference rejection. Second, the complexity of optimum integration filters used to calibrate INS errors should increase to account for the artificial correlations among successive GPS measurements. Calibration will require longer observation intervals and will not converge as tightly compared to operations without the selective availability (SA) degradation.

The unauthorized user *can* still use GPS to generate position resets that keep the INS position errors below the SA limit, which is expected to be less that 100 m as measured by the $2d_{rms}$ criterion. However, in the absence of INS sensor *calibrations*, the INS error growth during a GPS outage will be faster than that for a calibrated INS. Assigning numbers to these qualitative comments requires either access to classified information or to measured values of SA waveforms obtained at well-surveyed fixed observation sites. As this information becomes available to civilian users, we should expect to see numerous contributions to the technical literature on this subject.

A. Operation During Outages

A stand-alone GPS receiver typically incorporates current measurements to four or more satellites to update its most recent PVT solution. Dead reckoning that incorporates recent estimates of vehicle acceleration may be used to propagate the current PVT solution in-between measurement updates. A GPS outage occurs when fewer than four valid satellite measurements are available at each update. During a partial or complete outage, the software for a stand-alone receiver can continue to produce a navigation output if it mechanizes one of the following options, albeit with reduced accuracy:

1) Compute the least-squares solution with fewer measurements than there are unknowns.[7]

2) Constrain one or more navigation outputs to be fixed, such as the UE clock bias or the vehicle altitude, or constrain the navigation solution to lie along a great-circle path.[8,9]

3) Incorporate measurements from an external sensor. A barometric indication of altitude is commonly available in military UE, as are radar- or pilot-inserted position updates.[9,10]

During an outage, the navigation solution becomes less accurate the longer the outage and the greater the vehicle dynamics since the last full set of measurements. The key factor to be specified when deciding whether an auxiliary sensor is required is the maximum acceptable error growth during the outage. In a conservative design, maximum error growth is calculated under worst-case conditions of vehicle dynamics.

Outages may be a concern even for UE that track more than four satellites at a time. For example, a GPS antenna mounted on top of an aircraft will only see a limited number of satellites during a banked turn, and the dilution of precision (DOP) parameters for that visible constellation may be unacceptably high. In more extreme cases, a vehicle passing through a tunnel may see no satellites for an extended period, and a military UE may be jammed as it approaches its target.

Combining GPS with an independent navigation sensor (item 3 in the preceding list) is one means to maintain the quality of the navigation service during a GPS outage. In effect, the independent sensor can act as a flywheel to provide continuous, high-quality navigation outputs. Inertial navigators are commonly considered for this role because they are passive, self-contained, and widely available. Moreover, they are not subject to the causes of GPS outage. However, they are generally more expensive to buy and integrate than other radionavigation sensors such as Loran or Omega. Their use has generally been limited to military and commercial aircraft. However, low-cost, low-performance inertial sensors implemented using mass-production microelectronics technology are emerging from research laboratories. These may provide the technological basis for an economical solution to GPS outages in general aviation and commercial applications, such as trucks and automobiles.[11,12]

With respect to GPS/INS integration performance during outages, the key questions are the following:

1) What quality INS is required?

2) How complex is the integration required to exploit the inherent INS quality to achieve mission objectives?

The resulting performance must then be weighed against the cost to determine whether to implement the optimum integration or to accept a less expensive, lower-performance solution.

B. Providing All Required Navigation Outputs

GPS UEs routinely estimate only the translational motion of a point referenced to the GPS antenna. Interferometric processing of GPS signals received at multiple antennas can also provide rotational (attitude) information.[13-15] However, we assert that an inertial solution is preferable to interferometry for terrestrial users whenever it is available. This preference is based on three considerations:

1) The inertial system is self-contained and is not vulnerable to outages (except those caused by equipment failure).

2) Installation of an inertial system on an aircraft is less demanding than an interferometer, and it is probably less demanding on ships and vehicles, also.

3) The noise floor on the accuracy of a short-baseline (1–5 m) interferometer has not yet been achieved. It seems that multipath is the culprit, and that it is premature to expect that antimultipath techniques will be effective and practical.[16]

In addition to attitude indication, the inertial navigator is desirable because its accelerometers typically sense velocity changes at up to a 1.0-KHz rate, with a 200-Hz output rate being commonly available. Therefore, the INS routinely outputs navigation solutions one to two orders of magnitude more often than a GPS UE. This high output rate allows the INS to provide accurate inputs to vehicle control subsystems, platform-stabilization systems, pilot displays, and velocity-aiding inputs to GPS tracking loops.

It follows that an integration in which GPS is used to bound the error growth of an INS-based system navigation solution would be very effective whenever GPS was available, and the availability of a calibrated INS may be the only means to maintain nearly as good performance during an outage.

C. Reduced Noise in GPS Navigation Solutions

In a stand-alone GPS receiver, the navigation processor usually implements a linear filtering algorithm in which the previous navigation solution is propagated to the current measurement epoch. Because GPS does not directly sense acceleration, the propagated solution is sensitive to errors in the previous acceleration estimate or to jerk that changes the true acceleration during the propagation interval. In contrast, an inertial system measures position change very precisely in the interval between GPS updates. This property can be exploited by a well-tuned Kalman filter, using GPS measurements to estimate *errors* in the INS output. Because these errors change slowly, the filter can smooth its update over many GPS measurements, thereby reducing the effect of additive noise on any one update. The result is that the "integrated navigation" solution seems to be much "smoother" than the stand-alone GPS solution. (See Sec. III.A, and especially compare Figs. 3 and 6 for an illustration of this feature.)

D. Increased Tolerance to Dynamics and Interference

The INS velocity solution may be fed back to the GPS UE to reduce the apparent dynamics of the input to the GPS code and carrier loops. This has two effects[1,4,6]:

1) A fixed bandwidth-aided tracking loop can maintain lock on GPS signals in the presence of dynamics that would cause the unaided receiver to break lock.

2) The tracking loop bandwidths can be reduced to the minimum amount required to track the *errors* in the INS aiding signals. (As noted previously, this feature breaks down for unauthorized users in the presence of selective availability clock dither.) (INS position errors are mostly low frequency.)

The net result of these actions is that the INS-aided GPS receiver can maintain lock and provide GPS measurements over a much wider range of vehicle dynamics and radio frequency interference than the unaided, stand-alone receiver.

II. GPS Integration Architectures and Algorithms

The degree of complexity of the integration should reflect the mission requirements; it may also be limited by the investment that can be made to obtain those objectives. Integration strategies and mechanisms may be very simple (for example: choose the GPS UE position and velocity as the integrated solution when GPS is available with a given precision, otherwise choose INS position and velocity as the integrated solution) or relatively complex (for example: optimally combine GPS UE measurements with INS outputs, Doppler radar outputs, baroaltimeter signals, true airspeed, and other sensor data). However, in the following remarks, we limit our attention to alternatives involving only GPS integrated with an inertial system.

A. Integration Architectures

Figure 1 illustrates three generic functional architectures for GPS INS integration. The GPS receiver and the INS are treated as navigation *systems* in architectures a and b, with GPS supplying a position, velocity, and time solution, and the INS supplying a position, velocity, and attitude (P,V,θ) solution, respectively. In architecture c, the GPS and INS are treated as sensors producing line-of-sight measurements $(\rho,\dot{\rho})$ and accelerations and angular rates $(\Delta V, \Delta \theta)$, respectively. In addition to the GPS and INS units, each architecture includes various data paths and a processor unit that mechanizes the integration algorithm. These alternatives are distinguished by the data passed between the subsystem components. The proper interfacing and control of these components may incur the largest part of the cost of an integration project, but those concerns are not within the scope of architectural considerations.

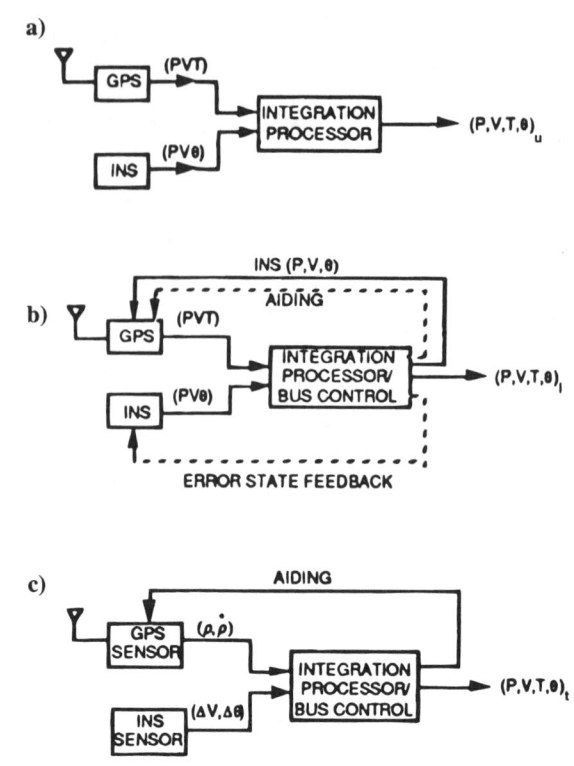

Fig. 1 Generic global positioning system/inertial navigation systems architectures: a) uncoupled mode; b) loosely coupled mode; c) tightly coupled mode.

1. Uncoupled Mode

Figure 1a illustrates the configuration in which GPS UE and an INS produce independent navigation solutions with no influence of one on the other. The integrated navigation solution is mechanized by an external integration processor that may be as simple as a selector or as complex as a multimode Kalman filter. All data busses are "simplex" (unidirectional). The characterization of Fig. 1a as an "uncoupled" mode is based on the independence of the GPS and INS navigation functions. Note that, in principle, the *hardware* could all be packaged in one physically integrated (embedded) unit; however, the functionality would still be that of uncoupled architecture.

The potential benefits of integrating the navigation solutions from uncoupled GPS and inertial navigators are:

1) It is the easiest, fastest, and potentially the cheapest approach when an INS and GPS are both available.

2) It provides some tolerance to failures of subsystem components (except in the embedded configuration, see Sec. II.C).

3) Using an integration processor as simple as a selection algorithm can provide en route navigation at least as accurate as available from an INS.

2. Loosely Coupled Mode

Figure 1b illustrates a configuration in which there are several data paths between the integration processor and the GPS and the INS equipment. Among these, the provision of the system navigation solution to the GPS UE is the most important for getting the maximum benefit from the integration filter. The inertial aiding of GPS tracking loops is of next greatest benefit, and feedback of error states to the INS is of second-order benefit. There may also be some improvement of system reliability to the extent that individual components are likely to be more mature and to have been better tested than emerging technology that features more highly integrated subsystems.

a. Reference navigation solution. GPS UE generally employs a Kalman filter mechanization to compute PVT updates based on current tracking loop measurements. A GPS UE does not directly sense acceleration; it must use relatively noisy acceleration estimates based on recent velocity measurements for a dead reckoning propagation of the previous navigation solution forward to the epoch of the current tracking-loop outputs. The situation changes dramatically when the system navigation solution is fed back to perform that propagation. In effect, the GPS measurements can now be used (within the UE navigation filter) to correct the system navigation solution. Over short periods of time, that solution is very accurate because it incorporates INS data based on acceleration sensing. The UE filter is then mechanized to estimate INS (or system) error states having relatively low dynamics and low uncertainty (process noise). The filter can be tuned to have a longer time constant (filter memory), thereby increasing the effective averaging of each noisy GPS measurement.

b. Inertial aiding of GPS tracking loops. As mentioned in Sec. I, the availability of a GPS navigation solution can be increased significantly when inertial aiding is used to reduce the vehicle dynamics tracked by the UE code and carrier loops. In principle, this aiding could be applied directly from the INS to the GPS UE, but it is shown as an output of the integration processor in Fig. 1b because of the following:

1) GPS tracking loops must be aided by the projection of vehicle velocity along the line-of-sight (LOS) to each satellite being tracked. The conversion from inertial coordinates to GPS LOS coordinates is most appropriately done in the integration processor or in the GPS UE itself. In either case, INS velocity information is available within the processor hence aiding can be part of the data flow to the UE. This avoids the expense and risk of developing a custom interface from the INS to the GPS UE.

2) Executing the coordinate transformation external to the INS retains flexibility in the selection of INS equipment and avoids the need to develop custom GPS/INS interfaces for each application. However, this raises a concern for "data latency" (i.e. feeding delayed data to the tracking loops) as mentioned in Sec. II.C.

c. Error-state feedback to the inertial navigation system. Most inertial navigation systems have the means to accept external inputs to reset their position and velocity solutions and to adjust the alignment of their stable platform. The adjustment may be executed by a mathematical correction in a "strap-down" inertial system, or it may be realized by torquing a gimballed platform. In either case, the use of feedback can maintain inertial navigation errors at a level for which their dynamics are accurately modeled by the error state propagation equations embodied in the integration filter. However, the impact of this feedback on error growth for a navigation grade INS is relatively small until the errors grow much larger than 10 km and 1 m/s, respectively.

3. Tightly Coupled Mode

Figure 1c illustrates the so-called tightly coupled integration mode. It differs from the loosely coupled mode in that both the GPS receiver and the inertial components are limited to their sensor functions. They are treated as sources of GPS code and carrier measurements and inertial indications of acceleration (velocity change) and angular rate, respectively. These sensor outputs are then combined in one navigation processor that may mechanize an appropriately high-order integration filter.[17-19]

In the tightly coupled mode, there is only one feedback from the navigation processor. Figure 1c illustrates the use of velocity aiding to the GPS tracking loops. Acceleration aiding could also be effectively used, but we are not aware of any particular mechanization using other than velocity aiding. The other paths used in loosely coupled architectures are not needed here because all computations involved in navigation processing are now internal to one processor.

The concept of tightly coupled integration is often raised in connection with embedded GPS receivers. These are not necessarily synonymous. However, it is reasonable that we would choose to mechanize a tightly coupled integration algorithm if we had already taken the effort to design a GPS receiver that is physically and electrically integrated with an inertial sensor or with a powerful navigation processor. We return to this point in Sec. II.C.

B. Integration Algorithms

The basic choices for GPS integration algorithms are 1) selection, with or without INS resets; 2) fixed-gain filter; and 3) time-varying filter. These are listed in order of increasing complexity and optimality. Each one can be used with any one of the architectures listed in Fig. 1, but the incremental payoff of a more complex filter is directly related to the quality of the input information.

1. Selection

A selection algorithm chooses the GPS indicated (PVT) as the system navigation solution whenever the GPS UE indicates that this solution is within acceptable bounds on its accuracy [via the GPS figure-of-merit (FOM)]. Inertial navigation systems data may be used to interpolate between successive GPS updates when a higher output rate is needed than can be provided by the UE. During GPS outages, the INS solution extrapolates from the last valid GPS solution. (The process of forcing the INS solution to equal the current GPS indicated velocity and/or position is known as a "reset" if that correction is actually fed back to the INS.)

2. Filtering

The general filtering problem involves trying to estimate time-varying states whose evolution is characterized by known laws of propagation, which usually are taken to be a coupled system of linear differential equations driven by white noise.

States usually cannot be measured directly, but they are inferred from measurable quantities to which they are related. These measurements may be made simultaneously or sequentially at a series of distinct points in time. The filter will generally incorporate knowledge of the statistics of the measurements.

Knowledge of the way the states change (propagate) in time, knowledge of the way the measurements are related to the states, measurement statistics, and measurement data are all used in each state update. The most common update algorithms use linear filters; e.g., ones in which the updated state is a linearly weighted sum of the measurements and the previous state value.

Position and velocity of an aircraft are examples of quantities that may be chosen as *states* in a filter (these are referred to as whole-value filter states). For whole-value position and velocity states, the propagation equations are simply the equations of motion of the aircraft. To make the whole-value filter propagation equations a better reflection of the real world, acceleration states could be added (otherwise, by its omission, acceleration must be treated as "noise," driving the derivative of velocity). GPS-indicated position and velocity are examples of *measurements* that might be processed by an integration filter with whole-value states. At one extreme, the integration filter could ignore everything except the GPS receiver position and use this as the integrated position. This degenerate case is the selection mode cited above in which the state propagation equations and any other available measurements would be ignored. For the degenerate case, the weight on the GPS UE position is one and the weight on the propagated state is zero. The weight on the measurement is referred to as the filter *gain*. In general, some rule must be used in order to determine how much weight should be put on a measurement and how much weight should be put on the propagated states.

Another choice of states are the *errors* in position and velocity indicated by the INS (these are referred to as error states). For a filter whose states are INS errors, accurate representations and linear approximations of the propagation equations are well known. As in the case of whole-value states, additional INS error states (for example, states for azimuth and tilt errors, accelerometer bias, and gyro drift) could be added to the filter in order to make the propagation equations a better model of the real world. Of course, the degree to which the filter must reflect the real world is a function of the estimation accuracy required, and that is a reflection of the mission requirements.

For a GPS/INS integration filter with INS error states, the measurements would actually be the differences between GPS position and INS position and the differences between GPS velocity and INS velocity. As with the case of whole-value states, some rule must be used in order to determine how much gain should be put on the measurements and how much weight should be put on the propagated states when computing state updates.

We should mention in passing that the optimum filter may require an impractically large number of states. Options to decompose a high-dimensional estimator into combinations of lower-dimensional filters have been described in the literature. Distributed filtering and federated filtering are the terms under which these options are usually cited, as in Refs. 20 and 21. Both are believed to be more robust than the optimum filter when the design must be tolerant to imperfect information about the estimation problem, and it is claimed that federated filters are more fault-tolerant. These details are beyond the scope of this survey.

a. Fixed-gain filters. In a fixed-gain filter, the propagated estimates are combined with new measurement data using predetermined gains. The gains are fixed in the sense that they have been loaded into computer memory a priori, so that the filter selects from a short list of gains, rather than computing them. Different gains may be used for different sensor status and operational status, reflecting the uncertainties in the propagated solution and in the measurements. In general, the gains in a fixed-gain filter can have any value (they should at least properly reflect the relationships among the measurements and the states).

If the state dynamics and their uncertainty are limited, and there is negligible variation of measurement noise during the interval of interest, it may be that the optimum filter gains will not vary much during the mission. In that case, the performance penalty of mechanizing one fixed set of gains (or a few selectable sets of gains) compared to optimum time-varying gains may be acceptably small. The benefit to the integrator is a vast decrease in computational burden and memory required to implement the filter. It may even be effective to precalculate an approximation to time-varying Kalman gains that can be stored for use during a mission.

b. Time-varying gain (Kalman filter). In the Kalman filter, new gains are computed every time measurements are available. The Kalman filter is a recursive implementation of the optimum least-squares error estimation algorithm. It is optimum in the sense that it strikes the correct balance between uncertainly in the presumed dynamics of the states being estimated (process noise), uncertainly in the measurements (measurement noise), and the observability of individual states (sensitivity) required to minimize the figure-of-merit. See Refs. 22–24 for a detailed discussion of Kalman filtering. In the present context, we note that the updating of N states by M measurements involves substantial matrix manipulations, propagation of difference equations, and memory to store the matrices. Current technology can handle updates of around 20 states at up to a few times per second in a reasonably cost-effective processor. Because upwards of 100 error sources may influence an integrated GPS INS solution, the brute force approach to real-time integration is not yet computationally feasible. Each designer must complete detailed design studies to determine the minimum number of states and the update rate that will result in an acceptable navigation error using the available processor resources and with acceptable design margin. Given the rapidly changing computational capabilities available to avionics integrators, questions of computational feasibility should be reconsidered every few years.

3. Discussion

The uncoupled mode is inferior (in performance) to the loosely coupled integration mode. The uncoupled and loosely coupled integration modes are inferior (in performance) to a tightly coupled mode because information inherent in the sensor measurements is lost in the receivers mechanization of the PVT solution; i.e., it is not always possible to backtrack from a PVT solution to the raw GPS measurements with sufficient bandwidth and precision to support a tightly coupled integration. The feedback of the system navigation solution to the INS (via resets) is a second-order improvement in the loosely coupled mode (and is inherent in the tightly coupled mode).

The current generation of military high-dynamics GPS UE (receiver 3A, MAGR) mechanizes an integration filter when it is operated in its "INS mode." This internal filter is tightly-coupled in the sense of Fig. 1c. However, its performance is suboptimal because it incorporates a very simplified model for the dynamics of its inertial error states, and because the filter is tuned very conservatively. Integrators who need better performance usually resort to a higher-order external integration filter that combines the RCVR-3A navigation output with the inertial navigation solution. This cascaded integration (e.g., filter-driving-filter) is often criticized by proponents of tightly-coupled integration. This criticism is valid, but it really addresses a cost tradeoff in which the integration filter in RCVR 3A was limited to 12 states, of which 9 represent very generic INS errors (P,V,θ). The decision for RCVR 3A was based on unit cost and the desire to produce a generic standard equipment that did not burden any user with features not justifiable in his or her application. In principle, a modified RCVR 3A operating in the INS mode with an expanded internal filter and appropriate software could perform as well as an externally mechanized tightly coupled integration.

In design studies of tightly coupled GPS/INS integration filters, as many as 80 inertial error states are modeled, in addition to GPS error states related to delay measurement bias, tracking loop errors, propagation errors, and user clock errors. In some ultraprecise systems, it may even be useful to incorporate additional states that model multipath effects. Nevertheless, many studies have shown that most of the benefit of expanded error state formulations is gained with 25–30 states, and that adequate performance can usually be obtained from 14–17 states.[25]

The tightly coupled mechanization does avoid one problem commonly attributed to loosely coupled integration, namely the possibility of instability (in state estimates) arising when the GPS navigation errors become highly correlated with INS navigation errors. This situation may occur at low input signal-to-noise ratios when GPS code loops remain in lock only because inertial aiding allows the loop bandwidth to be reduced, thereby reducing the effective levels of noise and interference. Now, the narrower the loop bandwidth, the more the loop error approximates the error of the aiding signals so that the correlation cited above becomes significant. See Ref. 26 for further discussion of this point.

C. Embedded Systems

As GPS approaches its operational status, there has been a massive increase in investment in civil GPS technology, which has led to smaller, lower power-consuming, higher-performance UE than were dreamed of as recently as the late 1980s. One consequence of this trend is that GPS UE can be packaged on a single card that can be embedded in other systems. As noted in Sec. II.B, the concept of GPS embedded in an INS is one such application that is being prominently discussed at present, with several efforts underway to demonstrate the concept.[27-30]

Setting aside the valid claims of savings in size, power, and weight that accrue from embedding, it is reasonable to ask whether there is any functional or performance payoff directly attributable to embedding. The answer is a qualified yes. There are potential performance improvements, but the system may be vulnerable to a single-point failure, such as a power supply or the processor.

1. Tight Coupling

There is no inherent reason to claim that embedding implies tightly coupled integration. An embedded receiver could be stand-alone, loosely coupled, or tightly coupled. However, developers of embedded systems have tended to mechanize tight coupling as a performance feature.

2. Carrier Loop Aiding

Standard military UE use inertial aiding only for code loops and only after carrier loops have lost lock. The decision to limit the INS aiding goes back to the late 1970s when it was argued that the latency (time delay) between the sensing of inertial velocity and its receipt at a GPS receiver could be as large as tens of milliseconds, even with the high-speed data busses that were available. With this much delay, it was argued that errors in the aiding signal during accelerations or turns could be large enough to drive the carrier loop out of lock.

There are at least two ways to mitigate this concern in an embedded system. The most common approach is to customize the data link between the INS and the GPS carrier loop to reduce the latency to a few tens of microseconds and to minimize the uncertainty in the latency. With that small a delay, the maximum error of the aiding signal is negligible, even for an aircraft rolling as fast as 1 rps, and moving toward a satellite with a relative velocity of 2000 ft/s. Under those conditions, the error caused by a 20-μs delay in attitude indication would be approximately V_ϵ, where

$$V_\epsilon < 2\pi (20 \times 10^{-6}) \text{ rad} \times 2000 \text{ ft/s} < 0.25 \text{ ft/s}$$

which is well within the acceptable range for GPS receivers. An alternative that has not yet been mechanized is to delay the GPS signals by an amount that matches the latency, before the aiding signal is applied. For modern precorrelation digital GPS receivers, we could store tens of milliseconds of GPS samples in a data buffer mechanized by a single memory chip.

3. Tracking Fewer Than Four GPS Satellites

The loose-coupling approach (Fig. 1b) integrates a GPS PVT solution with an inertial (P,V,θ) solution. When the GPS solution is unavailable, the integrated solution "flywheels" using the inertial solution as corrected at the start of the

GPS outage. In contrast, the tightly coupled solution uses raw GPS measurements, which are available as long as one or more satellites is being tracked. Thus, it is a more robust solution vs outages that could prevent a GPS navigation solution from being formed in the loosely coupled configuration. Reference 25 gives a good insight into the potential performance improvement (see also Sec. III).

However, there is one caveat. Conventional GPS UE, such as GPS RCVR 3A, can continue to provide a navigation solution (albeit degraded) with only two or three satellites. Thus, it is inappropriate to claim that a loosely coupled integration *must* convert to a free-running inertial solution in the presence of one or more satellite outages. The performance will depend on the details of the GPS UE mechanization (see Sec. III).

4. Quantization

All calculations within an embedded system are more likely to be executed as "full-precision" quantities than in a system wherein the GPS and INS and navigation processor are connected by data busses. These busses are usually so heavily used that data must be coarsely quantized for data transmission (compared to their internal precision) in order to satisfy communication bandwidth constraints and to conform to data transmission protocols. An alternative that has not been explored in GPS navigation data communications is to use data compression to increase the information content of the message structure. This would require reworking interfaces and message protocols, but the effort might be cost effective in high-precision applications.

III. Integration Case Studies

We consider three case studies that illustrate performance benefits of GPS/INS integration. The properties to be addressed include 1) noise quieting (reduced variance) with an INS error state filter; 2) in-flight INS alignment; and 3) reduced error growth of a "calibrated" INS during a GPS outage.

A. GPS/Inertial Navigation Systems Navigation Performance in a Low-Dynamics Aircraft

In 1985, a five-channel GPS UE (RCVR-3A), operating as a stand-alone system, was flown in a DeHaviland Twin-Otter aircraft that was also equipped with a laser-inertial integrated navigation system. The combination of this very stable inertial system with precise angle-angle-range pulsed optical measurements to surveyed [by the U.S. Geodetic Survey (USGS)] retroreflectors removed the long-term increase of inertial position errors attributable to drift, misalignment, and gravitational anomalies, and tied the reference solution to local geodetic coordinates.[31]

With postprocessing, the laser-inertial reference system located the aircraft to within 50 cm (position) and 5 mm/s (velocity) at any time during the flight, and to within 1.0 cm (position) and 0.3 mm/s during lock-on to a retroreflector. Therefore, it was at least one order of magnitude more accurate than GPS UE specifications and was, therefore, uniquely suited to score the inflight performance of the GPS navigation system. It was the most accurate reference system ever used in GPS flight testing.

The GPS UE was operated in the stand-alone PVA mode; however, each pseudorange and delta-range measurement was recorded in addition to the navigated UE solution and supporting data. Therefore, the performance of the receiver with different navigation processing algorithms could be evaluated by postflight emulations, and then compared to the stand-alone output and to the reference solution.

During the flight tests, the aircraft flew six circuits around five retroreflectors placed in suburban Boston (Fig. 2). The reference solution (altitude, latitude, longitude) is shown in Fig. 3.

1. GPS Point Positioning

The GPS pseudorange measurements were processed to form a point-positioning solution. In other words, the navigation equations were solved for each set of GPS measurements as a single set of four equations with four unknowns. In this case, there is no navigation filter; nor is any "memory" of previous navigation solutions used to smooth the results. If the noises on each satellite measurement were the same, then the position errors would be proportional to the position dilution of precision (PDOP) for the collection of satellites being tracked.

Figure 4 illustrates the resulting navigation errors, and Fig. 5 illustrates the DOP values that were current during the data collection. (In regions where Fig. 4 is multivalued, the smaller value is the minimum DOP, and the larger value corresponds to the DOP for the satellites actually selected.) The discontinuities

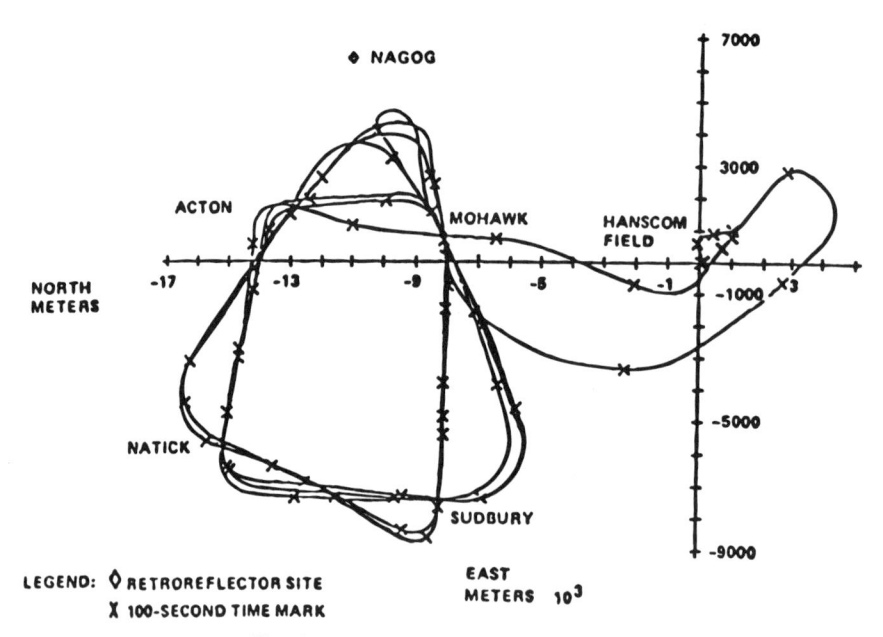

Fig. 2 Flight path on May 30, 1985.

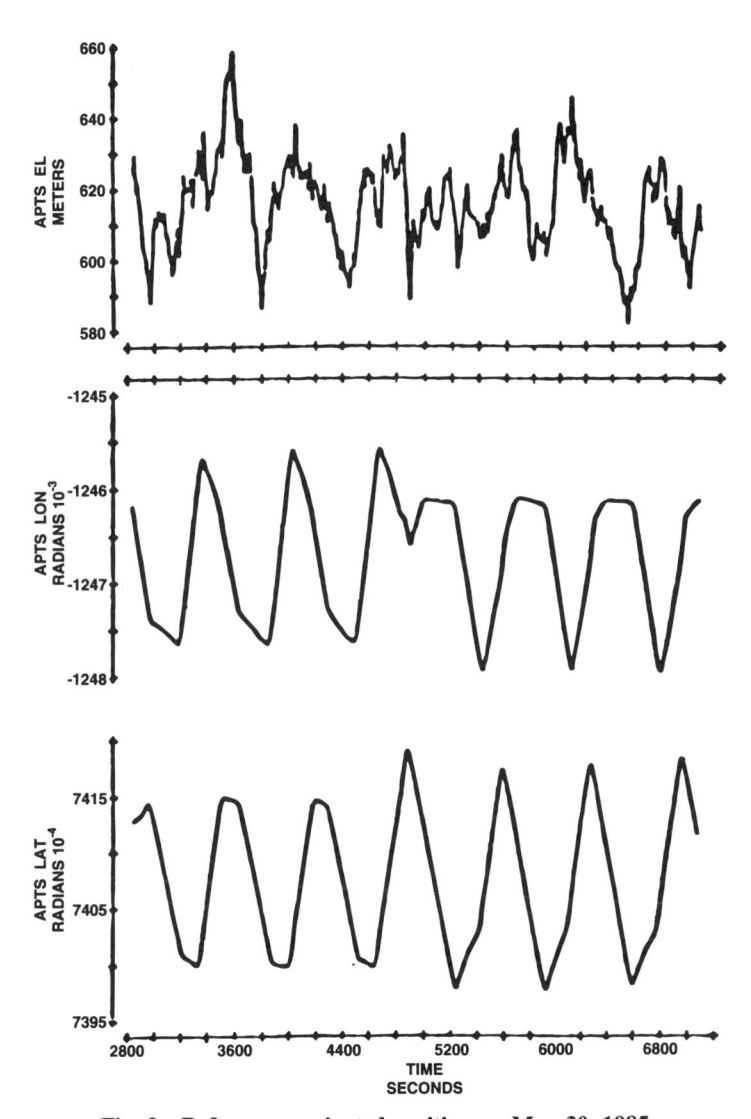

Fig. 3 Reference navigated position on May 30, 1985.

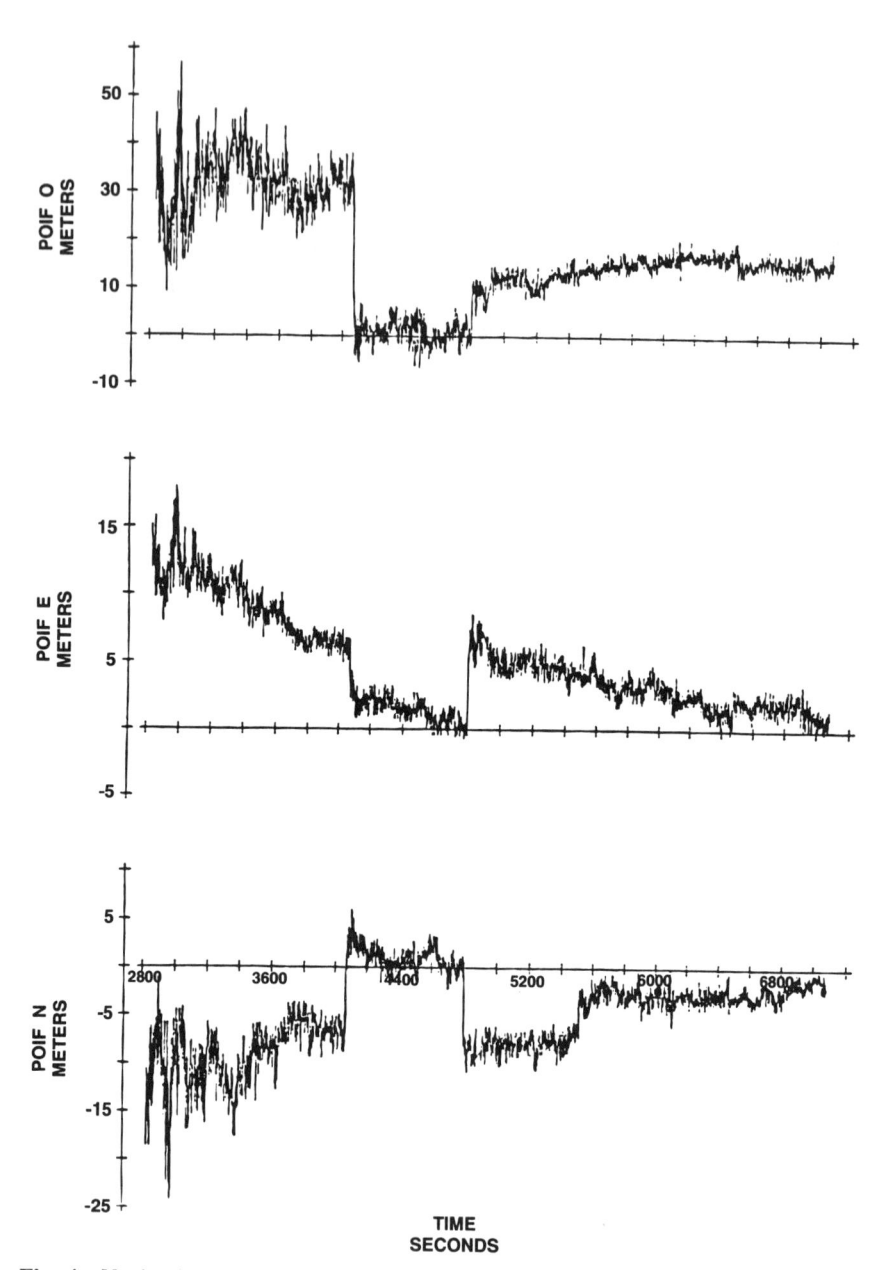

Fig. 4 Navigation errors (position) for point positioning with unaided five-channel receiver.

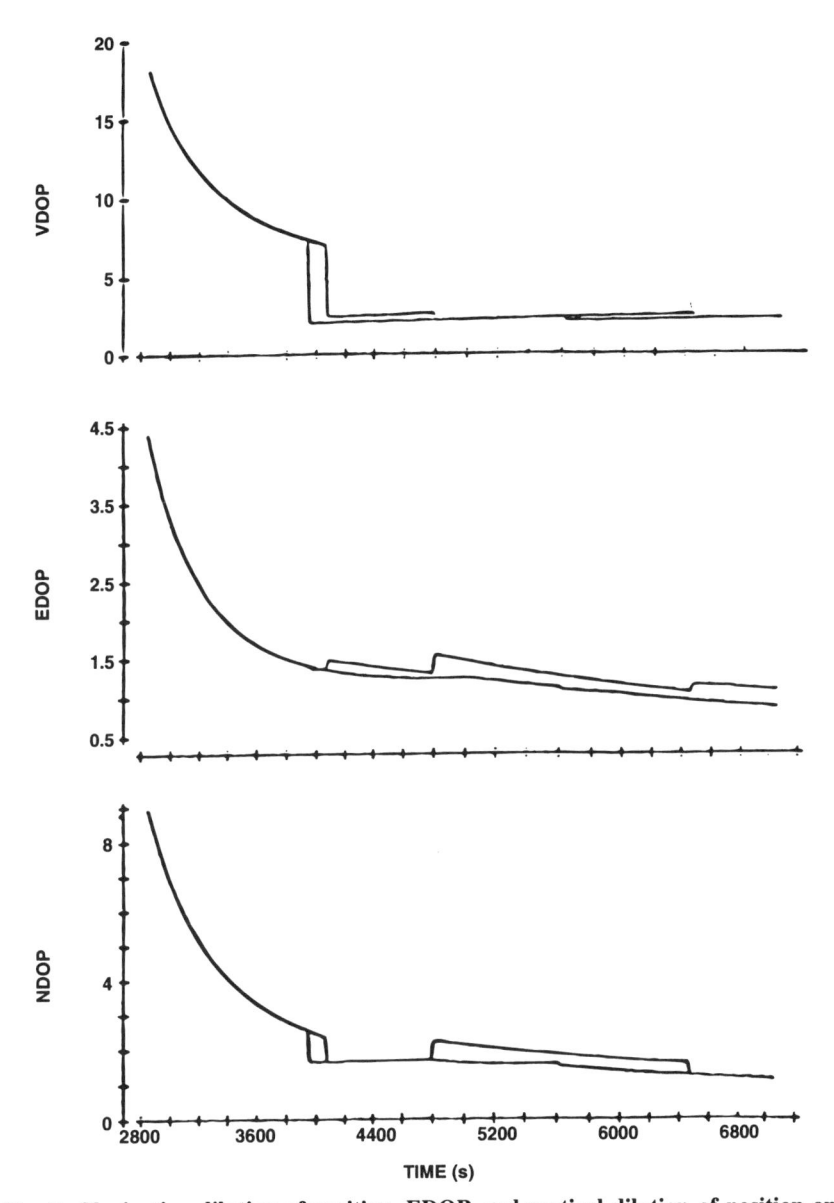

Fig. 5 Navigation dilution of position, EDOP, and vertical dilution of position on May 30, 1985.

in DOP signify satellite switches. The hypothesis that the point-positioning errors are proportional to DOP is supported by a comparison of these two figures.

2. GPS Internal Navigation Filter

Figure 6 illustrates the errors in the position solution output by the receiver, using its internal 12-state PVA-mode navigation filter. The errors are the difference between the RCVR 3A solution and the laser-inertial reference solution. The RCVR 3A navigation filter implements 12 error states including 9 for three-dimensional position, velocity, and acceleration (hence, the name PVA mode), and 2 for the internal receiver clock drift (seconds), and clock drift rate (frequency) offset from their nominal values. The 12th state calibrates bias in barometric altitude when that measurement is available. The baro output was not used in these flight tests, so only 11 states were updated by the filter.

Comparison of Figs. 4 and 6 show some of the features of the filtered solution compared to the point solution:

1) The influence of satellite constellation changes (e.g., at about 4100 s and 4800 s) is substantially reduced.

2) The noise standard deviation (estimated as one-half of the peak to peak variation of short-term errors) after "good" geometry was established (about $t = 4100$ s) is reduced by filtering from 2.5 m to about 1.0 m per coordinate.

3) Errors in the filtered solution show smaller values of bias, and they are more nearly a zero-mean process. We speculate that the reduction in bias errors is a consequence of the particular "tuning" of the receiver filter, which causes the biases to show up more in the clock error states than in the position error states.

Further analysis of the field test data showed large error spikes in the GPS velocity solution at each turn. These errors were consistent with an apparent lag in the GPS output relative to the reference solution. Indeed, it turns out that the RCVR 3A PVA filter output epoch is the *end* of the delta-range accumulation interval. This is 0.39 s later than the *middle* of that period, which is the effective epoch of the delta-range observable. This underscores the need for an integrator to be aware of all potential sources of time bias between GPS, inertial, and other sensors that are being integrated.

3. GPS Navigation in the Inertial Navigation Systems Mode

Inasmuch as the pseudorange and delta-range measurements from the receiver were available, as well as all the inertial outputs, it was possible to emulate the navigation solution that *would* have been produced by RCVR 3A operating in the INS mode. With respect to that particular receiver, we note that this procedure is valid because:

1) The receiver tracked in "State 5" throughout the flight tests. Therefore INS "aiding" would not have been applied to the receiver tracking loops, even if the receiver had been operated in the INS mode during the flight tests.

2) The bandwidth of the tracking loops is the same as would have been in effect in the INS mode. Therefore, the measurement noises input to the navigation filter emulation have the same standard deviation as the measurement noises for operation in the INS mode.

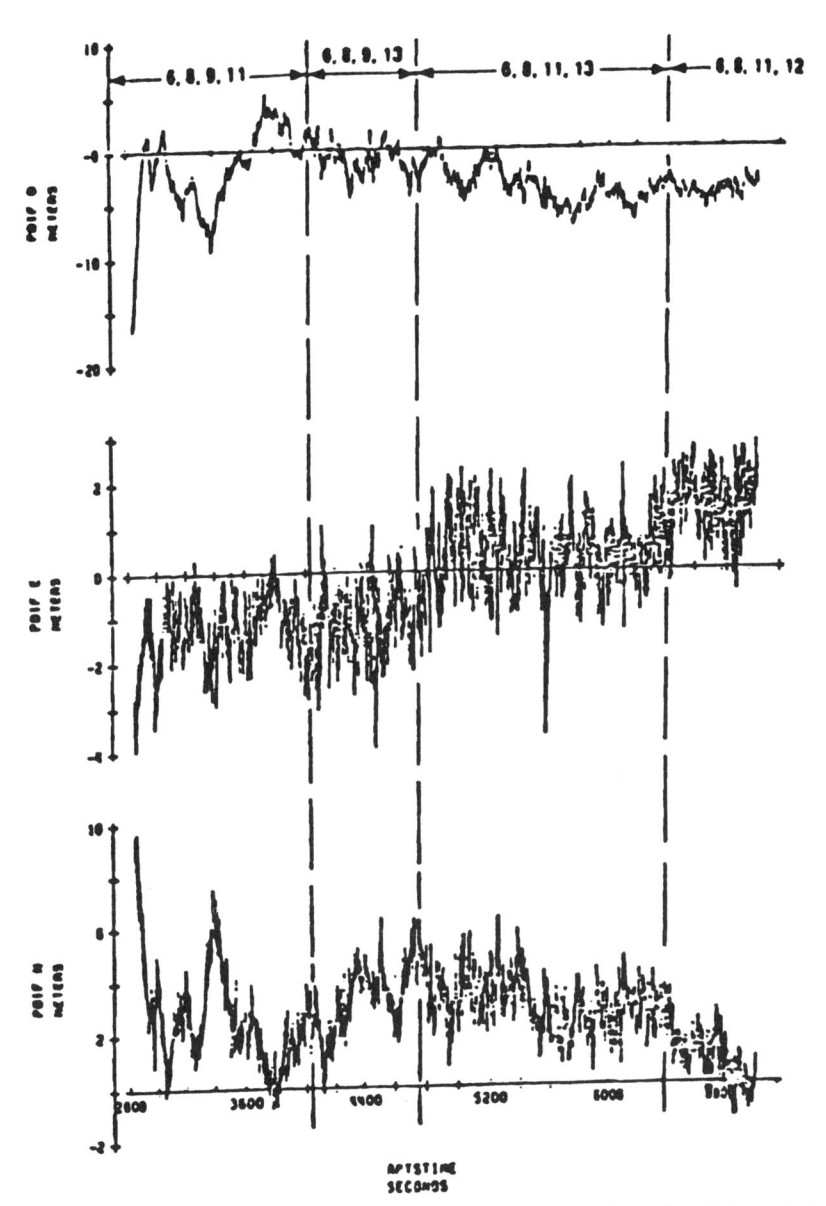

Fig. 6 Difference between global positioning system and laser-inertial navigated position on May 30, 1985.

Figure 7 illustrates the navigation solution output by a 17-state Kalman filter with measurements available once per second. The seventeen states are: position error (three states); velocity error (three states); UE clock error (two states); inertial platform misalignment (three states); gyro drift rate (three states); and accelerometer bias (three states). The 17-state filter adds 6 error states for gyro drift rate and accelerometer bias that are not available in the RCVR 3A navigation filter. Thus, the results are more suggestive of performance to be obtained with certain tightly coupled GPS/INS integrations than of the RCVR 3A performance.

The most striking feature of Fig. 7 is the suppression of additive noise. The navigation errors track satellite DOP variations with greatly reduced noise compared to the point solution shown in Fig. 4. The position errors are clearly dominated by slowly varying biases in the pseudorange measurements to each satellite. A secondary feature is the treatment of errors in the period from $t = 2800$ s to $t = 4100$ s when the satellite geometry is poor. The 17-state filter is clearly "tuned" differently than the 11-state filter in RCVR 3A. (Filter tuning is the process of adjusting filter parameters that model measurement noise and process noise in the computation of the filter weights at each update.) This tuning is probably closer to optimum because the pseudorange and delta-range residuals computed using those clock error states, were smaller than residuals computed using the receiver's clock error states. This example suggests the influence that filter tuning can have on navigation performance.[32]

Finally, we note that for "Navigation grade" inertial systems (commonly defined as having 1.0 nautical mph error growth attributable to gyro drift), the error growth of a GPS "calibrated" unit is dominated by the initial velocity error, accelerometer bias, and gravity anomalies for about the first 10 min after a GPS outage. Gyro bias begins to dominate around 20 min (about 0.25 Schuler period). The effectiveness of any "calibrations" on reducing errors at 20-min, or longer, intervals depends on the level of "random walk" gyro and accelerometer errors that accumulate in that interval.

B. Using GPS for In-flight Alignment

Alignment is the process that ties inertial platform coordinates to the geographic frame in which the host vehicle navigates. Alignment establishes the conditions necessary for the proper integration of accelerometer outputs to accurately estimate the changes in user position and velocity.

Gyrocompassing typically extends over a 10–15-min period prior to a standard aircraft takeoff. We need to look at the factors that contribute to alignment time in order to understand the opportunities for using GPS to speed up the process, or to allow it to proceed in-air in addition to on the ground.

1. Conventional Alignment Procedures

Alignment consists of platform leveling and establishing a reference bearing. The following discussion is based on the alignment of a gimballed platform INS, but the general conclusions and timing estimates are good approximations to the performance of a "strapdown" INS.

a. Leveling. Leveling is the process of establishing one plane of the inertial measurement unit (IMU) instrument package perpendicular to the local gravity

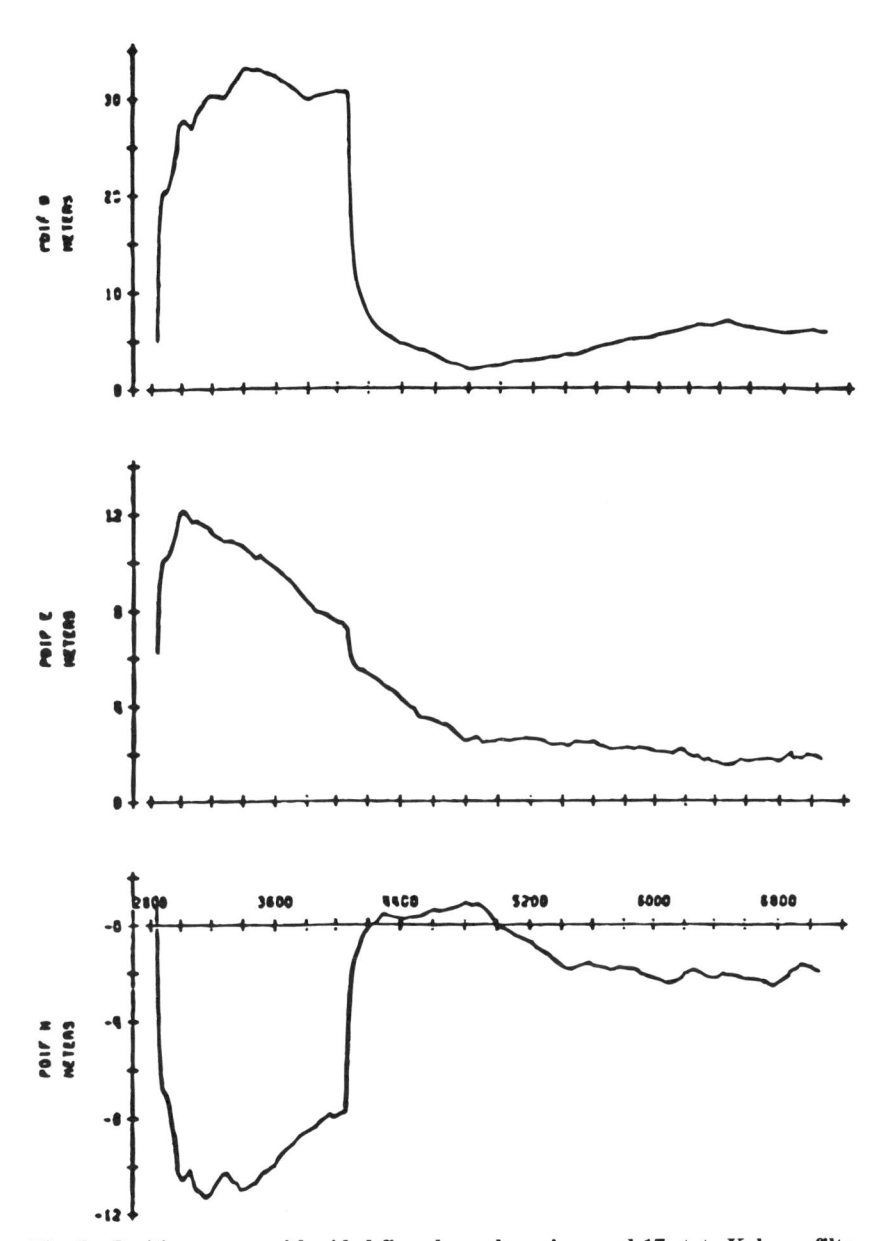

Fig. 7 Position errors with aided five-channel receiver and 17-state Kalman filter.

vector. This is typically accomplished by mounting two accelerometers on that plane so that their input axes are not co-linear (they are usually mounted at right angles), and tilting the plane to null the accelerometer outputs. (Alignment for "strap-down" inertial systems is a mathematical operation that does not involve physical motion of the sensor platform.) The perpendicular to this plane defines "UP" in an East, North, Up coordinate system.

The error signal that drives the leveling loop is proportional to the instantaneous value of the tilt error; i.e., to approximately 1.0 g times the alignment error (in rads). In mathematical notation, the output for each horizontal accelerometer is $\hat{g}$, where

$$\hat{g} = (1 + s_f)g\, e_\phi + b$$

where g is the magnitude of the local gravity vector; e_ϕ is the angular error of the platform normal to the accelerometer axis (i.e., the tilt error); b is the accelerometer bias; and s_f is the accelerometer scale factor error. For an aircraft at rest, $\hat{g}$ should be zero when the platform is level; however, the accelerometer bias causes the null to occur for a nonzero value of tilt error. For accelerometers used in contemporary fighter aircraft, b is about 150 μg, and s_f is about 500 ppm. For $g = 1.0$, the steady state tilt error e_ϕ is then approximately

$$|e_\phi| = b/(1 + s_f) \approx 150 \text{ μrad} = 30 \text{ arcsec} \tag{1}$$

where we assume that angle quantization effects in the leveling feedback loop are negligible. This 30 arcsec is a bias error; the random component of leveling errors is on the order of a few arcseconds.

b. North seeking. North seeking is the process of establishing a reference direction (azimuth) in the leveled plane containing the east and north accelerometers. The most widely used scheme for self-alignment of the platform is gyrocompassing.

Gyrocompassing exploits the following properties of gyroscopes. For a local level north-oriented system at latitude λ

1) If the gyro input axes are physically aligned with the geographic axes (North, East, Up), then the system will remain aligned if each gyro is individually torqued at a rate equal to the projection of the Earth rate on its input axis.

2) If the platform is level, but not aligned to north, it will rotate around the east axis, causing the north axis to rotate from horizontal. This rotation produces a level error than can be sensed by the north accelerometer and used to drive the azimuth error to zero.

The rate of rotation of the north axis in response to an azimuth misalignment of θ_A radians is given by $\dot{\theta}_N$

$$\dot{\theta}_N = \theta_A \Omega_e \cos \lambda$$

where Ω_e is the Earth rate. If this rate acts for t seconds, it produces net level error (tilt) of θ_N (radians)

$$\theta_N = \theta_A \Omega_e\, t \cos \lambda \tag{2}$$

From Eq. (2) we see that for gyrocompassing to convert a misalignment error

GPS AND INERTIAL INTEGRATION

θ_A into an equally large observable tilt error (i.e., for $\theta_N = \theta_A$), then the effect of the error must be integrated for at least t_o seconds, where

$$t_o = (\Omega_e \cos \lambda)^{-1}$$

At midlatitudes, (say $\lambda = 45$ deg) and for

$$\Omega_e = 2\Pi \text{ rad/day}$$

we have

$$t_o = \frac{1}{\sqrt{2\Pi}} \text{ days} = 5.4 \text{ h} \qquad (3)$$

In practice, the requirement for alignment accuracy is substantially less than the requirement on leveling accuracy, and this allows the alignment to proceed faster than indicated by Eq. (3). We can estimate this speedup using the approximation that the standard deviation of leveling errors is inversely proportional to the time spent in leveling. For medium accuracy inertial sets, which are found in contemporary aircraft, the random noise in leveling is on the order of a few arcsec, and the uncertainty in azimuth alignment after gyrocompassing is on the order of 160–200 arcsec. The ratio of these standard deviations is about 50. Substituting $\theta_A = 50\,\theta_N$ in Eq. (2) yields $t_o = t_o/50$, which is approximately 7 min. This estimate is consistent with Air Force specifications for the time to achieve standard accuracy alignment by gyrocompassing under favorable conditions.[33]

2. GPS Aided Alignment

Leveling is essentially an instantaneous process (especially in a strap-down system) because it seeks to drive two directly sensed gravity indications to be equal. This gives the clue as to the benefits of inflight alignment. Velocity changes sensed by the inertial system are compared to velocity changes sensed by GPS, and the alignment parameters are adjusted to drive the residuals to zero. Sensing INS alignment errors from velocity residuals rather than from position residuals (as in traditional gyrocompassing) is significant because it eliminates the delay incurred to integrate a velocity error to a position error of detectable magnitude. Now, combine this observability with the opportunity to use aircraft maneuvers (turns, climbs, dives) to apply acceleration in the horizontal navigation plane (North, East), in addition to the vertical plane (Up), and the potential use of GPS for inflight alignment is clear.

The issues for in-flight INS alignment using GPS are related to the noisiness of GPS velocity measurements compared to inertial measurements. Unless the INS misalignment is so large that, at the vehicle velocity, it produces a velocity error that exceeds the GPS noise level, the alignment will require several minutes, or more, of observations. Thus, the time to align is a function of the accuracy goals, the magnitude of acceleration that can be applied during maneuvers, and the noise level of GPS measurements.

Figures 8–10 were generated in a linear covariance simulation of INS alignment for a fighter aircraft using a standard (1.0 n.mi./h) inertial navigator. The integration filter mechanized 11 states (3 position, 3 velocity, 3 misalignment, 2 clock).

The measurements input to the filter are GPS pseudorange and delta-range and inertially indicated velocity change projected along the nominal line-of-sight toward each satellite being tracked. Although the filter is similar to the internal filter in RCVR 3A, its performance is superior because the inertial platform dynamics are more accurately modeled and the filter tuning (process–noise values) reflects the nominal INS error variances rather than worst-case values used in the RCVR 3A tuning. The results presented can be viewed as an upper bound on the performance available from RCVR 3A when the filter accurately models the integration environment. The RCVR 3A filter trades suboptimum performance for reduced sensitivity to any mismatch between actual INS performance and the INS as modeled in the filter. Two takeoff profiles were investigated.

a. Standard takeoff. 1) accelerate at 0.3 g to 160 knots indicated airspeed (KIAS) using MIL (standard) acceleration; 2) begin rotation to 5-deg pitch angle and accelerate to 170 KIAS takeoff speed; 3) climb at 5 deg, and accelerate to 385 KIAS, on 45-deg heading (northeast); and 4) Continue climbing at 5 deg and constant velocity to cruise altitude at 20,000 ft, and 385 KIAS.

b. Alert (scramble) takeoff. 1) accelerate at 0.4 g to 165 KIAS using maximum afterburner thrust; 2) rotate to 12-deg pitch angle and climb at 0.4 g to 415 kt at 2500 ft altitude; 3) climb from 2,500 to 20,000 ft altitude at constant speed and 12-deg pitch, arriving at 20,000 ft approximately 180 s after takeoff; and 4) level flight at 20,000 ft, then accelerate to Mach 0.85 (533 kt).

Navigation for standard takeoff assumes that the INS has been aligned by gyrocompassing and that GPS is continually available. Navigation for alert takeoff assumes that the INS is aligned by a stored heading reference and that GPS is not available until 5 min after takeoff. In both cases, different turn, climb, and dive maneuvers after takeoff were considered for their effectiveness in reducing alignment errors after takeoff.

The values cited in Table 1 were used to characterize the standard errors (1–σ levels) remaining after gyrocompassing and stored heading alignment respectively. These are the initial conditions for subsequent Kalman filtering of GPS plus inertial sensor outputs.

Three individual test cases are presented here. These are listed in Table 2.

c. Case 1. Case 1 illustrates the potential benefits of using GPS-aided alignment in a standard mission, where the INS is gyrocompassed prior to takeoff, and GPS is available throughout takeoff. A 2/3-g turn that produces a 45-deg heading change beginning at $t = 495$ s is included to illustrate the additional improvement available from in-flight maneuvers.

Figure 8 illustrates the history of rms alignment errors. As expected, the availability of GPS during the period of acceleration at takeoff reduces the

Table 1 Alignment errors at takeoff (arcsec)

	Level	Azimuth
Stored heading	40	450
Gyrocompassing	37	225

GPS AND INERTIAL INTEGRATION   211

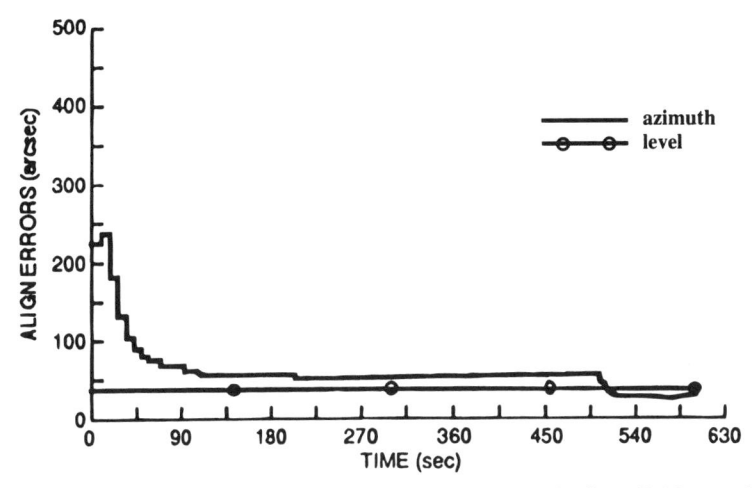

Fig. 8   Inertial navigation systems alignment errors with GPS available at takeoff and gyrocompassed alignment (standard takeoff).

azimuth alignment error; it seems to decrease asymptotically toward about 50 arcsec. The additional acceleration after $t = 495$ s produces a modest further improvement. The final azimuth error is approximately the same as the level error, which signifies that accelerometer bias is the limiting factor in improving alignment by this procedure. An analysis of the error sensitivity of the tilt estimates confirms this intuition, with nonorthogonality of the horizontal accelerometers with respect to the "up" direction being the only other significant error source.

   d.   *Case 2.*   Case 2 illustrates potential benefits of GPS for a "scramble" takeoff, when GPS is available only after 300 s. The aircraft takes off with a stored heading alignment having a nominal azimuth uncertainty of 450 arcsec. The aircraft executes a 1-g turn that produces a 45-deg heading change starting at $t = 495$ s. (The heading angle for Case 2 is essentially the same as for Case 1, but the turn is completed faster.)

   Figure 9 illustrates the history of RMS alignment errors. The error is constant until $t = 300$ s, when GPS becomes available. The availability of GPS makes a

Table 2   Test cases for in-flight alignment studies

Case	Takeoff mode	Alignment mode	GPS on at	Alignment maneuver
1	Standard	Gyrocompass	0 s	2/3-g turn
2	Alert	Stored heading	300 s	1-g turn
3	Alert	Stored heading	300 s	2-g S-turn

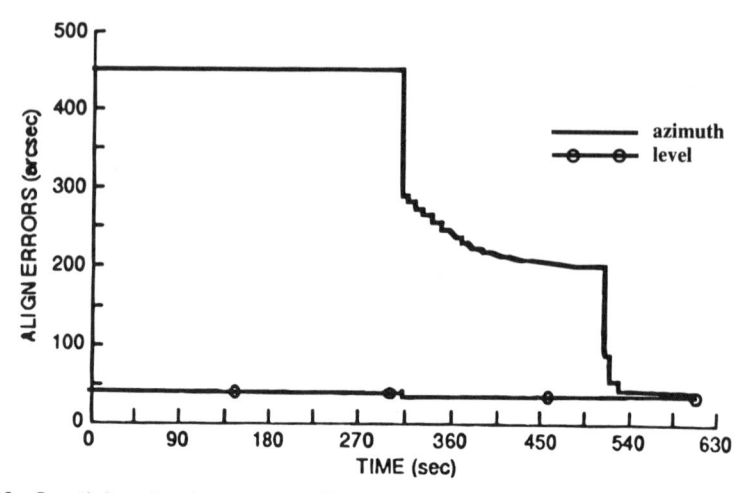

Fig. 9 Inertial navigation systems alignment errors with GPS available 5 min after an alert takeoff.

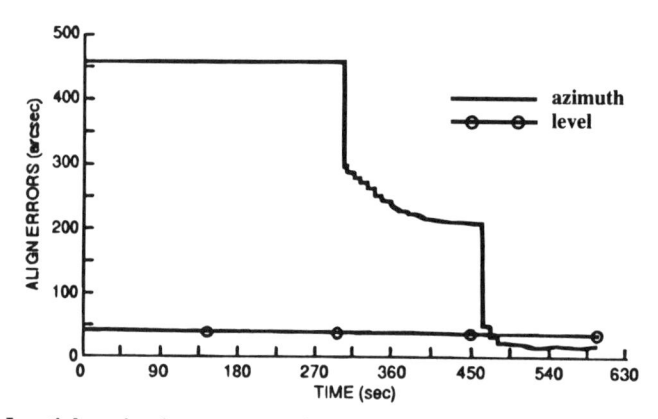

Fig. 10 Inertial navigation systems alignment errors with GPS available at 300 s after an alert takeoff with stored heading alignment.

dramatic improvement in alignment even without maneuvers because the position and velocity errors that have accumulated in the INS navigation results are highly correlated with misalignments. We see that, following an initial decrease to 280 arcsec, the azimuth errors fall to an asymptote of 200 arcsec, which is comparable to gyrocompass alignment accuracy. The execution of a 1-g lateral maneuver drives the azimuth errors to the level predicted by accelerometer bias.

  e. *Case 3.* Case 3 involves an "alert" takeoff with nominal stored heading azimuth errors and a 2-g "s" turn maneuver at $t = 460$ s. GPS is not available until $t = 300$ s (corresponding to a five-minute warm-up and acquisition time).
  Figure 10 illustrates the rms alignment error. As expected, Figs. 9 and 10 are identical until the maneuver begins, and then the higher g applied in Case 3 reduces the azimuth error below the level for Case 2.

3. Discussion

These simulations support the following performance conclusions:

1) In-flight GPS measurements during periods when lateral acceleration is on the order of 1 g can be used to reduce azimuthal alignment errors to the same magnitude as level errors. Larger accelerations lead to smaller errors.

2) Even in the absence of lateral acceleration, the availability of GPS measurements can be used to reduce the errors of a stored-heading alignment to the level of gyrocompassing. This reduction exploits the correlation between INS alignment errors and the INS navigation errors that are uncovered by comparison with GPS observations.

3) Simple maneuvers, such as 1-g coordinated turns over as little as 45 deg, and lasting for no more than 30 s, are adequate for realizing the benefits of in-flight alignment.

C. Integrated Navigation Solutions During a GPS Outage

The previous examples illustrated certain benefits of GPS/INS integration in a benign environment, with low-to-moderate vehicle dynamics. The use of integration to lengthen the interval that GPS data are available, and to ride out the period of outages caused by the interference or high dynamics is the focus of the following remarks.

1. Integration Case Study Overview

In a recent military aircraft avionics upgrade, RCVR 3A was to be integrated with a "navigation-grade" gimbaled-platform INS. The principal performance objective was to limit the error growth after 5 min of any GPS outage, provided that the INS had been calibrated by at least 7 min of GPS measurements prior to the outage. Based on extensive simulations, the following results were determined:

1) The horizontal circular error probable (CEP) after 5 min of GPS outage could be held to 30 m. (Recall that the free-running INS is specified as a 1.0 n.mi./h system.)

2) A 21-state filter mechanized as two independent "horizontal" and "vertical" filters would provide close enough to optimum performance to be a cost-effective, computationally effective approach.

These results were derived from a covariance analysis that incorporated a 73-state truth model for the INS, plus additional error states for GPS.

Table 3 lists these truth states. Table 4 characterizes the mission segments of the standard aircraft flight path used in the simulation.

2. Navigation Performance of the Integrated System

Figure 11a illustrates the growth in rms east-positioning navigation error vs time after the loss of GPS at approximately 2550 s. In this case, the GPS and INS position and velocity measurements were processed by a full optimal 73-state filter with dynamics, plant noise parameters, and measurement noise parameters that exactly match those in the truth model. In other words, this represents the best that any filter can do given this truth model and trajectory. Five minutes after the loss of GPS, the east component of horizontal position errors has grown to 80 ft (24.4-m rms), exclusive of the low-frequency GPS bias. Figure 12 shows the contribution of various error sources to the net navigation error. For at least 10 min, the accelerometer and gravity disturbance terms are dominant. Gyro errors do not become dominant until more than a quarter of a Schuler period (22 min) after the outage.

Figure 11b illustrates the growth in east errors for a suboptimum 21-state filter, whose states are listed in Table 5. These results are almost indistinguishable from Fig. 11a, for the "optimum filter." Finally, Fig. 11c gives the east error for a 15-state horizontal filter whose states are listed in Table 5. These results, too, are nearly indistinguishable from optimum for about the first 420 s of GPS outage.

Table 3 Truth model states

State number	Description	Comment
1,2	Horizontal position error	
3	Wander angle error	
4,5,6	Platform misalignment	100 arcsec initial horizontal 6 deg initial vertical
7,8,9	Velocity error	
10	Vertical position error	
11–13	Auxiliary baro-inertial states	Used in describing baro-inertial loop
14–16	Markov gyro bias	0.002 deg/h rms horizontal 0.005 deg/h rms vertical Correlation time = 5 min
17,18	Markov accelerometer bias	3 μg rms 10 min correlation time
19	Markov baro bias	100 ft rms 10 min correlation time
21–23	Gravity disturbance	35 μg rms 20 n.mi. correlation distance
24–26	Gyro bias	0.01 deg/h rms horizontal 0.022 deg/h rms vertical
27–29	Gyro scale factor error	0.0002 rms
30–32	Gyro misalignments about spin axes	3.3 arcsec rms
33–35	Remaining gyro misalignments	20 arcsec rms
36–41	Gyro g-sensitivity	0.015 deg/h/g rms
42–44	Gyro g-squared-sensitivity	0.02 deg/h/g^2 rms
45–47	Accelerometer bias	150 μg rms
48–50	Accelerometer scale factor error	0.0002 rms
51–53	Accelerometer scale factor asymmetry	0.0001 rms
54–56	Accelerometer nonlinear scale factor asymmetry	10 μg/g^2 rms
57–59	Accelerometer nonlinearity	10 μg/g^2 rms
60–65	Accelerometer orthogonal g-squared sensitivity	30 μg/g^2 rms
66–71	Accelerometer misalignments	20 arcsec rms
72	Baroaltimeter bias	300 ft rms
73	Baroaltimeter scale factor error	0.04 rms

GPS AND INERTIAL INTEGRATION

Table 4 Test trajectory

Time, s	Description
0–300	Gyrocompass at true heading = 120 deg
300–490	Taxi and turn left approximately 75 deg to prepare for takeoff at a heading of approximately 45 deg
490–540	Takeoff; speed = 434 ft/s, altitude approximately 1000 ft
540–840	Climb; accelerate to 655 ft/s, altitude approximately 9660 ft
840–1020	Level off at approximately 9660 ft; cruise at 655 ft/s
1020–1100	Descend to approximately 200 ft
1100–2720	Level-off at 200 ft; cruise at 655 ft/s
2720–2780	90 deg right turn.
2780–2806	Pop-up, climb to approximately 12,000 ft; accelerate to approximately 820 ft/s
2806–2838	Dive to 635 ft
2838–2858	Level-out at 635 ft; decelerate to 655 ft/s; weapons delivery
2858–2918	90-deg right turn
2918–3226	Low-altitude combat, 300 to 2700 ft; 485 to 655 ft/s
3226–3276	Turning climb to 9900 ft; speed = 655 ft/s
3276–3346	High-altitude combat
3346–5086	Turn toward home; high-altitude cruise at 9900 ft, 655 ft/s
5086–5446	Descend to 3000 ft; accelerate to 820 ft/s
5446–5876	Loiter; decelerate to 434 ft/s
5876–6126	Land

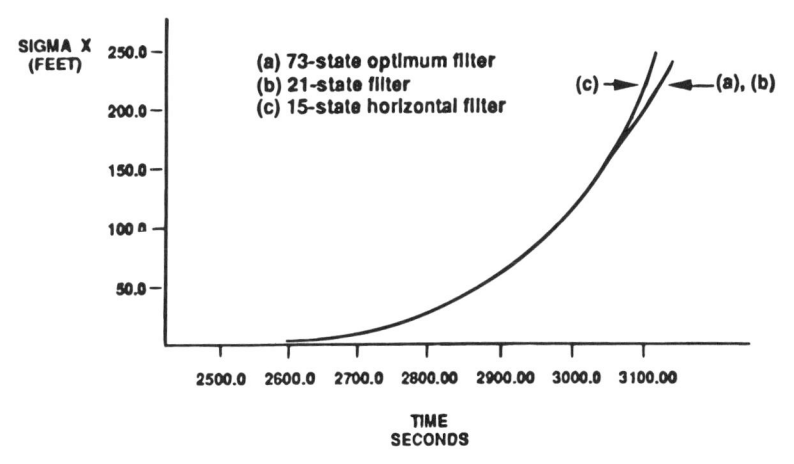

Fig. 11 Root mean square position error vs time. GPS measurements stop at 2550 s.

SIGMA X (FEET)

A: Accelerometer Errors and Horizontal Misalignments
B: Gravity Disturbance
C: Initial Position and Velocity Errors, Vertical Misalignment, Etc.
D: Gyro Errors

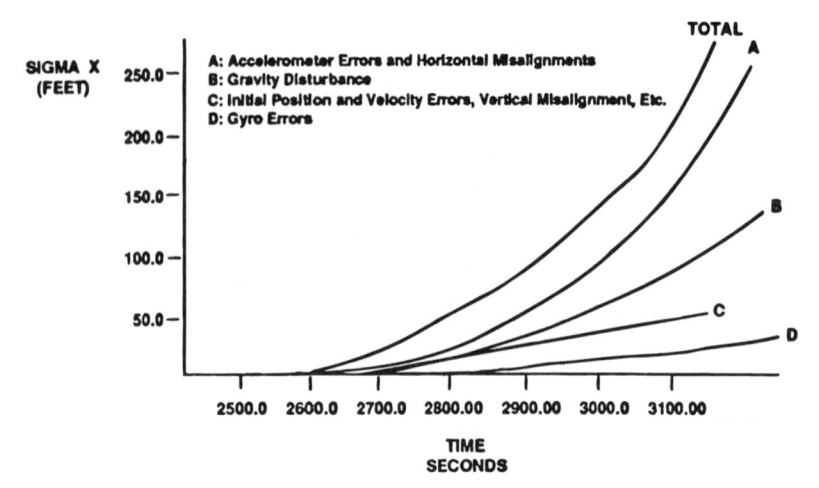

Fig. 12 Contributions to inertial navigation systems error growth during a GPS outage.

Table 5 States of the 15-state horizontal filter

State number	Description	Comment[a]
1,2	Horizontal position error	
3	Wander angle error[a]	
4,5,6	Platform misalignment (4 and 5 also absorb constant accelerometer bias)	Plant noise of 9.4E-17 rad^2/s (horizontal) and 5.9E-16 rad^2/s (vertical) used to account for unmodelled gyro errors
7,8	Horizontal velocity error	Plant noise of 9.E-12 ft^2/s^3 used to account for unmodeled accelerometer error
9–11	Gyro bias	
12,13	Horizontal gravity disturbance (also absorbs other accelerometer errors)	
14,15	Accelerometer scale factor error	

[a]In an actual implementation, wander angle error would be combined into a single state with vertical misalignment.

IV. Summary

We have presented a description of the processes whereby the combination of GPS with an inertial navigator can produce a system performance that is superior to either one acting alone. These benefits include the following:
  1) Smaller random errors than seen in stand-alone GPS navigation solutions.
  2) Improved availability of GPS operations during maneuvers and in the presence of radio frequency interference (RFI).
  3) A navigation solution whose position and velocity errors are bounded by the errors in the GPS navigation solution.
  4) A calibrated navigation solution whose errors grow slower than those of a free-running uncalibrated INS during GPS outages.

They are available because the long-term (low-frequency) content of GPS errors is negligible.

All four benefits are available to GPS users who are authorized to obtain the GPS Precise Positioning Service (PPS). Only the first and third are guaranteed to GPS users who are vulnerable to the selective availability clock dither that corrupts the GPS Standard Positioning Service (SPS). Clock dither prevents the tracking loop bandwidths from being reduced to the bandwidth of the INS errors, and corrupts the use of GPS measurements to calibrate the INS error states.

GPS receiver technology is evolving rapidly in response to pressure from the civilian market. This trend is evident in the miniaturization of full-capability receivers that can be physically embedded in a host system, such as an INS or a mission computer. This in turn makes it practical to obtain even higher performance levels by treating both GPS and the INS as sensors that produce measurements to be optimally combined by a navigation filter into an "optimum" navigation solution.

For many users, the primary benefit of embedded configurations is not improved performance. These users focus on savings in volume, weight, power consumption, and cost that are predicted for the embedded system. Some cost savings are nonrecurring (e.g., the one-time investment in developing integration software may be borne by the vendor of the embedded system rather than by the integrator), whereas others are recurring (e.g., reduced production cost of the system hardware).

In addition to these cost and performance benefits, the integrated system may be able to support functional capabilities that were previously not available to the user. For example, the use of in-flight INS alignment can make it possible for an aircraft to take off without waiting for the 5–10 min routinely reserved for INS alignment. This may be a life-saving capability for military aircraft that must take off immediately after an alert that hostile forces are incoming. Another example is the inclusion of error states for gravity anomalies in the integration filter. This is the basis for balloon-borne GPS/INS instrumentation that will allow for more extensive gravity mapping than has ever been possible.[34]

The design of a Kalman integration filter that mechanizes the navigation solution must also address the following standard questions:
  1) How many states should be estimated?
  2) How often should the filter be updated?
  3) How should correlated measurements be treated?

4) By how much can the computational burden of the filter be reduced by exploiting sparseness of the state dynamics matrix or the measurement covariance matrix?

5) How should the filter accommodate transient events such as changes in the constellation of satellites being tracked?

6) How should the filter be tuned to provide robust performance vs unknown aspects of the design problem?

7) How can the filter be made robust against variations in the error characteristics of individual INS or GPS units?

8) How can the filter be used to detect the onset of anomalous conditions that may indicate a failure in the GPS or the INS subsystems?

These design questions are addressed in more detail in the following chapters of this book.

References

[1]Cox, D. B., "Integration of GPS with Inertial Navigation Systems," reprinted in *Collected GPS Papers*, Vol. I, Institute of Navigation, Alexandria, VA, 1980, pp. 144–153.

[2]Johannessen, R., and Asbury, M. J. A., "Towards A Quantitative Assessment of Benefits INS/GPS Integration Can Offer to Civil Aviation," *Navigation*, Vol. 37, No. 4, 1990–1991, pp. 329–346.

[3]Greenspan, R. L., et al., "The GPS Users Integration Guide," *Proceedings of the ION National Technical Meeting* (Santa Barbara, CA), Institute of Navigation, Washington, DC, Jan. 26–29, 1988, pp. 104–112.

[4]Wiederholt, L., and Klein, D., "Phase III GPS Options for Aircraft Platforms," *Navigation*, Vol. 31, No. 2, 1984, pp. 129–151.

[5]Brown, A. K., and Bowles, W. M., "Interferometric Attitude Determination Using GPS," *Proceedings of the Third Geodetic Symposium on Satellite Doppler Positioning* (Las Cruces, NM), Feb. 1982, pp. 1289–1302.

[6]Eller, D., "GPS/IMU Navigation in a High Dynamics Environment," *Proceedings of the First International Symposium on Precise Positioning with GPS* (Rockville, MD), April 15–19, 1985, pp. 773–782.

[7]Bridges, P. D., "Influence of Satellite Geometry Range, Clock and Altimeter Errors, on Two-Satellite GPS Navigation," *Proceedings of the ION GPS '88* (Colorado Springs, CO), Institute of Navigation, Washington, DC, Sept. 19–23, 1988, pp. 253–258.

[8]Kalafus, R. M., and Knable, N., "Clock Coasting and Altimeter Error Analysis for GPS," *Navigation*, Vol. 31, No. 4, 1984–85, pp. 289–302.

[9]Bartholomew, R. G., et al., "Software Architecture of the Family of DOD Standard GPS Receivers," *Proceedings of the First Technical Meeting of ION Satellite Division* (Colorado Springs, CO), Institute of Navigation, Washington, DC, Sept. 1987, pp. 23–24.

[10]Dellicker, S. H., and Henckel, D., "F-16/GPS Integration Test," *Proceedings of ION GPS-89* (Colorado Spring, CO), Institute of Navigation, Washington, DC, Sept., 1989, pp. 295–303.

[11]Barbour, N., et al., "Inertial Instruments—Where to Now?," *Proceedings of the AIAA Guidance, Navigation, and Control Conference* (Hilton Head, SC), Washington, DC, Aug. 1992 (AIAA Paper 924414-CP).

[12]Boxenhorn, B., et al., "The Micromechanical Inertial Guidance System and Its Application," *Proceedings of the Fourteenth Biennial Guidance Test Symposium* (Holloman, AFB, NM), Oct. 3–5, 1989, Vol. 1, pp. 113–131.

[13]Ward, P., and Rath, J., "Attitude Estimation Using GPS," *Proceedings of the ION National Technical Meeting* (San Mateo, CA), Institute of Navigation, Washington, DC, Jan. 23–26, 1989, pp. 169–178.

[14]Kruczynski, L. R., and Li, P. C., "Using GPS to Determine Vehicle Attitude," *Proceedings of the ION GPS '88* (Colorado Springs, CO), Institute of Navigation, Washington, DC, Sept. 19–23, 1988, pp. 139–146.

[15]Jurgens, R., "Real Time GPS Azimuth Determining System," *Proceedings of the ION National Technical Conference* (San Diego, CA), Institute of Navigation, Washington, DC, Jan. 23–25, 1990, pp. 105–110.

[16]van Graas, F., and Braasch, M., "GPS Interferometric Attitude and Heading Determination; Flight Test Results," *Proceedings of the 47th Annual Meeting (ION)* (Williamsburg, VA), Institute of Navigation, Washington, DC, June 10–12, 1991, pp. 183–191.

[17]Graham, W. R., and Johnston, G. R., "Standard Integration Filter State Specification and Accuracy Predictions," *Navigation*, Vol. 33, No. 4, 1986–1987, pp. 295–313.

[18]Cunningham, J., and Lewantowicz, Z. H., "Dynamic Integration of Separate INS/GPS Kalman Filters," *Proceedings of the ION GPS '88* (Colorado Springs, CO), Institute of Navigation, Washington, DC, Sept. 19–23, 1988, pp. 273–282.

[19]Diesel, J. W., "Integration of GPS/INS for Maximum Velocity Accuracy," *Navigation*, Vol. 34, No. 3, 1987, pp. 190–211.

[20]Berman, G. J., and Belzer, M., "A Decentralized Square Root Information Filter/Smoother," of the *Proceedings 24th IEEE Conference on Decision and Control* (Ft. Lauderdale, FL), Institute of Electrical and Electronics Engineers, New York, Dec. 1985.

[21]Carlson, N. A., "Federated Square Filter for Decentralized Parallel Processes," *IEEE Transactions on Aerospace and Electronic Systems*, Vol. 26, No. 3, 1990, pp. 517–525.

[22]Gelb, A. (ed.), *Applied Optimal Estimation*, MIT Press, Cambridge, MA, 1974, Chap. 4, pp. 102–132.

[23]Maybeck, P. S., *Stochastic Models, Estimation and Control*, Vol. 1, Academic Press, Orlando, FL, 1979.

[24]Bletzacker, F., Eller, D., Forgette, T., Seibert, G., Vavrus, J., and Wade, M., "Kalman Filter Design for Integration of Phase III GPS with an Inertial Navigation System," *Proceedings of the ION National Technical Meeting* (Santa Barbara, CA), Institute of Navigation, Washington, DC, Jan. 26–29, 1988, pp. 113–129.

[25]Lewantowicz, Z. H., and Keen, D. W., "Graceful Degradation of GPS/INS Performance with Fewer than Four Satellites," *Proceedings of the ION National Technical Meeting* (Phoenix, AZ), Institute of Navigation, Washington, DC, Jan. 21–24, 1991, pp. 269–276.

[26]Widnall, W. S., "Stability of Alternate Designs for Rate Aiding of a Noncoherent Mode of a GPS Receiver," Intermetrics Corp., Cambridge, MA, Sept. 25, 1978.

[27]Buechler, D., and Foss, M., "Integration of GPS and Strapdown Inertial Subsystems into a Single Unit," *Navigation*, Vol. 34, No. 2, 1987, pp. 140–159.

[28]Franklin, M., and Pagnucco, S., "Development of Small Embedded GPS/INS RLG and FOG systems," *Proceedings of the ION National Technical Conference* (San Diego, CA), Institute of Navigation, Washington, DC, Jan. 27–29, 1992, pp. 3–12.

[29]Tazartes, D. A., and Mark, J. G., "Integration of GPS Receivers into Existing Inertial Navigation Systems," *Navigation*, Vol. 35, No. 1, 1988, pp. 105–120.

[30] Homer, W. C., "An Introduction to the GPS Guidance Package (GGP)," *Proceedings of the 15th Biennial Guidance Test Symposium, CIGTF Guidance Test Division*, Vol. 1 (Holloman AFB, NM), Sept. 24–26, 1991 pp. 12–15.

[31] Greenspan, R., and Donna, J., "Measurement Errors in GPS Observables," *Navigation*, Vol. 33, No. 4, 1986–1987, pp. 319–334.

[32] Soltz, J. A., Donna, J. I., and Greenspan, R. L., "An Option for Mechanizing Integrated GPS/INS Solutions," *Navigation*, Vol. 35, No. 4, 1988–1989, pp. 443–458.

[33] "Specification for USAF Standard Form, Fit and Function (F^3) Medium Accuracy Inertial Navigation Unit," SNU-84-1, Rev. 3, Amendment, Feb. 13, 1987, USAF Aeronautical Systems Division (now Aeronautical Systems Center).

[34] Jekeli, C., Doyle, T., Nicolaides, P., and Galdos, J., "Instrumentation Design and Analysis for Balloon-Borne Gravimetry and Attitude Determination Using GPS and INS," *Proceedings of the ION National Technical Meeting* (San Francisco, CA), Institute of Navigation, Washington, DC, Jan. 20–22, 1993, pp. 519–533.

Chapter 8

Receiver Autonomous Integrity Monitoring Availability for GPS Augmented with Barometric Altimeter Aiding and Clock Coasting

Young C. Lee*
MITRE Corporation, McLean, Virginia 22102

I. Introduction

ONE of the most important criteria for the operational approval of the use of Global Positioning System (GPS) for civil air navigation for instrument flight rules (IFR) operations is safety, which requires assuring the integrity of navigation solutions derived from GPS. Integrity monitoring detects erroneous information not detectable by other means, such as signal disappearances or failures identified through the satellite health bit flag. An integrity monitoring method for GPS that does not require an external monitoring system, and thus, is simple and practical to implement is receiver autonomous integrity monitoring (RAIM). RAIM, which is discussed in another chapter in this book, is to perform detection (detection of position error beyond protection limits) and identification (identification of failed satellite) functions.[1] For near term use of GPS as a supplemental system in the National Airspace System (NAS), the RAIM detection function is essential; on the other hand, the RAIM identification function, although not essential, is highly desirable. Without the identification function, a satellite failure would cause an outage of the position fixing function. High availability of the navigation and RAIM detection functions is important because when this function cannot be performed, the user must fall back on another approved navigation system. High availability of the navigation and RAIM functions is even more important for GPS to be authorized to provide required navigation performance (RNP), because a failure of a GPS-based system would have an impact over a large area, and thus, have a serious operational impact because of

---

Copyright © 1994 by the American Institute of Aeronautics and Astronautics, Inc. All rights reserved.
*Lead Engineer, Center for Advanced Aviation System Development, MS W307, 7525 Colshire Drive.

the lack of a backup system. (An RNP system is one that satisfies all requirements for flight in a given airspace.)

As a means of improving availability of the navigation and RAIM functions, GPS augmentations in the form of barometric altimeter aiding and clock coasting are often suggested. Implementation of these augmentations would be relatively simple and could be done in a timely manner. It is noteworthy that a requirement for barometric altimeter aiding was included in the FAA's Technical Standard Order (TSO)-C129,[2] which specifies requirements for GPS avionics equipment to be used as a supplemental means of navigation for en route (domestic and oceanic), terminal, and nonprecision approach phases of flight. This paper presents the technical analyses of availability and outage duration statistics of navigation and RAIM detection and identification functions for GPS with the altimeter aiding and/or clock coasting for en route, terminal, and nonprecision approach phases of flight.

The approach used in this paper is to compute temporal characteristics of both navigation and RAIM availabilities over three major domestic and oceanic routes. In addition, navigation and RAIM availabilities for users making a final approach was evaluated for five major airport locations in the conterminous United States (CONUS). The constellations considered include the Optimal 21 constellation (Opt21), 24 Primary constellation (24Pr), and the 24 Primary constellation with a typical set of 3 satellites failed (24Minus3). Because the Department of Defense (DOD) guarantees at least 21 operating satellites 98% of the time, but 24 operating satellites only 70%, it was considered important to determine availability for different constellations with less than 24 operating satellites. In particular, the 24Minus3 constellation is considered to give a reasonable lower bound in terms of the expected RAIM availability under published DOD policy.

Approaches for the augmentation are described first. This is followed by the analytical results. Finally, the findings of the analyses are summarized. The annex, which is the contribution of Rolf Johannessen and Charles Dixon, describes statistical distribution of the barometric error growth as a function of distance flown. This error growth rate has a significant impact on the effectiveness of barometric altimeter aiding, as discussed later.

II. Methods of Augmentations

This section describes a procedure for augmenting the basic GPS solution equation; first for the case of barometric altimeter aiding, and then for the case of clock coasting. One basic assumption in the following discussion is that GPS will be augmented with the baroaltitude or clock only when necessary; that is, if navigation or RAIM functions can be provided with GPS alone at a particular point in time, no augmentation is to be incorporated.

A. Augmented Geometry for Barometric Altimeter Aiding

In this method, barometric altimeter data provide an additional piece of information for navigation, RAIM detection, and RAIM identification functions. The barometric altimeter data are incorporated into the basic GPS position solution.

The basic equation for the set of GPS measurements, without any augmentation, is a linearized equation of the form

RECEIVER AUTONOMOUS INTEGRITY MONITORING 223

$$\begin{bmatrix} \Delta\rho_1 \\ \Delta\rho_2 \\ \ldots \\ \Delta\rho_n \end{bmatrix} = G \begin{bmatrix} \Delta x \\ \Delta y \\ \Delta z \\ \Delta t \end{bmatrix} + \begin{bmatrix} \epsilon_1 \\ \epsilon_2 \\ \ldots \\ \epsilon_n \end{bmatrix} \quad (1)$$

where $\Delta\rho_i$ is the difference between the measured pseudorange and predicted pseudorange based on the nominal GPS position and clock bias; G is $n \times 4$ geometrical connection matrix (directional cosines of line-of-sights to satellites) relating the measurements to the three components of position, and time; Δx, Δy, Δz, and Δt are perturbations in x, y, and z coordinates and time t from the nominal position and clock bias; and ϵ_i is range measurement error for the ith satellite.

In the case of barometric altimeter aiding, the above equation is augmented by a linear equation of the following form, making use of the baroaltitude measurement.

$$\Delta B = [0\ 0\ 1\ 0] \begin{bmatrix} \Delta x \\ \Delta y \\ \Delta z \\ \Delta t \end{bmatrix} + \epsilon_{baro} \quad (2)$$

where ΔB is the difference between the measured barometric altitude (calibrated with either GPS as described below or the local barometric setting) converted to WGS-84 altitude, and the predicted altitude based on the nominal GPS position; and ϵ_{baro} is the error in measured barometric altitude after it has been calibrated (either by GPS or local pressure setting) and converted to WGS-84 altitude.

In a conventional RAIM algorithm such as the least-squares residual RAIM method, it is assumed that all independent sensor measurement errors have the same variance. Therefore, in case the standard deviation of ϵ_{baro} is different from σ_{sv}, the above equation should be properly scaled before it is added to the basic GPS measurement equation. That is,

$$\frac{\Delta B}{(\sigma_{baro}/\sigma_{sv})} = \begin{bmatrix} 0 & 0 & \dfrac{1}{(\sigma_{baro}/\sigma_{sv})} & 0 \end{bmatrix} \begin{bmatrix} \Delta x \\ \Delta y \\ \Delta z \\ \Delta t \end{bmatrix} + \frac{\epsilon_{baro}}{(\sigma_{baro}/\sigma_{sv})} \quad (3)$$

where σ_{baro} is the standard deviation of ϵ_{baro}. Function availability is determined from the expanded coefficient matrix consisting of matrix G and the coefficient matrix in Eq. (3).

Two different ways of aiding GPS with barometric altimeter have been analyzed because of their utility for aviation applications. One way is to use pressure altitude data (altitude measurements with no correction for local temperature or pressure), but calibrated with GPS. The other way is to use altitude data calibrated with a local barometric setting.

B. Barometric Altimeter Aiding with GPS-Calibrated Pressure Altitude Data

This case uses altimeter data with a standard setting and no correction for temperature or local pressure, using these steps.

1) When navigation and RAIM detection (or identification) functions can be provided with GPS alone, baroaltitude is not used for these functions. At these times, if the geometry is good so that integrity can be assured to a certain level for the vertical position estimate from GPS, the barometric altimeter data are calibrated; that is, the offset between the altitude from the altimeter and altitude estimated from GPS is recorded. The geometric criterion that defines goodness for barometric altimeter data calibration is discussed below in conjunction with barometric altimeter calibration error.

2) If navigation and RAIM functions are not available with GPS alone, GPS is augmented by the measured barometric altimeter data calibrated with the offset obtained in step 1 at the most recent time at which geometry met the criterion for calibration. The augmented geometry for barometric altimeter aiding is used with the proper value of σ_{baro}.

In step 2, whatever value we use for σ_{baro} makes an impact on the availability analysis results. Most RAIM analyses to date for barometric altimeter aiding[3-5] have assumed a certain fixed value of σ_{baro} considered typical for each phase of flight. To model σ_{baro} more realistically, a different approach is taken in the present analysis. That is, the two sources of the error in the barometric altimeter data are separately evaluated before they are combined for the calculation of σ_{baro}: calibration error, and error caused by a change of the offset between the pressure altitude and geometric altitude. Each of these error sources is discussed below. Because these are independent sources of error, σ_{baro} can be obtained from their rss.

1. Barometric Altimeter Calibration Error Caused by GPS

This is the error associated with calibration of barometric altimeter data with GPS. There are two questions involved in this GPS calibration of barometric altimeter data. First, under what conditions can we calibrate barometric altitude? Second, how is the calibration error to be estimated?

To answer the first question, we can calibrate barometric altitude when the RAIM integrity algorithm ensures integrity (i.e., no alarm). If none of the satellites is failing with an abnormally large range error, the following should hold:

$$\sigma_h = \text{(all-in-view-set VDOP)} \times (1 \ \sigma \text{ pseudorange error}) \qquad (4)$$

where σ_h is the standard deviation of the GPS vertical position error. In this case, we can take σ_h as the standard deviation of barometric altimeter calibration error.

It should be noted, however, that before calibrating barometric altimeter data with GPS vertical position data, the validity of the GPS vertical data must be assured. Of course, the calibration will be done only when RAIM does not raise an alarm. However, we should remember that no alarm in RAIM does not mean every satellite has a nominal value of error; it only means that none of the satellite range errors has become large enough to cause the horizontal position error to violate the protection limit for the given phase of flight. Therefore, it is possible that one of the satellites is beginning to fail with a large error and has not yet been detected by the RAIM algorithm. In this case, barometric altimeter data would be calibrated with erroneous GPS vertical data. When this erroneous barometric altimeter data are subsequently used to augment GPS, the RAIM capability cannot be trusted. For this reason, a proper amount of calibration error

RECEIVER AUTONOMOUS INTEGRITY MONITORING 225

should be assigned that would reflect the level of integrity (i.e., protection limit) in the GPS vertical position data. The following is the approach used to account for calibration error.

Earlier analyses of RAIM performance show that the level of integrity (i.e., protection limit) that can be assured is strongly correlated with the worst subset horizontal dilution of precision (HDOP) and position dilution of precision (PDOP) (i.e., the maximum HDOP or PDOP among $n - 1$ out of the n satellites used for navigation). Specifically, it is shown in Refs. 6 and 7 that if the worst subset HDOP is less than or equal to three, then a horizontal error of greater than approximately 300 m can be protected against. This is under the assumption of the presence of selective availability (SA) with a standard deviation of 33 m for the pseudorange error. From this, it is observed that

$$\text{Horizontal protection limit} = 3 \times \text{(worst subset HDOP)}$$
$$\times (1 \, \sigma \text{ pseudorange error}) \quad (5)$$

By extrapolating the above to the case of vertical position integrity, we obtain

$$\text{Vertical protection limit} = 3 \times \text{(Worst subset VDOP)}$$
$$\times (1 \, \sigma \text{ pseudorange error}) \quad (6)$$

The Radio Technical Commission for Aeronautics (RTCA) specification[8] requires a missed detection probability of 0.001. This means that when there is a degradation of satellite ranging data, the probability that the protection limit is exceeded will be less than 0.001. Although it is difficult to characterize the statistical distribution of the error in the vertical position estimate used for calibration, we can assume that it has a normal distribution with a standard deviation of the error determined from the missed detection probability of 0.001. Under this assumption, the protection limit would correspond to 3.3 times the standard deviation of the error, which is approximated as 3 instead of 3.3. Therefore,

$$\sigma_{\text{baro calib}} = \text{worst subset vertical dilution of precision (VDOP)}$$
$$\times (1 \, \sigma \text{ pseudorange error}) \quad (7)$$

One thing to note is that although the barometric altimeter data should be calibrated whenever possible, it is better not to calibrate them when $\sigma_{\text{baro calib}}$ is too large. This is because the barometric altimeter calibration error may increase faster than the error caused by the pressure variation as a result of level or nonlevel flight. For this reason, it is assumed in the current analysis that barometric altimeter data are calibrated only when the worst subset VDOP is smaller than 5, which corresponds to a vertical protection limit of 500 m with SA on.

As a further measure to ensure that pressure altitude data are not calibrated by contaminated GPS data, TSO-C129 requires that calibration of pressure altitude data by GPS be performed only when the GPS test statistic is less than a threshold that corresponds to the 95th percentile point of the distribution of the test statistic in the presence of SA and that no other errors are present. This measure, along with the constraint on the maximum subset VDOP, would guarantee a probability of less than 0.001 of calibration with contaminated GPS data.

2. Error Caused by Change of Offset Between Pressure Altitude and Geometric Altitude

As the aircraft flies along after the barometric altimeter is calibrated with GPS, the offset between the altitude given by the barometric altimeter and the geometric altitude varies with changes in local pressure and temperature, as well as due to any change in aircraft altitude.[9,10] The appendix to this chapter describes statistical distribution of altitude offset vs distance flown. The growth of this uncertainty in barometric altimeter data varies widely, depending on the flight profile. In the current analysis, two different types of flight are considered, level flight and descending flight.

a. Level flight. Table A.1 in the appendix indicates that the standard deviation of the growth rate is approximately 0.5 m and 0.2 m per n.mi. of level flight at 250 and 850 mbars, respectively. A similar analysis in Ref. 10 determines that the worst growth rate of the uncertainty is 1 m per n.mi. of level flight. On the basis of these two sources, the standard deviation of the growth rate is estimated somewhat conservatively at 0.5 m per n.mi.; this value is used in the current analysis irrespective of the altitude. Although using the larger value of the rate may result in somewhat lower availability of functions, it should not affect integrity performance negatively.

b. Descending flight. For this case, a flight profile illustrated in Fig. 1 is assumed. The rates of growth of uncertainty shown in the figure are based on the results of an analysis reported in Ref. 10. (The rates shown may be considered

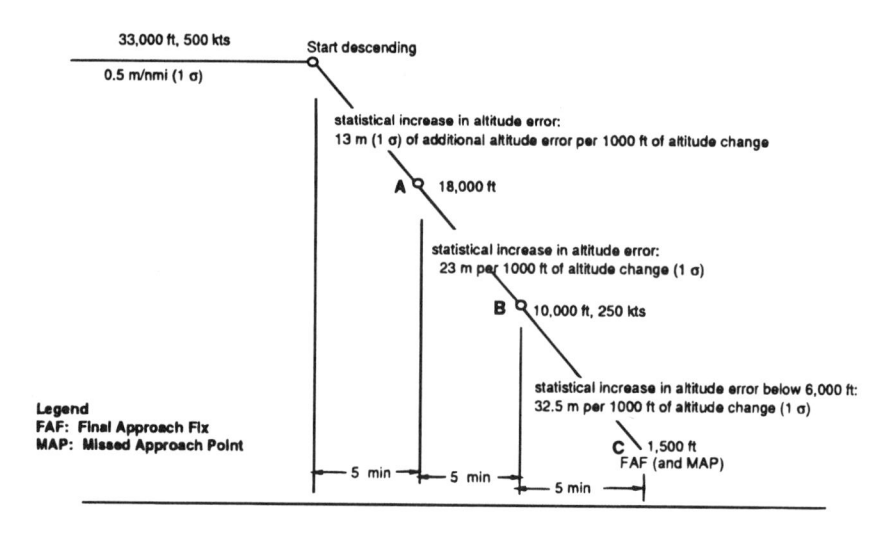

Fig. 1 Flight profile model.

representative for pressure altitude with no temperature correction. It might be possible to reduce the rates significantly with temperature correction of pressure altitude.)

3. Procedure to Calculate Function Availability

Figure 2 outlines the procedure to calculate function availability for barometric altimeter aiding. It is shown that the barometric altimeter is calibrated only if the detection function is available and the worst subset VDOP is less than a certain threshold.

C. Barometric Altimeter Aiding with Local Pressure Input

This is the case in which the barometric altimeter data with local barometric setting are used with no GPS calibration process. For σ_{baro}, the following values are used, based on Ref. 10: σ_{baro} = 290 m at 10,000 ft; and σ_{baro} = 49 m at a typical assumed minimum descent altitude of 1500 ft.

As defined earlier, σ_{baro} is the standard deviation of ϵ_{baro}, which is the error in measured barometric altitude after it has been calibrated (either using GPS altitude or local pressure setting) and converted to WGS-84 altitude. It is noted that there exists a finite amount of difference (e.g., maximum of about 50 m over CONUS) between the WGS-84 ellipsoid altitude and mean sea level altitude as a function of location. TSO-C129 (Ref. 2) requires that this difference be taken into account below 18,000 ft geometric altitude in calibrating the pressure altitude corrected for the local barometric pressure setting. It is believed that with this provision in TSO-C129, the above σ_{baro} values, especially the first, are conservative.

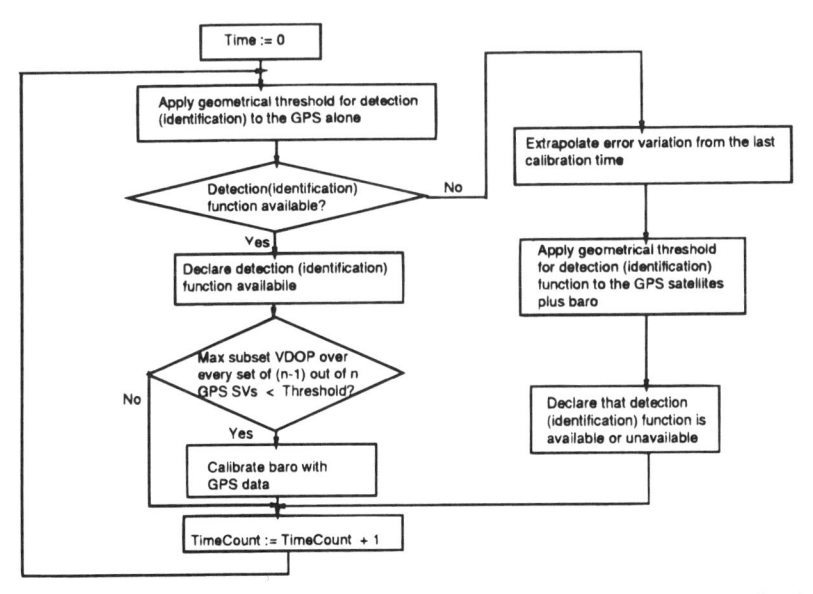

Fig. 2 Flow chart used to determine availabilities of detection and identification functions for the case of baro aiding with GPS calibration.

D. Augmented Geometry for Clock Coasting

The procedure for clock coasting is similar to that for barometric altimeter aiding with pressure altitude. That is, the user receiver clock is calibrated by GPS under good geometry defined below. When GPS satellite geometry is no longer good enough to provide reliable RAIM functions, GPS is augmented with a clock time measurement. The user receiver clock time is incorporated into the solution for navigation and RAIM functions as outlined below.

An additional equation to be included in Eq. (1) from the user receiver clock time measurement is expressed as:

$$\Delta T = [0 \quad 0 \quad 0 \quad 1] \begin{bmatrix} \Delta x \\ \Delta y \\ \Delta z \\ \Delta t \end{bmatrix} + \epsilon_{\text{clock}} \quad (8)$$

where ΔT is the difference between the measured user receiver clock time calibrated with GPS and the nominal GPS clock time, and ϵ_{clock} is the error in user receiver clock time (calibrated with GPS). If the standard deviation of of ϵ_{clock} is not equal to σ_{sv}, the above equation should be scaled before it is added to Eq. (1):

$$\frac{\Delta T}{(\sigma_{\text{clock}}/\sigma_{sv})} = \begin{bmatrix} 0 & 0 & 0 & \frac{1}{(\sigma_{\text{clock}}/\sigma_{sv})} \end{bmatrix} \begin{bmatrix} \Delta x \\ \Delta y \\ \Delta z \\ \Delta t \end{bmatrix} + \frac{\epsilon_{\text{clock}}}{(\sigma_{\text{clock}}/\sigma_{sv})} \quad (9)$$

where σ_{clock} is the standard deviation of ϵ_{clock}; σ_{clock} is determined by calibration error and error caused by receiver clock drift. Because these are independent sources of error, σ_{clock} can be obtained from their rss. Each of these error sources is discussed below.

1. Clock Calibration Error

As in the case for the barometric altimeter calibration error, calibration of the receiver clock requires integrity in the GPS time estimate. The level of integrity that can be assured in this time estimate is strongly correlated with the worst time dilution of precision (TDOP) among $(n - 1)$ out of n satellites used for navigation. Following logic similar to that for the barometric altimeter calibration,

$$\sigma_{\text{clock calib}} = (\text{worst subset TDOP}) \times (1 \ \sigma \ \text{pseudorange error}) \quad (10)$$

As in the case of barometric altimeter aiding, if calibration cannot be done with a certain tight integrity level, it is better not to calibrate the receiver clock. In the current analysis, the receiver clock was calibrated by GPS time only when the worst subset TDOP was less than or equal to three.

2. Error Because of Receiver Clock Drift

Uncertainty in the receiver clock time grows because of clock drift from GPS calibration time until clock aiding time. It is assumed that the growth rate is independent of the flight profile. A drift of 10^{-9} was assumed, which corresponds to a good temperature-controlled crystal oscillator. It is assumed here that there is no attempt to estimate the clock drift rate. Although estimation of the clock rate may potentially be promising, the presence of SA would limit the accuracy with which the clock rate can be estimated.

3. Procedure to Calculate Function Availability

The procedure to calculate function availability for clock coasting is similar to the procedure for barometric altimeter aiding shown in Fig. 2.

E. Simultaneous Use of Barometric Altimeter Aiding and Clock Coasting

If barometric altimeter aiding and clock coasting are used simultaneously, both equations of the form in Eqs. (3) and (9) are included in the augmented solution. The same set of model parameters and thresholds are used as that for the individual cases of barometric altimeter aiding and clock coasting.

III. Definitions of Function Availability

Availability is defined as the percentage of time that the system provides the required performance of the function for the phase of flight. In this analysis, availability is calculated on the basis of whether the user-to-satellite geometry (or an augmented geometry) is good enough to provide a satisfactory level of performance. The performance requirements for each of the functions are described below.

A. Navigation Function

Availability of the navigation function is based on the 2 drms (twice the distance root mean square) horizontal position radial error value calculated from the well-known equation

$$2 \text{ drms} = 2 \times \text{HDOP} \times \sigma_{sv} \quad (11)$$

where σ_{sv} is the standard deviation of satellite range error and HDOP is horizontal dilution of precision. That is, if the geometry gives a small enough HDOP for the all-in-view satellite solution to make 2 drms less than a given position error tolerance, then the navigation function is considered available for that geometry.

B. Receiver Autonomous Integrity Monitoring Detection Function

Availability of the RAIM detection function is based on the requirements adopted by RTCA SC-159 for the supplemental use of GPS[8]: presence of a malfunction of a satellite causing the position error protection limit to be violated shall be detected with a minimum probability of 0.999 given that the protection limit is violated; and the rate of alarms (false or true) will not be more than 0.002/h. An empirical relationship that determines, for any given geometry, availability/unavailability of RAIM detection function performing within the level specified above is established via extensive Monte Carlo simulations in

Ref. 11. This relationship, which is based on what is called approximate radial-error-protected (ARP), is used in the current availability analysis.

C. Receiver Autonomous Integrity Monitoring Identification Function

No formal requirements for the RAIM identification function have been established. For this reason, the following assumption is made. Upon occurrence of a malfunctioning satellite with an abnormal range error, RAIM will be able to detect the occurrence and also to correctly identify the satellite before the protection limit is violated. This detection and identification function must be accomplished with a probability of 0.999.

The geometric criteria that satisfy the above requirement are derived from a study reported in Ref. 12, which developed a new identification algorithm and showed, via simulation, that the above identification function performance requirement can be met if PDOP is smaller than a PDOP threshold for every set of $n - 2$ out of n satellites in view where

$$\text{PDOP threshold} = 25 \times [\text{protection limit (n.mi.)}] \times (33/\sigma_{sv}) \qquad (12)$$

This is the criterion used in the current analysis.

IV. Results

This section first describes the cases analyzed and parameters evaluated. Then, the results are presented and discussed. Many different cases were analyzed in terms of such factors as constellation, protection limits, and user locations. These are summarized in Table 1. As noted in the table, the user-to-satellite geometries have been sampled every 5 min.

A. Parameters of Interest

1. Availability

As stated earlier, availability is defined as the percentage of time that the system provides the required performance for the phase of flight. For the case of a user on one of the routes, availability is the average over the duration of flight and over six departure times. For the case of a user at an airport area, the availability is the average over a 24-h period.

2. Average Duration of Outage

This is the average duration of outages of the respective function (navigation, RAIM detection, or RAIM identification) over a flight or over 24-h period.

3. Maximum Duration of Outage

For an airport area, this is the maximum outage duration observed over a 24-h period. For flight along a route, this maximum duration of outage is the maximum of the maximum outage duration in each flight over flights with six different departure times.

RECEIVER AUTONOMOUS INTEGRITY MONITORING

Table 1 Summary of analysis scenarios

Constellation	User locations
Optimal 21[a]	User on a moving platform[d]
24 Primary[a]	New York (JFK)–Los Angeles (LAX)
24 Primary with 3 satellites failed	San Francisco (SFO)–Narita, Japan (NRT)
(satellites 1, 10, and 11)[b]	Dallas (DFW)–Paris (CDG)
Protection limits[c]	Airport locations[e]
0.3 n.mi.	San Francisco (SFO)
1 n.mi.	Dallas-Fort Worth (DFW)
2 n.mi.	Chicago (ORD)
12.6 n.mi.	New York (JFK)
Standard deviation	Atlanta (ATL)
of pseudorange error	
$\sigma = 33$ m (SA present)	
$\sigma = 10$ m (SA absent)	

[a]21 and 24 satellite positions defined in Ref. 14 but with repeating ground tracks.
[b]This is considered to be an "average" 21 satellite constellation in terms of coverage.[15]
[c]The first three protection limits are RTCA requirements[8] for the nonprecision approach, terminal, and en route phases of flight, respectively. The last protection limit of 12.6 n.mi. corresponds to the current requirement for the oceanic phase of flight.
[d]It is assumed that the user aircraft is flying at a speed of 500 n.mi./h over the great circle defined by each pair of locations. User-satellite geometries are sampled every 5 min. Availability is obtained as an average over six departure times, spaced four hours apart.
[e]These airport locations in CONUS were used to evaluate availability, for terminal and nonprecision approach phases of flight. Availability is obtained as an average over 24-h period with user-to-satellite geometries sampled every 5 min.

B. Discussion of Results

Referring to Tables 2–4, availabilities of navigation, detection, and identification functions, respectively, for protection limits of 1, 2, and 12.6 n.mi. are shown for the three cases of no augmentation, barometric altimeter aiding with GPS calibration, and clock coasting (The combined use of barometric altimeter aiding and clock coasting is treated later in Tables 6 and 7.) For barometric altimeter aiding, level flight was assumed, which is typical for the oceanic and the majority of the en route phases of flight. The results show the following:

1) For navigation, all three constellations have comparable very high availabilities. In fact, only 24Minus3 without an altimeter or clock augmentation suffers any outage at all and has average and maximum outage durations up to 15 and 25 min, respectively.

2) For detection, 24Pr and Opt21 have comparable availability, and 24Minus3 has significantly degraded availability. For the 24Minus3 constellation, average and maximum outage durations are mostly 15–40 min and 50–80 min, respectively. For the other two constellations, they are typically 10 and 25 min, respectively.

3) For identification, only augmented 24Pr has an availability that might be acceptable for a sole-means of navigation (i.e., without other augmentations, such as inertial systems). For the 24 Pr constellation, average and maximum outage durations are typically 5–10 min and 15–25 min, respectively. For the

Table 2 Availabilities of navigation function, SA on, mask angle = 7.5 deg

	24 Primary			Optimal 21			24 Pr minus 3		
	A	B	C	A	B	C	A	B	C
	Protection limit of 1 n.mi.								
GPS alone									
JFK–LAX	100	0	0	100	0	0	100	0	0
SFO–NRT	100	0	0	100	0	0	99.1	8	15
DFW–CDG	100	0	0	100	0	0	99.2	25	25
Baro aiding with GPS calibration in level flight									
JFK–LAX	100	0	0	100	0	0	100	0	0
SFO–NRT	100	0	0	100	0	0	99.8	5	5
DFW–CDG	100	0	0	100	0	0	100	0	0
Clock coasting (1E-9 drift)									
JFK–LAX	100	0	0	100	0	0	100	0	0
SFO–NRT	100	0	0	100	0	0	99.7	5	5
DFW–CDG	100	0	0	100	0	0	100	0	0
	Protection limit of 2 n.mi.								
GPS alone									
JFK–LAX	100	0	0	100	0	0	100	0	0
SFO–NRT	100	0	0	100	0	0	99.1	8	15
DFW–CDG	100	0	0	100	0	0	99.2	25	25
Baro aiding with GPS calibration in level flight									
JFK–LAX	100	0	0	100	0	0	100	0	0
SFO–NRT	100	0	0	100	0	0	100	0	0
DFW–CDG	100	0	0	100	0	0	100	0	0
Clock coasting (1E-9 drift)									
JFK–LAX	100	0	0	100	0	0	100	0	0
SFO–NRT	100	0	0	100	0	0	99.8	5	5
DFW–CDG	100	0	0	100	0	0	100	0	0
	Protection limit of 12.6 n.mi.								
GPS alone									
JFK–LAX	100	0	0	100	0	0	100	0	0
SFO–NRT	100	0	0	100	0	0	99.4	10	15
DFW–CDG	100	0	0	100	0	0	99.2	25	25
Baro aiding with GPS calibration in level flight									
JFK–LAX	100	0	0	100	0	0	100	0	0
SFO–NRT	100	0	0	100	0	0	100	0	0
DFW–CDG	100	0	0	100	0	0	100	0	0
Clock coasting (1E-9 drift)									
JFK–LAX	100	0	0	100	0	0	100	0	0
SFO–NRT	100	0	0	100	0	0	100	0	0
DFW–CDG	100	0	0	100	0	0	100	0	0

A = Availability, %.
B = Average outage duration, min.
C = Maximum outage duration, min.

RECEIVER AUTONOMOUS INTEGRITY MONITORING

Table 3 Availabilities of receiver autonomous integrity monitoring detection function, SA on, mask angle = 7.5 deg

	24 Primary			Optimal 21			24 Pr minus 3		
	A	B	C	A	B	C	A	B	C
	Protection limit of 1 n.mi.								
GPS alone									
JFK–LAX	100	0	0	99.7	5	5	88.5	14	80
SFO–NRT	99.4	5	5	94.1	10	25	91	17	85
DFW–CDG	99.5	8	10	94.1	8	15	92.9	12	55
Baro aiding with GPS calibration in level flight									
JFK–LAX	100	0	0	100	0	0	93.9	48	80
SFO–NRT	99.8	5	5	100	0	0	94.3	31	85
DFW–CDG	99.8	5	5	99.7	10	10	96.3	16	50
Clock coasting (1E-9 drift)									
JFK–LAX	100	0	0	100	0	0	92.6	23	80
SFO–NRT	99.8	5	5	97.7	9	20	92.7	21	85
DFW–CDG	99.7	5	5	97.4	11	15	94.2	12	55
	Protection limit of 2 n.mi.								
GPS alone									
JFK–LAX	100	0	0	100	0	0	91.3	17	80
SFO–NRT	99.8	5	5	96.9	7	15	93.7	16	80
DFW–CDG	99.8	5	5	96.8	6	10	95	12	55
Baro aiding with GPS calibration in level flight									
JFK–LAX	100	0	0	100	0	0	94.9	40	75
SFO–NRT	99.8	5	5	100	0	0	95.7	20	50
DFW–CDG	100	0	0	100	0	0	98.1	15	35
Clock coasting (1E-9 drift)									
JFK–LAX	100	0	0	100	0	0	93.6	33	80
SFO–NRT	99.8	5	5	99.8	5	5	94.6	22	80
DFW–CDG	100	0	0	99	6	10	96.5	16	55
	Protection limit of 12.6 n.mi.								
GPS alone									
JFK–LAX	100	0	0	100	0	0	92.9	22	80
SFO–NRT	100	0	0	99.4	7	10	94.9	24	80
DFW–CDG	100	0	0	99.2	5	5	97	16	50
Baro aiding with GPS calibration in level flight									
JFK–LAX	100	0	0	100	0	0	96.8	50	50
SFO–NRT	100	0	0	100	0	0	98.8	8	15
DFW–CDG	100	0	0	100	0	0	99.2	25	25
Clock coasting (1E-9 drift)									
JFK–LAX	100	0	0	100	0	0	96.8	25	35
SFO–NRT	100	0	0	100	0	0	98.6	23	30
DFW–CDG	100	0	0	100	0	0	98.4	25	40

A = Availability, %.
B = Average outage duration, min.
C = Maximum outage duration, min.

Table 4 Availabilities of receiver autonomous integrity monitoring identification function, SA on, mask angle = 7.5 deg

	24 Primary			Optimal 21			24 Pr minus 3		
	A	B	C	A	B	C	A	B	C
				Protection limit of 1 n.mi.					
GPS alone									
JFK–LAX	87.8	10	15	51	19	65	57.7	23	125
SFO–NRT	85	12	35	55.7	17	85	53.4	27	200
DFW–CDG	89.9	9	25	56.9	18	110	62.2	24	180
Baro aiding with GPS calibration in level flight									
JFK–LAX	99.4	5	5	94.2	10	20	80.4	24	100
SFO–NRT	98.9	6	10	90.9	11	35	83.3	20	145
DFW–CDG	99.2	6	10	92	8	20	88.1	17	125
Clock coasting (1E-9 drift)									
JFK–LAX	98.7	7	10	92	9	25	76.9	21	100
SFO–NRT	97.1	12	25	85.6	13	65	76.5	25	185
DFW–CDG	98.9	9	15	86.9	12	60	82.9	21	180
				Protection limit of 2 n.mi.					
GPS alone									
JFK–LAX	93.9	7	10	64.4	12	60	67	17	110
SFO–NRT	91.2	7	30	71.1	11	60	64.8	16	190
DFW–CDG	94.4	7	15	68.6	12	70	71.8	14	140
Baro aiding with GPS calibration in level flight									
JFK–LAX	99.7	5	5	97.1	9	15	86.5	19	95
SFO–NRT	99.1	6	10	95.7	9	20	87.2	18	125
DFW–CDG	99.4	5	5	96.3	6	15	91	14	75
Clock coasting (1E-9 drift)									
JFK–LAX	99.7	5	5	96.5	7	15	84	21	95
SFO–NRT	98.6	6	15	93.7	14	50	82.7	26	175
DFW–CDG	99.4	5	5	91.5	10	40	87.7	16	125
				Protection limit of 12.6 n.mi.					
GPS alone									
JFK–LAX	99	5	5	90.1	7	20	77.6	19	105
SFO–NRT	97.4	9	25	85.3	11	50	78.2	19	155
DFW–CDG	98.4	7	15	86.5	15	60	83.5	16	140
Baro aiding with GPS calibration in level flight									
JFK–LAX	99.7	5	5	99	5	5	92.3	17	80
SFO–NRT	99.7	5	5	98.9	6	10	93.5	15	85
DFW–CDG	100	0	0	99.2	5	5	96	10	50
Clock coasting (1E-9 drift)									
JFK–LAX	100	0	0	99.7	5	5	92.3	24	80
SFO–NRT	99.8	5	5	98.5	6	10	94	20	100
DFW–CDG	99.8	5	5	98.9	6	10	95	10	50

A = Availability, %.
B = Average outage duration, min.
C = Maximum outage duration, min.

other two constellations, outage durations are much longer; for example, maximum outage duration for 24Minus3 is typically 100–200 min.

It should be noted that although barometric altimeter aiding always improves availability, the improvement is most dramatic for the identification function, when availability is fairly low without barometric altimeter aiding (e.g., for identification function for a protection limit of 1 n.mi.). On the other hand, compared with GPS-calibrated barometric altimeter aiding in level flight, clock coasting does not improve availabilies as much, even with the assumption of a clock drift corresponding to a high-quality, temperature-controlled crystal oscillator. It should be noted, however, that a program is currently underway to develop an inexpensive microminiature atomic clock with a drift of 10^{-11} or less within the next five years.[13] If such an accurate clock is incorporated into a GPS receiver for aviation users, clock coasting would increase RAIM availability significantly because for a given error, the coasting time increases by a factor of 100 relative to that of a crystal oscillator. Also, RAIM performance and RAIM availability could be improved if the clock drift rate is also estimated along with the clock bias. In that case, however, the presence of SA would limit the accuracy with which the clock rate can be estimated.

For the 24Minus3 constellation, Table 5 compares availabilities at five major airport areas for three different augmentation cases: 1) GPS-calibrated barometric altimeter aiding in descending flight; 2) barometric altimeter aiding using local pressure input; and 3) clock coasting. It is observed that in descending flight at 10,000 ft altitude, barometric altimeter aiding using local pressure input brings more improvement in RAIM availability than the case of GPS-calibrated barometric altimeter aiding. This difference in improvement is much more pronounced at 1500 ft altitude. The table also shows that in descending flight, the availability improvement with GPS-calibrated barometric altimeter aiding is comparable to the case of clock coasting with a receiver clock having a drift assumed in the current analysis. For both cases, however, the results for RAIM identification show very low availabilities.

In Table 6, the availabilities for the case of SA off is compared with those for the case with SA on for the 24Minus3 constellations. As expected, turning SA off always brings some improvement in availability. According to results not reported in this paper, the improvement is much more significant for a protection limit of 0.3 n.mi. Also shown in the table are the case of simultaneous use of barometric altimeter aiding and clock coasting. It is observed that using both barometric altimeter aiding and clock coasting always improves the availability to at least the better of the individual availabilities and often somewhat more than that.

In Table 7, for the 24Minus3 constellation with SA off, the availability of the RAIM detection function for a mask angle of 2.5 deg is compared with the case of a mask angle of 7.5 deg. In general, lowering the mask angle significantly improves availability for the RAIM detection and identification functions.

V. Summary and Conclusions

On the basis of the results, the following summary/conclusions are drawn.
1) Even if GPS is used as a supplemental system, high availability of the

Table 5 Availabilities for terminal and nonprecision approach phases of flight, 24Minus3 constellation, SA on, mask angle = 7.5 deg

	At 10,000 ft with protection limit of 1 n.mi.									At 1,500 ft with protection limit of 0.3 n.mi.								
	Navigation			Detection			Identification			Navigation			Detection			Identification		
	A	B	C	A	B	C	A	B	C	A	B	C	A	B	C	A	B	C
GPS alone																		
SFO	100	0	0	89.9	18	75	54.5	30	175	100	0	0	72.2	27	115	28.8	64	480
DFW	100	0	0	85.8	17	55	50.7	32	150	100	0	0	68.1	35	125	25.7	82	460
ORD	100	0	0	86.5	20	95	54.9	31	265	100	0	0	70.1	43	135	31.3	58	515
JFK	97.9	15	25	82.3	32	175	58.3	26	285	97	15	25	68.8	50	230	30.2	53	310
ATL	96.9	45	45	86.1	18	100	56.9	34	285	96.5	25	45	73.3	55	235	32.3	54	460
Baro aiding with GPS calibration in descending flight																		
SFO	100	0	0	92.7	26	75	80.0	32	145	100	0	0	72.6	30	115	42	70	445
DFW	100	0	0	90.3	23	55	77.4	22	135	100	0	0	69.1	34	125	36.8	70	460
ORD	100	0	0	92	29	95	73.3	35	200	100	0	0	71.5	46	135	43.8	68	355
JFK	100	0	0	87.2	46	140	76.7	34	235	99	8	10	69.1	40	225	46.5	39	285
ATL	100	0	0	90.3	35	100	77.1	33	255	99.7	5	5	74	63	235	46.5	51	335
Baro aiding with local pressure input																		
SFO	100	0	0	93.1	33	75	80.0	32	145	100	0	0	86.5	28	95	68.8	35	155
DFW	100	0	0	91.0	26	55	78.5	21	135	100	0	0	83.7	34	55	66	35	135
ORD	100	0	0	93.8	30	75	75.3	24	95	100	0	0	89.2	17	90	67	37	185
JFK	100	0	0	92.0	19	55	78.1	35	225	100	0	0	85.1	36	80	68.4	41	230
ATL	100	0	0	92.7	21	65	78.8	28	130	100	0	0	88.2	24	70	68.4	50	285
Clock coasting (1E-9 drift)																		
SFO	100	0	0	92.7	26	75	78.1	24	110	100	0	0	73.6	27	115	53.1	45	315
DFW	100	0	0	86.8	15	55	71.2	32	135	100	0	0	70.8	38	125	46.5	70	295
ORD	100	0	0	87.9	25	95	70.5	39	135	100	0	0	71.5	51	135	53.1	56	325
JFK	100	0	0	84.7	44	175	72.9	39	230	97.2	13	25	69.8	54	230	57.3	38	285
ATL	100	0	0	86.5	18	100	74	54	200	99.7	5	5	74.7	73	230	57.6	51	295

A = Availability, %.
B = Average outage duration, min.
C = Maximum outage duration, min.

Table 6 Availabilities of receiver autonomous integrity monitoring functions with and without SA, 24Minus3 constellation, mask angle = 7.5 deg

	Protection limit of 1 n.mi.						Protection limit of 2 n.mi.					
	SA on			SA off			SA on			SA off		
	A	B	C	A	B	C	A	B	C	A	B	C
	Detection function											
Baro aiding with GPS calibration in level flight												
JFK–LAX	93.9	48	80	94.6	43	75	94.9	40	75	95.2	38	70
SFO–NRT	94.3	31	85	95.4	30	80	95.7	20	50	96.6	22	50
DFW–CDG	96.3	16	50	97.6	19	50	98.1	15	35	98.6	15	35
Clock coasting (1E-9 drift)												
JFK–LAX	92.6	23	80	93.3	26	80	93.6	33	80	94.2	45	80
SFO–NRT	92.7	21	85	94.8	21	80	94.6	22	80	95.4	30	80
DFW–CDG	94.2	12	55	96.2	12	55	96.5	16	55	97.4	27	50
Combined use of baro and clock coasting												
JFK–LAX	93.9	48	80	94.6	43	75	94.9	40	75	95.2	38	70
SFO–NRT	94.1	32	85	95.4	30	80	96	22	50	96.6	22	50
DFW–CDG	96.3	16	50	97.6	15	50	98.2	14	35	98.7	20	35
	Identification function											
Baro aiding with GPS calibration in level flight												
JFK–LAX	80.4	24	100	86.9	21	95	86.5	19	95	89.1	17	95
SFO–NRT	83.3	20	145	87.7	19	125	87.2	18	125	90.1	18	125
DFW–CDG	88.1	17	125	91.2	15	95	91	14	75	94.2	13	70
Clock coasting (1E-9 drift)												
JFK–LAX	76.9	21	100	81.1	25	100	84	21	95	85.9	28	95
SFO–NRT	76.5	25	185	81.5	29	175	82.7	26	175	84.6	29	150
DFW–CDG	82.9	21	180	86.4	17	140	87.7	16	125	89.7	15	125
Combined use of baro and clock coasting												
JFK–LAX	85.3	21	95	89.1	24	95	91	28	95	91.3	27	95
SFO–NRT	86.4	24	145	89	22	125	91.4	19	125	92.7	20	120
DFW–CDG	91.5	16	95	93.6	14	85	94.7	12	70	95.7	15	70

A = Availability, %.
B = Average outage duration, min.
C = Maximum outage duration, min.

RAIM detection function during normal operation (e.g., with a typical 21-satellite constellation) is important to avoid arriving at the final approach fix and finding that a GPS approach cannot be conducted because the satellite geometry does not provide the RAIM detection function. This is undesirable especially if no other approach is available at the airport. Because barometric altimeter aiding increases the availability significantly, the FAA decided to require this augmentation in TSO-C129.

2) In the approach mode, barometric aiding with local pressure input improves availability significantly more than barometric altimeter aiding with GPS calibration. This led the SOIT to require use of local pressure information in the approach mode in TSO-C129.

Table 7 Receiver autonomous integrity monitoring availabilities with two different mask angles, SA off, 24Minus3 constellation

	Protection limit of 1 n.mi.						Protection limit of 2 n.mi.					
	7.5 deg			2.5 deg			7.5 deg			2.5 deg		
	A	B	C	A	B	C	A	B	C	A	B	C
	Detection function											
Baro aiding with GPS calibration in level flight												
JFK–LAX	94.6	43	75	96.5	55	55	95.2	38	70	96.8	25	45
SFO–NRT	95.4	30	80	99.5	15	15	96.6	22	50	99.7	10	10
DFW–CDG	97.6	19	50	99.8	5	5	98.6	15	35	99.8	5	5
Clock coasting (1E-9 drift)												
JFK–LAX	93.3	26	80	96.5	55	55	94.2	45	80	96.5	55	55
SFO–NRT	94.8	21	80	99.4	10	15	95.4	30	80	99.4	10	15
DFW–CDG	96.2	12	55	99.8	5	5	97.4	27	50	99.8	5	5
Combined use of baro and clock coasting												
JFK–LAX	94.6	43	75	96.8	25	45	95.2	38	70	97.4	40	40
SFO–NRT	95.4	30	80	99.5	15	15	96.6	22	50	99.7	10	10
DFW–CDG	97.6	15	50	99.8	5	5	98.7	20	35	99.8	5	5
	Identification function											
Baro aiding with GPS calibration in level flight												
JFK–LAX	86.9	21	95	92	21	70	89.1	17	95	92.9	16	70
SFO–NRT	87.7	19	125	97.5	10	20	90.1	18	125	98	9	20
DFW–CDG	91.2	15	95	98.2	14	30	94.2	13	70	98.9	9	20
Clock coasting (1E-9 drift)												
JFK–LAX	81.1	25	100	90.7	24	85	85.9	28	95	92.6	58	85
SFO–NRT	81.5	29	175	96.6	16	50	84.6	29	150	97.4	14	50
DFW–CDG	86.4	17	140	96	14	45	89.7	15	125	97.4	20	40
Combined use of baro and clock coasting												
JFK–LAX	89.1	24	95	93.9	24	70	91.3	27	95	95.2	38	70
SFO–NRT	89	22	125	98.6	9	20	92.7	20	120	98.9	9	20
DFW–CDG	93.6	14	85	98.9	12	20	95.7	15	70	99.8	5	5

A = Availability, %.
B = Average outage duration, min.
C = Maximum outage duration, min.

3) However, even with the use of local pressure input, the availability of the detection function at the final approach fix is only 84–89% with the 24Minus3 constellation. According to FAA certification specialists, in order for a navigation system to be useable for nonprecision approach, the minimum availability of approach capability (which requires the RAIM detection function to exist) upon arrival should be at least 95%. Because this goal cannot be achieved, even with use of local pressure input for the 24Minus3 constellation, and because the majority of outage periods are predictable once the constellation is known, TSO-C129 requires GPS receivers to be able to predict RAIM availability at the destination airport at the estimated time of arrival.

4) For the other phases of flight (i.e., oceanic, en route, and terminal), the availability of the RAIM detection function is high enough (typically 95% or higher with the 24Minus3 constellation) so that GPS may successfully be used as a supplemental system. In order for GPS to be used as a sole-means (stand-alone) system, however, a very high availability (close to 100%) is required both for detection and identification functions. However, the analysis indicated that availability of the identification function is not high enough even for the oceanic phase of flight (i.e., about 95% with the 24Minus3 constellation). Therefore, for GPS is to be used as a sole-means system, a system such as an inertial system, or some other augmentation would be required.

5) Clock coasting with a user receiver clock with a drift of 10^{-9} is not as effective as barometric altimeter aiding, especially in level flight. As stated earlier, however, if a microminiature atomic clock with a drift of 10^{-11} or smaller becomes available in the future, clock coasting would significantly improve RAIM availability. Also, estimation of the clock rate may be promising, although the presence of SA would limit the accuracy with which the clock rate can be estimated.

Appendix: Statistical Distribution of the Height Gradients[†]

A. Background

An aircraft is assigned a flight level, which means it follows whatever height gives a constant reading on its barometric altimeter. A particular law of height vs pressure is then assumed in order to derive height. Even if the pilot maintains a constant height reading of his barometric altimeter, the aircraft is for most of the time actually altering its vertical distance from mean sea level because of deviations from the assumed pressure–height law and variations in true ground pressure.

This becomes important for the process of aiding the GPS receiver, because the height input the receiver can utilize is the height indicated by the barometric altimeter. Because that height is in error, it follows that there will be an error in the receiver's output. Three factors have an impact on the likely success in aiding the receiver from a barometric altimeter: 1) the rate at which this altitude error varies with distance flown; 2) the characteristics of the error arising when the GPS receiver with its erroneous height is used to calibrate the barometric altimeter; and 3) the geometric limitation inherent in treating the information from the barometric altimeter as a distance from the Earth's center (i.e., the dilution of precision when this sensor is used).

Feature 1) is specific to the atmosphere and has nothing to do with GPS. Feature 2) is specific to GPS and has nothing to do with the atmosphere, but is a result of range error distributions modified by GPS geometry. Feature 3) is a combination of GPS and baormetrically measured altitude; whereas the satellite moves about, altitude is always relative to the Earth's center. In each of these cases, we could work out a probability distribution for the consequences.

[†]R. Johannessen and C. Dixon, BNR Europe Limited, Harlow, England.

B. Technique Adopted

Meteorological measurements from aircraft and radiosonde measurements are fed into the model[16] developed by the UK Meteorological (MET) Office, which produces, among other parameters, values for height having a given barometric pressure, at fine intervals in latitude and longitude. This is done at regular intervals of time as part of the work of the Met Office. A selection of these values were stored on floppy disks and given to BNR Europe Limited, in Harlow, England for analysis, funded from the Chief Scientist's Division of the UK CAA in London. Geopotential height values were provided for both 250 mbar and for 850 mbar, corresponding approximately to heights of 33,000 ft and 4000 ft, respectively. These height values on a regular geographic interval, taking account of the horizontal separation between the sampling points, provided a large number of slopes expressed in meters height change per nautical miles horizontal distance flown when maintaining constant barometric pressure.

Samples were taken between 80.625W and 30E at intervals of 0.9375° long., and between 30N and 60N at intervals of 0.75° lat. 4879 points were available. At each of these points four gradients were computed so that some 19,000 gradients resulted for each day. Twelve different days (the first day of every month for one year) were sampled providing a data base of 225,000 gradients, all at 250 mbar. Additionally a subset of points were chosen at 850 mbar. These individually computed values for slope were then analyzed statistically and the probability distribution derived.

C. Results Obtained (Table A.1)

Taking a typical aircraft speed of 450 kt and assuming an integrity outage time of 10 min, the aircraft will have traveled 75 n.mi. leading to a height variation of less than $(0.92 \times 75) = 69$ m at 95% level of confidence (250 mbar). Likewise, if the integrity outage is 60 min, the corresponding height variation is 414 m. In this context, it should be noted that at least six satellites are needed to provide fault isolation, whereas at least five are needed for fault detection. The proportion of time that the requirements for fault isolation are met is, therefore, lower than the proportion of time that the requirements for fault detection are met. It follows that the duration of the periods when a receiver cannot isolate a faulty satellite will be longer than the duration of periods when a receiver cannot detect whether or not the integrity is good. Whereas the 69 m above may be appropriate for fault detection difficulties (as is of interest to supplemental means navigation), the longer 414 m is likely to be more relevant to the harder fault isolation cases (as is important to sole means navigation).

For comparison, it is noted that Ref. 17 has 0.7 m/n.mi. as "typical" for 40,000 ft altitude over the United States and that Ref. 10 has 1.0 m/n.mi. for 30,000 ft altitude, also over the United States. It seems from the latter paper that the 1.0 m/n.mi. is derived by looking at the most closely spaced pressure contours, thus providing a worst case value. Both these reference therefore, are, in broad agreement with Table A.1.

The probability distribution in Table A.1 can be used in two different ways. First, we can take a simulated flight proceeding along the track until there is an "outage"; i.e., when the RAIM or navigation functions cannot be performed

Table A.1 Height change/distance flown for different probabilities at 250 and 850 mbar[a]

Confidence level, %	Slope in m/n.mi.	
	250 mbar	850 mbar
5	0.04	0.04
15	0.07	0.04
25	0.13	0.06
35	0.18	0.07
45	0.23	0.09
55	0.31	0.13
65	0.38	0.16
75	0.49	0.20
85	0.65	0.25
95	0.92	0.36

[a]250 mbar and 850 mbar correspond approximately to heights of 33,000 ft and 4000 ft, respectively.

without aiding. At that stage, the barometric height is "calibrated," and the navigation or RAIM functions continue in height aiding mode with the height error changing because of the slope. Both the calibration error and the slope error are selected at random; the former from a distribution curve for height error in GPS and the latter from the distribution of Table A.1. This will most closely represent the procedure followed in the navigating aircraft.

Second, we can approximate the slope distribution to a standard deviation, and combine that error along with the pseudorange errors through normal matrix principles. The distribution was derived in order to allow the first option. On the other hand, the analysis contained earlier in the chapter used the second option.

References

[1]Lee, Y., "RAIM Availability for GPS Augmented with Barometric Altimeter Aiding and Clock Coasting," *Navigation,* Vol. 40, No. 2, 1993.

[2]"Airborne Supplemental Navigation Equipment Using the Global Positioning System (GPS)," Technical Standards Order (TSO)-C129, FAA Aircraft Certification Service, Washington, DC, Dec. 10, 1992.

[3]Brown, A., and Schmid, T., "Integrity Monitoring of the Global Positioning System Using a Barometric Altimeter," *Proceedings of the Institute of Navigation National Technical Meeting,* (Santa Barbara, CA), Institute of Navigation, Washington, DC, Jan. 26–29, 1988.

[4]Brown, R., Grover, G., Chin, Y., and Kraemer, J. H., "Update on GPS Integrity Requirements of the RTCA MOPS," *Proceedings of the Institute of Navigation Satellite Division's International Technical Meeting,* (Colorado Springs, CO), Institute of Navigation, Washington, DC, Sept. 11–13, 1991 (ION GPS-91).

[5]Van Dyke, K., "RAIM Availability for Supplemental GPS Navigation," *Navigation,* Vol. 39, No. 4, 1992–1993, pp. 429–43.

[6]Brown, R. G., and McBurney, P., "Self-Contained Integrity Check Using Maximum Solution Separation as the Test Statistic," *Navigation,* Vol. 35, No. 1, 1988, pp. 41–54.

[7]Lee, Y., "Performance Analysis of Self-Contained Methods for GPS Integrity Function," The MITRE Corporation, MTR-88W89, McLean, VA, Nov. 1988.

[8]"Minimum Operational Performance Standards for Airborne Supplemental Navigation Equipment Using Global Positioning System (GPS)," Radio Technical Commission for Aeronautics RTCA DO-208, Washington, DC, July 1991.

[9]Asbury, M. J. A., Forrester, D. A., Dixon, C. S., and Johannessen, R., "Probability Distributions That are Important When Assessing Barometric Aiding to GPS," *Proceedings of the Institute of Navigation Satellite Division's International Technical Meeting,* (Colorado Springs, CO), Institute of Navigation, Washington, DC, Sept. 19–21, 1990 (ION GPS-90).

[10]Dobyne, J., "The Accuracy of Barometric Altimeters with Respect to Geometric Altitude," *Proceedings of the Institute of Navigation Satellite Division's International Technical Meeting,* (Colorado Springs, CO), Institute of Navigation, Washington, DC, Sept. 19–23, 1988.

[11]Brown, G., "A Baseline RAIM Scheme and a Note on the Equivalence of Three RAIM Methods," *Navigation,* Vol. 39, No. 3, 1992, pp. 301–316.

[12]Lee, Y., "Receiver Autonomous Integrity Monitoring (RAIM) Capability for Sole-Means GPS Navigation in the Oceanic Phase of Flight," *IEEE Aerospace and Electronic Systems Magazine,* Vol. 7, No. 5, 1992, pp. 29–36.

[13]Winkler, G., "Briefing Presented to the DOD GPS Integrity Technical Tiger Team Meeting," Colorado Springs, CO, May 21, 1992.

[14]Green, G., et al., "The GPS 21 Primary Satellite Constellation," *Navigation,* Vol. 36, No. 1, 1989, pp. 9–24.

[15]Dobyne, J., "GPS Availability," Paper presented to the FAA Satellite Operational Implementation Team, Dec. 10–12, 1991.

[16]Bell, R. S., and Dickinson, A., "The Meteorological Office Operational Numerical Weather Prediction System," Met Office Scientific Paper 41, HMSO, London, 1987.

[17]Brown, A., "Integrity Monitoring of GPS using a Barometric Altimeter," (Preliminary report), RTCA Paper 405-87/SC159-117.

Chapter 9

GPS and Global Navigation Satellite System (GLONASS)

Peter Daly*
*University of Leeds,
Leeds, LS2 9JT, England, United Kingdom*
and
Pratap N. Misra†
*Massachusetts Institute of Technology,
Lexington, Massachusetts 02173*

I. Introduction to the Global Navigation Satellite System

A. History of Satellite Navigation Systems

BOTH the NAVSTAR Global Positioning System (GPS) and Global Navigation Satellite System (GLONASS) developed respectively by the United States and the (former) Union of Soviet Socialist Republics (USSR), now the Commonwealth of Independent States (CIS), are planned to become operational during the 1995–1996 time period. Known under the generic title of Global Navigation Satellite Systems (GNSS), they are intended to replace earlier satellite navigation systems (Transit and Cicada) also operated by the United States and the USSR. These two latter systems employ similar orbits with a small number of low-altitude (1100 km) polar-orbiting satellites transmitting information at dual frequencies around 150 and 400 MHz. The user waits for a single satellite (possibly as long as 2 h) and then makes a series of Doppler measurements during the short period (<16 min) when the satellite remains above the horizon. The satellites' position and velocity are transmitted in the navigation message and these, together with the Doppler measurements, are sufficient to allow the user to compute a position. Transmissions on two frequencies are used to allow an ionospheric group delay correction to be applied. The two systems have two major drawbacks; they are not available continuously and the user velocity must be known.

In an effort to overcome the difficulties associated with the earlier systems, both the United States and the CIS plan to introduce precise, global, continuous

---
Copyright © 1995 by the authors. Published by the American Institute of Aeronautics and Astronautics, Inc., with permission. Released to AIAA to publish in all forms.
*Director, CAA Institute of Satellite Navigation.
†Senior Staff Member, Lincoln Laboratory.

position-fixing capabilities by using navigation satellites transmitting dual-frequency spread-spectrum signals in L-band (1.2 and 1.6 GHz). In contrast to the earlier VHF systems, the primary navigation mode is based on range measurement rather than integrated Doppler. The two national systems, both of which possess a military and a civil role, are the CIS's GLONASS and the USA's NAVSTAR GPS,[1] designed to provide accurate position, velocity, and time information. At the end of 1993, the Initial Operating Capability (IOC) had been declared for GPS, while GLONASS was still in the preoperational stage, with a number of satellites already performing a navigation role.

The first release from the Soviet Union of detailed GLONASS information occurred at the International Civil Aviation Organization (ICAO) special committee meeting on Future Air Navigation Systems (FANS) in Montreal in May 1988.[2] At a later meeting of ICAO in September 1991, the concept of GNSS was adopted as providing for future air navigation requirements. The notion of GNSS encompasses GPS, GLONASS, and such alternative systems as the geostationary overlay to be provided by Inmarsat-3 satellites. GLONASS is intended to provide a navigation role for maritime and aviation interests; it offers many features in common with NAVSTAR GPS. In particular, the orbital plan foresees 24 satellites with 8 in each of three orbital planes separated by 120 deg with spacing of 45 deg within the plane. Clearly the orbital planning is such as to allow users anywhere access to at least the minimum number of satellites (four) required for navigation purposes. In practice, simple geometrical considerations tell us that, when fully operational, both GPS and GLONASS individually will allow access to eight or nine satellites for the greater part of each day. The combined resources of GPS and GLONASS together offer twice as many satellites as either system taken on its own. This doubling of available satellite numbers offers a level of independence, redundancy and cross-checking that enhances certain global applications of GNSS such as civil aviation, as discussed in detail in Sec. II.C.

GLONASS transmits two spread-spectrum signals in L-band at around the same power levels as NAVSTAR GPS. Satellites are distinguished by radio frequency channel rather than spread-spectrum code. A single narrow-band code is used of length 511 bits repeating every 1 ms. Information is differentially encoded in a return-to-zero (RZ) format with a final datarate of 50 baud. A separate broad-band code repeating every second is used to transmit differentially encoded data at 50 baud. The narrow-band code is to be found only at the higher of the two L-band transmit frequencies; the broad-band code is to be found on both the upper and lower L-band frequencies, and hence, offers the prospect of correction for the ionospheric delay effect. In this regard, the situation is entirely analogous to that of GPS. A plot of a typical GLONASS signal spectrum is shown in Fig. 1. The GLONASS C/A-code (coarse/acquisition) spectrum covering 1 MHz bandwidth is superimposed on the P-code (precise) signal transmitted in phase quadrature and covering 10 times the C/A-code bandwidth. On the ground, the spectrum is only reproducible in the first place by using a high-gain antenna (3-m dish or larger) to extract the spread-spectrum signal from the noisy background.

B. Orbits

For a given number of satellites in the final operational system, the choice of orbital planes and phases within the plane is constrained to ensure visibility of

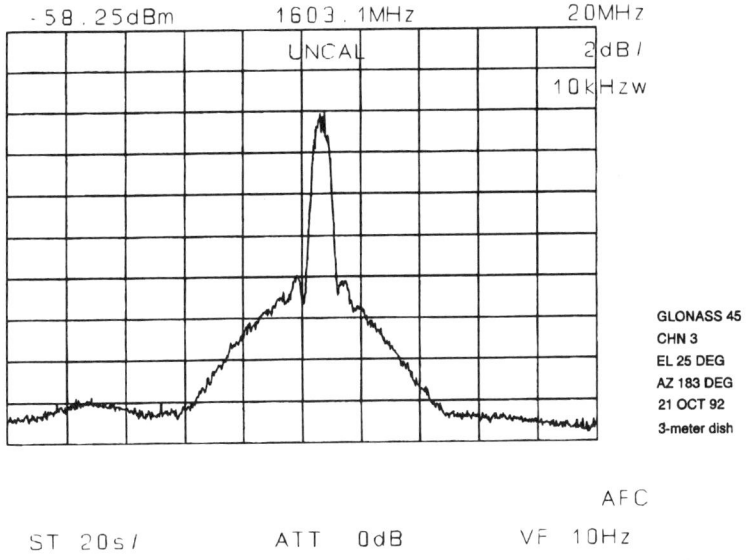

Fig. 1 Plot of typical GLONASS signal spectrum.

four well-located satellites on a continuous global basis and graceful degradation of the system during spacecraft failures. A common approach is to adopt a small number (three or six) equally separated inclined orbital planes with a number of satellites distributed equally in phase around each plane and with an offset in phase between planes. It is also possible to augment this approach with a number of satellites in the geostationary arc. The GLONASS satellite navigation system foresees an operational configuration of 24 satellites with 8 satellites in each of three orbital planes separated by 120 deg in Right Ascension of the Ascending Node (RAAN). RAAN may be thought of simply as equator-crossing longitude but with reference to an inertial (star-fixed) frame rather than a rotating (Earth-fixed) frame. The situation as of September 1994 is shown in Fig. 2 showing a separation in argument of latitude or orbital phase in the plane of 45 deg. There is also a displacement of +30 deg and −30 deg for satellites in planes two and three, respectively, with reference to plane one. This nomenclature follows that assumed by the GLONASS almanac format (described later in Sec. I.G). Relative positions of satellites remain very stable over long periods because they have very much the same, small rates of change of RAAN with time amounting to about −0.03 deg/day for near-circular GLONASS orbits.

All satellites have the same nominal orbital period of 675.73 min with longitude change of 169.41° W each orbit. This orbit produces a ground-track repeat every 17 orbits lasting 8 whole days less 32.56 min. The diurnal offset of $\Delta T = 4.07$ min from a full 24-h day coincides with that of NAVSTAR GPS and is very nearly the difference between a solar and sidereal day (3.93 min). This implies that each complete day less ΔT minutes a satellite performs 17/8 orbits or two whole revolutions plus an additional 1/8 revolution, equivalent to 45 deg. It follows that two satellites in the same plane but separated by 45 deg in orbital

246   P. DALY AND P. N. MISRA

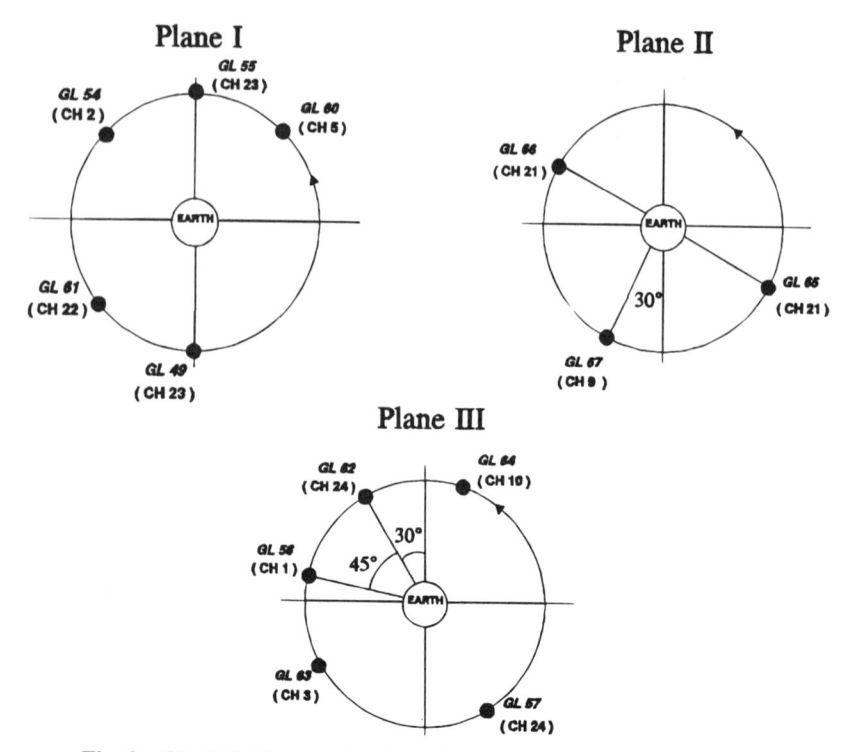

Fig. 2 GLONASS operational configuration as of September 1994.

phase, appear at precisely the same position on successive days less ΔT minutes. During that interval, the Earth has rotated very nearly 360 deg with the result that the ground-based observer sees both satellites at the same pointing azimuth and elevation. Over a ground-track repeat interval of eight days, then, all satellites in the same plane with separation of 45 deg appear in turn at the same position at intervals of 1 day less ΔT minutes. After 8 days, the whole cycle naturally repeats.

By examining the phases of satellites in the planes two and three, it becomes apparent that these satellites will also appear at the same position as the reference satellite in plane one within the same 8-day period. This arises because the time taken by the Earth to rotate through the angle 120 deg separating planes one and two is the same time taken by a satellite in that plane with phase $+255$ deg to travel around to the same position as the reference satellite. The Earth rotates through 120 deg in 478.69 min, very nearly 8 h, which corresponds almost exactly to 17/24 of a GLONASS orbit or $+255$ deg. The same argument holds for plane three at 240 deg separation for a satellite at phase $+150$ deg (or twice $+255$ less 360 deg). The angular separation of 45 deg within the plane together with the angular phase differences of ± 30 deg between planes assures that in an 8-day period, all 24 satellites will pass through the position with the reference subsatellite location.

C. History of Launches

Global satellite navigation systems have been under development by the United States and the former Soviet Union since the 1970s, although the nations already operated and, in fact, still operate a dual-frequency VHF satellite navigation system in near-polar, low Earth-orbit (U.S. Transit, USSR Cicada) providing limited time access but on a worldwide basis. The USA NAVSTAR GPS system saw its first launch in 1978; the USSR's GLONASS system was inaugurated 4 years later. GLONASS satellites are launched three at a time from the Tyuratam space center into near-circular orbits with period around 11.25 h at a height of 19,100 km and inclination of 64.8 deg. Before the demise of the Soviet Union, successful launches were followed by an announcement within a day or two in *Pravda*, which printed basic details of the mission. As of September 1994, there have been 23 launches since the first one in late 1982, all of them successful except those of April 1987 and February 1988, wherein the satellites failed to reach final orbit because of a malfunction of the fourth stage of the Proton launch vehicle. All launches but the most recent (launches 19–23) have taken place under the auspices of the USSR. Since the demise of the Union in 1991, the launches that took place were conducted by the CIS. Table 1 presents the international identifiers, Cosmos and GLONASS numbers of all known launches since

Table 1 Triple GLONASS launches history, September 1994

	International designator	Cosmos	GLONASS	Plane
1	1982—100A,D,E	1413 1414 1415	1 2 3	1
2	1983—84A,B,C	1490 1491 1492	4 5 6	1
3	1983—127A,B,C	1519 1520 1521	7 8 9	3
4	1984—47A,B,C	1554 1555 1556	10 11 12	3
5	1984—95A,B,C	1593 1594 1595	13 14 15	1
6	1985—37A,B,C	1650 1651 1652	16 17 18	1
7	1985—118A,B,C,	1710 1711 1712	19 20 21	3
8	1986—71A,B,C	1778 1779 1780	22 23 24	1
9	1987—36A,B,C	1838 1839 1840	25 26 27	3[a]
10	1987—79A,B,C	1883 1884 1885	28 29 30	3
11	1988—9A	1917 1918 1919	31 32 33	1[a]
12	1988—43A,B,C	1946 1947 1948	34 35 36	1
13	1988—85A,B,C	1970 1971 1972	37 38 39	3
14	1989—1A,B,C	1987 1988 1989	40 41[b]	1
15	1989—39A,B,C	2022 2023 2024	42 43[b]	3
16	1990—45A,B,C	2079 2080 2081	44 45 46	1
17	1990—110A,B,C	2109 2110 2111	47 48 49	3
18	1991—25A,B,C	2139 2140 2141	50 51 52	3
19	1992—5A,B,C	2177 2178 2179	53 54 55	1
20	1992—47A,B,C	2204 2205 2206	56 57 58	3
21	1993—10A,B,C	2234 2235 2236	59 60 61	1
22	1994—21A,B,C	2275 2276 2277	62 63 64	3
23	1994—50A,B,C	2287 2288 2289	65 66 67	2

[a] Failed to achieve final orbit.
[b] Etalon passive laser ranging satellite.

the first one of 1982; all triple launches, except the most recent, have taken place into one of two orbital planes (referred to in the table as planes 1 and 3) separated by 120 deg. The most significant novel feature of the most recent launch is that for the first time, plane 2 was chosen. The decision to begin to occupy the third plane can be interpreted as indicating serious intent to complete the entire 24-satellite GLONASS constellation.

Each launch aims to produce a final stable near-circular inclined orbit at a distance from the Earth's center of about four Earth radii. Of the first seven launches only two of the three launched satellites achieved the said stable orbit; the third satellite remained in an orbit without ground-track repeat and was not observed to transmit. Since then, launches 8, 10, 12–13, and 16–23 have all resulted in a stable orbit for all three satellites. They have also transmitted full navigation messages and can, therefore, be regarded as full-fledged members of the preoperational system. Launches 9 and 11 were failures caused by a malfunction of the fourth stage of the Proton SL-2 launcher. Launches 14 and 15 placed two (rather than the normal three) GLONASS satellites into stable orbit. In both cases, a third satellite was launched with the two GLONASS spacecraft. This third spacecraft (known as ETALON) was a passive spheroidal satellite provided with reflectors to act as a target for laser-ranging signals.

It was also a normal feature of the system in its early days for a launch to occur only when the number of functioning satellites had fallen or was about to fall below the number required for adequate testing of the system. This number cannot be stated with absolute precision because it depended on the orbital planes of the remaining satellites. However, a reduction of available units to any number less than four was likely to act as a precursor to a new launch. Given the number of satellites launched since the first one of 1982, we must presume that these have taken place to make up for the substantial number of in-orbit failures in the interim period. The attrition rate for GLONASS satellites is, indeed, very high (as compared with GPS), although it is not as easy as we might imagine for Western sources to determine whether or not a particular satellite has failed. On the face of it, the failure of a satellite is apparent when the satellite no longer transmits a navigation signal. However, it has been observed in the past that a satellite can remain in a passive, stand-by mode for 3 months before being returned to service. It is normally (but not always) quite correct to assume a satellite has failed when signals are no longer received over a period of several weeks.

D. Signal Design

A major aim of GLONASS is to provide high-precision position-fixing and time-reference capability on a worldwide and continuous basis to users on the Earth's surface, on land and at sea, in the air, and in space itself. Under the control of highly stable, onboard frequency references (clocks), timing signals (epochs), and data are transmitted simultaneously from a number of satellites. The data received from a particular satellite includes a precise ephemeris for that satellite allowing its position and velocity at a given time to be computed. In addition, each satellite provides information on the behavior of its own onboard clock. The observer makes time-of-arrival measurements to three satellites at the same time using the received data to compute the position of the satellites. Given a synchronized ground time reference, the observer can then, in principle, solve three range equations for three unknown position coordinates. In practice, the observer will not normally have a synchronized time reference and will, therefore, choose to "pseudorange" to four satellites instead of three, using the fourth measurement to compute the instantaneous time error.

Users of satellite navigation have the option to make measurements at the level of code phase or to enhance this with integration of Doppler frequency for averaging purposes or when in motion. In addition, the user may have access to carrier phase, which allows extremely accurate relative position fixing when used[3] in the differential mode. This particular method is of great use to the surveying community. In addition, a technique known as kinematic positioning has been derived[4] based on the notion of relative positioning between two observers, one of which is in motion, continuous, or in steps.

For the purposes of allowing computation of the user's own position, navigation satellites transmit details of their own positions and a time reference. In systems such as GLONASS and NAVSTAR GPS, whose purpose is primarily military, the user is expected to play a passive role, because any transmissions to a satellite might identify the user's position to an adversary. Similarly, the navigation message is protected against deliberate jamming by the use of spread-spectrum codes, which increases the bandwidth occupied by the signal and, hence, that of the intending jammer. It should be clear, however, that even were the system design to be exclusively for civil purposes, it is likely that spread spectrum would still be used for two reasons: 1) to minimize interference to others; and 2) to provide sufficient bandwidth and, hence, definition of the epoch timing edge. In the following discussion, the structure of the navigation signal from global navigation satellites to the user is treated.

The transmission carrier frequencies chosen for the new satellite navigation systems lie in L-band, around 1250 MHz (L_2) and 1600 MHz (L_1). Dual-frequency navigation transmissions at L_1 and L_2 allow the user to correct for ionospheric propagation effects and are incorporated into both NAVSTAR and GLONASS. A high-precision spread-spectrum code is modulated onto both carriers; whereas, the lower-precision civil code only appears at L_1. Russian publications on this subject often refer to the low-precision codes as providing the GLONASS SPS (standard positioning service) and to the higher-precision codes as providing the GLONASS "M" PPS (precise positioning service). Spread-spectrum techniques are primarily involved to reduce the effects of deliberate jamming of signals.

In stark contrast to the radio frequency transmission system chosen for GLONASS, which is FDM (frequency division multiplex), NAVSTAR GPS uses the same carrier frequency for all satellites that are distinguished one from another by the use of different spread-spectrum Gold codes, a form of CDM (code division multiplex). This difference between the two systems is of major significance in designing receivers capable of joint operation. It is perhaps worth noting in passing that a third global satellite navigation program proposed by the European Space Agency (ESA Navsat) is characterized by the use of TDM (time division multiplex) to distinguish individual satellites.

The GPS satellites use transmit frequencies at L_1 (1575.42 MHz) and L_2 (1227.60 MHz). Radio frequency carriers used by GLONASS are channelized within the bands 1240–1260 MHz and 1597–1617 MHz, the channel spacing being 7/16 or 0.4375 MHz at the lower frequencies and 9/16 or 0.5625 MHz at the higher frequencies. The carrier frequencies themselves are also multiples of channel spacing and the number of planned channels is 24. The relationship between channel number and the L_1 transmit frequency is simply $f(L_1) = f_0 + k$ × channel spacing, where $f_0 = 1602$ MHz, and k is the channel number (1–24). Transmit frequencies at the lower (L_2) frequencies are in the ratio 7/9. GLONASS satellites and channel numbers (designated GL and CHN) are presented in Table 2. All frequencies were measured at the University of Leeds, with the exception of that of GLONASS 5, which was obtained at the Haystack Laboratory in the United States. At the time of writing (Sept. 1994), there are 13 active GLONASS satellites. Note that toward the end of September 1993 following the regulations introduced at WARC 92, the technique of using the same channel for certain antipodal satellites was introduced.

As explained in a previous section, each satellite sends data at low speed from which its own position at any reference time may be calculated. This data, commonly sent at a 50-baud rate, is superimposed on a pseudorandom noise (PRN) code that is, in fact, periodic and very much longer than a single databit. The GLONASS low-precision code has length 511 bits as compared to NAVSTAR's 1023 bits for its equivalent code. A code sequence lasts only 1 ms, so each databit occupies 20 entire code sequences, the code itself or its inverse being sent depending on whether the databit is a "0" or a "1." In this manner, the information spectrum is spread across a wide bandwidth. On the assumption that, in transmission, the signal will be corrupted by Gaussian noise whose power level is proportional to bandwidth, then the signal will become immersed in the noise at the receiver's terminal and recoverable only be reversing the coding operation applied at the transmitter. This implies a knowledge of the PRN codes on the part of the receiver.

The code rate can be seen from the numbers already given to be 511 kbits/s and 1023 kbits/s for the civil GLONASS and NAVSTAR codes, respectively. Higher-rate codes are at 10 times these rates, and, of course, the sequence lengths are very much longer. To transmit the encoded data, a binary phase-shift keyed (BPSK) modulation technique is employed, the first nulls in the transmitted spectrum at ± the bit rate. Hence, bandwidths for the GLONASS transmission can be taken at 1 MHz and 10 MHz for the low-rate (C/A) and high-rate (P) codes, respectively. These figures compare with 2 MHz and 20 MHz for NAVSTAR's equivalent bandwidths. Both GLONASS codes are generated as maximum-length sequences. The corresponding generator polynomials are

C/A-code generator

$$g(X) = 1 + X^5 + X^9$$

and

P-code generator

$$g(X) = 1 + X^3 + X^{25}$$

the latter code[5] being truncated after every second. There is no need in GLONASS for a hand-over word (HOW) as with GPS to allow acquisition of the P-code. At the L_2 frequency, only the high-rate code is carried, but at L_1 both codes are transmitted on the same carrier, one in-phase and the other in-quadrature. This

Table 2 GLONASS channels

Sat ID	Cosmos	GL	Active	CHN
1982—100A	1413	1	——[a]	——[a]
1982—100D	1414	2	——[a]	——[a]
1982—100E	1415	3	——[a]	——[a]
1983—84A	1490	4	N	3
1983—84B	1491	5	N	1
1983—84C	1492	6	——[a]	——[a]
1983—127A	1519	7	N	2
1983—127B	1520	8	N	24
1983—127C	1521	9	——[a]	——[a]
1984—47A	1554	10	N	9
1984—47B	1555	11	N	18
1984—47C	1556	12	——[a]	——[a]
1984—95A	1593	13	——[a]	——[a]
1984—95B	1594	14	——[a]	——[a]
1984—95C	1595	15	N	17
1985—37A	1650	16	N	7
1985—37B	1651	17	N	10
1985—37C	1652	18	——[a]	——[a]
1985—118A	1710	19	N	4
1985—118B	1711	20	N	19
1985—118C	1712	21	——[a]	——[a]
1986—71A	1778	22	N	11
1986—71B	1779	23	N	20
1986—71C	1780	24	N	22
1987—79A	1883	28	N	14
1987—79B	1884	29	N	21
1987—79C	1885	30	N	5
1988—43A	1946	34	N	12
1988—43B	1947	35	N	23
1988—43C	1948	36	N	24
1988—85A	1970	37	N	18
1988—85B	1971	38	N	7
1988—85C	1972	39	N	10
1989—1A	1987	40	N	9
1989—1B	1988	41	N	6
1989—39A	2022	42	N	16
1989—39B	2023	43	N	17
1990—45A	2079	44	N	21
1990—45B	2080	45	N	3
1990—45C	2081	46	N	15
1990—110A	2109	47	N	4
1990—110B	2110	48	N	13
1990—110C	2111	49	Y	19
1991—25A	2139	50	N	20
1991—25B	2140	51	N	11
1991—25C	2141	52	N	14
1992—5A	2177	53	N	22
1992—5B	2178	54	Y	2

[a]Not yet observed.   (Continued on next page)

Table 2 GLONASS channels (continued)

Sat ID	Cosmos	GL	Active	CHN
1992—5C	2179	55	Y	17
1992—47A	2204	56	Y	1
1992—47B	2205	57	Y	24
1992—47C	2206	58	N	8
1993—10A	2234	59	N	12
1993—10B	2235	60	Y	5
1993—10C	2236	61	Y	23
1994—21A	2275	62	Y	24
1994—21B	2276	63	Y	3
1994—21C	2277	64	Y	10
1994—50A	2287	65	Y	21
1994—50A	2288	66	Y	21
1994—50C	2289	67	Y	9

results in a signal spectrum that superimposes the two individual spectra whose bandwidths differ by a factor of 10. Because both transmissions carry roughly the same power, a spectrum analyzer will display the narrower-band code at 10 times the strength of the wide-band code.

E. Message Content and Format

The data carried on transmissions from satellites are low bit rate at 50 baud essentially providing accurate positions for the transmitting satellite as well as information on its onboard frequency standard. In addition, data are given in the form of low-precision almanacs of all the other satellites currently available so as to allow the user to plan usage and to assist with signal acquisition. Data are sent in lines, subframes, and frames with preambles and parity checks at the end of each line. Essentially, each subframe of data lasts for 30 s and consists of 15 lines of 2-s duration. The 15-line subframe divides into a 5-line set of ephemeris data (including clock corrections) and a 10-line set of almanac data. The almanac data are subcommutated, each satellite being allocated 2 lines of almanac data. In this way, the subframe accommodates the almanacs of five satellites; the full constellation of 24 satellites occupies five complete subframes, the last of the 25 almanacs always being set to zero (spare). The subframe format is shown in Fig. 3 (the number of bits allocated to each parameter is found under the title of the parameter).

The start of each line marks the beginning of a 2-s subframe synchronized to GLONASS system time. The leading bit is always zero and followed by a line number, various databits, parity bits, and preamble. Some of the information bits are flags (for example, words $P1$-$P5$) and are not used by the navigator. The flag H on line 2 is most important, referring to the "health" of the space vehicle ("1" means unhealthy). The symbol TA gives the time of transmission in hours, minutes, and half-minutes Moscow Standard Time [MST (GMT + 3 h)]. The symbol TE gives the time of validity of the ephemeris in hours and quarter-hours

GPS AND GLOBAL NAVIGATION SATELLITE SYSTEM

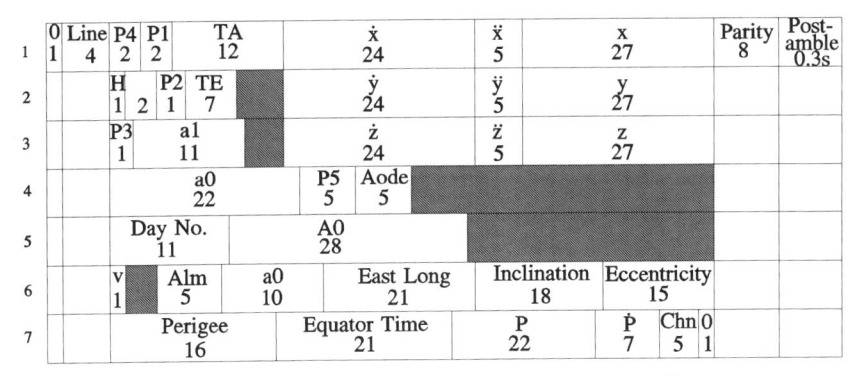

	0	Line	P4	P1	TA		$\dot{x}$	$\ddot{x}$	x	Parity	Post-
1	1	4	2	2	12		24	5	27	8	amble 0.3s
2		H 1 2	P2 1	TE 7			$\dot{y}$ 24	$\ddot{y}$ 5	y 27		
3		P3 1		a1 11			$\dot{z}$ 24	$\ddot{z}$ 5	z 27		
4			a0 22			P5 5	Aode 5				
5			Day No. 11			A0 28					
6		v 1	Alm 5		a0 10		East Long 21	Inclination 18	Eccentricity 15		
7			Perigee 16				Equator Time 21	P 22	$\dot{P}$ 7	Chn 5	0 1

Fig. 3 GLONASS C/A-code data message subframe format.

(MST). Usually the ephemeris is valid at odd quarter-hours, but occasionally (when a satellite is temporarily unhealthy) the ephemeris will be valid only on the half-hour. The remaining databits are described in the following sections. The reader is referred elsewhere for further details of the preamble and parity corrections.[6]

F. Satellite Ephemerides

The NAVSTAR GPS ephemeris data describe a Kepler ellipse with additional corrections that then allow the satellite's position to be calculated in an Earth-centered, Earth-fixed (ECEF) set of rectangular coordinates at any time during the period of validity of the data. In the GLONASS system, the satellite's instantaneous position and velocity are encoded at fixed time intervals (usually one half-hour) in an ECEF rectangular coordinate system. Positions and velocities at intermediate times are intended to be calculated using interpolation procedures and acceleration terms provided. The resolution in satellite position is 0.5 m and in velocity, 1 mm/s.

As far as timing is concerned, the global navigation satellites transmit satellite clock corrections to a GPS or GLONASS system time and, in addition, corrections from system time to a national time reference. NAVSTAR GPS represents satellite clock behavior in terms of a clock offset ($a0$), frequency offset ($a1$), and rate of change of frequency ($a2$). NAVSTAR GPS system time from Universal Coordinated Time at the U.S. Naval Observatory, UTC(USNO), is given as $A0$, and its first derivative as $A1$. In contrast, GLONASS only transmits in the ephemeris two parameters relating to the onboard clocks, $a0$ and $a1$. The first time offset (with resolution 1 ns) refers to the instantaneous time difference between space vehicle time and GLONASS system time. The second parameter (with resolution 1 in 10^{-12}) gives the rate of change of space vehicle time offset. Use of both parameters allows the user to establish individual space vehicle time offsets at any required instant of observation. Reference of GLONASS system time to Universal Coordinate Time/Soviet Union, UTC(SU) is by a single time offset parameter, $A0$ (with resolution 7 ns). An additional parameter called age-of-data of ephemeris (AODE) gives the integer number of days starting at the previous local midnight since the ephemeris data was updated. On a particular day, the parameter will normally be "0," increasing by "1" each successive midnight unless, in the interim, the ephemeris data are based on a current (same day) set of measurements. When the GLONASS system is operating normally, this parameter is always either "0" or "1". On occasion it has been observed to grow as large as 25.

G. Satellite Almanacs

There is greater similarity between NAVSTAR and GLONASS in the transmission of almanacs than in the transmission of ephemerides. Both systems transmit the basic elements of an osculating Kepler ellipse as Table 3 attempts to show.

In terms of using almanacs to predict satellite position from the reduced Kepler orbit, the two sets of data are quite similar, as we would expect.

Each 2-line almanac consists of a validity flag (V), an almanac number, a reduced-precision satellite clock phase offset, a set of orbital elements, and a satellite channel number. The set of orbital elements is represented as follows (all angular quantities are in semicircles and times are in seconds):- (i) equator-crossing longitude; inclination (offset 0.35); eccentricity; argument of the perigee; equator-crossing time; period P (offset from 12 h); and rate of change of period. The equator-crossing time of the reference orbit is always the first of the day. The reference day number itself occurs at the start of line 5. Day 1 corresponds to the first day of the year at the start of the 4-year leap cycle (currently 1 January 1992). The reference day number does not necessarily change each day—a set of almanacs is often allowed to remain unchanged for 2 days. The almanac number just referred to ranges from 1–25, each number in sequence (except the last) referring to a satellite location within one of the three reference planes. The first 8 almanac numbers refer to location within the orbital arc in plane one starting with phase "0" and working around clockwise in steps of 45 deg. The

Table 3 Satellite almanacs

NAVSTAR	GLONASS
Week of validity	Day of validity
Identifier	Channel number
Eccentricity	Eccentricity
Inclination	Inclination
Time of almanac	Equator time
Health	Validity of almanac
Right ascension (RA)	Equator longitude
Rate of change of RA	—
Root of semimajor axis	Orbital period
Argument of perigee	Argument of perigee
Mean anomaly	—
—	Luni–solar term
Time offset	Time offset
Frequency offset	—

GPS AND GLOBAL NAVIGATION SATELLITE SYSTEM 255

second set of 8 almanac numbers (9–16) refers to satellites in plane two, and the third set of 8 almanac numbers (17–24) to satellites in plane three.

Formally, we would expect the primary purpose of almanac data to be to allow the user to predict in fairly crude terms which satellites are above his local horizon at a given time and whether their geometry is favorable for navigation. This end would be achieved by almanacs giving a position of each satellite to within 100–200 m and is the case with the NAVSTAR almanac. However, the inclusion in the GLONASS almanac of a luni–solar correction term implies a position error perhaps an order of magnitude better than a NAVSTAR almanac. Almanacs are valid for several days in the case of NAVSTAR; they are usually, but not always changed, every day in GLONASS at local midnight. It is interesting to observe that the GLONASS almanacs differ from the earlier Cicada almanacs in one major respect—the earlier almanacs were based on an equinoctial Kepler set where eccentricity and argument of the perigee are transmitted as $h = e \sin \omega$ and $l = e \cos \omega$. The equinoctial set of elements is suitable for orbits with small eccentricity, because it leads to equations with no singularities when e tends to zero.

H. GPS/GLONASS Onboard Clocks

Both GPS and GLONASS will offer, independently, precise location and time transfer continuously anywhere in the world and, indeed, in space itself. Many potential users, in particular the civil aviation community, are keenly interested in a joint GPS/GLONASS operation, because it would offer substantial advantages in defining and maintaining the integrity of the navigation aid. The question arises of compatibility of GPS/GLONASS from the point of view of satellite onboard clocks, their system references, their national standards, and, ultimately, UTC. GLONASS provides worldwide time dissemination and time transfer services in the same manner as NAVSTAR GPS, with both exhibiting substantial advantages over other existing timing services. Time transfer is both efficient and economic in the sense that direct clock comparisons can be achieved via GLONASS between widely separated sites without the use of portable clocks. Event time tagging can be achieved with the minimum of effort, and users can reacquire GLONASS time at any instant because of the continuous nature of time aboard the satellites.

Time transfer from GPS/GLONASS is achieved in a straightforward manner. Each satellite transmits signals referenced to its own onboard clock. The GPS Control Segment monitors the satellite clocks and determines their offsets from the common GPS/GLONASS system time. The clock offsets are then uploaded to satellites as part of their transmitted data message. A user at a known location receives signals from a satellite and by decoding the datastream modulated on to the transmission, is able to obtain the position of the satellite, as well as the satellite's clock offset from the common system time. Hence, the signal propagation time can be calculated at any instant. The time at which the signals are transmitted is also contained in the data message; by combining this with the propagation time and correcting first for atmospheric effects and other delays, and then for the satellite's own clock offset, the user can effect transfer to GPS/GLONASS system time. Correction to an external time scale [such as

UTC(USNO) or UTC(SU)] is then possible because the relevant offset is one of the transmitted data parameters. Any other user who has the same satellite visible is also able to transfer to the same common time scale. Clearly, if two users access the same satellite simultaneously (known as common-view reception), the difference between the two users' measurements eliminates the systematic errors common to both, such as satellite ephemeris error. In this way, time transfer between users in common view offers increased accuracy. In fact, global networks of GPS stations currently exist for the purpose of comparative ranging For example, time transfer using GPS satellites in common view is organized according to an international, global schedule by the Bureau International des Poids et Mésures (BIPM) acting as the coordinating center in Paris. A similar schedule for GLONASS is under preparation; when ready, the capability of transferring international time standards via GPS or GLONASS will result in improved measures of international time. In turn, this improved coordination of time will lead to improvements in our capability to fix position using GNSS.

In an attempt to compare GPS and GLONASS system time references, a series of measurements was conducted of the difference between each reference and UTC(USNO) using a prototype single-channel GLONASS/NAVSTAR GPS receiver. Time comparisons between system times are referenced to a 1–pulse-per-second (1 pps) strobe synchronized to UTC(USNO). The NAVSTAR system time to UTC(USNO) comparison is used as a calibration of the measurement, because the offset between GPS time and UTC(USNO) is already known—it is transmitted as part of the GPS data message.

Table 4 shows a set of measurements over a typical 24-h period on 26 October 1990. Each individual measurement lasts 180 s; satellites are accessed many times in the course of the day. The data have been corrected for tropospheric,

Table 4 NAVSTAR GPS and GLONASS system time offset from UTC(USNO)

Date	Satellite	Readings, 1-s	Average offset/ns	Standard deviation/ns
26/10/90	NAVSTAR 2	2336	−194	76
26/10/90	NAVSTAR 3	2157	−195	49
26/10/90	NAVSTAR 6	3413	−242	54
26/10/90	NAVSTAR 9	4850	−228	72
26/10/90	NAVSTAR 11	4312	−185	55
26/10/90	NAVSTAR 12	3058	−214	51
26/10/90	NAVSTAR 13	4300	−192	58
26/10/90	NAVSTAR 19	538	−205	50
26/10/90	GLONASS 34	3224	5425	58
26/10/90	GLONASS 36	4291	5444	64
26/10/90	GLONASS 39	3052	5449	60
26/10/90	GLONASS 40	4111	5437	71
26/10/90	GLONASS 41	4484	5436	54
26/10/90	GLONASS 44	4130	5478	65
26/10/90	GLONASS 45	3585	5436	62
26/10/90	GLONASS 46	4828	5437	55

relativistic, and Earth rotation effects but not for ionospheric effects. Only two of the available GPS Block II satellites (NAVSTAR 2 & 19) were used; the absence of SA on both at this time is noticeable. Both sets of data are consistent in the sense that all eight satellites individually produce results that differ from the average by much less than the standard deviation. This validation is an important feature of the measurement, because most of the measurement equipment is common to both GPS and GLONASS. By means of the transmitted offsets $A0$, it is possible to deduce a value for UTC(USNO)–UTC(SU) obtained by the satellite navigation systems GPS and GLONASS with an uncertainty of less than 50 ns. Current research is aimed at reducing the uncertainty in these measurements to the order of 10 ns.

It is known that GPS satellites carry two cesium and two rubidium atomic clocks as frequency/time standards. GLONASS satellites carry three cesium standards. The question arises as to the comparative performance of onboard GPS and GLONASS clocks. Data on the performance of certain GLONASS satellites are available.[7] Over the years 1986–1989, a steady improvement in performance has been demonstrated with clocks on board spacecraft launched during 1989 showing the qualities of high-quality cesium standards of roughly the same level of performance as the GPS block I cesiums. Since 1989, the level of performance of onboard clocks has been consistently high. It is planned to use improved cesium clocks on future GLONASS satellites. These new clocks, known as "Malachite" atomic standards will offer long-term stabilities five times better than those currently in operation.

In determining the accuracy with which time (or its equivalent, pseudorange) can be measured, it is important to remember that two basic quantities are available: 1) code phase; and 2) carrier phase. In crude terms, we can measure code phase, meaning the time interval between local and transmitted code epochs, to an accuracy limited by the code frequency, the signal-to-noise ratio, and the bandwidth of the measurement. Ultimately, if all satellite onboard frequency sources are locked to the same frequency standard, the accuracy available is limited by the stability of that standard. In the case where onboard standards are synchronized atomic clocks, range, and hence, position can be established to a fraction of a wavelength at the carrier frequency. At L-band frequencies of 1.6 GHz, the free-space wavelength is around 19 cm. Because carrier-phase measurements are fractional, range to a satellite can only be found by inclusion of the integer number of carrier wavelengths between the satellite and observer. In practice, the integers may only be resolved if two or more receivers are operating simultaneously (differential operation) on the same satellite.

It is also important to keep in mind the rate at which ranging measurements can be made. Both GPS and GLONASS transmit a timing epoch every millisecond. Taken in conjunction with the fact that most modern receivers are designed as multichannel units, this implies an ultimate receiver capability of producing raw pseudoranges to all visible satellites simultaneously 1000 times a second. In most applications, such a high rate of raw measurement is not necessary. However, the point does demonstrate that the full capability of GNSS is far from being reached.

II. Performance of GLONASS and GPS + GLONASS

A. Introduction

In this section, we examine the performance capabilities of GLONASS and compare them with those of GPS. The GPS performance has been discussed fully in the companion volume, Part II. We include some additional results here for a side-by-side comparison with GLONASS, where appropriate. The context for our performance analysis is civil aviation. The interest in GLONASS stems from the recognition that GPS alone falls short of meeting the requirements of a global sole-means, or stand-alone, navigation system, as discussed below; GLONASS alone does the same. The two systems taken together, however, offer amply redundant measurements to all users, and seem capable of meeting these requirements globally. Using the combined set of measurements from GPS and GLONASS for positioning is referred to as an integrated use of the two systems, and denoted as GPS + GLONASS.

A user equipped with a GPS + GLONASS receiver may consider the two systems as augmenting each other. Other approaches to augmentation of either system are based on LORAN, inertial reference system, and baroaltimeter, as discussed in Chapters 6–8, this volume. Each approach brings about an improvement in the availability of a GNSS-based navigation service. Currently, the most important of these augmentations appears to be the FAA's Wide Area Augmentation System (WAAS), which is scheduled to provide a sole-means capability over the conterminous U.S. starting in 1998 (Chapter 4, this volume). If successful, WAAS would to expand to other countries and is expected to evolve into a seamless global augmentation of GPS.

The performance measures relevant to positioning are: distribution of the number of satellites in view and a characterization of their geometry; and the quality of the measurements. The performance capabilities of GLONASS, are substantially similar to those of GPS. The main difference between the two is SA. GLONASS, which has disavowed an SA-like feature, offers a significantly better positioning accuracy than does GPS with SA active. On the other hand, at this writing GPS is close to achieving operational-status, while the prospects of GLONASS are less clear.

We begin with a brief review of the requirements of a civil aviation navigation system, followed by a discussion of the technical issues related to an integrated use of the two autonomous systems. In Sec. II.C, we discuss the performance achievable from GLONASS and from integrated use of GPS and GLONASS via-à-vis the requirements of civil aviation. We focus exclusively on the signals planned to be available for civil use; namely, C/A-code on L_1. The data analysis results are based on nearly continuous measurements from the two systems since 1990 and have been reported in greater detail elsewhere.[8,9] While GPS receivers have been widely available, GLONASS receivers have remained rare. MIT's Lincoln Laboratory has had one or more of the following GLONASS receivers since 1990: a prototype receiver built by the Magnavox Corporation; ASR-16, an aviation model from the erstwhile Leningrad Radiotechnical Research Institute; and R-100, a GPS + GLONASS receiver under development at 3S Navigation.

B. Requirements of Civil Aviation

For a navigation system to be adopted for use in civil aviation, it must meet certain stringent criteria. The criteria are stated as standards and certification procedures for each piece of equipment installed in the cockpit or deployed at the airports or elsewhere for use in navigation. International civil aviation requires agreement on these standards and procedures among the national and regional regulatory agencies.

The requirements basically relate to three areas: coverage, accuracy, and integrity monitoring. The coverage of a navigation system deals with where and when can the system be used. A satellite-based navigation system is usable for three-dimensional positioning when four or more satellites are in view of the user. A global system must, therefore, deploy a large enough constellation so that all users at different locations and times see enough satellites. The requirement on accuracy refers to the positioning accuracy provided by the system. The accuracy requirements in civil aviation depend upon the phase of the flight, and they currently range from several kilometers for en route phase to several hundreds of meters for a nonprecision approach. The precision approaches, executed under poor visibility conditions, require that the navigation system guide an aircraft down to an altitude of 60 m or less. These approaches, currently executed on specially equipped runways, require much more precise position estimates. Although satellite-based navigation seems promising for precision approaches also, these are outside the scope of this chapter.

The requirement on integrity monitoring deals with an issue vital to civil aviation: the ability of the system and its users to detect a system malfunction in a timely manner. The main point is that the user must be able to rely on the position estimate provided by the system. A system may be certified as *supplemental* or *sole means*. A supplemental system must provide a position estimate of the required accuracy, when it can, and recognize a situation when it cannot. In the latter case, the system must warn the user, who may then switch to an alternate system for navigation. A sole-means system, as the name suggests, should require no other navigation system as a backup. Therefore, a sole-means system, or its users, must be able to recover from the possible malfunctions. Obviously, the idea of a sole-means system is economically attractive, and the integrated use of GPS and GLONASS is seen as a potential sole-means system. Indeed, if this promise could be shown to be met, there would be no need for any of the current ground-based navigation aids: VHF omnidirectional radio (VOR), distance-measuring equipment (DME), LORAN, and OMEGA. This was seen as particularly important because at present there are no ground-based navigation aids over long stretches in economically underdeveloped parts of the world and in sparsely populated areas (e. g., Alaska and parts of Russia and Canada).

Integrity monitoring is discussed in detail in Chapter 5, this volume. For completeness, the main ideas are reviewed below. The integrity-monitoring requirements for a satellite navigation system are typically stated as follows. If the error in a position estimate exceeds a certain threshold, the user must be notified within a certain time interval. Both the error threshold and the required response time depend upon the phase of the flight and can range widely. The system failure scenario for a satellite navigation system is defined as an erroneous or out-of-tolerance signal transmitted by one of the satellites in the constellation. The current view is that the constellations would be managed so that at any instant the probability of two or more satellites transmitting anomalous signals simultaneously while marked as healthy is considered negligible. Note, however, that even if a system is known to be operating to specifications, a user has to guard against an unacceptable position estimate obtained because of poor geometry or poor measurement quality.

The current accuracy and integrity-monitoring requirements for the various phases of flight are discussed in Chapters 12–14, this volume. For the purpose of evaluation of GPS and GLONASS performance, recall that: 1) in en route and terminal phases of flight and during a nonprecision approach, a navigation system is required to provide only a two-dimensional location of the aircraft; altitude is provided by a baroaltimeter; and 2) for nonprecision approach, the alarm limit for the horizontal error is 0.3 n.mi. (555 m), and time to alarm, 10 s.

C. Integrated Use of GPS and GLONASS

GPS and GLONASS are autonomous systems, each with its own time scale and coordinate frame in which to express a three-dimensional position. As discussed in Sec. I, the time scales adopted by the two systems are their national standards: UTC(USNO) and UTC(SU). The offset between the two time scales has been stable in recent years, but this stability cannot be taken for granted.

Clearly, a user interested in the integrated use of the two systems must be able to determine the instantaneous difference between the two time scales. The problem can be thought of as one of position estimation from two sets of pseudoranges, each with an unknown time bias, which makes five unknowns in all. Obviously, one or both systems could carry information on this bias as a part of their navigation messages. At worst, without the information on bias, we could solve for the additional unknown by "sacrificing" a measurement. As we shall see, the integrated use of GPS and GLONASS offers amply redundant measurements, and the additional unknown does not create a problem.

The two systems express the positions of their satellites, and, therefore, of their users, in different geocentric coordinate frames. In 1987, GPS adopted the WGS84 system.[10] GLONASS started with the SGS85 system[11] but switched in 1993 to PE-90. Our current knowledge of both SGS85 and PE-90 is quite limited; the differences between the two are believed to be insignificant for civil aviation purposes.

Combining measurements from GPS and GLONASS requires that we estimate a transformation between their coordinate frames. Estimation of the transformation is straightforward in principle; it requires positions of a set of points expressed in both coordinate frames. While a point on Earth can now be surveyed to centimeter-level accuracy in WGS84 using GPS measurements, the corresponding SGS85/PE-90 coordinates have been difficult to determine. The main difficulty has been the lack of precise and sturdy GLONASS receivers.

Access to GLONASS receivers and the facilities of the Deep-Space Tracking Network (DSTN) operated by MIT's Lincoln Laboratory gave us the resources to take a different approach to this problem in 1992. We took advantage of the fact that the positions of GLONASS satellites as defined in SGS85 were available to us as a part of the navigation messages broadcast by the satellites. The remaining task, then, was to obtain the corresponding coordinates in WGS84, and that is where the resources of the DSTN came into play. In experiments, we tracked several GLONASS satellites independently to characterize their ephemerides in WGS84 and compared these to the satellite positions in SGS85 as broadcast by the satellites themselves and recorded by the GLONASS receivers. The results showed that the coordinates of points on Earth as expressed in the two coordinate frames differ by no more than 20 m,[12] and that the two geocentric coordinate frames are brought substantially into coincidence by a small rotation (0.6") of the z-axis, and a small displacement of the origin. Figure 4 illustrates the process of gathering the position data in the two coordinate frames, and the resulting estimated transformation.

With the time and space reference standards reconciled, the design of a receiver to obtain measurements from both GPS and GLONASS poses no basic challenge. That such receivers remain rare is attributable mainly to the present uncertainty about the future of GLONASS.

D. Performance of GLONASS and GPS and GLONASS

We discuss next the level of performance achievable from GLONASS and from an integrated use of GPS and GLONASS. In particular, we review coverage, accuracy, and integrity-monitoring capabilities and compare the performance in each of these areas with the requirements of civil aviation.

There is uncertainty at present about the size and management of the GLONASS constellation. GLONASS was defined initially as a constellation of 21 active satellites plus up to three on-orbit spares. More recently, however, it seems that both GLONASS and GPS may be planning to maintain a 24-satellite constellation.[13] We present performance results for an average 21-satellite GLO-

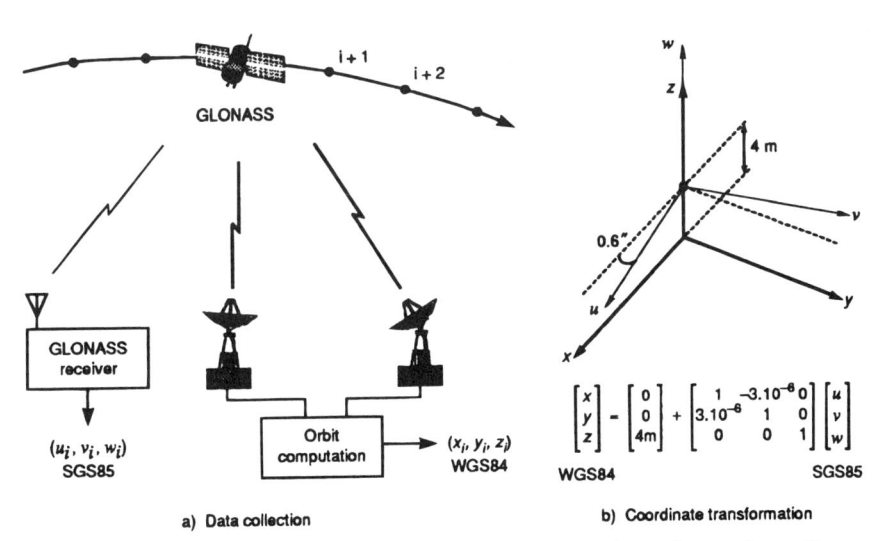

Fig. 4 Position data-gathering process and resulting estimated transformation.

NASS constellation (GLONASS-21) and for the average GPS and GLONASS constellations combined: GPS + GLONASS (2 × 21). These results are obtained in simulations with three randomly chosen satellites in each constellation declared as unavailable in each trial.

1. Coverage

Figure 5a shows the distribution of the number of satellites visible to a user at a random location on Earth at a random time, counting only those satellites that are well above the horizon (i.e., elevation > 7.5 deg).

The coverage results for GLONASS-21 are quite similar to those for GPS-21. About 0.5% of the users would see fewer than four satellites; nearly 20% fewer than six; and about 50% fewer than seven. With the combined constellations, however, all users see 7 or more satellites, 99% see 10 or more, and nearly half see 14 or more! Clearly, some users may not be able to estimate their positions using GLONASS (or, GPS) alone. With the combined constellation, however, *all* users will have abundantly redundant measurement sets on which to base a position estimate.

Figure 5b gives the cumulative distribution functions (cdf) of the horizontal and vertical dilutions of precision (HDOP and VDOP, respectively) available globally from GLONASS-21 and GPS + GLONASS (2 × 21). As discussed in Chapters 5 and 9, the companion volume, DOP parameters characterize the quality of the position estimates available from a constellation of satellites in view. The cdf in Fig. 5b define the availability of favorable satellite geometries for position estimation, and are to be interpreted in view of the relationship: rms position error = DOP. σ_{URE}, where σ_{URE} is the rms value of the user range error (URE). We discuss this relationship further in the next section, and indeed, use it to estimate σ_{URE} from position error data.

The distribution of VDOP has been included in Fig. 5b for completeness; we concern ourselves mainly with HDOP and horizontal accuracy for the reason noted earlier. The distributions of HDOP and VDOP for GLONASS are similar to those for GPS. With GLONASS-21, satellite geometries characterized by HDOP < 2 would be available to nearly 94% of the users; with GPS + GLONASS, such favorable geometries would be available to *every* user. Civil aviation, however, has no requirements related to DOPs, only to position accuracy, to which DOP is a contributor. The other contributor is measurement data quality, which we examine in the next section.

2. Accuracy

We turn next to the accuracy of the position estimates obtained from GLONASS and examine the quality of the position estimates based on measurements recorded in our laboratory. Figure 6a is typical of GLONASS. For comparison, we have also included a corresponding display of the quality of position estimates obtained from GPS (Fig. 6b). Figures 6a and b were both generated in the same way. Snapshots were taken at 1-min intervals of range measurements from the satellites in our view from each constellation over a period of 24 h on 31 December 1993 and position estimates computed when possible. The discrepancy in each position estimate was computed relative to the known, surveyed location of the

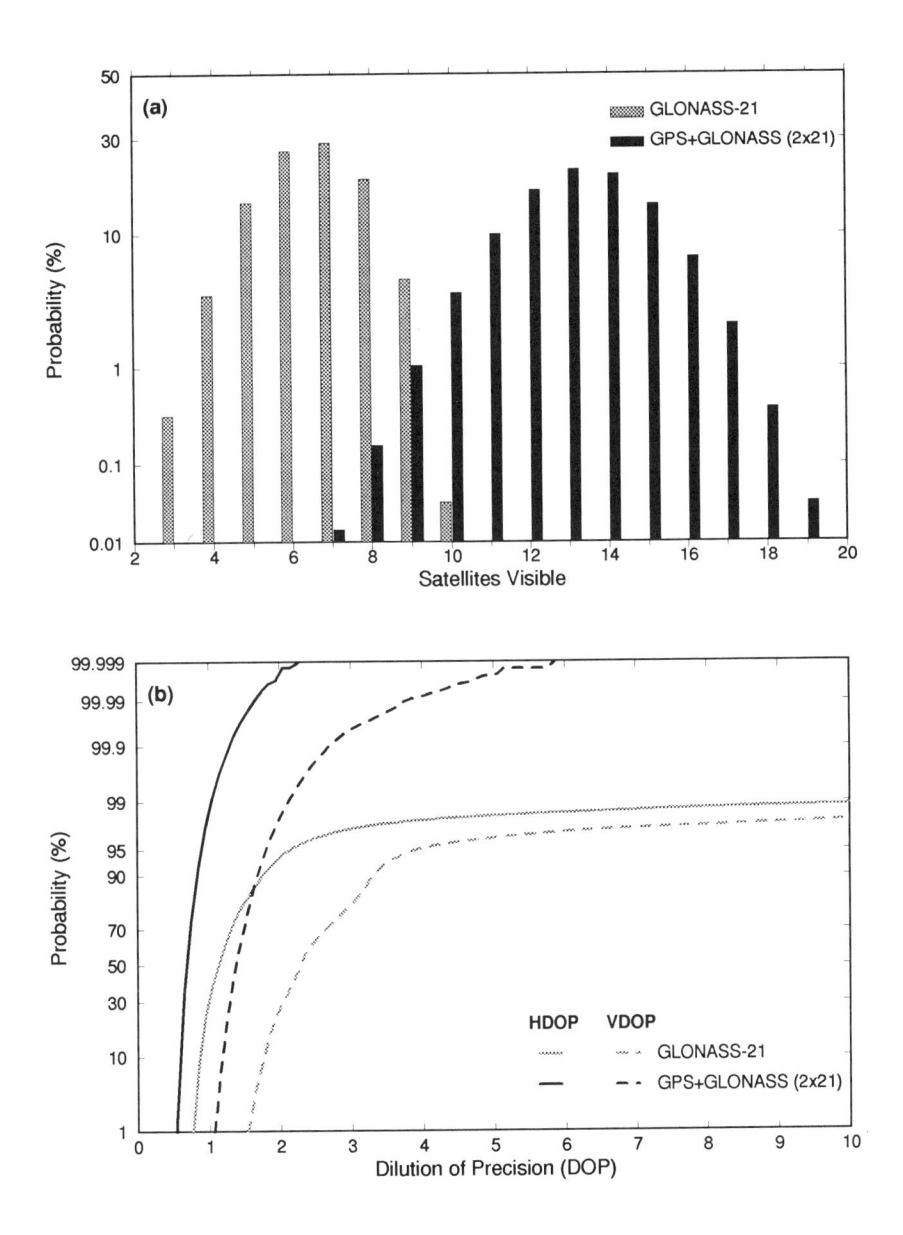

Fig. 5 Coverage provided by GLONASS and GPS + GLONASS.

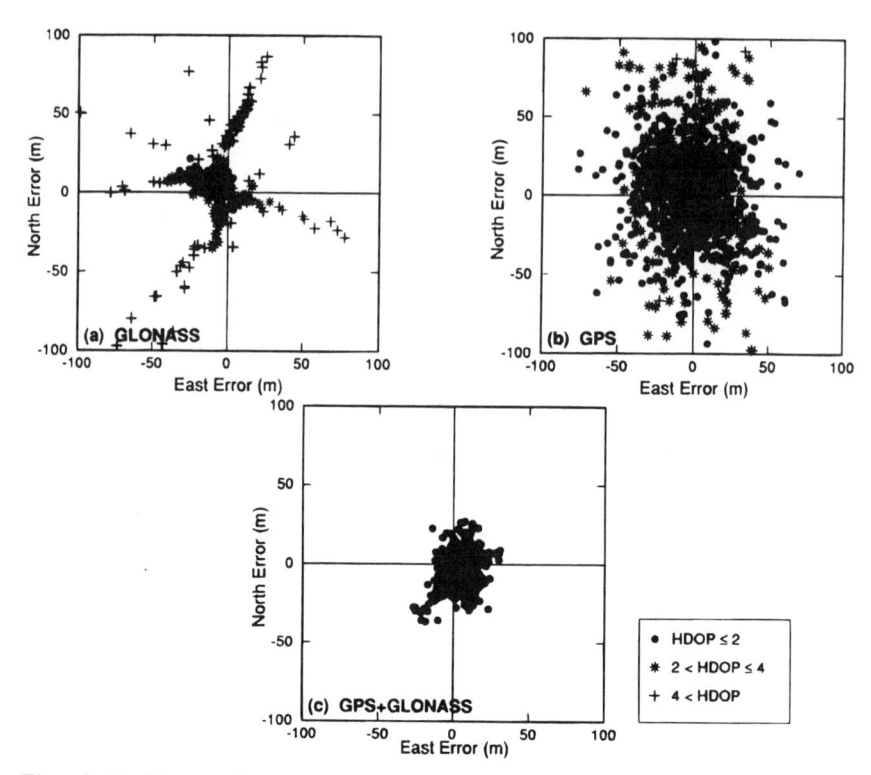

Fig. 6 Position estimates from GPS, GLONASS, and GPS + GLONASS (1-min samples, 31 December 1993).

antenna in the WGS84 coordinate frame, and the horizontal components of the discrepancy were plotted. Because the accuracy of a position estimate depends upon the satellite geometry at the time, each point is coded to reflect the corresponding HDOP. On 31 December 1993, GPS had 24 satellites in orbit, 20 of which were subject to SA; GLONASS had 12 satellites marked healthy. With measurements limited to satellites with SA active, position estimates could be computed from GPS nearly 90% of the time. The leaner GLONASS constellation provided the position estimates about half the time.

Figure 6a shows a tight cluster of position estimates corresponding to favorable satellite geometries (HDOP < 2). As noted earlier, such geometries would be available globally to 94% of the users with an average 21-satellite GLONASS constellation. The straggling position estimates shown correspond to poor satellite geometries (HDOP > 4). With the current sparse GLONASS constellation, the proportion of these points is relatively large, but with the operational constellation, fewer than 2% of the users would encounter such geometries. That the tight cluster in Fig. 6a is off-center was expected, because of the difference in the coordinate frames referred to earlier. The observed difference is consistent with our estimated transformation between SGS85 and WGS84.

In Fig. 6b, with SA on, the GPS position estimates appear to be widely scattered, as compared to Fig. 6a. These position estimates, however, are consistent with the GPS specifications on horizontal accuracy available to civil users. The strength of the constellation is reflected in the relatively small proportion of the symbols corresponding to HDOP > 2. Actually, HDOP was below 2 for nearly 90% of the measurement samples.

Finally, we look at the position estimates obtained from the combined set of measurements from GPS and GLONASS. The results, presented in Fig. 6c, illustrate the reason for our interest in GLONASS. Figure 6c combines the best features of Fig. 6a and b, and is indeed a distinct improvement over both. GPS contributes a larger satellite constellation, and GLONASS, measurements of better quality. The result is consistently good satellite geometries and mitigation of SA. Of course, with a full GLONASS constellation, the results would be better yet.

Figures 6a–6c are intended as a qualitative view of the positioning capabilities of GLONASS and GPS. To be able to predict the positioning accuracies achievable from these systems, when operational, we must take a closer look at the measurement error in the two systems. As noted earlier, we use the well-known relationship: rms position error = DOP. σ_{URE}, to estimate the rms value of the range measurement error for each system. From each snapshot of the measurements from a surveyed antenna location, we compute a DOP value for the measurement set and the error in the corresponding position estimate. Figures 7a and 7b give plots of horizontal error vs HDOP from measurement snapshots taken 3 minutes apart over a period of about three months in 1993. These scatter plots illuminate the nature of the often misunderstood relationship between DOP and position error. We divide the points in HDOP bins and compute the rms value of the position error in each bin. It is no surprise that the relationship between the rms position error and DOP is linear with slope σ_{URE}.

As seen in Figure 7a, the rms URE for GLONASS is approximately 10 m, and it has remained substantially at this level over the past 3 years. With SA on, the URE for GPS is apparently changeable, but it has been relatively stable during 1992–1993, with an rms value of 25–27 m. With SA off, σ_{URE} for GPS has remained at about 7 m. We have also analyzed the range measurements from the two systems over an extended period to characterize the cdf of the range error for each.[14] The difference in σ_{URE} between GLONASS and GPS (SA off) is attributed mainly to the fact that GPS transmits in its navigation message the values of parameters to compensate partially for the ionospheric delays on the basis of a model; GLONASS does not.

We now have all the elements necessary for a global view of the positioning accuracies achievable from the operational GPS and GLONASS systems, separately and together. To recapitulate, the error in a position estimate is determined by the spatial distribution of the satellites around the user (satellite geometry) and the error in the range measurements, and we now have a complete characterization of both. Table 5 summarizes the global projections for the quality of the position estimates available from "average" 21-satellite constellations of GPS and GLONASS, when operational, on the basis of their performance as observed over 1992–1993.

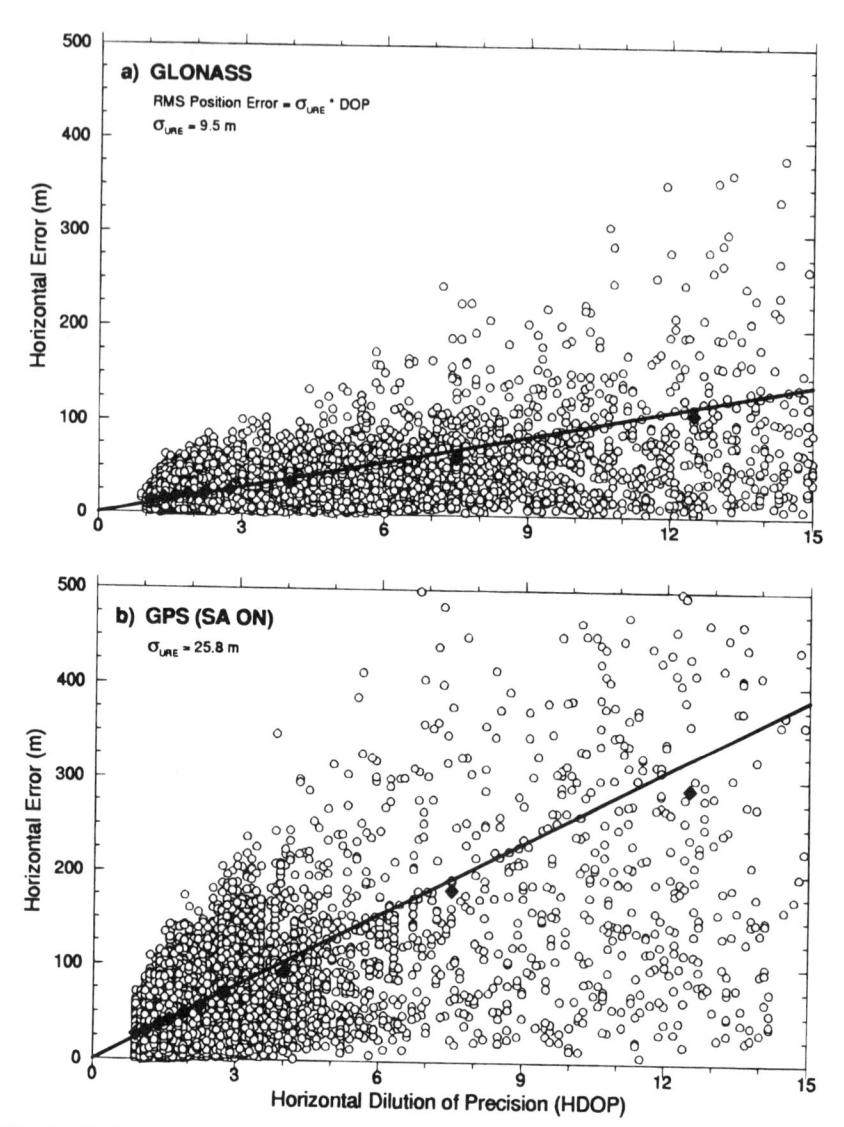

Fig. 7 Estimation of the rms value of the User Range Error (URE) from measurements taken over a 3-month period in 1993.

Table 5 The projected positioning accuracy of GPS and GLONASS

	Horizontal error		Vertical error, 95%
	50%	95%	
GPS (SA off)	7 m	18 m	34 m
GPS (SA on)	27	72	135
GLONASS	10	26	45
GPS + GLONASS	9	20	38

3. Integrity Monitoring

The navigation accuracy results of Table 5 assume that each system is operating to specifications. A user, however, cannot take this premise for granted. Indeed, both GPS and GLONASS have an extensive self-diagnosis capability on board the satellites, as well as monitoring facilities at the ground control stations. It is unclear, however, if an error can be detected and the appropriate flags set in the navigation message transmissions quickly enough to suit a pilot who is using the satellite signals in preparation for a landing. Basically, a critical demand of civil aviation is that the navigation system provide not only a position estimate but also an assurance that the estimate is "good" (i.e., the position error does not exceed a tolerable level). The idea of guarding against anomalous position estimates is called system integrity monitoring.

An approach to integrity monitoring of a satellite navigation system is to base an inference on the accuracy of a position estimate on the measurements themselves. The idea is to verify that the measurements are consistent with a model and to characterize the quality of a position estimate. This can be done if the measurement set is redundant (i.e., we have more measurements than the minimum needed for position estimation.) An important benefit of this approach, which is known as *receiver autonomous integrity monitoring* (RAIM), is that it eliminates the need for an external means of detecting system malfunction and a communication network to disseminate this information to the users.

The problem of detection and isolation of an anomalous range measurement may be thought of as one of detecting inconsistency in a set of linear equations and then identifying the anomalous equation. Obviously, at least one redundant equation is required to detect the presence of an anomaly via a consistency check. Similarly, at least two redundant equations are required to identify the anomalous equation. These, however, are only the necessary conditions, and do not guarantee an effective consistency check. The effectiveness of the check depends upon the conditioning of the set of equations and their subsets. Our task is complicated further because the equations are only approximate, being based on range measurements that include errors, the sources of which have been cited earlier. As an aside, note that the DOP parameter basically reflects the notion of linear independence of the direction vectors to the satellites and conditioning of a set of measurement equations.

Because at least four satellites are required to be in view to compute a three-dimensional position, users with five or more satellites in view may be able to use the system as supplemental. On the other hand, the real economic payoff will follow the adoption of a satellite navigation system as sole means. However, this would require that all users have six or more satellites in view. Based on the satellite visibility results for GPS and GLONASS (Fig. 5), each system by itself falls considerably short as a candidate for a sole-means system. Both systems taken together, however, offer amply redundant measurements and potential for RAIM and a sole-means system.

According to the integrity requirements, a supplemental navigation system must provide each user with a position estimate of the required accuracy, or an indication that it cannot. Obviously, the more often a system is usable, the better. If it is usable 100% of the time, we have a sole-means system. By definition, the users of a sole-means system must be capable of recovering from system failure.

A simple RAIM scheme could work as follows. Figure 8 is an azimuth-elevation sky map of the satellites, shown as stars, from the two constellations in view of a user. The total number of measurements available to a user is considerably larger than the minimum required. This is consistent with our previous results (Fig. 5). Suppose that the satellite shown as a dark star in Fig. 8 is providing anomalous measurements undetected by the system and unknown to the user. A user tracking the eight satellites inside the solid ellipse could compute a position estimate by using all of the measurements and, as a check on its quality, could compare it with the eight additional position estimates obtained when leaving out one measurement at a time. As noted earlier, the quality of a position estimate depends upon two factors: the error in the range measurements and the geometry of the satellites. Although unknown to the user, all eight satellites are performing to specifications, and the satellite geometries involved in all nine position estimates are uniformly good. The nine position estimates form a tight cluster, as shown on the right side of Fig. 8. This reassures the user of the consistency of the measurements and the quality of the position estimate.

On the other hand, suppose a user tracking the eight satellites inside the dashed ellipse in Fig. 8, including the faulty satellite, were to try this same check. The

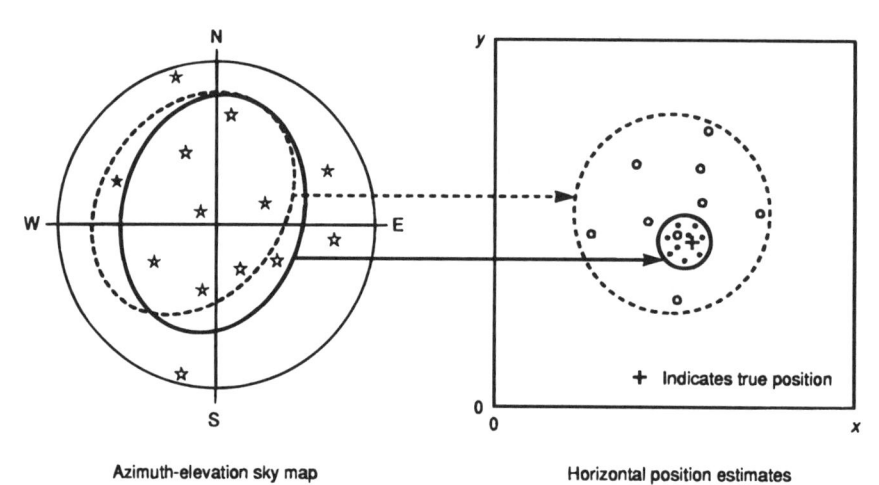

Fig. 8 Receiver autonomous integrity monitoring on the basis of scatter in the position estimates obtained from subsets of measurements.

resulting cluster would be larger; the actual size would depend primarily upon the size of the error in the faulty measurement. A user, if assured of good satellite geometries associated with the position estimates computed as a part of this check, could treat the size of the cluster as a predictor of the quality of the position estimate.

We have pursued this approach to RAIM, and have developed an algorithm for a position estimate and a measure of its quality, given the probable failure scenario discussed earlier. We define the measure of quality as a high-confidence estimate of an upper bound on the error in the position estimate and call it the *integrity level*. The integrity level is defined as $P\{\text{position error} < \text{integrity level}\} < \epsilon$, where ϵ is a suitably low, user-defined parameter. To be usable, the integrity level must be a "tight" error bound consistent with the required alarm limits. Two other essential points must be mentioned. First, the above relation is to be interpreted as a conditional probability, given that one of the satellite measurements could be in error by an indeterminate amount. The total probability that a position error could exceed its associated integrity level would be even lower; namely, it would be the probability of a system failure (expected to be quite rare) multiplied by ϵ. Second, the ability to compute the integrity level is predicated on the availability of measurements from n satellites (where $n \geq 5$), assuring good geometries for each subset of $(n - 1)$. To obtain a "tight" integrity level, we require at least five satellites satisfying the above requirements on geometry and operating to specifications.

Figure 9 illustrates the idea to be implemented. Given a set of measurements, the user computes a position estimate and its associated integrity level. The estimate is acceptable if the integrity level does not exceed the alarm limit for that phase of flight. If a system can assure all its users of the integrity levels they require at all times, then we have a sole-means system. The RAIM algorithm, described in greater detail in Ref. 14, consists of the following steps. Select n

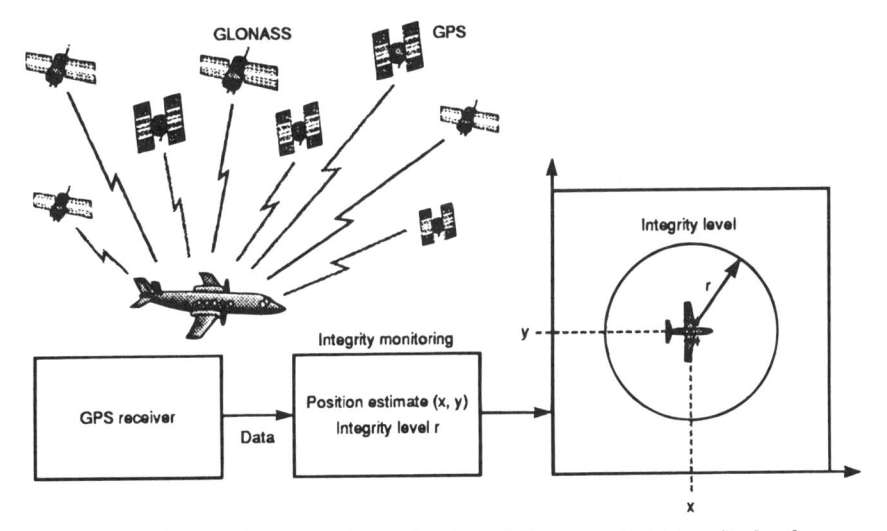

Fig. 9 Computing a position estimate and its associated integrity level.

satellites (where $n \geq 5$) among those visible, estimate positions from all n measurements, and from $(n - 1)$ measurements at a time, determine the size of the cluster (i.e., our RAIM statistic) formed by these position estimates, and obtain the corresponding integrity level from a precomputed table. If the integrity level is unsatisfactory to the user, then switch satellites for a better estimate, if possible. The computation of the table of scatter vs integrity level is at the heart of the algorithm. It requires computation of the conditional probability distribution function of the position error computed from a measurement set containing a faulty measurement, given the scatter.

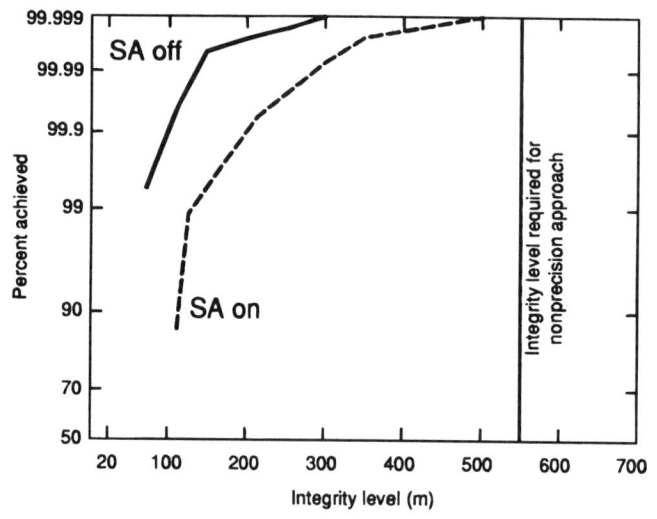

Figure 10 gives the distribution of integrity levels available to the users of GPS + GLONASS worldwide corresponding to the value of 10^{-5} for ϵ. The conclusion evident in this figure is that the combined measurements from the GPS and GLONASS systems offer a comfortable level of redundancy so that even if one of the measurements is anomalous, 99.9% of the users would be able to compute position estimates with an assurance that their position error does not exceed 200 m. Nearly all users would be able to obtain position estimates with an error below 500 m, meeting the requirements for a nonprecision approach. If SA were to be switched off in GPS, the estimates would be significantly better, as also shown in Fig. 10. This performance corresponds to the 2×21 constellation of GPS + GLONASS and reflects other assumptions on measurement quality and constellation availability that are believed to be on the safe side. By using a RAIM-based approach, therefore, GPS + GLONASS is expected to meet the requirements of a sole-means navigation system for en route and terminal phases of flight, and for nonprecision approaches.

We should note that although GLONASS, like GPS, falls short of meeting the requirements of a sole-means system, it can be used as a supplemental navigation system. GLONASS-21 can offer nearly 90% of the users the integrity level required for nonprecision approaches; GLONASS-24, 99%.

Fig. 10 Distribution of integrity levels available to the users of GPS + GLONASS.

III. Summary

The Global Navigation Satellite Systems GPS and GLONASS have much in common in terms of their signal designs, orbital plans, and ownership by military. A principal difference apparent in December 1994 is that GPS is very close to achieving operational status; GLONASS is planned by the Russian Federation to follow, but the development schedule remains unclear. This uncertainty is reflected in the current lack of ready availability of commercial GLONASS receivers anywhere, including Russia.

The performance of GLONASS in terms of coverage and positioning accuracy is potentially similar to that of GPS in the absence of SA. Like GPS, GLONASS is an attractive candidate as a supplemental navigation system for civil aviation. The combination of signals from the two systems, however, is a potential sole-means system for en route and terminal phases of flight, and for nonprecision approaches globally. The performance achievable in integrated use in coverage, accuracy, and integrity monitoring seems capable of meeting the requirements of a sole-means satellite navigation system.

While GLONASS deployment has lagged, the Wide Area Augmentation System (WAAS) has emerged as an attractive approach to achieving a sole-means capability over large geographic areas. The acquisition of WAAS is underway in the US, and several other countries have expressed an interest in expanding it beyond the conterminous United States (CONUS). This development notwithstanding, GLONASS remains potentially an attractive autonomous system of great value to the civil community.

Acknowledgments

The ISN Leeds program of research on global navigation satellite systems has been supported by the UK Civil Aviation Authority (CAA). The work on GLONASS performance at MIT Lincoln Laboratory was sponsored by the Federal Aviation Administration.

References

[1]Anon., "Global Positioning System," *Navigation*, Special Issue, Summer 1978.

[2]Anodina, T. G., "Global Positioning System GLONASS," Special committee on future air navigation systems (FANS/4), International Civil Aviation Organization, Montreal, May 2–20, 1988.

[3]Blackwell, E. G., "Overview of differential GPS Methods," *Navigation*, Vol. 32, No. 2, 1985, pp. 114–125.

[4]Remondi, B. W., "Kinematic and Pseudo-kinematic GPS," *Proceedings of the ION Satellite Division's International Technical Meeting*, (Colorado Springs, CO), Institute of Navigation, Washington, DC, Sept. 19–23, 1988, pp. 115–128.

[5]Lennen, G., "The USSR's GLONASS P-Code Determination and Initial Results," *Proceedings of the Institute of Navigation Satellite Division's International Technical Meeting*, (Colorado Springs, Co), Institute of Navigation, Washington, DC, Sept. 1989.

[6]Dale, S. A., Daly, P., and Kitching, I. D., "Understanding Signals from GLONASS Satellites," *International Journal of Satellite Communications*, Vol. 7, No. 1, 1989, pp. 11–22.

[7]Daly, P., Kitching, I. D., Allan, D., and Peppler, T., "Frequency and Time Stability of GPS and GLONASS Clocks," *International Journal of Satellite Communications*, Vol. 9, No. 1, 1991, pp. 11–22.

[8]Misra, P., et al., "GLONASS Data Analysis: Interim Results," *Navigation*, Vol. 39, Spring 1992, pp. 93–109.

[9]Misra, P., et al., "GLONASS Performance in 1992: A Review," *GPS World*, May 1993, pp. 28–38.

[10]Anon., "Department of Defense World Geodetic System 1984: Its Definition and Relationships with Local Geodetic Systems," Defense Mapping Agency, DMA TR 8350.2, 1987.

[11]Anon., "GLONASS Interface Control Document," GLAVCOSMOS, USSR, 1991.

[12]Misra, P., and Abbot, R. A., "SGS85–WGS84 Transformation," *Manuscripta Geodaetica*, Vol. 19, 1994, pp. 300–308.

[13]Anodina, T. G., "Status of and Prospects for the Development of the GLONASS Satellite Navigation System," FANS(II)/4-WP/47, International Civil Aviation Organization (ICAO), Montreal, 15 Sept.–1 Oct. 1993.

[14]Misra, P., et al., "Receiver Autonomous Integrity Monitoring (RAIM) of GPS and GLONASS," *Navigation*, Vol. 40, Spring 1993, pp. 87–104.

Part V. GPS Navigation Applications

Chapter 10

Land Vehicle Navigation and Tracking

Robert L. French*
R. L. French & Associates, Fort Worth, Texas 76107

I. Application Characteristics and Markets

LAND vehicle navigation and tracking systems represent one of the largest potential applications for Global Positioning System (GPS) receivers in terms of the sheer numbers that will be required over the next 20 years. Both vehicle navigation and tracking are important subsets of intelligent vehicle highway systems (IVHS),† a major worldwide movement to improve the efficiency, safety, and environmental aspects of road traffic through the application of information, communications, positioning, and control technologies.

The ability to determine vehicle location is the most fundamental requirement of both vehicle navigation and tracking systems. However, although navigation and tracking applications share certain common requirements and characteristics, they also are also distinctly different in many ways. Accordingly, this chapter starts with a brief characterization of tracking, navigation, and IVHS to establish a frame of reference for discussing these GPS applications.

A. Commercial Vehicle Tracking

Vehicle-tracking systems are often called by other names such as automatic vehicle location-monitoring systems or position-reporting systems. The term "commercial vehicle" as used here refers to public service vehicles such as those of ambulance, fire, police, and transit departments as well as to all classes of vehicles used in business and government service. With few exceptions (e.g., those that subscribe to commercial security or stolen vehicle tracking services), vehicles in ordinary consumer use are not considered to be part of the vehicle tracking market.

Copyright © 1995 by the American Institute of Aeronautics and Astronautics, Inc. All rights reserved.
*Principal.
†As this book goes to press, the term "Intelligent Vehicle Highway Systems" is generally being changed to "Intelligent Transportation Systems" (ITS) in the United States.

One of the most distinct differences between vehicle-tracking and navigation systems is that tracking systems do not necessarily require that vehicle location be known to the onboard component. Rather, vehicle location must be known at a central dispatch office or monitoring station, thus allowing tracking to be accomplished by multilateration using vehicle transmitter signals arriving at fixed receiver sites. However, it is common for vehicle location to be determined by onboard equipment and then transmitted to the dispatch office over a mobile data communication link.

Depending upon application, vehicle location may be communicated automatically at programed time or distance intervals, or in response to inquiries or "polling" by the dispatch office.

Thus, tracking systems typically incorporate, or are incorporated into, mobile data communications systems that report vehicle location and may also report sensed vehicle status information (e.g., engine temperature) and èxchange messages between the driver and the central dispatch office. Most vehicle-tracking systems are closely integrated with computer-aided dispatch (CAD), routing, and scheduling, or other forms of computerized fleet management systems using the same communication link.

A second major difference between vehicle navigation and tracking systems is the accuracy and continuity required of the positioning subsystem.[1] Automobile navigation and route guidance require continuous location information accurate to within 10–25 m in order to know always exactly which street is being traveled and which intersection is being approached. Vehicle tracking, depending upon the particular application (e.g., emergency response vehicles) may be equally or even more demanding in terms of location accuracy and continuity.

At the other extreme, accuracies of 500–1000 m may be acceptable for systems installed in intercity trucks because it is often adequate for the dispatcher to know on what major highway the truck is traveling or what city it is near. Tracking systems for many classes of commercial vehicles operating in urban areas may require only intermediate levels of accuracy (typically 100–300 m) depending upon the application.

Transit buses were among the first fleet operations to start using vehicle-tracking systems. Other early vehicle-tracking applications included police fleets. Major programs were sponsored by the Urban Mass Transit Administration (UMTA) during the 1970s to evaluate electronic signpost, radio frequency pulse trilateration, Loran-C, and other approaches.[2] In 1991, major transit fleets started turning to GPS for vehicle tracking.

During the 1980s, Loran-C became popular for tracking urban fleet vehicles and was used extensively with satellite communications in early intercity trucking applications. Other forms of satellite positioning such as Geostar's failed Radiodetermination Satellite Services (RDSS) endeavor[3] and Qualcomm's successful QASPR[4] (Qualcomm Automatic Satellite Position Reporting system) for intercity trucking also came on the scene during the 1980s. However, as the Navstar satellite constellation approached completion and GPS receiver prices started becoming more competitive in the early 1990s, many vehicle-tracking system developers turned to GPS.

Figure 1 illustrates the overall concept and principal subsystems for a representative vehicle-tracking system that uses an onboard radionavigation receiver (e.g.,

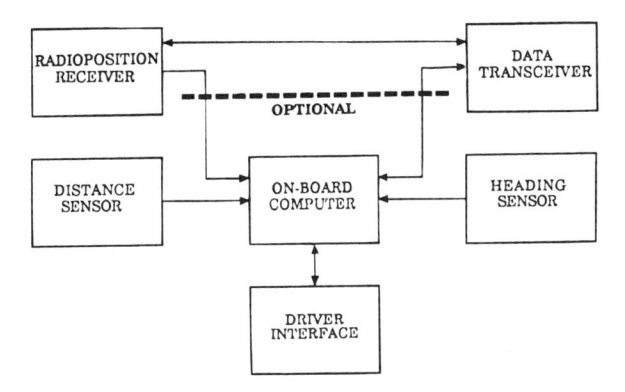

Fig. 1 Typical components and subsystems of vehicle tracking system.

GPS or Loran-C) to determine vehicle location for reporting over an arbitrary mobile data communication link. Satellite communications are typically used for tracking intercity trucking, whereas private land mobile radio or specialized mobile radio (SMR) are commonplace for urban fleet applications. (See Sec. III.E for a brief characterization of alternative mobile data communications approaches.)

Occasional discontinuities in vehicle location information, such as those encountered with GPS, Loran-C, and other radiolocation technologies in urban environments, may be a problem depending upon application, reporting protocol, and frequency of reporting. Thus, as indicated below the dashed line in Fig. 1, optional dead reckoning subsystems may be incorporated to keep track of vehicle location between valid radiolocation fixes. Although effective over short intervals, dead reckoning accuracy deteriorates with distance and/or time unless updated by GPS or other means, such as map matching in the case of automobile navigation. (See Sec. III.A for a brief description of dead reckoning approaches for vehicles.)

B. Automobile Navigation and Route Guidance

Automobile navigation systems develop and present to the driver various forms of navigation information useful in determining how to reach the desired destination.[5] Although navigation information may be perceived largely as a convenience by the individual driver, there are also broader societal needs according to a Federal Highway Administration study.[6] The study estimated that almost 7% of all distance traveled by noncommercial vehicles and over 12% of time spend driving is wasted because of poor navigation and route-following skills, thus contributing to congestion and accidents. The additional costs associated with this excess travel amount to $45 billion per year.

Most automobile navigation systems already on the market include an electronic road map display with icons indicating current vehicle location and the destination. More advanced systems also compute optimum routes and use simplified graphics and/or synthesized voice to provide real-time, step-by-step route guidance instructions for reaching the destination. Future systems may include "head-up" displays that project guidance or other information onto the windshield of the vehicle.

In addition to means for automatically and continuously determining vehicle location with sufficient accuracy to identify the road being traveled and each intersection approached, automobile navigation systems include digital road map databases giving the location, classification, and address ranges for each road in the areas where the vehicle operates. The map database may also include traffic regulations and typical travel times for individual road links for use in calculating optimum routes, as well as additional information such as the location and description of service stations, garages, parking, public buildings, hotels, restaurants, tourist attractions, and other types of directory information commonly used by travelers. Automobile navigation systems must also include man/machine interfaces for driver input and control and for presentation of navigation or route guidance information.

Figure 2 shows the major elements of a typical vehicular navigation system. Distance and heading (or heading change) sensors are invariably included for dead reckoning calculations, which, in combination with map matching, form the basic platform for keeping track of vehicle location. However, dead reckoning with map matching has the drawback of occasionally failing because of dead reckoning anomalies, extensive travel off mapped roads, ferry crossings, etc. Thus, the location sensor indicated by dashed lines is an optional means of providing absolute location to avoid occasional manual reinitialization when dead reckoning with map matching fails. Although electronic signposts or proximity beacons served to update vehicle location in some early developmental systems, most state-of-the-art systems include a GPS receiver for this purpose.

The data transceiver (see Fig. 2) is an option to permit navigation and route guidance systems to receive real-time traffic data from a traffic management center for use in determining optimum routes under prevailing conditions. The transceiver may also be used for transmitting measured link travel times to a traffic management center to maintain a real-time database of link travel times for transmittal to equipped vehicles. In-vehicle navigation and route guidance

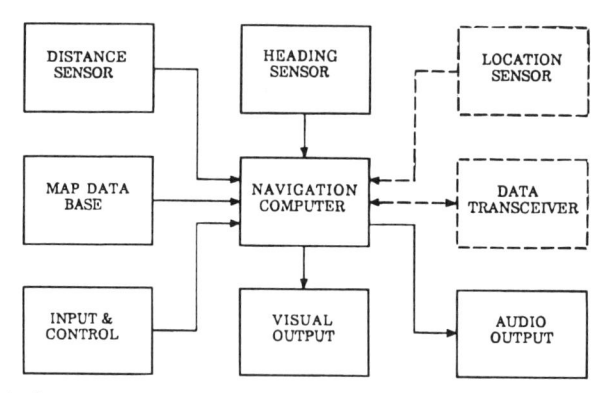

Fig. 2 Typical components and subsystems of vehicle navigation system. "Location sensor," if included, may be a GPS receiver or a proximity beacon receiver.[33]

systems coupled to traffic management centers by communication links are one of the most central aspects of IVHS.

C. Intelligent Vehicle Highway Systems

Intelligent vehicle highway systems apply computer, positioning, communications, and control technologies to integrate vehicles and highways in a coherent information network that facilitates the travel of individual vehicles, while optimizing traffic flow and increasing traffic capacity throughout the entire road system. Some basic concepts and components of IVHS are illustrated by Fig. 3.

Major efforts got underway in the United States, Europe, and Japan during the late 1980s to develop and apply IVHS to reduce congestion, improve mobility and road transportation efficiency, enhance safety, conserve energy, and protect the environment. IVHS AMERICA was formed in 1990 as a public–private educational and scientific organization to coordinate and promote the development of IVHS in the United States.

Although the taxonomy of IVHS is not yet fully consistent worldwide, the following six categories used for defining the United States' IVHS program encompass virtually all elements of IVHS approaches being pursued elsewhere as well:

1. Advanced Traffic Management Systems (ATMS)

Advanced traffic management systems extend real-time computer optimization of traffic signal timing to the urban road network level as opposed to individual

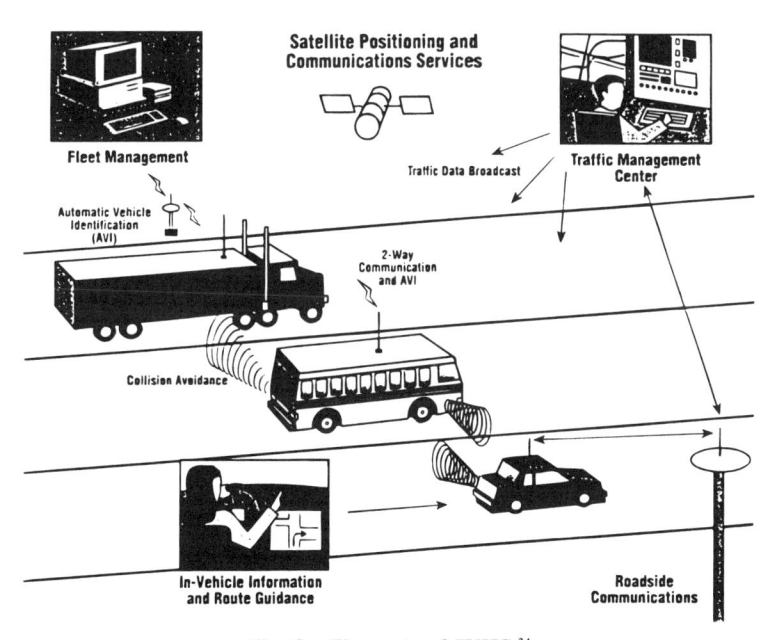

Fig. 3 Elements of IVHS.[34]

intersections or streets. This requires information on traffic conditions throughout the network in a real-time database that may also serve as an information source for dynamic route guidance in advanced traveler information systems-equipped vehicles.

2. Advanced Traveler Information Systems (ATIS)

Advanced traveler information systems keep drivers informed of their location and provide route guidance to selected destinations along with information on services such as lodging, food, fuel, repair, medical facilities, etc. ATIS permit communication between in-vehicle equipment and ATMS for data on traffic conditions, diversion routes, alternative modes of transportation, etc. Although ATIS concepts originally centered on vehicular navigation and route guidance for drivers, new ATIS concepts include portable versions for use by pedestrians and multimodal travelers also.

3. Commercial Vehicle Operations (CVO)

These include vehicle tracking and fleet management systems for commercial and emergency vehicles to improve operational efficiency and increase safety. They also include technologies such as automatic vehicle classification (AVC), weigh-in-motion (WIM), and communications among automated regulatory checkpoints so that intercity trucks may travel among different jurisdictions with minimal stopping.

4. Advanced Vehicle Control Systems (AVCS)

These apply additional technologies to vehicles to detect obstacles and adjacent vehicles, thus enhancing vehicle control by augmenting driver performance. Advanced vehicle control systems assist in the prevention of collisions for safer high-speed driving to increase roadway capacity, and they will eventually interact with fully developed ATMS to enable automatic vehicle operation.

5. Advanced Public Transportation Systems (APTS)

In addition to applying the above IVHS technologies to public transportation systems, APTS have a strong focus on customer interface. Examples include onboard displays (e.g., for next stop, transfer information, etc.), real-time displays at bus stops, and smart card fare systems as well as ride share and HOV (high-occupancy vehicle) information systems.

6. Advanced Rural Transportation Systems (ARTS)

These focus on issues and problems involving the development and application of IVHS to rural transportation. The major thrusts of ARTS include emergency communications (e.g., automated Mayday calls) and safety applications of IVHS technologies.

Intelligent vehicle highway systems are still in the early stages of development and, although numerous operational field trails are underway, relatively few actual applications of IVHS technology have been implemented. However, according to a General Accounting Office report,[7] available results indicate that travel times have been reduced by as much as 15%, and freeway speeds have been increased from 15 to 40 mph in congested areas. Other reported benefits include an 8–12% decrease in motor fuel consumption, an 8% decrease in hydrocarbon emissions, and an 18% decrease in roadway accidents.

Autonomous navigation systems started coming on the market in the late 1980s, and approximately 500,000 were already in use in Japan in 1993. ATIS versions with communication links to real-time traffic databases will start becoming available in the mid-1990s. An estimated 2.5 million vehicles per year will be sold with factory-installed ATIS by the year 2000, and 11 million per year will be sold by the year 2010, according IVHS AMERICA strategic planning projections.[8]

The IVHS movement also adds momentum to existing and emerging vehicle-tracking applications. Strategic planning for IVHS in the United States assumes that almost 300,000 CVO systems per year will be purchased by the year 2000. However, IVHS also has potential for modifying the architecture of vehicle-tracking systems when ATIS start becoming commonplace in all types of vehicles by the late 1990s. Because ATIS include navigation systems requiring accurate location information, it is likely that a common positioning subsystem will come to be used for both navigation and tracking.

II. Historical Background

Although vehicular navigation and tracking system concepts have become widely known only during the last few decades, their historical roots go much deeper.

A. Early Mechanical Systems

The world's first vehicular navigation system was the "south-pointing chariot," an automatic direction-keeping system developed by the Chinese around 200–300 AD (possibly earlier according to some legendary accounts), almost 1000 years before the magnetic compass was invented.[9] Its operation was based on the phenomenon that as a vehicle changes heading, the outer wheels travel farther than the inner wheels by a distance that is a simple mathematical function of the change in heading. When changing heading, a gear driven by the outer wheel of the south-pointing chariot automatically engaged and rotated a horizontal turntable to exactly offset the change in heading. Thus, a figure mounted on the turntable continuously pointed an outstretched arm in the same direction, like a compass needle, regardless of which way the chariot turned.

The differential odometer, the principle used in the south-pointing chariot, is now popular as a dead reckoning subsystem for modern automobile navigation systems because it is not subject to magnetic field disturbances as a magnetic compass is. It has the further advantage of not requiring additional hardware in automobiles equipped with wheel rotation sensors for antilock braking systems (ABS).

Mechanical route guides for automobiles began appearing in the United States around 1910 and were developed to aid drivers of early automobiles.[10] These pioneering devices incorporated the information of route maps in various forms including sequential instructions printed on a turntable, punched in a rotating disk, and printed on a moving tape, all being driven by an odometer shaft in synchronization with distance traveled along the route.

In fact, the odometer is itself a navigation device because it may be used in conjunction with road signs or road maps to estimate present position and monitor progress toward a destination. The mechanical route guides, however, went a step further to provide explicit real-time route instructions automatically at decision points along the way.

One of the earliest, and one of the most popular mechanical route guides to reach the market, was the Jones Live-Map, which was patented in 1909.[11] It consisted of a turntable slowly rotated by a gear train connected by flexible shaft to one of the vehicle wheels. Paper disks for individual routes with a scale of miles printed around their perimeters were mounted on the turntable beneath a glass cover with a fixed pointer. Detailed road directions keyed to specific distances from the beginning of a route came into view under the pointer at the time for execution. An advertisement for the Jones Live-Map claimed: "You take all the puzzling corners and forks with never a pause. You never stop to inquire...."

The far more sophisticated Chadwick Road Guide was introduced in 1910.[12] Like the Live-Map, the Chadwick device rotated a calibrated disk in synchronization with distance traveled. However, the metal disk contained holes spaced to coincide with decision points along the route represented by the disk. An array of spring-loaded pins behind the slowly rotating disk was normally depressed, but when a punched hole traversed a pin, the pin released and raised a signal arm bearing a color-coded symbol indicating the action to be taken. Simultaneously, a bell sounded to draw the driver's attention to the signal. Moreover, beneath a pointer, the same disk gave printed information regarding the location or the action to be taken.

B. Early Electronic Systems

As the expanding roadway system became better marked with standardized signs, and reliable road maps became available, the need for route guidance devices diminished, and only sporadic developments occurred between World War I and the late 1960s. Mechanical approaches faded into the background with the advent of electrical servomechanisms, electronic controls, and, eventually, the digital computer.

One of the earliest land vehicle navigation systems to incorporate electronic elements was the vehicular odograph developed for military vehicles during World War II.[13] The vehicular odograph was a dead reckoning system that included a light beam and photocell arrangement to read the output of a magnetic compass. The compass output drove a servomechanism to rotate a mechanical shaft corresponding to vehicle heading. The shaft was coupled to a mechanical computer that resolved distanced traveled from an odometer shaft into X and Y components, and drove a stylus to trace the vehicle's course automatically on a map of corresponding scale.

In one of the first steps toward applications now called IVHS, the U.S. Bureau of Public Roads (now the Federal Highway Administration), started researching means in the late 1960s for integrating in-vehicle route guidance with traffic management under a project called ERGS (Electronic Route Guidance System).[14] This system used proximity beacons in the form of short-range transmitters with inductive loop antennas buried beneath the roadway at strategic intersections. A dash-mounted console with thumbwheel switches permitted the driver to enter a selected destination code that was transmitted when the vehicle unit was triggered by a roadside unit upon nearing key intersections. The roadside unit immediately selected the optimum route to the destination considering the current traffic patterns and transmitted instructions for display on the vehicle's console.

Although technically sound, ERGS required expensive roadside infrastructure, and the development was canceled in 1970, following limited testing at the subsystem level. Similar approaches were carried through further stages of development and testing during the 1970s in Japan and Europe. Although a quarter of a century old, the basic ERGS concept is still representative of IVHS systems that use proximity-beacons for communications and/or position updates. In the meantime, new communications approaches, availability of inexpensive onboard computers, development of map-matching algorithms, and the promise of satellite positioning gave rise to a number of alternatives to the proximity-beacon approach.

Another early development toward IVHS was map-matching of the early 1970s to augment dead reckoning in vehicular navigation systems. Networks of roads and streets may be modeled as internodal vectors in a digital map database, and a particular route may be programmed as a unique sequence of mathematical vectors. An onboard computer may be used to analyze dead reckoning inputs and match the deduced vehicle path with programmed routes to remove position discrepancies automatically that otherwise accumulate. Similarly, map-matching may be used to trace the location of a vehicle traveling arbitrary paths within a network.

The first map-matching system, called the automatic route control system (ARCS), was developed in 1971 for a commercial delivery operation rather than IVHS.[15] It used the differential odometer principle to compute the vehicle's approximate path. A map-matching algorithm correlated each sequentially measured (i.e., dead reckoned) route vector and change in heading with its map database counterpart. Thus, the vehicle's location was confirmed by ARCS, and real-time route guidance (albeit without consideration of real-time traffic conditions) was issued at appropriate points using prerecorded voice messages. A second version issued route instructions visually, using simplified graphics on a plasma display panel.

III. Enabling/Supporting Technologies

Positioning technologies are fundamental requirements of both vehicle navigation systems and vehicle-tracking systems. Almost all vehicular navigation systems are heavily dependent upon dead reckoning with map matching as the main positioning technology, but most such systems now include GPS to good advantage. Map matching as well as route guidance requires digital road maps.

Another important supporting technology is mobile data communications. Although vehicle-tracking systems may not require digital road maps aboard the vehicle, they do require a mobile data communications link between the vehicle and dispatch office unless they use infrastructure-based integrated positioning and communications technologies. However, GPS and other radiopositioning technologies, especially Loran-C, are typically the sole or main technologies used for vehicle-tracking systems.

A. Dead Reckoning

Dead reckoning, the process of calculating location by integrating measured increments of distance and direction of travel relative to a known location, is used in many vehicle-tracking systems and in virtually all vehicle navigation systems.

1. Mathematical Formulation

The basic mathematical formulation for determining a vehicle's current coordinates, X_n and Y_n, relative to its initial coordinates, X_0 and Y_0, as depicted in Fig. 4 is as follows:

$$X_n = X_0 + \sum_1^n \Delta X_i = X_0 + \sum_1^n \Delta \ell_i \sin \phi_i$$

$$Y_n = Y_0 + \sum_1^n \Delta Y_i = Y_0 + \sum_1^n \Delta \ell_i \cos \phi_i$$

where ϕ_i is the heading associated with ℓ_i, the ith increment of travel, as illustrated. Thus, dead reckoning for vehicles requires a means for sensing distance traveled and heading (or change in heading).

2. Distance/Speed Sensors

Distance measurements for vehicle navigation systems are usually made with an electronic version of the odometer. Electronic odometers provide discrete

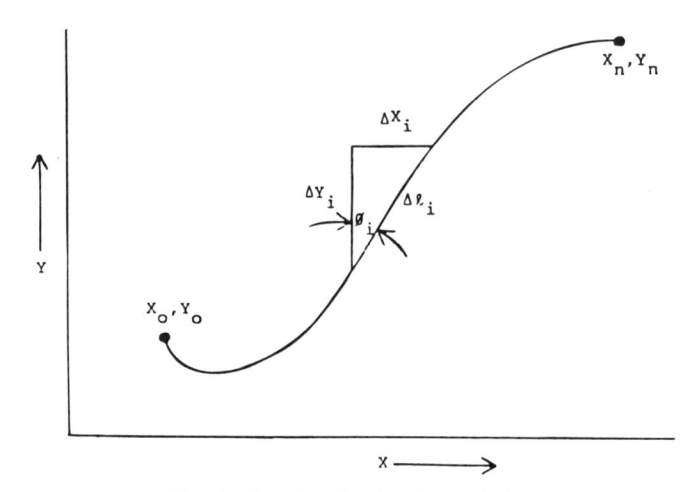

Fig. 4 Dead-reckoning formulation.

signals from a rotating shaft or wheel, and a conversion factor is applied to obtain the incremental distance associated with each signal. Magnetic, inductive, or capacitive sensors typically are mounted on a stationary member to detect the passage of closely spaced magnets or metallic protrusions attached to the hub or rim of a vehicle's wheel. Automobiles equipped with ABS already have wheel sensors that may also be used with navigation systems.

Odometer measurements are subject to a number of random and systematic errors, some of which can be defined and corrected in the distance conversion process. The difference in the diameter of a new tire and a well-worn tire, for example, can contribute distance errors as high as 3%. The error in distance measurements increases by approximately 0.1–0.7% when vehicle speed is increased by 25 mph because of the effect of centrifugal force on the tires, and a 10 psi change in tire pressure can induce an error of 0.25–1.1% percent.[16]

3. Heading/Heading-Change Sensors

Vehicle heading may be determined by direct measurement with a magnetic compass, or by keeping track of heading relative to an initial heading by accumulating individual changes in heading. A number of alternative means are available for measuring vehicle heading or heading changes. However, most have at least one drawback. As a result, it is a common practice for two different types of sensors to be used in combination to offset one another's weaknesses.

The magnetic compass's well-known accuracy problems caused by anomalies in the Earth's magnetic field are compounded when installed in a vehicle because of induced fields that depend upon vehicle heading. In addition, a vehicle may have a strong permanent magnetic field of its own, and subpermanent magnetism may be acquired or lost when hitting bumps. Buildings, bridges, and other structures external to a vehicle can also cause magnetic aberrations. Moreover, unless a compass is pendulum mounted, more errors are introduced when operating on an incline. A special consideration in installing magnetic compasses in trucks is the potential for disturbances from various types of cargo.

Nevertheless, some form of magnetic compass is used in most vehicular navigation systems. Compact, solid-state, flux-gate compass with software processes for compensating errors resulting from both permanent and induced magnetism of the vehicle[17] are especially popular in current systems. In many applications, both a flux-gate compass and a differential odometer or a gyroscopic device are used along with a software filtering process that combines the two outputs. Relatively more weight is placed on the differential odometer or gyro output for short-term changes in heading and on the compass for longer-term trends in absolute heading.

A differential odometer essentially is a pair of odometers, one each for the wheels on opposite ends of an axle. When a vehicle changes heading by an amount θ, as illustrated in Fig. 5, the outer wheel travels farther than the inner wheel by ΔD:

$$\Delta D = D_R - D_L$$

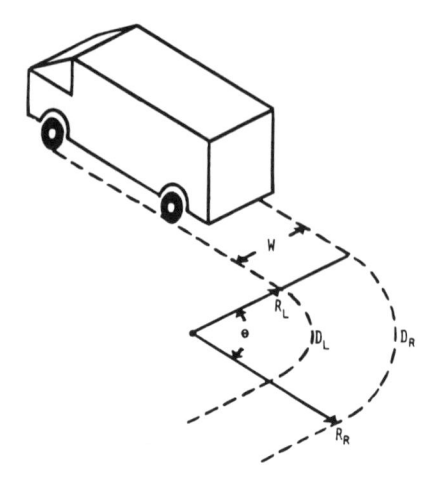

Fig. 5 Differential odometer principle.

Expressed in terms of heading change and vehicle width W, the difference in wheel travel is:

$$\Delta D = \theta(R_R - R_L) = \theta W$$

Thus, $\theta = \Delta D/W$.

Most other types of dead reckoning sensors used to detect heading changes for vehicle location and navigation systems use some form of the gyroscopic principle. These range from traditional spinning devices and gas-jet sensors to vibrating bars and fiberoptic gyros, and, although more expensive than differential odometer sensors, they are much simpler to install. The fiberoptic gyro started appearing in production automobile navigation systems in 1991.[18]

B. Digital Road Maps

An onboard digital road map database is an essential feature of vehicular navigation and route guidance systems. Digital maps also play important roles in vehicle-tracking systems for displaying vehicle location at the fleet dispatch office. The two basic approaches to digitizing maps are matrix encoding and vector encoding. Matrix encoding preserves map detail at a selected degree of resolution in the form of a digitized image, whereas vector encoding models roads in a mathematically abstracted form.

A matrix encoded map may be thought of as a digitized image in which each image element or pixel, as determined by an X–Y grid with arbitrary spacing, is defined by digital data giving characteristics such as shade or color. Thus, in addition to requiring more data storage, matrix encoding does not facilitate analytical treatment (such as map matching or route finding) of logical connections among the road elements. Nonetheless, matrix encoding is sometimes used in digitizing maps for display purposes such as tracking systems that superimpose vehicle location or other information.

The vector encoding approach applies mathematical modeling concepts to represent such geometrical features as roadways and boundaries with a minimum of data. By considering each road or street as a series of straight lines and each intersection as a node, a map may be viewed as a set of interrelated nodes, lines, and enclosed areas, as illustrated in Fig. 6.[19] Nodes may be identified by their coordinates (latitude and longitude). Additional nodes, typically called "shape points," are positioned along curves if the link between two intersections is not a straight line. Thus, curves are approximated by a series of vectors connecting shape points, whereas a single vector directly connects the node points representing successive intersections if there are no curves on the road segment in between.

The X-Y coordinates of node points may be encoded from maps or aerial photographs. In practice, they are usually encoded using special work stations that record the coordinates of a given point when the crosshair of an instrument is placed over the point and a button pressed. This process has been automated in varying degrees. In some cases, the printed map is scanned to obtain a matrix image, which is then converted to vector form by software.

Various combinations of attributes associated with the encoded road network are included in digital map databases. Of particular importance are roadway classifications, street names, and address ranges between nodes. Map databases used with systems that give turn-by-turn route guidance instructions also require traffic attributes such as turn restrictions and delineation of one-way streets. Directory and "yellow pages" information for selecting attractions, parking, restaurants, hotels, emergency facilities, etc. commonly are included.

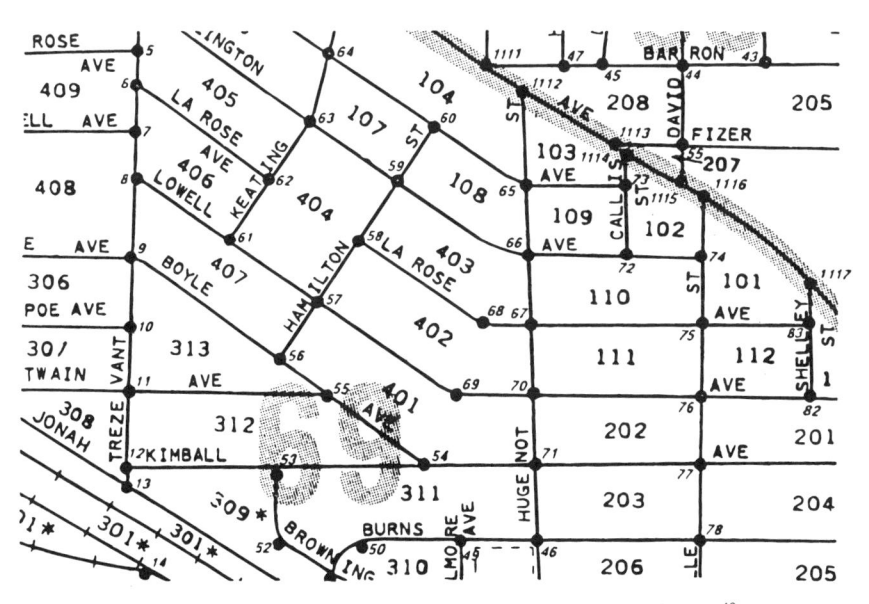

Fig. 6 Nodes and street segments of vector encoded map.[19]

C. Map Matching

The fact that motor vehicles are largely constrained to a finite network of streets and roads with only occasional excursions into driveways, parking lots, etc. makes it possible for computer algorithms utilizing road map information stored in vector encoded form to correlate a vehicle's path approximated by dead reckoning or other means with the digital map to maintain accurate knowledge of the vehicle's location within the defined road network. Known as "map matching" this process is used in virtually all vehicle navigation and route guidance systems. Most map matching algorithms may be classified as either semideterministic or probabilistic.[20]

1. Semideterministic

Semideterministic map-matching algorithms assume that the equipped vehicle is essentially confined to a defined route or road network, and thus are designed to determine where the vehicle is along a route or within the network. The basic concept can be illustrated by tracking the location of a vehicle over the simple route shown in Fig. 7a.

Figure 7b defines the route from node A through nodes B, C, D, E, and thence back to A in terms of instantaneous direction, ϕ, of travel vs cumulative distance, L, from the beginning. Locations of nodes where direction changes occur (or could occur) thus are defined in terms of distance L. The solid line is the plot of heading vs distance corresponding to the simple route. Alternative routes emanating from each node are indicated by dashed lines.

The kernel of a semideterministic map-matching algorithm is shown in highly simplified form in Fig. 8. Once initialized at a starting location (ϕ = 90 deg and L = 0 at Node A in the simple example of Fig. 7), the algorithm, in effect, repeatedly asks, "Is the vehicle still following the route?" and "What is the present location along the route?" The vehicle is confirmed on the route if certain tests are satisfied. The location along the route is estimated by odometry, and error in the estimate is automatically removed at each node where it is determined that an expected change in the vehicle heading actually occurs.

The simplified deterministic map-matching algorithm is driven by interrupts from differential odometer sensors installed on the left and right wheels. The distance L from the beginning of a route segment is updated by adding an increment ΔL for each left wheel interrupt, and the vehicle heading ϕ is updated by adding an increment $\Delta\phi$ calculated from the difference in travel by the left and right wheels occurring since the count N was last set to 0. As explained below, the N counter controls monitoring for unexpected heading changes occurring over relatively short distances.

Unless the turn flag is set to denote that the vehicle is approaching a distance L where a heading change should occur, count N is checked after each interrupt to determine if it has reached a limit C corresponding to an arbitrary amount of travel on the order of several meters. When the count limit C is reached, a test is made to determine if ϕ is within arbitrary limits (say ± 5 or 10 deg to allow for lane changes, slight road curvature, etc.). If so, ϕ is reset to ϕ_0 (the initial direction of the vector being traveled) and N is set to 0 to start another cycle of monitoring for unexpected heading changes.

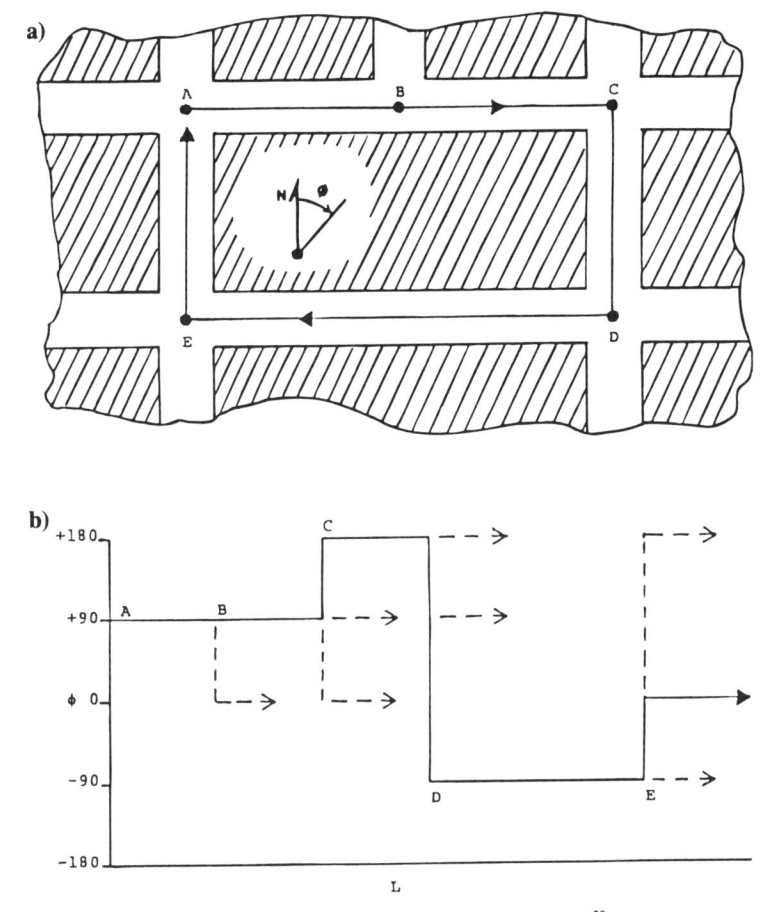

Fig. 7 Simplified route and vector model.[20]

Thus, in addition to verifying that the vehicle remains on the route while traveling between nodes, this process removes error in measured vehicle heading that accumulates while $0<N<C$. If the above test finds ϕ to be outside the limits, the vehicle is presumed to have turned off the route (perhaps into an unmapped driveway or parking lot), and other routines are called into play. For example, the driver could be informed of a route error and issued recovery instructions.

When the vehicle approaches within an arbitrary distance (say 75 m) of a node where a change in vehicle heading should occur, the turn flag is set and a route guidance instruction indicating the direction of the turn and, if appropriate, the name of the road to take is issued. The algorithm then continuously monitors changes in ϕ to confirm that the midpoint of the expected turn is reached within arbitrary limits (say 10 m) of the value of L specified for the node, and to confirm that the turn is completed.

When the midpoint is reached, the current value of L is adjusted to that specified, thus removing any error in the measured distance accumulated since

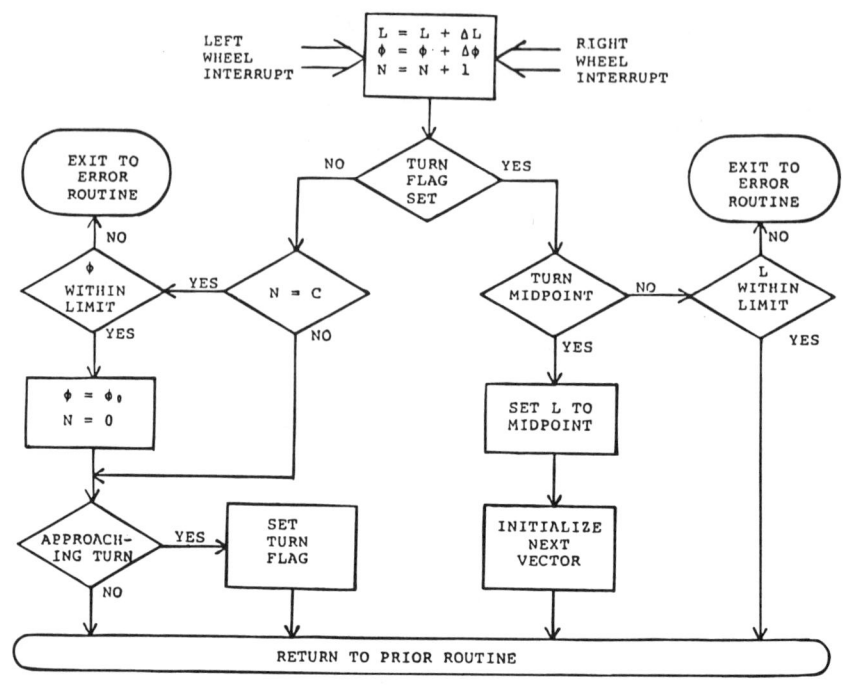

Fig. 8 Simplified map-matching algorithm.[20]

the last turn. If the expected turn is not observed within the allowed limits on distance L, the vehicle is assumed to have missed the turn or to have taken an alternate turn (see dashed lines in Fig. 7b), and other routines may be entered to identify the alternate route taken from the node.

The semideterministic algorithm concept outlined in the preceding paragraphs may be extended to tracking a vehicle's location as it moves over arbitrary routes within a road network rather than following a preplanned route. As long as the vehicle stays on roadways defined by a vector-encoded digital map, the vehicle must exit each node via some vector. Thus, a map-matching algorithm can identify successive vectors traveled by measuring the direction of vehicle travel as it leaves each node and comparing the vehicle direction with that of various vectors emanating from the node.

2. Probabilistic

Another type of map-matching routine is required for tracking vehicles not presumed to be constrained to the roads of a particular route or network. When the vehicle departs from the defined route or road network, (e.g., into a parking lot), or appears to depart as a result of dead reckoning error, the routine repeatedly compares the vehicle's dead reckoned coordinates with those of the links surrounding the off-road area that encompasses the vehicle location in order to recognize where the vehicle returns to the road network. Unlike travel on defined roadways, frequent map-matching adjustments do not prevent the accumulation of dead reckoning error. Thus, depending upon the distance traveled off road and the accuracy of the dead reckoning sensors, there may be considerable uncertainty in vehicle coordinates, which could produce misleading conclusions when tested against the surrounding links.

Probabilistic map-matching algorithms are used to minimize the potential of off-road errors by maintaining a running estimate of uncertainty in dead reckoned location, which is taken into consideration in determining whether the vehicle is on a street or not. The estimate of location uncertainty is reduced each time it is deemed that the vehicle is on a street, but the uncertainty resumes growth in proportion to further vehicle travel until the next match occurs. Thus, a probabilistic algorithm repeatedly asks, "Where is the vehicle?" with no a priori presumption that it is on a road.

D. Integration with GPS

Just as backing up GPS with dead reckoning may be highly useful in some vehicle-tracking applications, GPS may be integrated with map matching in others. Backing up dead reckoning and map matching in vehicular navigation systems with GPS eliminates the occasional failure of the former. Although failures may be infrequent (e.g., *Travelpilot*, Sec. IV.A), manual reinitialization can be a tedious and time-consuming task. Thus, GPS augmentation is commonly used in state-of-the-art automobile navigation systems.

The integration of GPS with dead reckoning as is often done for vehicle-tracking systems may be accomplished by a variety of approaches ranging from very simple to relatively complex. In the simplest case, dead reckoning may be used to fill in discontinuities in GPS position by initializing the dead reckoning position to the last GPS fix and incrementing the position as outlined in Sec. III.A. At the other extreme are sophisticated software filter approaches, such as the example outlined in Sec. IV.D.

GPS receivers may be integrated with dead reckoning and map matching in automobile navigation systems through relatively simple modifications of map-matching algorithms or through use of Kalman or other software filtering schemes. In the simplest approach, the GPS position may be ignored if it is within reasonable agreement with the dead-reckoning/map-matching position. However, when combining positions from the different techniques to determine the most likely location, the map-matching software must take into account the probable errors or uncertainties of each of the different techniques. This can be accomplished by incorporating filters in map-matching software.

The Kalman filter is the best known filter technology used for combining by developers of vehicle-tracking and navigation software. The basic mathematical models from which a Kalman filter equations are derived are f_k at time epoch t_k for map matching, $f_k + 1$ at time epoch $t_k + 1$ for GPS, and $g_k, k + 1$ for the interval between map-matching and GPS updates.[21] A Kalman filter uses the predicted state vectors and covariance matrices to compute the filtered state vectors and covariance matrices. This process is repeated for each time epoch.

E. Mobile Data Communications

Vehicle-tracking systems as well as navigation systems integrated with IVHS depend upon mobile communication for exchanging information between in-vehicle equipment and dispatch offices or traffic control centers. The principal communication systems identified for electronic transmission of information to and from vehicles are listed and broadly characterized in Table 1.[22]

Regulatory bodies allow commercial FM and TV broadcast stations to transmit inaudible ancillary information on sideband channels displaced 53–99 KHz from the central frequency. Called subcarrier authorization (SCA) in the United States, these sideband channels may carry analog or digital signals that are detected and decoded by special attachments or design features of ordinary receivers. The European Radio Data System (RDS) standardized by the European Broadcasting Union in 1984 operates on the same principle for transmitting station and programming data for automatic tuning along with other information in digital form.[23]

Table 1 Characteristics of alternative mobile communication approaches

Approach	Characteristics
Broadcast SCA	One-way only
	Voice and data
	Low data rates
	Extended area coverage
	Includes RDS
Proximity beacon	One-way or two-way
	Data only
	High data rates
	Spot-area coverage
Inductive loop	One-way or two-way
	Data only
	Low data rates
	Spot-Area coverage
Land-mobile radio (dedicated)	Two-way
	Voice and data
	Local-area coverage
Specialized mobile radio (SMR)	Two-way
	Voice and data
	Extended-area coverage
Cellular radio	Two-way
	Voice and data
	Local/extended-area coverage
Mobile satellite communications	One-way or two-way
	Voice or data
	Wide-area coverage
Meteor burst communications	Two-way
	Data only
	Wide-area coverage
	Involves time delays

Source: Ref. 22.

Use of the SCA approach for area broadcast of digitized information to vehicles requires relatively little additional equipment other than a minor enhancement of the FM radio receiver commonplace in vehicles. However, broadcast SCA has the disadvantage of being limited to one-way communication into vehicles, thus it cannot be used for vehicle tracking.

Infrared and microwave proximity beacons, which have the advantages of high data rates in combination with spot coverage (i.e., typically limited to tens of meters) that permit messages to be tailored to highly localized needs, are strong candidates for a mobile data communications role in route guidance systems being developed and tested in Europe and Japan. In these densely populated geographic areas, heavy infrastructure costs of installing beacons at close intervals throughout the road system and integrating them into overall communications networks do not seem as formidable as in the United States where, except for use as electronic signposts in vehicle-tracking systems for transit buses, proximity beacons have been used for few applications.

Inductive loops, which essentially are proximity beacons in the form of radio antennas buried beneath the roadway, have the low data rates characteristic of radio-frequency proximity beacons. They have the further disadvantage of being very expensive and awkward to install because traffic must be temporarily diverted from lanes being equipped. Maintenance is also expensive because of the wear and tear from traffic. Nonetheless, inductive loops were considered for communications in the early ERGS route guidance system research (see Sec. II.B), and they are being re-examined under current IVHS programs.

Although individually owned and operated land mobile radio systems for two-way voice communication have seen widespread use for fleet management within city areas, their spread has been severely limited by the amount of spectrum allocated for this purpose. However, great gains in dedicated land mobile throughput have resulted from increased use of data communications as an alternative to voice.

Specialized mobile radio is a class of land mobile radio service first authorized by the FCC in 1974. Specialized mobile radio quickly emerged to become the pre-eminent provider of private land mobile communications service, particularly in large metropolitan areas. Basically, SMR is a business and regulatory approach that permits different entities to share common transmission facilities and frequencies. With one type of trunked SMR service, a user vehicle unit monitors a number of frequency channels for a unique digital code identifying a transmission addressed to the individual unit and monitors channel use to select an available frequency for transmitting its own messages addressed to the user's dispatch station. With the second type of trunked system, the vehicle unit monitors a "control channel" that manages channel assignments.

Cellular radio telephones have provided a major step forward for voice communications in the vehicular environment, and are now seeing increased use for data communications. However, the present scheme of using cellular radio technology exclusively for full-duplex individually addressed communications will limit the potential use of cellular radio in providing traffic and routing data for vehicular navigation and information systems. Effective adaptation of cellular radio to the data communications requirements of navigation and route guidance would require the establishment of a dedicated channel for repeatedly broadcasting the map updates, traffic data, etc., by all cell transmitters in the local area served by a cellular system. Equipped vehicles could have special receive-only units to detect the data for transfer to the onboard equipment.

In the meantime, commercial fleet management applications of conventional cellular telephone are growing, and have particular attraction for small vehicle fleets with only infrequent requirements for voice communications within local cellular service areas. Cellular data links are also being considered by a number of vehicle location monitoring systems integrators for localized applications.

Unlike urban vehicle location monitoring, which can use short-range land-mobile radio or cellular telephone communication links, crosscountry truck location monitoring requires long-distance communication links such as those characteristic of mobile satellite services. Thus, satellite communications have long been viewed as having great potential for crosscountry trucking applications because of their wide-area coverage. A well-established example is the Omni-TRACS two-way satellite communication and position reporting service offered by Qualcomm, Inc. using geostationary satellites. Low-Earth-orbit (LEO) satellite systems such as Iridium proposed by Motorola, Inc. and ORBCOMM proposed by Orbital Communications Corp. also hold potential for extending mobile communications in rural and remote areas. However, the wide area covered by satellites is not an advantage for traffic data communications because traffic data are not useful outside the subject local area.

IV. Examples of Integrated Systems

The following examples of state-of-the-art systems collectively illustrate the recent evolution and present trends worldwide for vehicular navigation, route guidance, and tracking systems.

A. Etak Navigator™/Bosch Travelpilot™

The first commercially available automobile navigation system to include digitized road maps, dead reckoning with map matching, and an electronic map display was the Etak, Inc. Navigator introduced in California in the mid-1980s.[24] It used a flux-gate compass and differential odometer for dead reckoning. The equivalent of two printed city street maps were vector encoded and stored on 3.5-Mb digital cassettes for map matching and display purposes. Although sales were modest, the highly publicized Etak Navigator drew widespread attention to the concept of an electronic map display with icons showing current location and destination.

The Travelpilot essentially is a second generation of the Navigator jointly designed by Etak, Inc. and Bosch GmbH.[25] It was introduced in Germany in 1989 and in the United States two years later. One of its most conspicuous enhancements is the use of CD-ROM storage for digitized maps. The 640-Mb capacity permits the entire map of some countries to be stored on a single CD-ROM.

The primary function of the Travelpilot vehicle navigation and information system is to display to the driver a road map of the area around the vehicle, as illustrated in Fig. 9. The vehicle location and heading are indicated by an arrowhead icon below the center of the screen. The vertical bar at the right edge of the map indicates the display scale that can be zoomed in to 1/8 mile for complete street detail or out to 30 miles to show only major highways. The map is normally oriented so that the direction in which the vehicle is heading points straight up on the display, thus allowing the driver to relate the map display easily to the view outside. However, the map may be viewed in a north up orientation for reference purposes when parked.

Also when parked, a menu accessible through the "MEN" button permits use of soft-labeled buttons in a speller and scroller scheme to enter destinations by street address, intersection, etc. Travelpilot uses a process called "geocoding" to locate an input destination and display it as a flashing star on the map. As illustrated in Fig. 9, a destination geocoded by street address is bracketed by two flashing stars when the map is zoomed in. In this case, the stars mark the block whose address range includes the street number of the destination. A line of information across the top of the map display indicates the crow-flight distance and points the direction from the vehicle's current location to the destination. Up to 100 input destinations may be stored for future use.

A submenu provides several methods for the driver to reset the vehicle's position on the map if the Travelpilot gets off-track. The frequency with which the system requires reinitializing depends upon dead-reckoning anomalies and the completeness and accuracy of the map data for the area being driven. For example, map matching typically fails once in a thousand miles when operating in an environment such as greater Los Angeles or Dallas/Fort Worth. As for location accuracy the rest of the time (i.e., with map-matching operative), Travelpilot is claimed to have infinitesimal error relative to the map. The map-matching performance is compared to that of a servo-amplifier in which map-matching failure corresponds to loosing servolock to the map.

In addition to the CD-ROM player and vector-drawn 4.5-in. monochromatic display, the Travelpilot hardware includes a V50 processor, 1/2 Mb DRAM, 64-

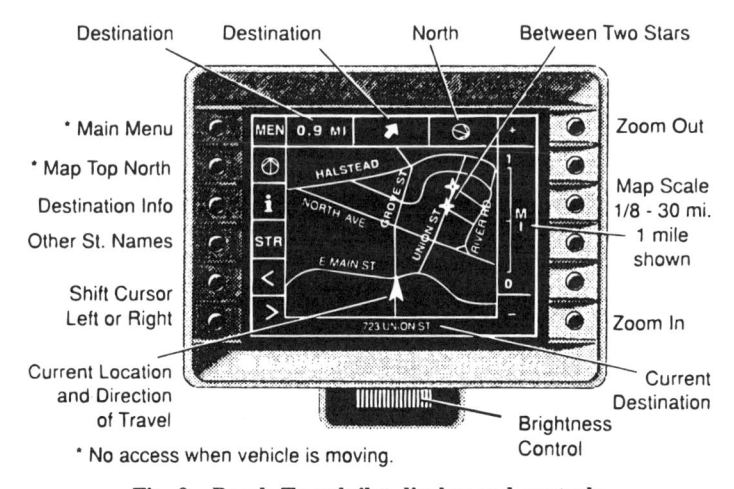

Fig. 9 Bosch Travelpilot display and controls.

Kb EPROM, and 8-Kb nonvolatile RAM. The nonvolatile RAM is used for storing vehicle location while the ignition is off, calibration factors, up to 100 saved destinations, etc.

The Travelpilot may interact with other devices through a RS-232 serial port and an expansion card slot. For example, Travelpilots in 400 Los Angeles fire trucks and ambulances are connected by digital packet radio to the city's emergency control center. The emergency operators can monitor each vehicle's location and status, and can send destinations directly to a vehicle's Travelpilot for emergency dispatch.

B. Toyota Electro-Multivision

Introduced in 1987 as the first sophisticated navigation system available as a factory option on automobiles sold in Japan, the Electro-Multivision has undergone numerous refinements including the recent addition of routing and voice guidance features. Except for a few features, it is representative of the more comprehensive models of navigation systems now available in Japan from almost all of the major automobile and electronics manufacturers.

The design and features of the original version of the Electro-Multivision can be summarized with reference to those of the Travelpilot. Both use dead reckoning and digitized maps stored on CD-ROM for display on a CRT screen with an icon representing present position, and are generally similar in their basic navigation features. However, a raster-scan color CRT rather than a vector-drawn monochromatic CRT is used in the Electro-Multivision. Also unlike Travelpilot, the Electro-Multivision map database includes "yellow pages" information such as the locations of facilities likely to be of interest to motorists.

The Electro-Multivision may also be used as a reference atlas. An initial display shows a color map of all Japan with 16 superimposed rectangles. Touching a particular rectangle causes the map area it encompasses to zoom and fill the entire screen, again with grid lines superimposed to form 16 rectangles. Thus, a few touches of the screen takes the driver from an overview of the entire country down to major roads and landmarks in some quarter of Tokyo. However, in spite of Electro-Multivision's sophisticated map-handling capabilities, map matching was not used in the first version because the digital maps then available for Japan did not contain sufficient detail at the city street level.

In addition to detailed digital maps and map matching, subsequent versions of Electro-Multivision include a GPS receiver and a color LCD rather than CRT display.[26] In 1991, a routing feature was added to calculate a suggested route to specified destinations and highlight the trace on the LCD map display.[27] The most recent version[28] adds synthesized voice route guidance instructions.

As with most other state-of-the-art automobile navigation systems in Japan, the Electro-Multivision navigation features are integrated with a full suite of entertainment features (e.g., AM-FM radio, tape cassette, audio CD player, color TV, etc.). In addition, Electro-Multivision includes a CCD camera for rear-vision on the LCD screen. Figure 10 shows the typical layout of Electro-Multivision components installed in an automobile.

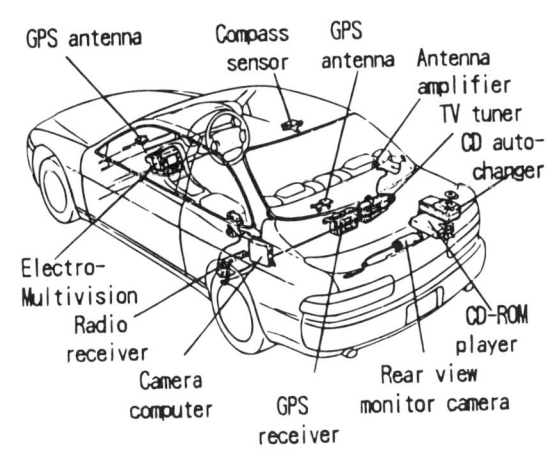

Fig. 10 Distribution of Toyota Electro-Multivision system elements in automobile.[27]

C. TravTek Driver Information System

Unlike Travelpilot and Electro-Multivision which are automobile systems already available in certain markets, TravTek is a functional prototype of a navigation-based in-vehicle traveler information system developed specifically for the TravTek IVHS operational field trial conducted for a one-year period ending March 31, 1993 in Orlando, Florida. The field trial was a joint public sector–private sector project with the primary objective of obtaining field data on the acceptance and use by drivers of navigation and other information provided by comprehensive in-vehicle systems linked with traffic operations and other data centers.

General Motors equipped a total of 100 cars with the system shown schematically in Fig. 11 to provide navigation, route selection and guidance, real-time traffic information, local "yellow pages" and tourist information, and cellular phone service.[29] Most of these vehicles were made available to Orlando visitors through Avis Rent A Car for short-term trials and the rest were assigned to local drivers for extended periods. The American Automobile Association selected the test subjects and operated a TravTek Information and Services Center that could be accessed via cellular telephone.

The City of Orlando, in conjunction with the Federal Highway Administration and the Florida Department of Transportation, operated a supporting Traffic Management Center that consolidated traffic data from various sources including "probe" data consisting of road segment travel times received from the equipped vehicles themselves. Data communications between the equipped vehicles and the Traffic Management Center were via specialized mobile radio (SMR).

TravTek navigation is based on a combination of dead reckoning and map matching with a GPS receiver playing a "watchdog" role. Although functionally realistic, the TravTek in-vehicle system design made extensive use of readily available modules that would not typically appear in a production system. For example, rather than consolidated databases stored on CD-ROM, TravTek used

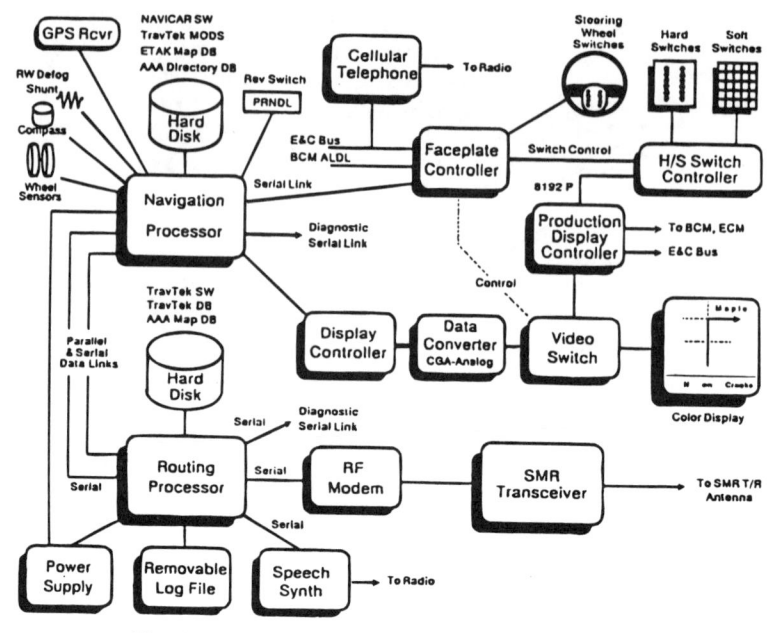

Fig. 11 Architecture of TravTek vehicle system.[29]

separate map databases stored on separate hard disk drives for navigation processing and route guidance processing.[30]

Similar to Travelpilot and Electro-Multivision, the TravTek navigation function superimposes vehicle location and destination on the map display screen and, like Electro-Multivision, highlights suggested routes on the map display and issues route guidance instructions via synthesized voice. In addition, turn-by-turn route guidance in the form of simplified graphics may be displayed, as indicated by Fig. 11.

Based on analysis of questionnaires completed by some 3000 test subjects, TravTek's effectiveness and benefits are rated very highly by drivers (5.1 on a scale of 1.0–6.0).

D. NavTrax™ Fleet Management System

NavTrax is a dispatch-type automatic vehicle location reporting system developed by Pulsearch Navigation Systems of Calgary, Alberta, Canada.[31] The positioning module is a robust GPS-based system integrated with dead-reckoning devices by a decentralized-federated filter making the module fault tolerant and suited for off-road as well as on-road use. The dispatch center subsystem provides map displays, means for selecting and polling (i.e., sending periodic requests for vehicles to report their positions), logging of fleet movements, etc. To date, the vehicle and the dispatch center have been linked by two-way UHF/VHF communications, although plans are underway to implement cellular technology as well.

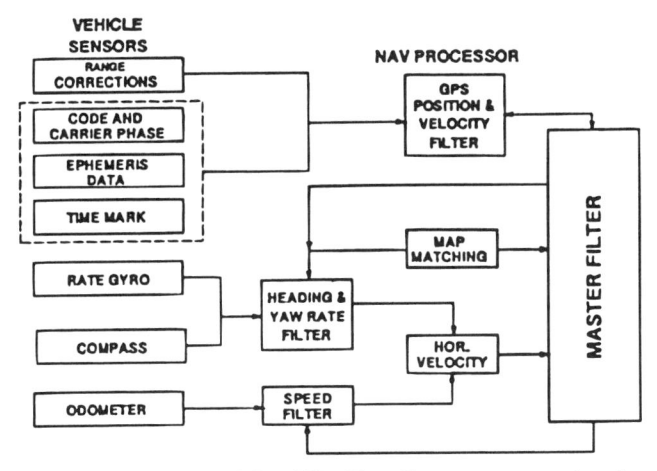

Fig. 12 Navigation module of NavTrax fleet management system.[32]

The major elements of the NavTrax positioning module are shown in Fig. 12.[32] Real-time position coordinates are continuously computed from sensor inputs including GPS pseudorange and carrier phase rate for position and velocity; rate gyro for azimuth change; compass for azimuth; and odometer for speed. With the decentralized-federated filtering approach, the reference filter is GPS based, while each dead reckoning sensor has its own local filter for determination of biases and fault detection. The reference filter and all local filters feed into the master filter where fusion of all position information takes place. Fusion feedback from the master filter to each local filter provides for frequent automatic calibration updates for each sensor.

The NavTrax filter scheme makes maximum use of short bursts of GPS information as it becomes available at intersections and openings in tree-covered areas. However, positioning can be continued for extended periods on the basis of dead-reckoning information alone. During periods of long lapses of GPS information, a limited form of map matching is carried out by taking the dead-reckoning coordinates and entering the road network database and exiting with location in the database frame of reference. The location is based on a probability function using the most probable road segment considering proximity, azimuth, and past connectivity. The associated map coordinates and azimuth of the road link are used in the filter, along with their associated variance information, to determine the best position of the vehicle.

NavTrax has been carried through several stages of Beta testing and is now being marketed. The trial applications included Calgary police vehicle dispatch, Amoco oil field operations, and vehicle location monitoring by the Royal Canadian Mounted Police.

References

[1]French, R. L., "Land Vehicle Navigation—A Worldwide Perspective," *Journal of Navigation*, Vol. 44, No. 1, 1991, pp. 25–29.

[2]"Automatic Vehicle Monitoring Program Digest," Urban Mass Transit Administration Report DOT-TSC-UMTA-81-11, April 1981.

[3]Rothblatt, M. A., *Radiodetermination Satellite Services and Standards,* Artech House, Boston, 1987.

[4]Sellers, D. L., and Bernard, T. J., "An Update on the OmniTRACS Two-Way Satellite Mobile Communications System and its Application to the Schneider National Truckload Fleet," *Proceedings of the 1992 International Congress on Transportation Electronics,* Society of Automotive Engineers, Dearborn, MI, 1992 (SAE P-260), pp. 351–356.

[5]French, R. L., "The Evolving Roles of Vehicular Navigation," *Navigation,* Vol. 34, No. 3, 1987, pp. 212–228.

[6]King, G. E., "Economic Assessment of Potential Solutions for Improving Motorist Route Following," Federal Highway Administration Rept. FHWA/RD-86/029, 1986.

[7]"Smart Highways: An Assessment of Their Potential to Improve Travel," U.S. General Accounting Office Rept. GAO/PEMD-91-18, May 1991.

[8]"Strategic Plan for Intelligent Vehicle-Highway Systems in the United States," IVHS America Rept. IVHS-AMER-92-3, May 1992.

[9]Needham, J., "Science and Civilization in China," *Mechanical Engineering,* Vol. 4, Part II, Cambridge University Press, New York, 1965.

[10]French, R. L., "U.S. Automobile Navigation: Early Mechanical Systems," *Navigation News,* Vol. 4, No. 3, 1989, pp. 6–7.

[11]Perry, H. W., "Some Remarkable Mechanical Road Guides," *Scientific American,* Vol. 104, No. 2, 1911, pp. 33, 47–48.

[12]Ellis, W. D., "Chadwick," *True's Automobile Yearbook,* No. 2, 1953, pp. 88–89, 132–133.

[13]Faustman, J. D., "Automatic Map Tracer for Land Navigation," *Electronics,* Vol. 17, No. 11, 1944, pp. 94–99.

[14]Rosen, D. A., Mammano, F. J., and Favout, R., "An Electronic Route Guidance System for Highway Vehicles," *IEEE Transactions on Vehicular Technology,* Vol. 19, 1970, pp. 143–152.

[15]French, R. L., and Lang, G. M., "Automatic Route Control System," *IEEE Transactions on Vehicular Technology,* Vol. 22, No. 2, 1973, pp. 35–41.

[16]Lezniak, T. W., Lewis, R. W., and McMillen, R. A., "A Dead-Reckoning/Map-Correlation System for Automatic Vehicle Tracking," *IEEE Transactions on Vehicular Technology,* Vol. 26, No. 1, 1977, pp. 47–60.

[17]Peters, T. J., "Automobile Navigation Using a Magnetic Flux-Gate Compass," *IEEE Transactions on Vehicular Technology,"* Vol. 35, No. 2, 1986, pp. 41–47.

[18]Ikeda, H., Kobayashi, Y., and Kawamura, S., "Sumitomo Electric's Navigation Systems for Private Automobiles," *Proceedings, VNIS '91—Vehicular Navigation & Information Systems Conference,* Vol. 1, Society of Automotive Engineers, Dearborn, MI, Oct. 20–23, 1991 (SAE Paper 912789), pp. 451–462.

[19]Silver, J., "The GBF/DIME System: Development, Design, and Use," Paper presented at the 1977 Joint Annual Meeting of American Society of Photogrammetry and American Congress on Surveying and Mapping, 1977.

[20]French, R. L., "Map Matching Origins, Approaches, and Applications," *Proceedings, Second International Symposium on Land Vehicle Navigation,* (Münster, Germany), July 4–7, 1989, 93–116.

[21]Karimi, H. A., private communication, Athabasca University, Edmonton, AB, Canada, March 16, 1993.

[22]French, R. L., "Mobile Communication," *Concise Encyclopedia of Traffic & Transportation Systems,* edited by M. Papageorgiou, Pergamon, New York, 1991, pp. 263-268.

[23]Shute, S., "RDS, The EBU Radio Data System," *International Broadcast Engineer,* May-July 1987.

[24]Honey, S. K., and Zavoli, W. B., "A Novel Approach to Automobile Navigation and Map Display," *Proceedings, NAV 85-Royal Institute of Navigation Conference on Land Navigation and Location for Mobile Applications,* Paper 27, York, England, Sept. 9-11, 1985.

[25]Buxton, J. L., Honey, S. K., Suchowerskyj, W. E., and Tempelhof, A., "The Travelpilot: A Second-Generation Automotive Navigation System," *IEEE Transactions on Vehicular Technology,* Vol. 40, No. 1, 1991, pp. 41-44.

[26]Ishikawa, K., Ogawa, M., Azuma, S., and Ito, T., "Map Navigation Software of the Electro-Multivision of the '91 Toyota Soarer," *Proceedings, VNIS '91—Vehicular Navigation & Information Systems Conference,* Vol. 1, Society of Automotive Engineers, Dearborn, MI, Oct. 20-23, 1991 (SAE Paper 912790), pp. 463-473.

[27]Umeda, Y., Morita, H., Azuma, S., and Itoh, T., "Development of the New Toyota Electro-Multivision," SAE Paper 920601, 1992.

[28]Ito, T., Azuma, S., and Sumiya, K., "Development of the New Navigation System—Voice Route Guidance," SAE Paper 930554, 1993.

[29]Krage, M. K., "The TravTek Driver Information System," *Proceedings, VNIS '91—Vehicular Navigation & Information Systems Conference,* Vol. 2, Society of Automotive Engineers, Dearborn, MI, Oct. 20-23, 1991 (SAE Paper 912820), pp. 739-748.

[30]Rillings, J. H., and Krage, M. K., "TravTek: An Operational Advanced Driver Information System," *Proceedings of the 1992 International Congress on Transportation Electronics,* Society of Automotive Engineers, Dearborn, MI, 1992 (SAE P-260), pp. 461-472.

[31]McLellan, J. F., Krakiwsky, E. J., and Huff, D. R., "Fleet Management Trials in Western Canada," *Proceedings, VNIS '91—Vehicular Navigation & Information Systems Conference,* Vol. 2, Society of Automotive Engineers, Dearborn, MI, Oct. 20-23, 1991 (SAE Paper 912826), pp. 797-806.

[32]McLellan, J. F., Krakiwsky, E. J., Schleppe, J. B., and Knapp, P. L., "The NavTrax™ Fleet Management System," *Proceedings, PLANS '92, IEEE Position, Location, and Navigation Symposium,* (Monterey, CA), IEEE, New York, March 25-27, 1992, pp. 509-515.

[33]French, R. L., "Automobile Navigation: Where is it Going?" *Navigation—Land, Sea, Air, & Space,* edited by M. Kayton, IEEE Press, New York, 1990, pp. 101-107.

[34]"IVHS Strategic Plan: Report to Congress," U.S. Department of Transportation, December 18, 1992.

Chapter 11

Marine Applications

Jim Sennott* and In-Soo Ahn†
Bradley University, Peoria, Illinois 61625
and
Dave Pietraszewski‡
*United States Coast Guard Research and Development Center,
Groton, Connecticut 06340*

I. Marine Navigation Phases and Requirements

NAVIGATION can be defined as the process of planning, recording, and controlling the movement of a craft or vehicle from one place to another. It is a process that looks ahead in an effort to determine how a safe arrival can be secured. Physical sensors can only measure what "just" happened, at best. The navigation process, including sensor systems and associated human and electronic controls and monitor algorithms, must take this "historic" information and convert it into rudder and thrust commands that affect future events. How well this is achieved can be judged by various cost and safety measures, such as average vessel passage time, safe passage probability, and ship footprint control deviation.

In an effort to differentiate and quantify marine navigation safety requirements in the United States, the departments of Transportation and Defense have defined marine navigation in terms of "phases." The four phases of marine navigation defined in the Federal Radionavigation Plan[1] are ocean, coastal, harbor/harbor approach (HHA), and inland waterway. Each phase of marine navigation is distinguished by a clearly different set of performance requirements. These requirements are based on safety and environmental concerns and support the desire to minimize marine collisions, rammings, and groundings.

The current technological characteristics and policy constraints on the Global Positioning System standard positioning service (GPS SPS) will allow GPS to

---

This paper is a work of the U.S. Government and is not subject to copyright protection in the United States.
*Professor of Electrical and Computer Engineering.
†Associate Professor of Electrical and Computer Engineering.
‡Senior Navigation Scientist.

satisfy many of the ocean and coastal phase performance requirements. The same characteristics and constraints [particularly selective availability (SA)] make it unacceptable for HHA (and inland waterway) navigation. The major distinction is accuracy. The accuracy required for the ocean phase is 1800–3700 m (2 drms) and 460 meters (2 drms) for the coastal phase. The accuracy required for the HHA phase is 8–20 m (2 drms).

In the HHA phase, a vessel pilot needs accurate, frequent, and timely verification of the vessel's position. Deviation from the desired vessel track drives the pilot's decision-making process. This is quite unlike the ocean phase where position updates on the order of minutes are quite satisfactory. The need for frequent position verification places additional burdens on the radionavigation service provider. Momentary unexpected signal outages can significantly jeopardize the safety of vessels executing sensitive maneuvers. Akin to aircraft precision approach and landing guidance, an availability specification of 0.997 is currently stated for HHA, and integrity requirements will ultimately be established.

These specifications must be used with some caution. The stated accuracy and update specifications lack specificity with respect to dynamic conditions under which a stated level of sensor performance shall be provided. Moreover, no explicit allowance is made for pilot error contributed by personnel skill, pathway geometry, ship dynamics, and disturbances. The familiar concept of flight technical error employed in aircraft guidance has yet to have been systematically exploited in marine navigation design. Clearly, radionavigation requirements ultimately depend on vessel size, maneuvering activity, and the geographic constraints of the operating area. Ship pilotage may be unsafe, even with perfect navigation sensors, if maneuvers are complicated by tight channel tolerances, unexpected traffic, poor helm dynamics, and strong wind and current disturbances.

In developing the role of GPS and differential GPS (DGPS) in the marine environment, particular attention will be given to vessel footprint steering performance, and the interplay between sensor and ship models. Other related functions such as hazard warning, risk assessment, and on-line dynamics modeling, are also discussed. Before turning to the development of these GPS applications, some background on the experimentation that led to deployment of a standardized marine DGPS service in the United States is in order.

II. Marine DGPS Background

In the early 1980s, the Department of Transportation (DOT) began studying the potential civil use of GPS. DOT quickly realized that many potential applications would require higher levels of accuracy and integrity than SPS would be able to provide.[2]

The feasibility of providing DGPS for marine navigation with an absolute accuracy of 10 m (2 drms) was demonstrated in 1987. In 1989 the Montauk Point, New York radio beacon was temporarily converted to provide a DGPS test broadcast. Given the success of these 1989 tests, and the desire to evaluate DGPS in operational environments more thoroughly, a prototype DGPS service was established on August 15, 1990. The Montauk Point marine radiobeacon is now part of the U.S. Coast Guard Northeast testbed prototype service, providing accurate navigation from Cape Hatteras, North Carolina to Canada. The U.S.

Coast Guard has announced plans to cover most coastal areas of the United States by 1996.[3]

An essential part of the coastal service is a radio link for transmission of differential corrections. An approach that has been extensively investigated, and adopted, is MSK digital transmission of data over the existing network of low-frequency marine radiobeacons. The chosen modulation format minimizes interference with the radio direction finding function presently provided by these beacons. Furthermore, forward error detection and correction features are employed for reducing the impact of Gaussian and impulsive noise. This results in a highly reliable link, even beyond the normal rated range of radiobeacon service.[4,5]

Paralleling DOT DGPS sensor research, the radionavigation accuracy requirements for HHA and inland confined waterways were under study in the 80s. In 1987, an assessment of DGPS, differential Loran-C, and other candidates for the Great Lakes and St. Lawrence Seaway was carried out under U.S. Maritime Administration sponsorship.[6] Bradley University explored the interplay between sensor dynamics/noise characteristics and ship control performance in an automated steering environment. The steering characteristics of DGPS and other radionavigation sensor combinations were integrated with an augmented state navigation filter, and an optimal steering controller was developed.

In 1988 the Coast Guard Research and Development Center conducted human factors simulations with experienced pilots operating with simulated radio navigation sensors to validate the Federal Radionavigation Plan requirement for 8–20 m (2 drms) accuracy for harbor and harbor approach.[7] The study concluded that the 8–20 m requirement was appropriate for large vessels in restricted waterways. The study also concluded that radio aids to navigation with this level of accuracy would enhance traditional visual and radar navigation if implemented and used properly and would also enhance navigation under restricted visual conditions down to 0.25 nautical miles (n.mi.). However, the study noted that "additional understanding needs to be gained, particularly on the proper design and utilization of radio aid (RA) devices for negotiating turns under 0.0 n.mi. visibility conditions, before concluding that an RA system with 8–20 m (2 drms) accuracy can be used safely to support an all weather navigation system." Clearly, the usage of DGPS for steering in confined waterways will require careful integration with risk assessment and hazard warning functions that collect information from other sources such as radars, geographic data bases, and from shore side vessel traffic systems (VTS).

III. Global Positioning Systems-Assisted Steering, Risk Assessment, and Hazard Warning Systems

The overall model for GPS steering and hazard warning is shown in Fig. 1. The system is partitioned as follows:
1) Ship
   Vessel steering hydrodynamics
   Wind forces model
   Water current model

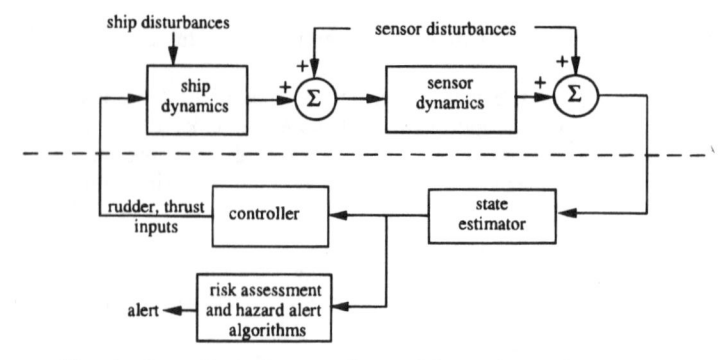

Fig. 1 Overall marine steering and hazard warning system.

2) Sensors
   Sensor lag dynamics
   Sensor noise disturbances
   User clock model
3) State estimation
   Nominal ship state variable model
   On-line vessel dynamics evaluator
   Navigation filter
4) Control
   Thrust control
   Rudder controller (human or automatic)
   Waypoint decision logic
5) Hazard warning and risk assessment
   Waypoint planning and risk evaluation
   Deviation and hazard warning algorithm

Figure 2 defines the coordinate systems and state variables that characterize the truth environment and form the basis for the navigation filter and rudder control design. Sway velocity, longitudinal velocity, and yaw rate are ship-referenced; heading angle error, crosstrack, and alongtrack position are waypoint referenced; current and wind components are east-north referenced.

1) Ship-fixed states
   x_{sway} = sway velocity
   x_{yaw} = yaw rate
   x_{vel} = longitudinal velocity
2) Waypoint referenced states
   x_{head} = heading angle error relative to waypoint segment
   x_{cross} = crosstrack error on a waypoint segment
   x_{along} = alongtrack position on a waypoint segment
3) East-north referenced states
   $x_{c,e}$ = crosstrack current bias
   $x_{c,n}$ = alongtrack current bias
   $x_{w,e}$ = crosstrack wind velocity component
   $x_{w,n}$ = alongtrack wind velocity component

MARINE APPLICATIONS 307

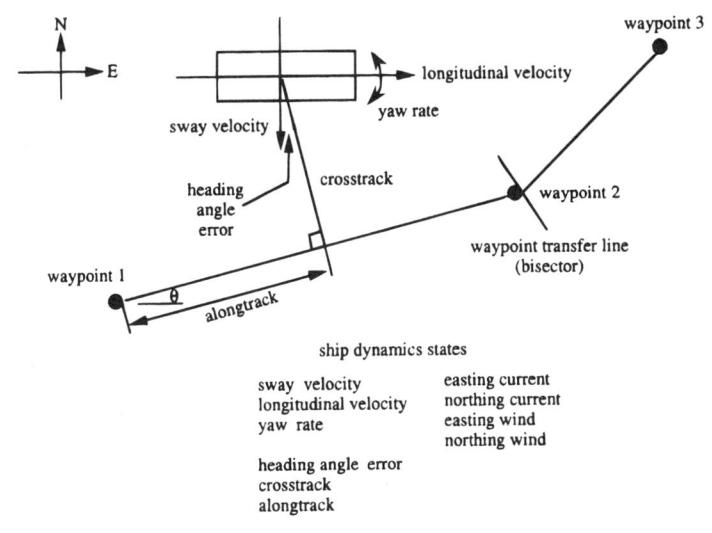

Fig. 2 Coordinate system and state definitions.

ship dynamics states

sway velocity
longitudinal velocity
yaw rate

heading angle error
crosstrack
alongtrack

easting current
northing current
easting wind
northing wind

Ship dynamics are nonlinear and unstable. To implement the rudder control and navigation filter, a procedure for on-line linearization of hydrodynamics yielding stability derivatives as well as on-line linearization of the coupled state equations will be carried out.

The sensor portion of the system typically includes differential DGPS, Loran-C, ship-heading reference, waterspeed indicator, and wind anemometer. Careful attention must be given to the widely different sensor dynamics. When designing the navigation filter, state augmentation techniques can be applied to compensate for lags.

The rudder and thrust controller may consist of the human pilot with appropriate situation display, or automatic steering may be employed. As a human pilot navigates through a series of waypoints, a combination of factors are considered. If the potential for traffic conflict is low, the pilot generally applies rudder control so as to (subjectively) minimize the probability of violating the channel boundaries during the waypoint system passage. In the simulations presented later in this section, an optimal control law acts as "stand in" for the human pilot, with rudder control and navigation filter tuned to achieve the best performance from each sensor combination.

Although, ideally, system design would be based upon a safe passage probability (SPP) performance index accounting for waterway geometry, ship maneuverability, ship footprint size, and navigation sensor quality, the rigorous design of steering control under this criteria is a difficult unsolved problem. Therefore, more tractable design and evaluation criteria, described below, are employed.

The visual aid literature describes ship crosstrack error[8-10] and relative risk factor[7,11-14] performance indexes. Relative risk factor (RRF) assesses the minimum clearance between the vessel footprint and the channel boundary, at a specified alongtrack station. Accounting for the vessel footprint orientation in the channel, the RRF measure is closely related to the probability of grounding and collision at a specific channel location.

A more comprehensive measure is the channel clearance width distribution (CCWD).[6,15,16] The channel clearance width (CCW) process from which this distribution is derived is the collection of ship clearance envelopes swept out during many passages through the waypoint structure. The CCWD is simply the first-order cumulative distribution of ship CCWs, over all sample stations along the channel. This distribution will be developed below for a representative ship steering system driven by DGPS.

The last component shown in Fig. 1 supports risk assessment and hazard alert functions. These functions may be performed aboard ship and/or at a shoreside vessel traffic management site. For planning prior to vessel passage, a safe passage probability (SPP) performance measure could be computed for a given set of waypoints and vessel subsystems. The Global Positioning System can play a central role in supporting these calculations. Also, real-time supervisory alerts for fixed hazards and poor steering may be desired as a backup to human or autopilot steering and thrust control. One such function is based upon a running computation of latest safe alarm time (LSAT) for known hazards along the waterway.

IV. Vessel and Sensor Modeling

A. Vessel Dynamics Model

The three body-fixed ship states, sway velocity, yaw rate, and longitudinal velocity, are described by nonlinear coupled differential equations. Just as in aerodynamic flight, hydrodynamic stability derivatives are employed to express body-fixed dynamics in a linear fashion. A set of stability derivatives is valid in the region about which the system is linearized. These coefficients will vary depending upon longitudinal and sway velocities and yaw rate, and on ship draft and waterway bottom clearance. The equations governing body-fixed states are given below. The two control variables are rudder angle and propeller thrust.

$$\dot{x}_{sway} = D_{s,s}x_{sway} + D_{s,y}x_{yaw} + D_{s,bw}\{[x_{w,n}\cos\theta + x_{w,e}\sin\theta]\sin x_{head} + x_{w,n}\sin\theta + x_{w,e}\cos\theta]\cos x_{head}\} + D_{s,\delta}\delta_{rudder}$$

$$\dot{x}_{yaw} = D_{y,s}x_{sway} + D_{y,y}x_{yaw} + D_{y,bw}\{[x_{w,n}\cos\theta + x_{w,e}\sin\theta]\sin x_{head} + x_{w,n}\sin\theta + x_{w,e}\cos\theta]\cos x_{head}\} + D_{y,\delta}\delta_{rudder}$$

$$\dot{x}_{vel} = D_{v,v}x_{vel} + D_{v,lw}\{[x_{w,n}\cos\theta + x_{w,e}\sin\theta]\cos x_{head} + [x_{w,n}\sin\theta + x_{w,e}\cos\theta]\sin x_{head}\} + D_{v,thrust}\delta_{thrust}$$

where D coefficients, defined below, are the hydrodynamic and wind force coefficients of the ship, and θ is the waypoint leg heading. Linearized dynamics provide information about the stability of the above nonlinear dynamics. In linearized dynamics, in states x_{sway} and x_{yaw} there is one positive eigenvalue. Hence, the rudder controller design must provide ship stability.

Table 1 Ship parameters

Length of ship	305 m
Width of ship	38 m
Hull type	Tanker
Speed	5.14 m/s (10 k)
Dynamics coefficients	
$D_{s,s} = -.0145$	
$D_{s,y} = -2.477$	
$D_{s,\delta} = -.01515$	
$D_{y,s} = -.00029$	
$D_{y,y} = -.0897$	
$D_{y,\delta} = .000392$	

Stability derivatives are obtained by taking first-order derivatives of nonlinear hydrodynamic equations at the nominal operating point. In simulation study, these derivatives are later matched to follow actual ship dynamics. For an example, see Table 1.

$D_{s,s}$ = sway dynamics coefficient
$D_{s,y}$ = yaw-to-sway coupling coefficient
$D_{s,\delta}$ = rudder-to-sway coupling coefficient
$D_{y,s}$ = sway-to-yaw coupling coefficient
$D_{y,y}$ = yaw dynamics coefficient
$D_{y,\delta}$ = rudder-to-yaw coupling coefficient
$D_{v,v}$ = longitudinal velocity dynamics coefficient
$D_{v,thrust}$ = thrust-to-longitudinal velocity coupling coefficient
$D_{s,bw}$ = beam wind-to-sway coupling coefficient
$D_{y,bw}$ = beam wind-to-yaw coupling coefficient
$D_{v,lw}$ = longitudinal wind-to-longitudinal velocity coupling coefficient

In addition to the hydrodynamic effects, wind forces play a very important part in the vessel motion. In the preceding list, coupling of winds into dynamics is treated as follows. Neglecting nonhomogeneous effects, the true wind vector is first resolved into the body-fixed coordinate frame. Then the relative beam and bow wind components are determined. It is assumed that these components induce torque and force terms linear in wind component magnitude. For example, the beam wind component generates beam force and torque values proportional to wind component magnitude.

The next three state equations describe ship position relative to the waypoint system. The coupling of longitudinal and sway velocity into alongtrack and crosstrack equations is nonlinear because of the rotation between waypoint and body-fixed frames.

$$\dot{x}_{\text{head}} = x_{\text{yaw}}$$

$$\dot{x}_{\text{cross}} = x_{\text{sway}} \cos x_{\text{head}} + x_{c,e} \cos \theta - x_{c,n} \sin \theta + x_{\text{vel}} \sin x_{\text{head}}$$

$$\dot{x}_{\text{along}} = -x_{\text{sway}} \sin x_{\text{head}} + x_{c,e} \sin \theta + x_{c,n} \cos \theta + x_{\text{vel}} \cos x_{\text{head}}$$

The current and wind states, in the easting-northing frame, are expressed with four linear state equations

$$\dot{x}_{c,e} = -\alpha_c x_{c,e} + w_{c,e}$$
$$\dot{x}_{c,n} = -\alpha_c x_{c,n} + w_{c,n}$$
$$\dot{x}_{w,e} = -\alpha_w x_{w,e} + w_{w,e}$$
$$\dot{x}_{w,n} = -\alpha_w x_{w,n} + w_{w,n}$$

where the forcing terms are chosen to model the random wind and water current terms.

A linear state space representation for the ship must be developed from these nonlinear state equations prior to obtaining the rudder controller and navigation filter. The state matrix coefficients of the continuous-time incremental linear model are derived by taking partial derivatives with respect to states and evaluating at an assumed operating point. Then the incremental continuous-time linear model may be written in the usual form as follows

$$\Delta \dot{x}^{\text{ship}} = A_{\text{ship}} \Delta x^{\text{ship}} + Bu$$

where

$$\Delta x^{\text{ship}} = (\Delta x_{\text{sway}} \Delta x_{\text{yaw}} \Delta x_{\text{vel}} \Delta x_{\text{head}}, \Delta x_{\text{cross}} \Delta x_{\text{along}} \Delta x_{c,e} \Delta x_{c,n} \Delta x_{w,e} \Delta x_{w,n})^t$$

and

$$u = (\delta_{\text{rudder}} \; \delta_{\text{thrust}})^t$$

with superscript t denoting transpose.

B. Standardized Sensor Model

The fundamental quantities observed in GPS or Loran-C systems are pseudorange and Doppler. These are corrupted by noise and dynamical errors. Noise errors are those observed at sensor outputs in the absence of any vehicle motion. It is often assumed that these are uncorrelated in time. Upon closer examination such jitter errors are found to be colored, reflecting both receiver front-end noise and signal tracking loop properties. Dynamical errors in pseudorange and Doppler result from the filtering actions of signal tracking loops. In critical vessel control problems these errors can be a very significant part of the overall error budget. Indeed, mariners using Loran-C are accustomed to lags between actual and reported position, compensating to a degree in their vessel pilotage. To compare different sensor combinations clearly for the precise navigation problems of interest, a linear equivalent sensor dynamics model is used to portray the tracking loop for each sensor channel.

Consider a generic sensor that tracks pseudorange. The navigator observes pseudorange via the navigation sensor hardware, in GPS from the delay-lock loop, and in Loran-C from the third cycle zero-crossing tracker. Both noise and dynamic lag terms can distort true pseudorange. Therefore, it is desirable to include these effects in the system design and evaluation. Let R_i be the pseudorange to the ith sensor. Then, the following equation results

$$R_i(t) = \sqrt{[e_{\text{ship}}(t) - e_{\text{sensor},i(t)}]^2 + [n_{\text{ship}}(t) - n_{\text{sensor},i}(t)]^2} + x_{\text{clock}}(t)$$

where $e_{\text{ship}}(t)$ and $n_{\text{ship}}(t)$ denote the ship's easting and northing positions at time t, respectively. Similarly, $e_{\text{sensor},i}(t)$ and $n_{\text{sensor},i}(t)$ denote the ith sensor's easting and northing positions, respectively. The $e_{\text{ship}}(t)$ and $n_{\text{ship}}(t)$ variable are nonlinear, through geometry, in ship along-track and cross-track states. The state $x_{\text{clock}}(t)$ is the offset between user and system time. Let R be the collection of true pseudorange, and Doppler values for the following sensor set

$$R = (R_1 \ R_2 \ \cdots \ R_{n-1} \ R_n)^t$$

A linear state equation representation portrays demodulator loops appropriate for both GPS and Loran-C tracking channels. Sensor states are the quantities actually seen as navigation filter inputs. The dimension of A_{sensor} is chosen to match the tracking loop order and number of channels employed. The noise power spectral density of w^{sensor} models equivalent atmospheric and/or receiver front-end sources. The forcing vector R is as defined above.

$$\dot{x}^{\text{sensor}} = A_{\text{sensor}} x^{\text{sensor}} + B_{\text{sensor}} R + w^{\text{sensor}}$$

C. Combined Ship and Sensor Model

To form a complete model the sensor state equations are combined with the previously derived ship state equations. To do so, the state nonlinearity in R is linearized about the nominal ship state solution. Periodically the nominal ship solution is updated and the incremental ship state is reset to zero

$$x_{\text{total,ship}} = x_{\text{nominal,ship}} + \Delta x^{\text{ship}}$$

where Δx^{ship} is the incremental solution defined above, and the nominal dynamics are treated as an unforced system with nonzero initial conditions.

Then, the sensor forcing terms in vector R are expanded about the nominal solution, the incremental term providing the desired connection between the ship dynamics and sensor dynamics. In continuous time, the combined sensor and incremental ship state equations are given by the following

$$\begin{bmatrix} \Delta \dot{x}^{\text{ship}} \\ \dot{x}^{\text{sensor}} \end{bmatrix} = \begin{bmatrix} A_{\text{ship}} & 0 \\ A_{\text{ship, sensor}} & A_{\text{sensor}} \end{bmatrix} \begin{bmatrix} \Delta x^{\text{ship}} \\ x^{\text{sensor}} \end{bmatrix} + \begin{bmatrix} B_{\text{ship}} & 0 \\ 0 & B_{\text{sensor}} \end{bmatrix} \begin{bmatrix} u^{\text{ship}} \\ u^{\text{sensor}} \end{bmatrix} + \begin{bmatrix} w^{\text{ship}} \\ w^{\text{sensor}} \end{bmatrix}$$

where

$$x^{\text{sensor}} = \text{navigation filter input}$$

with

$$u^{\text{ship}} = [\delta_{\text{rudder}} \ \delta_{\text{thrust}}]^t$$

and

$$u^{sensor} = [R_1\ R_2\ R_3\ \cdots\ R_n]^t_{nominal}$$

From the continuous-time system the usual discrete-time linear system is obtained (Ref. 17)

$$\Delta x(k+1) = \Phi \Delta x(k) + \Gamma u(k)$$

where the state transition matrix Φ is given by

$$\Phi = e^{A\Delta t}$$

and

$$\Gamma = \int_0^{\Delta t} e^{A\xi} B\,d\xi$$

In discrete time the complete model is given by the following

$$\begin{bmatrix} \Delta x^{ship}(k+1) \\ \Delta x^{sensor}(k+1) \end{bmatrix} = \begin{bmatrix} \Phi_{ship} & 0 \\ \Phi_{ship,sensor} & \Phi_{sensor} \end{bmatrix} \begin{bmatrix} \Delta x^{ship}(k) \\ \Delta x^{sensor}(k) \end{bmatrix} + \begin{bmatrix} \Gamma_{ship} & 0 \\ 0 & \Gamma_{sensor} \end{bmatrix} \begin{bmatrix} u^{ship} \\ u^{sensor} \end{bmatrix} + \text{noise}$$

It is important to note the ship state transition matrix is not static. While the ship is underway its hydrodynamics vary, because of changes in channel bottom and side clearance and large variations in vessel side-slip, yaw rate, and longitudinal velocity. Therefore, it is desirable to monitor and update the hydrodynamic coefficients by on-line identification techniques.[18-20] GPS-derived position, velocity, and attitude data are an excellent basis for this on-line hydrodynamic modeling. In any event, the ship state transition matrix is impacted by rotations between body-fixed and waypoint-fixed systems; therefore, updating of the above is performed whenever a significant change in ship heading is detected. In so doing, the nominal ship solution is set to the present estimate of total ship state, and the ship incremental state estimate is reset to zero. These updates impact both the navigation filter and rudder controller portions of the system.

V. Waypoint Steering Functions

Vessel steering in confined waterways is at present conducted manually, with the great majority of steering performance studies focused upon human factors simulations and shipboard observations of pilot performance. However, DGPS offers the possibility of a transition to a more automated system. In anticipation of such systems, and with an additional goal of understanding how well human pilots might perform with DGPS under ideal conditions, a simulator was developed encompassing the above ship and sensor models.[6,15,16] In this work, an automatic steering system acts as stand-in for the human pilot, a simplification that sets aside the admittedly complex issue of pilot display configuration and man-machine interface. The discrete-time ship and sensor model derived above provides a foundation for the autopilot and navigation filter design. The most promising configurations can later be evaluated by full-fledged human factors simulation.

A. Filter and Controller Design

The ship navigation filter developed for this study integrates a variety of radionavigation sensors, each properly characterized in terms of equivalent noise sources and signal tracking loop dynamics. In the filter, a reduced-order ship model with constant thrust is assumed. Five ship states are included: two in the body-fixed frame, sway velocity and yaw rate, and three in the local waypoint coordinate system frame, alongtrack position, crosstrack error and heading. Two water current velocity states, easting and northing, are also estimated. Random wind disturbances are introduced on sway and yaw rate states. In the navigator, these seven states portray the ship motion more accurately than a simpler easting–northing or Earth-centered, Earth-fixed (ECEF) based filter. This filter model entails nonlinearities in the system dynamics matrix, requiring periodic linearization about the estimated ship trajectory.

Within the navigation filter, seven additional states are associated with the sensor model. As shown previously, these may be modeled together with five ship incremental states as one larger linear system. Five of the seven sensor states portray sensor smoothing lags, as encountered in the pseudorange and heading sensor elements of the navigation system. The two remaining sensor states model the navigation receiver clock bias and bias rate internal to the DGPS and differential Loran-C receiver.

Navigation filter estimates of all ship states are passed on to the rudder controller, whose job is to position and orient the ship footprint relative to the specified channel boundaries. When steering large ships in confined waterways, attention must be given to both crosstrack steering and yaw errors. Specifically, control seeks to minimize the CCW, defined as the minimum channel width that will clear the ship footprint. To be viable, such a control must tolerate random disturbances of current, wind, and model mismatch, making best use of the complete navigation filter state vector. In turning maneuvers, optimality under the CCW criteria is an unsolved problem. However, in straight track-keeping segments, CCW may be formulated as a stochastic regulator problem. To handle both straight and turning segments, the implemented steering algorithm consists of 1) an "inner" steering regulator designed to maintain tight control of ship cross-track and yaw error on each leg of the waypoint structure, and 2) an "outer" decision algorithm for transitioning between waypoint segments.

The inner control is a stochastic regulator whose goal is to drive waypoint-referenced error states to zero. The following quadratic performance index is utilized

$$J_i = \frac{1}{2} \sum_{k=i}^{N-1} \{q_h \Delta x_{\text{head}}^2 + q_c \Delta x_{\text{cross}}^2 + r_{11} \delta_{\text{rudder}}^2 + r_{22} \delta_{\text{thrust}}^2\}$$

$$= \frac{1}{2} \sum_{k=i}^{N-1} \{\Delta x^t(k) \, Q \, \Delta x(k) + u^t(k) \, r \, u(k)\}$$

The weighting matrix Q consists of diagonal elements q_h and q_c, and other arbitrarily small positive elements along the diagonal to make Q positive semidefinite. This choice of Q implies emphasis on heading angle error and crosstrack error, possibly with some weighting of sway velocity and yaw rate. The matrix r for the control input is positive definite with r_{11} for the rudder control and r_{22} for the thrust control. Using the above Q and r matrices the optimal control u is obtained.[21]

B. Sensor/Ship Bandwidth Ratio and Straight-Course Steering Performance

A very simple straight track-keeping scenario is first explored. This gives an insight into the interaction of GPS and LORAN signal-tracking parameters with vessel dynamics parameters. Consider for the moment, lateral motion control and minimization of crosstrack error. The ship bandwidth on this axis was adjusted to a typical value for a medium-sized ocean vessel, 0.05 rad/s. Three different random vessel disturbance levels were considered, resulting in open-loop crosstrack standard deviations of 50, 15, and 5 m, respectively.

In this simplified model, an equivalent one-dimensional sensor was employed for estimating crosstrack displacement. First-order sensor dynamics were modeled, with sensor state process noise adjusted to maintain a constant crosstrack sensor error variance for each bandwidth setting considered. The navigation filter included a sensor lag state properly matched to this model. Then, an optimum linear quadratic Gaussian (LQG) controller was derived and the steady state crosstrack performance index for the combined filter and regulator was developed analytically.

The achieved crosstrack steering performance was strongly influenced by the sensor/vessel bandwidth ratio. Very poor control was obtained with a sensor bandwidth narrower than the vessel bandwidth. To underscore this effect, the steering error statistics were plotted in terms of the sensor/vessel bandwidth ratio, shown in Fig 3. In all cases, the sensor output standard deviation was held to

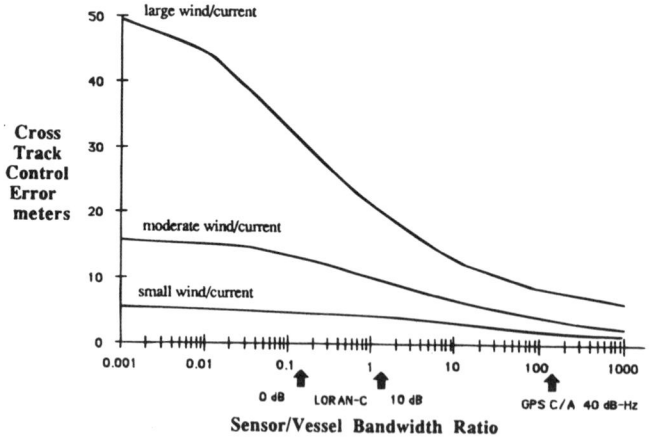

Fig. 3 Vessel steering performance vs bandwidth ratio.

5 m. At very small sensor/vessel bandwidth ratios, the controller is essentially operating without a sensor, and the crosstrack errors are the open-loop values of 50, 15, and 5 m. At unity bandwidth ratio, errors decreased to 21, 10, and 4.5 m, the largest improvement for the high wind/current disturbance scenario.

The above generic results offer insight into comparative DGPS and Loran-C performance. For a 20-kHz signal-to-noise ratio (S/N) of + 0 dB, a typical Loran-C receiver operates with a bandwidth of 0.007 rad/s. At +10 dB S/N the receiver loop bandwidth may be widened to 0.07 rad/s. Loran-C falls between 0.14 and 1.4 on the sensor/vessel bandwidth axis of Fig. 3. For a coherent delay-lock GPS tracking loop operating at a carrier-to-noise ratio of 40 dBHz and with the assumed 5 m standard deviation, the bandwidth is considerably broader, approximately 6 rad/s. Global positioning systems C/A reception at 40 dBHz carrier-to-noise ratio (C/N) achieves a bandwidth ratio of 125, permitting considerably better vessel control performance.

For the large wind/current example, the control performances for the different sensors fall between 33 and 9 m, crosstrack error. Although both Loran-C and GPS receivers can be designed to have identical noise jitter, five meters in the above analysis, the much wider bandwidth of GPS affords far better vessel control. Global positioning systems Doppler observables, or carrier-smoothed pseudoranges, would further enhance this performance. Human pilots tend to compensate for lags in guidance sensors. This subjective behavior has been formalized in the above by inclusion of lag states in the navigation filter design. We now return to the complete waypoint simulation.

C. Comparative Footprint Channel Clearance Width Distributions

Prior to closed loop operation with the sensor system and navigation filter, ship hydrodynamic coefficients were calibrated against published at-sea test data. The general vessel parameters considered are as shown in Table 1. An important test of ship dynamics is the 20/20 Z maneuver test.[8] From a straight line path with rudder amidships, the helm (rudder command) is deflected 20 deg to the right. When the heading changes 20 deg to the right of the initial heading, the helm is reversed to 20 deg left. When the heading changes to 20 deg left of the initial heading, the helm is reversed to 20 deg right. Beginning with available hydrodynamic coefficients,[19] coefficients were trimmed to give 20/20 Z maneuver dynamics similar to reported at-sea tests.[8] In an operational system, these model coefficients could be continuously refined by on-line identification techniques.

The basic test scenario consisted of a strait approach segment of 1500 m, a 35-deg/course change to port, and another 2500 m of course keeping. A diminishing southwesterly current of 1.5 k was applied early in each run, and to simulate the effect of wind and unmodeled hydrodynamic forces, random disturbances were introduced on the ship sway velocity and yaw rate. The higher of the two disturbance levels was selected to model storm conditions likely to be encountered only infrequently in river and harbor areas. The low-disturbance scenario is more typical of everyday piloting.

Even with perfect knowledge of ship position, velocity, heading, and yaw rate states, some deviation from the desired waypoint trajectory is unavoidable. Thus, before examining GPS and other sensor systems, it was important to determine steering errors contributed by ship controllability factors and wind and water disturbances. To this end, the optimal ship controller was driven with a perfect navigation state vector under a variety of disturbance conditions, and baseline values for sway and yaw rate disturbances were established. Also, the "outer" decision control for course leg switching points was optimized to obtain best CCW performance.

Figure 4a shows typical clearance widths swept out by the ship footprint for two disturbance levels in the region just after the turn. The graph ordinate is accumulated alongtrack position, and the abscissa is the minimum channel width needed to just clear the extremity of the ship. With small ship disturbance, the turn approach and postturn recovery clearance values are nearly the same as the ship half width, 19 m. Not shown is the turn region, where the right clearance width must be increased to 150 m and the left clearance width must be increased to 80 m to avoid boundary contact. The high disturbance CCW values superimposed on the plot show substantial increases in needed channel width for safe passage.

To further quantifying this behavior, the cumulative probability distribution for CCW samples at alongtrack points was estimated by simulation. Each distribution is the result of several runs and thousands of data points. Because CCW distributions are likely to differ in approach, turn, and recovery regions, they should be computed separately for each region. Figure 4b compares the CCW distributions in the postturn region, for both low- and high-disturbance conditions. For the low-disturbance runs, 95% of the clearance width samples are under 30 m. For the high-disturbance group, the 95% clearance half width is 48 m. An analytical derivation of these distributions is under development, with the ultimate goal of determining safe passage probability.

As discussed earlier in this book, GPS user and reference station equipment can operate in either code mode or integrated Doppler mode, with the most accurate results obtained when both sites are in the integrated Doppler mode. Delta range is useful in estimating velocity states during the required "warm-up" interval of the integrated Doppler mode. In the following simulations, perfect reference station corrections are assumed, together with a shipboard DGPS receiver operating with unsmoothed code pseudorange. In this GPS configuration, the sensor error budget is dominated by code-tracking loop noise, multipath, and tracking loop dynamical errors. Commensurate with a received C/N of 40 dBHz, and a tracking loop jitter of 5 m, the simulated code tracking loop time constant was set at 1.0 s. The navigation filter also processes a heading sensor good to 1 deg, 1-sigma. All data are sampled at a 1 Hz rate.

Fig. 5a shows typical GPS CCW plots obtained under low- and high-disturbance conditions, for the postturn region. At the low-disturbance level, CCW is very close to that obtained with perfect ship state knowledge. It is instructive to compare CCW cumulative distributions against the perfect state knowledge case. These are shown in Fig. 5b. The DGPS CCW distribution closely tracks the perfect-state-knowledge case. At the 95% level, an increase of about 2 m in CCW is contributed by the DGPS sensor and heading reference system. The sensor system contribution is greater with large disturbance, about 16 m. This

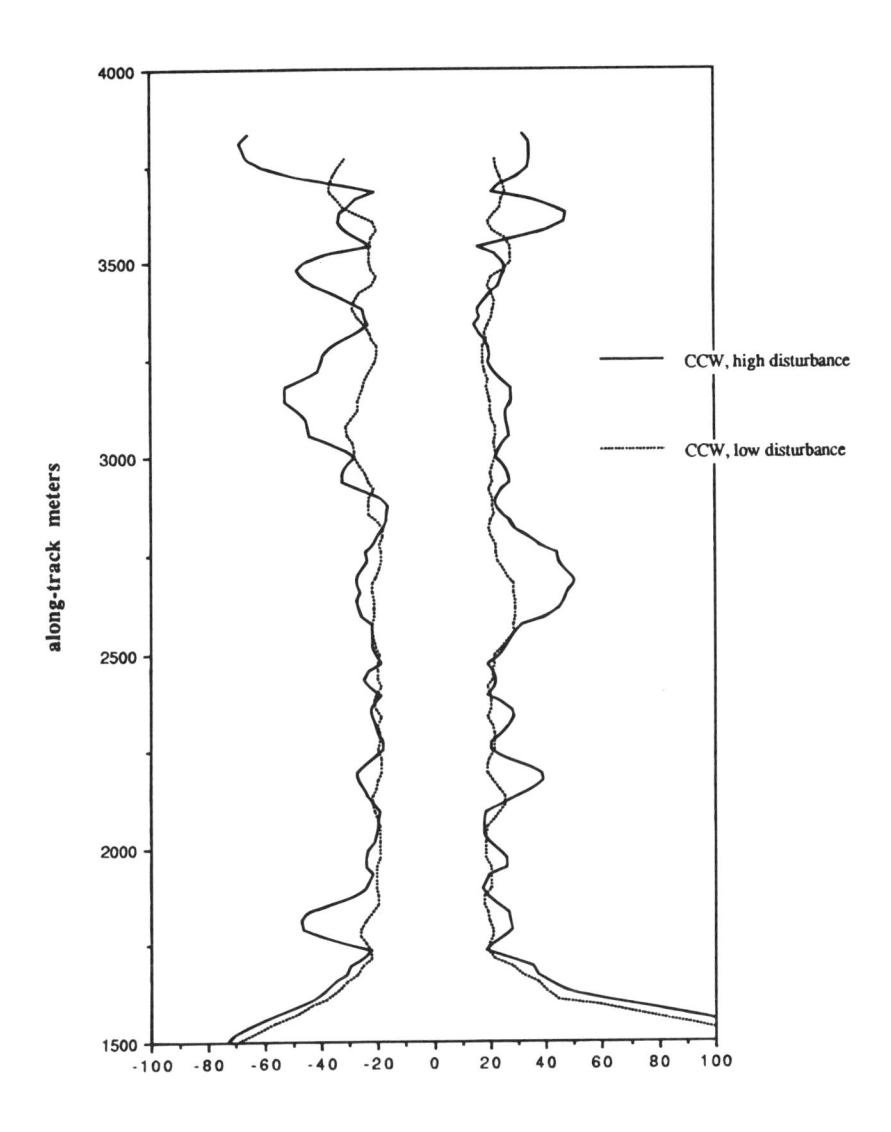

channel width, meters

Fig. 4a Channel clearance width with perfect navigation sensors.

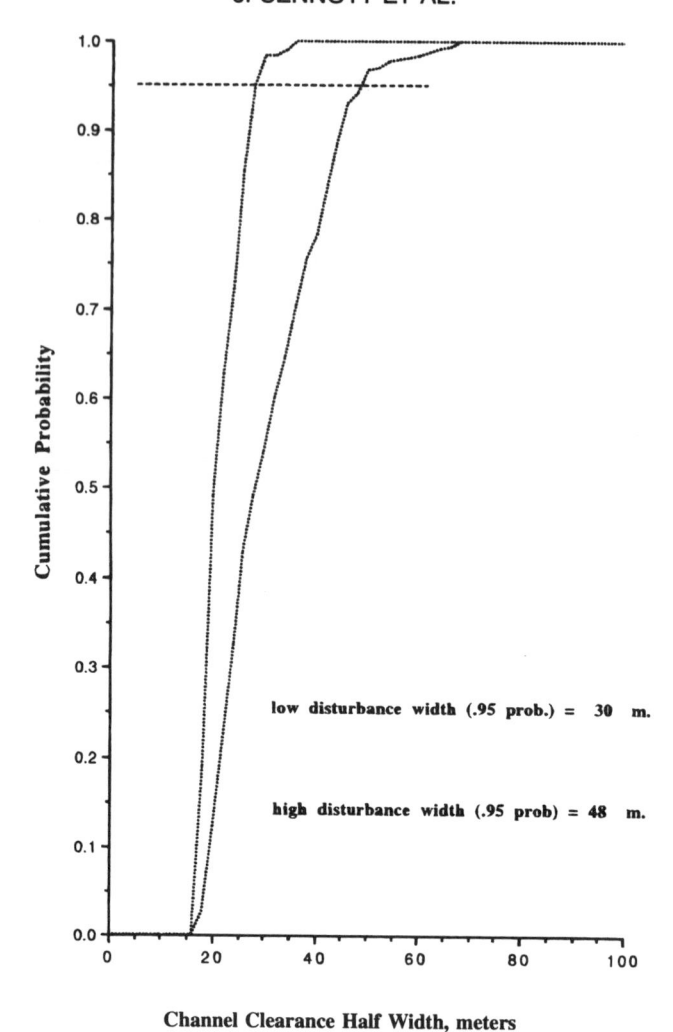

Fig. 4b Channel clearance width distributions, perfect nav.

(Graph shows cumulative probability vs. Channel Clearance Half Width, meters.
low disturbance width (.95 prob.) = 30 m.
high disturbance width (.95 prob) = 48 m.)

could be significantly improved with delta range velocity or smoothed pseudorange observables.

Loran-C may be operated differentially to achieve improved accuracy. Calibration is problematical, however, given local grid warp conditions in river and harbor areas. Furthermore, from the earlier discussion, signal tracking loop dynamics will significantly degrade steering performance. To understand better its fundamental performance limits, a perfectly calibrated Loran grid was assumed. Then, a simulated tracking loop jitter of 5 m, identical to the GPS pseudorange jitter above, was introduced. For an assumed 20-kHz S/N of +10 dB this corresponds to a tracking loop time constant of about 14 s. Clock and heading reference parameters and rudder data quality were the same as for GPS.

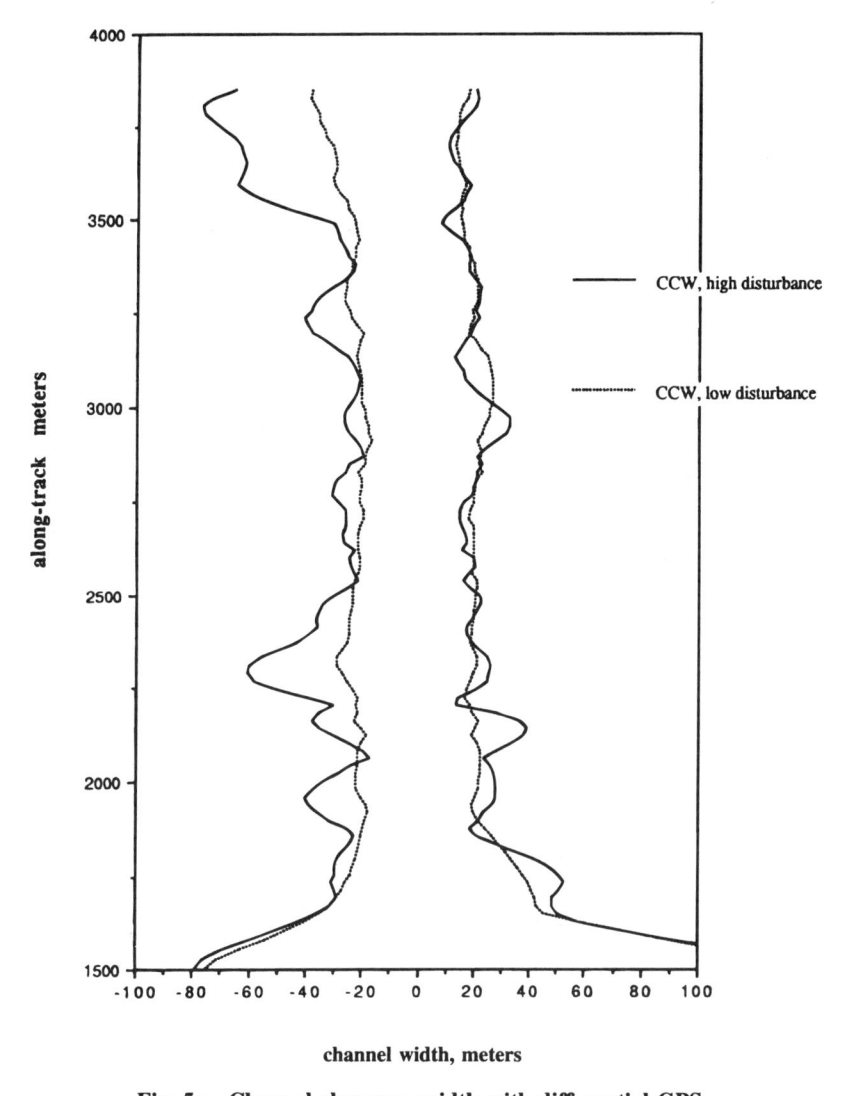

Fig. 5a Channel clearance width with differential GPS.

Figure 6a shows typical CCW results for high- and low-disturbance conditions in the turn recovery region. The plot exhibits a good deal of steering drift, indicative of poorer velocity estimates within the navigation filter, and a consequent loosening of the ship control loop. Turning to the cumulative distributions for channel clearance (Fig. 6b), a substantial loss in performance over the perfect-nav case is observed. At the 95% level, an increase of about 37 m in CCW is observed. This is with a five meter pseudorange jitter, and with sensor lag augmentation in the nav filter.

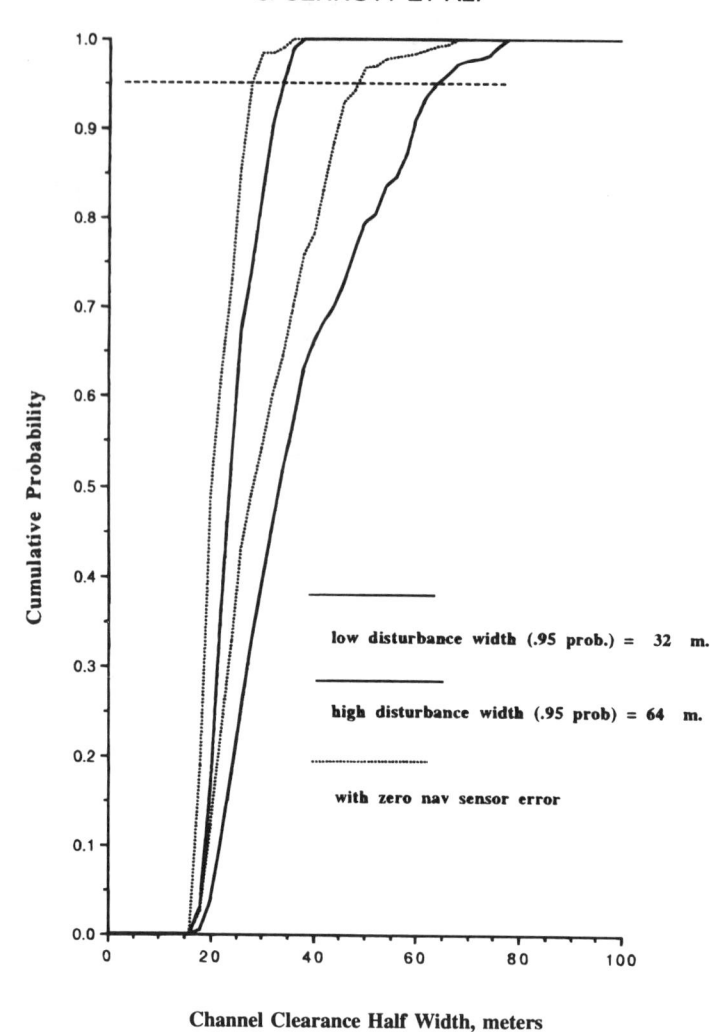

Fig. 5b Channel clearance width distributions, differential GPS.

low disturbance width (.95 prob.) = 32 m.

high disturbance width (.95 prob) = 64 m.

with zero nav sensor error

In summary, although tracking loop output standard deviation characteristics and sensor sample rates were identical for both DGPS and Loran-C tests, implying the same static accuracy for both sensors, achieved control performances differed markedly. As these results clearly show, specification of receiver jitter and update rate alone is insufficient for predicting ultimate closed-loop steering performance. Sufficiency requires characterization of both sensor dynamics and sensor noise sources. In the aviation environment, this issue is of even greater significance.

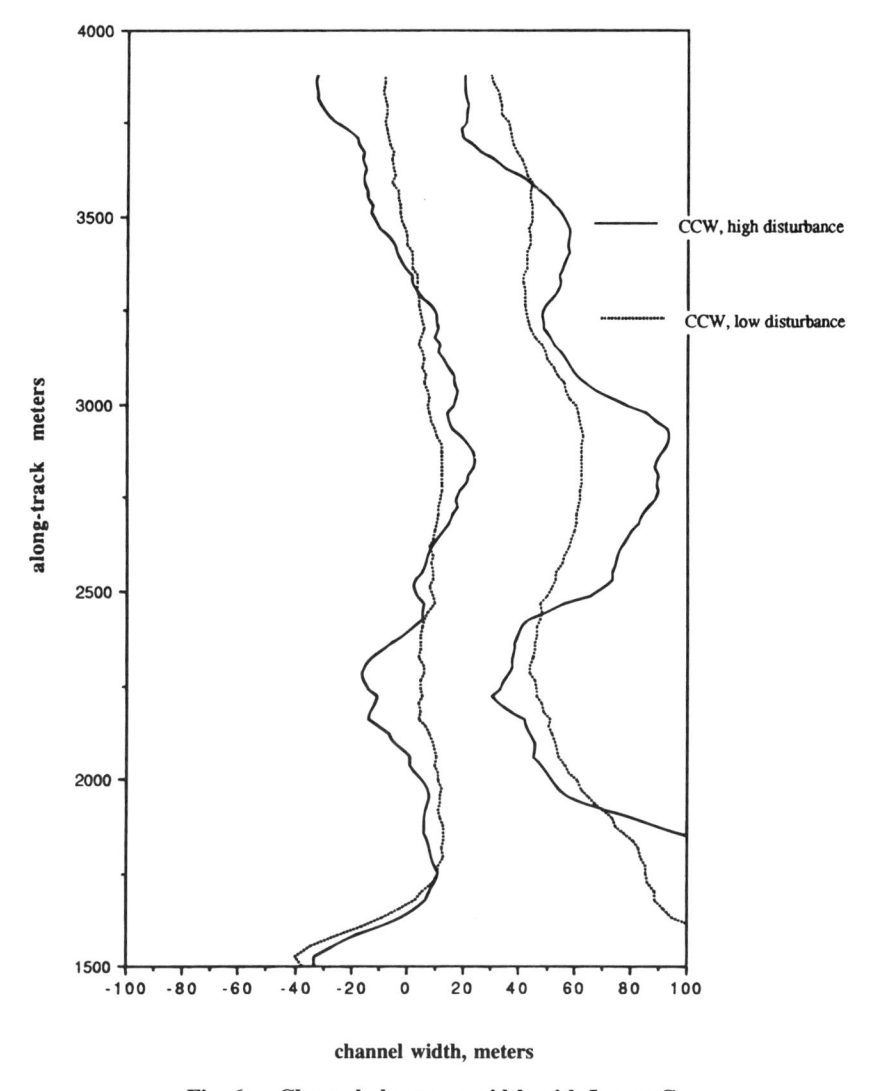

Fig. 6a Channel clearance width with Loran-C.

VI. Hazard Warning and Risk Assessment Functions
A. Risk Assessment

A preview of risk in advance of vessel passage through a waterway system, considering waterway geometry, anticipated steering quality, and navigation system anomalies can be performed aboard ship or at the shore-based vessel traffic system. A well-designed graphics interface would enhance the usefulness of this tool for testing "what if" scenarios. A safe passage probability figure of merit could be used to rank alternative waypoint steering sequences prior to passage.

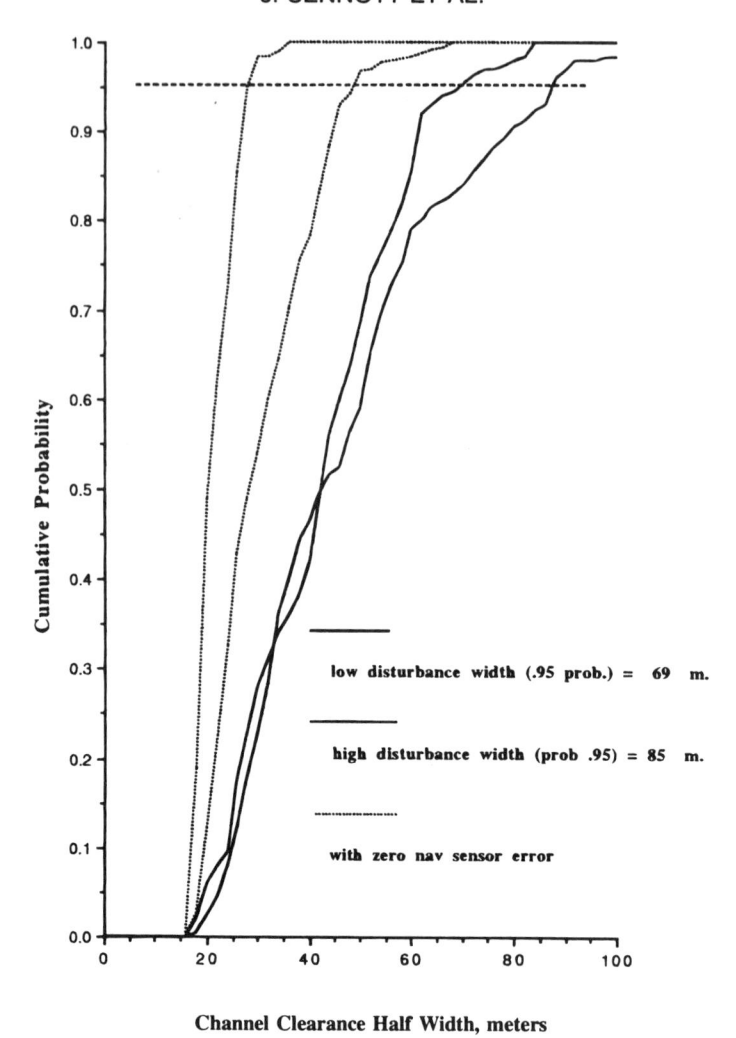

Fig. 6b Channel clearance width distributions, Loran-C.

low disturbance width (.95 prob.) = 69 m.

high disturbance width (prob .95) = 85 m.

with zero nav sensor error

The computation of risk depends upon knowledge of ship-specific hydrodynamic coefficients obtained over a history of runs in the same, or similar, confined waterway geometries. Differential GPS could be the basis for estimation of these ship coefficients, which would be entered into a closed-loop steering simulation similar to that discussed in Sec. V. Operators would input desired waypoints either by keyboard or by light-pen on a map display. The following computations would then be carried out: 1) optimization of waypoint segment switch points; 2) generation of the disturbance-free nominal-ship/nominal-sensor trajectory; 3) introduction of the ship disturbance model; 4) introduction of the sensor disturbance model; 5) computation of first- and second-order CCWDs; and 6) evaluation of SPP.

Computations can be computed for a specified waypoint sequence in several seconds with today's processors. In addition to statistical data, graphical outputs overlaid on the electronic chart would include points of waypoint turn initiation and the resulting ship footprint sweeps.

B. Hazard Warning

In contrast to the planning aspect of risk assessment, hazard warning addresses the immediate threat from nearby traffic and fixed hazards. Warning of conflict with other vessels requires continuous updates of relative position and the exchange of maneuver information confirming rules-of-road procedures. Warning of fixed hazards requires integration of the geographic database with ship and environmental states. GPS plays a vital role here.

Present-day vessel collision avoidance systems acquire relative positions from short-range marine radar. Automatic exchange of GPS position, velocity, and identity via a dedicated vhf channel would greatly improve the quality of this collision avoidance data. The beacon device would consist of a modified vhf transceiver connected to an existing GPS receiver. A similar GPS position reporting beacon could be affixed by authorities to new or critical hazards. At a very low incremental cost, smaller vessels that frequent busy commercial shipping lanes could also participate by equipping themselves with a GPS reporting beacon.

State estimates derived from GPS can be used to support a warning algorithm with the following properties. First, it does not generate an alarm (false) when the hazard miss distance would be acceptable in the absence of an alarm-induced maneuver. Second, it does generates an alarm (successful) in sufficient time for avoidance when the miss distance would otherwise be unacceptable in the absence of an alarm-induced maneuver.

The trade between successful and false alarms has been evaluated for an algorithm that estimates the Latest Safe Alarm Time (LSAT). Making use of the present state vector estimate, the vehicle maneuver model, and current rudder activity if available, the algorithm continuously probes for the latest time a maneuver may be undertaken to avoid the given hazard boundaries successfully. It has been shown that GPS velocity data can greatly improved the trade between the false and successful alarms.[22]

VII. Summary

The phase-in of more automated control and warning systems must include a period where their information is used in an advisory capacity. This provides an opportunity to identify and evaluate the remaining problems with the implementation without risking the safety of the crew and vessel involved. The phases of navigation clearly represent distinct levels of risk. An integrated approach using a technology such as DGPS combined with other sensors and the actual maneuvering characteristics of the vessel will be needed. In conclusion, GPS/DGPS means much more to marine transportation than "where the vessel is located." The investigation of other potential benefits and associated technologies as yet unforeseen will advance transportation safety and economic benefits in the future.

References

[1]"Federal Radionavigation Plan," U.S. Department of Transportation.

[2]Hartberger, A., "Introduction to the U.S. Coast Guard Differential GPS Program," *Proceedings, IEEE PLANS '92, IEEE Position, Location, and Navigation Symposium*, (Monterey, CA), IEEE, New York, March 25–27, 1992.

[3]Spalding, J., and Krammes, S., "Impact of USCG Differential GPS Service on User Equipment Technology," *Proceedings GPS-92*, Sept. 1992, pp. 573–578.

[4]Enge, P. K., Ruane, M., and Sheynblatt, L., "Marine Radiobeacons for the Broadcast of Differential GPS Data," *Proceedings of IEEE PLANS*, IEEE, New York, Nov. 1986, pp. 368–376.

[5]Enge, P. K., Kalafus, R. M., and Ruane, M. F., "Differential Operation of the Global Positioning System," *IEEE Communications Magazine*, July 1988, pp. 48–60.

[6]Sennott, J. W., and Ahn, I. S., "Assessment of Candidate Navigation Systems for Great Lakes St. Lawrence Seaway," Rept. of U.S. Department of Transportation Contract DTRS-57-85-C-0090, TTD 3, March 1988.

[7]Gynther, J. W., and Smith, M. W., "Radio Aids to Navigation Requirements: The 1988 Simulator Experiment," U.S. Coast Guard Rept. CG-D-08-90, Dec. 1989.

[8]Bertsche, W. R., Atkins, D. A., and Smith, M. W., "Aids to Navigation Principal Findings Report on the Ship Variables Experiment: The Effect of Ship Characteristics and Related Variables on Piloting Performance," U.S. Coast Guard Rept. CG-D-55-81, Nov. 1981.

[9]Bertsche, W. R., and Smith, M. W., "Aids to Navigation Principal Findings on the CAORF Experiment—The Performance of Visual Aids to Navigation as Evaluated by Simulation," U.S. Coast Guard Rept. CG-D-51-81, Feb. 1981.

[10]Bertsche, W. R., Mirino, K. L., and Smith, M. W., "Aids to Navigation Principal Findings Report: The Effect of One-Sided Channel Marking and Related Conditions on Piloting Performance," U.S. Coast Guard Rept. CG-D-56-81, Dec. 1981.

[11]Cooper, R. B., Cook R. C., and Marino, K. L., "At-Sea Data Collection for the Validation of Piloting Simulation," U.S. Coast Guard Rept. CG-D-60-81, Dec. 1981.

[12]Marino, K. L., Smith, M. W., and Moynehan, J. D., "Aids to Navigation SRA Supplemental Experiment Principal Findings: Performance of Short Range Aids Under Varied Shiphandling Conditions," U.S. Coast Guard Rept. CG-D-03-84, Sept. 1984.

[13]Moynehan, J. D., and Smith, M. W., "Aids to Navigation and Meeting Traffic," U.S. Coast Guard Rept. CG-D-19-85, June 1985.

[14]Smith, M. W., Marino, K. L., and Multer, J., "Short Range Aids to Navigation Systems Design Manual for Restricted Waterways," U.S. Coast Guard Rept. CG-D-18-85, March 1985.

[15]Sennott, J. W., and Ahn, I. S., "Simulation of Optimal Marine Waypoint Steering with GPS, LORAN-C, and RACON Sensor Options," Fifth Annual Technical Meeting Institute of Navigation, Jan. 1988.

[16]Sennott, J. W., and Ahn, I. S., "Design of a State Estimator and Regulator for Marine Steering Incorporating Sensor Model Dynamics," *Proceedings of Twenty-First Annual Pittsburgh Conference on Modeling and Simulation*, May 1990.

[17]Gelb, A. (ed.), *Applied Optimal Estimation*, MIT Press, Cambridge, MA, 1974.

[18]Amerongen, J., et al., "Model Reference Adaptive Autopilots for Ships," *Automatica*, Vol. 11, 1975.

[19]Astrom, K. J., and Kallstrom, C. G., "Identification of Ship Steering Dynamics," *Automatica,* Vol. 12, 1976, p. 9.

[20]Fung, P., and Grimble M. J., "Dynamic Ship Positioning Using a Self-Tuning Kalman Filter," *IEEE Transactions on Automatic Control,* Vol. 28, No. 3, March 1983, pp. 339–349.

[21]Lewis, F. *Optimal Control,* Wiley, New York, 1986.

[22]Sennott, J. W., and Ahn, I. S., "Evaluation of Sensor Performance and Supervisory Control for Terminal Air Traffic," *Proceedings of Twenty-Second Annual Pittsburgh Simulation Conference,* May 1991, pp. 2135–2143.

Chapter 12

Applications of the GPS to Air Traffic Control

Ronald Braff*
MITRE Corporation, McLean, Virginia 22102
J. David Powell†
Stanford University, Stanford, California 94305
and
Joseph Dorfler‡
Federal Aviation Administration, Washington, DC 20591

I. Introduction

THIS chapter identifies and discusses applications of GPS to air traffic control (ATC). The first section provides a very brief overview of ATC for the purpose of providing the reader with sufficient operational context for understanding why and how the GPS will be integrated into the ATC system. The second section discusses basic operational requirements considerations with respect to GPS implementation into the ATC system. The third and fourth sections contain descriptions and discussions of the potential specific applications of GPS in providing navigation and surveillance services, respectively, to ATC.

II. Air Traffic Control System

Air traffic control is "a service provided by the appropriate authority to promote the safe, orderly, and expeditious flow of air traffic" (Ref. 1, p. 232). In the United States, the authority for ATC is the Federal Aviation Administration (FAA), which operates and regulates the National Airspace System (NAS). The NAS encompasses "the common network of airspace, airports, navigation aids, and air traffic control equipment across the United States" (Ref. 1, p. 536). Control of air traffic involves the following functions: 1) *procedures* and *regulations* by which the ATC system operates and the *organization of airspace* in the form of

---

Copyright © 1994 by the authors. Published by the American Institute of Aeronautics and Astronautics, Inc., with permission. Released to AIAA to publish in all forms.
*Principal Engineer, Center for Advanced Aviation System Development.
†Professor of Aeronautics and Astronautics and Mechanical Engineering.
‡Program Manager, Satellite Navigation Program.

routes for departures, en route airways, and arrivals; 2) *Human air traffic controllers* who are responsible for providing the ATC service; 3) *Automation systems* (e.g., computers and displays) providing information to the controllers on the status, location, and separation of aircraft in the system; 4) *Communications systems* providing air–ground and ATC interfacility voice and data communications; 5) *Surveillance systems* (e.g., radar) providing real-time positional information to ATC for tracking aircraft and hazardous weather; and 6) *Navigation systems* providing real-time positional information for aircraft navigation.

Today's continental airspace is mainly organized into airways defined by the VHF omnidirectional range (VOR) navigation aids. A VOR provides directional information to aircraft. The directional information is the angular parameter (referenced to magnetic north) of a polar coordinate system with the origin at the VOR antenna. The resulting radials of constant angle define the centerlines of VOR airways. To complete the polar coordinate positioning capability, distance-measuring equipment (DME) is collocated at most VOR facilities. An aircraft with only VOR capability (guidance on the selected radial) is constrained to fly VOR airways from one VOR to another, which extend into the terminal area. In most high-density terminal areas, the airways end at the approach feeder fixes, and air traffic controllers provide vectors (heading change commands) to line aircraft up on the final approach segment. Before the advent of small low-cost digital computers, the VOR was the best method for lateral guidance because the demodulation of its signals provides a direct read-out of the radial the aircraft is traversing.

The Instrument Landing System (ILS) provides both lateral guidance with respect to the extended runway centerline and vertical guidance for a fixed glide path of usually 3 deg. It also provides "marker beacons" at fixed points along the approach course to provide along-course information. The ILS is used in conjunction with lighting systems and other visual aids for final approaches during low-visibility conditions.

Area navigation (RNAV) permits aircraft to fly from waypoint to waypoint, where the waypoints can be defined at any location in two or three dimensions. A waypoint is a point in space that defines the beginning and end of a desired flight path. Radionavigation-based RNAV requires a computer to transform at least two signals into suitable guidance information. The essential benefit of RNAV is to allow users to choose, along a path defined by the waypoints, the best route they determine with respect to such performance criteria, as minimum time or fuel burn to destination. Examples of systems that provide RNAV capability are VOR/DME, DME/DME, Loran-C, Omega, the GPS, and the inertial navigation system (INS).

The present ATC system would have problems coping with the wide scale use of RNAV. The VOR airway system orders the flow of traffic, minimizing the number of route intersection points. With RNAV, aircraft would be flying user-preferred trajectories that would yield random routes, resulting in more complexity of the intersection of routes. This would be further compounded by the inevitable increase in traffic over the years. Hence, additional automation is needed to help controllers and traffic managers predict and resolve conflicts in such an environment; otherwise, the wide-scale introduction of RNAV would be impeded. The FAA will be implementing enhancements to the NAS Advanced

Automation System (AAS) that are designed to handle greater numbers of aircraft flying random routes. This set of enhancements is called automated en route ATC (AERA). An FAA benefit/cost study for AAS with AERA indicated that the accommodation of user-preferred trajectories accounted for nearly 70% of the benefits of AAS to the users.[2]

Figure 1[3] is a schematic diagram of the present NAS infrastructure of facilities. In addition, there are organizational entities providing standards and certification of user equipment and operation, and maintenance of the thousands of ATC facilities that provide the functions displayed in Fig. 1.

The foregoing description of the ATC system is the background to provide a context for the subsequent discussion of the application of GPS to ATC. An up-to-date detailed description of the ATC system is given in Ref. 1.

III. General Considerations

The role of the GPS in ATC involves providing highly accurate position, velocity, and time for the navigation and surveillance functions for all phases of flight and ground movement of aircraft. The GPS can be used for air navigation in three basic roles:

1) A *required navigation performance* (RNP) or *primary navigation* system is a navigation system that meets all requirements to use certain procedures or to fly in certain airspace without the need for any other navigation system onboard the aircraft, except, of course, compass and airspeed indicator. An RNP system may include one or more integrated navigation sensors in its definition (e.g., the GPS with an inertial reference system).

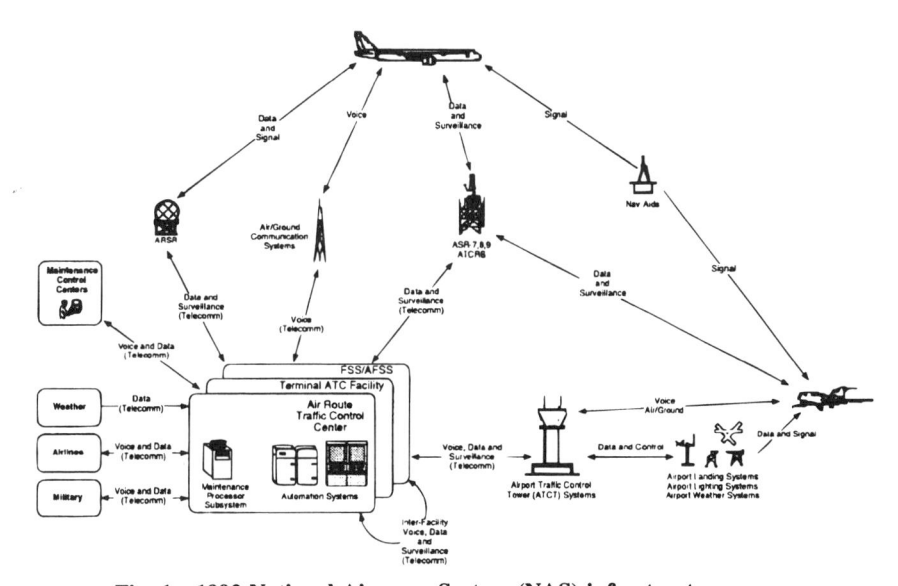

Fig. 1 1992 National Airspace System (NAS) infrastructure.

2) A *supplemental system* is one that can be used alone without comparison to another system; however, an RNP system that could be used in the event that the supplemental system is not available must be on board the aircraft.

3) A *multisensor navigation system* is one that can be used for navigation, but only after it has been compared for integrity with an RNP system in the aircraft.

When employed in surveillance, a navigation system can provide the sensor function that is used in conjunction with a data link to transmit positional reports to ATC. This type of surveillance is called "automatic dependent surveillance" (ADS).

Why is there such a great interest in use of the GPS in ATC? The basic benefits of the GPS are its higher accuracy and worldwide coverage of airspace. These benefits provide the user with the potential for minimum avionics that provides worldwide navigation capability to fly user-preferred routes, rather than airways, for all phases of flight, and the government an opportunity to reduce its vast infrastructure of thousands of ground-based transmitters dedicated to navigation and surveillance. Only a satellite navigation system such as the GPS provides all of these benefits. Specific user benefits include: capability to fly preferred routes, landing system capability to any runway, and reduced separation in nonradar airspace in conjunction with reliable air-to-ground datalinks for ADS-position reports. The benefits of satellite navigation and ADS are discussed in detail in Refs. 4 and 5.

Figure 2[3] illustrates a possible transition phase to GPS-based navigation and ADS. There is a mix of GPS-based and the ground-based systems where GPS is used in all phases of flight. Further transitions could involve complete GPS

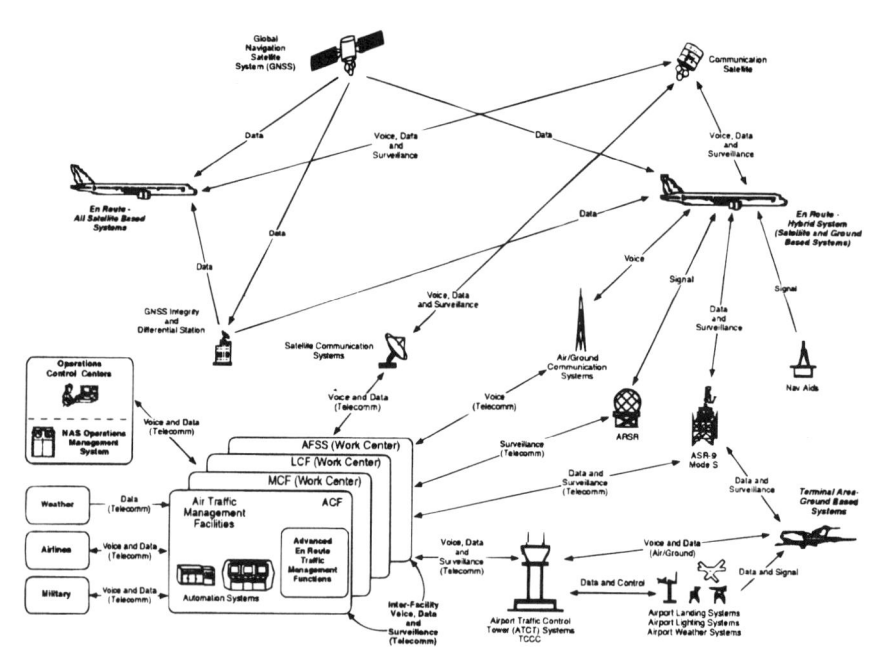

Fig. 2 2010 National Airspace System (NAS) infrastructure.

replacement of the ground-based navigation aids and much of the surveillance infrastructure. Such a stepped transition will be necessary because not all of the users can equip with the GPS in a short period of time, and ATC automation must evolve to accommodate and benefit from the increased capabilities provided by GPS-based navigation. As indicated in Fig. 3,[3] the approach taken here is to explore the role of satellites for navigation and ADS applications

A. **Operational Requirements**

The major operational considerations in the introduction of a new system concern its *accuracy, integrity, availability, continuity of service,* and the *procedures.*

Accuracy is the degree of conformance of estimated position with true position. It is usually expressed in statistical terms, such as the 95th percentile error. The most common accuracy metric for horizontal error is 2 drms, twice the rms radial (distance) error. If the navigation sensor errors are normally distributed then 2 drms is a circle about the true position containing approximately 95–98% of the position determinations, depending upon the eccentricity of the resulting bivariate error ellipse. Vertical accuracy is usually expressed at the 95 percentile or 2 standard deviation (2σ). Accuracy is usually the first parameter of consideration in the evaluation of a navigation system because it represents a physical limitation of a system, and it is straightforward to estimate by analysis or measurement.

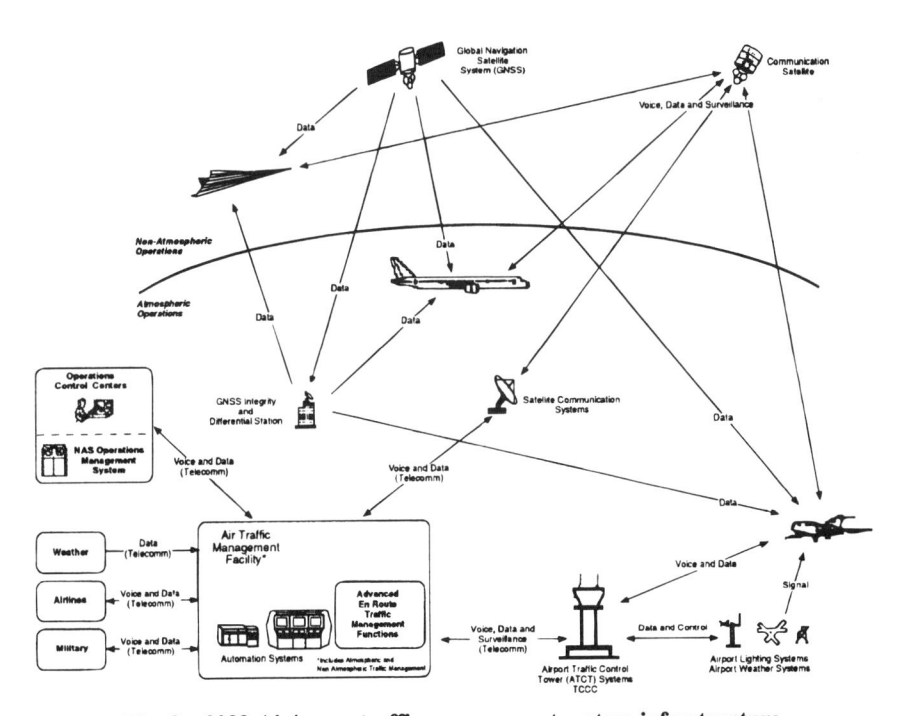

Fig. 3   20?? Air/space traffic management system infrastructure.

"*Integrity* is the ability of a system to provide timely warnings to users when the system should not be used for navigation."[6] The integrity function of a navigation system involves monitoring of the system's errors, and if specified protection levels are estimated to be exceeded, a warning is given to the pilot that the system cannot be used for navigation, or the system shuts itself off.

The integrity requirements and solutions with respect to a combined use of the GPS as the sensor for both navigation and ADS has not yet been addressed. In the present radar-covered airspace where the navigation and surveillance functions are completely independent, there is a very high level of integrity in the sense that although the ground-based radionavigation systems have an excellent integrity function, any failure of that function would be caught by the independent radar surveillance system. Furthermore, if an aircraft loses its radionavigation capability when in radar airspace, then its dead-reckoning navigation can be monitored by radar, and speed and heading corrections ("radar vectors") communicated by the air traffic controller to the pilot via voice radio transmissions. The issue here is whether more demanding integrity requirements should be put on the GPS when it is both a sensor for navigation and ADS. The introduction of more stringent integrity requirements could entail an increased alarm rate, thereby decreasing availability. Any time there would be an integrity alarm, there could be an absence of both navigation and surveillance to the affected users.

"The *availability* of a system is the percentage of the time that the services of the system are usable."[6] The FAA requires an availability of 99.999%[7] for RNP enroute and terminal navigation, and for surveillance. This only applies to the services provided by the FAA and does not include the airborne equipment. When serving high-density airspace, this high availability is attained through redundant coverage of navigation and surveillance ground facilities, and redundant subsystems within each facility. The 99.999% requirement has been used as a guideline in estimating augmentations to increase the GPS availability.

The separation of routes in nonradar covered airspace is much greater than in radar airspace (e.g., 60 n.mi. vs 8 n.mi.). Thus, the first combined application of GPS navigation and GPS-based ADS with adequate datalinks will be in oceanic airspace where there is a great desire to decrease the large required separation between aircraft tracks. Using qualitative arguments, the formerly Radio Technical Commission for Aeronautics (RTCA) Task Force report (Ref. 4, p.17) was quite optimistic about the feasibility of adequate integrity and availability of GPS applications for all airspace. GPS availability requirements should not be as stringent in nonradar airspace as in radar airspace because there is a wide margin for decreasing aircraft separation standards, but still keeping them larger than in radar airspace.

Continuity of service is the ability of a navigation system to provide required service over a specified period of time without interruption. Continuity is particularly important in the approach and landing phase of flight. "The level of continuity is expressed in terms of the probability of not losing the radiated guidance signals."[8]

Procedures are based on criteria that have to do with where and how the system can be certified for operation. This is determined by the regulatory authority (FAA) based on the system's capabilities. For example, at the time of writing, GPS has been certified as a supplemental system for en route and nonprecision approach navigation.[9] A nonprecision approach procedure is that wherein a radionavigation system provides only lateral guidance for the approach; whereas in a precision approach procedure, the radionavigation system provides both lateral and vertical guidance.

When a system is implemented with new and better capabilities, the users would like to use it to its full capabilities to derive the most benefits. However, mainly for safety reasons, it takes regulatory authorities time to approve use of a system to its fullest capabilities. For example, the GPS standard positioning service (SPS) accuracy (100 m, horizontal) is such that it could be used to provide a level of accuracy for nonprecision approaches that is at least equivalent to, or, in most cases, better than that provided by today's standard, the VOR.[6] However, to take full advantage of this capability, new criteria must be developed so that procedures can account for smaller ground obstacle clearance criteria for a GPS-based system. Therefore, as a transition strategy, the FAA is allowing suitably equipped aircraft[9] to use the GPS to fly all present nonprecision approaches, except those where lateral guidance is provided by the localizer subsystem of the ILS.

B. Government Activities

The following description of government activities concerning the GPS's application to ATC is presented to indicate the seriousness of the commitment to lay the ground work for the GPS to assume a major role in the NAS and internationally.

The FAA Satellite Navigation Program Plan,[10] updated annually, provides the scope, objectives, schedules, and other requisite planning information for implementing satellite navigation in the NAS. The FAA program covers all required activities for implementation of satellite navigation in all phases of flight, including precision approaches.

The FAA Satellite Operational Implementation Team (SOIT) was formed on 19 August 1991 to facilitate the introduction of satellite navigation and communications into the NAS. The team consists of FAA experts in aviation and flight standards, avionics certification, instrument flight procedures, and other operational areas. At the time of writing, the team has developed and approved the process, procedures, and standards for operational use of the GPS for all phases of flight down to nonprecision approaches: on 25 February 1991 (and revised 20 July 1992) for using the GPS as an additional sensor input to an approved multisensor navigation system; and on 10 December 1992 a Technical Standard Order (TSO)[9] was written describing the required capabilities of GPS receivers to be used as a supplemental system of navigation. Furthermore, on 23 April 1992, the FAA published the following notices developed by the SOIT: Notice 8110.47, "Airworthiness Approval of GPS Navigation Equipment for Use as a VFR and IFR Supplemental Navigation System," and Notice 8110.48, "Airworthiness Approval of Navigation or Flight Management Systems Integrating Multiple Navigation Sensors." On 9 June 1993 the FAA authorized supplemental navigation approval for use of GPS equipment (TSO-C129) to conduct oceanic, domestic en route, and terminal instrument flight rules (IFR) operations, as well as nonprecision approaches with certain limitations.

RTCA activities relating to the GPS have been underway since the establishment of RTCA Special Committee 159 (SC-159), at the request of FAA, on 20 September 1985. The RTCA provides an organizational framework for interested parties representing airspace users, avionics manufacturers, and government organizations who volunteer to develop, by consensus, minimum operational performance standards (MOPS) for avionic systems, and more broad-based consensus on defining communication, navigation, and surveillance (CNS) systems for aviation use and determining their benefits.[4] The SC-159 documents are recommended guidelines used by the FAA and other parties in developing TSOs and technical programs. The SC-159 MOPS for supplemental GPS[11] navigation equipment was used as a major technical input for the TSO-C129 on the GPS for supplemental navigation. At the request of the FAA, the RTCA also formed a task force to develop a consensus strategy with recommendations regarding early implementation of an operational Global Navigation Satellite System (GNSS) capability in the United States. The task force was composed of high-level representation from commercial, business, and general aviation users, industry, the U.S. Department of Defense, and the FAA. They reached a solid consensus that the user community wants, needs, and is ready to implement GNSS-based operations, and that the benefits will apply to virtually all aspects of aviation operations.[4] GNSS is a concept for "a worldwide position and time determination system. GNSS includes one or more satellite constellations, end user receiver equipment, and a system integrity monitoring function. GNSS will be augmented as necessary to support the RNP concept for a wide range of specific operations" (Ref. 4, p. 9).

The *International Aviation Civil Aviation Organization* (ICAO) Special Committee on Future Air Navigation Systems (FANS) has defined a future ATC system where satellites play the major role in providing the communications, navigation, and surveillance (ADS based on GNSS) infrastructure.[12] The ICAO is an organization of the United Nations responsible for promulgating standards and recommended practices (SARPS) that have the status of international treaties. The purpose of SARPS is to ensure the international interoperability of CNS systems. Because the GPS is an integral part of the envisioned GNSS, it receives attention at ICAO. In 1992, the U.S. government stated to ICAO its intention to provide GPS signals for the foreseeable future with no direct user charges (Ref. 4, p. 57). It was also stated that ICAO be afforded at least 6 years' advance notice prior to the termination of GPS signals.

The *Federal Radionavigation Plan* (FRP) "delineates policies and plans for federally provided radionavigation services."[6] It has a biannual update. The FRP includes the Federal government's policy on GPS and information on the retention of other radionavigation systems that may be impacted by GPS. It is a collaborative effort by DOD and DOT; therefore, it includes information relevant to aviation and marine- and land-based users, both in the civilian and military communities.

IV. Air Navigation Applications

Applications of the GPS to air navigation are best partitioned into 1) en route, terminal, and nonprecision approach phases of flight; and 2) precision approach phase of flight. This is a natural partition for historical and practical reasons.

The 100-m (2-drms) accuracy specification for the GPS standard positioning service is based on civil aviation's need for a level of accuracy that is as good or better than present approved navigation systems for all phases of flight down to and including nonprecision approaches. Therefore, in the early 1980s, this SPS accuracy was recommended by the FAA and accepted by the Department of Defense. At that time, receiver and differential GPS (DGPS) technology and projected system performance precluded GPS application to precision approaches where very accurate vertical guidance is required.[13] However, within the last few years flight tests of DGPS, with state-of-the-art receiver developments, have indicated the accuracy feasibility of GPS for precision approaches. From the users' point of view, the GPS may be looked upon as a "seamless" potential replacement of inertial navigation and Omega for oceanic en route navigation, VOR/DME for domestic navigation through nonprecision approaches, and ILS at least up to Category I (CAT I) approaches. At the time of writing, application of the GPS to the more stringent CAT II and CAT III categories of approach are under intense investigation.[14]

The following discussion stresses the navigation operational considerations for GPS. This discussion should provide the reader with some insight as to what is involved in implementing the GPS as an approved air navigation system.

A. En Route, Terminal, and Nonprecision Approach Operational Considerations and Augmentations

1. Accuracy

The accuracy of an air navigation system usually considers three basic error sources; namely, sensor error, course-centering error, and flight technical error (FTE). The sum of these three errors is called total system error (or system use error). For clarity, these errors are defined here by the way they are measured.

1) *Sensor error* is the difference between the navigation receiver position determination and a truth source of position (such as a surveyed point, theodolite angle, or a laser tracker position). Sensor error consists of the sum of the nominal GPS error components (e.g., signal, atmospheric delays, receiver noise, multipath, and coordinate conversion) (Ref. 11, p. 15). Sensor error is usually expressed as twice the rms error (or the error not exceeded 95% of the time).

2) The *FTE* is the measure of how well a human pilot or an autopilot can follow the guidance commands derived from the navigation position determination. It is measured as the negative of the guidance command. It is usually expressed as twice the rms error or a 95% error. It considers pilot performance in specified wind environments. The FTE must be such that it will not cause an unacceptable total system error when combined with sensor accuracy.

3) *Course-centering error* is the measure of how accurately the navigation sensor position is transformed into guidance commands, where the guidance commands are relative to the desired flight path. It is the difference between the displayed cross-track guidance command and the computed guidance command.

4) *Total system error* has been traditionally calculated as the root-sum-square of the 2-rms sensor error, course-centering error and FTE. More recently, for approach flight tests, the FAA has also been estimating total system error as the difference between a laser tracker truth position and the assigned flight path.

There should be, and it is generally observed, that there is no significant difference in the results obtained by the two methods.

The cross-track accuracy requirements for each of the three phases of flight through nonprecision approach have been agreed upon (Ref. 11, p. 31). The 95% sensor error requirement for GPS RNAV is 0.124 n.mi. (230 m) for all phases of flight through nonprecision approach. The course-centering error requirement for the en route and terminal phases of flight are 0.2 n.mi. (370 m), and 0.1 n.mi. (185 m) for nonpecision approach. No FTE requirements are included in Ref. 11 because it is stated that FTE is beyond the control of the equipment manufacturer or installer. Estimates of manual FTE are 1.0, 1.0, and 0.5 n.mi. for domestic en route, terminal, and nonprecision approach, respectively. For coupled automatic flight control the FTE estimates are 0.25, 0.25, and 0.125 n.mi. for domestic en route, terminal and nonprecision approach, respectively.[11] Root-sum-square (rss) combining of the aforementioned error budget components yields estimates of 95% total system error contained in Table 1.

The aforementioned accuracy values are expected to hold for both supplemental and RNP GPS navigation.

2. Integrity

Integrity of the GPS as a navigation system in the NAS has been addressed and recognized as an issue for many years (Ref. 15, pp. 1214–1223 and Ref. 11, Appendix B). The problem is that GPS as implemented today does not have the capability to notify users of a signal malfunction in a timely manner. For some types of signal malfunctions, it can take on the order of an hour for notification; whereas, the integrity monitoring response times required for flight operations are on the order of seconds. Integrity solutions have been found and are described elsewhere in this volume.

The provision of integrity for supplemental satellite navigation requires only the detection of a navigation sensor malfunction; whereas, the integrity of an RNP system requires both detection and correction of the malfunction. The latter requirement is needed for RNP because a faulty satellite needs to be identified so it can be removed from the position determination solution, or a subset of satellites that do not cause an alarm can be found. There are two basic ways for providing integrity: ground-based monitoring (the GPS Integrity Channel or GIC), and airborne monitoring by using redundant measurements (Receiver Autonomous Integrity Monitor or RAIM).

The GIC will consist of a network of ground-based GPS signal-monitoring stations located at known reference points that cover a wide geographical area

Table 1 Total system cross-track error estimates for GPS

Phase of flight	Steering	95% Total system error, n.mi.
Domestic en route	Manual	1.0
	Coupled	0.3
Nonprecision approach	Manual	0.52
	Coupled	0.2

over which signal integrity is guaranteed by a navigation provider, such as the FAA. These monitors will be connected to a central control station where the integrity decisions will be made and messages composed. The integrity messages will be broadcast through geostationary satellite relays. Recently the FAA has called its implementation of the GIC the GPS Integrity Broadcast (GIB).

Receiver autonomous integrity monitoring (RAIM) (see Chapter 5, this volume) is essentially various algorithmic techniques for integrity monitoring that use redundant pseudorange measurements (i.e., $n - 4$ satellites when $n > 4$ satellites are visible) or aiding from another sensor (e.g., barometric altimeter). Receiver autonomous integrity monitoring without aiding requires at least five satellites in view with good geometry to permit detection of a violation of GPS position error tolerance. At least six satellites with good geometry are required for the identification of a faulty satellite or the determination of a useful subset of satellites.

With the planned GPS constellation and no augmentations to GPS, the availability of RAIM identification for RNP-nonprecision approach is clearly not sufficient for operations. Because identification is required for RNP, the FAA is considering the implementation of a GIC to provide the integrity monitoring function.[16] Not only is GIC expected to provide enhanced integrity monitoring, it will also increase availability and continuity of service because navigation can be conducted with only four satellites in view. The requirements for integrity monitoring are usually stated in terms of four parameters.[11]

a. Protection level. The positional error magnitude that cannot be exceeded, and for which the integrity monitoring system provides protection by warning the pilot.

b. Alarm rate. An alarm must be annunciated when position errors exceed the protection level. The alarm rate should not be so excessive that it becomes a nuisance during operations.

c. Time to alarm. The maximum allowable time from the onset of a failure to the annunciation of the failure.

d. Missed detection probability. The probability that a failure occurred and was not detected and displayed to the pilot.

Table 2, derived from Ref. 11, contains the RTCA SC-159 recommended requirements for RAIM integrity monitoring for supplemental navigation in the en route, terminal, and nonprecision approach phases of flight.

The RAIM performance requirements for RNP have not been specified at the time of writing, but they may contain a much lower maximum allowable alarm rate to ensure the high performance required for an RNP system. When RAIM identification is possible, no alarm need be annunciated if a fault is identified, and sufficient navigation capability is provided by the other satellites in view. Integrity monitoring requirements must be specified for the GIC, a point that is discussed later with respect to the FAA's proposed GIC implementation.

3. Availability

The availability of GPS for navigation has been addressed for many years.[16] Assessments of GPS availability for air navigation indicate that use of the planned

Table 2 GPS RAIM performance requirements for supplemental navigation as a function of phase of flight

Phase of flight	Protection level, n.mi.	Maximum allowable alarm rate, h^{-1}	Time to alarm, s	Minimum detection probability
En route	2.0	0.002	30	0.999
Terminal	1.0	0.002	10	0.999
Nonprecision approach	0.3	0.002	10	0.999

constellation, without any augmentations from other navigation sensors, may fall short of providing sufficient availability in the context of GPS as an RNP system, especially for the approach phase of flight.[17,18] These referenced studies used average availability over continental airspace as the measure of GPS availability. However, because the geometry of GPS satellites is varying in both time and space, these variations must also be considered. Figure 4[19] illustrates the diurnal variation of unavailability (1—availability) of a 24-satellite GPS constellation augmented with Inmarsat-3 geostationary satellites, where both the geometric constraints on satellite–user geometry and satellite failures are considered. Note the significant order-of-magnitude changes in unavailability during the day. Clearly, this variation indicates that average availability is not a complete measure. The times where the spikes indicate poor availability could be eliminated through further augmentations. Figure 5[19] illustrates the spatial variability of GPS (augmented by Inmarsat-3 satellites) unavailability using average unavailability as the measure. Again, significant variation can be seen.

The availability of a satellite-based system is of more concern than that of a ground-based system because a loss of signal coverage in the former could involve a very wide area; whereas, a ground-based system outage would involve a facility that covers a much smaller area. For example, the impact of an outage of ILS could be minimized by diverting aircraft to another ILS runway on the airport, if feasible, or to a nearby alternate airport. However, with a satellite-based system, the alternate airport could also have lost the signal coverage. On the other hand, the ILS outage persists until repaired, which could take several hours from intitial equipment shutdown; whereas, a GPS outage caused by lack of satellite coverage or sufficient postion fix geometry could be alleviated within 20 minutes or so when another satellite(s) comes into view. Also, the predictability of satellite availability at the destination (made at departure time) would ease the impact of periodic satellite coverage holes caused by failed satellites.

A predicted outage does not compromise safety, but it would have an impact on the efficiency of the ATC system if GPS-RNP is to replace existing navigation aids. These are the reasons why augmentations to the GPS are under consideration, such as hybrid combination with other systems (e.g., GLONASS and Loran-C), and additional satellites with GPS-like signals.[10,20–22] These augmentations of the GPS to increase availability are discussed below.

The requirements for availability depend upon whether the GPS is to be used as a supplemental or an RNP system. Reference 9 contains no quantitative signal

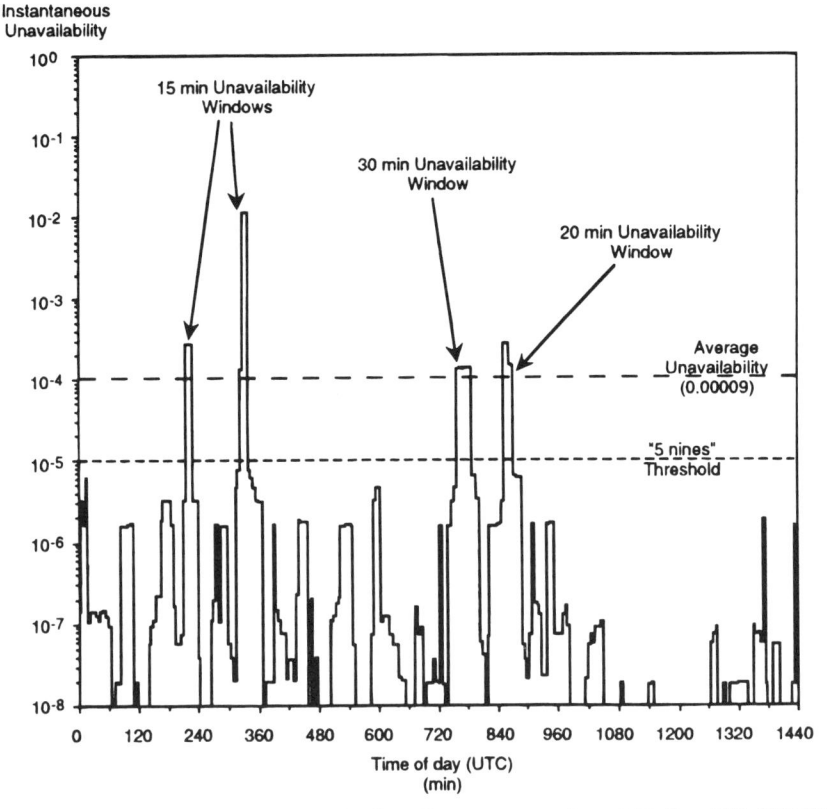

Fig. 4 Temporal unavailability: GPS + Inmarsat overlay at Dallas, TX (VDOP > 4.5).

availability requirements for supplemental GPS. However, availability of the GPS for supplemental use must to have a reasonable value (e.g., 95%); otherwise, it could not be relied upon to provide adequate area navigation service, and could be disruptive to air traffic control when aircraft have to transition from area navigation to VOR radial navigation during a GPS signal outage. Stated availability requirements for supplemental use tend to be concerned with predicting availability of RAIM (sufficient number of satellites and geometry), and providing enhancements to increase it. For instance, there is a requirement for the GPS navigation set to provide the pilot with information to determine whether RAIM will be available at the planned destination.[9] To enhance availability, there is also a requirement for barometric altimeter aiding of the RAIM function.

At this time, there seems to be no stated availability requirement for GPS as an RNP system. However, it seems that such a requirement may have to consider the 99.999% requirement stated in Sec. II. Augmentations to GPS that have the potential to achieve this level of availability are discussed below.

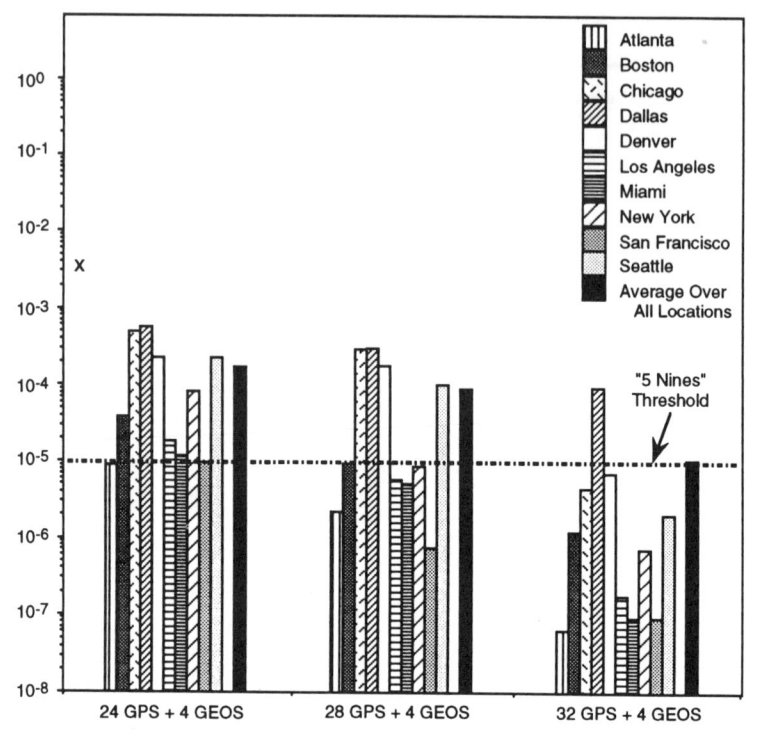

Fig. 5 Spatial unavailability: Various Constellation Sizes of GPS + Inmarsat Overlay (VDOP > 4.5).

4. Continuity

For GPS supplemental continuity, the traditional requirements tend to be somewhat relaxed. For instance, during the final approach segment of a nonprecision approach, with the GPS navigation function in operation, the warning flag is not displayed until the RAIM detection function is lost for more than 5 min.[9] Also, if RAIM is augmented by altimeter or receiver clock coasting, the warning flag may be delayed for a period of time consistent with the worst case drifts of these aiding sensors.

There are no existing quantitative or qualitative requirements for RNP continuity for nonprecision approach; however, a sometimes stated rule of thumb is no more than 1 out of 1000 approaches should be broken off because of loss of navigational signals.

5. Augmentations

The following are discussions of some augmentations for increasing GPS availability with respect to an adequate integrity function. The first three augmentations, barometric altimeter, Loran-C, and GLONASS, increase the availability of RAIM. The GIC with geostationary satellite overlay, the last discussed augmentation, provides an external ground monitoring system and communication of GPS signal integrity to users. Relying on sources available in the navigation literature, an estimate of the availability of the integrity function attributable to each augmentation is provided. These availabilities are not compared in a quantitative sense because they were estimated by different parties using somewhat different assumptions and methods. However, as shown, the availability estimates seem to indicate that GIC with geostationary satellite overlay and GPS/GLONASS would provide sufficient availability of the integrity function to provide RNP nonprecision approach.

a. Barometric altimeter. Barometric altimeter augmentation is a relatively inexpensive method of augmentation. In the FAA TSO-C129 for supplemental navigation,[9] it is required to be used in conjunction with RAIM for the purpose of increasing RAIM availability. A digital altimeter read-out is the input to the navigation solution, along with the pseudoranges. The altimeter may be looked upon as a ranging source from the center of the Earth when the altitude readout is converted into a range from the center of the Earth using the mean sea level relationship to the ellipsoid. The barometric altitude read-out can be calibrated by either of two methods.

One method is to calibrate the altimeter with the local pressure correction that is transmitted from an airport. This ties the altitude to the surface of the airport. The other method is to calibrate the altimeter with GPS-derived altitude while geometry is adequate to ensure integrity of the GPS vertical position. Present approved practice is to use the altimeter input only as a source for RAIM when RAIM is not available; e.g., only four satellites are in view, or the RAIM geometry is inadequate.[23] Reference 23 contains the equations for analyzing altimeter inputs.

Table 3 presents some results for en route availability of RAIM for horizontal position error protection levels of 1 and 2 n.mi.[23] These results assume selective availability (SA) is in place (pseudorange error standard deviation = 33 m); 21 operating satellites in the the GPS constellation; and satellites are used if they have an elevation angle above 7.5 deg (or 7.5 deg mask angle). It is seen that even with a typical set of 21 satellites up, the RAIM availability for en route

Table 3 21-satellite constellation RAIM detection availabilities over some air routes, %

Augmentation	Route	Protection level = 1 n.mi.	Protection level = 2 n.mi.
None	New York–Los Angeles	88.5	91.3
None	San Francisco–Japan	91.0	93.7
None	Dallas–Paris	92.9	95.0
Altimeter via GPS	New York–Los Angeles	93.9	94.9
Altimeter via GPS	San Francisco–Japan	94.3	95.7
Altimeter via GPS	Dallas–Paris	96.0	98.1

supplemental navigation is generally adequate for a protection level of 2 n.mi. For that protection level, the availability is generally in the neighborhood of 95%.

Table 4 presents the results for nonprecision approach extracted from Ref. 23. These results assume SA is in place (pseudorange error standard deviation = 33 m); barometric altitude correction standard deviation = 49 m; a typical set of 21 operating satellites in the GPS constellation; and 7.5 deg mask angle. It should be recognized that the absolute availability values are strongly dependent on the foregoing parameter assumptions, particularly the mask angle. However, the results provide an insight into the relative benefits of the two methods of calibrating a barometric altimeter. Referring to the fourth column of Table 4, it is seen that for nonprecision approach GPS calibration increases availability for navigation, but has negligible effect in increasing RAIM detection availability. However, local calibration does provide a significant increase in RAIM availability.

b. Loran-C augmentation. Loran-C is a low-freqency navigation aid that covers the conterminous U.S and is certified as a supplemental air navigation system. It is normally operated in the hyperbolic mode, where each line-of-position is derived as the difference between the time-of-arrival of two signals. A detailed description of Loran-C is given in Ref. 24.

It has been recognized that combining Loran-C with the GPS can provide an availability for navigation that is significantly greater than that of either system

Table 4 21-satellite constellation RAIM availabilities for nonprecision approach at some major airports, % (protection limit 0.3 n.mi.)

Augmentation	Airport	Navigation only	RAIM detection of position error	RAIM identification of a malfunctioning satellite
No augmentations	SFO[a]	100	72.2	28.8
	DFW[b]	100	68.1	25.7
	ORD[c]	100	70.1	31.3
	JFK[d]	97.0	68.8	30.2
	ATL[e]	96.5	73.3	32.3
GPS calibration of altimeter	SFO	100	72.6	42.0
	DFW	100	69.1	36.8
	ORD	100	71.5	43.8
	JFK	99.0	69.1	46.5
	ATL	99.7	74.0	46.5
Local calibration of altimeter	SFO	100	86.5	68.8
	DFW	100	83.7	66.0
	ORD	100	89.2	67.0
	JFK	100	85.1	68.4
	ATL	100	88.2	68.4

[a]San Francisco.
[b]Dallas-Forth Worth.
[c]Chicago.
[d]Kennedy-New York.
[e]Atlanta.

alone, but more importantly, it will increase the availability of RAIM. In Ref. 25, descriptions are given of how the GPS could be augmented with Loran-C signals. The essence of this concept is to treat a Loran-C signal as a pseudorange, and combine the Loran-C pseudoranges with the GPS pseudoranges. However, the minimum number of pseudoranges required depends upon whether Loran-C system time is accurately calibrated with the GPS system time. If time is not accurately calibrated, then one extra pseudorange measurement is required in the position determination solution. It is estimated that such a combined system would have access to at least 9 and an average of 11 pseudorange measurements.[15]

Reference 25 contains curves showing the unavailability of GPS/Loran-C hybrid RAIM (21-satellite constellation) as a function of a position determination geometry parameter. For the geometries of the New England/New York area, the unavailabilities are essentially constant, independent of that parameter. The results indicate the availabies for combined GPS/Loran-C fault detection and isolation increase to the orders of 99.99 and 99.9%, respectively. This is a very significant improvement over GPS-RAIM (see Table 5). Thus, this hybrid combination shows great promise for increasing RAIM availability.

c. GLONASS augmentation. GLONASS is a Russian navigation system that is planned to be a 24-satellite constellation. A description of GLONASS and its performance are given in Ref. 26, where it is stated that there seems to be no problem in bringing the spatial and time coordinates of the two systems into coincidence.

With a combined GPS/GLONASS, each containing 21 operating satellites, 99% of the users would see 10 or more satellites and almost all would see at least 8 satellites.[27] Thus, augmentation with GLONASS would provide a great increase in RAIM availability. Evaluation of a RAIM algorithm, assuming 21 operating satellites in each constellation, indicated that for a critical-system availability of 99.999%, a horizontal protection level of approximately 350 m (0.2 n.mi.) could be achieved. Thus, a combined GPS/GLONASS satellite navigation system shows great promise in providing an RNP nonprecision approach capability.

d. GPS integrity channel. Reference to column 3 in Table 4 indicates that for nonprecision approach, there is near 100% availability for navigation if RAIM were not required. With the implementation of the GIC, these levels of availability will be achievable and perhaps surpassed, as explained below.

When the the GIC was first envisioned, the health of each GPS satellite signal was determined by the ground integrity monitoring segment, and the GIC broadcast indicated the healthy satellites.[28] If a satellite were to be indicated unhealthy, it was not used in the position solution. The broadcast would be via planned mobile communication satellites. Later, however, the RTCA Special Committee SC-159[29] developed a concept where the GIC would broadcast the satellite errors that were estimated by the ground integrity monitoring segment.

The integrity message, consisting of satellite error estimates, would be broadcast as part of the navigation message of the GPS-like signals transmitted from geostationary satellites, and the signals also would provide additional sources of pseudoranges.[21] The quantization of the broadcasted errors would be coarse (e.g., preset 25-m quantization levels). The airborne receiver would then use these errors as an input to a real-time position error estimation process that modeled the satellite geometry being used. If the estimated position error remained within the horizontal position error protection level, integrity of the signals used in the GPS position determination would be assumed. If not within the protection level, an attempt would be made to find a combination of satellites yielding a solution that was within the level; otherwise, an alarm flag would be raised.

The advantage of the RTCA approach is that it would produce a much lower alarm rate because it tailored the integrity decision to the user's satellite geometry and integrity protection based on phase of flight. During this period, it was decided not to use the GIC-estimated errors as differential corrections because of the traditional reluctance of mixing postion determination with integrity monitoring.

Presently, the bold step of using the GIC error estimates as corrections is under serious consideration. This step may be justified by realizing that if effective independent ground monitoring of the corrections can be achieved, then the corrections would be guaranteed to provide a truncation of any large signal-in-space errors. In this sense, the corrections can be considered as integrity-monitored "digital navigation signals" rather than as integrity messages so that the tradition of separating position determination from integrity is not violated. The effective integrity monitoring of the corrections can be achieved by independent ground monitors that compare their estimates of the pseudorange error components (e.g., satellite clock, orbital data, and ionospheric errors) with those received from the GIC broadcast.

The bounding of signal-in-space errors bounds the resulting position determination errors. It was shown in Ref. 30 that there is a factor, HMAX, so that $\delta r <$ HMAX$_n \cdot E$, $|e_i| < E$, $i = 1, 2, \ldots, n$, where δr is the horizontal radial error; E is a known bound on the range errors; $|e_i|$ are the magnitudes of the pseudorange measurement errors; and n is the number of satellite pseudoranges used in the position solution. Assuming the errors attributable to the quantization of the coarse corrections are the major source of pseudorange errors attributable to signal-in-space errors (within $\pm$ 12.5 m for a 25-m quantization), the signal-in-space measurement errors can be assumed to be bounded by the half-quantization interval. It then follows that E represents the half-quantization interval.

Also, for a four-satellite position determination solution, HMAX$_4 <$ 2 HDOP$_4$,[30] where HDOP is horizontal dilution of precision (a measure of the amplification of position error attributable to satellite geometry), $\delta r <$ 2 HDOP$_4 \cdot E$. The utility of this bound is that most accuracy analyses of the GPS are given in terms of HDOP. For example, if HDOP$_4 <$ 4 (a large upper bound), and for 25-m quantization, the resulting position error upper bound (attributable to the signal-in-space) is 100 m. Reference 31 contains results of an analysis indicating that 0.99999 availability would be obtained for HDOP $<$ 4 if GPS is used in conjunction with geostationary satellite signal sources.

B. Precision Approach Operational Considerations and Augmentations

Differential GPS (DGPS) will make it possible to conduct precision approaches at any runway within the coverage area. As pointed out in Ref. 10, important safety benefits will result because both horizontal and vertical centerline guidance will be provided throughout the approach, requiring no alignment or transition. Moreover, if aircraft can be directed to all available runways during instrument meteorological conditions (IMC), significant improvements in capacity and terminal airspace capacity will be possible. Differential GPS may also be used for airport surface position determination and situational awareness.

Precision approach is the most demanding application of GPS to air navigation. First, the sensor accuracy requirements are of the order of 20 times more stringent as for nonprecision approach, and it involves the need for very accurate vertical guidance. It is well known that GPS vertical accuracy is usually significantly less than horizontal accuracy (e.g., the average ratio of VDOP/HDOP = 1.4^6). Second, the position update rate requirement is much higher, especially for autopilot coupled approaches. Third, the integrity and continuity requirements are the most stringent. Thus, for precision approach some type of differential augmentation is needed to achieve very accurate position and velocity determinations.

1. Accuracy

The lateral and vertical sensor accuracies for precision approach traditionally have been based on the three categories of approach: Category I, II, or III (or CAT I, etc.) The operational definitions of these categories are based on visibility or runway visual range (RVR) and decision height (DH). There is equipment at airports (called RVR system) that measures the visibility along the runway, and this measurement is reported to pilots approaching the airport. The CAT III requirements also include very stringent equipment redundancies, lateral guidance in rollout for CAT IIIb and CAT IIIc, and other requirements.[32] The DH is "a specified height at which a missed approach must be initiated if the required visual reference to continue the approach to land has not been established."[32] Table 5 contains a summary of the ILS visibility and DH requirements. However, it should be recognized that the widespread use of DGPS for precision approaches would probably lead to the definition of new landing minima criteria, such as the tunnel concept discussed below.

The microwave landing system (MLS), presently proposed in the International Civic Aviation Organization (ICAO) as the replacement system for ILS, is specified to have essentially the same sensor accuracies as CAT III ILS. Because DGPS also is now considered a serious replacement for ILS, the ILS sensor accuracy requirements are used here, where applicable, for specifying estimates

Table 5 Traditional categories of precision approach

Category	Visibility or runway visual range	Decision height
CAT I[a]	0.5 mile visibility or 2,400–1,800 ft	200 ft
CAT II[a]	1,200 ft	100 ft
CAT IIIa[b]	> 700 ft	DH < 100 ft
CAT IIIb[b]	150 < RVR < 700 ft	DH < 50 ft
CAT IIIc[b]	RVR < 150 ft	0

[a]Ref. 1.
[b]Ref. 32.

of DGPS accuracy requirements for the various categories of approach. The technique used is that of Ref. 13 where the requirements are derived from ICAO Annex 10.[8] In this technique, the sensor error components are categorized as is done in MLS.

Both ILS and MLS are systems that primarily measure the *angle* to the approaching aircraft; therefore, the errors translate into varying aircraft displacements depending upon the range to the aircraft from the ground antennas. On the other hand, GPS errors have no significant spatial variation through the approach path. This fundamental difference in the two types of systems makes it complex to translate the accuracy requirements established for ILS and MLS into similar ones that are applicable for DGPS.

In Annex 10, the MLS sensor error is broken down into three components.

1) Path following error (PFE) is "that portion of the guidance signal error which could cause aircraft displacement from the desired course line or desired glide path."

2) Path following noise (PFN) is "that portion of the guidance signal error which could cause aircraft displacement from the mean course line or mean glide path."

3) Control motion noise (CMN) is "that portion of the guidance signal error which causes control surface, wheel, and column motion and could affect aircraft attitude angle during coupled flight, but does not cause aircraft displacement from the desired course and/or glide path."

In terms of DGPS errors, PFE is interpreted as any errors that remain constant during an approach plus any slow varying errors (e.g., waypoint coordinate error + multipath at DGPS reference station). Path following noise, a component of PFE, is interpreted as an error that varies slowly during the approach (e.g., multipath at DGPS reference station), and CMN is interpreted as an error that varies fairly rapidly so such that the aircraft body may respond to it by attitude changes only (e.g., multipath at the aircraft that is reflected from the terrain). As derived from Annex 10 error tolerances, Table 6 contains estimates of the above three portions of errors as a function of approach category. In developing Table 6, PFE is derived from the RSS of ILS alignment and beam bend error tolerances; PFN is ILS beam bend error tolerance; and CMN is based on MLS (because

Table 6 Estimated precision approach 95% sensor accuracy requirements for DGPS at the decision height (9000-ft runway)

Approach category	Error direction	Path following error, ILS, m	Path following noise, ILS, m	Control motion noise, MLS, m
CAT I, DH = 200 ft	Lateral	16.0	13.3	4.3
	Vertical	4.1	2.9	1.2
CAT II, DH = 100 ft	Lateral	5.0	3.8	3.6
	Vertical	1.8	1.0	0.6
CAT III, DH = 50 ft	Lateral	4.0	3.5	3.2
	Vertical	0.6	0.5	0.3

APPLICATIONS OF GPS TO AIR TRAFFIC CONTROL

there are no CMN requirements for ILS). Also, the Annex 10 alignment error is considered a 3σ error because it is an "adjust and maintain" tolerance.

In the ICAO Annex 10 method of assessing PFE and CMN, measured sensor errors are passed through filters. One filter is a second-order filter whose output represents PFE, and the other is a high-pass filter whose output represents CMN. The filter outputs are compared to the sensor accuracy requirements. The procedure and the filter parameters can be found in Ref. 8, and the results of their application to DGPS guidance errors measured during flight test can be found in Ref. 33.

The FAA's SOIT has developed a tunnel-in-space (tunnel) concept that focuses on total system error rather than sensor error when specifying RNP.[34] The tunnel emphasizes the continuous containment of total system error from the final approach fix through rollout. The lateral and vertical dimensions of the tunnel decrease with decreasing distance from the runway threshold. Unlike the conventional approach of emphasizing sensor errors, it gives credit to the users who attain small FTE and allows a tradeoff between sensor accuracy and FTE. The tunnel concept is defined by an inner and outer tunnel, as shown in Fig. 6.[34] The inner tunnel is a surface within which the guidance reference point on the aircraft must remain with a probability of 95%. The outer tunnel is an outer containment surface where any part of the aircraft must not penetrate. The maximum penetration probability is 1×10^{-7} for vertical or lateral penetration per approach because of navigation errors. The basic total system accuracy performance is based on the inner surface because it is directly measurable through flight testing. The outer tunnel performance is mainly evaluated by analysis and simulation. The definition of the tunnel surfaces are continuous; therefore, in principle, it could allow users to attain a range of landing minimums, depending upon equipage for RNP and the local obstacles about the approach path (e.g., towers and high terrain). Table 7 contains tunnel total system requirements at various DH points along the final approach path.[34] In contrast to the sensor errors shown in Table 6, there is no specification for 50-ft DH because, at this point, it is assumed that for a CAT III approach, vertical guidance is mainly provided by a radar altimeter.

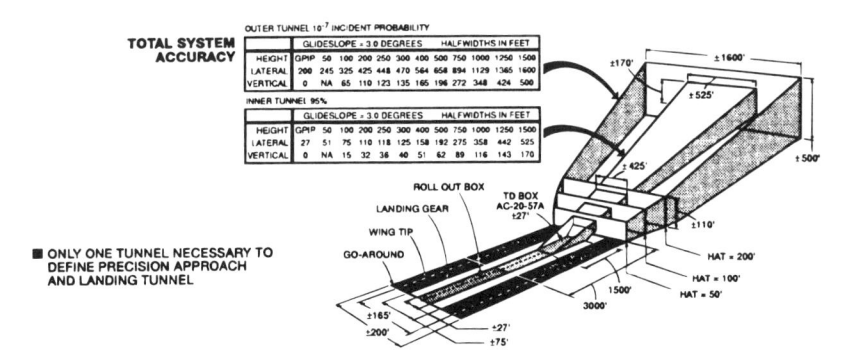

Fig. 6 Definition of inner and outer surfaces of the RNP tunnel.[34]

Table 7 Precision approach total system accuracy requirements based on the tunnel concept[34]

Tunnel surface	Decision height, ft	Vertical 1/2 width, ft	Lateral 1/2 width, ft
Inner, 95%	200	32	110
	100	15	75
	50	NA	51
	Runway	NA	27
Outer, 10^{-7}	200	110	425
	100	65	325
	50	NA	245
	Runway	NA	200

Note: aircraft dimensions must be subtracted from outer tunnel limits.

2. Integrity

Two basic alternatives are discussed for ensuring signal integrity for precision approach. One approach is similar to the monitoring of the ILS or MLS. In ILS and MLS, the transmitted signal alignment error is monitored. If the monitor alarm limits are exceeded for some specified period of time the system is shut down or there is a transfer to a standby transmitter. In a similar manner, the DGPS data transmitted to the aircraft can be independently monitored on the ground to ensure that the transmitted differential data have integrity. As noted in the discussion on nonprecision approach, monitored differential data ensures the truncation of large correction errors, and provides, in a sense, monitored navigation signals. Based on an ICAO Annex 10 standard[8] for ILS signal monitoring (shift of mean course line or glide path angle), Table 8 contains estimated precision approach monitor protection levels for each category of approach. These are only estimated protection levels for errors in differential corrections because they are based on ILS. However, they could be conservative levels given that GPS has less unobservable errors than ILS with respect to errors that cannot be monitored on the ground, such as multipath at the aircraft.

It should be noted that in comparing the monitor limit values in Table 8 to the PFE accuracy requirements in Table 6, in some cases, the monitor limits are smaller than the PFE sensor accuracy (CAT I) or comparable to them. This is explained by recalling that the PFE is composed of both alignment and PFN, and PFN is not monitored because it is a local error at the aircraft position. For a ground-based landing system, the aircraft antenna is pointed toward the ground and moving toward the source of radiation in such a way that any multipath error would be of low frequency (course bend). This results in an aircraft displacement. With DGPS, the most likely source of low-frequency multipath would be at the ground reference station antenna. Therefore, this antenna must be sited very carefully.

Another approach is where integrity monitoring is performed onboard the aircraft. In this method, the current flight technical error (FTE) is subtracted from the outer tunnel surface (Table 7). The result is compared to the current estimate of sensor error. The sensor error estimate includes estimated errors in

Table 8 Estimated precision approach monitor protection levels to ensure ILS equivalent signal error integrity at decision height, 9000-ft runway

Approach category	Protection level direction	Protection level, m	Allowable duration out of protection level, s	Probability of undetected error per landing
CAT I, DH = 200 ft	Lateral	13.5	10	1×10^{-7}[a]
	Vertical	4.8	6	1×10^{-7}[a]
	Total			2×10^{-7}
CAT II, DH = 100 ft	Lateral	8.2	5	1×10^{-7}
	Vertical	2.3	2	1×10^{-7}
	Total			2×10^{-7}
CAT III, DH = 50 ft	Lateral	6.1	2	0.5×10^{-9}
	Vertical[b]	2.3	2	0.5×10^{-9}
	Total			1×10^{-9}

[a]Not specified, but recommended.
[b]Assumes radar altimeter provides vertical guidance below 100 ft HAT; therefore, protection level is set for 100-ft HAT.

the differential data, where these errors are transmitted as part of the differential message. Because no part of the aircraft is allowed to penetrate the outer tunnel surface, relevant aircraft dimensions must be subtracted from the tunnel surfaces (e.g., wingspan and main landing gear).

In any integrity method, the time to alarm must be specified. This is the allowable duration that the the signals may be out of the protection level. Based on ICAO Annex 10 standards, Table 8 contains the durations for each category of approach. Because with GPS, the vertical and lateral position solutions are based on the same data, the more stringent vertical times to alarm should be the requirement.

Also of interest is the probability of undetected error, which includes only those errors that could place an aircraft in a hazardous position (e.g., very large positional error). Based on ICAO Annex 10 standards, the probabilities for each category of approach are listed. It should be noted that Annex 10 does not explicitly specify a probability of undetected error for CAT I and only states that the CAT II requirement is recommended. Also, as noted, the probabilities are given separately for lateral and vertical. These are combined to take GPS into account. Because the undetected error probabilities are rather small, they would be evaluated by analysis and simulation.

3. Availability and Continuity of Service

The discussion here assumes that for precision approach, the GPS is used as an RNP system. According to ICAO Annex 10, the continuity requirements are of the order of 10^{-6} for loss of the navigation signals for a period of time. Specifically for CAT II, the probability of continuity is $1 - (8 \times 10^{-6})$ in any period of 15 s, and for CAT III it is $1 - (4 \times 10^{-6})$ for any period of 30 s. There is no explicit standard for CAT I, but the CAT II value is recommended. These values for continuity are based on combining the values for localizer and glideslope.

For the case of DGPS, the ground reference, monitor, and the satellite signals are included in these requirements. Furthermore, a loss of continuity could occur because of unfavorable geometry; e.g., if the vertical dilution of precision (VDOP) falls above a required maximum value. The overall availability requirement could be as high as 0.99999.[7] Therefore, it seems that the required signal redundancy in terms of the satellite constellation size and the need for pseudolites (ground-based transmitters of GPS-like signals) will be driven by these precision approach requirements.

With the tunnel concept, it would be possible to adjust the landing minima based on the estimated accuracy of DGPS caused by variations of geometric dilution of precision (GDOP). For example, an increase in VDOP above the value required for a CAT I-minima approach could lead to a DH higher than 200 ft. Such a concept could lead to an increase in operational availability.

4. Augmentations

As stated previously, some form of DGPS is needed to attain the required performance for precision approach. The augmentations discussed here are local area DGPS, wide area DGPS (WADGPS or WGPS), and pseudolites (PL). At the time of writing, some significant flight test results are available for local area DGPS and preliminary results are available for pseudolites. Wide area is in a much earlier developmental phase; therefore, there are no available flight test results for its performance.

a. Local area differential GPS. In local-area DGPS, the reference station is located at or nearby the airport. Besides a reference station and data link, a basic local area DGPS configuration will need to include a monitor station to ensure the integrity of the transmitted differential data. Both code phase and carrier phase (real-time kinematic GPS) implementations are being considered; the latter mainly for CAT III type implementation. The FAA Technical Center in Atlantic City, New Jersey has been engaged in an extensive cooperative effort with industry to flight test various DGPS implementations for precision approach.[35] Summary accuracy results of the test campaigns are presented, and their implications are discussed.

Table 9 is a tabulation of the results of flight tests. The accuracy results are presented in terms of the magnitude of the mean + two-standard deviation ($|\mu|$ + 2σ). This parameter has been used by the FAA and industry as the summary parameter of the tests because it is assumed to provide an upper bound of the 95 percentile error magnitudes. Data in the table for the first five entries are from Ref. 35, and the data for the last entry are from Ref. 36. The sensor accuracies are based mostly on measurements of the real-time sensor outputs as compared to the position determinations of a laser tracker, the exception being the last entry, which is explained below. The accuracy results of Table 9 represent a progression of improving sensor accuracies. This improvement is mainly because of the state-of-the art of the receiver under test, whether the receiver

Table 9 Summary of precision approach flight test results

Receiver type	Vertical sensor accuracy, m	Lateral sensor accuracy, m	DGPS data update rate, Hz	Average VDOP	Number of approaches
2-Channel DGPS	11.2	5.9	0.5	2.0	33
2-Channel DGPS/ INS	8.4	4.3	0.5	2.0	33
5-Channel carrier aided, CAID	5.1	4.7	1.0	?	18
10-Channel narrow correlator, NCOR	2.1	1.2	1.0	1.7	35
Carrier tracking, dual frequency, DLFR	1.0	1.0	0.5	1.7	18
Carrier tracking,	0.36	0.15	1.0	?	6
pseudolites, PL	0.16[a]	?	1.0	5.0	16

[a]This value of vertical sensor accuracy was determined by calculating the mean and standard deviation of the sensed vertical position of the wheels from the runway while the aircraft was on the runway following the 16 approaches.[36]

used the L_2 frequency, or whether a pseudolite was used. The methodology of using both the L_1 and L_2 frequency to resolve the integer ambiguities while tracking the carrier wave is used extensively by the surveying community and is discussed in Chapter 13, this volume. The technique of using pseudolites under the approach to determine the integer ambiguities is discussed in Chapter 2, this volume.

The 2-channel sequencing receiver implementation was representative of a first-generation receiver designed for airline service in the en route phase of flight. The receiver was certainly not designed for precision approach service, and as the receiver for a precision approach research project, it was modified to accept DGPS corrections. Furthermore, this receiver type was also used as the ground reference receiver. Because it did not have the enhancing features of the other receivers tested, the combined airborne receiver and ground reference receiver lower signal-to-noise ratios was a main contributor to the relatively large observed sensor error when operating in the differential mode. Another significant contributor to errors, when in the differential mode, may have been multipath caused by the siting of the reference antenna. This antenna was mounted high on the roof of a large hangar in the presence of potential sources of reradiation, such as the roof surface and protruding objects.

It is also noted that the 2-channel system was an implementation where the DGPS or DGPS/inertial position and velocity output was the input to a flight management computer (FMC). The FMC provided guidance signals to the aircraft's autopilot for complete automatic, fixed turn radius, precision approaches.

The other implementations used techniques that significantly enhanced accuracy: carrier aiding to smooth the pseudorange measurements (CAID), narrow correlator spacing (NCOR), kinematic carrier tracking using a dual-frequency reciever to resolve the cycle ambiguities (DLFR), and the use of 2 pseudolites (PL) under the approach path to provide sufficient information for near instantaneous resolution of the cycle ambiguities. In these implementations, the DGPS vertical and lateral position outputs were converted into analog ILS look-alike signals for driving existing aircraft course deviation and glide slope indicators for hand flying. In some implementations, DGPS-derived velocity was used to interpolate between position updates to provide a higher update to the indicators.

The most important accuracy result of these tests was the establishment by the NCOR system that CAT I accuracy sensor requirements can be consistently met within a wide margin when the VDOP is reasonably low. This is noted by comparing the requirements of Table 6 with the sensor accuracy results of Table 9. To achieve the vertical sensor error requirement of 4 m from Table 6, the vertical sensor accuracy of 2 m for a VDOP of 1.7 (NCOR) in Table 9 leads to a constraint that the VDOP must be less than 3.4. Furthermore, the NCOR system seems to meet the FRP Category III lateral accuracy requirement with a very wide margin.

The tested NCOR system vertical sensor accuracy comes very close to meeting CAT II requirements. Again, this was an early prototype, and more accuracy could possibly be achieved out of this receiver type. Furthermore, as discussed in Ref. 35, if the tunnel concept is used to achieve RNP, then NCOR would also be a viable system for CAT III consideration. In fact, narrow correlator technology has been demonstrated to be sufficient to support a series of automatic landings that achieved CAT III accuracy requirements based on the tunnel.[37] The system used for the landings included a radar altimeter and an INS. In such an implementation, the lateral error accuracy of the landing system is most critical for limiting the touchdown and rollout dispersion. It is expected that any CAT III implementation would include an inertial reference system to ensure continuity and to aid the autoland performance. The determination of the feasibility of using DGPS for CAT III.[14] is a major part of the FAA's satellite navigation program. The challenge in using the NCOR technology to meet the CAT III requirements will be to establish acceptable levels of integrity and availability.

With respect to vertical errors, the tested DLFR system seems to meet the CAT II vertical sensor accuracy requirement and almost meets the CAT III vertical sensor requirement (1 m vs 0.6 m). Taking into account measurement errors, the DLFR system may have been much closer. It must be noted that the DLFR system was a very early prototype system (maybe the first) that provided real-time kinematic GPS guidance for a landing system. Because the sensor accuracy of kinematic GPS is well known to be less than 0.1 m, further testing should indicate even more accurate results. This system is attractive because one set of receivers at the airport site will service all runway ends at that airport. The challenge in using the DLFR technology to meet CAT III requirements will be to ensure sufficient integrity and availability, and to address the issue of a possible lack of reliability of the civilian use of the encrypted code associated with using the L_2 frequency.

Pseudolites also seem to offer promise for CAT III performance. The satellite-like signals emanating from the ground transmitters under the approach path provide extra ranging signals with large changes in geometry as the aircraft pass overhead. These data are sufficient to resolve the carrier cycle ambiguities reliably and rapidly; therefore, the accuracy of the method should be on the order of a few centimeters. The six flight tests of the PL system[38] that were performed in a laser test range indicate accuracies that are approximately that of the laser range itself; therefore, the accuracies shown in the table probably represent those of the laser rather than the PL system. The early system used in the trials precluded acquiring data for a larger number of flights that would have been desirable for obtaining reliable statistical data. A subsequent system has been evaluated at the Palo Alto Airport (CA);[36] however, no laser range data were acquired. In these tests, 16 consecutive landings were carried out where the sensed vertical distance from the runway at touchdown was assumed to be the vertical error. The $|\mu|$ + 2σ of these trials are also reported in Table 9. Further development of this approach is underway; however, preliminary results indicate that it will easily meet the sensor requirements for CAT III landings. An important attribute of this system is that the extra two ranging signals under the approach not only provide for improved accuracy, they also provide sufficient redundancy for a reliable onboard integrity monitor[39] through RAIM and enhance the availability. The key issue to be addressed is whether the expense of placing two pseudolites under each runway is justifiable.

b. Wide-area augmentation system (WAAS). For the purposes of minimizing the number of DGPS ground sites and to provide precision approach benefits to virtually all airports, including those not currently instrumented, the concept of wide area DGPS has been introduced.[40,41] Wide area DGPS is a major part of the FAA's satellite navigation program.[10,42] The differential data are derived from a network of relatively few ground reference stations, and the broadcast to aircraft contains separate clock bias, orbital data, and ionospheric correction data. The required number of ground reference stations is determined mainly by the spatial variations of the ionospheric delay and degree of redundant coverage against station failures. Reference 42 contains a discussion of the concepts and requirements of the FAA proposed wide area DGPS.

Based on experimentation with a simple algorithm for estimating ionospheric delay at any location, it has been estimated that a minimum of 15 uniformly distributed ground reference stations would be required to provide wide area DGPS coverage over the conterminous U.S.[43] The diurnal delay estimation errors are not stationary during the day. The 95 percentile errors for a pseudorange ionospheric correction varied between 1.5–4.0 m. Even with a VDOP of 2, the pseudorange 95% error has to be 2 m to meet the CAT I vertical sensor accuracy requirement of 4 m. Thus, at this time, the accuracy of correction of ionospheric errors indicates the potential of this method to come near to the CAT I sensor accuracy requirement for vertical guidance. It is expected that a more sophisticated algorithm would decrease the upper bound on these sensor errors. However, a requirement for low VDOPs would lead to low availability.

Another approach to the problem of ionospheric error is to estimate the delay with a two-frequency user receiver,[44] and broadcast only the orbital data and satellite clock corrections. This method would require significantly fewer ground monitors for the clock bias and orbital parameter estimations. With this method, the experimental results indicated 2 rms vertical position errors between 4 to 6 m.

From the foregoing discussion of the accuracy potential of wide area DGPS, it seems that it may not be completely possible to achieve CAT I vertical sensor accuracy within a significant margin, as was achieved with local area DGPS. If this is the case, then a possible implementation strategy would be to provide for local area DGPS at, say, 100 major airports in the U.S., and use the wide area DGPS to provide the outstanding benefits of precision approach capability (discussed in the introduction of this section) to the remaining airports.

However, the adoption of the proposed tunnel RNP total system accuracy requirements would alleviate the tight sensor accuracy requirements. For example, according to Fig. 6,[34] at the 200 ft DH, the 95% tunnel half height is 9.8 m (32 ft). Assuming a hand-flying FTE error of 6 m (2σ), the resulting sensor vertical accuracy would be $\sqrt{[(9.8)^2 - 6^2]} = 7.7$ m rather than 4 m. Thus, the proposed RNP requirements provide a wider margin for sensor accuracy. Furthermore, the tunnel RNP allows precision approaches of less accuracy, as indicated in Fig. 6. With the RNP concept, aircraft with autopilots capable of coupling to the approach path and providing lower FTE values will be able to accept larger sensor errors and will, therefore, achieve a higher availability than aircraft not so equipped. With today's technology, coupled autopilots are available to virtually all aircraft, including much of general aviation.

The FAA may implement wide area DGPS as an augmentation to the GIC.[42] This system is being referred to as the Wide Area Augmentation System (WAAS). The GIC would be implemented with, say, a network of 8–10 monitor stations. Then the network would be expanded to include more stations, including additional ones to provide redundant coverage. This concept is a natural evolution because the GIC provides a message format that lends itself to differential data and a satellite broadcast capability. More detail on wide area DGPS is in Chapter 3 of this volume.

There is a strong likelihood that the WAAS concept will be able to provide CAT I (or near CAT I) approach capability at all airports with no radionavigation equipment required at the airport site.

C. Other Navigation Operational Considerations

There are other operational considerations and roles of the GPS beyond what has been discussed previously. Several of these are discussed briefly here to illustrate the versatility and additional benefits of satellite-based navigation. The subjects considered are parallel runways, reduced obstacle clearance areas, curved approaches, remote area operations, enhanced ATC conflict prediction, and runway surface operations.

1. Parallel Runways

The use of parallel runways for simultaneous operations during instrument meteorological conditions makes it possible to increase greatly the capacity of major airports. A goal of the FAA is to decrease the separation between parallel runways for simultaneous operations to 3000 ft or less. A discussion of this problem and alternative solutions are contained in Ref. 45. Figure 7 is an illustration of the geometry. There is a no transgression zone (NTZ) that neither aircraft is allowed to enter once their 1000 ft altitude separation is lost on the final approach segment. If an aircraft does enter the NTZ, then the threatened aircraft must be broken out of its approach. If the aircraft entering the NTZ is really navigating normally and returns to the correct approach without controller intervention, then the break-out is termed a nuisance break-out.[45] The objective is for each aircraft to remain within its assigned normal operating zone (NOZ). To achieve this close runway spacing, each runway will be monitored by a controller, and specialized radar and display equipment will be used (precision runway monitor (PRM).[46]

The upper part of Fig. 7 shows the 95% accuracy requirements of CAT I and CAT III PFE (Table 6) extrapolated along the extended runway centerline out to 10 n.mi. and the measured lateral accuracy of DGPS (NCOR in Table 9). It is readily seen that a CAT I landing system sensor accuracy of several hundred feet has a good chance of causing nuisance break-outs. The situation is not nearly as bad with CAT III capability; however, the vast majority of airports and users in the United States do not have CAT III equipment. Because its accuracy is independent of distance along the extended centerline, DGPS is vastly superior to a ground-based, angular-landing system that has increasing lateral error propor-

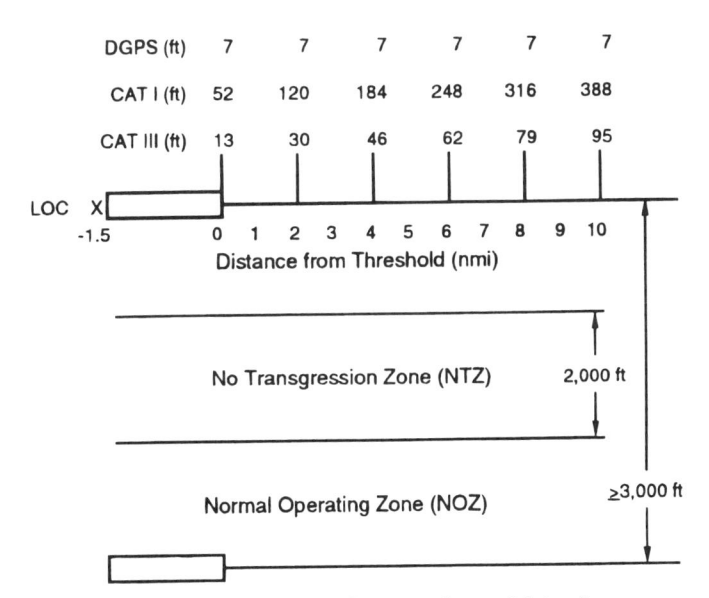

Fig. 7 Parallel approach geometry and comparison of lateral sensor accuracies (GPS vs ILS).

tional to distance from its transmitter antenna. With the high DGPS accuracy, the magnitude of the FTE would be virtually the sole determinant in keeping an aircraft in its NOZ. Parallel runway separation requirements are fairly large today because the angle-based ILS produces errors that increase with distance from the runway. By changing to a landing system based on the GPS with error characteristics that do not grow with distance from the runway, the potential to decrease the separation standards of parallel runways is achieved. If accomplished, this will increase the capacity of many existing airports whose runways do not meet the standards for parallel operation in bad weather. Therefore, the GPS has a potential to decrease the pressures to expand the land area of existing airports and the pressures to build new airports in more remote locations.

2. Reduction of Obstacle Clearance Areas

For each approach procedure, the FAA sets altitude minima based on the height of obstacles within an obstacle clearance area. For example, in a nonprecision approach, the lowest minimum descent altitude (MDA) is 250 ft above any obstacles. There are standard practices for setting up approach procedures.[47] For those receivers complying with the standards for GPS-based approaches (TSO-C129),[9] the FAA will allow equipped aircraft to fly VOR nonprecision approaches with GPS. In this case, the present procedures and obstacle clearance areas are the same as for the VOR. This will allow a rapid use of GPS for nonprecision approaches. However, the GPS has the potential to decrease obstacle clearance areas significantly on any VOR nonprecision approach independent of where the VOR is located relative to the airport. A VOR can be used for nonprecision approach when it is within 30 n.mi. of the airport.

The application of GPS to VOR-based nonprecision approaches has been analyzed in detail.[48] Figure 8, taken from Ref. 48 illustrates the difference of required primary obstacle clearance areas between GPS and VOR. The VOR is located at an airport and 30 n.mi. from an airport. In Fig. 8, it is assumed that the FTE for GPS nonprecision approaches is the same as that for VOR; thus the minimum half-width of a primary obstacle clearance area for a GPS approach need be no wider than 1 n.mi. For a VOR located at an airport, the primary areas are equivalent, and at 30 n.mi. from an airport, the VOR area is over twice as large as the GPS area. The GPS potential for reducing the size of obstacle clearance areas is important because it could lead to the reduction of MDAs and accommodation to new man-made structures as well as natural structures such as tree growth.

3. Curved Precision Approaches

One major benefits of the GPS will be its ability to provide accurate guidance for curved precision approaches. A flight path of such an approach is illustrated in Fig. 9. In this profile, an aircraft flies level at 4000 ft, and then performs a fixed radius 180-deg turn onto the runway centerline at 4 miles from threshold. This type of approach capability introduces great flexibility in the terminal maneuvering area for sequencing traffic and for aircraft noise abatement flight paths. In fact, the wide sector coverage of MLS is for the purpose of permitting curved

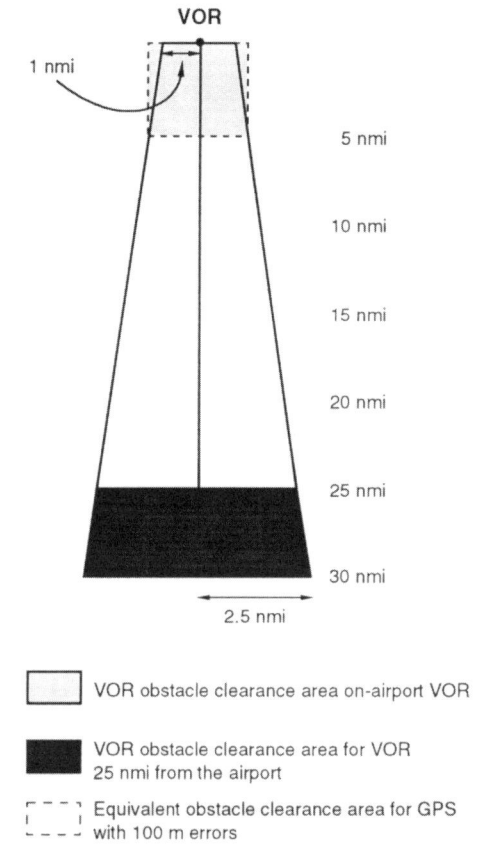

Fig. 8 Comparison of nonprecision final approach primary obstacle clearance areas.

approaches. An experimental system has been built and tested to illustrate how DGPS can provide guidance for curved approaches.

Extensive flight tests were conducted for a flight profile similar to that of Fig. 9. The results reported in Ref. 49 indicate for 32 approaches, using a DGPS/inertial hybrid for guidance, the total system lateral errors were within 40 m and the vertical errors within 15 m during the curved portion of the approach. The experimental system showed how the GPS can be integrated into an advanced flight management system (FMS) to provide a curved approach capability from anywhere in the terminal area.

4. Remote Area Operations

Because the GPS signal coverage is available worldwide, the system can be used in place of or in conjunction with traditional navigation aids for all phases of navigation. This coverage also allows the planned wide area DGPS to provide precision approach benefits to all airports covered by its network of ground

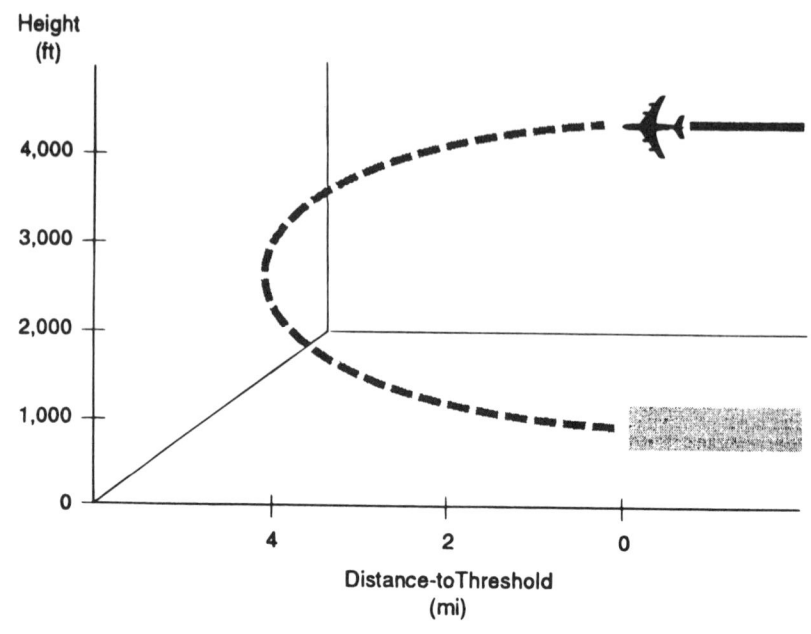

Fig. 9 Illustration of curved precision approach flight path.

reference monitors, including remote ones that do not have a precision approach capability. The benefits are in terms of increased safety and availability. As discussed previously, the safety benefit comes from the accurate three-dimensional guidance. The increased availability comes from a decrease in the landing minima. This benefit is illustrated in Fig. 10.[50] The figure contains a comparison between the unavailabilities of typical nonprecision approach minima and near CAT I minima. An approximate CAT I (near 200 ft/0.5 miles) capability provides a relatively large incremental reduction of minima as compared to the nonprecision approach in those areas where there are significant weather problems. For example, throughout the northwest corner of the United States, as illustrated by data from Spokane, WA, it is estimated that a nonprecision approach can only be made about 94.7% of the time, whereas a wide area DGPS based near CAT I approach could be made approximately 97% of the time. This is a significant increase in availability.

5. Prediction of Aircraft Conflicts

In ATC, the control system looks ahead in time to predict potential violations of aircraft separation minima (conflicts). The further away in time that a conflict can be predicted accurately allows increasingly more flexibility in changing the aircraft clearance(s) to avoid a conflict. This allows a greater probability of choosing maneuvers that have less impact on a users' preferred flight path. For example, if sufficient time is available, speed changes may be given to aircraft rather than path changes. The parameter to be predicted is the distance of closest

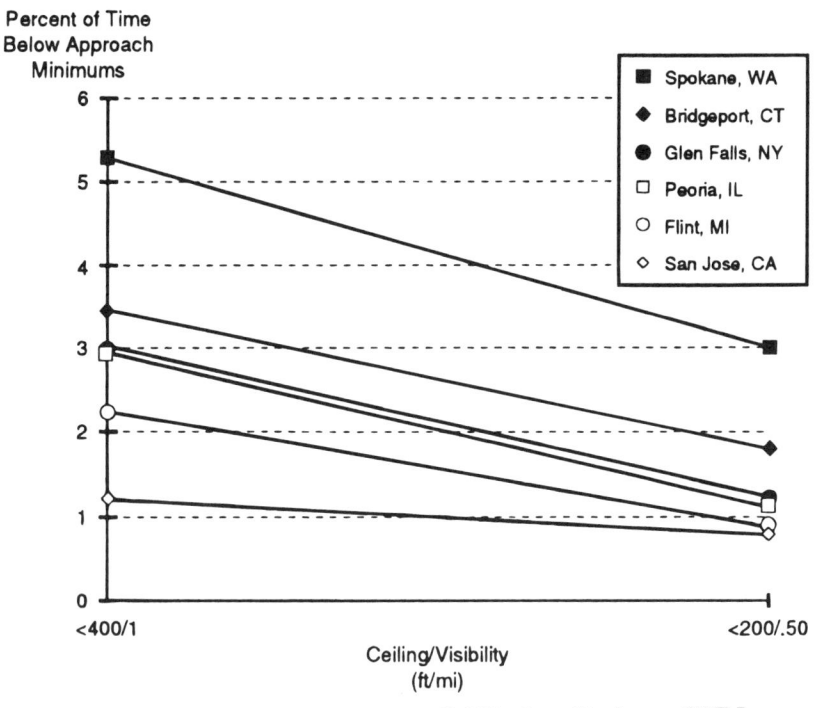

Fig. 10 Illustration of potential availability benefit of near CAT I.

approach. The accuracy of the estimation of this distance depends upon total system navigation accuracy, encounter geometry, and spatial decorrelation of wind estimations. Using a simple encounter geometry, the role of more accurate GPS navigation in the predictability of aircraft conflicts has been analyzed.[51] The encounter geometry is illustrated in Fig. 11, which shows a crossing encounter of two aircraft at the same altitude. In predicting a conflict, a threshold miss distance is compared to the estimated distance of closest approach. If the threshold miss distance is violated, then a change of aircraft clearance(s) is needed. The threshold miss distance is set to avoid the distance of closest approach becoming less than 5 n.mi., which is the present en route radar separation distance used in ATC.

The analysis of Ref. 51 assumed the 95% lateral total system error = 0.26 n.mi. (GPS sensor = 0.05 n.mi. and FTE = 0.25 n.mi.). For four-dimensional navigation (see Sec. IV.D), the along-track total system error = 1.3 n.mi. (GPS sensor = 0.05 n.mi. and 10-s control error). Clearly the FTE and control error dominate the navigation error budget, which will always be true in GPS applications. Based on the total system errors and an assumed automated conflict prediction capability, Table 10 contains the results of the analysis with respect to calculated threshold miss distances for 15-min predictions. Two cases are shown: 1) area navigation (GPS used for lateral guidance only); and 2) four-dimensional navigation (GPS used for both lateral guidance, and along-track guidance to meet

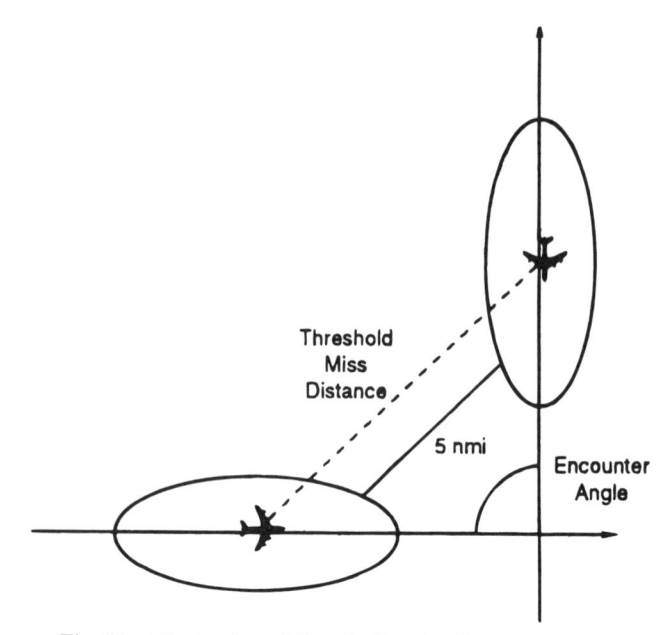

Fig. 11 Illustration of threshold miss distance concept.

a track/time profile). With larger total system errors, the thresholds would be made larger. To gauge the effectiveness of the magnitude of the calculated thresholds, they should be compared to the 7-n.mi. threshold presently used by radar controllers to ensure distance-of-closest approaches are not less than 5 n.mi., where these predictions are assumed to be made within a look-ahead time of 5 min. Clearly, with the present system, for a look-ahead time of 15 min, the threshold miss distance would have to be substantially larger than 7 n.mi. The larger the threshold, the greater the intervention rate. Thus, the four-dimensional results shown in Table 10 seem to indicate the potential of GPS accuracy in supporting a long-range conflict prediction capability that could be equivalent to the short-term predictions of today's system. It should be noted that four-dimensional control would not be the only means of assuring separation; a contingency plan may be needed to ensure separation in the event that the

Table 10 Threshold miss distances for 15-min encounter predictions

Encounter angle, deg	Intervention threshold, n.mi.	
	Area navigation	4-dimensional navigation
30	9.9	7.1
90	8.6	6.6
120	6.4	5.7

preferred maneuver does not exist. Such a significant increase in conflict prediction capability may be needed as more aircraft operators demand user-preferred routes that will result in many more crossing flight path encounters between aircraft.

6. Airport Surface Application

The navigation applications of DGPS on an airport surface are for guidance and situational awareness of both aircraft and surface vehicles during poor visibility. An airport surface application would need to include the use of a map display with aircraft position and alarms to alert the pilot of any incursions onto wrong runways or taxiways. Such a prototype system has been developed and tested with reported accuracies of 2–5 m on the airport surface.[52] Furthermore, at the time of writing, there has been an ongoing airport/airline test of the accuracy of DGPS position on the surface of Chicago's O'Hare International Airport. In this realistic environment, it was reported that 2 drms accuracies of 3 m were achieved, and the shadowing of GPS signals in gate areas was minimal.[53]

D. Area and Four-Dimensional Navigation

The current NAS consists of fixed routes that are defined by the ground-based VOR navigation aids. Although the ability to navigate without flying on outbound or inbound radials to VORs (RNAV) has been installed in a large percentage of aircraft for many years, the current FAA ATC system has not been structured to allow widespread use of this capability. Aircraft equipped with INS, Loran, or GPS are all capable of flying directly to their destination without flying over any intermediate VORs.

The current basic philosophy is to constrain aircraft to the VOR radials with RNAV flight on an exception basis.[54] To take full advantage of satellite technology, this needs to be reversed so that ATC expects RNAV flight except if an aircraft is not so equipped. Under this concept, conflicts between protected airspace around each aircraft should decrease.

The system might even go one step further. Aircraft could be given a four-dimensional (three-dimensional position and *time*) trajectory to be followed that ATC had determined was free of conflicts from any other assigned four-dimensional trajectory. If winds or aircraft airspeeds deviated making the time schedule difficult to meet, the time would be adjusted on an exception basis, and the assigned four-dimensional trajectory changed. Flight management systems in large carriers could be easily structured to provide a four-dimensional guidance.[55] In fact, the latest Boeing 737s and 747s all have four-dimensional FMS capability; the problem is that the restrictive ATC procedures allow little use of this feature. Even general aviation GPS navigators could easily be structured to provide the pilot information on what speed is required to arrive at the next waypoint at the prescribed time (or tell the pilot that it is not easily met).

To achieve the greatest benefit from satellite navigation, the vast majority of all users must switch to such a system. The current system is angular and has increasing errors as the distance to the VOR increases. The GPS is a linear system with consistent errors everywhere. This difference and the wider selection of routes are the fundamental reasons for improved airspace capacity for GPS.

The ATC system must be structured to take advantage of the improved error characteristics and routings so that we will be able to achieve the improved capacity.

V. Surveillance

Surveillance is the function that provides the current location of aircraft to air traffic controllers. This is accomplished today operationally in one of three ways: 1) pilot reporting of the aircraft's position and altitude at regular intervals via voice radio to the air traffic control centers; 2) primary surveillance radar, consisting of a rotating antenna on the ground that sends out signals that are reflected off the metal skin of the aircraft ("skin tracking"); and 3) secondary surveillance radar (SSR), consisting of a rotating antenna on the ground that sends out interrogations that trigger a transponder in the aircraft to send a reply transmission on a different frequency.

The emergence of computer and datalink technology has made it possible to automate option 1 above. This is now being evaluated in trials around the world and is referred to as automatic dependent surveillance (ADS) where *dependent* means that the surveillance *depends* on a position fix from equipment onboard the aircraft. The emergence of such satellite navigation as the GPS that provide worldwide position fix coverage makes it possible to build a relatively simple ADS device that requires input from a navigation system (GPS) and a barometric altimeter. This concept has the potential to reduce the need for radars, to increase the quality and extent of surveillance coverage including vehicles on the ground, and to reduce the cost of the total surveillance system. As a side benefit, the automatic broadcast of each aircraft's position creates the potential for any aircraft to acquire the capability to know the whereabouts of neighboring traffic at a cost that is affordable by general aviation.

In summary, surveillance based on the broadcast of navigation satellite derived position has the potential to provide increased coverage at lower overall cost. With this approach, each aircraft gains the knowledge of the whereabouts of all other aircraft; therefore, the scheme provides increased safety as well.

A. Current Surveillance Methods

There are two types of flights: those under visual flight rules (VFR) and instrument flight rules (IFR). In VFR flight, a pilot must be able to remain outside clouds at all times during the flight and is responsible for separation from all other aircraft by maintaining a constant visual scan of the surrounding sky. No contact with air traffic controllers is necessary for VFR flights, although it is recommended that VFR pilots file a flight plan, report the takeoff time to an FAA Flight Service Station (FSS), and report destination arrival to an FSS. The purpose of a VFR flight plan is to enable a timely search in the event of a crash. Air traffic control centers, to the extent possible, do carry out surveillance of VFR flights so they can warn IFR pilots of their existence. Furthermore, many VFR pilots call up ATC and request "flight following" so they can have the benefit of ATC surveillance for all the other traffic in their area. Controllers will provide this service to VFR pilots only if their workload permits.

APPLICATIONS OF GPS TO AIR TRAFFIC CONTROL

Instrument flight rules flight is required to penetrate any clouds or if the aircraft flies above 18,000 ft. Commercial air carriers using turbojet aircraft usually fly above 18,000 ft and, therefore, always fly under IFR. Aircraft of any type below 18,000 ft must fly IFR if in clouds, but it is also common for pilots to fly IFR without clouds present in order to have ATC help them maintain separation from other aircraft. An aircraft on an IFR flight plan must follow a precise path in the sky that has been filed with ATC and the controllers monitor the progress; hence, the need for good surveillance methods. When an aircraft is in clouds (called instrument meteorological conditions, or IMC), ATC is responsible for providing separation from other aircraft; when not in clouds, each aircraft is technically responsible for separation from others by visual means; however, IFR aircraft receive a lot of help from ATC because they are under surveillance by ATC and have an established voice communication link with the controllers at the radar screens. Hence, we see that ATC has an absolute requirement to know the whereabouts of all IFR aircraft and a strong desire to know the whereabouts of all VFR aircraft.

1. Pilot Reporting of Position

While flying an IFR flight plan that is not being tracked on radar, pilots are required to report their position to the ATC center at designated points on the charts roughly 160 km apart while over land and 1000 km apart while over the ocean. The report consists of the aircraft identification, position, time, altitude, estimated time of arrival (ETA), and name of the next reporting point and the name of the next succeeding reporting point. Once a pilot has been told by ATC that radar contact has been established, no further reports are made. When told that radar contact is lost, pilot reports must commence.

2. Primary Radar

A rotating, narrow-beam antenna on the ground sends out high-energy pulses at either 1300 or 2800 MHz that are reflected off the metal skin of the aircraft and returned to the rotating antenna. The time for the round trip of the pulse is measured and multiplied by the speed of light to find the distance of the target from the antenna. The azimuth of the antenna at the time of the return is also known so that a two-dimensional measure of the aircraft's position is found. The information is usually presented on a large cathode ray tube (CRT) so that the positions of all of the aircraft show up as if on a map of the area surrounding the radar. Radars typically take about 10 s (en route) or 5 s (terminal area) for a complete revolution (scan) of the antenna; therefore, each aircraft's position is updated on the CRT at that interval. No information on the altitude of the aircraft is obtained with this type of radar. Therefore, pilots are required to report altitude to ATC as they are handed off from controller to controller, even while under radar surveillance.

3. Secondary Surveillance Radar (SSR)

Secondary radar also consists of a rotating antenna. It sends out interrogations at 1030 MHz, which are lower power than primary radar because they are designed only to trigger a "transponder" in each aircraft. The transponder, after a short delay, transmits a return reply at 1090 MHz back to the rotating antenna. The return consists of digital data with certain quantities encoded in the reply. Different types of return data formats are referred to as "Modes." This system (air plus ground components) is referred to as the Air Traffic Control Radar Beacon System (ATCRBS).

a. Mode 3/A. In the case of "Mode 3/A," the reply includes a 4-digit identification code that is assigned by ATC and dialed in by the pilot. Visual flight rules aircraft without radar following transmit 1200, while other aircraft are assigned a unique code so that the controllers can see which radar return belongs to which aircraft. This allows the controllers to address a specific aircraft by name over the radio. There are special codes reserved for emergencies (7700) and lost communications (7600). The controllers have no altitude information with Mode A transponders.

b. Mode C. In the case of Mode C, the reply includes the altitude as well as the 4-digit identification code, as in Mode 3/A. The altitude is automatically determined by an encoding altimeter and sent to the transponder to be transmitted as a digital word in addition to the identification code. The resolution of the altitude reporting is 100 ft, and the accuracy must be checked every 2 years. It takes 13 bits to transmit a message and, at 1090 MHz, it is possible to have a data rate of about 1 Mb/s. Thus, it takes 20 μs for one reply. Because of the width of the interrogating beam, one scan of an aircraft will cause 15–30 replies of the transponder. The multiple replies cause unneeded congestion on the frequency and are eliminated in the Mode S system described next. When two aircraft are lined up on the same ray from the radar beacon and at similar slant ranges, their transponder returns often interfere with one another (called "garble"), and the condition will persist until the aircraft separate. In high-density areas, there are often several radar beacons within range of aircraft; therefore, transponders typically are sending replies to several beacons, and usage of the frequency increases. The reliability of the system improves with more radar beacons, however, because it is unlikely to have garble at all the radars. It is likely that the 1090 MHz channel garble would eventually become excessive as air traffic grows in the high-density environments, if a substantial portion of the fleet retain Mode C transponders.

Most aircraft are required to have a Mode C transponder. Specifically, Mode C is required to fly above 10,000 ft or within 30 miles of busy airports. Although Mode C transponders report the altitude automatically, and it shows up on the controllers radar CRT, it is still required that IFR pilots report the altitude by voice radio when being handed off from one controller to another or when changing altitudes. This altitude reporting requirement is a carryover from the days when only Mode 3/A transponders existed, and today it provides an extra layer of safety by allowing the controllers to check the Mode C transponder's altitude reporting accuracy and the pilot's attentiveness in holding the proper altitude.

c. Mode S

Mode S is a relatively new system[56,57] that has been designed to increase the capacity for data exchange between air and ground while simultaneously decreasing the channel occupancy at 1090 MHz and the resulting garble. Each aircraft is permanently assigned a 24-bit identification number that can be selectively interrogated. The key design feature of the Mode S that allows for a large increase in data exchange capacity without saturating the capacity of that frequency is that the ground uplink message at 1030 MHz influences what is transmitted by the aircraft. The ground radar asks a specific aircraft by its identification number for a reply in one of 256 56-bit message formats. As with Mode C, it will include an aircraft identification and altitude; however, the variable format makes possible much more data. Also, Mode S transponders reply only once per interrogation compared to the 15–30 replies from continuous interrogations for Mode C, thus reducing channel garble. Furthermore, Mode S transponders spontaneously broadcast a short message with their identification about once per second, whether in range of radar beacons or not. This spontaneous message is referred to as a "squitter," and is an integral part of an associated collision avoidance system.

Airliners with more than 30 seats were required to be equipped with a collision avoidance system (see next paragraph) which included Mode S by the end of 1993. Turbine-equipped aircraft with 10–30 seats are required to be equipped with a collision avoidance system (including Mode S) by 1995. For general aviation, it was originally required that all new transponder installations be Mode S after July 1992. However, this requirement was officially rescinded in November 1992 because of the lack of Mode S radar beacons that could make use of the Mode S transponders. Individual light aircraft owners felt that a requirement for them to spend extra funds for equipment that did not provide them or the overall system with any benefit was not logical and, through the political process, encouraged the FAA to rescind the requirement. As of the end of 1993, there were about 20,000 Mode S transponders, almost exclusively on large airplanes.

4. Traffic Alert and Collision Avoidance Systems (TCAS)

The TCAS systems[58] are based on the Mode S transponder technology in that each aircraft listens for other aircraft transponder broadcasts. It determines the range to another aircraft by sending out an interrogation and waiting for the reply. It determines the bearing to other aircraft by using a directional antenna for receiving the replies. The bearing measurement is not accurate enough to support use of that information for horizontal conflict resolution. Instead, all conflicts are resolved by advising the pilots to climb or descend. The accuracy of relative altitude is much greater than bearing because it is being reported in the transponder message by each aircraft's encoding altimeter. Although aircraft that carry a TCAS system must also have a Mode S transponder, the TCAS system will locate other traffic equipped with a Mode C transponder. The timetable for the installation of TCAS is identical to that shown above for the installation of Mode S; most airliners (more than 30 seats) were equipped by the end of 1993, while those with 10–30 seats must be equipped by February 1995. There is no requirement for general aviation to install TCAS or Mode S at the present time.

B. Surveillance via GPS

Satellite navigation provides a three-dimensional position that can be broadcast by each aircraft to be received by ATC on the ground and by other aircraft. For unaided GPS, the accuracy of the vertical determination is not as good as that available from the barometric altimeter. Therefore, it is usually conjectured that a system to broadcast position and altitude would utilize inputs from both the GPS and a barometric altimeter. Overland, this kind of system has the potential to replace the SSR ground station in moderate-to-low-density airspace. The antennas could be replaced by fixed omnidirectional or broad sector beam antennas located approximately at the old SSR sites for receiving the aircrafts' broadcasts. Alternatively, the concept could consist of broadcasts to satellites for relay to ATC on the ground. As a side benefit, each aircraft can also receive the transmissions of the other aircraft in the vicinity, thus providing a less expensive and more accurate system for traffic alerts and collision avoidance. It has been proposed to use both VHF frequencies and the current transponder frequencies (1030 and 1090 MHz) to transmit an aircraft's position directly to ATC on the ground.

A large attraction of replacing the rotating ground SSR antennas with non-rotating antennas having omni-directional coverage is the reduced cost. It has been projected that the life cycle costs would be about half for the latter, producing a savings of about $0.5 billion to the FAA.[59] In addition, the low cost of the receiving antennas would probably result in the installation at sites not currently covered by SSR, therefore resulting in increased coverage and safety.

In the very long term, one could speculate that it would be possible to eliminate all ground receivers. Instead, the position broadcasts might be received by low earth orbit satellites for relay to ATC on the ground. Whether such a system is capable of handling the large number of aircraft that would be in view of one satellite and/or whether the large number of satellites required would be cost effective remain to be seen.

The discussion in this section on the application of GPS to surveillance is not as mature as that for navigation since the state of the art for the surveillance application is just beginning to develop. Therefore, much of the literature from which the discussion is drawn may be speculative. Today, the navigation and surveillance functions are totally independent of one another: radar provides surveillance, and the aircraft individually navigate using various equipment, none of which relies on the ground radar. Using GPS for surveillance means that one system (GPS) would be a part of both the navigation and surveillance functions. The integrity, continuity, and availability issues discussed in Sec. III should be readdressed for this situation for a comprehensive treatment of the subject; however, this has not been done to date.

1. Swedish System

A system designed and tested in Sweden uses a 25 KHz bandwidth VHF channel and a coordinated transmission protocol.[60,61] The system transmits the aircraft identification, position, speed, heading, and altitude every 10–13 s. When more than one aircraft is active within the same geographic area, the rates of ADS transmissions are automatically coordinated and increased. The communica-

APPLICATIONS OF GPS TO AIR TRAFFIC CONTROL

tions are synchronized to GPS time (the atomic clocks in the satellites), thus providing a time-base accuracy of about 100 ns. This makes it possible for each aircraft to transmit their position during allocated time slots with reasonably small guard times between slots to account for timing errors and the signal transmission time (at the speed of light) over the coverage area.

The complete message requires about 128 bits, including parity checks, thus transmission time slots of 27 ms allows for the transmission at the 9,600 b/s rate that is easily accomplished with a 25 KHz VHF channel. This means that 37 transmissions per second are possible, and the system has a capacity of 370 aircraft if there are 10 s between position reports. This capacity applies to the region that is within range of a transmission. The aircraft capacity per square kilometer depends on the size of the area within range, which depends on the transmission power level. As the power level is reduced, the number of aircraft that can be sensed per km^2 per channel increases, and the number of ground receivers and system cost goes up. For example, if the transmission power level was set for a range of 100 km, the circular area within range is about 30,000 km^2, and the achievable density is about 0.01 aircraft per km^2. In a high-density area; e.g., the Los Angeles basin, a higher capacity density would be necessary. This could be achieved with a lower transmission power, and thus more ground receivers or more VHF channels allocated, thus increasing the transmission rate.

When within range of a ground receiver station, ATC would coordinate and assign the time slots for broadcast for each aircraft. When out of range of ground stations, the aircraft would transmit autonomously with each user transmitting in an available empty time slot determined by receiving all transmissions on the channel. The time slot being used would occasionally be changed so as not to interfere with an aircraft coming into range that had randomly selected that same time slot.

The system has been installed in several aircraft and has operated successfully in Sweden for several years. The Swedish CAA has proposed that ICAO coordinate activities to specify a format and to allocate a frequency for the implementation of such a system for all.

2. GPS–Squitter

M.I.T.'s Lincoln Laboratory has proposed an ADS system to augment and possibly replace the current SSR in the United States.[62] In addition, the system is capable of tracking the position of aircraft and vehicles on the airport surface. The basic idea is the same as the Swedish system in that each aircraft transmits its position. However, rather than use a VHF frequency, the idea here is to transmit the data utilizing the Mode S transponder that is already in many aircraft. Because these transponders already have the capability to transmit data with a digital format, a minor modification will provide the capability to transmit the GPS-derived position. Barometric altitude is already being transmitted by the Mode S tansponders as it is on Mode C transponders.

With the Lincoln Laboratory design, the transponders would be serving a dual purpose: they would be responding to interrogations from the rotating SSR antennas as they are currently designed to do, and they would spontaneously be transmitting the aircraft's position (called a "squitter") on a random schedule roughly 0.5 s apart. Because of the necessity to respond to the SSR interrogations, it is impossible to coordinate the aircraft transmissions into a systematic time schedule. A completely random schedule has been shown to have sufficient capacity to carry the required traffic in the less dense areas, but it does not seem to be capable of handling large airport surface areas or the busy air corridors.

The transponder downlink frequency (1090 MHz) is capable of a much higher data rate than the 9.6 Kb/s rate possible with VHF. Therefore, the time required for a position transmission (112 bits) is about 120 μs. It has been calculated[62] that a single, nonrotating, omnidirectional antenna can accommodate 140 aircraft with a probability of 99.5% that an aircraft will be detected in 5 s. The capacity can be increased to 350 aircraft by using six collocated antennas, each designed to receive signals from a limited sector. This analysis is conservative in that it assumes a lot of activity on the transponder frequency from non-Mode S transponders. This other activity is triggered by SSR beacons as well as military radar. This activity would decline when and if the GPS–Squitter became the standard throughout the entire fleet.

The GPS–Squitter has been successfully tested on the airport surface at Hanscom Field, a medium sized airport in Bedford, Massachusetts. Installation is complete, and testing is underway for a demonstration of the GPS–Squitter on the surface at Boston's Logan Airport in the spring of 1994.

This system extends the use of the Mode S signal formats, which are already covered by national and international standards. The ICAO panel that developed the international standards for Mode S and TCAS has accepted a task to develop the standards modifications required to add GPS–Squitter to the Mode S standards.

3. Oceanic ADS with the GPS

Trials of an ADS system have been underway since September 1990 over the Pacific Ocean.[63,64] Regularly scheduled airliners (United and Northwest Airlines) have been participating in an FAA program to gain early experience and to assess the benefits of using a geostationary satellite datalink for automatic position reporting.

The present system for aircraft tracking over the oceanic airspace consists of the aircraft manually reporting its position by voice over a high-frequency radio link to ATC on land. Reports are made every 10 deg of longitude (approximately once per hour). Based on the inertial navigators in practically all transoceanic aircraft today, pilots can maintain a track with a maximum drift error of about 0.5–1.0 n.mi./h. This results in requirements that aircraft stay nominally separated by 60 n.mi. laterally and 10 min in-trail.

In the ADS trials by satellite link, reports are made automatically at a rate that is about every 1 deg of longitude (approximately 10 per hour). The aircraft transmit their positions reports via an Inmarsat geostationary satellite, which then relays it to the oceanic ATC centers via existing land lines. The transmissions have proved to be very reliable and match closely with the voice reports. A two-way data link is also being investigated so that ATC and the aircraft can send and receive messages to one another. Unlike the ADS systems described in the previous two sections where the position reports are broadcast for all to hear, the oceanic ADS is a point-to-point transfer of data between the aircraft and ATC.

Although few commercial aircraft are GPS-equipped today, it is projected that this will change soon because it provides a significant improvement in oceanic (and other remote areas) navigation accuracy over that achievable with inertial navigation. The improved navigation accuracy, coupled with some kind of ADS system such as that being evaluated now, will result in significant improvements in the surveillance accuracy. Thus, reductions in the required spacing of aircraft will result. If using an ADS system with a reporting rate of 20 per hour and GPS to determine position, it has been estimated that the lane separation could be reduced from 60 n.mi. to 15 n.mi.[65] Even further reductions may be possible if each aircraft is aware of the position of its neighbors with some kind of traffic alert system or TCAS.[66]

These improvements would not only allow dramatic improvements in the capacity of the oceanic airspace, it would also allow for more optimal routings, thus saving time and fuel.

4. TCAS with the GPS

Another key benefit of an ADS system is the collision avoidance function that results as a byproduct. With each aircraft broadcasting its own position, that information becomes available to other aircraft who can then compute the relative position of potential threats.[67] In low-end general aviation craft, this information could be displayed on a liquid crystal display (LCD) that states something such as: "10, +500, D, L2R" which translated is: traffic, 10 o'clock, 500' above, descending, and moving left to right. Similar capability in terms of computation and display is now being sold in aviation navigators that cost about $1000. For more affluent owners, a map display would likely be the format to display threats.

Current commercial aircraft with more than 30 seats are already required to have a TCAS that operates by processing and interacting with the Mode S returns from neighboring aircraft and display the information on a CRT. The least expensive unit available is approximately $50,000, which is beyond the means of most general aviation owners. The current TCAS is fundamentally less accurate than a GPS ADS system because it determines the bearing to a threat by finding the direction at which the transponder signals arrived. Furthermore, the current TCAS determines the range to a threat by initiating a two-way transmission. Two-way transmissions for every pair of neighboring aircraft take large slices of the channel capacity, and, therefore, the technique is fundamentally limited as to the number of aircraft it can accommodate. The GPS ADS is a lower cost, more accurate system that makes much more efficient use of the 1090 MHz channel. Therefore, it allows room for growth to all aviation, providing safety at an affordable cost. The TCAS systems currently installed in aircraft can be converted to the GPS ADS format with minor modifications.[62]

As described earlier, there was a requirement for all new aircraft transponder installations to be Mode S after July 1992. The requirement was designed to alleviate congestion on the transponder frequencies (1030 and 1090 MHz) and to allow for increased capacity of the surveillance system over the coming decades. The FAA's goal to convert the entire fleet to Mode S has been, at least temporarily, thwarted by the elimination of the requirement for general aviation to equip with Mode S. However, congestion of the 1030 and 1090 MHz frequencies will now progressively become worse, and the reliability of the current SSR surveillance and TCAS systems will degrade somewhat over the next few decades without some action to fix the fundamental problem. If the ADS Mode S concept was adopted by the FAA, thus giving an affordable TCAS system to general aviation, the small aircraft owner would realize an important benefit and would be more likely to support the transition to Mode S.

VI. Summary of Key Benefits

As research, development, and implementations of GPS-based ATC systems continue, important benefits for civil aircraft operators, the FAA, and the traveling public will likely result. To ensure integrity and to enhance availability, an augmentation system is required. A viable candidate seems to be the WAAS currently being tested by the FAA. Important benefits that have been established to date are as follows.

1) A seamless area navigation (RNAV) capability at a reasonable cost; thus, improving efficiency by more optimal routing, increasing the capacity of the airspace, and decreasing the cost of avionics equipage.

2) Elimination of the need for the FAA to maintain numerous ground-based navigation aids [VORs, DMEs, NDBs (nondirectional beacons), and ILSs] with their associated high cost.

3) Instrument landings with a decision height of about 200 ft (CAT I or near CAT I) at *all* airports without the need for any radionavigation equipment at the airport site. (This is a substantial increase in the number of airports capable of supporting all-weather operation.)

4) A decrease in the required separation of aircraft in oceanic airspace, thus increasing the capacity and allowing for more optimal routing. (This will require a satellite communications link.)

Although unproved to date, research suggests that the GPS will also allow for the following.

1) An increase in the capacity of existing airports with parallel runways by a reduction in the separation standards of runways for simultaneous operation in low-visibility conditions. (This will reduce the pressure to expand existing airports and to add new airports at remote locations.)

2) Automatic landings in very-low-visibility conditions with a landing system substantially cheaper than the current CAT III ILS or the projected MLS.

3) A traffic alert system that is more accurate and less expensive than the current TCAS. (This will provide increased safety for all of aviation.)

4) The minimization of ground-based, rotating radar beacons and replacement with beacons with omnidirectional coverage. (This will be a considerable expense reduction for the FAA and will provide improved surveillance coverage of aircraft for ATC.)

References

[1]Nolan, M. S., *Fundamentals of Air Traffic Control,* Wadsworth, Belmont, CA, 1990, pp. 232, 536.

[2]McFarland, A. L., "Indications of the Need for AAS From the AAS Benefit/Cost Study," The MITRE Corporation, MP-88W23, McLean, VA Oct. 1988.

APPLICATIONS OF GPS TO AIR TRAFFIC CONTROL

[3]"Airway Facilities Strategic Plan," 1993 ed., Airway Facilities Executive Board, Federal Aviation Administration.

[4]Anon., "RTCA Task Force Report on the Global Navigation Satellite System (GNSS) Transition and Implementation Strategy," Prepared by RTCA/TF-1, Sept. 1992.

[5]Anon., "User Requirements for Future Communications, Navigation and Surveillance, Including Space Technology Applications," RTCA/DO-193, Sept. 1986.

[6]Anon., "1992 Federal Radionavigation Plan," U.S. Departments of Transportation and Defense, DOT-VNTSC-RSPA-92-2/DOD-4650.5, Washington, DC, 1992.

[7]Anon., "National Airspace System, System Requirements Specification," NAS-SR-1000, Federal Aviation Administration, March 1985.

[8]Anon., "International Standards and Recommended Practices, Aeronautical Telecommunications, Annex 10, to the Convention on International Civil Aviation," Vol. 1, International Civil Aviation Organization (ICAO), Oct. 22, 1987.

[9]Anon., "Airborne Supplemental Navigation Equipment Using the Global Positioning System (GPS)," Federal Aviation Administration, TSO C-129, Dec. 10, 1992

[10]Loh, R., Dorfler, J., and Braff, R., "The Federal Aviation Administration's (FAA) Satellite Navigation Program," Symposium on Worldwide Communications, Navigation, and Surveillance, sponsored by The Federal Aviation Administration, Transport Canada, and ARINC, Reston, VA, April 1993.

[11]Anon., "Minimum Operational Performance Standards for Airborne Supplemental Navigation Equipment Using Global Positioning System (GPS)," RTCA/DO-208, July 1991.

[12]Anon., "Report of the Tenth Air Navigation Conference," International Civil Aviation Organization Doc. 9583, AN-CONF/10, Montreal, Canada, Sept. 1991.

[13]Hogle, L., "Investigation of Potential Applications of GPS for Precision Approaches," *Navigation,* Vol. 35, No. 3, 1988.

[14]Swider, R., Loh, R., and Shively, C., "Overview of the FAA's Differential GPS CAT III Program," *Proceedings of the Symposium on Worldwide Communications, Navigation, and Surveillance,* Reston, VA, April 1993.

[15]Braff, R., Shively, C. A., and Zeltser, M. J., "Radionavigation System Integrity and Reliability," *Proceedings of the IEEE,* Vol. 71, No. 10, 1983.

[16]Anon., "Satellite Navigation Concepts for Early Implementation: The Federal Aviation Administration's (FAA) View," Associate Administrator for System Engineering and Development, Federal Aviation Administration, Washington, DC, Aug. 1992.

[17]Durand, J.-M., Michal, T., and Bouchard, J., "GPS Availability, Part I: Availability Achievable for Different Categories of Civil Users," *Navigation,* Vol. 37, No. 2, 1990.

[18]Van Dyke, K. L. "RAIM Availability for Supplemental Navigation," *Navigation,* Vol. 39, No. 4, 1992–93.

[19]W. A. Poor, Availability Estimates For GNSS, Proceedings of The Institute of Navigation Technical Meeting, San Francisco, California, Jan. 1993.

[20]Enge, P., Vicksell, F. B., Goddard, R. B., and van Graas, F., "Combining Pseudoranges from GPS and Loran-C for Air Navigation," *Navigation,* Vol. 37, No. 1, 1990.

[21]Kinal, G. V., and Singh, J. P., "An International Overlay for GPS and GLONASS," *Navigation,* Vol. 37, No. 1, Spring 1990.

[22]Denisov, V., Gorev, V., and Silantyev, Y., "The GLONASS Space Navigation System Features and GLONASS/GPS Joint Operations Considerations," Symposium on Worldwide Communications, Navigation, and Surveillance, sponsored by the Federal Aviation Administration, Transport Canada, and ARINC, Reston, VA, April 1993.

[23]Lee, Y., "RAIM Availability for GPS Augmented with Barometric Altimeter Aiding and Clock Coasting," *Navigation*, Vol. 40, No. 2, 1993.

[24]Anon., "U.S. Coast Guard, Loran-C User Handbook 1992," Commandant Publication COMDTPUB P16562.6, Nov. 1992.

[25]Enge, P., Vicksell, F. B., Goddard, R. B., and van Graas, F., "Combining Pseudoranges From GPS And Loran-C For Air Navigation," *Navigation*, Vol. 37, No. 1, 1990.

[26]Misra, P., Bayliss, E., Lafey, R., Pratt, M., and Hogaboom, R., "GLONASS Data Analysis: Interim Results," *Navigation*, Vol. 39, No. 1, 1992.

[27]Misra, P., Bayliss, E., Lafey, R., Pratt, M., and Muchnik, R., "Receiver Autonomous Integrity Monitoring (RAIM) of GPS And GLONASS," *Navigation*, Vol. 40, No. 1, 1993.

[28]Braff, R., and Bradley, J., "Global Positioning System as a Sole Means for Civil Aviation," *Proceedings, IEEE PLANS '84, Position, Location, and Navigation Symposium* (San Diego, CA), IEEE, New York, Nov. 1984.

[29]Kalafus, R. M., "GPS Integrity Channel RTCA Working Group," *Navigation*, Vol. 36, No. 1, 1989.

[30]Lee, Y., "New Concept for Independent GPS Integrity Monitoring," *Navigation*, Vol. 35, No. 2, 1988.

[31]Phlong, W. S., and Elrod, B. D., "Availability Characteristics of GPS and Augmentation Alternative," *Proceedings of the 1993 National Technical Meeting of The Institute of Navigation* (San Francisco, CA), ION, Washington, DC, Jan. 1993.

[32]Anon., "Criteria for Approval of Category III Landing Weather Minima," AC No. 120-28 C, Federal Aviation Administration, March 19, 1984.

[33]Hundley, W., Rowson, S., Courtney, G., Wullschleger, V., Velez, R., and O'Donnell, P., "Flight Evaluation of a Basic C/A-Code Differential GPS Landing System for Category I Precision Approach," *Navigation*, Vol. 40, No. 2, 1993.

[34]Davis, J., and Kelly, R., "Required Navigation Performance (RNP) for Precision Approach," *Navigation*, Vol. 41, No. 1, 1994.

[35]Till, R. D., Wullschleger, V., and Braff, R., "GPS for Precision Approaches: Flight Testing Results," *Proceedings of the Institute of Navigation Annual Meeting* (Cambridge, MA), ION, Washington, DC, June 1993.

[36]Cohen, C. E., Pervan, B. S., Cobb, H. S., Lawrence, D. G., Powell, J. D., and Parkinson, B. W., "Achieving Required Navigation Performance Using GNSS for Category III Precision Landing," Paper presented at the Differential Satellite Navigation Systems Conference, London, April 18–22, 1994.

[37]Rowson, S. E., Courtney, G. R., and Hueschen, R. M., "Performance of Category IIIB Automatic Landings Using C/A Code Tracking Differential GPS," Paper presented at ION National Technical Meeting, San Diego, CA, Jan. 24–26, 1994.

[38]Cohen, C. E., Pervan, B. S., Lawrence, D. G., Cobb, H. S., Powell, J. D., and Parkinson, B. W., "Real-Time Flight Test Evaluation of the GPS Marker Beacon Concept for Category III Kinematic GPS Precision Landing," *Proceedings of ION GPS-93* (Salt Lake City, UT), ION, Washington, DC, Sept. 22–24, 1993.

[39]Pervan, B. S., Cohen, C. E., and Parkinson, B. W., "Autonomous Integrity Monitoring for Precision Approach Using DGPS and a Ground-Based Pseudolite," *Proceedings of ION GPS-93* (Salt Lake City, UT) ION, Washington, DC, Sept. 22–24, 1993.

[40]Brown, A., "Extended Differential GPS," *Navigation*, Vol. 36, No. 3, 1989.

[41]Kee, C., Parkinson, B. W., and Axelrad, P., "Wide Area Differential GPS," *Navigation*, Vol. 38, No. 2, 1991.

⁴²Loh, R., "FAA Wide Area Integrity and Differential GPS Program," *Proceedings of the Second International Symposium on Differential Satellite Navigation Systems* (Amsterdam, The Netherlands) April 1993.

⁴³El-Arini, B. M., Klobachar, J. A., Wisser, T. C., and Doherty, P. H., "The FAA Wide Area Differential GPS (WADGPS) Static Ionospheric Experiment," *Proceedings of the 1993 National Technical Meeting of the Institute of Navigation*, ION, Washington, DC, Jan. 1993.

⁴⁴Kee, C., and Parkinson, B. W., "Algorithms and Implementation of Wide Area Differential GPS," *Proceedings of ION GPS-92, Fifth International Technical Meeting of the Satellite Division of The Institute of Navigation* (Albuquerque, NM), ION, Washington, DC, Sept. 2, 1992.

⁴⁵Toma, N. E., and Wroblewski, P., "An Evaluation of Methods to Reduce Nuisance Breakout Rates for Operations to Closely Spaced Parallel Runways," WP 92W0000135, The MITRE Corp., McLean, VA, Sept. 1992.

⁴⁶Anon., "Precision Runway Monitor Demonstration Report," Federal Aviation Administration, ARD-300, DOT/FAA/RD-91/15, Feb. 1991.

⁴⁷Anon., "United States Standard For Terminal Instrument Procedures (TERPS)," Federal Aviation Administration handbook 8260.3B, July 1976, CHG 11, May 7, 1992.

⁴⁸Loh, R., "GPS Monitor Alarm Limits for Nonprecision Approaches," *Navigation*, Vol. 36, No. 3, 1989.

⁴⁹Wullschleger, V., et al., "Curved Precision Approach Flight Tests with GPS Integrated into State-of-Art Avionic Suite," *Proceedings of ION GPS-92*, (Alburquerque, NM), ION, Washington, DC, Sept. 16–18, 1992.

⁵⁰Braff, R., and Loh, R., "Analysis Of Stand-Alone Differential GPS for Precision Approach," *Journal of Navigation*, Vol. 45, No. 2, May 1992.

⁵¹Rockman, M., and Braff, R., "Impact of Navigation Accuracy on the Intervention Rate in a Highly Automated Air Traffic Control System," *Proceedings, IEEE PLANS '92, Position Location and Navigation Symposium* (Las Vegas, NV), IEEE, New York, March, 1990.

⁵²Pilley, H. R., and Pilley, L. V., "Collision Prediction and Avoidance Using Enhanced GPS," *Proceedings of ION GPS-92*, (Albuquerque, NM), ION, Washington, DC, Sept. 16–18, 1992.

⁵³Hoffelt, R., Hogg, P., Stern, R., and Langone, C., "Initial Results and Revised Objectives of the Chicago GNSS Trials," Symposium on Worldwide Communications, Navigation, and Surveillance, Federal Aviation Administration, Transport Canada, and ARINC, Reston, VA, April 1993.

⁵⁴Davis, J., "A CNS Paradigm Shift," Foxfire, Inc., Aug. 2, 1993.

⁵⁵Sorensen, J. A., "4-D Flight Planning," American Control Conference, Seattle, WA, 1986.

⁵⁶Orlando, V. A., and Drouilhet, P. R., "Discrete Address Beacon system Functional Description," Rep. FAA-RD-80-41, Lincoln Laboratory, Lexington, MA, April 1980.

⁵⁷Chapman, C., and Brady, J. J., "Mode S System Accuracy," DOT/FAA/RD-81/90, FAA Tech. Center, Atlantic City, NJ, Feb. 1982.

⁵⁸Anon., "Minimum Operational Performance Standards for Traffic Alert and Collision Avoidance System (TCAS)—Final Draft," RTCA Paper 106-83/SC K7-RO, RTCA, Washington, DC, March 1983.

⁵⁹Hodgins, D., "ATCBI Replacement Alternatives Study," FAA ASE-300 Presentation.

[60]Nilsson, J., "Time-Augmented GPS Aviation and Airport Applications in Sweden," *GPS World,* April 1992.

[61]Nilsson, J., "The GNSS Transponder—A Low Cost Effective Worldwide GNSS-Based Civil Aviation CNS/ATM, ATC, ATN Data Link and Collision Avoidance System Concept," Global Navcom '93, Washington State Convention and Trade Center, Seattle, WA, July 1, 1993.

[62]Bayliss, E. T., Boisvert, R. E., and Knittel, G. H., "Demonstration of GPS ADS of Aircraft Using Spontaneous Mode S Beacon Reports," *Proceedings of ION GPS-93* (Salt Lake City, UT), ION, Washington, DC, Sept. 22–24, 1993.

[63]Massoglia, P. L., and Till, R. D., "Automatic Dependent Surveillance (ADS) Pacific Engineering Trials (PET)," *Proceedings, IEEE PLANS '92, Position, Location and Navigation Symposium, 500 Years after Columbus—Navigation Challenges of Tomorrow* (Monterey, CA), IEEE, New York, March 23–27, 1992, pp. 167–172.

[64]Massoglia, P. L., Pozesky, M. T., and Germana, G. T., "The Use of Satellite Technology for Oceanic Air Traffic Control," *Proceedings of the IEEE,* Vol. 77, No. 11, 1989.

[65]Rome, H. J., and Krishnan, V., "Causal Probabilistic Model for Evaluating Future Transoceanic Airlane Separations," *IEEE Transactions on Aerospace and Electronic Systems,* Vol. 26, No. 5, 1990.

[66]Deckert, J. A., "Integrating TCAS into the Airspace Management System," IEEE PLANS '92, Position, Location and, Navigation Symposium, *500 Years after Columbus—Navigation Challenges of Tomorrow* (Monterey, CA), IEEE, New York, March 23–27 1992, pp. 167–172.

[67]Livack, G. S., Miller, J. A., and Powell, J. D., "Collision Avoidance: Can GPS Reduce the Risk of Mid-Airs?" *The Southern Aviator,* Jan. 1994.

Chapter 13

GPS Applications in General Aviation

Ralph Eschenbach*
Trimble Navigation, Sunnyvale, California 94088

THE Global Positioning System (GPS) will have a profound impact on general aviation. Many observers believe that by the end of the decade, GPS will play an important role in the three principal aspects of flying required to allow a flight between two airports without outside intervention or assistance: navigation, collision avoidance, and landing. We look at these three in more detail; however, first we look at the market served and current solutions to these problems.

I. Market Demographics

The term *general aviation* (GA) usually is applied to all noncommercial aircraft applications. This includes all private, corporate, and business aircraft, but excludes commercial airline aircraft.

A. Airplanes

The United States dominates the general aviation airplane market. More than 75% of the GA fleet is located in the United States. As shown in Table 1, there are about 265,000 registered aircraft in the United States, of which 212,000 are active.

About 60% of these aircraft are primarily for personal use, and 40% are for business use. In addition to these GA aircraft, there are about 6000 air carrier aircraft. In all, GA accounts for about 67% of all hours flown, 25% of all passengers flown, and 45% of all miles flown. General aviation does this while burning only about 7% of the fuel consumed.[1] It plays a big role in transportation and interstate commerce, and GPS, by enhancing safety and reducing costs, will make this role even more important.

B. Pilots

In 1990, there were about 702,000 active pilots. They are broken into the categories shown in Table 2.[2]

Copyright © 1994 by the American Institute of Aeronautics and Astronautics, Inc. All rights reserved.
*Vice President, Navigation.

Table 1 Number of active general aviation aircraft in 1990 by type and primary use

Aircraft type	Active GA aircraft	Corporate	Business	Personal	Instructional	Aerial Application	Aerial observation	Other work	Commuter air carrier	Air taxi	Other	Inactive
Total all aircraft	212,229	10,906	35,496	120,636	19,889	6687	5302	1525	1242	6188	4,358	54,115
Piston, total	187,773	3933	33,863	113,429	18,603	5402	4,011	1041	643	3853	2,995	46,285
One-engine	165,073	1412	25,615	106,868	17,686	5152	3779	951	303	928	2,380	42,311
Two-engine	22,606	2521	8248	6559	915	234	228	90	284	2925	603	3,886
Other piston	94	0	0	3	3	16	4	0	56	0	12	88
Turboprop, total	5652	2861	847	262	38	220	23	16	466	640	280	759
Two-engine	5257	2856	834	224	38	44	22	13	439	547	240	655
Other turboprop	395	5	12	38	0	176	1	3	28	93	39	104
Turbojet, total	4374	3204	340	115	4	0	17	0	0	374	321	517
Two-engine	3950	2938	329	113	1	0	17	0	0	343	209	355
Other turbojet	425	266	11	3	3	0	0	0	0	31	112	161
Rotocraft, total	7397	863	393	1369	877	1,065	995	224	126	1,132	355	3025
Piston	3459	45	133	1174	798	723	412	65	2	0	108	2343
Turbine	3938	818	260	195	79	342	583	159	124	1,132	247	682
Other, total	7032	45	55	5459	367	0	256	245	7	190	408	3530

Source: FAA.
Note: Row and column summation may differ from printed totals because of estimation procedures, or because some active aircraft did not report use.

Table 2 Estimated active pilots

Students	128,663
Private	299,111
Commercial	149,666
Airline transport	107,732
Miscellaneous	17,487
Total	702,659
Flight instructor	63,775

Despite the general decline in the GA market, which has seen new aircraft sales drop from about 17,000 in 1978 to about 1200 in 1990,[3] the number of pilots and the number of hours flown has remained constant. During this time, the general proficiency of pilots has improved dramatically. Instrument ratings (which allow a pilot to fly in clouds) have increased as a percentage of active pilots from 38% in 1976 to 52% in 1990.[2] Here again the capabilities of GPS will greatly accelerate this trend.

During the last 20 years, GA safety has improved also. The accident rate has declined by 60%, and the fatal accident rate has been more than halved. This reflects both pilot proficiency and aircraft instrumentation. The GPS will greatly enhance the pilot's situation awareness, and this will improve the safety record even more.

C. Airports

The area of airports and their usage is where the GPS will have its most dramatic impact. Currently, about 17,500 airports are in use in the United States, and about 5200 of them are in public use.[2] Of these, only about 1100 (or 6%) have Instrument Landing Systems (ILS).[2] GPS will allow an all-weather approach to be made at virtually all of the airports. Thus, GPS may expand the airport landing capacity by an order of magnitude.

II. Existing Navigation and Landing Aids (Non-GPS)

People have been flying airplanes for many years before GPS arrived. How did they navigate? The history of navigation is a long one, and here we look only at the era of flight and, in particular, at radio navigation systems.

A. Nondirectional Beacons (NDB)

The first radio navigation systems were nondirectional beacons. For the most part, they are in the 200–400 kHz band, though AM broadcast stations from 550 to 1600 kHz can also be used as NDB. By using a pair of directional antennas, a receiver can be designed to display the relative bearing between the aircraft heading and the station. Such a receiver is called an automatic direction finder (ADF). These are still found in most cockpits today, but they are seldom used by the modern navigator. Automatic direction finders frequently are used today in less developed parts of the world.

By keeping the relative bearing zero degrees, a pilot can use the ADF to fly to the station. Note that in the presence of winds, this does not result in a direct path to the station, but it will get the pilot there nonetheless.

The major problem with this system is that the pilot cannot determine position unless the pilot crosses directly over the station. With the use of a compass, however, the pilot can determine the radial from an NDB, as shown in Fig. 1.

Because an NDB does not take winds into account, it is very difficult to use an NDB to fly a given radial to or from a station. This problem is solved by the advent of the vhf omnidirectional radio (VOR) system.

B. Very High Frequency Omnidirectional Radio

The VOR is the backbone of the current air traffic control (ATC) system. At present, there are some 1000 VORs in service. The VOR system allows a pilot to determine the radial from a station *independent* of the aircraft heading. Each VOR station transmits in the band from 108 to 118 MHz. A 9.96-kHz tone is FM modulated ± 480 Hz by a CW signal. The antenna is rotated to create an amplitude-modulated signal in the receiver as the antenna sweeps by. The FM is synchronized to the rotation so that there is zero phase difference between the FM and AM, when the antenna points due north. The phase angle between the AM and FM signals is the radial from the station.

Although this system does not yet determine aircraft position directly, two VORs can be used to get two radials, and the aircraft position is estimated to be at the intersection of these two lines. An alternate rule of thumb used by many pilots to determine the range from a single VOR is to fly perpendicular to the VOR and measure the time in minutes it takes to change the radial by one deg. The distance from the VOR in nautical miles (n.mi.) is equal to the pilot's speed in knots (kts) times the time in minutes to change 1 deg. Thus, if it takes 30 s to move 1 deg at 140 kts, then the pilot is 70 n.mi. from the station. A more

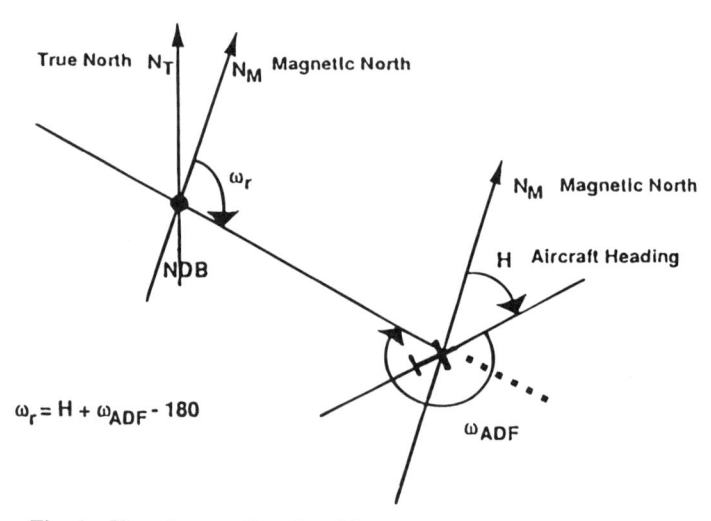

$$\omega_r = H + \omega_{ADF} - 180$$

Fig. 1 Use of a nondirectional beacon to determine a radial.

accurate and quicker method to determine range from a station came about with the development of distance-measuring equipment (DME).

C. Distance-Measuring Equipment

Distance-measuring equipment is based on measuring the propagation time of a signal from the airplane to the DME station, and back. DME is collocated with VORs, thus enabling the pilot to get both range and radial from the same point to quickly determine position. DME operates in the band from 960–1215 MHz. DME is linked to VORs so that the VOR frequency is used to identify the DME for that station.

An interesting problem with DME is that of determining which reply from a given station is the reply triggered by the transmission from the pilot's aircraft. Remember that all airplanes using the same DME are transmitting on the same frequency. Each radio transmits a sequence of pulses at a random interval from 5/s to 150/s. The receiver then gates a sliding window after each pulse. When replies are received *repeatedly* in the same window, those replies must be in response to that radio's interrogations. Thus, the time shift of the window is a measure of the distance to the station.[4]

The VOR/DME system is the basis for most overland aircraft navigation used in the world today. It is also used as the core of nonprecision approaches used throughout the world. One problem with VOR/DME is that it is a "line-of-sight" system; thus, the maximum useful range is about 70–100 miles. If a mountain range comes between the aircraft and the station, the system will not work, and the range could decrease substantially. The long-range radio navigation (LORAN) system attempts to solve that problem.

D. Long-Range Radio Navigation

The LORAN system was originally developed as a marine system, and it is still used primarily on the coastlines of the Northern Hemisphere. A LORAN chain is made up of three to six transmitters linked together as a master and multiple secondaries. They all transmit a series of pulses with a common repetition rate called a group repetition interval (GRI). By measuring the time difference (TD) between the arrival of the master pulses, and a given secondary's pulses, a line of position (LOP) can be formed, which results in a hyperbola with the two transmitters as foci. LOPs from two master–secondary pairs allows a point position to be determined, as shown in Fig. 2. Ambiguous solutions can arise, as seen in Fig. 2. When this occurs, an additional LOP can be used to resolve the ambiguity. If an additional LOP is not available, some other information must be used for resolution. In the worst case, the equipment notifies the user of an ambiguous solution, and the user is asked to determine the correct solution.

There are currently 17 different chains (called GRIs) located around the world transmitting on 100 kHz. The GRI designates the group repetition interval of the pulses for that particular chain. The low frequency gives the system a range of about 1000 n.mi. For many years, only U.S. coastlines were covered, but recent expansion of the system closed the "midcontinent gap," and we now get good coverage in the contiguous states. Depending upon the geometry of the transmitters, LORAN has an absolute accuracy of about 0.25 n.mi. and a repeatable

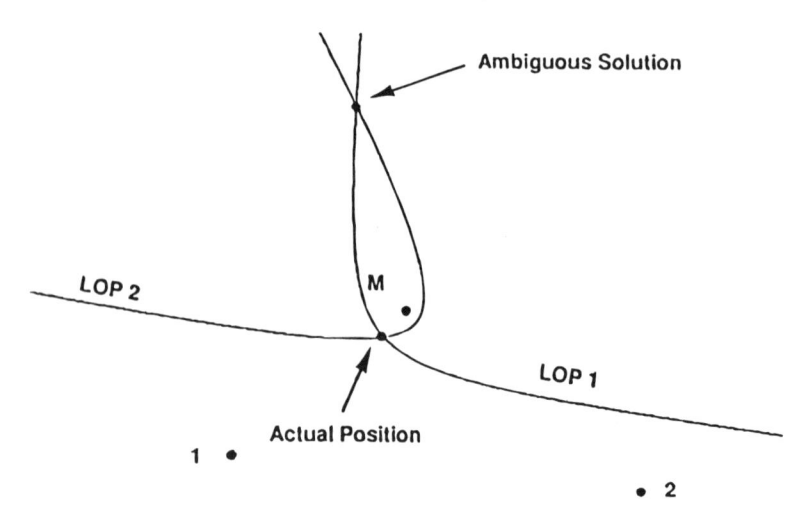

Fig. 2 LORAN is a hyperbolic system.

accuracy between 18 and 90 m.[5] Currently there are about 600,000 users worldwide.

One difficulty of the LORAN system is that it is susceptible to low-frequency noise. There are many man-made sources, including power lines, high-power Navy communication transmitters, street cars, and more, but the most detrimental to the GA usage is that caused by electrical storms. In extreme cases, this can cause a complete inability of the receiver to track the signal. Another limitation of LORAN is geographic. There is no LORAN coverage more that 1000 n.mi. from any coast over water, and there is no coverage in the southern hemisphere at all. The coverage problem is solved by Omega.

E. Omega

Omega is also a hyperbolic system, but provides worldwide coverage by transmitting at an even lower frequency than LORAN (on four frequencies from 10.2 to 13.4 kHz). It is currently the only radionavigation system certified for extended over water flights. Eight stations give worldwide coverage. Accuracy of the Omega system is usually assumed to be about 4 n.mi. It is currently estimated that there are about 27,000 users.[6]

F. Approaches

Approaches are categorized as either precision or nonprecision. A nonprecision approach is one that does not provide glide path guidance (see FAA Document 7110.65G). Precision approaches are further categorized as Category I, II, or III. Table 3 shows the accuracies required for each.

Whereas VORs are usually named with a three letter identifier (e.g., SJC, SFO), approaches are usually named for the type of navigational aid used.

Table 3 Accuracy requirements for different types of approaches

Precision approach type	Accuracy, m
Category I	
Horizontal	16.5
Vertical	3.4
Category II	
Horizontal	6.5
Vertical	1.6
Categoty III	
Horizontal	4.1
Vertical	0.5

Note: All values are 95% limits.

1. Nonprecision

There are four different nonprecision approaches—NDB, VOR, VOR/DME, and localizer back course. In all cases, the equipment required for the approach is included in the name. Thus, to do a VOR/DME approach, both a VOR and a DME must be in the airplane and in working condition. In a nonprecision approach, there is no vertical guidance from the navigation aid. Instead, the pilot uses the altimeter and descends in steps to specific minimum altitudes. In the example shown in Fig. 3, the pilot would maintain 2900 ft until crossing the Los Angeles International Airport (LAX) VOR, then he or she must descend to 1200 ft and remain at that altitude until LASKE intersection. After LASKE intersection, the pilot must descend to 980 ft until reaching the missed approach point (MAP).

Notice that one VOR (LAX) is used for guidance, and a second, SLI, is used as a cross-radial for position determination. In this example, it is used for both the final approach fix (FAF), and missed approach hold point (MAHP). Also, in a nonprecision approach, the MAP is frequently determined by time and velocity. This is seen in the table at the lower right of the approach plate. In a nonprecision approach, after the FAF (here LASKE INT), the pilot descends to the minimum descent altitude (MDA) (here 980 ft MSL). This altitude is held until the time has elapsed from the FAF to the MAP (here 1 min 12 s for a speed of 120 kts). If the field is not in sight at that time, a missed approach is executed. The sole altitude reference for this approach is a baro-altimeter. The GPS altitude cannot be used for the approach.

2. Precision Approaches

Precision approaches, and ILS in particular, are the standard civil landing system used in the United States and abroad. Whereas nonprecision approaches use the altimeter for vertical guidance, precision approaches use a radio signal, and the altimeter is used only to determine the decision height (DH). This is the altitude at which a decision must be made whether to execute a missed approach or not. A typical ILS approach is shown in Fig. 4.

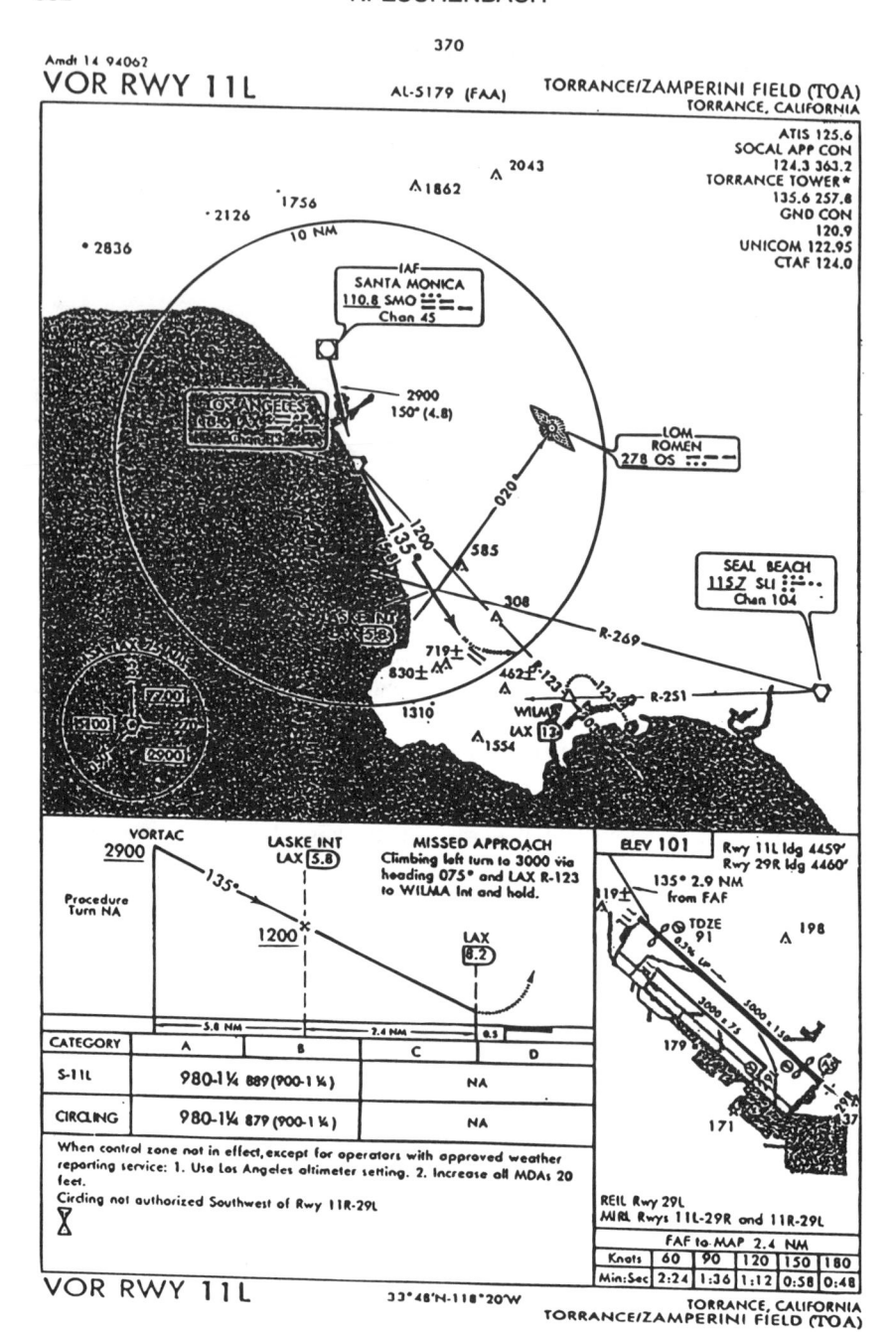

Fig. 3 A VHF omnidirectional radio system approach in Torrance, California.

GPS APPLICATIONS IN GENERAL AVIATION

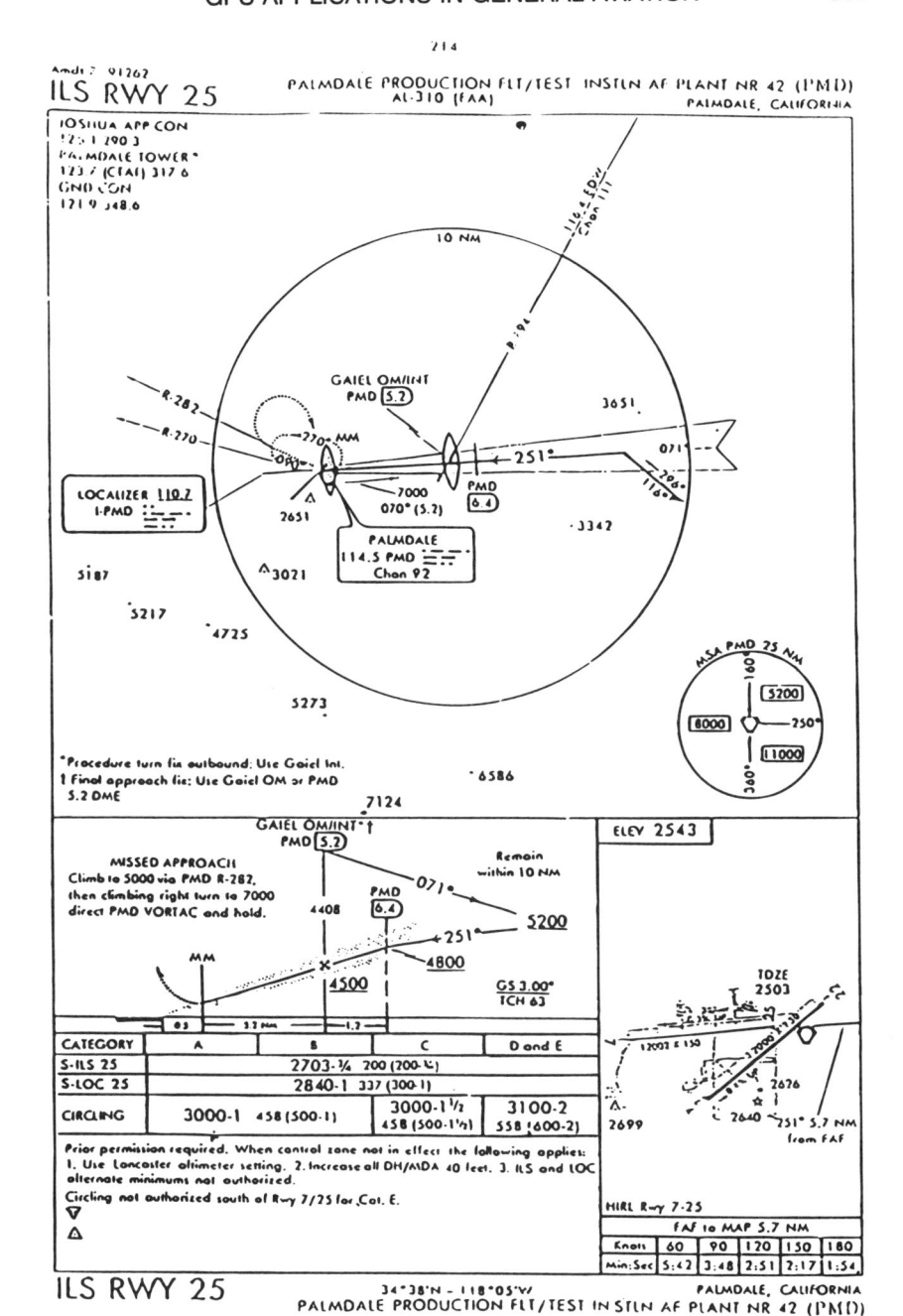

Fig. 4 Typical instrument landing system approach.

There are several things to note about this type of approach. From about 6 miles out, the vertical guidance is provided by the ILS. When the pilot reaches 2703 ft MSL (200 ft above the ground), a decision must be made. If the runway is not clearly visible, a missed approach must be executed.

The accuracy of ILS is sufficient for Category I, II, and III approaches, however it has limitations including siting, cost, frequency allocation, and performance. These are some of the reasons that there are only 1100 ILS approaches in over 5000 public airports. The siting problem occurs because the ILS requires a long, straight approach that is clear of obstructions. The microwave landing system (MLS) attempts to overcome some of these problems.

3. Microwave Landing System

The MLS is being developed by DOT, DOD, and NASA to replace the ILS. Because this system allows curved and steep approaches, it will allow closer spacing of aircraft on the same approach, as well as allowing approaches in more difficult terrain. The MLS operates by transmitting multiple signals to determine azimuth, elevation angle, and range to the end of the runway. The angles are determined by using scanning beams operating in the 5.25 GHz band. Range measurements are made in the DME band. The FAA has only recently begun to phase in the MLS. The transition will be slow. Some segments of the aeronautics industry doubt that MLS will ever be fully implemented because of the potential of satellite systems like GPS and GLONASS. The Air Transport Association has stated that its members believe that satellite systems will be able to demonstrate Category I approach capability by 1994, and Category III by 1997[5] (see also Chapter 12, this volume).

III. Requirements for GPS in General Aviation

To understand GPS in aviation, we must understand the certification environment. Specifically, we must understand Technical Standards Orders (TSO), Supplemental Type Certifications (STCs), and Form 337s. A TSO is a document put out by the FAA that outlines the specifications a piece of equipment must meet in order to comply with the FAA requirements. These include, but are not limited to, environmental, performance, user interface, and system interface requirements. An STC is a document that describes a typical installation in a particular airplane type. An avionics installer would use this document to comply with the installation requirements. If a TSO and an STC are not available for a given product, an installer may use a Form 337, but this is very difficult, because it requires local FAA approval.

The first receivers specifically designed for the general aviation market became available in 1990. These units were very similar to the LORANs available at that time in that they contained complete databases, and performed such typical area navigation functions as great circle range and bearing, ground speed, ETA, ETE, and much more. With GPS, however, the pilot got improved accuracy, quicker response to dynamics, much more accurate velocity measurements, and no geographic gaps in the coverage. These first units were installed with FAA Form 337 approvals. The STCs were first received about a year later. This allowed

GPS APPLICATIONS IN GENERAL AVIATION

more general installations, and allowed the units to be connected to other aircraft systems. The first TSOs were accomplished with the help of other approved systems in a multisensor application. GPS/LORAN and GPS/Omega TSOs were received in the spring of 1992. On December 10, 1992, a GPS TSO was issued under the number TSO-C129. The first equipment was approved under this TSO in August, 1993, in Category AII. Category AII applies to a stand-alone unit operating in the Terminal and Enroute phases of flight. The next approvals for GPS are for the "approach overlay" program. This is described as TSO-C129 Category AI and allows GPS approaches to be flown "over" existing nonprecision approaches. Although GPS is more accurate than the other systems, the minimum descent altitude will not be improved at this time.

A. Dynamics

In GA, the dynamics are relative benign. Typical accelerations are in the range of 1–3 g with acrobatic applications up to 4–6 g. This presents little challenge for most GPS receivers. Roll rates are typically around 10 deg/s, with maximum rates of 60 deg/s. In nonacrobatic applications, these rates do not continue past 60–70 deg of bank. A standard rate turn is 3 deg/s. Once again, this is not difficult for a normal GPS receiver.

B. Functionality

The primary function of a GPS receiver in a cockpit is to enhance the pilot's position awareness and to provide information for navigational guidance. For a pilot, position awareness is in terms of position relative to a known ground navigation aid (such as VORs, NDBs, or airports) A pilot will not say, "I'm at latitude 37 23.6 N and longitude 122 2.3W," but will instead say "I'm on the 268 degree radial, 4.6 miles from San Jose VOR."

In terms of track guidance, the pilot is used to using a course deviation indicator (CDI). This is an analog instrument that shows the angular error between the desired track and the current position as seen from the destination. In most avionics GPS receivers, the CDI is used to display the cross track error (XTE). The usual scale is about 1 nm per dot with a five-dot range right and left.

C. Accuracy

Avionics accuracy requirements are different for each of the different phases of flight. These are usually separated into *en route, terminal, approach,* and *landing*. Table 3 shows the accuracy requirements for landing, and Table 4a shows those for the other phases of flight.

Table 4a   Navigation accuracy requirements

Phase of flight	Accuracy
En route	2 n.mi.
Terminal	1 n.mi.
Approach	0.3 n.mi.

Table 4b GPS accuracy levels

Standard positioning service (SPS)	100 m
SPS without selective availability (SA)	25 m
Differential GPS	2–5 m
Kinematic carrier tracking	10 cm

The GPS has four different levels of accuracy. These are shown in Figure 4b. If we compare these levels with the requirements of the different phases of flight, we see that GPS standard positioning service (SPS) accuracies are sufficient for all phases of flight except landing, and for landing, we require differential GPS (DGPS). An interesting consequence of this is that selective availability (SA) is of no consequence for normal avionics applications. For en route and terminal applications, the expected level of SA has an insignificant effect, and for landings, differential GPS must be used in any case.

D. Availability, Reliability, and Integrity

For landing applications in particular, three other critical items must be addressed: availability, reliability, and integrity.

1. GPS Availability

Will the system be made available to the worldwide flight user community? The United States has stated that GPS will be made available. After KAL flight 007 was shot down over Russia, Larry Speakes, then deputy press secretary to President Reagan, stated that "the President has determined that the United States is prepared to make available to civilian aircraft the facilities of its Global Positioning System. . . ."[7] In addition, the United States has stated to the ICAO that the system will be available to users without charge for a minimum of 10 years.

Availability is not only a political question, but also a technical one. How many outages and of what duration will there be? The outages could be either planned (maintenance or improvements) or unplanned (equipment failures or lack of satellite visibility). For landing systems, for instance, a total system availability of 95% is expected. This must include the ground system, the user equipment, the datalink, and the satellite system.

2. GPS Reliability

How reliable will the system be? Most simulations show that GPS by itself would not meet the reliability requirements. Under specific scenarios of satellite failures, there would not be enough satellites in view to *guarantee* a reliable solution. Most solutions to this usually involve additional satellites. This includes GLONASS (the Russian equivalent of GPS), INMARSAT satellites, and pseudo-satellites (pseudolites). Pseudolites are ground transmitters that transmit on the GPS frequency, and the user can both range and get differential corrections from this link. Most recently, the FAA is proposing to use synchronous satellites that have ranging capability and, thus, increase both reliability and integrity (see Chapter 3, this volume).

3. GPS Integrity

How can we be certain that if an answer is displayed it is correct? The requirement here is both the ability to detect errors and the ability to report them to the pilot in a timely manner. If there were always five or more satellites available, and good geometry, this could be solved by receiver autonomous integrity monitoring (RAIM). In general, this is the ability of the receiver to determine when the solution can be trusted for the intended application. For the landing phase of flight, GPS alone cannot deliver the necessary integrity. About an additional 12 satellites would be necessary to meet the integrity requirement.

Another proposed solution to this problem is the use of a GPS integrity channel (GIC) (see Chapter 4, this volume). In this solution, ground monitors are used in combination with a communication link to detect and communicate a problem with the system to the pilot within two seconds. The assumed communication channel would be a number (1–4) of geostationary satellites that could provide ranging information, as well.

In summary, solutions exist that will allow GPS to be used as a precision landing system to open up virtually all the world's airports to all weather landing capability.[8] It is not clear which will be the eventual system of choice, but at this time the systems appear as shown in Table 5.

Once a communication link is required for the DGPS landing systems, two other important applications become available—tracking, and clearance delivery. A cooperative aircraft can transmit its position so that a base station, be that a private base station or an ATC center, can monitor the progress of the aircraft in flight without radar or voice contact. In addition, clearances, weather, or other traffic information could be transmitted to the pilot for improved situation awareness.

IV. Pilot Interface

Because of other cockpit demands, it is mandatory that the receivers be designed to facilitate the flow of information from the pilot to the navigation system and return. Both input and output have evolved substantially over the years.

A. Input

In almost all of the earlier navigation systems, the only input required by the pilot was to tune the receiver to a particular frequency. With the advent of long-range systems, LORAN, Omega, and now GPS, it became necessary to input destinations, and routes. In some of the earlier systems, this was accomplished with complete alphanumeric keypads. This was especially true in high-end sys-

Table 5 Landing type systems

Landing type	System
Nonprecision	SPS
Category I	Differential GPS
Category II, III	Real-time kinematic with pseudolite differential GPS

tems where panel space was available. In GA, however, the most common method now in use is done with two concentric knobs. One controls the cursor position, and the other scrolls the alphabet and the numerals. This method compromises panel space, and flexibility. Software has eased this transition by only allowing the letters which are possible to show and by completing the spelling with only what is available. For example, when searching for Albuquerque, only the first four letters need to be entered, and when going from "Alba" to "Albu," there are only three steps: "e," "i," and "u." The others are not possible combinations.

B. Output

In older navigation systems, the only output was the CDI. This output has been retained in newer systems, but much additional information is presented in both alphanumeric and graphic form. Information such as Desired Course, Actual Track, Range to Destination, Speed over the Ground, Time Enroute, Time of Arrival, and Cross Track Error are all available on most GPS receivers today. In addition, databases in the receiver contain information about the Airports, VORs, NDBs, and Intersections. Airport information such as City Name, Airport Name, Runway lengths, Field elevation, and Fuel availability are all available at the touch of a button. Frequencies such as ATIS, Tower, Ground, CTAF, and Unicom are also readily available.

Graphic information is also becoming available in the cockpit. The most common of these is the moving map. In the most simple form, this is a plan view of the surface of the Earth with relevant aviation data presented on it and the user is at the center of the picture. Typically, VORs and airports are shown. In some moving maps, the airways are shown, as well. Here we can also have access to the airport information listed above. Because of the three-dimensional aspect of GPS, we will soon see "tunnel in the sky" presentations to guide the pilot. This will give the pilot much more ability to guide the plane than the current dual needle system.

V. GPS Hardware and Integration

A. Installation Considerations

1. Antenna Siting

The basic tenet of GPS antenna placement is that the antenna must have a clear view of the sky. Usually the only thing that could shadow the antenna is the tail, and thus, the antenna is usually placed over the cockpit well forward of the tail structure. Shadowing by the wings and fuselage will also occur during turns, but nothing can be done about this short of complex dual antenna structures. Again, the software can help by using sophisticated algorithms for reacquisition after loss of lock. One of the better algorithms is to use a Doppler predictor based on current position and velocity. This requires at least a three-satellite solution. If a new satellite can be reaquired before an old one is lost, this assumption can be maintained.

2. VHF Communications (Comm) Interference

Because of an unfortunate choice of frequencies, the 12th and 13th harmonics of the Aircraft VHF Communication Band (118–136 MHz) lie directly in the GPS band (1575.42 ± 1 MHz). The 12 communication channels between 121.125–121.250 and 131.225–131.350 (25 MHz spacing) produce harmonics directly in the GPS band. Because they are directly in the band, there is nothing that can be done in the GPS receiver to filter them out. The only option is to filter them out before the comm antenna, and hope that there is no leakage directly out of the comm transceiver. Because the comm radios were there first, the STC for TSO-C129 requires that a test be performed to see if the particular installation causes interference with the GPS, and if it does, a filter is recommended to filter the comm radio output. If this problem cannot be solved, an IFR installation is not allowed.

B. Number of Channels

The problem above (loss of signals during turns) is simplified and minimized by having more channels. Thus, the adage that "more is better." Because there are rarely more than nine satellites in view at once, more than nine channels are of little value. With a sequencing receiver, this problem is exacerbated. Thus, for good performance in turns, a minimum of four channels is necessary, while six or more improves performance.

Another place where more channels helps is in *time to first fix* (TTFF). Because it usually takes about 5 min from power turn on until navigation information is necessary, TTFF is not an important issue in aviation.

C. Cockpit Equipment

As of this writing (1994) there are three types of GPS equipment available to the GA pilot. These are handheld, panel mount, and Dzus mount. All three types have gained rapid market acceptance as the constellation has filled. The market segments are described below, but more specific information can be found in Ref. 9.

D. Hand-held

The hand-held receiver is the lowest priced way to get GPS capability. This is also the best solution for the nonowner pilot. There are two entries into this market—the Garmin 55 AVD and the Trimble Flightmate. Both of these units have avionics databases containing the location of most airports and VORs. They are both battery powered and run for about 4 h on two AA cells. Inside an aircraft, they both work better with a remote antenna. A remote antenna can be mounted easily on the windshield for better satellite visibility. They both have a street price of about $1100. The units differ in the user interface and the display. These units have an RS-232 output, which can be connected to computers for data recording and processing.

E. Panel Mounts

This market has several entries. The principal products are made by Garmin, Trimble, King, Arnav, Narco, and IIMorrow. All have databases with complete information about airports, VORs, NDBs, and intersections. In addition, typically 100 user waypoints can be added. Output can drive moving maps, CDIs, HSIs, and more. Some units interface to air data computers and can automatically compute winds aloft, density altitude, and true airspeed. Interfaces to fuel sensors allow computation and display fuel consumption rates, fuel on board, fuel consumed, and fuel remaining at destination. Warnings are provided when fuel reserves are too low.

Although many similarities exist, there are some substantial differences. The number of GPS channels varies from Garmin with 1, to Trimble with 6, to Arnav with options from 5 to 12. The low end of this market uses LCD with most of the products using LEDs. King has a CRT display that gives added flexibility to the user interface.

Several of the high-end units interface to PCs, which allows for the ability to upload and download waypoints, flight plans, search patterns, and more. Commercial flight-planning software can now download flight plans directly into the unit or into a datacard that can be carried to the plane and loaded into the receiver.

F. Dzus Mount

The Dzus mount market, as the panel mount market, describes the installation method. Dzus refers to a 5-1/4 in.-wide package that usually mounts in the console between the two pilots on larger aircraft. This market is made up of the high-end twins, jets, and helicopters.

The primary participants in this market are Global, Trimble and Universal Navigation. This market requires FAA certification, usually in the form of a TSO. At present, GPS can only be TSOed in combination with other such systems as LORAN or Omega. Trimble is the only manufacturer that supplies certified equipment combining both GPS/LORAN and GPS/Omega. Interfacing to other aircraft systems is mandatory in this market. Roll steering and ARINC 429 interfaces are common. The digital interface protocol, ARINC 429, is used to communicate between different pieces of avionics equipment. Roll steering is an analog output that commands a roll angle to an autopilot. Roll in degrees is computed using the following formula:

$$ROLL = -(0.00281*XTE + 0.002*GS*TKE)$$

where ROLL is bank angle in deg; XTE is cross track error in ft; GS is ground speed in ft/s; and TKE is track angle error in deg.

VI. Differential GPS

As pointed out earlier, the only way approach accuracies can be achieved is through the use of differential GPS. The three components of a DGPS system are the Reference Station, the Communication Link, and the Airborne Receiver. For Special Category I approaches (SCAT I), these components are defined in RTCA Document DO-217.[10]

A. Operational Characteristics

To fly a SCAT I, an ATC clearance is required. Obviously, SCAT I compliant equipment must be in the aircraft and on the ground. A DGPS status indicator must indicate that the unit is receiving and using differential corrections. Upon activation of the approach, both vertical and horizontal guidance will be based upon angular deviation from the desired course. Once the approach is selected, detection of failures or losses of integrity will be annunciated by flags in the guidance indicators. Once a flag is present, the pilot must not continue the specified approach under instrument meteorological conditions (IMC) using DGPS. If a missed approach is initiated before crossing the threshold waypoint, guidance will be given to that waypoint, and then the pilot is expected to execute the published missed approach.

It should be noted that this initial use of DGPS for precision instrument approaches will be "supplemental" only. That is, when operating under instrument conditions, use of DGPS can only be done if other appropriate landing systems are available.[11]

B. Ground Stations

The ground station is made up of four components: a DGPS reference receiver, a data-processing function, a DGPS signal integrity monitoring function, and a data transmitter. The reference receiver must compute pseudorange corrections with an accuracy of better than 1.1 m rms. Over a 2.5-min approach, it must have a failure probability (attributable to hardware failure or integrity alarm) of less than 3.8×10^{-5}. The integrity of the system will be designed to meet the requirement that the probability that any part of an aircraft penetrates the outer tunnel without warning will be less than 1 in 10^7 approaches.[11]

The data-processing requirements will be such as to compute and format the differential correction messages in the proper manner. A user differential range error (UDRE) is also computed as a measure of the pseudorange corrections generated by the reference station. The confidence in this limit shall be at least 99.5%.

The integrity monitoring function must be completely independent from the rest of the ground equipment. It must monitor the integrity of all the data generated by the ground equipment before it gets transmitted over the data link. If the monitor itself fails, the data link must be shut down immediately. In all cases, a failure must be communicated to the pilot within 3 s.

There are currently two proposals for the data link. One uses the VHF aircraft navigation (VOR) band, and the other uses Mode S. Reference 11 contains complete implementation details of these two approaches. In either case, the frequencies will be selected automatically once an approach has been initiated by the pilot. The pilot will not be able to select the wrong frequency, as is now possible with an ILS approach.

C. Airborne Equipment Features

The primary functions of the airborne equipment are to receive the differential correction, receive the GPS signals, compute the corrected GPS position and navigation solution, and manage the navigation database.

To ensure that position accuracy requirements are satisfied, navigational information used for display must be updated at a 5-Hz rate or more. In addition, the latency must be 0.2 s or less. Accuracy and alarms will be determined by the tunnel concept. This involves both an inner and an outer tunnel. The total system accuracy must be within the outer tunnel always (less than 1×10^{-7} incident probability) and within the inner tunnel 95% of the time.[11] Integrity warnings must have a latency of no more than 3 s. This is so that the overall system can maintain a 6-s warning of any malfunction to the pilot because the ground segment can also have a 3-s latency.

The navigation database also requires some modifications from the usual avionics database. In addition to some new waypoints, the resolution must be increased to 0.0001 min for latitude and longitude and 0.1 ft in altitude. The equipment will at a minimum store the *glidepath intercept waypoint* (GPIWP), the *threshold crossing waypoint* (TCWP), and the *threshold crossing height* (TCH). The GPIWP and the TCWP are used to define the *final approach segment* (FAS), and the TCH is used to define the containment tunnels.

To get complete system integrity, pilot error must also be reduced as much as possible. Pilots will not be allowed to manipulate individual waypoints, as is now possible for en route navigation. When an approach is selected by the pilot, all the appropriate waypoints will be automatically concatenated into a "route" to be flown by the pilot. The waypoints will be sequenced automatically, and missed approach guidance will be given.

VII. Integrated Systems

Until *initial operational capability* (IOC) was declared in December 1993, the availability of GPS was at times so poor that GPS could not be relied upon for a navigation system. Primarily for this reason, GPS was integrated with other systems to provide the reliability and availability needed for commercial navigation. These systems achieved credibility through the Multisensor TSO C-115a. In general, this TSO requires that if the two independent sensors disagree, GPS must be ignored. Also, because of integrity, the two systems should be kept as independent as possible. This requirement eliminated the option of integrating pseudorange measurements from different systems into one navigation solution.

A. GPS LORAN

This combination provided TSOed capability in the domestic en route structure. This allowed GPS to be used by GA pilots about 2 years before it would otherwise have been available. The main drawback to this system was LORAN's susceptibility to electromagnetic interference. Just when navigation becomes very important, when the weather is poor, LORAN would become unavailable.

B. GPS/Omega

GPS was first combined with Omega in 1991. This allowed the use of GPS for worldwide en route navigation. By adding a GPS sensor to an Omega navigation system, a seamless path was provided to the pilot to get GPS accuracies without additional training.

Now that GPS has achieved operational status, these integrated combinations will be less important to the aviation community.

VIII. Future Implementations

We have only begun to tap the potential that GPS brings to aviation. If the last decade is any indication, the next decade will see spectacular results. Navigation, collision avoidance, and landing systems will, of course, be improved, but there will be completely new applications, as well. A very exciting area is in the use of GPS for an attitude and heading reference system (AHRS).[10]

A. Attitude and Heading Reference System

Because we can measure differential phase between two antennas from a single satellite to an accuracy of about 1 mm, by tracking four or more satellites simultaneously, we can determine the relative position of two antennas to a few millimeters. If two antennas are placed about 2 m apart, we can then determine the vector between them with an angular resolution of 1 mrad. By adding another antenna perpendicular to the first, we can then determine the complete three-dimensional attitude of a body. If we now mount these three antennas on the two wings and tail of an airplane, we can use the system as an AHRS. This will allow the complete determination of the attitude without a gyroscope. At present, the update rates are too slow to replace the gyro in auto pilot applications, but a gyro replacement will certainly come in this decade. For a more detailed treatment of this, see Chapter 19, this volume.

B. Approach Certification

By using the tracking capabilities of GPS, approach certification and landing pattern determination will be made substantially easier. By tracking an airplane during a normal visual flight rules (VFR) approach a safe landing pattern can be determined. At a later time, that exact path could be uploaded to the pilot for an instrument flight rules (IFR) approach.

By using a computer to add the flight technical error (the error induced by the man–machine combination) to the desired approach pattern, a box could be flown in VFR that would be the limits of expected errors from the desired path. Again, in VFR, we could quickly determine if the approach with the expected errors was a safe one.

We can now imagine that the equivalent of an automated terminal information system (ATIS) would upload the current approach in use so that the pilot would only need to acknowledge the approach in use. This would eliminate the cumbersome job of waypoint entry during the most stressful part of an IFR flight.

C. Collision Avoidance

Another area where GPS will be exploited is the area of collision avoidance. Passive Traffic Alert and Collision Avoidance Systems (Passive TCAS) rely heavily on knowledge of position. GPS could provide this information very easily.

An even better solution involves broadcasting an exact time-tagged position. This could be done with a Mode S transponder transmitter. A cooperating aircraft could then receive these transmissions, and compute the projected paths of both the "own" and "other" aircraft, and warn the pilot or recommend an evasive maneuver if a collision were imminent. This could reduce the need for en route traffic advisories.

We could envision a modular system, where the simplest block is a low-cost GPS combined with a uhf transmitter. This combination would transmit the exact position, velocity, time, and ID of the host aircraft about once a second. This would have to be low cost because it would have to be mandated and carried on all aircraft to be effective. These transmissions could be listened to by the ATC system, and aircraft could be presented on a screen in a manner similar to that currently provided by radars. By phasing out radars, huge maintenance costs could be eliminated. Two additional modules would enhance this system: a database and display for navigation, and a receiver and display for collision avoidance (see also Chapter 12, this volume).

The Navigation Module (NM) would use the GPS signals to drive a navigation management system (NMS) similar to that provided by the current GPS receivers. A database containing waypoints, frequencies, airport information (runways, elevations, location), airways, minimum en route altitudes (MEA), etc., would eliminate the need for charts and their cumbersome usage in the cockpit.

The collision avoidance module (CAM) would receive the transmissions from other aircraft in the area, and would compute and display the relative position and velocity and, thus, the collision threat possibility. In a similar manner, transmitter modules could be placed at or near tall obstacles. The CAM would then display their positions, as well as those of nearby aircraft.

As can be seen, GPS will play a very important role in the three principal aspects of flight: navigation, landing, and collision avoidance.

D. Autonomous Flight

With the systems described above (navigation, collision avoidance and landing) we can easily imagine what might be called autonomous flight. In much the same way a person gets in a car and safely drives to a destination without having to file a "drive plan" and being in constant communication with a "Land Traffic Control Center," we can now imagine a person being able to get in a plane and safely flying to a destination without having to file a "flight plan" and being in constant control of an "Air Traffic Control Center". The ATC role would become one more of coordination than of control and separation.

Because of the tracking capabilities of GPS, the ATC will know the position and velocity of all aircraft in flight. This will allow them to monitor congested routes and airports and recommend alternates when necessary.

IX. Summary

As has been shown, GPS has already made a profound impact on general aviation. GPS has brought precise navigation to much of the world where little or no capability existed before. Collision avoidance and landing systems will soon be using GPS. When all three are in common usage, the impact will be even greater allowing autonomous flight for many applications. Even so, we've only just begun.

References

[1] Anon., "AOPA 1992 Aviation Fact Card," AOPA, 421 Aviation Way, Frederick, MD, 21701, 1992.

[2] Anon., *General Aviation Statistical Databook,* General Aviation Manufacturers Association, 1400 K Street NW, Suite 801, Washington, DC, 1992.

[3] Anon., "1991 Avionics Retrofit Market Analysis," Aircraft Electronics Association.

[4] Kayton, M., and Fried, W. R., *Aviation Navigation System,* John Wiley, New York, 1969, pp. 181–192.

[5] Langley, R., "The Federal Radionavigation Plan," *GPS World,* March 1992, pp. 50–53.

[6] Anon., "1990 Federal Navigation Plan," Copies available from NTIS, 5285 Port Royal Road, Springfield, MA 22161 as Document DOT-VNTSC-RSPA-90-3/DOD-4650.4.

[7] Montgomery, H., "Uncommon Ground," *GPS World,* June 1992, pp. 16–19.

[8] Lechner, W., "The Potential of Global Satellite Systems for Precision Aircraft Navigation," *GPS World,* June 1992, pp. 40, 41.

[9] Connes, K., *The Loran, GPS, & NAV/COMM Guide,* Butterfield Press, 1992, pp. 98–99.

[10] Cohen, C., and Parkinson, B., "Aircraft Applications of GPS-Based Attitude Determinations," pp. 775–782; also *Proceedings of the ION GPS—92* (Albuquerque, NM), Institute of Navigation, Washington, DC, Sept. 16–19, 1992.

[11] Anon., "Minimum Aviation System Performance Standards DGNSS Instrument Approach System: Special Category I (SCAT I)," Radio Technical Commission for Aeronautics, RTCA DO-217 Washington DC, Aug. 27, 1993.

Chapter 14

Aircraft Automatic Approach and Landing Using GPS

Bradford W. Parkinson* and Michael L. O'Connor†
Stanford University, Stanford, California 94305
and
Kevin T. Fitzgibbon‡
São Jose Dos Campos, Brazil

I. Introduction
A. Autolanding Conventionally and with GPS

MOST conventional aircraft automatic landing systems use the instrument or microwave landing systems (ILS or MLS) in their terminal approach phases. These systems supply the autopilot with the aircraft's *angular* deviation from a desired flight path, which essentially corresponds to the measurement of vertical and lateral positions. Basic ILS can only satisfy the FAA's nonprecision and Category I landing requirements; and aircraft using the MLS or an improved ILS system can land with Category III required accuracy. In these autopilots, velocity is typically calculated by filtering and differentiating position, or by integrating the acceleration outputs of an Inertial Measurement Unit. Some disadvantages of these integrated systems include their high user costs and dependence on expensive ground equipment. Also, the noisy ILS and MLS signals are typically processed with a smoothing filter. This causes lags that must be compensated for by the landing autopilot.

Unlike most other navigation aids, a *GPS receiver directly measures three-dimensional velocity* with extreme accuracy (for DGPS and CDGPS,§ better than 5 cm/s). For the autopilot designer, this is of great value. A *direct* measurement of true ground speed not only assists in normal landings, it gives important advanced knowledge of wind gusts and shears. In addition to velocity, a single, state-of-the-art, GPS receiver can provide accurate three-dimensional position

---
Copyright © 1995 by the authors. Published by the American Institute of Aeronautics and Astronautics, Inc., with permission. Released to AIAA to publish in all forms.
*Professor, Department of Aeronautics and Astronautics; Director of the GPS Program.
†Research Assistant, Department of Aeronautics and Astronautics.
‡Consultant and Professor of Aeronautics.
§DGPS is differential GPS, which is covered in Chapter 1 of this volume. CDGPS is carrier-phase differential GPS.

and attitude measurements. *The power of GPS is that a single electronic box can measure three-dimensional position, velocity, and attitude for all phases of flight, including precision landing.* Traditionally, full autolanding requires an expensive inertial navigator. The results presented herein suggest that a simple system of rate gyros may be all that is required for a Category II or possibly a Category III landing. Of course, further effort is required to ensure that integrity specifications are met, which would include determining the level of redundancy and reliability required.

B. Simulations Results Presented

In this chapter, an automatic landing system is designated and simulated for a Boeing 747 using a discrete-time controller and an optimal estimator, which both rely on GPS sensors. The technique for designing the autopilot is fully described; the appendices include the parameters used. Four alternative sets of sensors are included in the simulations:
1) standard GPS;
2) standard GPS augmented with a radar altimeter;
3) code differential GPS (DGPS) without radar altimeter; and
4) carrier-phase differential GPS (CDGPS).

A block diagram of the landing system is presented in Fig. 1, where the radar altimeter and differential aids are optional.

The autopilot controller is implemented in two ways. The first is a fairly standard linear quadratic Gaussian (LQG) regulator, which estimates wind disturbances directly. The second uses an integral control law (ICL), which does not directly estimate the winds, but includes integral states that "soak up" the output errors. Optimal estimator theory, LQG regulator theory, and representative aircraft models are presented in many textbooks.[1-4] Holley and Bryson[5] presented a modified integral control design for multi-input multi-output *continuous* autolanding systems. They applied their results to an aircraft lateral mode in the presence of constant crosswinds.

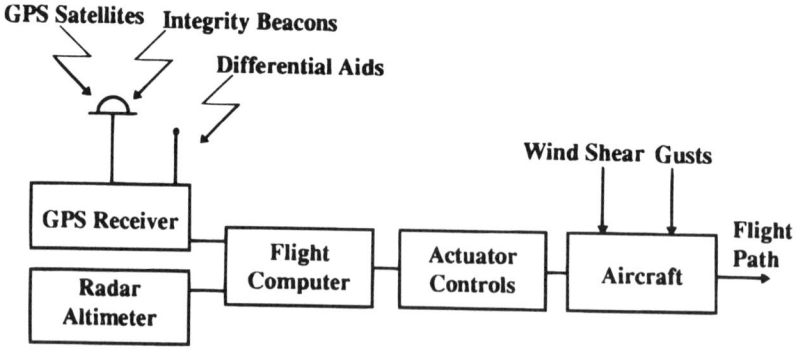

Fig. 1 The GPS autoland system block diagram.

AIRCRAFT AUTOMATIC APPROACH AND LANDING USING GPS

The results of the simulations presented here are for the *discrete* design version of this ICL, and are applied to a *complete* aircraft model with both lateral and longitudinal modes subjected to wind shear and gust disturbances. Disturbance models for the simulation and their numerical values were extracted from Holley and Bryson,[5] Roskam[6] and Bryson.[2]

The simulation results show that under normal wind conditions and typical satellite geometries, GPS and DGPS can easily meet the FAA navigation system accuracy required for a nonprecision approach. Augmenting stand-alone GPS with a radar altimeter meets the accuracy requirements for a precision Category I approach. Utilizing carrier-phase measurements with CDGPS meets the required navigation system accuracy for a precision Category III approach *without an inertial navigation system*. Comparison of the two control laws suggests that the ICL controller is more robust to unexpected variations in disturbance inputs than the LQG controller.

II. Landing Approach Procedures

A. Instrument and Microwave Landing Systems

The most common landing system currently in use (1994) is ILS. MLS is the most recently developed system with improved accuracy over ILS. Both systems provide a reference path to the aircraft in terms of an azimuth and an elevation. Because of the nature of these angular measurements, position errors with these systems vary with distance from the ground-based transmitters. Very often these systems are aided by onboard inertial navigation systems that provide additional information about attitude, position, and velocity. The aircraft is controlled in order to keep its path within a reference cone that guarantees the position and velocity accuracies required at touchdown. Table 1 shows FAA landing requirements at the time of this writing. GPS is providing a continuum of accuracy that has led to the development of new types of specifications called the *tunnel concept*.

Figure 2 presents a sketch of the typical flight phases for an ILS landing. These are: 1) the initial approach; 2) the glide-slope phase; and 3) the flare phase. This nomenclature belongs to the ILS/MLS systems and is preserved and used in this work.

Table 1 FAA navigation system accuracy standards

Operational phase	Minimum altitude	Accuracy lateral, 2 drms	Vertical, rms
En route terminal	152 m	7400 m	500 m
Approach landing			
Nonprecision	76.2 m	3700 m	100 m
Precision Category I	30.5 m	9.1 m	3.0 m
Precision Category II	15.2 m	4.6 m	1.4 m
Precision Category III	0 m	4.1 m	0.5 m

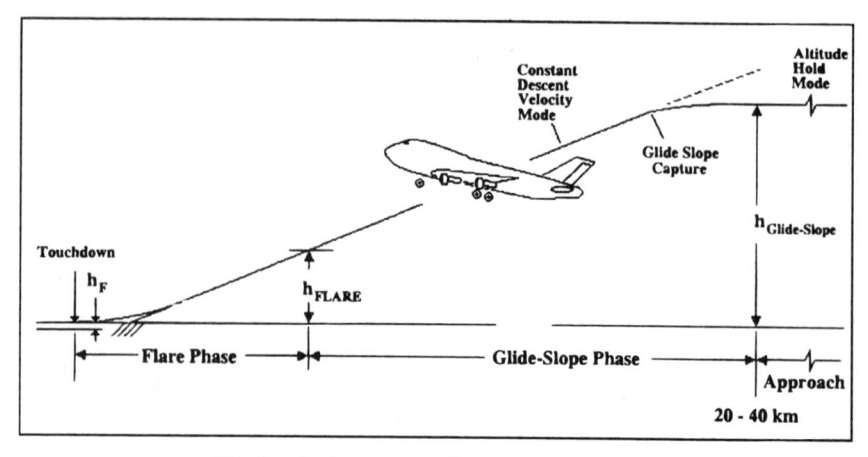

Fig. 2 Typical longitudinal approach path.

1. Initial Approach

During the *initial approach* phase, the aircraft starts at cruising altitude and descends to a lower altitude between 500 and 1500 m, which occurs at a distance of less than 40 km from the runway. After this descent, the aircraft enters an *altitude hold* mode. From this condition, the aircraft is able to capture the ILS/MLS radio signals and follow an accurate path toward the runway. Existing navigation equipment and autopilots can bring an aircraft within 150 m in position and less than 10 degs in azimuth accuracy with respect to the runway's position and azimuth during initial approach, even in the presence of disturbing winds.

2. Glide Slope

When the initial approach path intersects the desired glide path, the aircraft enters a constant-descent or *glide-slope* mode in which the altitude rate is kept between -2 and -3 m/s. This typically results in a path inclination between 2 and 3 degs. The transition maneuvers are designed to be safe and comfortable to passengers with accelerations not exceeding 0.15 g. During the glide slope, the autopilot keeps the aircraft deviation from the center of the ILS/MLS radio beam as small as possible.

3. Flare

The *flare phase* starts at a switching altitude h_{FLARE}, which depends upon the glide-slope altitude rate and the desired altitude rate for aircraft touchdown on the runway This last phase ends with touchdown where the altitude rate should be about -0.5 m/s. The autopilot performs this maneuver by flying an asymptotic approach toward a final altitude (h_F) chosen to be slightly below the runway. An approach to the exact runway altitude is not desirable, because it would greatly magnify small positioning errors at touchdown. Also, we see in the simulation that a sensor bias greater than h_F can cause the aircraft never to reach the runway.

B. GPS Approach

The GPS system is an independent position, velocity, and attitude sensor with no ground aiding equipment in the sense that no reference beam is provided by the ground equipment to the autopilot. Unlike ILS and MLS, *any* convenient reference path can be created with a GPS system based upon the runway's known position. Integrity beacons are used with CDGPS to calibrate satellite integer ambiguities and provide an additional carrier-phase reference. These simple, inexpensive devices also provide integrity: calculations show that they can meet the FAA requirement of less than one misleading position in a billion landings. Although the aircraft is required to fly in the vicinity of these transmitters to reliably achieve centimeter-level accuracies, they do *not* restrict aircraft motion to a particular reference path for landing.

In the examples presented, we have used a path similar to the ILS and MLS systems for landing with GPS. Other approach paths could be used. For example, a parabolic, continuous-arc descent, combining phases 2 and 3 of the conventional approach, may afford advantages and would be very easy for the GPS-equipped aircraft to fly.

III. Aircraft Dynamics and Linear Model

In the simulation, the aircraft is modeled with six degrees of freedom in small perturbations around a stable equilibrium point. The particular steady-state equilibrium point is the landing configuration at sea level for a Boeing 747 as derived from Bryson.[2] The components of the state vector are the aircraft position, velocity, attitude, attitude rate, and thrust specific force. The components of the control vector are the elevator, aileron, and rudder deflections, as well as the thrust specific force command. The disturbances are the longitudinal, vertical, and lateral winds.

The aircraft controller is assumed to have a perfect model of plant dynamics and sensor characteristics. However, the wind model for the simulation is inexact for both the LQG and ICL control law designs. The differences are described below.

A. State Vector

The simulations are performed in the state-space domain and the components of the state vector $X(t)$ are as follows:

$$X(t)^T = [u(t)\ w(t)\ q(t)\ \theta(t)\ d(t)\ h(t)\ x(t)\ \delta_T(t)\ U_0(t)$$
$$|\ v(t)\ r(t)\ p(t)\ \phi(t)\ \psi(t)\ y(t)] \qquad (1)$$

where,

Longitudinal mode:

u = longitudinal groundspeed
w = vertical groundspeed
q = pitch rate
θ = pitch attitude angle

d = vertical deviation from glide-slope
h = altitude
x = longitudinal displacement
δ_T = thrust specific force
U_o = nominal forward speed (doesn't change—used in calculation of d)

Lateral mode:

v = lateral groundspeed
r = yaw rate
p = roll rate
ϕ = roll attitude angle
Ψ = yaw attitude angle
y = lateral displacement

B. Control Vector

The components of the control vector $U(t)$ are as follows:

$$U(t)^T = [\delta_E(t) \; \delta_{TC}(t) \, | \, \delta_A(t) \; \delta_R(t)] \quad (2)$$

where,

Longitudinal mode:

δ_E = elevator deflection
δ_{TC} = commanded thrust specific force

Lateral mode:

δ_A = aileron deflection
δ_R = rudder deflection

C. Disturbance Vector

The components of the disturbance (wind) vector $W(t)$ are as follows:

$$W(t)^T = [W_U(t) \; W_W(t) \, | \, W_V(t)] \quad (3)$$

where,

Longitudinal mode:

W_U = longitudinal wind
W_W = vertical wind

Lateral mode:

W_V = lateral wind

D. Measurement Vector

During the glide-slope phase, the components of the measurement vector $Z(t)$ are as follows:

$$Z(t)^T = [u(t)\ w(t)\ \theta(t)\ d(t) \mid v(t)\ \phi(t)\ \psi(t)\ y(t)] \quad (4)$$

E. Equations of Motion

The first-order differential equations of motion of the aircraft model are put into the continuous state-space representation,

$$dX(t)/dt = AX(t) + BU(t) + B_W W(t) \quad (5)$$

with the following output and measurement equations,

$$Y(t) = CX(t) \quad (6)$$

$$Z(t) = HX(t) + \mu(t) \quad (7)$$

$W(t)$ and $\mu(t)$ are assumed to be uncorrelated white Gaussian noises with given means and variances in the ICL controller design. In the LQG controller, the estimated wind vector $We(t)$ uses an exponentially correlated model represented by the following state-space equation:

$$dWe(t)/dt = -1/\tau_W\ We(t) + r(t) \quad (8)$$

The time constant τ_w chosen is large compared to the time constants of the aircraft, so the wind estimate is very close to the actual wind bias. $We(t)$ is appended to the estimated state vector $Xe(t)$ for estimator design, creating the following augmented state vector:

$$Xe_A(t) = \begin{bmatrix} Xe(t) \\ We(t) \end{bmatrix} \quad (9)$$

F. Wind Model

The true wind disturbances are modeled in three dimensions with respect to the airframe of the aircraft. In each direction, the wind is composed of a random but correlated gust component and a steady shear component. Table 2 presents the correlation times used for each wind gust component. For the nonsteady wind disturbances, the following exponentially correlated model is used:

$$dW(t)/dt = -1/\tau\ W(t) + v(t) \quad (10)$$

Table 2 Wind gust disturbance correlation times

Disturbance	$1/\tau\ (s^{-1})$
Longitudinal wind	0.43
Vertical wind	1.06
Lateral wind	0.14

where

$W = W_U, W_v$ or W_W
v = white Gaussian noise
τ = true correlation time

The continuous disturbance covariance matrix $W_C = E\{v^2\}$ is chosen so that after discretization, the wind covariance $E\{W^2\} = 0.7$ m/s in all three axes.[2]

Steady winds in the simulation are modeled as a function of altitude. The wind intensity in each direction varies linearly from altitude zero (runway's altitude) up to a steady constant value. Two parameters characterize each wind profile:
1) intensity W_{SAT} (constant saturation value); and
2) intensity gradient D_H

or mathematically, as follows:

$$W_{BIAS}(t, h) = D_H h(t), \qquad h(t) \leq W_{SAT}/D_H \tag{11}$$

$$W_{BIAS}(t, h) = W_{SAT}, \qquad h(t) > W_{SAT}/D_H \tag{12}$$

G. Throttle Control Lag

The following equation was used to model the lag in throttle control as a first-order process with a 4-s time constant. This has an appreciable effect, especially during the flare phase of landing.

$$d\delta_T/dt = -0.25(\delta_T - \delta_{TC}) \tag{13}$$

H. Glide-Slope Deviation

Typically, the equations of motion for an aircraft are given for a vehicle in straight and level flight. However, during the glide-slope phase we are interested in controlling the deviation from a nominal trajectory that is not horizontal. Implementing a controller that follows a ramp input in altitude h is one solution, but it turns out to be simpler to define a new variable d, which is the perpendicular distance from the glide slope. The equations derived in Bryson's text[2] give a differential equation for h:

$$d(h)/dt = -w \cos \theta + (U_0 + u)\sin \theta \tag{14}$$

or

$$d(h)/dt \simeq -w + U_0 \theta \tag{15}$$

Using this, we can take the glide-slope angle γ into account and compute the differential equation for d:

$$d(d)/dt = [-w \cos \theta + (U_0 + u)\sin \theta]\cos \gamma$$
$$+ [-w \sin \theta + (U_0 + u)\cos \theta]\sin \gamma \tag{16}$$

or

$$d(d)/dt \simeq -w \cos \gamma + U_0 \theta \cos \gamma - u \sin \gamma + U_0 \sin \gamma \tag{17}$$

In the simulation truth model, both h and d are computed. The controller operates on d in the glide-slope phase and h in the flare phase. This causes some difficulty in controller design. For example, care must be taken in transition from glide-slope to flare phase, because the state variable changes (see Appendix A). Also, the new state variable (U_0), which does not change with time, must be added to the dynamic equations to account for the constant fourth term of Eq. (17).

IV. Autopilot Controller

A. Linear Quadratic Gaussian and Integral Control Law Controllers

Two different types of autopilot controllers were used in these simulations. The first is based on a standard LQG regulator that has extra states for direct estimates of wind disturbances. The second is based on an ICL regulator that does not directly estimate the winds but includes added states for the integral of the output errors.

During the glide-slope phase, the controller works to correct the perpendicular deviations from the desired flight path. During flare, it tries to keep the aircraft altitude on the exponential path described above while driving the lateral displacement to zero. For this reason, both controllers were actually designed with the capability of adjusting to nonzero set points. A detailed description of how this was done can be found in Appendix A.

Each of the two methods of controller design has its advantages and limitations. The LQG regulator is truly optimal in minimizing the quadratic cost function (described below) when the plant, control, and disturbance models are known exactly. If the controller model differs significantly from the truth model, rms performance is degraded, and nonzero disturbances and set-points can lead to a nonzero steady output offset. The ICL controller slightly degrades the closed-loop system performance compared to the LQG controller when the controller model matches the truth model. However, the ICL control law has the ability to compensate for nonzero steady disturbances of (possibly) unknown origin, so modeling errors do not result in a steady output error.

B. Regulator Synthesis

During the landing phases, the aircraft's attitude, position, velocity, and controls are limited either because of physical constraints, such as the maximum available rudder deflection, or constraints that ensure structural safety and passenger comfort. The latter constraints usually include accelerations and attitude angle limits. Table 3 presents the typical maximum values for a Boeing 747 aircraft. These maximum values affect the optimal controller gains for the glide-slope and flare phases of flight.

The continuous system presented in Eqs. (5–7) can be discretized and represented by the following state, output, and measurement equations:

$$X_{K+1} = A_D X_K + B_D U_K + B_{WD} W_K \qquad (18)$$

$$Y_K = C X_K \qquad (19)$$

$$Z_K = H X_K + \mu_K \qquad (20)$$

Table 3 Maximum limits for states and controls for the B747 on landing

Variable	Units	Glide slope	Flare
Pitch rate	deg/s	*	*
Pitch attitude	deg	*	5
Roll rate	deg/s	5	5
Roll attitude	deg	15	5
Yaw rate	deg/s	*	*
Yaw attitude	deg	10	5
Lat. displacement	m	15	8
Lat. velocity	m/s	*	*
Long. displacement	m	*	*
Long. velocity	m/s	*	*
Vert. displacement	m	8	1.5
Sink rate	m/s	3	0.6
Elev. deflection	deg	5	5
Aileron deflection	deg	5	5
Rudder deflection	deg	5	5
Throttle specific force	m/s²	1	1

Constraints were not placed on the asterisk quantities.

where the time index K refers to the time $t = KT$ (T = sampling period), and A_D, B_D, B_{WD}, C, and H are the discrete transition, control, disturbance, output, and measurement distribution matrices, respectively. The optimal regulator is designed to minimize the following cost function:

$$J = \Sigma(X_K^T Q_D X_K + 2 X_K^T N_D U_K + U_K^T R_D U_K) \tag{21}$$

where Q_D, R_D, and N_D are the discrete weighting matrices of the states, controls, and their correlated terms. The discrete weighting matrices can be obtained by discretizing the continuous weighting matrices Q and R. N_D is caused by the coupling resulting from the discretization and is generally nonzero, even if there is no coupling in the continuous case. The following relations are used:

$$Q_D = \int_0^T [F^T(t)QF(t)]dt \tag{22}$$

$$R_D = RT + \int_0^T [G^T(t)QG(t)]dt \tag{23}$$

$$N_D = \int_0^T [F^T(t)QG(t)]dt \tag{24}$$

where

$$F(t) = \exp(At)$$

$$G(t) = \int_0^t F(t - \tau)Bd\tau$$

The Q and R matrices can be defined as diagonal matrices, and the diagonal elements are defined using the following rule-of-the-thumb method (sometimes called Bryson's rule):

$$Q = \text{diag}[1/(X_{1\text{max}})^2 \cdots 1/(X_{N\text{max}})^2] \quad (25)$$

and

$$R = \text{diag}[1/(U_{1\text{max}})^2 \cdots 1/(U_{M\text{max}})^2] \quad (26)$$

where $X_{i\text{max}}$ and $U_{j\text{max}}$ are the maximum values that each variable is allowed to reach in the dynamic or steady environment. These values are usually determined by such physical limitations as available power, limited angular deflections, and safety and structural failure requirements. The following steady-state control law results:

$$U_K = -C_X X_K \quad (27)$$

V. GPS Measurements

A key factor in the design of the autolanding system is a realistic set of GPS biases and noise. These errors can vary considerably among the various equipment manufacturers. The values used to evaluate these quantities are the current (1994) state-of-the-art in accuracy. The rms measurement error and biases of typical GPS [without selective availability (SA)], GPS aided by a radar altimeter, DGPS, and carrier-phase DGPS systems are presented in Table 4. GPS was also used for attitude determination in the simulation, with a 0 deg bias error and 0.2 deg standard deviation.

Figure 3 details the landing system as implemented with the integral control law, including the GPS receiver, the estimator, and the digital controller. The measurements provided by the receiver are sampled at discrete time intervals and sent to the estimator. The function of the estimator is to combine the limited sensor information with the known plant model to generate estimates of all the

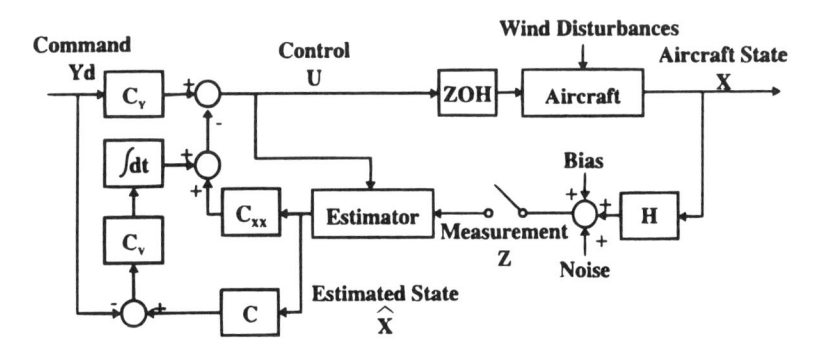

Fig. 3 Integral control law simulated block diagram.

Table 4 1σ measurement error and biases of typical GPS and DGPS

System	GDOP[a] HDOP[b]	VDOP[c]	UERE,[d] m Bias	σ	UERRE[e] m/s Bias	σ	Horizontal pos. error, m Bias	σ	Vertical pos. error, m Bias	σ	Horizontal vel. error, m/s Bias	σ	Vertical vel. error, m/s Bias	σ
GPS (no SA)	3.0	5.0	3.0	0.5	0.02	0.01	9.0	1.5	15.0	2.5	0.06	0.03	0.10	0.05
GPS + Alt.	3.0	N/A	3.0	0.5	0.02	0.01	9.0	1.5	0.2	0.1	0.06	0.03	0.05	0.025
DGPS	3.0	5.0	1.5	0.5	0.01	0.005	4.5	1.5	7.5	2.5	0.03	0.015	0.05	0.025
CDGPS	3.0	5.0	0.01	0.002	0.005	0.005	0.03	0.006	0.05	0.010	0.015	0.015	0.025	0.025

[a]GDOP = geometric dilution of precision.
[b]HDOP = horizontal dilution of precision.
[c]VDOP = vertical dilution of precision.
[d]UERE = user equivalent range error.
[e]UERRE = user equivalent range rate error.

AIRCRAFT AUTOMATIC APPROACH AND LANDING USING GPS

state variables (see Appendix B). The controller calculates the control commands based on the estimated states using full state feedback. The control signals are sent to a zero-order-hold (ZOH) digital-to-analog converter and into the aircraft as a continuous signal.

VI. Results

A. Cases Simulated

The following cases were simulated:
1) standard GPS, GPS with altimeter, DGPS, and CDGPS;
2) linear quadratic Gaussian and integral control law regulators; and
3) Glide-slope and flare phases.

The numerical results of the simulation are presented in tables with the statistical mean and standard deviation of the flight path errors. Plots of the altitude, lateral displacements, and altitude rates during a typical CDGPS landing are also presented.

Tables 5–10 give the statistics of the lateral and vertical errors for GPS alone, GPS with altimeter, DGPS, and CDGPS configurations. Tables 5 and 6 show the total system error during the approach, while Tables 7–10 break this error down into the *navigation system error* (the difference between actual and estimated position), and the *flight technical error* (the difference between estimated and desired position). The altitude rates at touchdown and landing success rates are also presented in the Tables 5 and 6. The landing success rate is important because in some simulation runs, the stand-alone GPS sensor bias was large enough to keep the aircraft from ever touching down.

To create the tables, 60 landings were simulated for each configuration with a glide-slope angle of 2.5 deg, $h_{FLARE} = 15$ meters, $h_F = 1.2$ m, and $1/a = 5.7$ s (see equation A10). The measurement sample frequency was 10 hertz. Sensor bias errors were recalculated along with random errors for each run. The tables show the mean and standard deviation values for all runs combined.

Glide-slope acquisition was performed at the low altitude of 200 m to keep a reasonably short simulation length. There was a significant transient at the

Table 5 Total system error for linear quadratic Gaussian controller

Displacement	GPS alone (no SA) Mean	σ	GPS + altimeter Mean	σ	DGPS Mean	σ	CDGPS Mean	σ
Glide-slope phase								
Vertical, m	4.90	14.76	0.40	1.54	2.11	7.58	0.30	0.95
Lateral, m	2.86	8.56	2.81	8.46	1.55	4.50	0.37	1.62
Flare phase								
Vertical, m	9.58	11.33	1.58	1.24	4.89	6.17	0.50	0.75
Lateral, m	1.12	8.87	0.09	8.06	0.01	4.64	−1.85	1.13
Alt. rate, m/s	−1.81	1.64	−0.91	0.43	−1.33	1.00	−0.71	0.34
% Successful landings	48		100		52		100	

Table 6 Total system error for integral control law controller

Displacement	GPS alone (no SA) Mean	σ	GPS + altimeter Mean	σ	DGPS Mean	σ	CDGPS Mean	σ
Glide-slope phase								
Vertical, m	0.78	14.54	−0.25	0.89	1.19	7.49	0.29	0.73
Lateral, m	3.00	8.38	2.83	8.29	1.20	4.35	−0.13	1.39
Flare phase								
Vertical, m	6.66	11.39	−0.14	0.72	3.68	6.25	−0.11	0.58
Lateral, m	4.44	9.02	3.40	8.40	1.74	4.78	−0.28	1.62
Alt Rate, m/s	−2.50	1.95	−0.61	0.30	−1.19	1.09	−0.56	0.27
Successful landings	57		100		63		100	

beginning of each simulation while the wind bias compensation built up. In the LQG case, the initial estimates of wind biases were zero, whereas, in the ICL case, the initial integral error terms were zero. Tables 5–10 show the results of data taken after the effects of this initial transient have settled out.

B. Landing with GPS Alone

For autoland using nondifferential GPS, there is little difference in performance between the LQG and ICL controllers. Tables 5 and 6 show that the vertical rms position errors lie between 10 and 15 m, while lateral rms position errors are around 8 or 9 m. Tables 7 and 8 show that these errors are primarily caused by navigation system error, not flight technical error. In other words, the imprecise GPS measurements are the main cause of the total system errors, not the autopilot controller. From these results, we see that the GPS system in the absence of SA clearly *meets the FAA nonprecision approach requirements* shown in Table 1, but the vertical position errors exceed the Category I precision landing specification.

C. Landing with GPS plus Altimeter

Specifications for precision landing approach are more stringent in the vertical dimension than in the horizontal dimension. Unfortunately, GPS horizontal measurements are typically more accurate than vertical measurements because of satellite geometry. One solution to improve vertical navigation accuracy is to augment GPS with a radar altimeter. Although this sensor does nothing to improve lateral accuracies, Tables 5 and 6 show that vertical rms position errors are reduced to around 1 m. Once again, the total system error is dominated by the navigation system error for both the LQG and ICL controllers. This combined sensor system *meets the accuracy specifications for the FAA precision Category I approach,* with extremely good vertical navigation system error compared to GPS alone.

D. Landing with Differential GPS

Another method for improving the accuracy of the stand-alone GPS signal is to use code-differential corrections from a nearby reference station. This has the advantage of improving both lateral and vertical measurements of position and velocity. Tables 7 and 8 show that DGPS offers a significant improvement in the navigation system accuracy of both autopilots, as expected. It is interesting to note that the addition of reference corrections also improves the flight technical error, because wind gusts and biases are now better estimated (LQG) or otherwise accounted for (ICL). Tables 7 and 8 show that lateral and vertical position estimate errors during flare were reduced from 12 or 13 m with GPS to around 7 m with DGPS. The results suggest that under the wind conditions described for this simulation, *DGPS alone meets the FAA Category I precision landing lateral error requirement, but exceeds the vertical requirement of 3 m.* The vertical navigation error bias was responsible for many unsuccessful landings, because the flare asymptote was only 1.2 m below the runway in these simulations.

E. Landing with Carrier-Phase

A third method for improving sensor accuracy during aircraft landing is to perform real-time carrier-phase differential GPS. As shown in Table 4, this method

Table 7 Navigation system error for linear quadratic Gaussian controller

Displacement	GPS alone Mean	σ	GPS + altimiter Mean	σ	DGPS Mean	σ	CDGPS Mean	σ
Glide-slope phase								
Vertical, m	4.51	14.76	0.08	0.36	1.80	7.52	0.02	0.08
Lateral, m	2.31	8.22	2.23	8.13	1.14	4.10	0.01	0.03
Flare phase								
Vertical, m	9.05	12.57	0.33	0.32	4.49	6.74	0.03	0.07
Lateral, m	2.44	8.67	2.37	7.96	1.34	4.42	0.00	0.03

Table 8 Navigation system error for integral control law controller

Displacement	GPS alone Mean	σ	GPS + altimeter Mean	σ	DGPS Mean	σ	CDGPS Mean	σ
Glide-slope phase								
Vertical, m	0.30	14.59	−0.69	0.31	0.78	7.49	−0.11	0.06
Lateral, m	3.19	8.21	3.01	8.12	1.37	4.10	0.05	0.03
Flare phase								
Vertical, m	7.34	12.77	−0.07	0.31	4.08	6.86	−0.01	0.05
Lateral, m	4.14	8.70	3.66	8.12	1.72	4.44	0.02	0.03

offers the tremendous advantage of raw measurement errors that are much smaller than stand-alone GPS. The total position error for an aircraft landing with GPS, GPS with altimeter, or DGPS was primarily dominated by navigation system error. *The simulation shows that the system error of a Boeing 747 landing with CDGPS is dominated by the aircraft flight technical error;* i.e. the ability of the aircraft to follow a known trajectory in the presence of external physical disturbances. This means the sensor measurements of position, velocity, and attitude are so accurate that autopilot performance is basically determined by actuator control authority and passenger safety and comfort. The navigation system error for CDGPS shown in Tables 9 and 10 *easily meets the FAA precision Category III accuracy requirements.*

Figure 4 shows a typical landing using CDGPS in the presence of wind gusts and steady wind disturbances. Both altitude plots show the transient at the beginning of the simulation where the wind bias compensation is building up. The lateral displacement plots show the performance of each controller in the presence of wind gusts alone and in the presence of wind gusts and a wind bias. The initial 7.5 m/s "step" in wind bias leads to a transient with about a 5 m maximum lateral error. All four simulations were run with the same initial conditions and the same random errors for sensors and disturbances.

Figure 5 shows the altitude and altitude rate for typical CDGPS landings using the LQG and ICL control laws. A landing with no winds is compared to a landing

Table 9 Flight technical error for linear quadratic Gaussian controller

Displacement	GPS alone		GPS + altimeter		DGPS		CDGPS	
	Mean	σ	Mean	σ	Mean	σ	Mean	σ
Glide-slope phase								
Vertical, m	0.39	1.51	0.32	1.28	0.32	1.12	0.28	0.93
Lateral, m	0.55	2.10	0.57	2.13	0.41	1.75	0.36	1.62
Flare phase								
Vertical, m	0.54	2.07	1.25	1.13	0.40	1.24	0.47	0.76
Lateral, m	−1.32	1.93	−2.27	1.57	−1.33	1.51	−1.85	1.13

Table 10 Flight technical error for integral control law controller

Displacement	GPS alone		GPS + altimeter		DGPS		CDGPS	
	Mean	σ	Mean	σ	Mean	σ	Mean	σ
Glide-slope phase								
Vertical, m	0.47	0.79	0.44	0.75	0.41	0.72	0.39	0.71
Lateral, m	−0.19	1.59	−0.18	1.58	−0.17	1.40	−0.18	1.38
Flare phase								
Vertical, m	−0.68	1.74	−0.07	0.59	−0.40	0.97	−0.10	0.57
Lateral, m	0.30	2.22	−0.27	1.85	0.02	1.78	−0.30	1.62

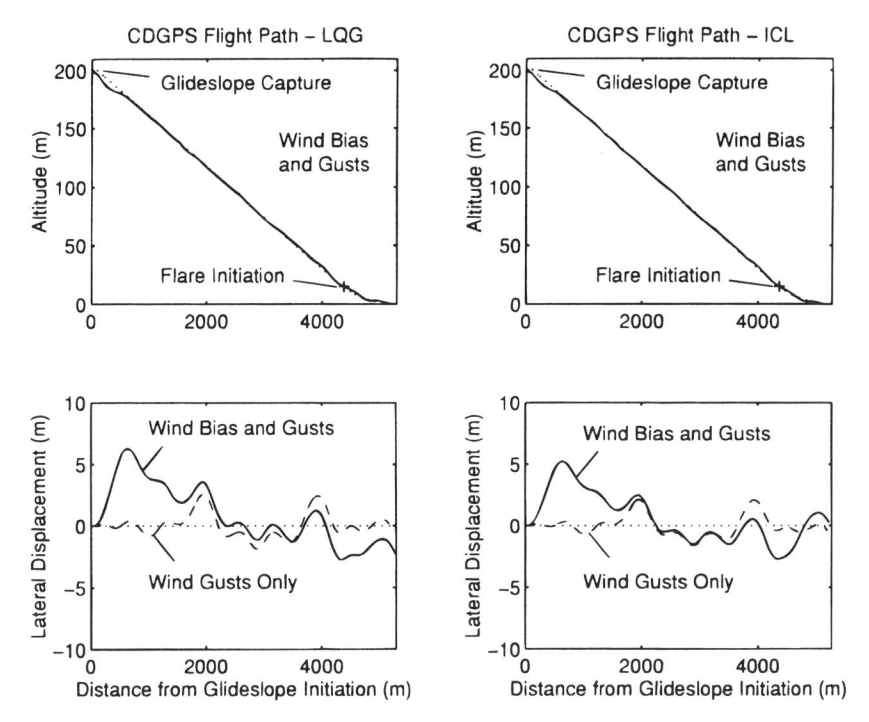

Fig. 4 Typical CDGPS approach.

with both wind biases and gusts. The goal to land with an altitude rate of -0.5 m/s is easily met with no winds present. As seen in Tables 5 and 6, this goal is slightly exceeded in the presence of winds. The plots of Fig. 5 suggest that errors in altitude rate are primarily caused by the wind gusts rather than the wind bias, which has fallen off appreciably at this low altitude.

F. Linear Quadratic Gaussian vs Integral Control Law

As expected, *the navigation system errors for both controller types are very similar.* The primary difference between the LQG and ICL controllers is seen in the flight technical error.

From the simulation results shown in Tables 5 and 6, we can see that the LQG and ICL controllers have approximately the same total system error *standard deviations.* For example, using DGPS, Table 5 shows the LQG vertical accuracy standard deviation is 7.58 m during glide slope and 6.17 m during flare. These are similar to the ICL values in Table 6, which are 7.49 m and 6.25 m, respectively. Although the LQG controller was the "optimal" design for the given cost matrices, the ICL controller was able to achieve comparable performance *in response to random zero-mean inputs.*

In fact, the ICL controller actually performed better than the LQG controller in response to *nonzero wind biases.* From Tables 9 and 10 we see that the LQG *mean* flight technical error is relatively high, especially during flare. For example,

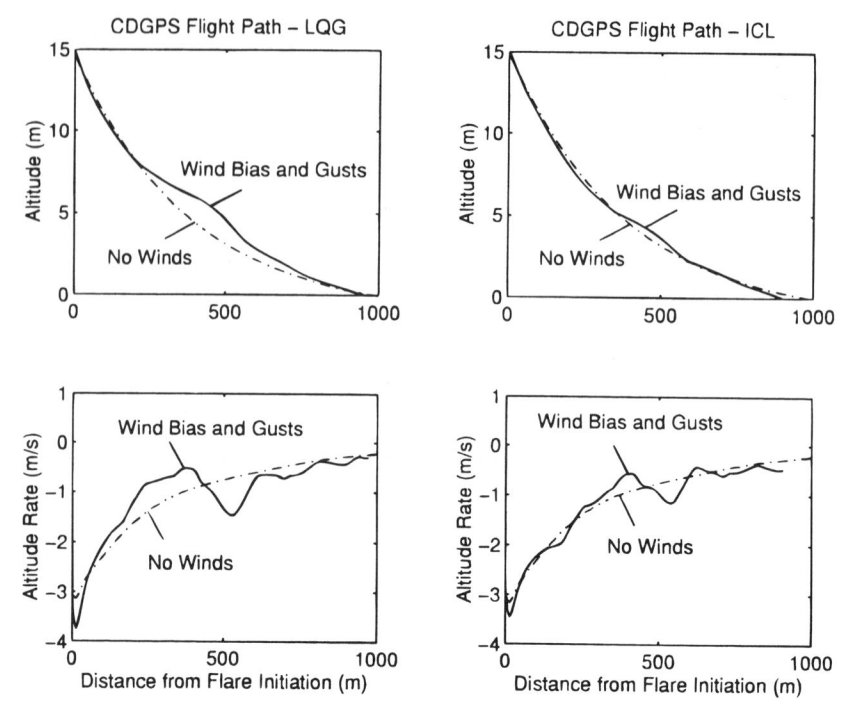

Fig. 5 Typical CDGPS landing.

using CDGPS during flare, the LQG bias errors are 0.47 m vertically and −1.85 m laterally. The ICL controller results in mean errors of −0.10 m vertically and −0.30 m laterally. During flare, the wind bias is varying linearly with altitude. This suggests that the ICL control law is better adapted than the LQG controller to compensate for a changing, nonzero wind bias. Figure 4 also suggests that the LQG controller is more susceptible to variations in wind bias disturbances, because the initial lateral position transient is larger than for the ICL controller.

VII. Conclusions and Comments

Autopilot design based on GPS has several advantages over ILS- and MLS-based systems. GPS is clearly less expensive because it does not rely on costly equipment on the ground or in the air. It is also more flexible, because approaches are not confined to take place within a narrow radar beam. *Most importantly, the GPS sensor measurements are fundamentally better suited for use by an autopilot.* ILS and MLS measurements are based on angular deviation from a desired flight path. They, therefore, have changing sensitivity to position errors as the aircraft comes closer to the transmitter. The GPS measures position in three dimensions with no real degradation in accuracy nor any change in sensitivity through touchdown. GPS also offers the enormous advantage of highly accurate velocity measurements. The estimator can use this information to improve

AIRCRAFT AUTOMATIC APPROACH AND LANDING USING GPS

its position estimates and to better determine the magnitude of wind disturbances. This is particularly helpful in the event of wind shears near the ground.

As with any computer simulation, there are some limitations to the study presented here. For example, the simulation assumes approximately average wind conditions and typical GPS sensor errors. More severe winds or poorer GPS performance (such as a higher GDOP or a less accurate receiver) would produce worse results, although the conclusions about CDGPS would probably still be valid. Also, we have assumed that the aircraft dynamics are linear in the region of operation with no significant cross coupling between the lateral and longitudinal modes. Finally, we have assumed that the computer has perfect knowledge of the linear plant model and sensor characteristics and that corrections from the ground reference station when applicable, are continuously available. All of these assumptions are reasonable for the initial autopilot design. To ensure robustness, further analysis, backed by flight tests must be undertaken.

Simulation results suggest the following conclusions based upon the previous assumptions:

1) The ICL control law seems to be more robust to variations in design parameters than a standard LQG controller. In particular, the ICL is less sensitive to variations in nonzero wind disturbances.

2) Nondifferential GPS, even in the absence of SA, can only satisfy the FAA nonprecision landing requirement.

3) Nondifferential GPS augmented with a radar altimeter meets the FAA Category I precision landing requirement.

4) Differential GPS (code phase) meets the lateral accuracy but exceeds the vertical accuracy requirement for FAA Category I precision landing.

5) Carrier-phase differential GPS meets the accuracy specification for FAA Category III precision landing. Unlike the other three cases, position errors for CDGPS are dominated by autopilot error, not sensor uncertainty.

The only four sensor configurations examined in this study were GPS alone, GPS with radar altimeter, DGPS alone, and CDGPS alone. In all cases, system accuracy (and integrity) could be improved by using additional sensors. A simple inertial measurement unit would provide redundancy and improve estimates of aircraft position and heading. Also, because GPS is a highly accurate sensor of *ground* speed, the addition of an *air* speed sensor could greatly improve the estimates of wind speed—especially wind shear.

Appendix A: Discrete Controllers

Linear Quadratic Gaussian Controller for Constant Set Points and Steady Wind Disturbances

If we want the output vector Y_K to take on specific values (Y_D), we can often look at the linear state equations as perturbations about a nonzero steady value. When doing this, we must be careful that the linearized equations still hold in the new regime.

The steady-state values are defined as follows:

$$X_{SS} = X_{SS}(K) = X_{SS}(K+1) \quad (A1)$$

$$W_{SS} = W_{SS}(K) = W_{SS}(K+1) \quad (A2)$$

$$Y_{SS} = Y_D \quad (A3)$$

Substitute into Eqs. (18) and (19) to obtain the following:

$$-B_{WD}W_{SS} = [A_D - I]X_{SS} + [B_D]U_{SS} \quad (A4)$$

$$Y_D = [C]X_{SS} + [0]U_{SS} \quad (A5)$$

where I is the identity matrix. For arbitrary nonzero W_{SS} and Y_D, the system has a solution if the rank of M, a matrix whose elements are defined by,

$$\begin{bmatrix} X_{SS} \\ U_{SS} \end{bmatrix} = \begin{bmatrix} M_{XW} & M_{XY} \\ M_{UW} & M_{UY} \end{bmatrix} \begin{bmatrix} -B_{WD}W_{SS} \\ Y_D \end{bmatrix} \quad (A6)$$

is equal to the rank of the column-augmented matrix. Also, the number of outputs must be less than or equal to the number of controls. If the number of outputs equals the number of controls and M is nonsingular, the solution is unique and is given by the following:

$$\begin{bmatrix} M_{XW} & M_{XY} \\ M_{UW} & M_{UY} \end{bmatrix} = \begin{bmatrix} (A_D - I) & B_D \\ C & 0 \end{bmatrix}^{-1} \quad (A7)$$

The controller Eq. (27) can now be rewritten to account for the steady offsets:

$$U_K = U_{SS} - C_X[X_K - X_{SS}] \quad (A8)$$

which becomes the following:

$$U_K = -C_X X_K + C_Y Y_D - C_W W_{SS} \quad (A9)$$

where $C_Y = C_X M_{XY} + M_{UY}$; and $C_W = [C_X M_{XW} + M_{UW}]B_{WD}$. Note that so far we have assumed X_K and W_{SS} are known exactly. The method for actually estimating these quantities is discussed in Appendix B.

Linear Quadratic Gaussian Controller for Exponential Set Points

It is sometimes desirable to follow a reference input that is decaying exponentially in time. For example, during flare, the aircraft is asymptotically approaching an elevation below the runway. To create a controller for this situation, we simply modify the design procedure just described to follow a continuous desired output

$$Y_D(t) = Y_i e^{-at} \quad (A10)$$

which corresponds the following discrete reference command:

$$Y_{DK} = Y_i r^K \quad (A11)$$

where $r = e^{-aT}$, and T is the sample time.

AIRCRAFT AUTOMATIC APPROACH AND LANDING USING GPS

Nominally, while tracking this reference command input,

$$X_K = X_i r^K \tag{A12}$$

$$U_K = U_i r^K \tag{A13}$$

Combining this with Eqs. (18) and (19), we have the following:

$$X_i r^{K+1} = A_D X_i r^K + B_D U_i r^K + B_{WD} W_K \tag{A14}$$

We simplify as before to find the following:

$$-B_{WD} W_K r^{-K} = [A_D - rI]X_i + [B_D]U_i \tag{A15}$$

$$Y_i r^K = [C]X_i + [0]U_i \tag{A16}$$

and

$$\begin{bmatrix} X_i \\ U_i \end{bmatrix} = \begin{bmatrix} N_{XW} & N_{XY} \\ N_{UW} & N_{UY} \end{bmatrix} \begin{bmatrix} -B_{WD} W_K r^{-K} \\ Y_D \end{bmatrix} \tag{A17}$$

where

$$\begin{bmatrix} N_{XW} & X_{XY} \\ N_{UW} & N_{UY} \end{bmatrix} = \begin{bmatrix} (A_D - rI) & B_D \\ C & 0 \end{bmatrix}^{-1} \tag{A18}$$

The controller equation can finally be written to account for exponential reference inputs:

$$U_K = U_{SS} - C_X[X_K - X_{SS}] \tag{A19}$$

which becomes

$$U_K = -C_X X_K + C_Y Y_{DK} - C_W W_K \tag{A20}$$

where

$$C_Y = C_X N_{XY} + N_{UY}$$

$$C_W = [C_X N_{XW} + N_{UW}]B_{WD}$$

We have again assumed X_K and W_K are known exactly. The method for estimating these quantities is discussed in Appendix B.

Integral Control Law Controller for Constant Set Points and Steady Wind Disturbances

The continuous form of the modified integral control was developed and presented by Holley and Bryson.[5] The same logic is followed to develop the discrete-time version.

One way to develop a controller that allows for nonzero set points and steady disturbance inputs is to define a general control law that will eventually lead to a form of integral control:

$$U_K = -C_X X_K + C_Y Y_0 \tag{A21}$$

Combining this with Eqs. (18) and (19), it turns out that if $[I - A_D + B_D C_X]$ is nonsingular, the steady-state solution is as follows:

$$X_{SS} = [I - A_D + B_D C_X]^{-1}[B_D C_Y Y_0 + B_{WD} W_{SS}] \qquad (A22)$$

$$Y_D - TB_{WD} W_{SS} = TB_D C_Y Y_0 \qquad (A23)$$

where $T \equiv C[I - A_D + B_D C_X]$.
Equation (A23) is satisfied for all Y_D and W_{SS} if

$$TB_D C_Y = I \qquad (A24)$$

$$Y_0 = Y_D - TB_{WD} W_{SS} \qquad (A25)$$

Solving for C_Y is straightforward when the number of outputs is equal to the number of controls and the square matrix (TB_D) is full rank. In order to compute Y_0, i.e., the implicit influence of the disturbances W_K in the control law, define a new state as follows:

$$Y_{WK} \equiv TB_{WD} W_K \qquad (A26)$$

and define a new estimator for this state with gain L_2 (which is chosen empirically), so that

$$Ye_{W(K+1)} = Ye_{WK} - L_2[Ye_{WK} - Y_{WK}] \qquad (A27)$$

Using Eq. (18)

$$Ye_{W(K+1)} = Ye_{WK} - L_2 TB_{WD} We_K + L_2 T[X_{K+1} - A_D X_K - B_D U_K] \qquad (A28)$$

By applying Eqs. (A21), (A24), and (A25), Eq. (A28) can eventually be simplified to the following useful form:

$$Ye_{W(K+1)} - L_2 T X_{K+1} = Ye_{WK} - L_2 T X_K - L_2[Y_D - CX_K] \qquad (A29)$$

This can be simplified even further by defining V_K so that

$$L_2 V_K = Ye_{WK} - L_2 T X_K \qquad (A30)$$

Equations (A29) and (A30) become the following:

$$V_{K+1} = V_K - [Y_D - CX_K] \qquad (A31)$$

We now have an expression that represents the integral of the error between desired output and actual (or measured) output. The control law can be determined by expanding Eq. (A21):

$$U_K = -C_X X_K + C_Y [Y_D - Ye_{WK}] \qquad (A32)$$

$$U_K = -C_X X_K + C_Y [Y_D - L_2 V_K + L_2 T X_K] \qquad (A33)$$

and finally

$$U_K = -[C_{XX}] X_K + [C_Y] Y_D - [C_V] V_K \qquad (A34)$$

where

$$C_{XX} = C_X + C_Y L_2 T$$
$$C_Y = (TB_D)^{-1}$$
$$C_V = C_Y L_2$$

Note that the discrete form of the integral control law is similar to the continuous form described by Holley and Bryson.[5]

The size of L_2 is linearly related to the amount of control used to zero the integral error. The selection of L_2 for these simulations was done empirically using the following considerations: it must be large enough to produce a satisfactory response to disturbances and unmodeled errors, but cannot be so large that unacceptable control authority is required.

Integral Control Law Controller for Exponential Set Points and Wind Disturbances

The previous ICL controller design assumes reference inputs are constant and disturbances have a constant bias combined with zero-mean noise. To account for the exponential decaying input associated with the flare phase, the previous controller design is modified with the following assumptions:

$$Y_{DK} = Y_i r^K \tag{A35}$$

and nominally

$$X_K = X_i r^K \tag{A36}$$

With these assumptions made, the design for the new integral control law is almost the same as for the old one. We begin with the general control law:

$$U_K = -C_X X_K + C_Y Y_0 \tag{A37}$$

Combining this with Eqs. (18) and (19), it turns out that if $[rI - A_D + B_D C_X]$ is nonsingular, the flare solution is as follows:

$$X_i = [rI - A_D + B_D C_X]^{-1} [B_D C_Y Y_0 + B_{WD} W_K] \tag{A38}$$

$$Y_D - TB_{WD} W_K = TB_D C_Y Y_0 \tag{A39}$$

where

$$T \equiv C[rI - A_D + B_D C_X]^{-1}$$

Note that this T is slightly different from the T defined for the constant reference input case because of the r term.

Equation (A39) is satisfied for all Y_D and W_K if

$$TB_D C_Y = I \tag{A40}$$

$$Y_0 = Y_D - TB_{WD} W_K \tag{A41}$$

Solving for C_Y is straightforward when the number of outputs is equal to the number of controls and the square matrix (TB_D) is full rank. To compute Y_0, i.e., the implicit influence of the disturbances in the control law, define a new state as follows:

$$Y_{WK} \equiv TB_{WD}W_K \tag{A42}$$

and define a new estimator for this state with gain L_2 so that

$$Ye_{W(K+1)} = rYe_{WK} - L_2[Ye_{WK} - Y_{WK}] \tag{A43}$$

Using Eq. (18)

$$Ye_{W(K+1)} = rYe_{WK} - L_2TB_{WD}We_K + L_2T[X_{K+1} - A_DX_K - B_DU_K] \tag{A44}$$

By applying Eqs. (A37), (A40), and (A41), Eq. (A44) can eventually be simplified to the following useful form:

$$Ye_{W(K+1)} - L_2TX_{K+1} = r[Ye_{WK} - L_2TX_K] - L_2[Y_D - CX_K] \tag{A45}$$

This can be simplified even further by defining V_K so that

$$L_2V_K = Ye_{WK} - L_2TX_K \tag{A46}$$

Equations (A45) and (A46) become the following:

$$V_{K+1} = rV_K - [Y_D - CX_K] \tag{A47}$$

We now have an expression that represents the integral of the error between desired output and actual (or estimated) output. The control law can be determined by expanding Eq. (A37):

$$U_K = -C_XX_K + C_Y[Y_D - Ye_{WK}] \tag{A48}$$

$$U_K = -C_XX_K + C_Y[Y_D - L_2V_K + L_2TX_K] \tag{A49}$$

and finally

$$U_K = -[C_{XX}]X_K + [C_Y]Y_D - [C_V]V_K \tag{A50}$$

where $C_{XX} = C_X + C_YL_2T$; $C_Y = (TB_D)^{-1}$; and $C_V = C_YL_2$.

Transition from Glide-Slope Phase to Flare Phase

In the LQG controller the phase transition from glide-slope to flare is relatively straightforward. The winds are estimated along the glide slope, so these state variables are held and used during flare. The controller switches an internal state variable from d to h, updates the internal model to include this new state, and recomputes controller and estimator gains. The new controller gains account for flare error specifications (see Table 3).

The procedure is basically the same for the ICL controller, however, because the winds are not directly being estimated, care must be taken to update the integral states. The governing equations for the integral state for glide-slope phase and flare phase are, respectively:

$$V_{gsK} = L_{2gs}^{-1}T_{gs}B_{WD}We_K - T_{gs}Xe_{gsK} \tag{A51}$$

$$V_{flK} = L_{2fl}^{-1}T_{fl}B_{WD}We_i r^K - T_{fl}Xe_{flK} \tag{A52}$$

AIRCRAFT AUTOMATIC APPROACH AND LANDING USING GPS

To match correctly at transition, V_{f0} should be chosen so that the state and wind estimates match those reflected in V_{gsKf}. Equation (A51) can be expressed with the unknown quantity (We_K) as a function of the "known" quantities $(V_{gsK}$ and $Xe_{gsK})$,

$$T_{gs} B_{WD} We_K = L_{2gs}[V_{gsK} + T_{gs} Xe_{gsK}] \tag{A53}$$

Because the solution for $B_{WD} We_K$ is underdetermined, a least-norm solution can be found by using the pseudoinverse of T_{gs}:

$$B_{WD} We_K = T_{gs}^T (T_{gs} T_{gs}^T)^{-1} L_{2gs}[V_{gsK} + T_{gs} Xe_{gsK}] \tag{A54}$$

Substituting this into Eq. (A52), we get a solution for the integral error term at the beginning of flare in terms of the integral error term at the end of the glide-slope phase and other known quantities:

$$V_{f0} = L_{2fl}^{-1} T_{fl} T_{gs}^T (T_{gs} T_{gs}^T)^{-1} L_{2gs}[V_{gsKf} + T_{gs} Xe_{gsKf}] - T_{fl} Xe_{f0} \tag{A55}$$

Appendix B: Discrete Time Optimal Estimator

Estimator Synthesis

In our design of the ICL and LQG controllers, we have assumed that the quantitites X_K and possibly W_K are available. In truth, these quantities must be estimated from the plant model, control history, and measurements. These estimates are the values actually used in the control law.

Integral Control Law Estimator

One way to implement the ICL is to assume the wind disturbances and sensor measurements are Gaussian, zero mean, and uncorrelated. We expect any biases caused by modeling errors and wind biases to be accounted for with the integral error term. Referring to Eqs. (18) and (20), we can write these assumptions in the following way:

$$E\{W_K\} = 0 \quad \text{for all } K \tag{B1a}$$

$$E\{W_J W_K^T\} = W_D \quad \text{for } J = K \tag{B1b}$$

$$= 0 \quad \text{for } J \neq K \tag{B1c}$$

$$E\{\mu_K\} = 0 \quad \text{for all } K \tag{B2a}$$

$$E\{\mu_J \mu_K^T\} = V_D \quad \text{for } J = K \tag{B2b}$$

$$= 0 \quad \text{for } J \neq K \tag{B2c}$$

$$E\{W_J \mu_K^T\} = 0 \quad \text{for all } J, K \tag{B3}$$

The discrete covariance of the disturbance noise is W_D, and V_D is the discrete covariance of the measurement noise.

We can compute the prediction estimator gain L given W_D and V_D using standard linear quadratic estimation techniques.[1] We then generate state estimates using the following equations:

Measurement update:

$$X_K^* = Xe_K + L[Z_K - HXe_K] \tag{B4}$$

Time update:

$$Xe_{K+1} = A_D X_K^* + B_D U_K \tag{B5}$$

Control law:

$$U_K = -C_{XX}Xe_K + C_Y Y_{DK} - C_V V_K \tag{B6}$$

Linear Quadratic Gaussian Estimator

In the LQG case, we generate estimates of W_K in an attempt to directly compensate for wind disturbances. The straightforward approach of modeling these disturbances as constants is not helpful with a constant gain estimator. The optimal result is $We_{K+1} = We_K$, so the initial estimate is never changed as measurements are taken.

A more useful model is to assume the disturbance is exponentially correlated with time constant τ_w, which is long compared to the characteristic times of the aircraft. The continuous disturbance model is given by the following:

$$dW(t)/dt = -1/\tau_W W(t) + 1/\tau_W r(t) \tag{B7}$$

where

$$E\{r(t)\} = 0 \tag{B8}$$

$$E\{r(t)r^T(t)\} = W_C \delta(t) \tag{B9}$$

The size of W_C and τ_W determine how quickly the estimator responds to changes in wind. For a small W_C, the estimator reacts sluggishly in the presence of gusts; whereas, for a large W_C, the estimator reacts quickly, but performance is degraded because of measurement noise.

The disturbance vector is appended to the state vector before discretization. An estimator is designed using linear quadratic techniques on the augmented state vectors and matrices to find an estimator for the plant states and disturbance states simultaneously.

Appendix C: Numerical Values for Continuous System

This appendix contains the continuous truth matrices used in the autopilot simulation. The numerical values were derived from Bryson[2], and are in units of feet, seconds, and centiradians.

AIRCRAFT AUTOMATIC APPROACH AND LANDING USING GPS

Automatic Landing System—Longitudinal Mode

Matrix A:

−0.0210	0.1220	0.0000	−0.3220	0.0000	0.0000	0.0000	1.0000	0.0000
−0.2090	−0.5300	2.2100	0.0000	0.0000	0.0000	0.0000	−0.0440	0.0000
0.0170	−0.1640	−0.4120	0.0000	0.0000	0.0000	0.0000	0.5440	0.0000
0.0000	0.0000	1.0000	0.0000	0.0000	0.0000	0.0000	0.0000	0.0000
0.0349	−0.9994	0.0000	2.2087	0.0000	0.0000	0.0000	0.0000	0.0349
0.0000	−1.0000	0.0000	2.2100	0.0000	0.0000	0.0000	0.0000	0.0000
1.0000	0.0000	0.0000	0.0000	0.0000	0.0000	0.0000	0.0000	0.0000
0.0000	0.0000	0.0000	0.0000	0.0000	0.0000	0.0000	−0.2500	0.0000
0.0000	0.0000	0.0000	0.0000	0.0000	0.0000	0.0000	0.0000	0.0000

Matrix B:

0.0100	0.0000
−0.0640	0.0000
−0.3780	0.0000
0.0000	0.0000
0.0000	0.0000
0.0000	0.0000
0.0000	0.0000
0.0000	0.2500
0.0000	0.0000

Matrix B_W:

0.0210	−0.1220
0.2090	0.5300
−0.0170	0.1640
0.0000	0.0000
0.0000	0.0000
0.0000	0.0000
0.0000	0.0000
0.0000	0.0000
0.0000	0.0000

Glide-slope output distribution matrix C:

0.0000	0.0000	0.0000	0.0000	1.0000	0.0000	0.0000	0.0000	0.0000
0.0000	0.0000	0.0000	0.0000	0.0000	0.0000	1.0000	0.0000	0.0000

Glide-slope measurement distribution matrix H:

1.0000	0.0000	0.0000	0.0000	0.0000	0.0000	0.0000	0.0000	0.0000
0.0000	1.0000	0.0000	0.0000	0.0000	0.0000	0.0000	0.0000	0.0000
0.0000	0.0000	0.0000	1.0000	0.0000	0.0000	0.0000	0.0000	0.0000
0.0000	0.0000	0.0000	0.0000	1.0000	0.0000	0.0000	0.0000	0.0000
0.0000	0.0000	0.0000	0.0000	0.0000	0.0000	1.0000	0.0000	0.0000
0.0000	0.0000	0.0000	0.0000	0.0000	0.0000	0.0000	0.0000	1.0000

Wind disturbances:

Wind gradient:
$$0.1000 \text{ (ft/s)/ft}$$

Steady wind:
$$25.000 \text{ ft/s}$$

Automatic Landing System—Lateral Mode

Matrix A:

−0.0890	−2.1900	0.0000	0.3190	0.0000	0.0000
0.0760	−0.2170	−0.1660	0.0000	0.0000	0.0000
−0.6020	0.3270	−0.9750	0.0000	0.0000	0.0000
0.0000	0.1375	1.0000	0.0000	0.0000	0.0000
0.0000	1.0094	0.0000	0.0000	0.0000	0.0000
1.0000	0.0000	0.0000	0.3010	2.1894	0.0000

Matrix B:

0.0000	0.0327
0.0264	−0.1510
0.2270	0.0636
0.0000	0.0000
0.0000	0.0000
0.0000	0.0000

Matrix B_W:

0.0890
−0.0760
0.6020
0.0000
0.0000
0.0000

Glide-slope output distribution matrix C:

0.0000	0.0000	0.0000	0.0000	1.0000	0.0000
0.0000	0.0000	0.0000	0.0000	0.0000	1.0000

Glide-slope measurement distribution matrix H:

1.0000	0.0000	0.0000	0.0000	0.0000	0.0000
0.0000	0.0000	0.0000	1.0000	0.0000	0.0000
0.0000	0.0000	0.0000	0.0000	1.0000	0.0000
0.0000	0.0000	0.0000	0.0000	0.0000	1.0000

Wind disturbances:

Wind gradient:
0.10000 (ft/s)/ft

Steady wind:
25.000 ft/s

Bibliography

Beser, J., and Parkinson, B. W., "The Application of NAVSTAR Differential GPS in the Civilian Community," *Navigation*, Vol. 29, No. 2, 1982.

Etkin, B., "Dynamics of Atmospheric Flight," Wiley, New York, 1972.

Parkinson, B. W., and Fitzgibbon, K. T., "Aircraft Automatic Landing System Using GPS," *Navigation*, Vol. 42, No. 1, Jan. 1989.

Stengel, R. F., *Stochastic Optimal Control*, Wiley, New York, 1986.

References

[1]Bryson, A. E., and Ho, Y. C., *Applied Optimal Control*, Hemisphere, Bristol, PA, 1975.

[2]Bryson, A. E., *Control of Spacecraft and Aircraft*, Princeton University Press, Princeton, NJ, 1994.

[3]Bryson, A. E., and Henrikson, L. J., "Estimation using Sampled Data Containing Sequentially Correlated Noise," *Journal of Spacecraft and Rockets*, Vol. 5, No. 6, 1982, pp. 662–665.

[4]Gelb, A., *Applied Optimal Estimation*, MIT Press, Cambridge, MA, 1974.

[5]Holley, W. E., and Bryson, A. E., "MIMO Regulator Design for Constant Disturbances and Non-zero Set Points with Applications to Automatic Landing in a Crosswind," Stanford University, SUDAAR No. 465, Aug. 1973.

[6]Roskam, J., *Airplane Flight Dynamics and Automatic Flight Control*, The University of Kansas, Lawrence, KS, 1979.

Chapter 15

Precision Landing of Aircraft Using Integrity Beacons

Clark E. Cohen,* Boris S. Pervan,† H. Stewart Cobb,†
David G. Lawrence,† J. David Powell,‡ and Bradford W. Parkinson§
Stanford University, Stanford, California 94305

LANDING aircraft in poor visibility imposes the very highest standards of performance for a navigation system. Required to work under extreme weather conditions and at life-critical levels of performance, a Category III (lowest visibility) landing system must meet a vertical position accuracy requirement of 2 ft (95%) with extremely demanding integrity. For each approach, the probability of missed detection of failure cannot exceed 5×10^{-9}. This chapter explores the augmentation of GPS with *Integrity Beacons*—a special type of pseudolite—to achieve the required navigation performance (RNP) for precision landing of aircraft.

I. Overview of the Integrity Beacon Landing System

The Integrity Beacon Landing System (IBLS)[1-3] is illustrated in Fig. 1. It is founded on using GPS augmented with Integrity Beacons—compact, low-power, ground-based marker beacon "pseudolites" (transmitters used as pseudo-GPS satellites). Integrity Beacons are nominally situated in pairs on either side of the approach path to a runway. The power is set low so that the broadcast signal is measurable only inside of the "bubble" shown in Fig. 1. The bubble radius (determined by the broadcast signal power) is adequate when it is only a few times larger than the nominal altitude of approach. A conventional differential GPS (DGPS) reference station is located at the airport tower. This station broadcasts GPS reference information to all aircraft in the vicinity of the airport, both on the ground and in flight. Flying through the integrity bubbles, an aircraft is capable of tracking enough ranging sources to initialize DGPS to centimeter-level accuracy with a high degree of integrity. The aircraft can then maintain this initialization from bubble exit through touchdown and rollout.

Copyright © 1995 by the authors. Published by the American Institute of Aeronautics and Astronautics, Inc., with permission. Released to AIAA to publish in all forms.
*Research Associate, Department of Aeronautics and Astronautics; Manager, GPS Precision Landing.
†Ph.D. Candidate, Department of Aeronautics and Astronautics.
‡Professor, Department of Aeronautics and Astronautics.
§Professor, Department of Aeronautics and Astronautics; Director of GPS Program.

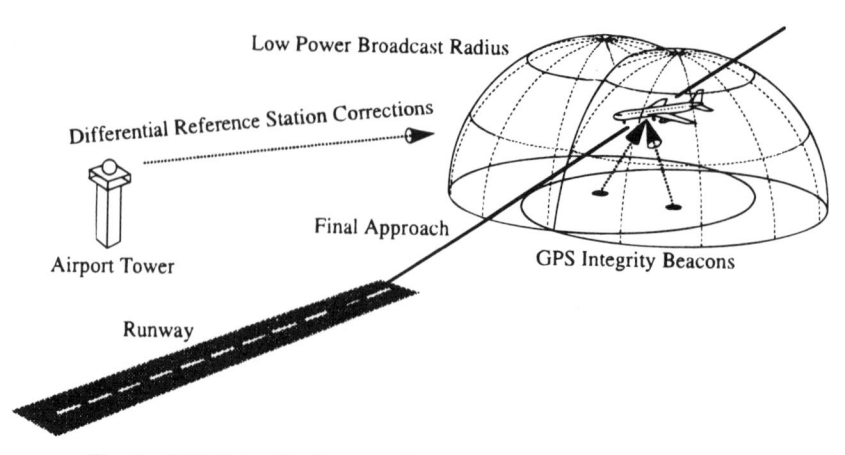

Fig. 1 GPS Integrity Beacons for Category III precision landing.

A. Centimeter-Level Positioning

Kinematic aircraft positioning is based on precise measurements of the GPS carrier phase. The GPS L_1 carrier wavelength is 19 cm, and a state-of-the-art GPS receiver can measure the carrier phase in real-time to a small fraction of a wavelength (i.e., subcentimeter precision). Precision positioning is accomplished by measuring the carrier phase difference between an antenna at a surveyed location on the ground and the aircraft antenna. By resolving the carrier-phase cycle ambiguities (the number of integer wavelengths between each given pair of antennas in the direction of each given GPS satellite), a receiver can determine its position to centimeter-level accuracy.

Historically, cycle ambiguities have been resolved by using integer search techniques based on redundant ranging measurements with the optional use of the dual-frequency GPS signal.[4,5] In contrast, IBLS resolves integers by using the ranging information from the Integrity Beacons over a large change in geometry caused by aircraft motion. This allows the user to solve explicitly and analytically for the exact numerical values of the integers.

Initially flying in on traditional differential GPS, an IBLS-equipped aircraft flies over a pair of Integrity Beacons to resolve the cycle ambiguities. Thereafter, centimeter-level positioning accuracy is achieved, all the way through landing and taxi. The system utilizes the single-frequency C/A-code and carrier signal that has been explicitly provided for civilian use. Real-time operation is essentially independent of selective availability (SA).

By allowing the aircraft to use the precision of GPS carrier phase reliably Integrity Beacons yield centimeter-level sensor accuracy for the aircraft. The aircraft receiver can convert this accuracy into a high-level of onboard integrity. As it flies through the bubble, the aircraft obtains GPS ranging information from every direction—both from the GPS satellites in the sky above and from the Integrity Beacons below. If any element of the system is not performing to specification, the inconsistencies between measurements (precise to the centimeter level) make the problem clear, and the system issues an integrity alarm.

B. History of the Integrity Beacon Landing System

IBLS is a spin-off of NASA-sponsored research at Stanford University directed toward a satellite test of Einstein's General Theory Relativity. On this spacecraft, called Gravity Probe B, GPS will be used for both precise orbit determination and spacecraft attitude determination. A new high-performance attitude determination system based on GPS carrier phase was developed and flight tested on both spacecraft and aircraft.[6] Many of the kinematic positioning techniques pioneered in the attitude system laid the groundwork for the landing system. Under FAA sponsorship, the IBLS "Pathfinder" was developed as a feasibility test bed for Category III precision landing.

C. Doppler Shift and Geometry Change

The principle by which cycle ambiguities are resolved is similar to the familiar changing pitch of a passing locomotive whistle as heard by a stationary listener. As it flies overhead, the moving aircraft measures the carrier phase (the derivative of which is the Doppler shift) of the stationary ground-based transmitter signal. A large change in angular geometry occurs on a time scale of seconds. When referenced to the slowly changing satellite geometry, the Integrity Beacon carrier phase range measurements coupled with the large change in angular geometry quickly provide enough information to pinpoint the cycle ambiguities for each satellite.

A *single* Integrity Beacon below provides enough information to resolve altitude and along-track position (as described further in Sec. IV.A). Flying between a *pair* of Integrity Beacons placed on either side of the ground track (as in Fig. 1A) provides enough information to initialize all three components of aircraft position to high accuracy and also provide a crosscheck.

The code phase component of the GPS signal is not explicitly required for IBLS positioning. The only indirect requirements for code modulation are to enable the receiver to distinguish between different GPS satellite carrier signals and to provide a coarse position initialization for the algorithm. Therefore, because code-based ranging can be considered optional, it can provide an *additional, independent* layer of integrity checking. Of course the modulation can also include data communications in such a way that is similar to the satellite data messages.

II. Required Navigation Performance

Required navigation performance for precision landing is being quantified by the parameters *accuracy, integrity, availability,* and *continuity*. The IBLS performance in the context of these required performance parameters is discussed in the following subsections.

A. Accuracy

The consistent and dependable centimeter-level accuracy provided by the use of the GPS Integrity Beacons exceeds both the ICAO Annex 10[7] and Federal Radionavigation Plan (FRP)[8] system specification of 2-ft vertical (95%) navigation sensor error (NSE) for Category III landings with a substantial margin.

Another proposed specification—the RNP tunnel concept[9]—sets a 15-ft (95%) requirement at 100-ft altitude for total system error (TSE). At any given instant, TSE is the sum of NSE and the pilot or autopilot's flight technical error (FTE). In flight testing with different autopilots, IBLS has also met this specification.

The accuracy provided by the Integrity Beacon architecture has important implications in engineering the landing system to be resistant to adverse conditions. An important overall contributor to the utility of the Integrity Beacon may be its very low NSE. As discussed in Sec. V, Integrity Beacon positioning is largely insensitive to position dilution of precision (PDOP). The system can easily handle worst-case satellite failures and still maintain excellent NSE. Newer pilot displays may allow landing systems to meet the total accuracy specification at lower cost if larger pilot-in-the-loop FTE is tolerable. When NSE is small, the TSE becomes insensitive to NSE. On average, TSE is the root-sum-square of the two 95% components:

$$\text{TSE} = \sqrt{\text{NSE}^2 + \text{FTE}^2} \cong \text{FTE}$$

High sensor accuracy and the largest possible allowance for FTE also translates into more margin to safely reject unpredictable wind gust disturbances on final approach.

Very importantly, this ample margin for navigation sensor accuracy can be used to improve system integrity. As described next, the high accuracy means that extremely tight thresholds can be set for Receiver Autonomous Integrity Monitoring (RAIM).

B. Integrity

Integrity is the measure of trust that can be placed on the correctness of the navigation system output. The requirement for Category III integrity[7] is given as a probability of missed failure detection per approach of 5×10^{-9}.

Perhaps the most powerful benefit provided by Integrity Beacons is the capacity for RAIM during precision approach and landing. A precision approach position solution based on GPS Integrity Beacons is overdetermined. Because of the redundancy of information and the centimeter-level precision of the measurements, tight thresholds on the solution rms residual (typically on the order of tens of centimeters) can be set for the detection of anomalous conditions.

Receiver autonomous integrity monitoring provides an important improvement over the traditional Ground Monitor Station used for integrity. Perhaps the most direct benefit of RAIM is that it covers failure modes in *all* segments of the system, including the aircraft segment. Ground monitors have no way of resolving these types of failures. Another problem with traditional ground integrity monitoring schemes alone [even with a landing system as mature as instrument landing systems (ILS)] is that there are still some nonaircraft error modes that can still slip through undetected. In the case of the GPS, one such potentially dangerous error mode is that recently associated with space vehicle pseudo random noise code (PRN) 19.[10] In the case of PRN 19, a GPS signal that was somewhat abnormal affected receivers from different manufacturers in different ways over a period of a few months. Some receivers experienced range biases that differed by several meters. Suppose that both the ground reference receiver and the ground integrity monitor were affected identically, while the aircraft receiver was affected differently. No integrity warning would be issued. However, a potentially dangerous situation for the aircraft could exist.

The IBLS (which provides both ground monitoring *and* RAIM) is immune to the class of PRN 19-type anomalies and other failure modes, because it employs the GPS carrier to solve explicitly for all range biases (see Sec. IV). An integrity detection scheme that emphasizes RAIM enables the ultimate integrity decision to be made by the aircraft, not the ground. This autonomous decision capability ensures that, regardless of the state of the system ground components (including any monitoring equipment), there is always more than enough information for the aircraft to make an independent assessment of integrity.

Overall system integrity is analyzed quantitatively in Sec. VII. Analysis indicates that a probability of missed detection of 10^{-9} in actual flight operations will be achievable through the use of Integrity Beacons.

C. Availability

Availability is the fraction of time that the complete landing system will be able to carry out its function at the initiation of the intended approach. Note that in this case, *availability* also implies the satellite constellation geometry for the availability of *integrity*.

Use of the GPS Integrity Beacons significantly augments the availability of GPS landing capability by providing additional ranging measurements. IBLS requires only four GPS satellites for full performance when an inertial reference unit (IRU) is employed (see Sec. VI). Therefore, the availability of IBLS is projected to be *better than that of enroute GPS navigation,* which requires more than four satellites for the redundancy needed to carry out RAIM. IBLS also provides significant margin against high PDOP (see Sec. V.C). These advantages provide important protection in scenarios where satellite failures reduce coverage over a large geographical area. In such scenarios, flying to an adjacent airport is not a practical alternative.

D. Continuity

Continuity is a measure of interruptions in the system operation once an approach has been initiated. Because of the high accuracy afforded by the GPS Integrity Beacon concept, the probability of RAIM integrity false alarms can be brought to 10^{-7} (the continuity of service requirement) *or lower* if desired. Section VII provides a further discussion on the issues involved.

It may be possible to rely simply on redundant GPS satellites to satisfy continuity of integrity, especially if direct ILS receiver replacement is adopted for retrofitting existing aircraft. In the long run, however, it may be more beneficial to optimize the system to take full advantage of the IRU that is typically employed to drive the control surfaces of the aircraft. In this case, the GPS Integrity Beacon can be used to initialize accurately the three-axis position biases of the inertial unit at a safe altitude, allowing the aircraft to continue on even in the event of jamming or a complete failure of GPS.

III. Integrity Beacon Architecture

The centerpiece of IBLS is the Integrity Beacon itself. Working in conjunction with a traditional local area differential GPS ground station, this low-power transmitter provides the required ground augmentation of GPS for an aircraft to carry out a high-integrity landing. In this section, two types of Integrity Beacons are described: the *Doppler Marker* and the *Omni Marker*.

A. Doppler Marker

The *Doppler Marker* Integrity Beacon is an independent, low-power GPS signal transmitter that interfaces directly to the differential reference station. The Doppler marker circuit board is shown in Fig. 2. Designed to be the size of a business card, this transmitter is capable of running for more than half a day on an ordinary 9-volt battery. For a 300-m radius bubble, the transmitted power is on the order of a microwatt (-30 dBm). This type of GPS Integrity Beacon was used for the flight test results presented in Sec. V.

Figure 3 shows a block diagram of the complete system using the Doppler Marker Integrity Beacon. The signals from the beacons are fed directly into the differential reference station. The reference station measures the carrier phase of both the Integrity Beacon signals and the GPS satellite signals. Both sets of measurements are transmitted as a group up to the aircraft via the traditional differential data link.

B. Omni Marker

The Omni Marker is a more advanced version of the Integrity Beacon that offers several improvements over the Doppler Marker. When applied to IBLS, the Omni Marker eliminates most of the need for cabling to connect the various components of the ground system. The concept is illustrated in Fig. 4. In one continuous signal-processing chain, the Omni Marker locks onto the GPS signal from the receive antenna, strips off the satellite PRN code, reapplies a new pseudolite PRN code to the carrier, and rebroadcasts the signal. The outgoing code and carrier are kept phase coherent with respect to their incoming counter-

Fig. 2 GPS Integrity Beacon hardware.

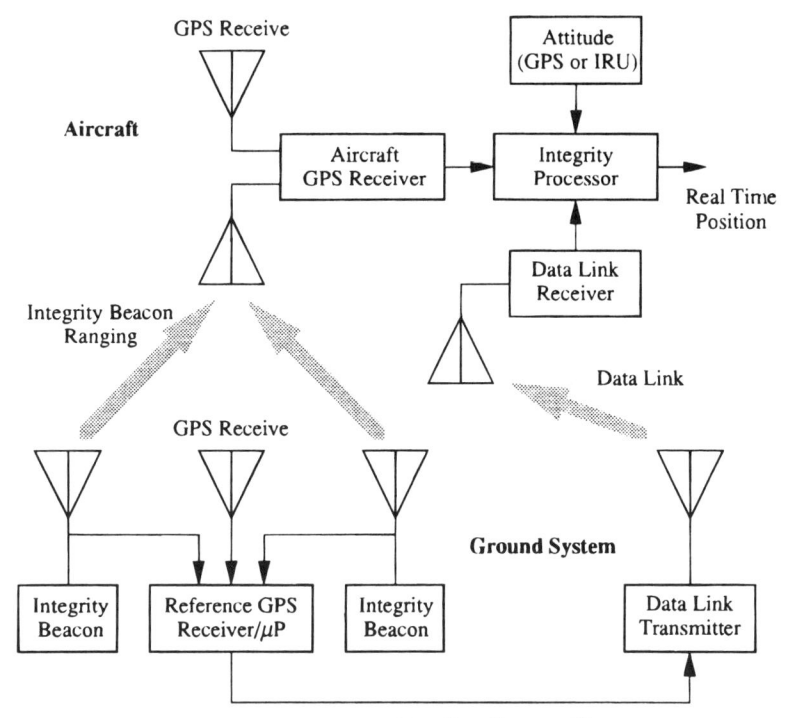

AIRCRAFT PRECISION LANDING USING INTEGRITY BEACONS 433

Fig. 3 Integrity Beacon Landing System diagram.

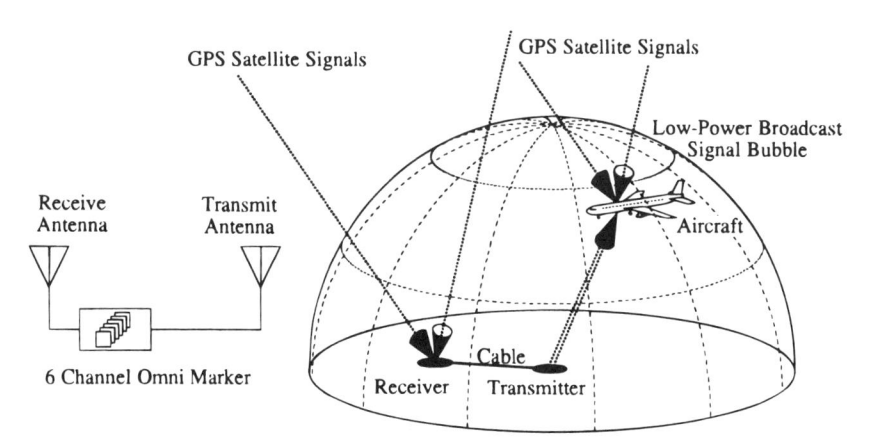

Fig. 4 Omni Marker Integrity Beacon concept.

parts. Ideally, the transmit and receive antennas would be collocated, but in practice some separation may be required for radio frequency isolation. Each Omni Marker Integrity Beacon is autonomous and independent. The result is a reliable, federated architecture that is much less vulnerable to individual component failures.

In addition to being the key to dependably resolving cycle ambiguities, the omni marker can also serve to relay the ground reference measurements to the aircraft. The digital data link traditionally used in DGPS (usually situated centrally at the airport) can be replaced by a coherent rebroadcast of all the received GPS signals. This application is further described in Ref. 71.

IV. Mathematics of Cycle Resolution

This section presents the mathematics of cycle ambiguity resolution using Integrity Beacons. Section IV.A on observability analysis offers a qualitative description of how Integrity Beacons are used and a look at what components of position can be resolved. Section IV.B on matrix formulation offers a quantitative description of how IBLS is able to carry out precision positioning and achieve high integrity.

A. Observability Analysis

A simplified analysis illustrates how a single Integrity Beacon provides both altitude (radial) and along-track position. Figure 5 shows a simple linear trajectory directly over the Integrity Beacon, located at the origin. The aircraft coordinates are along-track position x, cross-track (lateral, into the page) position y, and constant altitude (radial) z. The magnitude of the aircraft position vector (x,y,z) is the range to the pseudolite r. The measured range φ is the single-difference carrier phase measured between the ground reference receiver and the aircraft receiver via its belly-mounted antenna:

$$\varphi \equiv \varphi_{\text{aircraft}} - \varphi_{\text{reference}} = r + b - \Delta t_{\text{aircraft}} + \Delta t_{\text{reference}}$$

where b represents the sum of the cycle ambiguity and all system and cable biases and delays for the Integrity Beacon ranging link, and each Δt represents each receiver clock bias. Initialized with a trial trajectory from differential GPS, conventional kinematic positioning is used to eliminate the relative clock bias ($\Delta t_{\text{aircraft}} - \Delta t_{\text{reference}}$) between the two receivers. The Integrity Beacon range can then be measured directly, subject to the additive bias b.

$$\varphi' \equiv \varphi + (\Delta t_{\text{aircraft}} - \Delta t_{\text{reference}}) = r + b = \sqrt{x^2 + y^2 + z^2} + b$$

Linearizing the measured phase about the nominal trajectory, which (nominally)

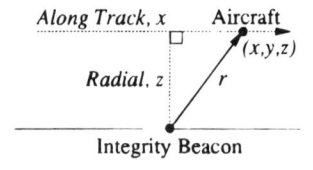

Fig. 5 Overflight geometry.

runs directly over the pseudolite ($y = 0$), it can be shown that the nominal observation matrix is given by the following:

$$d\varphi'|_{y=0} = \begin{bmatrix} \dfrac{x}{r} & 0 & \dfrac{z}{r} & 1 \end{bmatrix} \begin{bmatrix} dx \\ dy \\ dz \\ db \end{bmatrix}$$

For reference, these observation functions (i.e., the row vector in the above equation) are plotted in Fig. 6 (normalized by the radius of closest approach) as a function of along track position. As long as the ranging signal is observed over a large enough arc, each observable component of the Integrity Beacon ranging signal is clearly distinguishable, including the most important parameter, altitude.

Note that the cross-track (lateral) component of position y is unobservable with a simple, linear trajectory over the Integrity Beacon. *For this reason, the dual Integrity Beacon configuration of Fig. 1 is used.* With two transmitters on either side of the glide slope, all three components of position are directly observable. In many cases, it may also be possible to use the same information from another Integrity Beacon placed under the glide slope of a parallel runway.

B. Matrix Formulation

To provide further insight as to how IBLS is able to provide such high performance accuracy and integrity using the GPS, the matrix formulation of cycle ambiguity resolution is presented here. The mathematical development of the system is defined with respect to Fig. 2, which shows a block diagram of the flight test system. The development of cycle ambiguity resolution algorithms is most easily done within the context of conventional carrier-based differential ranging with the Doppler marker, although the development is readily adaptable to the omni marker Integrity Beacon. Figure 7 serves as a guide for the vector definitions employed herein.

Single differencing of raw carrier phase measurements obtained at airborne and reference station receivers yields for the space vehicle (SV) i at epoch k the following:

$$\varphi_{ik} = -\hat{s}_{ik}^T x_k + \tau_k + N_i^s + \epsilon_{ik}^s$$

where φ_{ik} is the single-differenced (aircraft minus reference) SV phase; $\hat{s}_{ik}$ is the

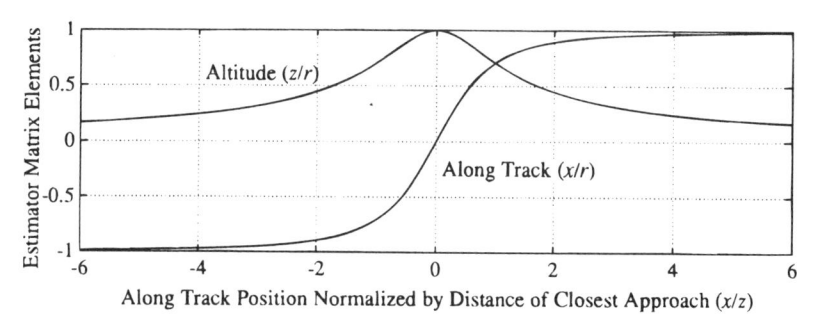

Fig. 6 Error profile for a pseudolite pass.

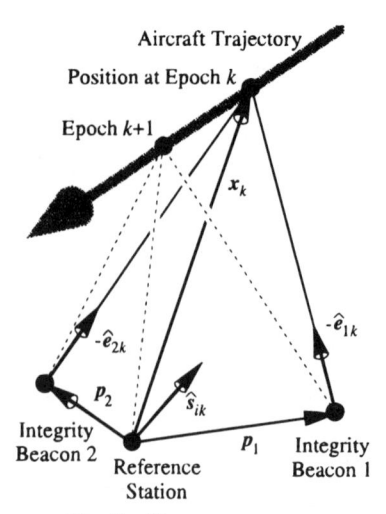

![Fig. 7 Vector geometry showing aircraft trajectory, position at Epoch k and k+1, integrity beacons 1 and 2, reference station, and vectors x_k, p_1, p_2, $\hat{s}_{ik}$, $-\hat{e}_{1k}$, $-\hat{e}_{2k}$.]

Fig. 7 Vector geometry.

line-of-sight unit vector to the SV; x_k is the displacement vector from the differential station GPS receive antenna to the top-mounted aircraft GPS antenna; τ_k is the difference in the aircraft and reference receiver clock biases; N_i^s is the satellite integer cycle ambiguity; and ϵ_{ik}^s is the satellite range measurement error caused by multipath and receiver noise. Similarly for Integrity Beacon j at epoch k, we have the following:

$$\phi_{jk} = |p_j - x_k| - |p_j| + \tau_k + N_j^p + \epsilon_{jk}^p$$

where ϕ_{jk} is the single-differenced Integrity Beacon phase, and p_j is the vector from the differential station to Integrity Beacon j. Because the transmitter is quite close, the formulation for satellites (whose wave fronts are essentially planar) is not appropriate. Instead, the use of range magnitude is necessary.

Given an approximate trajectory $\bar{x}_k$ obtained from code-based DGPS, the equations above can be expressed in terms of the deviation from the approximate trajectory: $\delta x_k \equiv x_k - \bar{x}_k$. Keeping first-order terms only, the result is as follows:

$$\delta\varphi_{ik} \equiv \varphi_{ik} + \hat{s}_{ik}^T \bar{x}_k = -\hat{s}_{ik}^T \delta x_k + \tau_k + N_i^s + \epsilon_{ik}^s$$

and

$$\delta\phi_{jk} \equiv \phi_{jk} - |p_j - \bar{x}_k| + |p_j| = -\hat{e}_{jk}^T \delta x_k + \tau_k + N_j^p + \epsilon_{jk}^p$$

where $\hat{e}_{jk}^T \equiv (p_j - \bar{x}_k)/|p_j - \bar{x}_k|$. To resolve cycle ambiguities, the value of one integer must be specified because of the existence of the clock bias τ_k, which is common to all measurements at epoch k. For simplicity, we choose $N_1^s = 0$. Defining $\delta\Phi_k$ to be the vector of m SV and two Integrity Beacon measurements at epoch k

AIRCRAFT PRECISION LANDING USING INTEGRITY BEACONS 437

$$\delta\Phi_k \equiv \begin{bmatrix} \delta\varphi_{1k} \\ \vdots \\ \delta\varphi_{mk} \\ \delta\phi_{1k} \\ \delta\phi_{2k} \end{bmatrix}, \quad \text{and} \quad \hat{S}_k \quad \text{as} \quad \hat{S}_k \equiv \begin{bmatrix} -\hat{s}_{1k}^T & 1 \\ \vdots & \vdots \\ -\hat{s}_{mk}^T & 1 \\ -\hat{e}_{1k}^T & 1 \\ -\hat{e}_{2k}^T & 1 \end{bmatrix}$$

we stack all n measurements collected during Integrity Beacon overpass to obtain the following:

$$\begin{bmatrix} \delta\Phi_1 \\ \vdots \\ \delta\Phi_k \\ \vdots \\ \delta\Phi_n \end{bmatrix} = \begin{bmatrix} \hat{S}_1 & 0 & \cdots & 0 & 0 & \bar{I} \\ 0 & \ddots & 0 & \ddots & 0 & \vdots \\ \vdots & \ddots & \hat{S}_k & \ddots & \vdots & \bar{I} \\ 0 & \ddots & 0 & \ddots & 0 & \vdots \\ 0 & 0 & \cdots & 0 & \hat{S}_n & \bar{I} \end{bmatrix} \begin{bmatrix} \delta x_1^* \\ \vdots \\ \delta x_k^* \\ \vdots \\ \delta x_n^* \\ N \end{bmatrix} + \epsilon$$

where

$$\bar{I} = \begin{bmatrix} 0 & \cdots & 0 \\ 1 & \cdots & 0 \\ \vdots & \ddots & \vdots \\ 0 & \cdots & 1 \end{bmatrix}$$

$$\delta x_k^* = \begin{bmatrix} \delta x_k \\ \tau_k \end{bmatrix}$$

and

$$N = [N_2^s \quad \cdots \quad N_m^s \quad N_1^p \quad N_2^p]^T$$

The least-squares solution to the above can be obtained efficiently by sparse matrix batch algorithms or equivalently by sequential forward–backward smoothing. Because of the nonlinear nature of the problem, the "solution" δx_k is not the final answer. Instead, the approximate trajectory and observation matrix must be improved by the computed estimate of δx_k, and the process above repeated through convergence (i.e., until the update δx_k becomes negligible). Computation time for convergence takes considerably less than a second on a 25-MHz 486 PC. Experience has shown that the solution converges in 3–10 iterations. The current algorithm has been tested in simulation and always converged when presented with an initial condition within 300m of the correct value for a 100m altitude bubble pass. In repetitive flight trials, presented in the following section, the algorithm has converged on every approach. In the unlikely event of convergence failure, the signal to the pilot would be a continuity alarm at 200 ft, *not* an integrity problem.

V. Experimental Flight Testing

This section quantifies the centimeter-level accuracy of the landing system and describes a sampling of the real-time flight testing that has occurred. Most of this section covers flight tests that use a laser tracker as the means for establishing navigation sensor accuracy. For the purposes of Category III precision landing, the approximate 1-ft accuracy of the laser tracker is satisfactory. However, it is claimed throughout this chapter that IBLS is capable of providing centimeter-level accuracy—an order of magnitude better than the laser tracker. The following introduction is provided to quantify these claims.

A. Quantification of Centimeter-Level Accuracy

Quantification of the centimeter-level accuracy of the system is indirect, because there is no other independent positioning sensor (including a laser tracker) known that is practical enough to facilitate comparison to the required level of performance. Therefore, a partial list of indirect means for establishing the centimeter accuracy of IBLS are given as follows:
1) Position checks against independent GPS static surveying.
2) Dynamic comparison of attitude determination using GPS against an IRU.

1. Position Checks Against Independent GPS Static Surveying

This method of comparison checks kinematic GPS initialized with Integrity Beacons against static survey with the GPS after the airplane lands. Each of these techniques measures position with respect to the defined reference runway coordinate frame. As presented in Sec. V.A, the quantitative agreement between these two independent means of positioning support absolute positioning accuracy to the centimeter level. In fact, centimeter-level static results are routinely obtained in surveying with satellite range rates in excess of 1 km/s.

2. Dynamic Comparison of Attitude Determination Using GPS Against an Inertial Reference Unit

This method of comparison checks relative positioning of antennas on an aircraft using GPS (employed primarily as a means of attitude determination) and the same quantity determined from an independent IRU. Chapter 19, Fig. 11 (this volume) shows an example of such relative positioning. Translating the 0.05 deg angular error of the IRU through a baseline of 16 m, the resulting dynamic position error is 1.4 cm rms. Translational experiments used to check GPS against an IRU position provide similar results over the short term, such as in Ref. 12.

These two comparison techniques—one static and absolute; the other dynamic and relative—combine to support the centimeter-level positioning accuracy using IBLS. Static survey results indicate the absolute accuracy of IBLS. Then, inertial comparison is one way to validate that kinematic survey techniques are just as accurate as static survey.

B. Piper Dakota Experimental Flight Trials

For the flight trials, the landing system hardware shown in Fig. 2 was set up in a single-engine Piper Dakota. Onboard the aircraft, signals from a Trimble TANS Quadrex receiver (specially modified for precision landing at Stanford University) were fed into the flight computer. A second GPS receiver, a TANS Vector, was used for attitude determination to supply the lever-arm correction for the positions of the belly-mounted Integrity Beacon receive antenna, landing gear, laser altimeter, and laser tracker retro reflector. Figure 8 is a photograph of the aircraft in flight. The fig. shows the four GPS antennas used for attitude determination mounted on the fuselage, tail, and each wingtip. Figure 9 shows a close-up of the integrity antenna mounted on the underside.

The relative positions of the ground station reference and Integrity Beacon transmit antennas are known to the subcentimeter level using standard GPS static surveying techniques. This procedure emphasizes one of the operational advantages of IBLS. The ground antennas may be placed wherever convenient. Then, they are self-surveyed with GPS. The system has proved it is ready to support the first flight inspection landing within an hour of the initial antenna placement.

By employing the simple hemispherical, upward-looking antenna pattern, site-specific multipath is of negligible consquence. Carrier multipath (which is approximately 1000 times smaller than code phase multipath) typically accounts for less than 0.5 cm of ranging error. Therefore, the flight inspection process is actually required only as a simple check of the installation database parameters and for obstacle clearance. Because Integrity Beacon antennas are upward-looking rather than side-looking, as with ILS, site-specific multipath is not an operational issue with GPS augmented with Integrity Beacons. Flight inspection should be simplified.

1. Independent GPS Survey Results

To demonstrate the centimeter-level accuracy of positioning using Integrity Beacons, position fixes from IBLS were compared with those from an independent GPS static survey. The cycle ambiguities were resolved in flight using Integrity Beacons broadcasting from the approach path at Palo Alto (CA) airport 1-km out. The aircraft flew an approach and landed without losing lock on the integers.

Fig. 8 Piper Dakota flight test aircraft.

Fig. 9 Belly-mounted integrity antenna.

After the aircraft was secured at the tie-downs, the final real-time position fix output by IBLS was recorded and compared to an independent estimate of this final position obtained using standard commercial static GPS survey receivers and software. These results are shown in Table 1. The quantity marked Δ is the difference between IBLS and the static survey. The quantity σ is the estimated standard deviation of Δ based on the covariance derived from the measurement geometry of the bubble pass. Repeated comparison experiments of this sort have shown that the static agreement is consistently on the centimeter level. Based on the discussion at the start of this section, it is believed that this same level of accuracy is available on a point-by-point basis throughout the entire portion of the flight following the bubble pass.

Table 1 Comparison of the Integrity Beacon Landing System and static survey

cm	Altitude	In-track	Cross-track
Δ	−0.1 cm	1.2 cm	1.4 cm
σ	1.7 cm	0.6 cm	1.1 cm

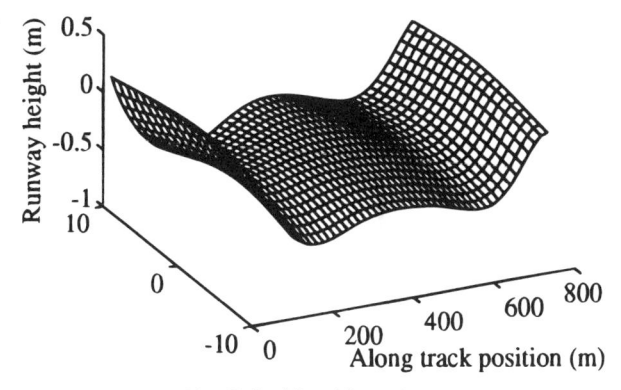

Fig. 10 Palo Alto Airport runway.

2. Laser Altimeter Comparison

For the laser altimeter tests, a laser rangefinder was installed in the aircraft.[2] To compare the laser altimeter data with IBLS position fixes, an accurate model of the Palo Alto Airport runway height was needed. Over a half-hour period, a comprehensive kinematic GPS survey was performed by driving a golf cart up and down the runway. A three-dimensional surface was fitted to these data to give a precise computerized model of runway height as a function of horizontal position.

Using 15 coefficients, the surface shown in Fig. 10 was generated. Very little memory storage is required to represent a runway in this manner range of the coefficients of 6 m, the total required data storage is only 15 bytes using fixed-point storage and a resolution of 5 cm. With this simple model, IBLS is able to output accurate height above the runway without the need for an extensive runway data base. The total measured discrepancy between the GPS and the laser altimeter for a series of seven touch-and-gos is summarized in Table 2.

3. Laser Trackers

To obtain real-time confirmation of IBLS accuracy in all three axes, a number of flight tests using laser trackers have been carried out. These include approaches with the Piper Dakota at the NASA Ames Crows Landing facility in August, 1993,[2] autocoupled approaches with an FAA Beech King Air at the FAA Technical Center in July and August, 1994,[11] and automatic landings of a United Airlines Boeing 737 at NASA Crows Landing in October, 1994.[13] The laser tracker

Table 2 Integrity Beacon Landing System and laser altimeter differences on Piper Dakota (7 approaches)

Mean	Standard deviation	Total estimated measures Error
3 cm	11 cm	< 30 cm

Table 3 Navigation sensor error at 50ft on King Air (49 approaches)

NSE, m	Vertical	Cross-track	Along-track
Sigma (σ)	0.2 m	0.1 m	0.2 m
Mean (μ)	−0.1 m	0.1 m	−0.2 m
$\|\mu\| + 2\sigma$ (95% error)	0.5 m	0.3 m	0.6 m

measures the azimuth, elevation, and range to the retroreflector mounted on the aircraft. The range accuracy is specified at ±0.3 m. Azimuth and elevation accuracy at the two facilities are specified at better than ±0.2 mrad 1σ. The laser tracker results have proved to be nearly identical for each flight trial, essentially bounding the accuracy of IBLS by that of the laser tracker. Furthermore, the accuracy is effectively independent of the type of aircraft employed. Representative data from the flight trials detail the laser-tracking results in the following subsections.

C. Federal Aviation Administration Beech King Air Autocoupled Approaches

In July and August, 1994, a series of 49 autocoupled approaches were carried out in an FAA Beech King Air at the FAA Technical Center. Initially the aircraft was guided using conventional code-based DGPS to bring it down the approach path over the Integrity Beacons at 600 ft. Upon bubble exit at roughly 500 ft, the system performed its cycle ambiguity calculation (in roughly 0.2 s using a Pentium processor) and assumed its precise-positioning mode. With the safety pilot monitoring the ILS, the pilot had the option to disengage the autopilot at 100 ft. However, because of the steady guidance being displayed and the smooth descent of the aircraft, the pilots typically left the autopilot engaged down to 50 ft or lower. On all 49 approaches, IBLS successfully resolved the cycle ambiguities, performed its internal onboard integrity checks, provided navigation output to within the accuracy of the laser tracker (or better), and autocoupled into the flight controls to guide the aircraft through the approach.

1. Navigation Sensor Error

Ensemble statistics for the NSE at 50-ft altitude are assembled in Table 3. Based on previous calibration experiments (see Independent GPS Survey Results, above), it is believed that most of the error is attributable to the laser tracker. The vertical NSE of 0.5 m meets the 0.6 m, 95% error requirement found in both ICAO Annex 10[7] and the Federal Radionavigation Plan (FRP). The ensemble statistics for cross-track NSE of 0.3 m exceeds both the ICAO 95% requirement of 4.4 m and the FRP 95% requirement of 4.1 m by a wide margin.

Table 4 shows statistics that summarize the accuracy achieved with respect to requirements.

2. Total System Error

Figure 11 shows the vertical TSE for the approaches. For comparison, the 95% inner tunnel boundaries[9] are superimposed on the plot. The TSE is plotted

Table 4 Integrity Beacon navigation sensor error performance on King Air (49 approaches)

95% Error, m	Vertical	Cross-track
Integrity beacon	0.5 m	0.3 m
ICAO Annex 10[7]	0.6 m	4.4 m
Federal Radionavigation Plan	0.6 m	4.1 m
Meets requirement	YES	YES

as a function of altitude for a nominal 3-deg glide slope. In other words, the along-track component of position is scaled to the nominal 3-deg glide slope altitude where the aircraft should be for each particular along-track position.

The plots begin at the bubble exit point, so the first part of the plot shows the small transient that occurs as the sensor accuracy increases. At the extreme right, the pilot breaks off the pass near the aim point and resumes manual control. Using a Category I autopilot, at no point does the TSE approach the 95% Category III inner-tunnel boundary. In other words, with the near-perfect sensor accuracy of IBLS, a Category I autopilot was capable of achieving Category III vertical TSE specifications. In the long run, the nearly perfect NSE of the Integrity Beacon may enable a future generation of safe, low-cost landing systems based on less expensive autopilots or pilot-in-the-loop graphical displays. These systems may be able to meet the TSE requirements by allowing for larger FTE.

The TSE results are summarized in Table 5 at 50-ft altitude. In spite of a significant cross-track bias in the autopilot, this Category I autopilot meets Category III specifications.

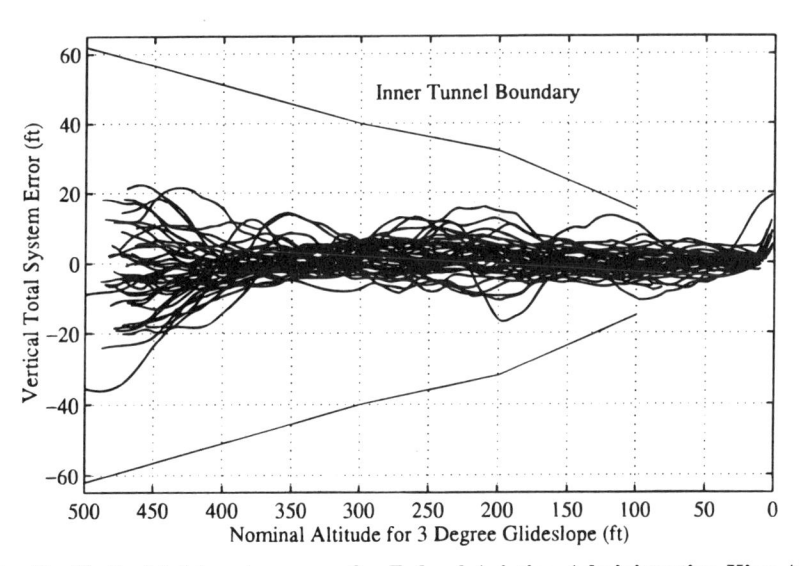

Fig. 11 Vertical total system error for Federal Aviation Administration King Air approaches.

Table 5 Integrity Beacon total system error at 50-ft altitude on King Air

Total system error, m	Vertical	Cross-track
Sigma (σ)	1.0 m	2.7 m
Mean (μ)	−0.5 m	5.3 m
$|\mu| + 2\sigma$ (95% error)	2.5 m	10.7 m
Required navigation performance tunnel	4.5 m (TBD)	15.5 m
Meets requirement	YES	YES

D. Automatic Landings of a United Boeing 737

In October, 1994, a United Airlines Boeing 737-300 was modified to accept IBLS guidance and was used to carry out 110 successful automatic landings using the GPS. Figure 12 shows the aircraft just following touchdown during one of its 110 autolands at NASA's Crows Landing research facility in California's Central Valley. These flight trials were sponsored by the FAA to help establish the feasibility of Category III precision landing using GPS.

The landing system configuration is shown in Fig. 13. A pair of standard ARINC 743 GPS antennas were mounted on fuselage—one on the top and and one on the underside of the aircraft. The Trimble TANS GPS receiver sent raw carrier-phase measurements to the navigation processor. These phase measurements—combined with the data link messages received through a VHF blade antenna on the top of the aircraft and attitude measurements from the inertial unit—provided the raw information for the IBLS flight computer to calculate precise aircraft position and glide path deviation. From the single-channel navigation processor, a dual-channel analog interface provided ILS localizer and glide slope signals to the autopilot. The 737-300 is equipped with a dual-channel flight control system designed for Category IIIA landings. The autolands were performed through touchdown without roll-out guidance.

A total of 111 approaches were attempted with 110 resulting in successful autolands. When the aircraft was at about 300 ft of altitude on the 37th approach (following the bubble pass), a U.S. Air Force upload transmission temporarily brought down the signal of one of the GPS satellites. As intended in the design, the landing system responded by raising a flag and calling off the approach

Fig. 12 One of 110 United Airlines Boeing 737 autolands.

AIRCRAFT PRECISION LANDING USING INTEGRITY BEACONS

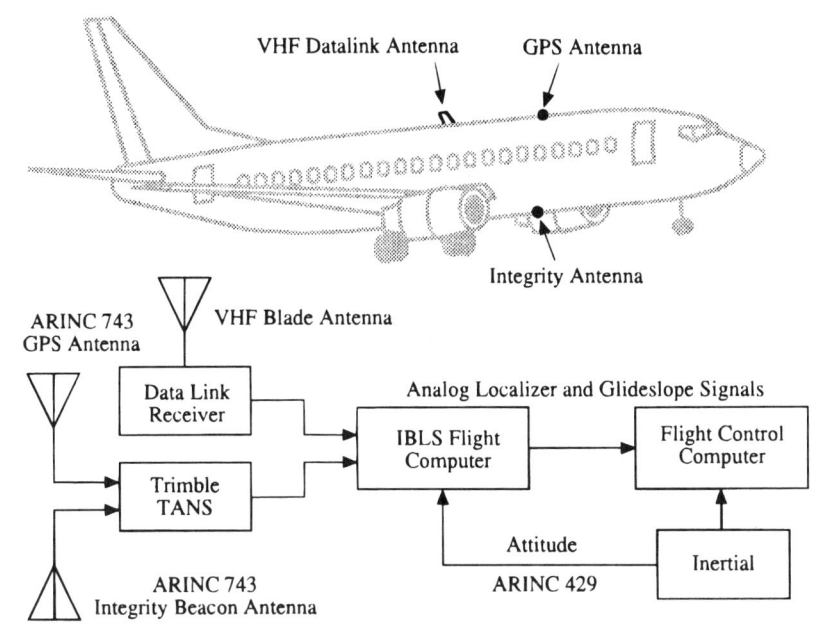

Fig. 13 Airborne configuration of United Boeing 737-300 Autoland Tests.

within 1/4 s of the event. The Category III specification for time-to-alarm caused by a system fault is 2 s. A second-generation system now allows landing to continue past such rare events using redundant GPS satellites or an inertial unit. Out of the total of 111 approaches flown, there were no false alarms and no missed detections.

1. Navigation Sensor Error

Figure 14 shows the vertical (most challenging) NSE for 100 of the autolands using the laser tracker as a reference. The plot is given as a function of distance to the aim point, converted into units of altitude assuming the standard 3-degree glide slope. To ensure that the approaches represented a true basis for operational evaluation, the plot shows only those 100 autolands for which the cycle ambiguities were intentionally reset (cleared) as a matter of procedure upon rollout onto final approach. (For experimentation purposes during some of the other autolands, it was demonstrated that the integers from a previous touch-and-go could be successfully carried around the pattern through to the next bubble pass.)

Again, it is believed that the error shown in Fig. 14 is dominated by the laser tracker. The standard error signature of the angular-based laser tracker is readily apparent in the plot as the spread on the vertical error increases with range. Prior to the advent of GPS, laser trackers have been traditionally considered the most accurate and convenient means of independently establishing position. It is interesting that GPS can be credited with finding new sources of error in laser trackers not previously considered nor encountered in this application. In the flight trials

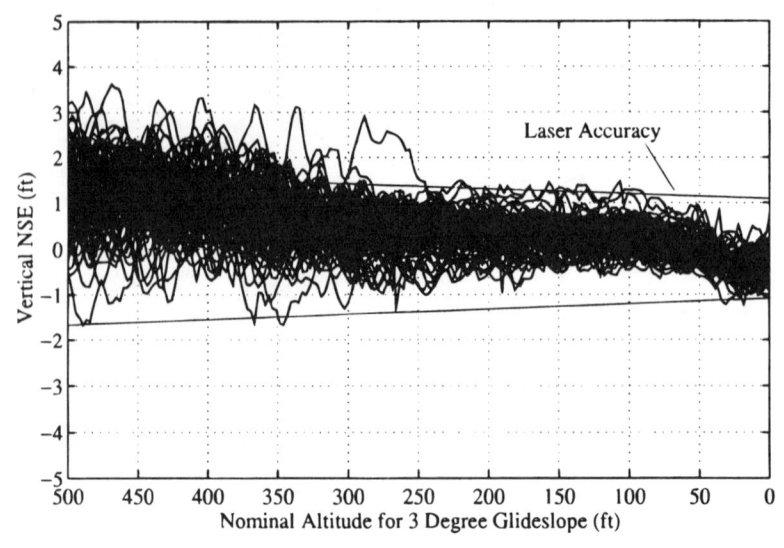

Fig. 14 Vertical navigation sensor error for 100 autolands.

at different test ranges, systematic errors found during data analysis on the order of tenths of a milliradian were traced to the omission of a correction for tropospheric refraction of the laser beam attributable to the gradient of atmospheric density with altitude. It also seems that other small systematic errors may still remain, such as harmonic noise in the elevation resolvers.

2. Safety Margin Attributable to Enhanced Availability

As a demonstration of the capacity of IBLS to provide enhanced availability, many of the approaches shown used a satellite selection algorithm picking the four highest elevation satellites in the sky. Occasionally this algorithm yields a PDOP greater than 10 for the four satellites. Interestingly, the positioning error in these high PDOP approaches is unnoticeable, because it is still outweighed by the laser error. In Fig. 14, even three autolands with satellite PDOP in the range of 17–18 do not stand out from the rest. In aircraft navigation applications with less-challenging performance requirements than Category III precision landing, a PDOP this large would be considered unusable. However, with IBLS, satellite PDOP of 18 resulted in no more than 1 ft of vertical error.

The statistical results confirm the high accuracy of the system. As expected, the statistics for NSE are comparable to those in Table 3. Table 6 summarizes

Table 6 Integrity Beacon flight technical error on the United Boeing 737-300

Flight technical error, m	Vertical, 100 ft	Cross-track, 100 ft	Vertical, 50 ft	Cross-track, 50 ft
Sigma (σ)	1.1 m	2.2 m	1.0 m	2.1 m
Mean (μ)	0.1 m	0.2 m	0.1 m	0.1 m
$|\mu| + 2\sigma$ (95%)	2.3 m	4.6 m	2.1 m	4.3 m

AIRCRAFT PRECISION LANDING USING INTEGRITY BEACONS 447

Table 7 Integrity Beacon flight technical error (total system error) performance on United Boeing 737 at 50 ft

95% Error, m	Vertical	Cross-track
Integrity beacon	2.1 m	4.3 m
Required navigation performance tunnel	4.5 m (TBD)	15.5 m
Meets requirement	Yes	Yes

the vertical and cross-track FTE (essentially the same as TSE for IBLS) at 50- and 100-ft altitude for the 110 autolands.

Again, using the RNP tunnel as the basis for comparison, Table 7 is constructed showing that the near-perfect NSE of IBLS allows the autopilot to reach its theoretical maximum of performance. Note that the crosswind component for approximately 20% of the landings exceeded the autopilot specification of 10 knots, sometimes by almost a factor of two. Despite this adverse condition, the system was still able to outperform any known proposed specification for TSE.

E. Flight Test Summary and Observations

The IBLS was shown to be sufficient in meeting accuracy requirements for Category III automatic landings. Additionally, the integrity of the cycle ambiguity resolution process, throughout the total of 160 test approaches (49 on the FAA King Air and 111 on the United Boeing 737), was 100% successful. No false alarms were issued, and no missed detections were registered. Note that to match the ILS sensor characteristics (which have limited dynamic range), the GPS signal had to be delayed and filtered. This extra lag is unnecessary for accuracy and suggests an avenue for autopilot improvements using the greater dynamic sensitivity of IBLS.

The consensus among all the test pilots was that the aircraft flew smooth descents with a "solid" guidance signal. FAA Administrator David R. Hinson, who flew as pilot in the left seat of the FAA King Air for approaches 46 and 47, observed that the system "seemed to be much more stable than an ILS approach". On the King Air flights that ran on a Category III ILS runway, many of the test crew observed the "scalloping" of the ILS needles running in parallel, when compared to the GPS display, which held absolutely steady.

This benign response to both high PDOP and crosswinds, which significantly exceeded specifications, is representative of how the IBLS architecture has margin to handle stresses arising out of adverse operating conditions. In operational terms, insensitivity to high satellite PDOP translates into a significant safety margin of availability. It is equivalent to a GPS satellite failure scenario, where suddenly only a suboptimal satellite geometry is available to an aircraft on final approach. With the centimeter-level precision of IBLS ranging, PDOP less than 20 is of little consequence. That small ranging error multiplied by a geometric factor of 20 still yields a small position error.

VI. Operations Using Integrity Beacons

Depending upon the required minimums, an operational system can be supplemented with other existing sensors, such as an inertial reference unit (IRU) and a radar altimeter. The IBLS architecture is designed to satisfy operational requirements, performance specifications, and institutional issues in a way that minimizes cost and maximizes performance and efficiency.

At the airport, a pair of Integrity Beacons would be situated on either side of the approach path at approximately the range of the ILS middle marker or farther. Taking after the simple, low-cost, rugged prototype shown in Fig. 2, operational Integrity Beacons are likely to be just as routine to install and maintain as ordinary light bulbs placed around the airport. Each site can easily have redundant units broadcasting on different codes in the unlikely event of a ground failure, because Integrity Beacons do not interfere with one another.

The airborne component of operational IBLS does not differ much from that shown in Fig. 13. A benefit of the second GPS antenna (in addition to the improved level of safety on final approach) is that the aircraft has nearly 4π sr visibility of the sky and GPS. With this additional antenna, the aircraft is able to maintain lock on GPS, even during curved approaches, steep banks, and turns. The optimized airborne component also includes a loosely coupled IRU for operating at Category IIIB and IIIC minimums. For direct retrofit compatibility with existing autoland systems (either digital or analog), the GPS can emulate a traditional ILS output. The GPS receiver can also use attitude measurements from the existing IRU to calculate the lever-arm (relative position) correction between the upper and lower GPS antennas.

A. Integrity Beacon Landing System Landing Sequence

There are at least two assumed means of implementing IBLS on board aircraft: as an ILS retrofit or as a fully optimized GPS/IRU package. The ILS retrofit uses the GPS to emulate the standard ILS signal fed into the autopilot. The autopilot also makes use of the IRU measurements, but not with any coupling back to the ILS receiver. In the optimized package, GPS and IRU data are filtered together in Cartesian coordinates (not localizer and glide slope coordinates), so that the landing system has full benefit of three-axis position (and velocity) from the GPS. For an aircraft on final approach, there is not much difference between the two implementations until the conclusion of the integrity bubble pass.

The chronology of approach and landing is shown graphically in Fig. 15. Upon initiation of the approach, DGPS is used to navigate the aircraft to the integrity bubble. There—at a safe altitude—the Integrity Beacon serves as the final checkpoint before landing and defines the transition point to the high-

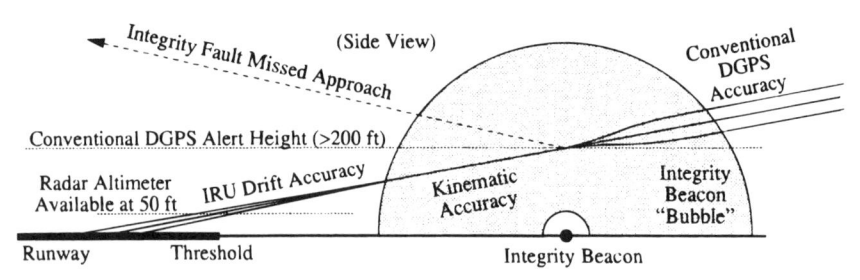

Fig. 15 Phases of Integrity Beacon system precision landing.

integrity GPS operation. Once well inside the signal bubble, the aircraft receiver has enough information to resolve cycle ambiguities and initialize its output to centimeter-level accuracy. Toward the end of the bubble pass, the GPS positioning has the highest integrity of any regime of flight. An important requirement of the design is that it preserve this level of integrity all the way through touchdown. The preferred methods for carrying this out are different for each airborne implementation of IBLS described in the following subsections.

1. Retrofit Instrument Landing Systems

With stand-alone GPS being used to emulate ILS (in a system topology similar to that in Fig. 13), the simplest means of preserving performance after the bubble exit is to maintain kinematic centimeter-level positioning all through touchdown. If five or more GPS satellites are visible (the same minimum requirements for enroute GPS navigation), RAIM can be continued with nearly the same effectiveness as that achieved inside the bubble. Once the cycle ambiguities are properly initialized inside the bubble, the same tight thresholds for integrity alarms applied within the bubble (as described in Sec. VII) can be applied to the carrier-phase positioning residuals. If a pseudolite is used at the airport to service the vicinity of touchdown with a modulation scheme to ensure adequate reception,[14,15] it is possible that its signal may be useable as a redundant measurement. Then, only four satellites are required for RAIM.

2. Optimized GPS/IRU

Given that the IRU velocity and scale factors are calibrated in flight using DGPS prior to reaching the bubble, the IRU position and velocity can be updated to kinematic accuracy at the time of the bubble pass. Thereafter, the IRU will preserve the required position accuracy (and integrity when checked against kinematic DGPS or another IRU) during the 15–20 s between the 200-ft alert height and the landing. Assuming that the system passes its internal integrity checks at the altitude of the bubble pass (at or above 200 ft), the aircraft continues its descent, navigating directly from the IRU as initialized by IBLS. Kinematic GPS continues to serve as an integrity "safety net," but even if there is any subsequent GPS failure or radio jamming of any sort, the aircraft can continue the landing, because the IRU has already been initialized. At roughly 50-ft altitude, the aircraft is over the threshold, and the radar altimeter (backed up by kinematic GPS, as demonstrated in Sec. V.A) can be employed for the flare maneuver.

Landing sequence is summarized as follows.
1) Initiate the approach using traditional differential GPS (accuracy: 2–5 m).
2) Acquire Integrity Beacon above 200-ft DGPS alert altitude.
3) Perform positioning/RAIM (integrity: missed detection probability 10^{-9}).
4) Execute go/no go decision above 200-ft alert altitude.
5) Final update of IRU position and velocity (accuracy: 2–5 cm rms).
6) Execute landing with IRU, radar altimeter, GPS; maintain 10^{-9} integrity.

VII. Integrity Beacon Landing System Navigation Integrity

The requirements on accuracy, integrity, and continuity for Category III precision approach demand the highest level of the GPS navigation performance. Specifically, combined navigation and flight control accuracy on the order of a few meters must be maintained, continuity of function preserved for all but one in 10 million (10^{-7}) approaches, and loss of integrity limited to *one in a billion* (10^{-9}) approaches.[9] Although high-accuracy navigation is possible using either differential high-performance C/A-code or differential L_1 carrier phase, the *centimeter-level* precision afforded by carrier phase provides two clear advantages. First, as previously stated, NSE represents a nearly negligible contribution to TSE. This leads to maximum margin in FTE and, therefore, maximum flexibility in flight control system design. Second, the high precision of carrier phase provides a foundation for a high level of RAIM performance.

The high performance of carrier phase can only be achieved, however, if the integer cycle ambiguities can be reliably resolved for each space vehicle (SV). IBLS is a *high-integrity* solution to real-time cycle ambiguity resolution for Category III precision approach, because the capacity for RAIM is built-in.[16] In this regard, two important observations can be made:

1) The centimeter-level precision of carrier-phase measurements provide maximum benefit from RAIM in the sense that extremely tight detection thresholds may be set without incurring unacceptably high false alarm rates. Therefore, both *high integrity* and *high continuity* are ensured.

2) The redundant ranging measurements obtained from ground-based pseudolites ensure the *availability* of RAIM.

Thus, the traditional limitations associated with high-performance navigation using RAIM-based fault detection (pseudorange measurement accuracy and low SV availability) do not exist when carrier-phase measurements are used and Integrity Beacons are present. Consequently, the integrity of IBLS-based cycle ambiguity resolution and positioning inside the IBLS bubble can be ensured through RAIM. Even after cycle ambiguity resolution, when the aircraft exits the bubble, the high precision of carrier phase is still available for kinematic positioning and RAIM, although the availability of RAIM will be somewhat degraded (to a lesser degree if a geostationary overlay is implemented). Supplementing RAIM with independent monitoring by an IRU (initialized with the high-integrity carrier-phase positioning available inside the bubble) may be beneficial in this regard and will also ensure navigation continuity even in the unlikely event of the GPS signal jamming. Ground monitoring can, of course, also be present in a supplementary role throughout the approach.

A. Receiver Autonomous Integrity Monitoring

A mathematical description of cycle ambiguity resolution is given in Sec. IV.B, and the basic RAIM theory pertinent to IBLS may be found in Ref. 16. Recall that the linearized observation equation is given by $\phi = Hu + \delta\phi$, where ϕ is the $n \times 1$ vector of stacked single-difference phase measurements (aircraft minus reference) collected at the aircraft during pseudolite overflight. The vector $\delta\phi$ ($n \times 1$) is the single-difference phase error. The observation matrix, H ($n \times m$, $n > m$), contains the geometric information associated with the overpass. The

AIRCRAFT PRECISION LANDING USING INTEGRITY BEACONS

state vector, u ($m \times 1$), contains the cycle ambiguities and position fixes at each measurement epoch in the bubble. The position vector at an arbitrary epoch during the bubble pass x (3×1) is a vector element of u:

$$u = \begin{bmatrix} \vdots \\ x \\ \vdots \end{bmatrix}$$

The least-squares state estimate error is as follows:

$$\delta u = \begin{bmatrix} \vdots \\ \delta x \\ \vdots \end{bmatrix} \approx H^+(\delta\phi - \delta H u)$$

where $H^+ = (H^T H)^{-1} H^T$, and δH is the error in the airborne user's knowledge of the observation matrix H. The measurement residual vector ($n \times 1$) is $r \approx (I - HH^+)(\delta\phi - \delta H u)$.

Under normal conditions (NC)—no system failures—δH is negligible, and $\delta\phi$ is normally distributed with zero mean and standard deviation of $\sigma_\phi = 1$ cm: $\delta\phi = N(0, I_n \sigma_\phi^2)$. Under these circumstances, it can be shown[17] that the norm of the residual vector is a χ^2 distributed random variable with $n-m$ DOF. A residual threshold R can then be obtained analytically to achieve any desired probability of false alarm under normal error conditions,[18] where the false alarm event is defined by the following:

$$FA \equiv (\|r\| > R \mid NC)$$

If a is defined to be the desired navigation system accuracy specification, then the missed detection event is given by the following:

$$MD = (\|r\| < R, \|\delta x\| > a)$$

Figure 16a is a conceptual plot of position error vs residual. The probability "ellipse" nearest the origin represents the case of such normal condition errors as multipath and receiver noise (represented by the Gaussian measurement error model given above). For a given failure mode, the ellipse will slide up the failure mode axis a distance proportional to the magnitude of the failure. In Fig. 16b a line constraint is drawn to represent the navigation system accuracy specification (a). Note that it is possible, for small failure magnitudes, that the accuracy specification will not be breached. Also shown in the figure is a threshold set on the measurement residual (R). The resulting RAIM fault detection algorithm is a simple one. Check the residual statistic to see if it is larger than the threshold. If so, a system failure is declared. Although there is complete freedom in the selection of detection thresholds, false alarms will increase as the threshold approaches zero. However, the detection threshold can always be chosen to produce a low false alarm rate under normal error conditions. In the case of GPS carrier phase measurements, the overall result is shown conceptually in Fig. 16c. A hypothetical failure mode penetrating the narrow missed detection region is mathematically possible to construct, but given that such a mode must be related physically to a *real system failure,* the likelihood of its occurrence will be low.

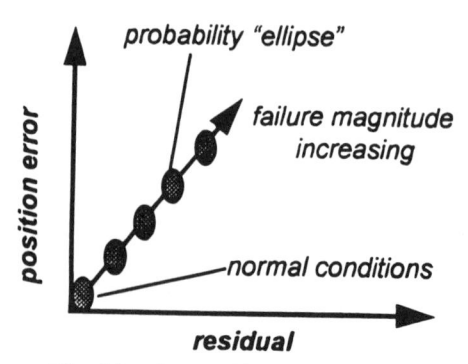

Fig. 16a State estimate vs residual.

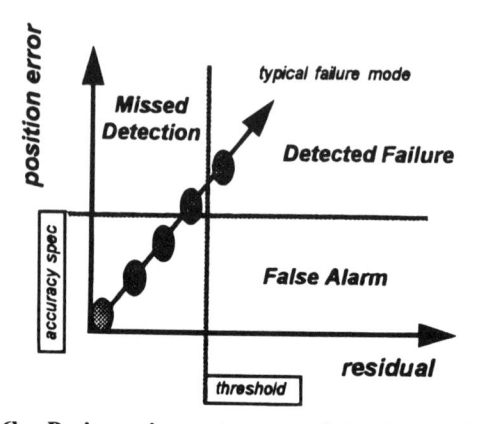

Fig. 16b Basic receiver autonomous integrity monitoring.

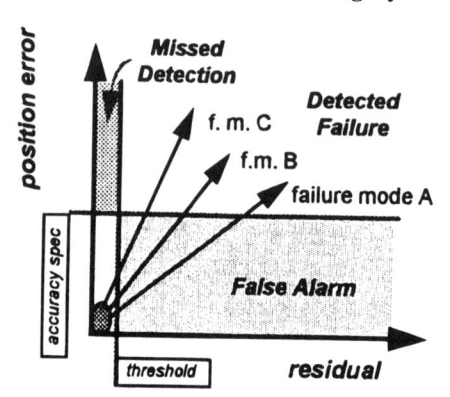

Fig. 16c High-performance receiver autonomous integrity monitoring with Integrity Beacon Landing System.

AIRCRAFT PRECISION LANDING USING INTEGRITY BEACONS

Clearly, the *quantitative* verification of RAIM fault detection capability depends upon the nature and likelihood of navigation system fault modes.

B. System Failure Modes

A loss of integrity event occurs when a navigation system failure, or an unusually large ranging error attributable to a familiar source such as multipath, causes a large position error that is undetectable by any form of monitoring. A top-level fault tree illustrating a number of integrity-threat failure classes associated with the space, airborne, and ground segments of IBLS is shown in Fig. 17. For this example, three diverse failure modes are chosen, one from each of the three segments of IBLS.

1. Airborne Segment: Cycle Slips

Cycle slips are most often associated with the airborne receiver; however, once differencing (or differential correction) is done, a cycle slip in the reference receiver will result in the same overall ranging error as a cycle slip in the aircraft receiver and must, therefore, be detected. The probability of cycle slip occurrence is dependent upon the particular receiver and antenna used. In general, however, the probability of a cycle slip event increases with increasing phase–locked-loop bandwidth, increasing time, and decreasing signal strength. As a first layer of the safety net, the low signal strength conditions under which cycle slips are likely to occur are identified at the signal processing (phase–locked-loop) level. Thereafter, RAIM provides an important additional layer of cycle slip monitoring.

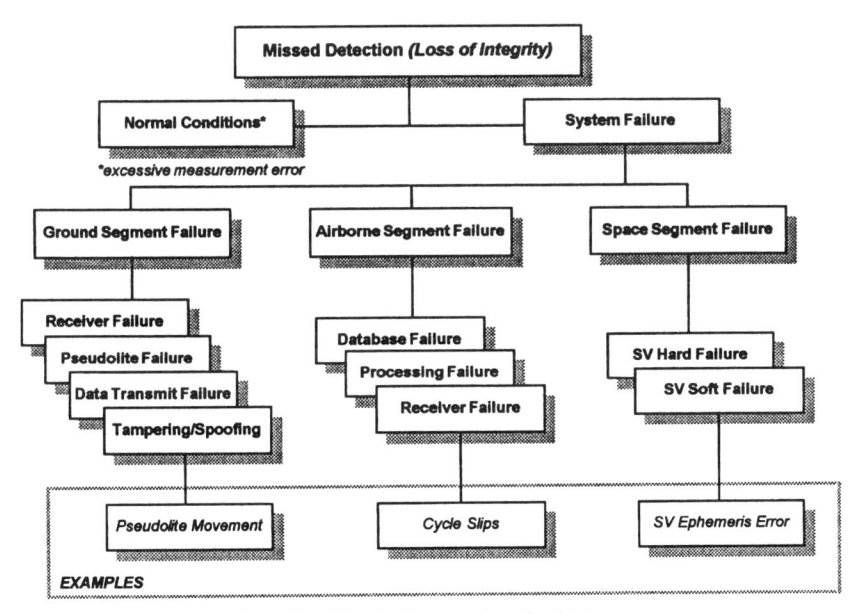

Fig. 17 Navigation system fault tree.

2. Ground Segment: Movement of Ground Hardware

This failure mode may be expressed as a mismatch between the actual ground hardware location and the location given in the airborne database. The question has been raised that perhaps a differential-based landing system such as IBLS, because of the existence of ground hardware, may be susceptible to either intentional tampering or unintentional errors in antenna siting (perhaps the result of maintenance work). Such discrepancies can also result from errors in the airborne IBLS ground survey database or, possibly, as errors in flight inspection. Movement of the reference station to an unsurveyed location before the approach begins, while leaving the pseudolites untouched, will not affect cycle ambiguity resolution or centimeter-level positioning.[19] However, movement larger than the code measurement noise, will generally trigger an alarm caused by the discrepancy between code and carrier. Furthermore, RAIM ensures that IBLS is also robust to movement of Integrity Beacon pseudolites.

3. Space Segment: Spacecraft Ephemeris Errors

Spacecraft soft failures—those spacecraft failures unknown to the user—fall into two basic categories: SV clock errors and SV ephemeris errors. Although the effects of clock errors originating at the spacecraft are almost completely eliminated through differential positioning, those of spacecraft ephemeris errors are not. Rather large ephemeris errors (> 500 m) are required to produce noticeable positioning errors. Among the possible origins of such an error are intentionally induced SA errors of unusually large magnitude, orbit determination error, and errors in ephemeris upload. The resulting user position estimate error, and RAIM measurement residual will both scale linearly with the displacement between the aircraft and the reference station.

An exception to the differential cancellation of ranging errors was recently exhibited by SV 19.[10] The symptom was a pseudorange bias of up to several meters when nonidentical receivers were used at the reference station and aircraft. Note that a code-ranging error of this type may or may not be detectable by ground monitoring, depending upon the actual receivers used at the reference station, monitor station, and aircraft. It is noteworthy, however, that IBLS tests during the occurrence of this ranging anomaly *demonstrated that IBLS carrier-phase tracking of SV 19 was not affected.*

C. Quantifying Integrity

The statistical significance necessary to demonstrate integrity $P(MD)$ on the order of 10^{-9} cannot, of course, be attained through flight test or other experimental means. The large total number of sample approaches and wide range of system failures can only be achieved through mathematical models and computer simulation. A valid method for quantifying integrity through simulation is discussed in Ref. 19 and can be applied both to normal error conditions and the three representative types of system failure already chosen. Considering the case of normal system errors (receiver noise and multipath) first, Fig. 18 shows a plot of $P(\|\delta x\| > a)$—a conservative measure of integrity under normal error conditions. For comparison, the equivalent result for high-precision C/A-code ranging

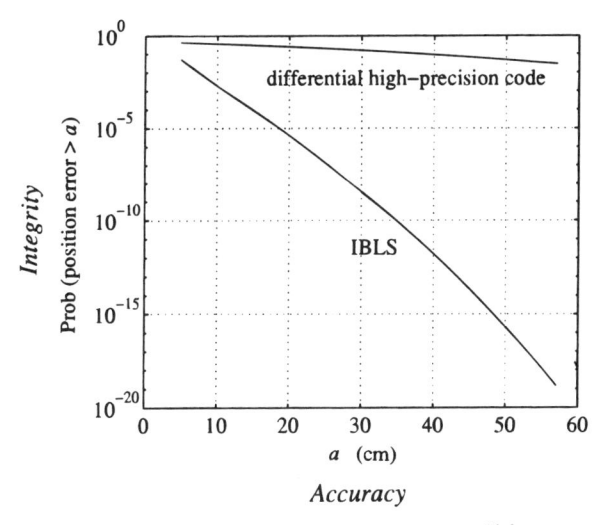

Fig. 18 Integrity under normal error conditions.

is also included. The level of navigation system integrity under normal error conditions is better than 10^{-10} even for accuracies of 35 cm with IBLS; whereas, for the equivalent accuracy the integrity of a system based on high precision code is roughly only 10^{-1}.

In addition, the effectiveness of the RAIM-based fault detection capability built into IBLS (Sec. VII.A) is demonstrated by the results of over 25 million simulated approaches using representative models for the three fault modes considered above.[20] Figure 19 shows the surface relating system integrity $[\log_{10}P(MD)]$, continuity $[\log_{10}P(FA)]$, and accuracy (a). As intuitively expected,

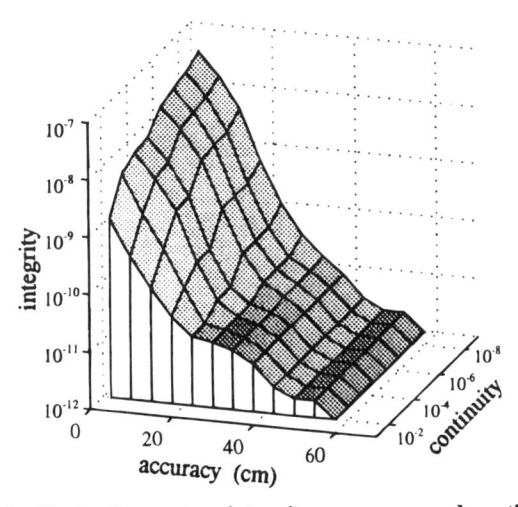

Fig. 19 Navigation system integrity, accuracy, and continuity.

integrity improves as the requirements on navigation system accuracy and continuity are relaxed. The resulting surface can be interpreted in an absolute sense as well. Integrity in the IBLS bubble to the three failures simulated is better than 10^{-10}, even for an accuracy requirement of 35 cm and continuity (false alarm probability under normal error conditions) of better than 10^{-7}. In summary, in the presence of normal error conditions and the three failure modes considered, Category III levels of integrity and continuity are achievable with submeter navigation system accuracy.

D. Signal Interference

This subsection addresses the issue of interference to the radio signals used on final approach and landing. The section deals with the potential effects of hostile jammers and spoofers, and it discusses the impact of the so-called "near–far" problem with respect to the Integrity Beacon pseudolites. As with any radionavigation system, spoofing (intentional hostile generation of a false radio signal in an effort to mislead an aircraft into an unsafe condition) is a more important issue than jamming.

1. Jamming

Fortunately, jamming is a readily detectable condition (either through receiver loss of lock, or by monitoring cross-correlation or AGC levels). Therefore, jamming is generally only a nuisance rather than a life-threatening situation. In a hostile jamming campaign scenario, pilots and ground controllers have a number of options with which to respond, including diverting the aircraft to another airport. Using the same kinematic techniques used for landing, equipment could be developed to locate a hostile jammer rapidly. To assess the likelihood of the jamming scenario, it is useful to consider that, although portable, hand-held aviation-band VHF radios are readily available to the general public for just a few hundred dollars, the instances where air-to-tower voice communications have been intentionally obstructed are extremely rare.

2. Spoofing

The GPS Integrity Beacon concept provides a significant barrier to intentional, hostile tampering with the ground system. Although the authors are unaware of any reported instances of hostile tampering with the ILS equipment, this potential vulnerability of any radionavigation aid must be considered. Unlike conventional DGPS, the built-in redundancy of IBLS signals makes it nearly impossible to corrupt the signal with anything that spoofs the aircraft into thinking it is anywhere other than where it really is.

An example of this robustness against spoofing is shown in Fig. 20. In the scenario shown, the spoofer attempts to direct the aircraft into the ground by broadcasting false GPS differential corrections. These corrections assume a lower altitude for the reference station. Conventional RAIM algorithms would not be able to detect this condition, because the residuals would all be self-consistent. With the Integrity Beacon, however, the aircraft immediately registers that the pseudolite phase signature is completely inconsistent with its conventional posi-

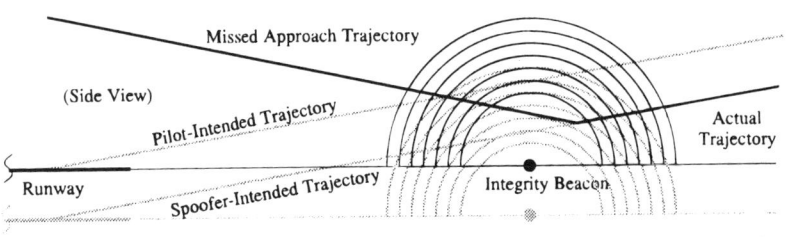

AIRCRAFT PRECISION LANDING USING INTEGRITY BEACONS 457

Fig. 20 Integrity Beacon spherical wavefronts make spoofing extremely difficult.

tion fixes. The system issues an alarm to the pilot, who executes a missed approach.

Spoofing scenarios in which simulated GPS Integrity Beacon signals are broadcast are extremely difficult to execute, given that the spoofer must know the aircraft position and attitude to high accuracy in order to calculate what signal to transmit. The key point is that such spoofing would be so difficult that any would-be tamperer will most likely direct his efforts elsewhere.

3. Near–Far Problem

One perceived shortcoming that is frequently pointed out in applications employing pseudolites is the so-called near–far problem. For the special case of the *low-power* Integrity Beacon, the architecture has been intentionally designed so that this phenomenon is of little consequence. The cross-correlation (isolation) between the GPS gold codes in two separate 1023-bit PRN sequences may be no better than 21.6 dB worst case.[21] The power level of -130 dBm received from GPS satellites is relatively constant, because a terrestrial user's range to the satellite does not vary much as a percentage of the average range. The problem is that when a user is operating at close range to pseudolites, there is a possibility that the pseudolite power may exceed the ambient GPS power by the cross-correlation threshold.

Figure 21 shows a side view of an Integrity Beacon and how the bubble geometry readily sidesteps this issue. The effective dynamic range of the signal is indicated by the shaded region. Based on the nominal glide slope shown, it is clear that there is significant margin for the approach altitude to be very much higher or lower than nominal while still staying within the correct power level. If the aircraft receiver registers a power level that becomes much too high, it is a clear indication that the aircraft is flying too close to the ground and should execute a missed approach.

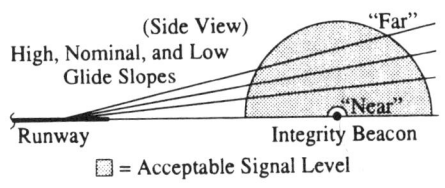

Fig. 21 Integrity Beacon Landing System sidesteps the near–far problem.

In future implementations, some thought is being directed toward increasing the bubble size for a larger coverage volume and perhaps extending the boundary to encompass the touchdown zone. Another concept is to augment the Integrity Beacons with a pseudolite back at the airport to provide redundant carrier ranging and a built-in data link (as mentioned in Sec. VI.A). Of course, these enhanced implementations would be subject to the near–far problem (as well as antenna pattern and placement considerations) and, therefore, might employ pulsing or other modulation schemes[14,15] to reduce interference with the GPS signals.

VIII. Conclusion

Analysis and flight testing are both demonstrating that IBLS is capable of meeting and exceeding the stringent requirements of Category III precision landing, especially in the most challenging regimes of accuracy and integrity. The centimeter-level precision of GPS carrier phase and the built-in crosschecks and redundancy provided by Integrity Beacons translate into robust, dependable performance and continued assurance to the pilot that it is safe to land. The end result is an architecture that is safe for passengers and crew.

References

[1]Cohen, C. E., Pervan, B. S., Cobb, H. S., Lawrence, D. G., Powell, J. D., and Parkinson, B. W., "Real-Time Cycle Ambiguity Resolution using a Pseudolite for Precision Landing of Aircraft with GPS," DSNS '93, Amsterdam, The Netherlands, March 30–April 2, 1993.

[2]Cohen, C. E., Pervan, B. S., Lawrence, D. G., Cobb, H. S., Powell, J. D., and Parkinson, B. W., "Real-Time Flight Test Evaluation of the GPS Marker Beacon Concept for Category III Kinematic GPS Precision Landing," *Proceedings of ION GPS-93*, (Salt Lake City, UT), Institute of Navigation, Washington, DC, Sept. 22–24, 1993.

[3]Cohen, C. E., Pervan, B. S., Cobb, H. S., Lawrence, D. G., Powell, J. D., and Parkinson, B. W., "Achieving Required Navigation Performance using GNSS for Category III Precision Landing," DSNS-94, London, UK, April, 1994.

[4]Hatch, R., "Instantaneous Ambiguity Resolution," KIS Symposium 1990, Banff, Canada, Sept. 1990.

[5]Lachapelle, G., Cannon, M. E., and Lu, G., "Ambiguity Resolution on-the-Fly: A Comparison of P-Code and High Performance C/A-Code Receiver Technologies," *Proceedings of ION GPS-92*, (Albuquerque, NM), Institute of Navigation, Washington, DC, Sept. 16–18, 1992.

[6]Cohen, C. E., "Attitude Determination Using GPS," Ph.D. Dissertation, Stanford University, Stanford, CA Dec., 1992.

[7]Anon., "*International Standards, Recommended Practices and Procedures for Air Navigation Services—Annex 10*," International Civic Aviation Organization, April, 1985.

[8]"U.S. Federal Radionavigation Plan," U.S. Departments of Transportation and Defense, 1992.

[9]Kelley, R. J., and Davis, J. M., "Required Navigation Performance (RNP) for Precision Approach and Landing with GNSS Application," *Navigation*, Vol. 41, No. 1, 1994.

[10]Nordwall, B. D., "Filter Center Column," *Aviation Week and Space Technology*, Aug. 30, 1993.

[11]Cohen, C. E., Lawrence, D. G., Pervan, B. S., Cobb, H. S., Barrows, A. K., Powell, J. D., Parkinson, B. W., Wullschleger, V., and Kalinowski, S., "Flight Test Results of Autocoupled Approaches Using GPS and Integrity Beacons," *Proceedings of ION GPS-94*, (Salt Lake City, UT), Institute of Navigation, Washington, DC, Sept. 20–24, 1994.

[12]Paielli, R., Bach, R., and McNally, D., work in progress, initially reported as "Carrier Phase Differential GPS for Approach and Landing: DGPS/INS Integration and Flight Test Validation," ION National Technical Meeting, Anaheim, CA, Jan. 1995.

[13]Cohen, C. E., Lawrence, D. G., Cobb, H. S., Pervan, B. S., Powell, J. D., Parkinson, B. W., Aubrey, G. J., Loewe, W., Ormiston, D., McNally, B. D., Kaufmann, D. N., Wullschleger, V., and Swider, R., "Preliminary Results of Category III Precision Landing with 110 Automatic Landings of a United Boeing 737 using GNSS Integrity Beacons," ION National Technical Meeting, Anaheim, CA, Jan. 1995.

[14]Stansell, T. A., "RTCM SC-104 Recommended Pseudolite Signal Specification," *Navigation*, Vol. III, 1986.

[15]Elrod, B., Barltrop, K., and Dierondorck, A. J., Van "Testing of GPS Augmented with Pseudolites for Precision Approach Applications," *Proceedings of ION GPS-94*, (Salt Lake City, UT), Institute of Navigation, Washington, DC, Sept. 20–24, 1994.

[16]Pervan, B. S., Cohen, C. E., and Parkinson, B. W., "Integrity Monitoring for Precision Approach Using Kinematic GPS and a Ground-Based Pseudolite," *Navigation*, Vol. 41, No. 2, 1994.

[17]Parkinson, B. W., and Axelrad, P., "Autonomous GPS Integrity Monitoring Using the Pseudorange Residual," *Navigation*, Vol. 35, No. 2, 1988.

[18]Sturza, M. A., "Navigation System Integrity Monitoring Using Redundant Measurements," *Navigation*, Vol. 35, No. 4, 1988–89.

[19]Pervan, B. S., Cohen, C. E., and Parkinson, B. W., "Integrity in Cycle Ambiguity Resolution for GPS-Based Precision Landing," DSNS '94, London, UK, April 18–22, 1994.

[20]Pervan, B. S., Cohen, C. E., Lawrence, D. G., Cobb, H. S., Powell, J. D., and Parkinson, B. W., "Autonomous Integrity Monitoring for Precision Landing Using Ground-Based Integrity Beacon Pseudolites," *Proceeding of ION GPS-94*, (Salt Lake City, UT), Institute of Navigation, Washington, DC, Sept. 20–24, 1994.

[21]Spilker, J. J., Jr., "GPS Signal Structure and Performance Characteristics," *Navigation*, Vol. I, 1980, pp. 29–54.

Chapter 16

Spacecraft Attitude Control Using GPS Carrier Phase

E. Glenn Lightsey*
NASA Goddard Space Flight Center, Greenbelt, Maryland 20771

I. Introduction

VIRTUALLY all but the simplest spacecraft employ some means of active attitude control using such actuators as control momentum gyros, reaction wheels, offset thrusters, and magnetic torque rods. Attitude control is almost always performed by a closed-loop system onboard the spacecraft; especially for low Earth orbit (LEO) spacecraft, where ground contacts are often limited to a few minutes per day. Only unusual events such as a momentum unload or spacecraft slew maneuver are commanded from the ground, and even then, the onboard system is usually responsible for some level of automatic control. Closed-loop attitude control, of course, requires sensor feedback of the vehicle orientation. This has traditionally been provided by such low-cost sensors as magnetometers, horizon sensors, and sun sensors, or more expensive high-performance instruments including gyroscopes and star trackers. Recently developed GPS attitude determination systems provide an opportunity to use this new technology in attitude control system designs.

As described in Chapter 19, this volume, precise measurement of differential carrier phase between multiple antennas may be used to determine the attitude of a vehicle. Thus, installing a GPS receiver onboard a spacecraft affords the opportunity to use a single lightweight, low-cost sensor for a multitude of functions: position, velocity, attitude, attitude rate, and time. This consolidation of resources is likely to lead to an overall savings in cost, power, weight, and complexity for spacecraft. Furthermore, the elimination of many different sensor devices and their interfaces can yield a substantial benefit in system reliability.

The key performance issues regarding the utility of GPS as a closed-loop attitude sensor are bandwidth, accuracy, and antenna placement. Typical low-

---

Copyright © 1995 by the American Institute of Aeronautics and Astronautics, Inc. No copyright is asserted in the United States under Title 17, U.S. Code. The U.S. Government has a royalty-free license to exercise all rights under the copyright claimed herein for Governmental purposes. All other rights are reserved by the copyright owner.

*Engineer, Guidance and Control Branch; also Ph.D. Candidate, Department of Aeronautics and Astronautics, Stanford University, Stanford, CA 94305.

precision attitude control systems using such sensors as magnetometers, Earth sensors, and Sun presence detectors, have bandwidths on the order of a few times orbit rate and pointing accuracies of 1–5 deg (example: the Solar Anomalous and Magnetospheric Particle Explorer, SAMPEX). For high-performance applications using gyros and star trackers, bandwidths range up to a few Hz with pointing accuracies of 0.1 deg or better (example: the Hubble Space Telescope, HST). The update rate of the currently designed GPS attitude determination system is in the range of 0.1–10 Hz, although the theoretical defined limit identified by Cohen[1] is 1 kHz. The accuracy of GPS attitude determination is strongly dependent on antenna placement and data-processing techniques. For a one meter antenna separation, the point solution accuracy is approximately 0.3 deg, with possible improvements to better than 0.1 deg using dynamic filtering. Both accuracy and bandwidth performance can be extended by combining the GPS sensor with gyroscopes.

The operational capability of the GPS attitude sensor was demonstrated in space in June 1993 when the Air Force sponsored RADCAL satellite was launched into an 800-km polar orbit. This gravity-gradient stabilized satellite contained two cross-strapped GPS receivers from which differential carrier phase measurements were used to obtain attitude solutions in postprocessing (see Fig. 1 and Refs. 2 and 3). One receiver failed after six months in orbit, but the two receivers together have provided more than 18 months of measurement data since launch.

Further plans are in progress to use a GPS receiver as a real-time attitude sensor on a host of other spacecraft including the OAST Flyer (discussed below), Gemstar, REX-II, Orbcomm, Globalstar, SSTI Lewis, SSTI Clark, and others. If the attitude determination performance using GPS is found to be acceptable in terms of both accuracy and bandwidth, a wide variety of control schemes can be used based on the type of actuators available. This development is likely to have tremendous benefit for LEO spacecraft missions, including remote sensing and mobile communications. The remainder of this chapter provides further details on GPS attitude sensing for spacecraft and a specific case study of the design of a control system using GPS as the attitude sensor.

II. Design Case Study

One of the first demonstrations of closed-loop attitude control in space using GPS will be the GADACS experiment (GPS Attitude Determination and Control System, pronounced "gay-dax"), which is manifested for flight onboard Space Shuttle mission STS-69 in Autumn 1995. The satellite bus is the rectangular SPARTAN "OAST Flyer" payload, measuring approximately $1 \times 1.25 \times 1.5$ m, as shown in Fig. 2. It is deployed from the Space Shuttle cargo bay as a free-flying spacecraft for approximately 40 h of operation before retrieval. The vehicle orientation is controlled in three axes through the actuation of small cold nitrogen gas jets. The OAST Flyer mission will be carrying at least two additional experiments, the Return Flux Experiment (REFLEX) and the Solar Exposure to Laser Ordnance Devices Experiment (SELODE).

The REFLEX and SELODE experiments will be conducted for approximately the first two-thirds of the 2-day mission. During this time, the SPARTAN attitude

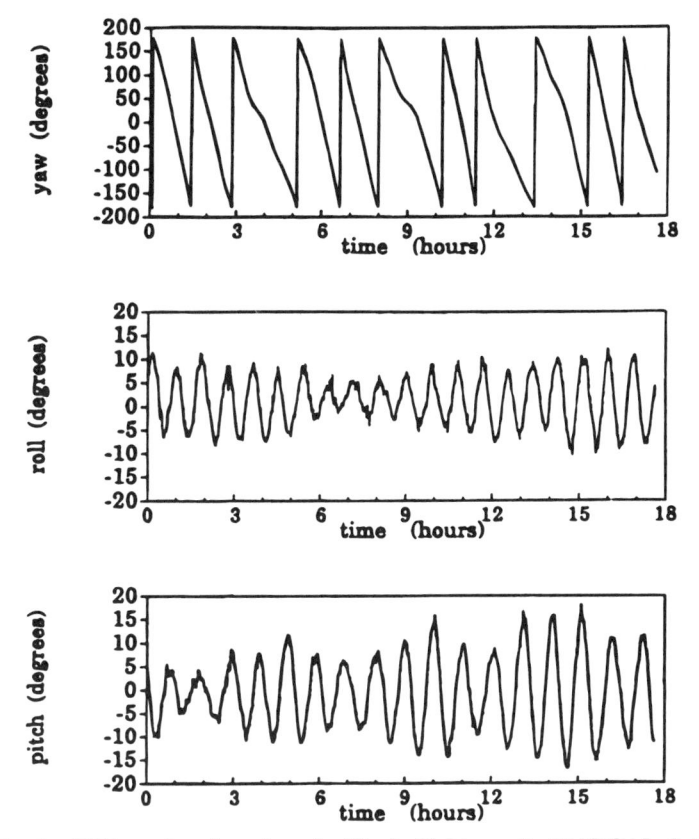

Fig. 1 GPS carrier-phase-based attitude flight results (RADCAL data, unfiltered).

control system will maneuver the spacecraft into a preset series of pointing profiles while these experiments are operating.

The attitude determination part of the GADACS experiment will be operational the entire time the OAST Flyer is free flying. During the REFLEX and SELODE portions of the mission, GADACS will be collecting attitude determination data using its two GPS attitude receivers. Control of the spacecraft will be maintained by the nominal OAST Flyer controller. GPS data will be recorded on tape along with gyro measurements that will be used to verify and calibrate the real-time GPS attitude measurements. Star tracker updates will be performed every other orbit throughout the mission to calibrate the onboard gyros.

For the last portion of the mission, GADACS will assume control of the vehicle, using only GPS-sensed attitude for closed-loop attitude control. During this time, GADACS will control the spacecraft in a series of Earth-pointing and inertial profiles and execute a series of test slews in order to test the performance of the GPS attitude-based control for a variety of mission types. The changeover from one receiver/antenna set to another will also be exercised by these maneu-

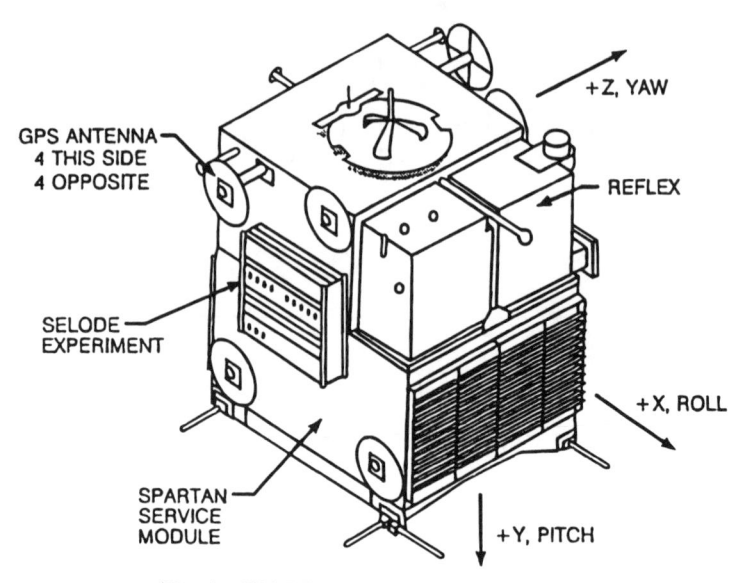

Fig. 2 SPARTAN OAST-flyer spacecraft.

vers. A detailed discussion of the GADACS mission and hardware is given in Ref. 4.

The three nonlinear single-axis control systems that GADACS will employ were based on the GPS sensor behavior, the expected environmental disturbance torques, and the performance and mission requirements imposed on the experiment by the spacecraft. Of the latter, the dominant consideration was conservation of an extremely limited actuation fuel budget. The final design is presented as a case study along with a discussion of the issues leading to its selection and expected performance.

III. Sensor Characteristics

The fundamental performance characteristics of any sensor should be understood before attempting to use it in a control application. In the case of a GPS attitude receiver, the main determinants of system performance are antenna separation (also known as baseline length) and noncommon mode error sources; the dominant component of the latter being multipath interference.

Because relative position error is roughly independent of antenna separation, greater baseline length leads to more precise attitude measurements. For a given antenna separation, as shown in Fig. 3, the baseline vector between the master antenna position a_o and the slave antenna position a_i is defined as follows:

$$b_i = a_i - a_o \qquad (1)$$

This baseline vector is assumed to be known in the vehicle body fixed frame B.

SPACECRAFT ATTITUDE CONTROL USING GPS CARRIER PHASE

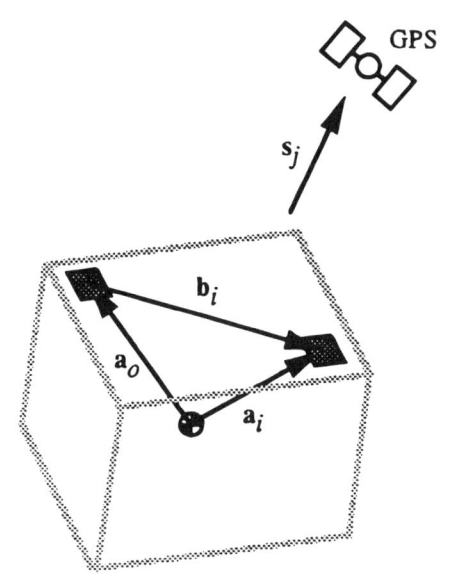

Fig. 3 Attitude observability geometry.

The carrier phase difference measured by the GPS receiver provides a measure of the projection of the baseline vector b_i onto the line of sight from the vehicle to the GPS satellite s_j:

$$\Delta\varphi_{ij} = b_i \cdot s_j \tag{2}$$

where the integer cycle ambiguity has been neglected. This subject is discussed in detail in Chapter 19, this volume.

Because the line-of-sight vector is known in the orbit local frame, and the baseline vector is known as the body frame, the phase difference can be expressed as a function of the attitude of the vehicle:

$$\Delta\varphi_{ij} = (b_i^B)^T A s_j^L \tag{3}$$

where the superscripts indicate the coordinate frame in which the vector is expressed. A is the attitude matrix of the vehicle, or in other words, the rotation matrix from the orbit local to the body fixed frame.

Equation (3) can be linearized about a nominal or previously estimate attitude as follows:

$$\Delta\varphi_{ij} = (b_i^B)^T (\delta A A_o) s_j^L \tag{4}$$

In this case, A_o is the nominal attitude matrix. Chapter 14, this volume, provides details on how measurement equations from several satellites and baselines can be solved for the best attitude estimate.

This relation shows that for given $\Delta\varphi$, A_o, and s, there is a direct relationship between δA and b. To achieve smaller resolution in attitude δA a larger baseline vector b is needed, as expected.

A. Antenna Placement

For most satellites, the payload size is required to be minimized. The maximum antenna separation should, therefore, be sought to achieve the best possible attitude resolution. This usually results in antenna placement on the corners of the zenith facing side of the satellite, if possible. Even for the relatively small baselines of many spacecraft (often <1 m), subdegree accuracy is still possible using GPS.

If the satellite is not Earth-pointed (for example, an inertially pointed platform), then potential blockage of GPS signals by the Earth must be considered. If continuous attitude information is required in this case, it may be necessary to have two independent GPS receivers with antennas installed on opposite faces of the spacecraft. At least three antennas (preferably four or more) must share the same sky view to make the differential carrier phase measurements necessary for attitude determination.

Another antenna placement issue is the location of the antenna array relative to the key control requirements. For example, if the antennas are mounted on the solar panels, and the pointing requirements are driven by an instrument mounted on the main body of the spacecraft, there will be a loss in accuracy. Furthermore, antennas mounted on flexible or moving parts may reduce accuracy. A design tradeoff exists between geometric resolution (increasing antenna separation) and sensor collocation with the point of interest (decreasing antenna separation). For most small spacecraft, such as GADACS, the additional geometric resolution gained by maximizing the antenna separation on nonmoving spacecraft parts results in an overall improvement in sensor accuracy.

B. Sensor Calibration

The theoretically achievable attitude accuracy is limited by mechanical system knowledge. Mechanical system knowledge is defined as the precision to which the antenna phase center locations are known on the spacecraft and the alignment of the spacecraft body axes to the antenna axes. These are important quantities that should be determined as accurately as possible during ground testing. Many GPS receivers measure the antenna locations autonomously by placing the receiver into a self-survey mode and allowing the GPS satellites to pass overhead for several hours. This test requires the spacecraft truss to have a sky view, and it should be planned in advance. The axis alignment usually must be measured by some independent means (a theodolite, for example).

For the GADACS experiment, a significant effort was undertaken to measure the antenna baseline lengths and the axis alignment precisely. Five 7-hour self-survey tests were performed over a 4-day period using the spacecraft truss in an open area selected to minimize multipath reflections. The antenna baselines were found to be repeatable to within 2.2 mm (roughly 1 part in 500). During this time, the antenna axis alignment was validated optically using a theodolite and an antenna-mounted optical reflection cube to within 1 arc minute. A detailed description of the calibration test and results is given in Ref. 5.

In the future, it is expected that an alternative approach to an outdoor test on the spacecraft truss will be developed. Some methods that have been considered to determine the same information are indoor calibration using simulated GPS signals,[6] and on-orbit sensor calibration performed in postprocessing[7] or near real time.[8]

C. Multipath

Another possible error source is multipath. Multipath occurs when the signal arrives at the antenna from reflected surfaces in addition to the line-of-sight source. The reflected signal is phase shifted with respect to the original transmission and appears as additive noise at the antenna. Because the antenna locations are different, the multipath signature at each antenna is unique and the error is not common mode.

Multipath is the dominant error source in many spacecraft applications, accounting for more than 90% of the total error budget in carrier phase measurement. Unfortunately, its presence is pervasive, although steps can be taken to minimize it. Isolating the hemispherical patch antennas, providing unobstructed fields of view, and adding ground planes will all reduce multipath. Canting the antennas away from reflective surfaces may also improve the measurement quality, although this technique also reduces the field of view common to all antennas. The options are usually limited on a typical mission with many conflicting design constraints. Empirical tests have shown that for complex reflective surfaces, such as spacecraft, a conservative value for carrier phase error caused by multipath is approximately 5 mm rms (see Refs. 1 and 9).

D. Sensor Accuracy

These and other error sources are discussed more rigorously in other works.[9,10] If the mechanical system is well known, the total system performance is characterized by the multipath environment. The relationship between rms attitude error, range error (carrier phase), and baseline length is approximately as follows:

$$\sigma_\theta = \sigma_r/b \qquad (5)$$

A plot of this empirically verified relationship along with datapoints for RADCAL and GADACS baselines is shown in Fig. 4. From this chart, about 0.5 deg attitude accuracy could be expected for a single RADCAL measurement. If the three simultaneous measurements are assumed to be statistically independent (not strictly true, but a useful approximation), a single, more accurate estimate is produced with the following covariance:

$$1/\sigma_m^2 = 1/\sigma_1^2 + 1/\sigma_2^2 + 1/\sigma_3^2 = 9.1 \text{ deg}^{-2} \qquad (6)$$

or

$$\sigma_m \sim 0.3 \text{ deg} \qquad (7)$$

Preliminary flight results from RADCAL (Fig. 1) indicate that the accuracy of the attitude solution is in the shaded region shown on Fig. 4 with σ_m approximately 0.5–1.0 deg. The reason for the difference from the theoretical performance is that the mechanical system is not well known. Accuracy better than 0.5 deg rms is expected when the data reduction is finished. However, because there is no other independent attitude sensor on RADCAL (other than a magnetometer,

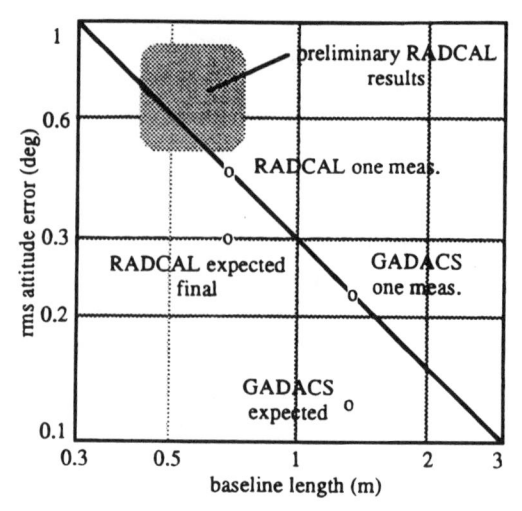

Fig. 4 Attitude determination accuracy.

which does not provide better than degree accuracy), the actual GPS attitude sensor accuracy is not precisely known.

The GADACS experiment, which has antenna baselines that are approximately twice as long as RADCAL (1.20 m vs 0.67 m), can expect $\sigma_m \sim 0.2$ deg. This is an important design parameter that will be used to size the thruster dead band in the attitude control loop.

E. Dynamic Filtering

Improved attitude estimates can be derived by using a Kalman filter to include knowledge of the vehicle dynamics in the estimation process. Fig. 5 illustrates the results for the same dataset shown in Fig. 1 with an extended Kalman filter. The filter states consist of three elements of a correction quaternion, three angular velocities, and three line biases. If the highest accuracy estimates are required for the controller, it is probably best to use a filter. However, if the dynamic model is not correct, the filter will produce unreliable and possibly divergent results.

IV. Vehicle Dynamics

The control system design begins with a study of the uncontrolled vehicle dynamics. To size the body-derived environmental torques, it is convenient to use an "orbit" reference frame to express vehicle attitude, as shown in Fig. 6. The frame is formed by taking the geodetic nadir as the z_o axis, crossing it into the inertial velocity vector to form the y_o axis, and completing the orthogonal set by computing x_o as the cross product of y_o and z_o. The x_o axis is then aligned with the local horizontal component of the velocity vector, which approximately coincides with the velocity vector for near-circular orbits.

The vehicle body axes (shown in Figs. 2 and 6) are assumed to be aligned with the principal axes of inertia. In fact, they are misaligned by about 10 deg,

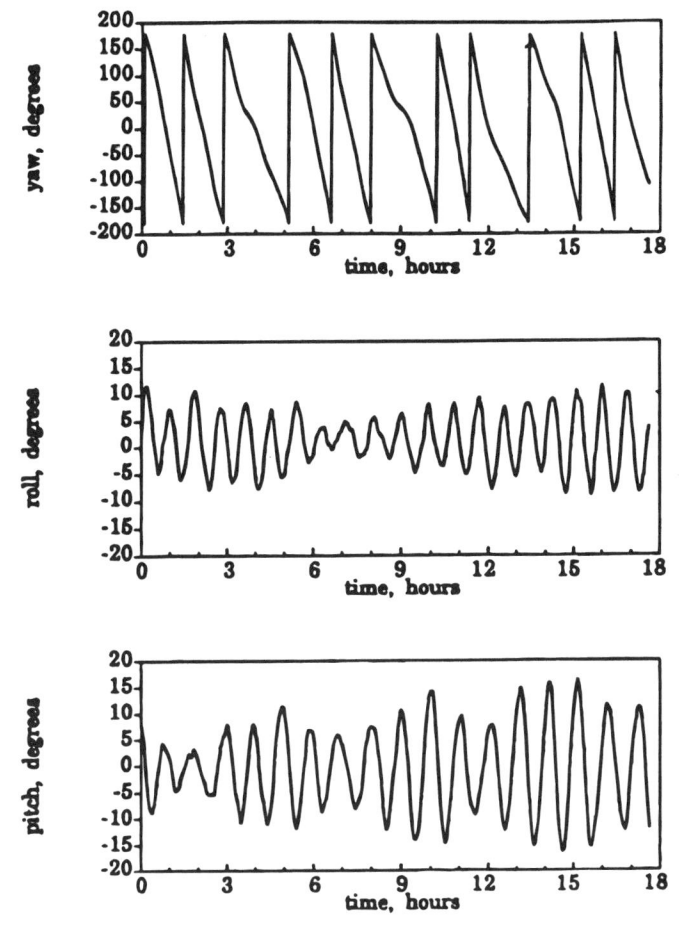

Fig. 5 GPS carrier-phase-based attitude flight results (RADCAL data, with Kalman filter).

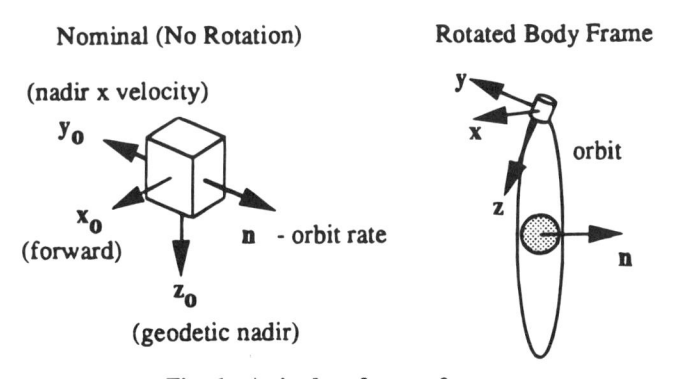

Fig. 6 Attitude reference frames.

but that effect is neglected in this simplified analysis. The vehicle attitude is then expressed as a 321 (yaw, pitch, roll) Euler rotation sequence from the orbit reference frame.

With the vehicle frame defined, the next step is to model the significant disturbance torques in terms of the vehicle attitude. At the Space Shuttle altitude of 380 km, aerodynamic drag is usually the dominant environmental torque. For small, relatively symmetric satellites, however, where the deviation between the center of pressure and center of mass is not large, the gravity gradient and drag torques may be equally important. In fact, the peak gravity gradient torque may be determined from the SPARTAN inertia listed in Table 1 and compared to the previously observed maximum aerodynamic torque for the Spartan payload[11]:

$$\max |M_{gg}| = 5.77e{-}5 \text{ N m} \tag{8}$$

$$\max |M_a| = 3.39e{-}4 \text{ N m} \tag{9}$$

The aerodynamic torque is seen to be about 5.9 times greater than the gravity gradient torque. Because this is less than an order of magnitude, a thorough analysis of the vehicle dynamics should include both effects.

A. Gravity Gradient Moment

The gravity gradient torque is examined first. Other references (e.g., Ref. 12) have demonstrated that for the 321 Euler sequence of a vehicle in a near-circular orbit whose body axes coincide with the principal axes of inertia (all true for GADACS), the gravity gradient torque may be linearized for small ψ (yaw), θ (pitch), and ϕ (roll) as:

$$M_{gg} = -3n^2[(J_y - J_z)\phi \mathbf{i} + (J_x - J_z)\theta \mathbf{j}] \tag{10}$$

where the [$\mathbf{i}, \mathbf{j}, \mathbf{k}$] unit vector set defines the [x, y, z] vehicle reference frame. The variable n is orbit rate.

A useful parameterization first performed by DeBra[13] combines the principal inertia into two dependent variables α and β as follows:

$$\alpha \equiv (J_y - J_z)/J_x = -0.282 \tag{11}$$

$$\beta \equiv (J_y - J_x)/J_z = -0.135 \tag{12}$$

The stability properties of the gravity gradient torque may then be conveniently represented by the plot in Fig. 7. Without going into detail, mechanical systems with inertia properties in the lightly shaded "Lagrange Region" are neutrally stable; i.e., the minor axis of inertia will align with the geodetic nadir vector and precess with it. The dark "DeBra-Delp Region" is stable only if the system has

Table 1 GADACS inertia properties

$J_x = 305.3$ kg m^2
$J_y = 258.7$ kg m^2
$J_z = 344.7$ kg m^2

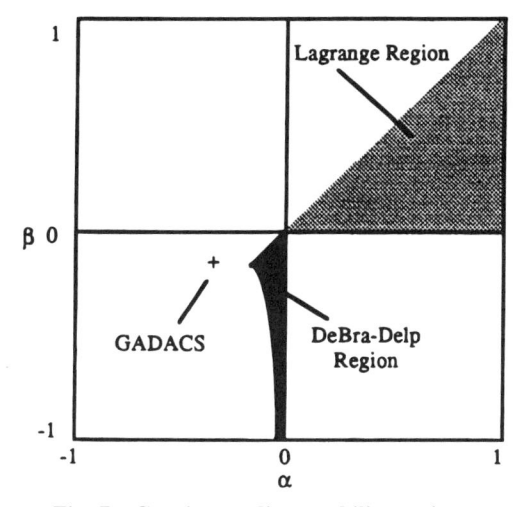

Fig. 7 Gravity gradient stability regions.

no energy damping. Because the SPARTAN mass properties lie outside these regions, it is concluded that gravity gradient torques alone are destabilizing and would reorient the spacecraft with the GPS antenna array looking along the local horizontal without active control.

B. Aerodynamic Moment

The aerodynamic torque is represented in Fig. 8. It is caused by a difference between the location of the center of pressure, where the resultant drag force acts along the negative velocity ("ram") vector and the center of mass, about

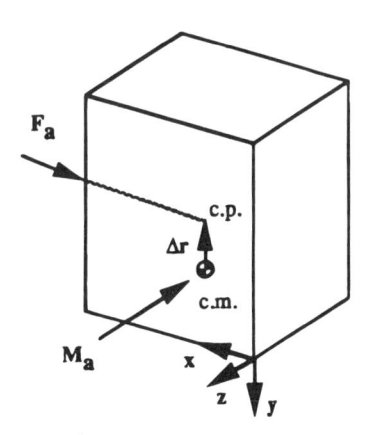

Fig. 8 Aerodynamic moment.

which moments are taken. The GADACS center of pressure (the center of area) is almost directly along the vehicle z axis:

$$\Delta r = 18.8\,k \text{ cm} \equiv z_b \tag{13}$$

$$M_a = \Delta r \times F_a = M_a \hat{z}_b \times (-\hat{v}_b) \tag{14}$$

where

$$M_a \equiv \max|M_a| = 3.39e - 4 \text{ N m} \tag{15}$$

and the ($\hat{}$) notation signifies the unit vector operation.

The velocity of the satellite is easily represented in the orbit reference frame for the circular GADACS orbit as follows:

$$v_o = |v|i_o \tag{16}$$

Then, following the 321 Euler sequence transformation

$$\hat{v}_b = (T_{bo})^T \hat{v}_o = (c\psi c\theta)i + (c\psi s\theta s\phi - s\psi c\phi)j + (c\psi s\theta c\phi + s\psi s\phi)k \tag{17}$$

where c and s represent cosine and sine, respectively. Thus, the aerodynamic torque in the body frame is:

$$M_a = -M_a[(c\psi s\theta s\phi - s\psi c\phi)i + (c\psi c\theta)j] \tag{18}$$

$$M_a \approx -M_a[\psi i + j] \tag{19}$$

The linearized vehicle equations of motion may then be determined by constraining force equilibrium

$$\dot{H}_c - \Sigma M_{\text{ext}} = 0 \tag{20}$$

which yields the following roll/yaw coupled equations:

$$\begin{array}{l} J_x\ddot{\phi} + 4n^2(J_y - J_z)\phi + (-J_x + J_y - J_z)n\dot{\psi} + M_a\psi = M_{xc} \\ J_z\ddot{\psi} + n^2(J_y - J_x)\psi + (J_x - J_y + J_z)n\dot{\phi} = M_{zc} \end{array} \tag{21}$$

and in pitch:

$$J_y\ddot{\theta} + 3n^2(J_x - J_z)\theta = -M_a + M_{yc} \tag{22}$$

where M_{xc}, M_{yc}, and M_{zc} are the control torques that will be applied.

C. System Natural Response

Before progressing to the design of the control system, the dynamic behavior of the uncontrolled equations of motion should be examined. It is interesting to note that the pitch equation is decoupled from the roll/yaw system. While this result is expected for a gravity gradient disturbance, the decoupling holds in the presence of aerodynamic drag (for small motions). This property arises from the fact that the aerodynamic moment arm lies along the z-axis of the vehicle reference frame.

By completing the Laplace transform of the roll/yaw equations of motion (assuming zero initial conditions and small deflections), the following representation is obtained:

$$\begin{bmatrix} s^2 + 4\alpha n^2 & (\alpha - 1)ns + \rho \\ (1 - \beta)ns & s^2 + \beta n^2 \end{bmatrix} \begin{bmatrix} \phi \\ \psi \end{bmatrix} = \begin{bmatrix} M_{xc}/J_x \\ M_{zc}/J_z \end{bmatrix} \quad (23)$$

where α and β are defined above and $\rho \equiv M_a/J_x$.

The effect of aerodynamic torque on the uncontrolled system may be examined by placing the characteristic equation into Evans form vs ρ:

$$s^4 + (3\alpha + \alpha\beta + n^2)n^2 s^2 + 4\alpha\beta n^4/(1 - \beta)ns = \rho \quad (24)$$

When J_y is the maximum moment of inertia, the numerator may be factored as follows:

$$(s^2 + \omega_1^2)(s^2 + \omega_2^2)/(1 - \beta)ns = \rho \quad (25)$$

This relation has the root locus shown in Fig. 9. Without the aerodynamic torque ($\rho = 0$), the system is characterized by the unstable gravity gradient poles. For any $\rho > 0$, the system is unstable. The roots given by the worst case aerodynamic torque of M_a are represented by the asterisks. The dominant unstable pole for this system has a time constant of 770 seconds. Therefore, it takes about 13 minutes for a small error to grow to 2.7 times its initial value. For more typical values of aerodynamic torque ($<M_a$), the time constant is longer.

Because the growth of the instability is slow compared to the designed controller bandwidth (~0.01 rad/s, or approximately 0.1 times orbit rate $\approx$ 9 min), the instability may be represented as a disturbance torque, and each axis of the controller may be treated as a simple double integrator plant. It is desirable to keep the plant model as simple as possible because of the nonlinear elements of the control system described in the following section.

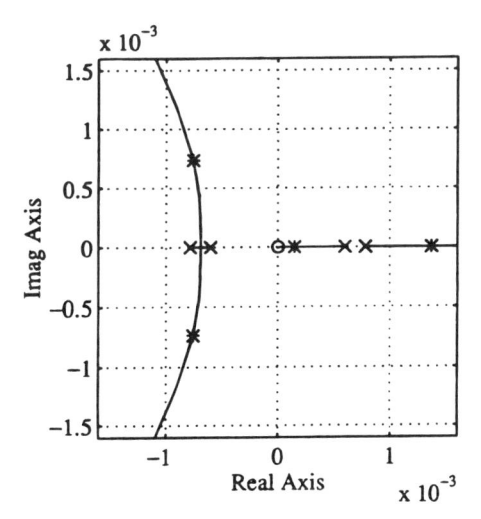

Fig. 9 Root locus for aerodynamic moment.

V. Control Design

The GADACS mission will be one of the first flight tests to quantify the accuracy of the GPS attitude solutions from experimental measurements. The expected in-flight accuracy of this device for a 1-m antenna separation is believed to be less than 0.5 deg at a 1-Hz sample rate. During the last third of the mission, the GPS solutions will be used as sensor inputs to the attitude control system; because this part of the mission is a demonstration of GPS-based attitude control, the control system will approximate several typical spacecraft pointing applications.

Because the GPS attitude solutions are relatively noisy, and the noise characteristics are not well known, and because of a very limited fuel budget, the controller requirements are to achieve reasonable pointing performance without excessive actuation in the presence of noise. Furthermore, in the event of loss of GPS attitude or significant discrepancies between the GPS attitude solutions and the inertially derived measurements, provisions have been made to use the gyro-sensed attitude for control.

A. Control Loop Description

A conceptual block diagram of the GADACS control system is shown in Fig. 10. The continuous plant is sampled by the GPS attitude receiver at an update rate of 1 Hz. Failure detection and correction logic (described in Ref. 3), chooses between the GPS solution and the inertially derived measurement of the vehicle attitude as an input to the plant estimator. This prediction estimator is designed to have settling qualities of the same natural frequency as the controller (~0.01 rad/s), and estimates both the vehicle attitude and Euler rates in roll, pitch, and yaw. These quantities are expressed in quaternion format and compared to the commanded attitude to produce an error quaternion, which is then converted into single-axis error signals that are provided to the controller. The controller is a single-axis position and velocity state feedback controller with a bandwidth of 0.01 rad/s. The command signal (with nonlinearities) is also provided as an estimator input (not shown).

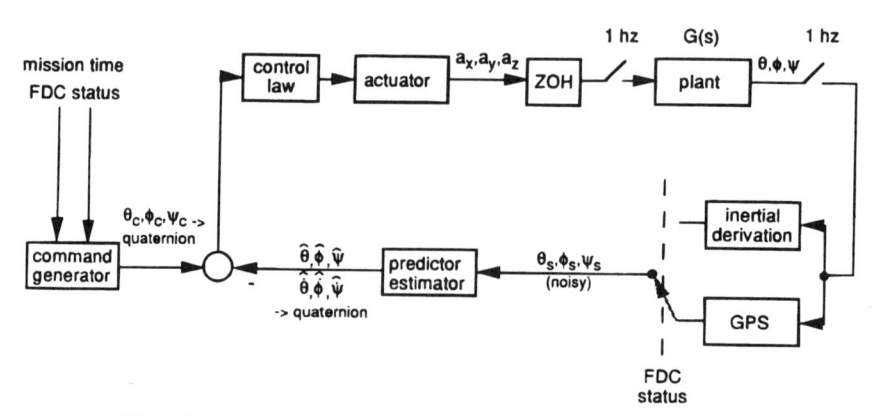

Fig. 10 Conceptual block diagram of GADACS control loop.

The control system has four nonlinearities that must be considered in the design. The control signal drives a nonlinear gas jet actuator, which is a pulse frequency-modulated system with dead zone, saturation, and rate limit. The pulse frequency modulation provides 25 ms impulses of gas calibrated to provided 0.17 deg/s^2 acceleration in a single body axis. The dead zone is designed to provide the proper fuel consumption limit cycle in the presence of sensor noise. Saturation occurs when the pulse frequency modulation reaches a full on state, which provides a constant (100% duty cycle) acceleration of 0.17 deg/s^2. A rate limit of 0.5 deg/s is implemented to prevent saturation of the DRIRU-II gyro.

For design and simulation purposes, this nonlinear system may be modeled as a linear gain with dead zone, saturation, and rate limit. Figure 11 demonstrates that this approximation is reasonably accurate when compared to the actual full nonlinear system. The effective acceleration of pulse-modulated system distributed over 1 s is compared to the simplified nonlinear model.

B. Simulation Results

Single-axis simulations were developed to demonstrate the performance of the control system. The GPS attitude solution noise was modeled as a Gaussian random process with statistically uncorrelated samples of 0.3 deg rms standard deviation. This noise level is approximately what was experienced on the RADCAL mission[2]; because of longer antenna baselines and a more favorable multipath environment, the actual noise experienced by GADACS should be less, but this level was used in the design to be conservative. In actuality, the noise is spectrally colored by multipath, but this effect was not modeled.

Environmental torques were modeled as worst-case steady-state torques acting in the same orientation. Because the controller bandwidth is high relative to the destabilizing time constant of the disturbances, no attempt was made to estimate the disturbances.

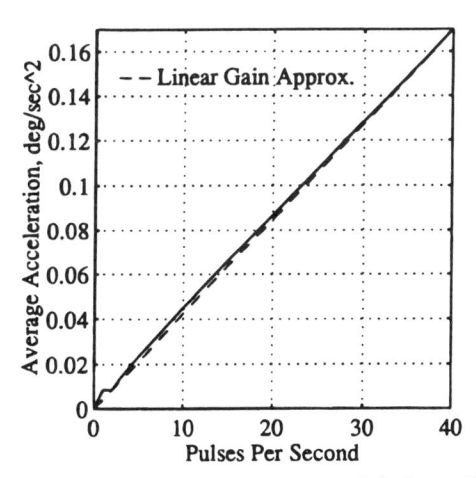

Fig. 11 Comparison of nonlinear pulse frequency modulation to linear gain approximation.

The main determinants of performance that were used in evaluating the design were controller bandwidth, steady-state error, dead zone size, and limit cycle frequency. Of these, the latter proved to be the main requirement on the design: the limit cycle frequency had to be very low, with a period greater than one-tenth orbit rate, for the design to remain within its limited actuation budget. Furthermore, the limit cycle frequency needed to be conservatively selected, given the uncertain noise characteristics of the sensor.

During the GADACS control portion of the mission, the controller will employ two types of motion, inertial hold and rate control. There will also be a limited number of inertial step and settle commands. The dead zone size was designed to achieve the proper limit cycle frequency and to reject spurious actuation caused by sensor noise. Partially because of the conservative noise characteristics used and mainly as a result of the limited fuel budget, a dead zone size for this system was selected as plus or minus 2 deg. Figure 12 demonstrates that over the course of a simulated orbit, no false actuation was produced with this deadband size for the input sensor noise. This result may be compared with a smaller deadzone of 0.5 deg in Fig. 13. No improvement in accuracy is achieved while fuel is wasted as the controller responds to the sensor noise. The dead zone is needed to keep the high-bandwidth actuator from responding to the noise in the lower-bandwidth sensor.

Figure 14 shows the design with no input, but with nonzero initial conditions to excite the limit cycle. The pitch plot shows the command input, and the sampled and estimated states. The limit cycle frequency of about six revolutions per orbit is clearly evident. The control effort required to maintain the limit cycle is reasonably small. Studies were performed to demonstrate that the limit cycle remained within the overall gas budget for up to twice the expected sensor noise.

The pitch axis, nadir-pointing profile is shown in Fig. 15. This is a rate-controlled mode with the pitch rate input equal to orbit rate (360 deg/5500 s).

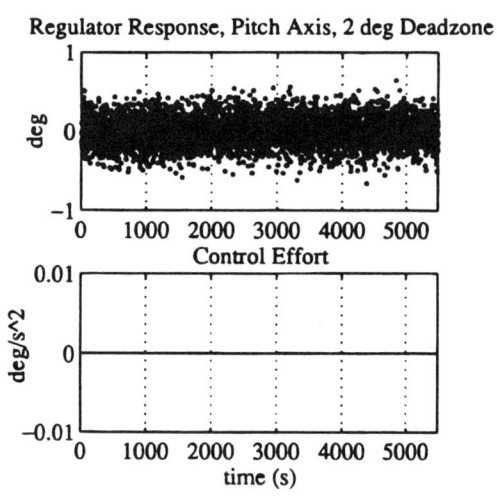

Fig. 12 2-deg dead zone reduces false actuation caused by sensor noise (estimator included).

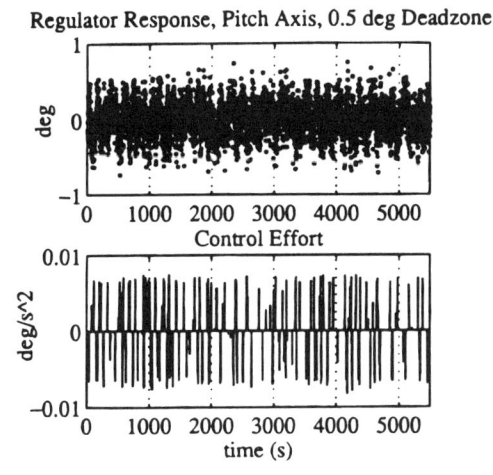

Fig. 13 0.5-deg dead zone responds to sensor noise without improving performance (estimator included).

The design is seen to accelerate to orbit rate and remain there without excess actuation once the response has settled.

The system response to a large (45 deg) step command is shown in Fig. 16. The 0.5-deg/s rate limit is seen during the first 80 s of the step. The saturation of the control system at the rate limit causes a slightly slower and more lightly damped response than in the unsaturated case, nonetheless, the performance is satisfactory.

VI. Conclusion

The design of an attitude-control system for the GADACS mission has been presented as an example of how GPS technology can be applied to spacecraft

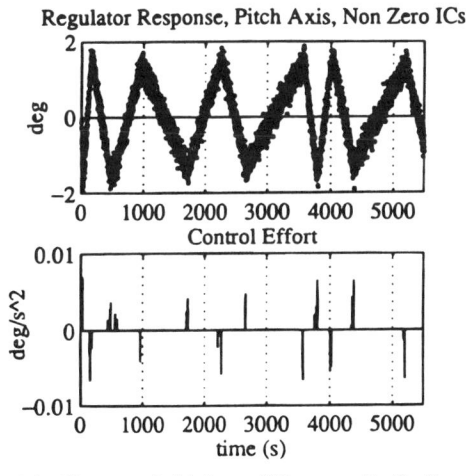

Fig. 14 Nonzero initial conditions excite limit cycle.

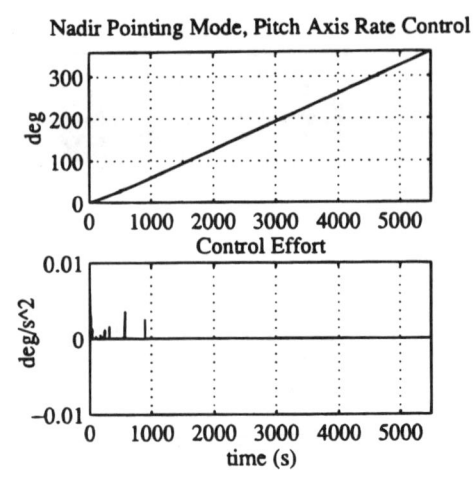

Fig. 15 Nadir-pointing mode using pitch orbit rate control.

closed-loop control. Its performance has been shown to be satisfactory with respect to fuel consumption and limit cycle frequency in the presence of sensor noise. The attitude determination accuracy of this system is expected to be less than 0.5 deg rms, and the controller accuracy is expected to be about 2 deg. Provisions have been made to allow sensor inputs to the controller from the precision inertial system in the event that GPS measurements are unavailable or unacceptable for spacecraft attitude control.

Every attempt has been made to produce a reliable control system that has good performance; however, it is important also to provide a conservative design, given the lack of flight experience for the GPS sensor. Until more space flight data are obtained, the best design is one that is fairly robust in the face of system

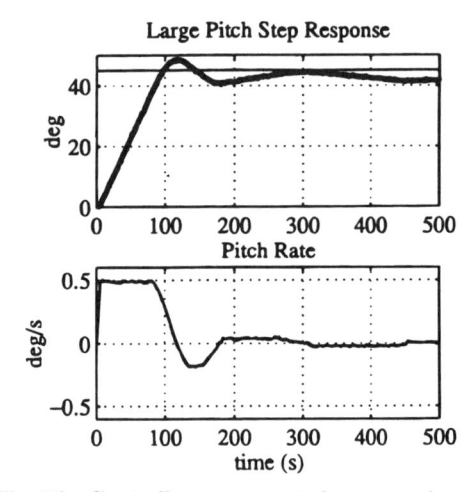

Fig. 16 Controller response to large step input.

uncertainty. As the heritage of GPS attitude determination grows, more aggressive designs can be implemented that provide the best possible performance for the system.

Cost effective subdegree attitude control is now possible in space using GPS carrier–phase-based attitude determination. This technology may soon provide an acceptable attitude sensor for many types of LEO missions. As applications grow in number, fabrication costs for space-qualified GPS receivers will be further reduced and manpower costs associated with custom designed spacecraft will be decreased. Attitude determination and control in space using GPS is an exciting new technology development that should experience substantial activity and growth over the next several years.

Acknowledgments

The author gratefully acknowledges the assistance of Penina Axelrad of the University of Colorado. The many helpful contributions of Clark Cohen and Bradford Parkinson of Stanford University; Trimble Navigation, Ltd.; and NASA Goddard Space Flight Center, including Frank Bauer, Jon McCullough, Jim O'Donnell, and the rest of the GADACS and SPARTAN design team have greatly aided in this research.

References

[1]Cohen, C. E., and Parkinson, B. W., "Expanding the Performance Envelope of GPS-Based Attitude Determination," *Proceedings of ION GPS-91* (Albuquerque, NM), Institute of Navigation, Washington, DC, Sept. 16–18, 1991, pp. 1001–1012.

[2]Cohen, C. E., Lightsey, E. G., Parkinson, B. W., and Feess, W. A., "Space Flight Tests of Attitude Determination Using GPS: Preliminary Results," *Proceedings of ION GPS-93* (Salt Lake City, UT), Institute of Navigation, Washington, DC, Sept. 22–24, 1993, pp. 625–632.

[3]Axelrad, P., Chesley, P. B., Comp, C. J., and Ward, L. M., "GPS Based Attitude Determination," FY 93–94 TR II to the Naval Research Laboratory, Oct. 1994.

[4]Bauer, F. H., Lightsey, E. G., McCullough, J., O'Donnell, J., and Schnurr, R., "GADACS: A GPS Attitude Determination and Control Experiment on a SPARTAN Spacecraft," IFAC Conference, Palo Alto, CA, Sept. 1994.

[5]Bauer, F. H., Lightsey, E. G., McCullough, J., O'Donnell, J., Schnurr, R., Class, B. F., Jackson, L., and Leiter, S., "Pre-Flight Testing of the SPARTAN GADACS Experiment," *Proceedings of ION GPS-94* (Salt Lake City, UT), Institute of Navigation, Washington, DC, Sept. 1994, pp. 1233–1241.

[6]Uematsu, H., and Parkinson, B. W., "Antenna Baseline and Line Bias Estimation Using Pseudolites for GPS-Based Attitude Determination," *Proceedings of ION GPS-94* (Salt Lake City, UT), Institute of Navigation, Washington, DC, Sept. 1994, pp. 717–726.

[7]Axelrad, P., Ward, L. M., "On-Orbit GPS Based Attitude and Antenna Baseline Estimation," ION Technical Meeting, San Diego, CA, Jan. 1994, pp. 441–450.

[8]Lightsey, E. G., Cohen, C. E., Feess, W. A., and Parkinson, B. W., "Analysis of Spacecraft Attitude Measurements Using Onboard GPS," 17th Annual AAS Guidance and Control Conference, Keystone, CO, Feb. 1994, pp. 521–532.

[9]Lightsey, E. G., Cohen, C. E., and Parkinson, B. W., "Application of GPS Attitude Determination to Gravity Gradient Stabilized Spacecraft," AIAA GNC, AIAA Paper 93-3788, Monterey, CA, Aug. 1993.

[10]Kruczynski, L. R., Li, P. C., Evans, A. G., and Hermann, B. R., "Using GPS to Determine Vehicle Attitude," *Proceedings of ION GPS-89* (Colorado Springs, CO), Institute of Navigation, Washington, DC, Sept. 1989, pp. 163–171.

[11]Schuler, B., "Impulse Budget: SPARTAN 204," TM to the SPARTAN Project, NASA/GSFC, Code 740, Oct. 1994.

[12]Hughes, P. C., *Spacecraft Attitude Dynamics,* Wiley, New York, 1986, pp. 281–346.

[13]DeBra, D. B., and Delp, R. H., "Rigid Body Attitude Stability and Natural Frequencies in a Circular Orbit," *Journal of Astronautical Sciences,* 1961.

Part VI. Special Applications

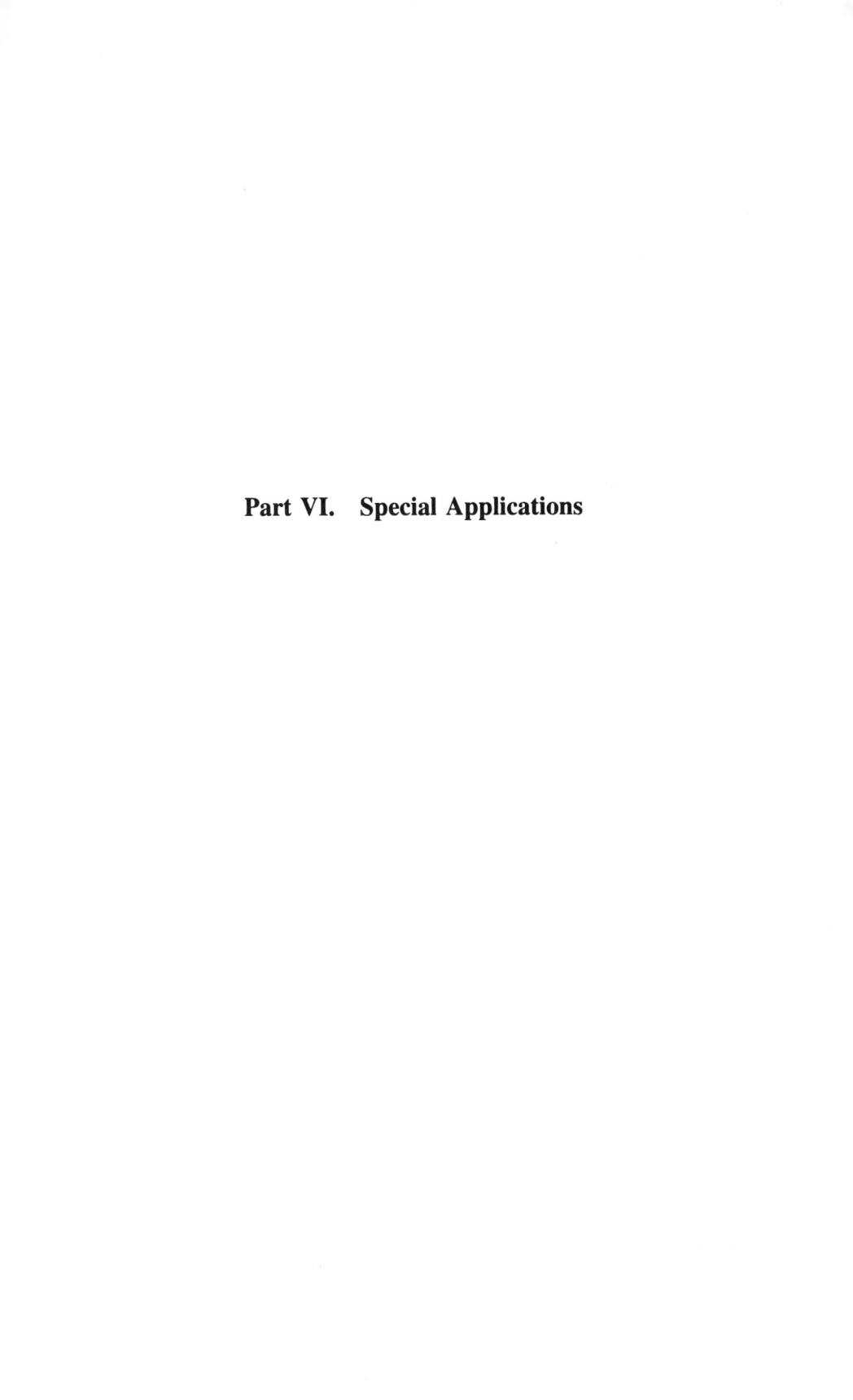

Chapter 17

GPS for Precise Time and Time Interval Measurement

William J. Klepczynski*
United States Naval Observatory, Washington, DC 20392

Introduction

THE Global Positioning System (GPS) has quickly evolved into the primary system for the distribution of Precise Time and Time Interval (PTTI). This is true not only within the Department of Defense (DOD) but also within the civilian community, both national and international. The users of PTTI are those who maintain and distribute time (epoch) to better than one-millisecond (1 ms) precision and/or accuracy and time interval (frequency) to better than one part in ten to the ninth (1×10^{-9}). The GPS is very effective not only in meeting these modest requirements of the PTTI community but also meeting more stringent ones, such as synchronizing clocks to tens of nanoseconds over large distances.

It is not surprising that this is the case. As with all navigation systems, the heart of the GPS is a clock. In the GPS, it controls the transmission of the navigation signals from each satellite and is an integral part of the ground monitor stations. This relationship between clocks and navigation is not unique. It goes back to the eighteenth century when John Harrison (1693–1776) developed his famous clock.[1] Harrison's clock solved the longitude problem for the Royal Navy by allowing a ship to carry Greenwich time with it to sea. The navigator then determined his own local time. The difference between the navigator's local time and the Greenwich time, which he was carrying with him, was his longitude difference from Greenwich. The GPS NAVSTAR satellites are similar to the Royal Navy H.M.S. Deptford. They carry a standard reference time onboard. The navigator then uses the difference between his local time and the reference time onboard the satellite to help him determine his position.

The importance of the GPS to the PTTI community can be neither understated nor underestimated. The GPS is and will be the primary means by which time, that is Universal Coordinated Time, U.S. Naval Observatory [UTC(USNO)], the

---

This paper is declared a work of the U.S. Government and is not subject to copyright protection in the United States.
*Department Head, Time Service Department.

time scale maintained at the U.S. Naval Observatory and the reference for all timed DOD systems, will be distributed within the DOD. The GPS provides time in the one-way mode (OWM) easily to a precision and accuracy of 100 ns in real-time. With a modest amount of care, it is possible to reach 25 ns. In the OWM, the GPS is considered to be akin to a clock on the wall. The output from the receiver provides time as if looking at a clock on the wall. In addition, the OWM also allows the user to determine the difference between a local clock and UTC(USNO) or GPS time. Corrections can be applied to the local clock, in real time or after the fact, so that it can be set on time to UTC(USNO) within the specifications of the system.

Through the GPS, PTTI users can also compare clocks in the common-view mode (CVM) over large distances to a precision and accuracy better than 10 nanoseconds. In the CVM, two users make measurements of their local clock with respect to the same GPS satellite at the same instant of time. If a user differences the values obtained at each site, he or she can determine the offset between the clocks at each site. However, this method requires the exchange of data by at least one of the participants.

The melting-pot method (MPM), which is similar to the OWM and requires an exchange of data as with the CVM, also allows clocks at remote sites to be synchronized and, more importantly, to be steered. In the MPM, a control station determines both the remote clock offset and rate from GPS time or UTC(USNO) and its own clock offset and rate from GPS time or UTC(USNO) by some form of regression to the observations of as many satellites as possible during the day. By comparing the two clock offsets and rates with respect to GPS time or UTC(USNO), corrections to the remote clock can be estimated. Then, corrections to the remotely located clock can be sent via a dial-up modem at any desired time. This last mode has the advantage of allowing automatic operation,[2] and it is not dependent upon any one satellite.

The ability to use the GPS in different modes to derive timing information ensures its prominence as a critical contributor to all timed systems. However, a word of caution is necessary. Prudent systems engineering requires that adequate and alternate back-up systems for PTTI be factored into the overall design of the system. *This point must be emphasized.*

I. Universal Coordinated Time

Universal Time (UT) represents a family of time scales based on the Earth's rotation on its axis. It is an important reference for navigation.

Early forms of time keeping employed time as indicated by the sundial or the apparent solar time. Ptolemy (c. 100–178 A.D.) noted the irregularity of the solar day and defined mean solar time by assuming a mean movement of the sun relative to an observer on Earth. In this way, a clock and the mean solar time were in approximate agreement. The difference between apparent solar time and mean solar time varies with the seasons and is approximated by the "Equation of Time," which in centuries past, was sometimes printed on sundials. Because the Earth is tilted on its axis by 23°.45 with respect to the ecliptic (plane of revolution of the Earth about the sun), the apparent rotation rate of the sun about the Earth is not constant throughout the year. Furthermore, the Earth's orbit around the sun is not perfectly circular. The time offset from the "Equation of Time" varies by roughly 16 min, and the two times are approximately equal four times a year—at the middle of April and June and at the end of August and December. The slight ellipticity of the Earth's orbit also causes a variation of the apparent solar day by approximately 4.7 s. We can make the appropriate correction for these effects to obtain mean solar time, and if this correction is made at the Greenwich meridian in England, we have UT0 the first UT scale. More precisely, UT0 is based on a mathematical expression for the right ascension of the fictitious mean sun and the clock time-of-transit of any celestial object with known position by an observatory yields UT0 after correction for longitude, aberration, parallax, nutation, and precession.[3] However, UT0 is not strictly uniform.

The coordinates of an observatory used to generate UT0 are subject to small changes caused by slow movements of the Earth rotation axis, called polar variation. Universal Time One (UT1) is a true navigator's time scale, which has been corrected for polar variation (PV). Universal Time One is not uniform because of small changes both seasonal (SV) and irregular and some unpredictable changes in the Earth's rotation rate. A further smoothing is performed to remove the SV to form UT2. Universal Time Two (UT2) still has small, unpredictable (10^{-8}) variation and a long-term drift (10^{-10}/year) effect. Universal Times One and Two are stable to within approximately 3 ms in a day.

With the availability of a large number and different kinds of high-precision atomic clocks, atomic time scales now exist with time stabilities better than 1 part in 10^{-14} from year to year. The accuracy of time measurement now exceeds that of any other physical measurement. Universal Coordinated Time is an atomic clock time scale coordinated by the Bureau International de Poids et Mésures (BIPM). Prior to 1982, it was coordinated by the Bureau International de l'Heure (BIH). Universal Coordinated Time differs from a pure atomic clock time, in that it occasionally introduces leap seconds because its epoch is set to astronomical time, while its rate is set to atomic time. These leap seconds are introduced to keep an atomic time scale in approximate step with the Earth's rotation. The leap second adjustment can cause a particular minute to have 59 or 61 s instead of 60. Universal Coordinated Time is, by international agreement, kept to within 0.9 s of the navigator's time scale, UT1. Leap seconds are *usually* added or deleted on June 30 or December 31. Leap seconds have been implemented since 1972.

II. Role of Time in the GPS

The practicality of using atomic clocks in space for navigation was proven with the Navigation Technology Satellites.[4] A very stable frequency source in the satellite can ensure the stability of transmissions over several revolutions of the spacecraft. This ensures adequate tracking to update the satellite's orbit and affords sufficient predictability in the clock's performance.

The GPS Block I developmental satellites contained three rubidium clocks and one cesium clock. These atomic clocks were launched into orbit to help evaluate their long-term performance in space and their effective contribution to overall operations. Rubidium clocks normally exhibit better short-term stability than cesium clocks. They also are much cheaper. However, they are subject to larger frequency variations caused by changes in environmental conditions, and they exhibit a large drift in frequency. On the other hand, cesium clocks have better long-term stability, which tends to have a greater favorable impact on operations. Experience gained with the early Block I launches helped formulate the planning for the operational Block II satellites, which were configured to have two cesium clocks and two rubidium clocks. After reaching orbit, the Master Control Station (MCS) designates the primary clock. Usually, a cesium clock has been so designated; however, in order to get operational experience with all the types of clock in orbit, a rubidium is sometimes chosen.

The orbit determination process also determines the phase offset of the satellite clock with respect to a clock or a system of clocks that has been designated as the GPS master clock. This difference is transmitted in the navigation message as the coefficients of a quadratic expression.

The GPS control segment (CS) maintains a pair of cesium beam atomic clocks at the five monitor station (MS) sites. These clocks constitute a reasonably accurate reference for the GPS orbit determination. Presently, the entire ensemble of clocks, both those in space and on the ground, form the basis of the GPS time. Thus, the GPS time is an atomic clock time similar, but not the same as, UTC time. One marked difference is that the GPS time does not introduce any leap seconds. To do so would throw the GPS P(Y)-code receivers using the system out of lock. Thus, introducing leap seconds is out of the question for GPS. Please note that the UTC leap seconds will cause the GPS time and UTC to differ by an integral and known number of cumulative leap seconds as they are introduced. Other than the leap second effect, however, the GPS CS attempts to keep the GPS time to within 1 μs of UTC time (modulo 1s).

Historically, GPS time was kept at one of the MS. The MCS had the ability to designate any MS clock as the reference for GPS time. The orbit determination process then kept track of all satellite clocks with respect to that GPS master clock. Occasionally, when something happened to that clock, then another clock was designated as the GPS master clock and the Kalman filter was re-initialized with the states of the new clock. The MS clock plays a very important role in the GPS. The determination of each of the satellite orbits is intimately tied to the MS clock. In the orbit determination process, the measured pseudoranges of each satellite are compared to and time tagged by the MS clock. Unfortunately, the orbit determination process can not separate an error in estimated range to the satellite from an error in the clock. Therefore, in order to get a good estimate of the orbit, the MS clocks have to be very stable during the estimation period.

Because environmental conditions at the MS sites were not ideal, there were frequent jumps in frequency of the MS cesium clock designated as the GPS master clock. To improve this situation, an hardware clock ensemble was installed at the MCS.[5] This was done to demonstrate the capability of a rather stable clock system at the MCS that could give a very stable reference to the GPS. This set of clocks is sometimes referred to as the "Navy clock ensemble."

At the same time that the hardware clock ensemble was being developed, a GPS composite clock was developed[6] and put into operation. This is a software clock that averages all the clocks in the system, the ground clocks and the satellite clocks. Because of the intrinsic performance of the rubidium clocks now in orbit, they have been de-weighted in their contribution to the composite clock.

III. Translation of GPS Time to Universal Coordinated Time

The GPS time is based on atomic time, and the time broadcast from the satellite is continuous (modulo 1 s) without the leap seconds of UTC. The introduction of leap seconds would throw the P-code receivers out of lock. Because the time reference for GPS and all DOD timed systems is UTC(USNO), there must be a way to relate the GPS time to UTC(USNO). This is accomplished through the use of the coefficients A_0 and A_1, also transmitted in the navigation message. These coefficients are determined through monitoring the GPS satellites at the USNO. Data used for this purpose are transmitted over a secure line to the GPS MCS at Falcon Air Force Base, CO where the coefficients are determined and transmitted to each satellite. These coefficients give the difference between GPS time and UTC(USNO). The user navigation set or timing receiver can then easily compute the difference between the local clock driving the receiver and UTC(USNO). The algorithm defining the relationship between GPS time and UTC using the navigation data in Subframe 4 as quoted in ICD-GPS-200 is as follows:

1) Whenever the effective time indicated by the WN_{LSF} (week number) and the DN (day number) values are not in the past (relative to the user's present time), and the user's present time does not fall in the timespan which starts at DN + 3/4 and ends at DN + 5/4, the UTC/GPS time relationship is given by the following:

$$t_{UTC} = (t_E - \Delta t_{UTC}) \text{ (modulo 86,400 s)}$$

where t_{UTC} is in seconds and

$$\Delta t_{UTC} = \Delta t_{LS} + A_0 + A_1 [t_E - t_{ot} + 604,800 \text{ (WN–WN}_t)]$$

seconds; t_E = GPS time as estimated by the user on the basis of correcting t_{sv} for factors described in paragraph 20.3.3.3.3 as well as for ionospheric and SA (dither) effects

Δt_{LS} = delta time attributable to leap seconds

A_0 and A_1 = constant and first-order terms of polynomial

t_{ot} = reference time for UTC data (reference 20.3.4.5)

2) Whenever the user's current time falls within the timespan of DN + 3/4 to DN + 5/4, proper accommodation of the leap second event with a possible week number transition is provided by the following expression for UTC:

$$t_{UTC} = W \text{ [Modulo } (86,400 + \Delta t_{LSF} - \Delta t_{LS})], \text{ seconds}$$

where

$$W = (t_E - \Delta t_{UTC}) - 43,200) \text{ (modulo 86,400)} + 43,200, \text{ seconds}$$

and the definition of Δt_{UTC} (as given in paragraph 1) above) applies throughout the transition period. Note that when a leap second is added, unconventional time values of the form 23:59:60.xxx are encountered. Some user equipment may be designed to approximate UTC by decrementing the running count of time within several seconds after the event; thereby, promptly returning to a proper time indication. Whenever a leap second event is encountered, the user equipment must consistently implement carries or borrows into any year/week/day counts. The correction parameters to convert the GPS time broadcast by the satellite to UTC time are contained in the 24 most significant bits (MSBs) of words six through nine plus the eight MSBs of word ten in page 18 or subframe.

Performance of the GPS time reference vs UTC(USNO) is shown in Fig. 1. One year of data are plotted here. The abscissa is time in units of one day. The ordinate shows the difference, in nanoseconds, between GPS time and UTC(USNO) for that day. The GPS time can be converted to UTC(USNO) by using the transmitted coefficients. We can then difference UTC(USNO) as derived from GPS, called UTC' (USNO), and UTC(USNO) as kept at the Naval Observatory. These differences are shown in Fig. 2 for the same period of time as that shown in Fig. 1. It is important to note that the daily averages of the difference between GPS time and UTC(USNO) did not exceed 30 ns throughout the year.

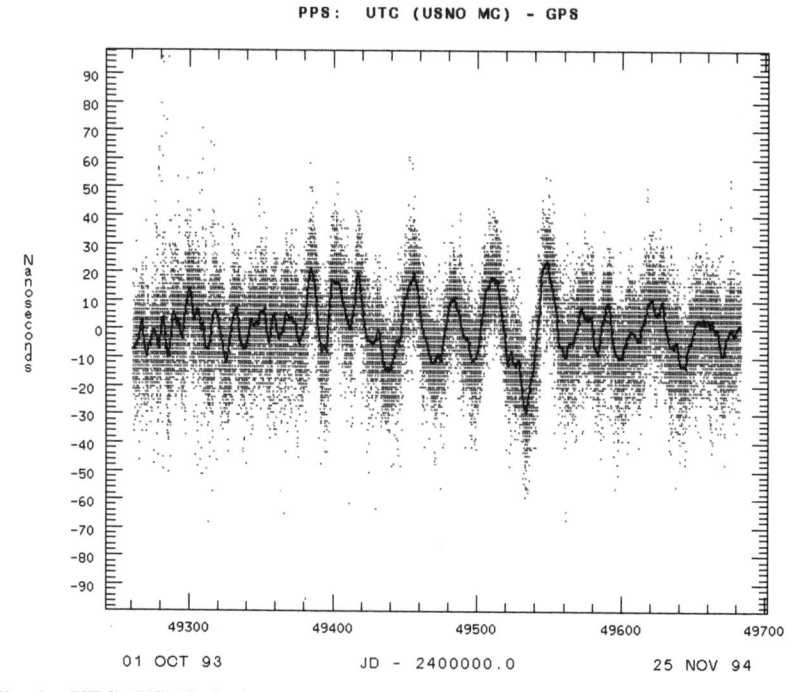

Fig. 1 UTC (USNO, MC)–GPS time over a 1-year period. One large division on the ordinate corresponds to 10 ns and on the abscissa to 100 days. The dots represent a 13-min averaged data point. A line has been drawn through the daily average of the 13-min data points.

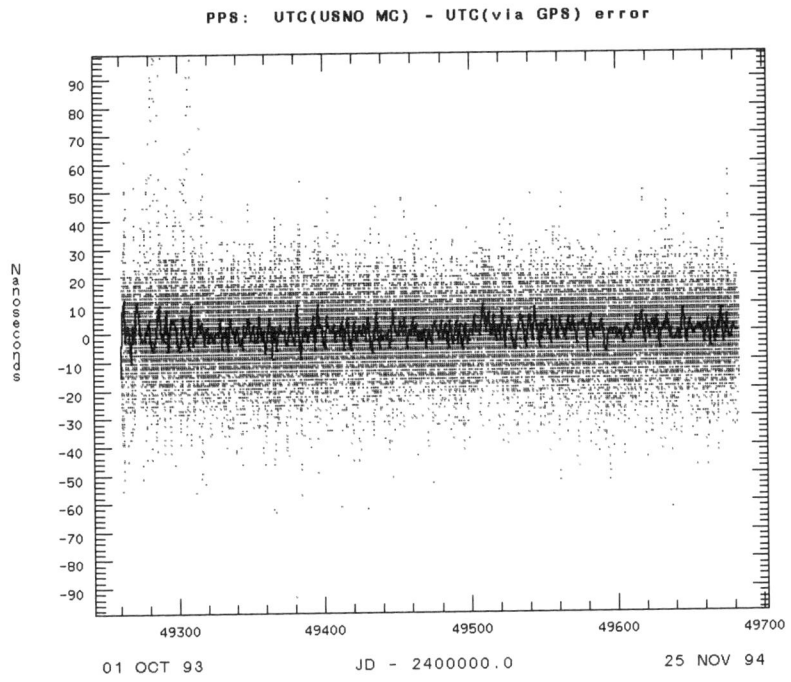

PRECISE TIME AND TIME INTERVAL MEASUREMENT 489

PPS: UTC(USNO MC) - UTC(via GPS) error

Fig. 2 UTC(USNO, MC)–UTC′ (USNO, MC via GPS) over the same 1-year period as shown in Fig. 1. UTC′ (USNO, MC via GPS) means UTC (USNO) as derived from the data contained in the navigation message. The dots represent a 13-min averaged data point. A line has been drawn through the daily average of the 13-min data points.

In addition, daily averages of the differences between UTC′ (USNO) and UTC(USNO) kept within 15 ns over the same period.

IV. GPS as a Clock in the One-Way Mode

In the computation of position, the user navigation set initially determines the difference between the local clock in the navigation receiver and each of the satellite clocks used in the fix. By application of the appropriate set of coefficients selected from the navigation message, the user can compute the difference between his local clock and GPS time or UTC (USNO). To make use of this knowledge, the receiver must then generate a timed output signal. This can be done in several ways, depending on your needs.

In most instances, a platform that has a GPS navigation unit will also be relying on that unit for time to be passed to other systems, such as, a keyed communication transceiver. Figure 3 depicts such a typical system. The timing output port of the GPS receiver is usually a serial or parallel output port, which sends the time to the system in question. The time sent is usually BCD-coded in a way that the system can understand. This, in fact, is proving to be a problem. Because of a lack of coordination, many systems have developed their own time

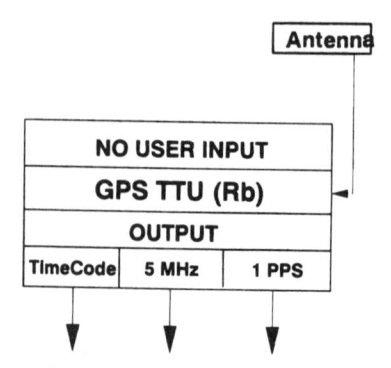

```
                    ┌─────────┐
                    │ Antenna │
                    └────┬────┘
         ┌───────────────┴──────────┐
         │      NO USER INPUT       │
         ├──────────────────────────┤
         │       GPS TTU (Rb)       │◄─
         ├──────────────────────────┤
         │         OUTPUT           │
         ├──────────┬───────┬───────┤
         │ TimeCode │ 5 MHz │ 1 PPS │
         └────┬─────┴───┬───┴───┬───┘
              ▼         ▼       ▼
```

Fig. 3 Block diagram for a typical GPS time transfer unit (receiver).

Because of a lack of coordination, many systems have developed their own time code specifications. Because there is usually only one time code output port on a GPS receiver, great care must be taken in selecting which time code is ordered with the set. If time is needed for more than one system on a platform, then care must be taken in designing a distribution system that passes on the necessary codes to the other systems.

It is possible to synchronize a local clock to the output of a GPS time transfer unit (TTU) to UTC(USNO). A GPS time transfer unit is the type of timing receiver commonly used by the PTTI community. The synchronization is usually accomplished through a feedback loop of some kind and appropriate filter. A small computer keeps track of the differences between UTC(USNO) and the local clock. This information is then used by the computer to set the local clock on time. Figure 4 depicts the block diagram for such a system. In it, the user has chosen as the input to his GPS TTU a local clock driven by his local frequency standard. The rate of the local frequency standard can be adjusted, for example, by a phase microstepper, which is under computer control. The adjusted rate is then fed into a clock that can also be stepped in time by the computer. The GPS TTU sends to the computer, through a serial port, the measured difference between the local clock and UTC(USNO). The computer can then either offset the clock to bring the local time closer to UTC(USNO), or it can adjust the phase microstepper so that the adjusted rate is closer to UTC(USNO). By continually measuring and controlling this process, the local clock will eventually be set to UTC(USNO).

V. Common-View Mode of GPS

Another way in which the GPS can be used to synchronize clocks is in the CVM. This mode of measurement offers some advantages over using the GPS in the previously discussed one-way mode. In this mode, two different observers, separated by large distances, observe the same GPS satellite at the same instant of time. By taking differences of the observations made at each site, the difference between the clocks at the two sites can be obtained. However, data must be exchanged and shared between the two users. Sometimes, this can be a concern because one of the sites may not be able to transmit or share its data over a convenient link.

PRECISE TIME AND TIME INTERVAL MEASUREMENT

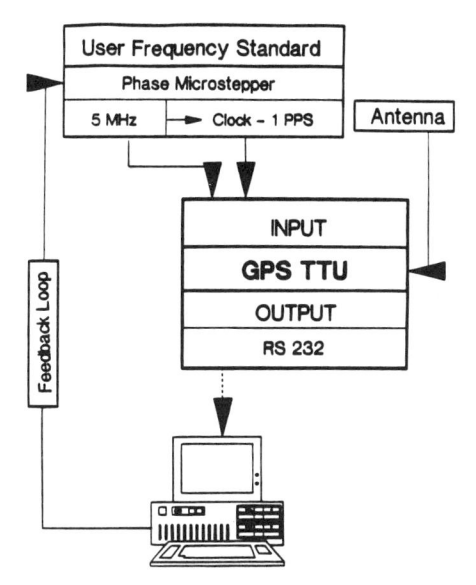

Fig. 4 Block diagram for a more sophisticated GPS time transfer unit block diagram.

The advantage of this technique is that it minimizes certain errors that might be present. Satellite clock errors are totally eliminated,[7] because they are common to both receivers. Ephemeris errors in the transmitted data are not cancelled but minimized. The amount depends upon the geometry between the two sites. Other disadvantages are that it is dependent on a few satellites, and data must be exchanged between the users. The mathematics used in this technique are very simple. We have only to subtract the values obtained at each site to obtain the differences between the two clocks at each site.

This technique is not only useful for synchronizing clocks, it could also prove to be a valuable tool in investigating ionospheric fluctuations over the two sites, as a study done in 1984 between the Tokyo Astronomical Observatory (TAO) and the U.S. Naval Observatory has shown. In this experiment, there were about 2 h of common view time between Washington and Tokyo. Observations were made as the morning terminator line crossed over between the two sites. Figure 5 exhibits the data between the two sites obtained on Feb. 15, 1984. The abscissa is time in hours. The ordinate represents the difference, in nanoseconds, between the clocks used as a reference for the local GPS time transfer units at the Tokyo Astronomical Observatory (now the National Optical Observatory of Japan) and the U.S. Naval Observatory. No discontinuities can be seen in the data between the observations that were made when both sites were in darkness and when one of the sites was in darkness and the other in daylight.

With a good amount of data, it is possible to provide for the steering of remotely located timed systems.[2] Because of the large number of satellites, adequate data can be obtained throughout the day. By smoothing over 2-day intervals, many of the fluctuations apparent in GPS data can be minimized. With the capabilities of today's rubidium and cesium clocks, this should prove more

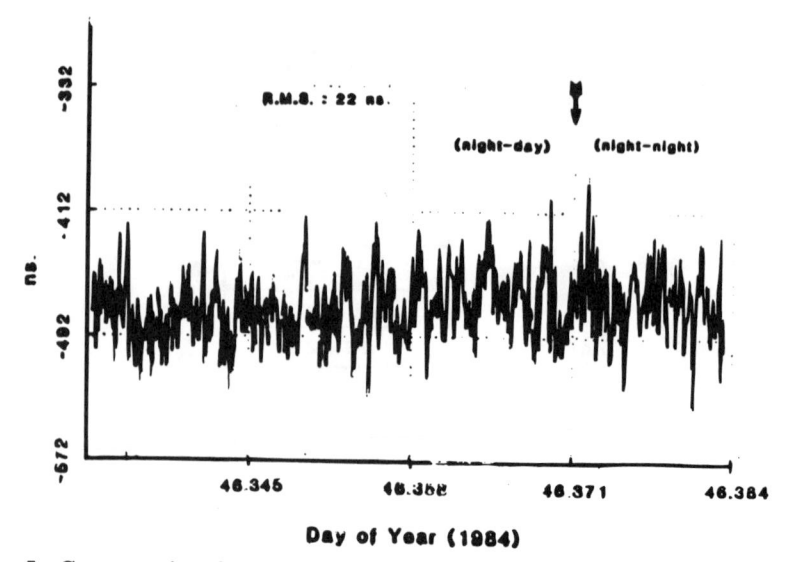

Fig. 5 Common-view GPS data, taken in 1984, between U.S. Naval Observatory (USNO) and the Tokyo Astronomical Observatory (TAO), now called the National Optical Observatory.

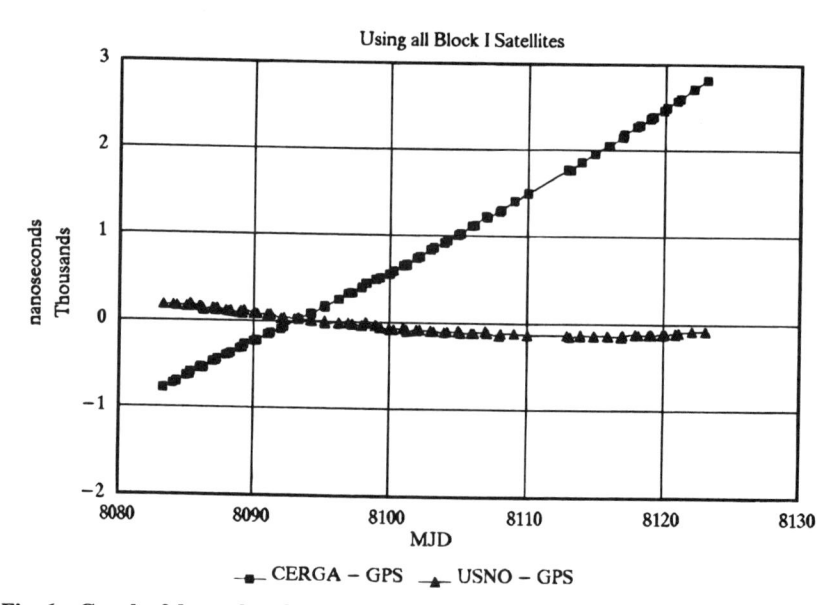

Fig. 6 Graph of data taken from Table 1, which shows the raw values of USNO–GPS and OCA/CERGA–GPS, prior to combination for GPS common-view analysis.

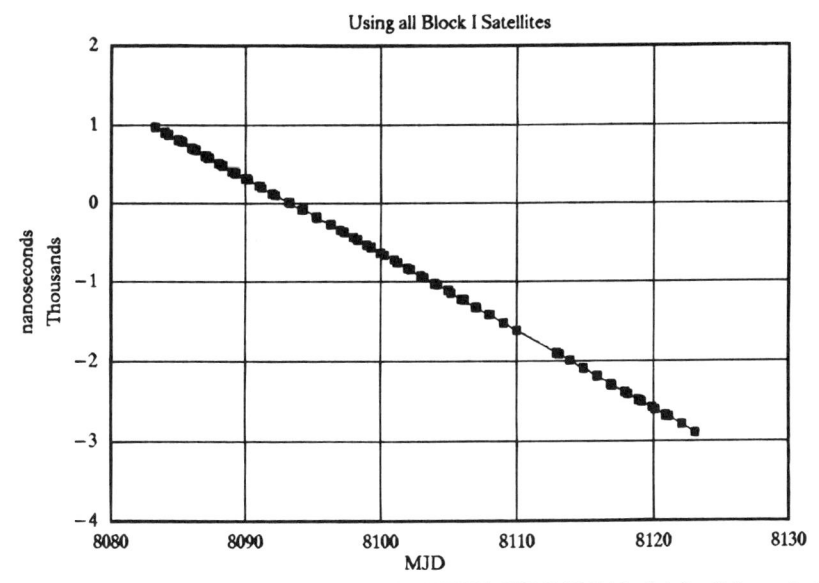

Fig. 7 Graph of the raw values of UTC (USNO)–UTC (OCA) obtained from data in Table 1 used in GPS common-view analysis. Values obtained by subtracting the data shown in Fig. 6.

than adequate to easily maintain a local time scale good to about one part in ten to the thirteenth (1×10^{-13}).

As an example of the practical application of the CVM, Table I gives a sample of GPS data obtained at two sites, the U.S. Naval Observatory in Washington, D.C. and Observatoire de la Cote d'Azure (OCA) in Grasse, France, from July 11–August 20 in 1990. Figure 6 is a plot of the differences between each station's reference clock and the GPS time. Some coordination between the two sites is required in order to ensure the greatest accuracy and precision. First, a sequence of satellites to be observed must be preselected. Factors that would come into play are geometry between the two sites, the starting time of the observations, and the length of tracking time. By selecting a satellite that passes midway between the two stations, any ephemeris errors can be minimized. However, this is not always possible to arrange. It is very important to ensure that the two sites start observing at the exact same second. In this way, all perturbations in the system are measured identically at both stations. This helps to minimize certain errors. Also, it is important to check that both stations are keeping the same time; i.e., UTC or GPS time. When one either averages or does a linear regression through the raw data, it is important that both stations have data over the exact same length of time. Some errors can bias the results because an average of data over different intervals can be different. If one differences the two sets of data, one obtains the difference between the two local reference clocks at each site. A plot of these differences is shown in Fig. 7. Figure 8 is a plot of the residuals after fitting a simple linear regression through the data exhibited in Fig. 7. The rms of the spread is about 6 ns.

Table 1 GPS Common-view data between U.S. Naval Observatory and Observatoire de l'Cote d'Azure (11 July '90–20 Aug. '90), Block I satellites

SVN[a]	MJD[b]	USNO–GPS	OCA–GPS	USNO–OCA	Res to lin. reg.[c]
12	8083.288	−781	195	976	−2
12	8083.310	−780	189	969	−7
13	8083.985	−725	187	912	2
13	8084.007	−723	181	904	−4
12	8084.285	−705	176	881	0
12	8084.307	−699	177	876	−3
13	8084.982	−642	173	815	2
13	8085.004	−637	172	809	−2
6	8085.204	−632	165	797	5
12	8085.282	−593	199	792	8
12	8085.304	−595	193	788	6
13	8085.979	−551	164	715	−1
3	8086.090	−531	165	696	−10
6	8086.201	−541	153	694	−1
12	8086.279	−552	135	687	0
12	8086.301	−551	128	679	−6
13	8086.976	−470	146	616	−4
3	8087.088	−465	145	610	1
6	8087.199	−463	127	590	−8
12	8087.299	−438	146	584	−4
13	8087.974	−392	123	515	−8
6	8088.196	−383	115	498	−3
12	8088.274	−367	124	491	−3
12	8088.296	−370	109	479	−13
13	8088.971	−313	101	414	−12
13	8088.993	−312	101	413	−11
3	8089.082	−303	114	417	2
6	8089.193	−306	86	392	−12
12	8089.271	−269	127	396	−1
12	8089.293	−262	120	382	−13
13	8089.968	−229	96	325	−4
13	8089.990	−230	90	320	−7
3	8090.079	−230	91	321	3
6	8090.190	−215	87	302	−6
13	8090.965	−152	75	227	−5
13	8090.988	−149	82	231	1
3	8091.076	−147	78	225	3
6	8091.188	−127	77	204	−7
13	8091.985	−87	36	123	−10
6	8092.185	−57	49	106	−8
6	8093.182	25	40	15	−2
12	8093.260	22	44	22	12
12	8093.282	24	25	1	−6
6	8094.179	99	16	−83	−3
12	8094.257	90	10	−80	7
12	8094.279	88	8	−80	9
6	8095.176	182	8	−174	2

(continued on next page)

Table 1 GPS Common-view data between U.S. Naval Observatory and Observatoire de l'Cote d'Azure (11 July '90–20 Aug. '90), Block I satellites (continued)

SVN[a]	MJD'[b]	USNO–GPS	OCA–GPS	USNO–OCA	Res to lin. reg.[c]
12	8095.254	172	−7	−179	5
12	8095.276	173	−20	−193	−7
12	8096.251	266	−5	−271	10
12	8096.274	268	−12	−280	3
13	8096.971	336	−11	−347	4
12	8097.249	342	−20	−362	16
12	8097.271	346	−27	−373	7
13	8097.946	402	−34	−436	9
12	8098.246	459	−1	−460	14
12	8098.268	461	−10	−471	6
13	8098.943	498	−36	−534	8
13	8098.965	503	−32	−535	9
12	8099.243	517	−45	−562	9
12	8099.265	515	−56	−571	2
13	8099.940	558	−73	−631	8
13	8099.963	562	−76	−638	3
12	8100.263	594	−75	−669	1
13	8100.938	662	−58	−720	16
6	8101.160	680	−76	−756	1
12	8101.260	669	−94	−763	4
13	8101.935	744	−84	−828	5
13	8101.957	750	−78	−828	7
6	8102.157	761	−82	−843	11
13	8102.932	842	−95	−937	−8
13	8102.954	843	−84	−927	5
6	8103.154	865	−91	−956	−5
13	8103.929	920	−103	−1023	3
13	8103.951	925	−101	−1026	2
6	8104.151	945	−98	−1043	5
13	8104.926	1011	−97	−1108	15
13	8104.949	1018	−90	−1108	17
6	8105.149	1030	−113	−1143	2
13	8105.924	1116	−109	−1225	−5
3	8106.035	1116	−110	−1226	5
6	8106.146	1122	−110	−1232	9
13	8106.921	1227	−96	−1323	−6
13	8106.943	1229	−90	−1319	0
3	8107.032	1212	−114	−1326	1
13	8107.918	1277	−140	−1417	−4
13	8107.940	1280	−137	−1417	−1
3	8108.029	1301	−115	−1416	8
3	8109.026	1401	−117	−1518	3
3	8110.024	1490	−127	−1617	1
13	8112.904	1768	−127	−1895	2
13	8112.926	1773	−129	−1902	−2
6	8113.126	1763	−148	−1911	8
13	8113.901	1858	−135	−1993	1
13	8113.924	1863	−125	−1988	8

(continued on next page)

Table 1   GPS Common-view data between U.S. Naval Observatory and Observatoire de l'Cote d'Azure (11 July '90–20 Aug. '90), Block I satellites (continued)

SVN[a]	MJD'[b]	USNO–GPS	OCA–GPS	USNO–OCA	Res to lin. reg.[c]
13	8114.899	1958	−136	−2094	−3
13	8114.921	1958	−131	−2089	4
13	8115.896	2050	−134	−2184	4
13	8115.918	2056	−137	−2193	−3
13	8116.893	2145	−142	−2287	−2
13	8116.915	2148	−146	−2294	−7
3	8117.004	2180	−124	−2304	−9
13	8117.890	2270	−116	−2386	−5
13	8117.913	2269	−119	−2388	−4
6	8118.113	2289	−121	−2410	−7
13	8118.888	2363	−121	−2484	−6
13	8118.910	2366	−123	−2489	−9
3	8118.999	2379	−111	−2490	−1
6	8119.110	2389	−118	−2507	−7
13	8119.885	2462	−116	−2578	−3
13	8119.907	2459	−122	−2581	−4
6	8120.107	2488	−117	−2605	−8
13	8120.882	2568	−108	−2676	−4
13	8120.904	2571	−114	−2685	−11
6	8121.104	2598	−95	−2693	0
6	8122.101	2712	−81	−2793	−3
6	8123.099	2813	−81	−2894	−7

Regression Output
Constant	785689.4
Standard error of Y Estimate	6.811978
R squared	0.999965
Number of Observations	117
Degrees of Freedom	115
X coefficient(s)	−97.0783
Standard error of coefficients	0.053525

[a] SVN: Space vehicle number.
[b] MJD': Modified Julian date.
[c] Residuals after linear regression.

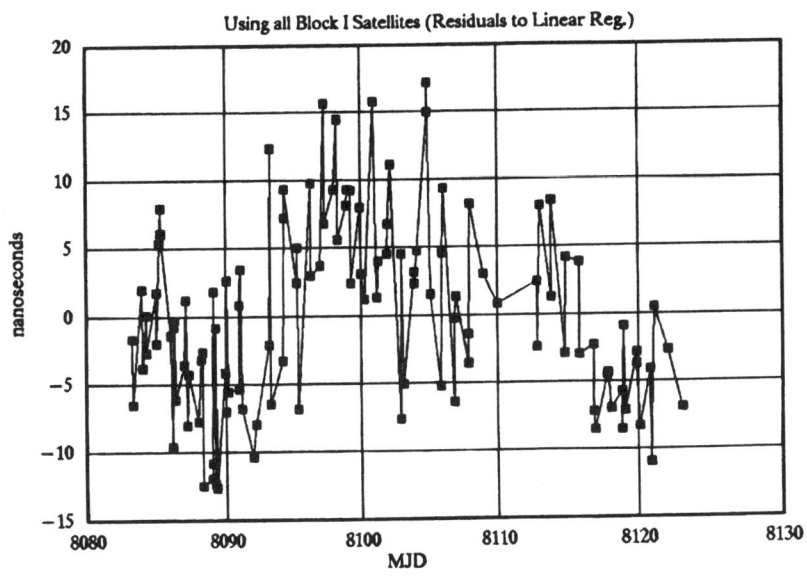

Fig. 8   Residuals with regard to a linear regression of the data shown in Fig. 7.

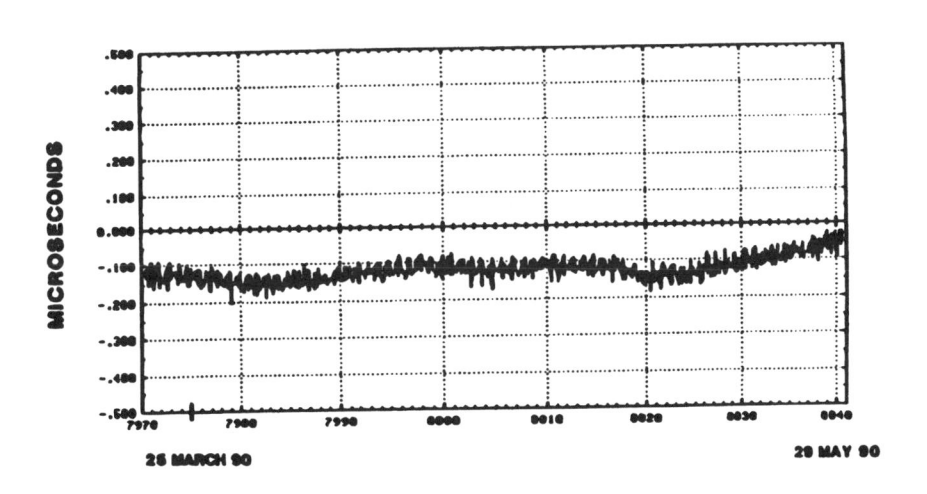

Fig. 9   UTC(USNO, MC)–GPS time over a 2-month period from which all effects of SA have been removed.

STel 502 RECEIVER / SPS UNCLASSIFIED

BLOCK II ONLY

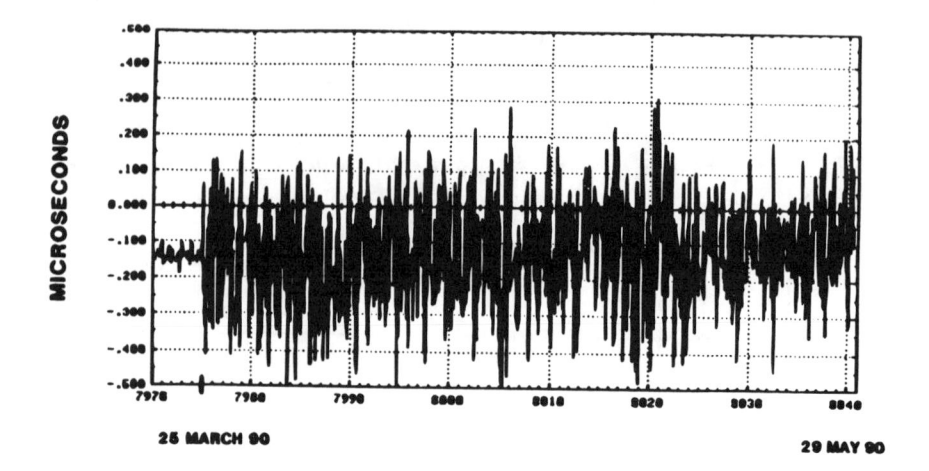

Fig. 10 UTC(USNO, MC)–GPS time over the same 2-month period as shown in Fig. 9, but the effects of SA have not been removed. It is easy to determine when SA was turned on.

VI. Melting-Pot Method

This method can also be used to synchronize clocks over widely separated distances. Unlike the CVM, which requires simultaneity of observations by both stations, the MPM only requires that each station observe as many satellites during the day that its receiver can track. This method is more robust than the CVM, because it observes significantly more satellites during the day. Therefore, it is more readily suitable to allowing the automation of steering a remotely located clock because it will not be affected by occasional gaps in data. The offset of the local clock with respect to GPS time can be ascertained by a simple regression. The offset in time and rate can than be compared with similar data obtained at another station to correct one of the clocks. In fact, it is possible to automate the process so that control of the remote clock can be done automatically when a set of prescribed limits are exceeded.

VII. Problem of Selective Availability

Selective availability (SA) will affect use of the GPS in the OWM of time transfer unless the TTU is an authorized, keyed receiver. Figure 9 shows the results of monitoring GPS with a dual-frequency TTU that has been keyed. The data are presented in a form similar to that used in earlier figures. Figure 10 shows the same data taken at the same time when SA was partially on. The

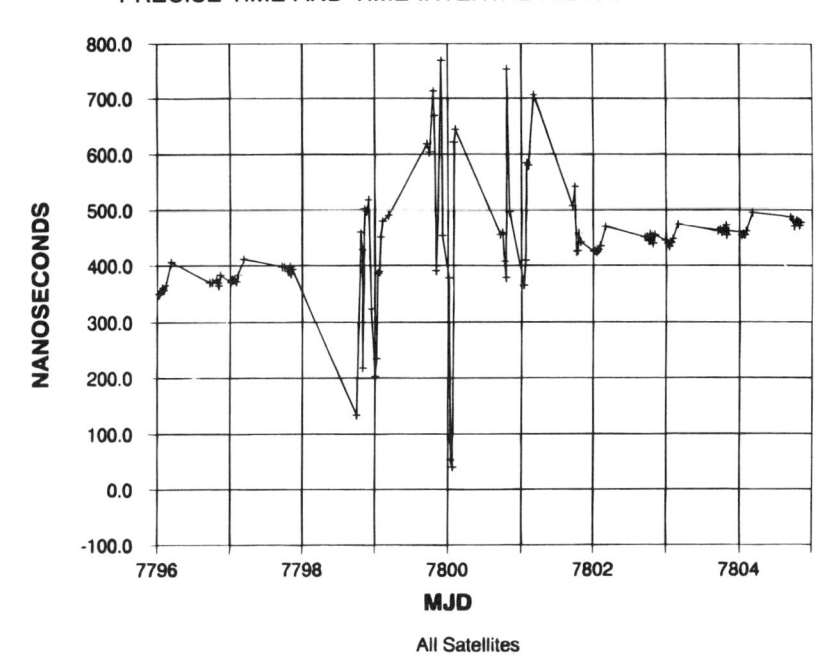

Fig. 11 Common-view comparison of two clocks during a period when SA was turned on.

degradation effects of SA on timing data is easily evident, as well as the time at which it was turned on.

The error induced in a clock that is tracking UTC derived from GPS will depend upon the level of SA. By employing some averaging techniques it is possible to minimize the affects of SA. Many of the more recent receivers on the market now track more than six satellites, thus allowing many different ways to average and smooth the data. This implies that the user clock being steered to UTC via GPS, is not updated in real time but only after a sufficient period of time has elapsed in order to smooth out the amounts of SA that are being applied.

The common-view mode of GPS can also minimize some of the effects of SA. Figure 11 shows a comparison between two different clocks via CVM GPS when SA was on. In this case, averaging will improve the results by minimizing the sudden deviations apparent in the data. The use of averaging in common-view depends upon what kind of SA is being applied. If the only form of SA being applied is clock dither, then no averaging will be necessary.[7]

IX. Future Developments

The GPS is a bright star within the PTTI community. It is now the workhorse time transfer system. It promises to be so for the next several decades. The GPS offers the user community great flexibility on how it can be used. Because of this flexibility, it has become a widely accepted and successful tool.

However, it must be cautioned that for this to continue to be so, improvements to the GPS must take place along with improvements in the PTTI field. It is obvious that a new generation of clocks will soon be developed. These clocks may have performance in the parts to the sixteenth region of stability and accuracy. There are also new techniques for time transfer over large distances that promise to exceed the current capabilities of the GPS. Therefore, for the GPS to maintain its position of pre-eminence then it, too, will have to show some progress.

There are several areas where some of the GPS subsystems can be improved, consequently improving overall system performance. The clocks at the monitor stations can be augmented with a small ensemble of cesium clocks, clock-averaging software, and independent means for comparing the clocks with those of other MSs. This would allow for the independent determination of the frequency and phase offset of the monitor station clocks from the Kalman filter process used at the MCS. In addition, newer receiver hardware can be installed at the monitor stations. There have been many improvements in receivers since the original monitor stations went into operation some 15 years ago. Because of receiver miniaturization and automation, it is now not unreasonable to think of augmenting the original set of five monitor stations to improve orbit determination.

References

[1] Gould, R. T., *John Harrison and His Timekeepers,* National Maritime Museum, 1958.

[2] Miranian, M., and Klepczynski, W. J., "Time Transfer via GPS at USNO," *Proceedings of ION GPS-91,* Institute of Navigation, Washington, DC, Sept. 1991.

[3] Winkler, G. M. R., "Timekeeping and Its Applications," *Advances in Electronics and Electronphysics,* Vol. 44, Academic Press, New York, 1972.

[4] White, J., Danzy, F., Falncy, S., Frank, A., and Marshall, J., "NTS-2 Cesium Beam Frequency Standard for GPS" *Proceedings 8th Annual Precise Time and Time Interval Applications and Planning Meeting,* Paper X-814-77-149, 1976, p. 637.

[5] Stein, S., "Improvement of CSOC Clock Ensemble Algorithm for Use at USNO," *Ball Aerospace Systems Group,* B7170-89-001, 1989.

[6] Brown, K., "The Theory of the GPS Composite Clock," *Proceedings of ION GPS-91,* Institute of Navigation, Washington, DC, Sept. 1991.

[7] Allan, D. W., and Weiss, M. A., "Accurate Time and Frequency Transfer during Common-View of a GPS Satellite," *Proceedings 34th Annual Frequency Control Symposium,* 1980, p. 334.

Chapter 18

Surveying with the Global Positioning System

Clyde Goad*
Ohio State University, Columbus, Ohio 43210

THE geodesy/geophysics and surveying communities are fortunate to have the Global Positioning System (GPS) satellites transmitting its pseudorange code on dual-frequency bands using very stable oscillators. As was discussed in Chapter 3 of the companion volume, the pseudorange code is affected by changing the phase state of the carrier by 0.5 cycles (180° or 200 gons). That is, at prescribed times, the state of the carrier is changed by 0.5 cycles, if the binary code is to be switched from a 0 to a 1 or from a 1 to a 0. These code change epochs, called chips, occur at a rate of 10.23 MHz for the P1 and P2 codes, and at a rate of 1.023 MHz for the C/A code. Should the receiver make available the difference between the phase of the transmitted signal and the phase generated by the receiver's oscillator, then these differences can be used together with the same information from other satellites being tracked by a receiver and the same satellites being tracked by other receivers. The goal of combining the one-way measures between satellites and stations is to determine, very precisely, the geometric vector (baseline) between electrical centers of two receivers' antennas. The technique used is similar to that used by electronic distance meters (EDMs), which use the (fractional) phase of the reflected signal difference to infer distance. However, unlike EDMs, the GPS signals are based on incoherent phase measures (one-way, not reflected signals) using only one or both of the two available frequencies. EDMs usually use five or so frequencies so as to determine the cycle of the reflected fractional phase without ambiguity.

Because the GPS satellite receivers can utilize only one or at most two available frequencies, some additional effort is required to determine the cycle ambiguity between oscillators, because they are not aligned. To accomplish the task of rendering the measurements in terms of quantities that will allow them to be of use, only two mathematical models are required: a model for distance traveled by an electromagnetic wave in a vacuum, and the phase change of an oscillator running with constant frequency.

Copyright © 1995 by the author. Published by the American Institute of Aeronautics and Astronautics, Inc., with permission. Released to AIAA to publish in all forms.
*Department of Geodetic Science and Surveying.

I. Measurement Modeling

Let us first look at the physical situation, as shown in Fig. 1. The GPS satellites transmit the carrier signal continuously. Suppose the phase of the carrier that was transmitted from a GPS satellite at time t_T arrives at the ground receiver antennas at time t_R. Assume, also, that this signal travels at the speed of light through a vacuum. Thus, the distance traveled equals the time interval multiplied by the speed:

$$r = c(t_R - t_T) \tag{1}$$

This is the true range if t_R and t_T are measured by the same clock. It is pseudorange if they are measured by different clocks. The phase front of the satellite generated at transmission time, then, arrives at the receiver later at receiver sample time. This is the same assumption used to process pseudorange measurements. The actual measurement here is the difference between the satellite phase at transmission time and the receiver phase at receipt time or the following:

$$\phi_R^S(t_R) = \phi^S(t_T) - \phi_R(t_R) \tag{2}$$

where t_T is the time the carrier signal left the satellite base; t_R is the time this same signal arrived at the receiver antenna based on the receiver's clock. Superscripts refer to satellite ID; subscripts refer to receiver ID.

The left side of Eq. (2) is given as a function of receiver time here, but either t_T or t_R could be used. The choice will be dictated by how the receiver manufacturer chooses to implement Eq. (2) in hardware. This is discussed shortly.

Regardless of whether t_R or t_T is used as the reference time on the left side of Eq. (2), both t_T and t_R also appear on the right side. Now we must substitute for one of the times to obtain a corresponding phase at the time of the other.

Here, all times are expressed in terms of t_R, which is chosen because of the way data from most GPS receivers are tagged. That is, most manufacturers choose to sample the incoming phase values from all satellites being tracked at the same (received) time. This allows the manufacturers to use inexpensive oscillators. The t_R, then, represents the current state of the receiver clock at the instant phases are sampled and compared.

The connection between t_R and t_T is known from Eq. (1):

$$t_T = t_R - r/c \tag{3}$$

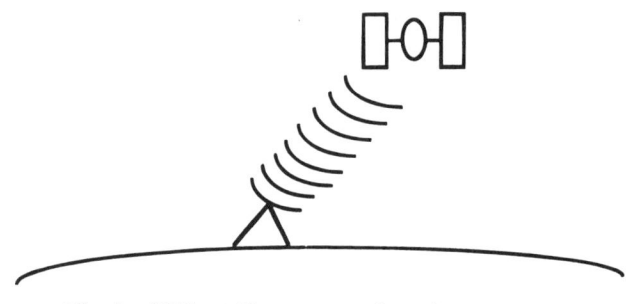

Fig. 1 GPS satellite-to-ground receiver geometry.

Substituting Eq. (3) into Eq. (2) yields the following:

$$\phi_R^S(t_R) = \phi^S(t_R - r/c) - \phi_R(t_R) \tag{4}$$

The first term on the right side of Eq. (4) can now be expanded using an ideal oscillator relation:

$$\phi(t + \Delta t) = \phi(t) + f \cdot \Delta t \tag{5}$$

Equation (5) is the same model for determining time intervals in quartz watches used by most persons today. The symbol f stands for the phase rate or frequency of the oscillator (in the satellite). We notice that the $-r/c$ in Eq. (4) is the Δt in Eq. (5). Thus, after substituting Eq. (5) into Eq. (4), we get the desired relation:

$$\phi_R^S(t_R) = \phi^S(t_R) - f/c\, r - \phi_R(t_R) + N \tag{6}$$

where $\phi^S(t_R)$ is the phase in the satellite oscillator at time t_R under the assumption of a constant running oscillator of frequency f; N is an (unknown) integer reflecting the fact that $\phi_R^S(t_R)$ measures only the fractional phase difference at time t_R or that the phase difference counter has an arbitrary integer value (i.e., only the fractional part is actually measured on the first measurement after signal acquisition).

It should be emphasized that the N is required only on the first measurement after signal lock is achieved. After lock is achieved, the phase (difference) counter counts the total (integer plus fractional) phase change from sample epoch to sample epoch. This total change in phase (integrated Doppler) continues until a loss of lock occurs.

Actually, we can directly process the phase measurement as given in Eq. (6); however, here, the generation of differenced measurements is given to show explicitly the removal of those terms not of interest to those needing position information. That is, the $\phi^S(t_R)$ and the $\phi_R(t_R)$ are of no direct interest to us. Thus, one way to remove them from the data used for positioning purposes is to generate differenced data types where the phase measurements participating in the differencing process are chosen to remove these "nuisance" variables.

Assume for now that the higher frequency L_1 from one satellite, number 6, is tracked by receivers 9 and 12. The mathematical representation of these two one-way measurements from Eq. (6) is given as follows:

$$\phi_9^6(t_K) = \phi^6(t_K) - f/c\, r_9^6(t_K) - \phi_9(t_K) + N_9^6 \tag{7a}$$

$$\phi_{12}^6(t_K) = \phi^6(t_K) - f/c\, r_{12}^6(t_K) - \phi_{12}(t_K) + N_{12}^6 \tag{7b}$$

In Eqs. (7a) and (7b), it is assumed that the two receivers sample the phase-tracking channels at exactly the same (received) time t_K. Note that $\phi^6(t_K)$ appears in both the equations.

If the two equations are differenced (7a) $-$ (7b), we get $\phi_9^6(t_K) - \phi_{12}^6(t_K) = -f/c[r_9^6(t_K) - r_{12}^6(t_K)] - [\phi_9(t_K) - \phi_{12}(t_K)] + N_9^6 - N_{12}^6$ or more simply:

$$\phi_{9,12}^6(t_K) = -f/c[r_9^6(t_K) - r_{12}^6(t_K)] - [\phi_9(t_K) - \phi_{12}(t_K)] + N_{9,12}^6 \tag{8}$$

Equation (8) gives the mathematical representation of the (between station) single difference. This type can also be used to estimate station coordinates. However, in addition to receiver (or more precisely, antenna) coordinates, we must also

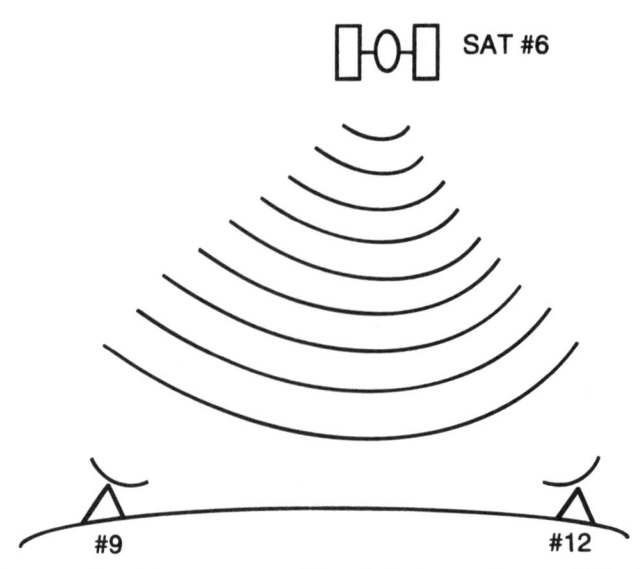

Fig. 2 Geometry involving two ground-based stations and one satellite. The figure shows the geometry of the situation. By looking at the figure, we see that the only common element between the two sampled phase measures is satellite #6.

solve for the receiver phase difference and the integer ambiguity $[\phi_9(t_k) - \phi_{12}(t_k)] + N^6_{9,12}$. This is not done here. The technique is quite similar to that used to process pseudoranges collected by a receiver during periods of no motion. If the reader is interested, the technique used to process single-difference phase measurements is given by Ref. 1.

However, generally, because of the presence of the receiver oscillator phase differences, the single differences between receiver oscillators are not of primary interest, so another differencing operation is used to remove these undesirable terms. Now we introduce another satellite (say, #18) that is also tracked by stations #9 and #12 (Fig. 3).

With the addition of satellite 18, we can generate an additional single-difference measurement. Here, the two available single differences are listed for ease of discussion.

$$\phi^6_{9,12}(t_K) = -f/c[r^6_9(t_K) - r^6_{12}(t_K)] - [\phi_9(t_K) - \phi_{12}(t_K)] + N^6_{9,12} \quad (9a)$$

$$\phi^{18}_{9,12}(t_K) = -f/c[r^{18}_9(t_K) - r^{18}_{12}(t_K)] - [\phi_9(t_K) - \phi_{12}(t_K)] + N^{18}_{9,12} \quad (9b)$$

Looking at Eq. (9a) and (9b), we see that $\phi_9(t_K) - \phi_{12}(t_K)$ is common to each. These oscillator differences between receivers can be removed through another difference operation. Thus, let us difference Eq. (9b) from Eq. (9a) to obtain the following:

$$\phi^6_{9,12}(t_K) - \phi^{18}_{9,12}(t_K) = -f/c[r^6_9(t_K) - r^6_{12}(t_K) - r^{18}_9(t_K) + r^{18}_{12}(t_K)] + (N^6_{9,12} - N^{18}_{9,12}) \quad (10)$$

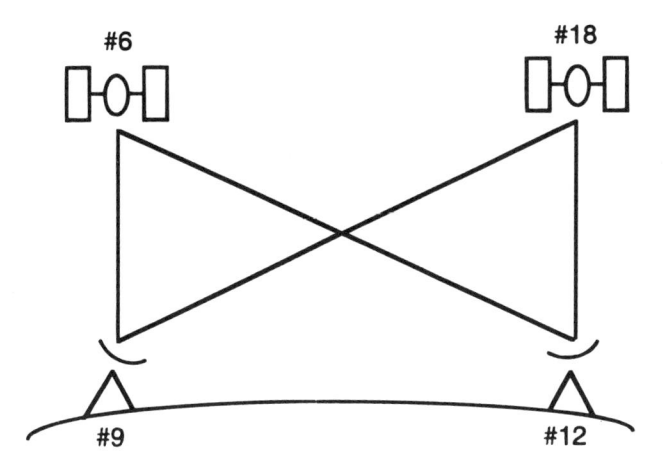

SURVEYING WITH THE GLOBAL POSITIONING SYSTEM   505

Fig. 3 Geometry involving two ground-based stations and two satellites.

Simplifying the notation as before, Eq. (10) is rewritten as follows:

$$\phi_{9,12}^{6,18}(t_K) = -f/c(r_9^6 - r_{12}^6 - r_9^{18} + r_{12}^{18}) + N_{9,12}^{6,18} \tag{11}$$

In Eq. (11), the t_K has been dropped from the r terms, because the time dependence is obvious from the t_K on the left side. The notation $\phi_{9,12}^{6,18}$ and $N_{9,12}^{6,18}$ implies that two stations (9 and 12) and two satellites (6 and 18) are involved in this "double difference" operation. $N_{9,12}^{6,18} = (N_9^6 - N_9^{18} - N_{12}^6 + N_{12}^{18})$. Because each N value on the right is an integer, then $N_{9,12}^{6,18}$ is also an integer. The reader is now reminded of the assumptions made in the deviation of the double difference observable given in Eq. (11). They are simultaneity of reception times at receivers, perfectly constant and equal oscillator rates (frequencies) in the satellites, and that the signals from the satellites travel at the vacuum speed of light. Although none of these conditions is ever achieved exactly, small corrections can be made to the one-way measurements to achieve a high degree of compliance.

Because of the subtraction of the many common elements in Eq. (11), the double difference is not very sensitive to the (absolute) position of either receiver location, but it is sensitive to the position of one receiver relative to the other (i.e., the baseline vector). Thus, double differences are very similar to distance and angle measurements used commonly by the surveying community.

To be of use, however, we must be able to collect sufficient data to allow for the separation of geometry (baselines) and the ambiguities. For example, Fig. 4 depicts a possible history of one double difference configuration (2 satellites, 2 stations). The lower curve represents the actual geometrical contribution to the double difference phase history given by the r terms in Eq. (11). The upper curve depicts what is actually measured. Thus, the difference between the two curves is the integer ambiguity. It should be obvious that with very little data (in time) and no a priori knowledge of the baseline, there is no way to discriminate between the baseline and ambiguity. But, as time passes, there will be only one baseline that satisfies the shape or change in time of all double difference histories. Also obvious is that the greater the number of satellites

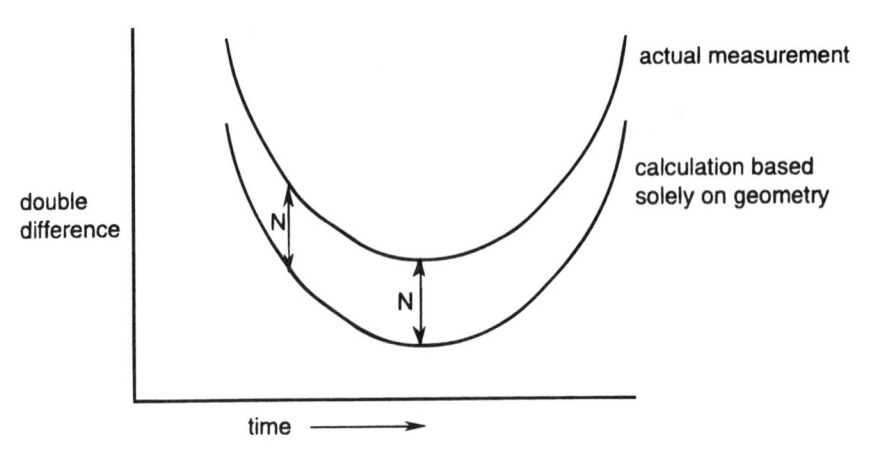

Fig. 4 A depiction of a possible double-difference measurement scenario vs what would be calculated based solely on geometry of the satellites and receivers.

tracked, the sooner the actual baseline can be identified. Once the baseline is determined unambiguously to within the order of 0.25 cycles, the N values can be constrained to integer values, which allows for the most desirable use of the double difference data.

Normally the baseline (vector) and ambiguities are estimated using the technique of least squares. That is, the best guess of the ambiguities and baseline are those values that minimize the sum of squares of measurement discrepancies once the estimated quantities' contributions are removed. In such implementations, we generally treat the ambiguities as real-valued parameters. These estimates, then, take on a (real) value that makes the measurement residual sum of squares a minimum. To the extent that common mode contributions to the measurements cancel, then the real-valued estimates of the ambiguities tend toward integer values. The classic case for such easy identification of integer-valued ambiguity estimates is when the baseline is short. That is, over short baselines, it is usual for those (error) sources not included in the original representation [Eq. (6)] to be removed through the differencing process [Eqs. (8) and (11)]. Such physical contributions usually canceling over short distances are errors in the refraction (tropospheric and ionospheric) and orbital errors.

Defining the concept of "short" baselines is not so easy, however. Let us consider more carefully the ionosphere, for example. The activity of the ionosphere is known to depend greatly on the 11-year cycle of sunspot activity. Therefore, when sunspot activity is low, the ionosphere is not as active, and the effect on microwave signals from GPS satellites is similar over a wider area than when the sunspot activity is greater. In 1983, when the sunspot activity was low, newly introduced single-frequency phase-measuring GPS receivers provided phase measurements that allowed for integer identification up to distances of 60 km. At the maximum of the most recent sunspot activity in 1990–1991, integers were difficult to identify, at times, over 10-km distances.

Not all possible difference combinations should be generated, however. Theoretically, only those combinations of double differences that are linearly independent offer new information to a data reduction. A linearly dependent combination is one that can be obtained by linearly combining previously used double differences. For example, consider the following possible double differences: $\phi_{9,12}^{6,18}$; $\phi_{9,12}^{6,20}$; and $\phi_{9,12}^{18,20}$. The last double difference can be obtained by a combination of the first two, as follows: $\phi_{9,12}^{18,20} = \phi_{9,12}^{6,18} - \phi_{9,12}^{6,20}$.

In other words, once $\phi_{9,12}^{6,18}$ and $\phi_{9,12}^{6,20}$ have been used, no new information is contained in $\phi_{9,12}^{18,20}$. Thus, such linearly dependent data should not be considered. If n represents the number of receivers and s the number of satellites being tracked at a data-sampling epoch, the maximum number of linearly independent combinations is $(n-1)(s-1)$. For the simple case of only two receivers, the generation of linearly independent data is not so difficult. However, when the number of receivers is greater than two, the task of generating the maximum number of linearly independent measurements in order to gain the maximum amount of information possible is not so trivial. Reference 2 has addressed this problem in detail.

Because there are usually several ways to combine data to form independent observables, there may be advantages of some schemes over others. Distance between receivers is one such consideration. Let us consider the case of three ground receivers (A, B, C) collecting data simultaneously as given in Fig. 5.

Here there are three possible baselines, only two of them linearly independent. Which two should be chosen? Now it is appropriate to discuss those contributions ignored in the generation of Eq. (11). These include such items as tropospheric and ionospheric refraction, multipathing, arrival time differences, orbit error, etc. Two of these unmodeled contributions are known to have errors that increase with increasing distance between receivers—orbit error and ionospheric refraction. (Tropospheric refraction does also, but only to a limit of, say, 15–50 km). Now, back to Fig. 5. Because we now realize that a more complete cancellation of unmodeled errors occurs for the shorter baseline, and thus, the use of Eq. (11) is more justified, we definitely should choose the baseline BC as one of the two independent lines. Although not so drastically different, we might as well choose AB as the other independent line, because it is slightly shorter than the line AC.

Although both orbit error and ionospheric influences are baseline length-dependent, the ionosphere causes the major degradation. Solar storms, traveling ionospheric disturbances (TIDs), day/night variations, etc. can cause large disturbances in the GPS signals. This is especially bothersome at equatorial and auroral latitudes. Moreover, these disturbances are especially prominent at the peaks of the 11-year solar cycle. Fortunately, there is one "fix" and that is to use the dispersive character of the ionospheric effect. Dispersion means frequency dependent. So, two signals transmitted at different frequencies will exhibit different

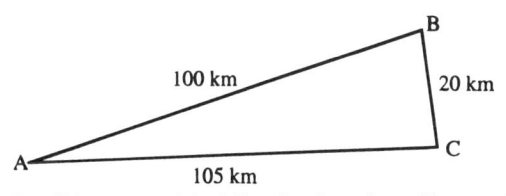

Fig. 5 Possible geographical distribution of satellite receivers.

ionospheric signatures. Actually, a better mathematical model for the GPS signals is one that recognizes a retarding of the code (group delay) and an advance of the phase—both inversely proportional to the square of the transmission frequency when expressed in distance units (say, meters).

Thus, having two measurements each at a different frequency, allows us to combine them in such a way that the ionospheric effect can be cancelled. Fortunately the GPS system was designed to transmit two different frequencies—L_1 (1575.42 MHz) and L_2 (1227.6 MHz). However, these two frequencies are transmitted only on the P codes and not on the C/A code, which is available to the civil sector. Not only is a dual-frequency receiver more expensive, it might be unable to track the dual-frequency P-code signals if antispoofing (AS) is being used.

Because the ionosphere contribution is inversely proportional to the square of frequency in range, then it is equivalently inversely proportional to frequency (to the first power) in angular units (cycles). Thus, we can now modify Eq. (6) to incorporate the ionospheric contribution as follows:

$$\phi_R^s(t_R) = \phi^s(t_R) - (f/c)r - \phi_R(t_R) + N + I(t)/f \tag{12}$$

Let us assume that we want to combine phase measurements at the L_1 and L_2 frequencies so as to remove the ionospheric terms I/f_i (where f_i stands for either the L_1 or L_2 frequency; i.e., $i = 1$ or 2); then, ϕ (no ion) $= \alpha_1\phi(L_1) + \alpha_2\phi(L_2)$ represents the "ion-free" combination. The I/f_i terms will cancel if the condition $\alpha_1/f_1 + \alpha_2/f_2 = 0$ is satisfied. In addition, another condition can be imposed, so the resulting combination is usually chosen so as to look like the original L_1 equation, but without the ionospheric term. This additional condition is given as $\alpha_1 f_1 + \alpha_2 f_2 = f_1$. These two conditions allow for a unique solution as follows:

$$\alpha_1 = f_1^2/(f_1^2 - f_2^2) = 2.5457$$

$$\alpha_2 = -f_1 f_2/(f_1^2 - f_2^2) = -1.9837$$

For the case of an "ion-free" measure, Eq. (12) should be written as follows:

$$\phi_R^S(t_R) = (\alpha_1 + \alpha_2)\phi^S(t_R) - (f_1/c)r - (\alpha_1 + \alpha_2)\phi_R(t_R) + \alpha_1 N_1 + \alpha_2 N_2 \tag{12'}$$

Here note that the ionosphere term I is time dependent. We have combined the L_1 and L_2 measurements to eliminate this ionospheric term—but at a price. We must now work with increased noise. Furthermore, the ambiguities are no longer integers, because the coefficients needed in the no-ion combination are not themselves integers, which destroys the integer nature of the resulting ambiguities.

Because of increased processing requirements and increased costs associated with receiver purchases, many surveyors choose to purchase or rent the less expensive single-frequency receivers (L_1 only) and try to counter the detrimental impact of the ionosphere by observing only over short baselines, as discussed earlier.

II. Dilution of Precision

Because the mathematical model of the double difference observables can be generated even before a survey is undertaken, with the anticipated data collection start/stop times and recent satellite ephemeris information (almanac), the least-squares adjustment process can be simulated that allows for a predetermination of the geometrical strength in the planned data. This has been useful in deciding the amount of time needed for a survey session. These recovered dilution of precision (DOP) values are, in essence, the very same as the position dilution of precision (PDOP) or geometric dilution of precision (GDOP) values used in navigation and described in the companion volume, but now based on the accumulated data over a survey session rather than on data collected at an instant, as is done for navigation purposes.

Two such measures can be calculated, one based on the ability to fix the ambiguities to their integer values (the "fixed" DOP values) and one that assumes that the integer ambiguities cannot be determined and must be estimated as real numbers along with baseline components (the "float" DOP value). Experience with a particular set of hardware and software techniques along with the calculated DOP values allows us to estimate the amount of data required to identify double difference integer ambiguities. Based on the techniques discussed up to this point, 0.5–1.0 h per baseline are typical.

III. Ambiguity Search

With the rapid improvement of personal (low-cost) computers, a technique introduced by Ref. 3 is now being pursued by some investigators. In essence, it is a search technique that requires baseline solutions to have integer ambiguities. Two techniques have evolved; one that searches arbitrarily many locations in a volume and one that restricts the search points to those locations associated with integer ambiguities. To put it another way, one "loops" over all locations in a volume, or one loops over possible integer ambiguity values that yield solutions within a given volume. The explanation of this technique requires only the use of Eq. (11). A sample location in space is chosen (arbitrarily). It can then be used to calculate the distances (r terms) in Eq. (11), and if it is the actual location of the antenna, then that which is left after removing the r terms should be an integer. All measurements to all satellites at all epochs will exhibit this behavior. Locations that do not satisfy this requirement cannot be legitimate baselines. The beauty of this search technique is that cycle slips (losses of lock) are not a consideration; that is, even if the ambiguity changes its integer value, such an occurrence has no impact on the measure of deviation from an integer.

The volume search technique is the easiest to envision and the most robust. A suitable search cube, say one meter on a side, is chosen and each location in a grid is tested. Initial search step sizes of 2–3 cm are reasonable. Once the best search point is found, a finer search can be performed to isolate the best fitting baseline to, say, the mm level. Although the most robust, this volume search can be time consuming. An alternative is to choose the four satellites with the best PDOP, and test only those locations found from assuming that their ambiguities are integers. That is, "loop" on a range of ambiguities rather than all locations in the test cube. Such a scheme is much faster, but can suffer if the implied test locations are in error because of unmodeled contributions to the measurements used to seed the search. Effects that can cause such errors are multipathing, ionosphere, etc.

In either case, the key to minimizing computer time is to restrict the search volume. One such way is to use differential pseudorange solutions if the pseudoranges are of sufficient quality. Here P-code receiver measurements are usually superior to those that track only the C/A codes. However, some manufacturers are now claiming to have C/A-code receivers with pseudorange precision approaching 10 cm. Of course, success can only be obtained if the initial search volume contains the location of the antenna within it. So one now must contend with competing factors. The search volume needs to be as large as possible to increase the probability that the true location can be found. However, the search volume must be small enough to obtain the estimate in a reasonable amount of time. Clearly, the better the available pseudoranges and the greater the number of satellites being tracked, the better such a search algorithm will work.

These search techniques can also be used even when the antenna is moving. However, in this case, one must assume that no loss of lock occurs for a brief time so that the search can be performed on ambiguities. This, then, allows for the different epochs to be linked through a common ambiguity value, because there is no common location between epochs of a moving antenna (unless the change in position is known, which could be the case if inertial platforms are used). As computers become even more powerful and if receivers can track pseudoranges with sufficient precision and orbits are known well enough, even baselines over rather long distances can be determined using these techniques. In the end, because of the required computer time, one probably would not use these search techniques to determine the entire path of an airplane or other moving structure, but they could be very useful in providing estimates of integer ambiguities in start-up or loss-of-lock situations.

IV. Use of Both Pseudoranges and Phase

It should now be obvious that for the most precise surveying applications, recovery of the ambiguities is required. Using the approach discussed earlier, the separation of the geometrical part (baselines) and the ambiguities requires some time to pass in order to utilize the accumulated Doppler. One major consequence of this approach is that the integer ambiguities are more difficult to identify with increasing baseline length because of the decoupling of unmodeled error sources such as tropospheric refraction and orbital errors. The same is true for the ambiguity search. With the introduction of affordable receivers collecting both dual-frequency pseudoranges and phases, this laborious approach might be laid to rest if sufficient noise reduction can occur with the tracking of the precise pseudoranges. Techniques utilizing the P-code pseudoranges are now discussed.

For some time, the ability to use readily the pseudoranges in addition to dual-frequency phase measurements provided by the ROGUE receivers designed at the California Institute of Technology/Jet Propulsion Laboratory to recover widelane phase biases has been well known.[4,5]

Here we examine the simultaneous use of all four measurements (phases and pseudoranges from both L_1 and L_2 frequencies). It is shown that the four-measurement filter/smoother can be generated numerically from the average of two three-measurement filters/smoothers. Each of the three-measurement algorithms can be used to provide estimates of the widelane ambiguities, provided that some preprocessing can be performed to reduce the magnitude of the L_1 and L_2 ambiguities to within a few cycles of zero. However, such a restriction is not required for the four-measurement algorithm.

A. Review

To aid in understanding these techniques, a review is presented using the notation of Ref. 5. First, the set of measurements available to users of receivers tracking pseudoranges and phases on both the L_1 and L_2 frequency channels at an epoch is given mathematically as follows:

$$\rho_1 = r^* + I/f_1^2 + \epsilon_{R_1} \tag{13a}$$

$$\Phi_1 = r^* - I/f_1^2 + N_1\lambda_1 + \epsilon_{\Phi_1} \tag{13b}$$

$$\rho_2 = r^* + I/f_2^2 + \epsilon_{R_2} \tag{13c}$$

$$\Phi_2 = r^* - I/f_2^2 + N_2\lambda_2 + \epsilon_{\Phi_2} \tag{13d}$$

Note, here that all the measurements in Eqs. (13a–13d) are expressed in linear units, which is different from that given earlier in Eq. (12) for phases. A simple scaling by λ_1 or λ_2 accomplishes this transformation; $\Phi = \lambda \times \phi$.

In Eqs. (13a–13d), the r^* represents the combination of all nondispersive clock-based terms; or, in other words, the ideal pseudorange; the dispersive ionospheric contribution is the I/f_i^2 term (theoretically a positive quantity) with group delays associated with pseudoranges and phase advances associated with the phases. The two phases (range) measurements include the well-known integer ambiguity contribution when combined in double-difference combinations. Finally, all measurements have noise or error terms ϵ.

Eqs. (13a–13d) can be expressed in the more desirable matrix formulation as follows:

$$\begin{bmatrix} \rho_1 \\ \Phi_1 \\ \rho_2 \\ \Phi_2 \end{bmatrix} = \begin{bmatrix} 1 & 1 & 0 & 0 \\ 1 & -1 & \lambda_1 & 0 \\ 1 & (f_1/f_2)^2 & 0 & 0 \\ 1 & -(f_1/f_2)^2 & 0 & \lambda_2 \end{bmatrix} \begin{bmatrix} r^* \\ I/f_1^2 \\ N_1 \\ N_2 \end{bmatrix} + \begin{bmatrix} \epsilon_{\rho_1} \\ \epsilon_{\Phi_1} \\ \epsilon_{\rho_2} \\ \epsilon_{\Phi_2} \end{bmatrix} \tag{14}$$

In Eq. (14), it is readily apparent that in the absence of noise, one would solve the four equations in four unknowns to recover ideal pseudorange, instantaneous ionospheric perturbations, and the ambiguities. Although the noise values on phase measurements are of the order of a millimeter or less, the pseudorange noises vary greatly from receiver to receiver. L_1 C/A-code pseudoranges have the largest noise values, possibly as high as 2–3 m. This is because of the relatively slow chip rate of 1.023 MHz. P-code chip rates are 10 times more frequent, which suggests noises possibly as low as 10–30 cm. Obviously, to determine ambiguities at the L_1 and L_2 carrier frequencies ($\lambda_1 \cong 19$ cm, $\lambda_2 \cong 24$ cm), low pseudorange noise values play a critical role in the time required to isolate either N_1 or N_2, or some linear combination of them. In a least-squares smoothing algorithm, Ref. 5 showed that the worst and best combinations of L_1 and L_2 ambiguities are the narrow-lane ($N_1 + N_2$) and wide-lane ($N_1 - N_2$)

combinations, respectively. With 20-cm pseudorange uncertainties, the wide-lane estimate uncertainty approaches 0.01 cycles; whereas, narrow-lane uncertainties are at about 0.5 cycles. These should be considered as limiting values, because certain contributions to Eqs. (13) and (14) were not included, such as multipath and higher-order ionosphere terms, with multipath being the more dominant of the two, by far.

The beauty of Eq. (14) lies in its simplicity and the ease of implementing a least-squares algorithm to obtain the wide-lane ambiguity values. Once the wide-lane ambiguity is obtained, the usual ion-free combination of Eqs. (13b) and (13d) yield biases that can be expressed as a linear combination of the unknown L_1 ambiguity and the known wide-lane ambiguity. Knowing the values of the wide-lane ambiguity makes it much easier to recover the L_1 ambiguity. However, not knowing either ambiguity, and even knowing the baseline exactly is a situation wherein it is possible the analyst will be unable to recover the integer values for N_1 and N_2.

Other factors, in addition to multipath, that could negatively influence the use of Eq. (13) would be the nonsimultaneity of sampling of pseudorange and phase measurements within the receiver or a smoothing of the pseudoranges using the phase (or Doppler) information that attempts to drive down the pseudorange noise but then destroys the relations (13a–13d). Note that theoretically no large ionosphere variations or arbitrary motions of a receiver's antenna negate the use of Eqs. (13) or (14). Thus, after sufficient averaging, wide-lane integer ambiguities can be determined for a receiver/antenna, say, involved in aircraft tracking or tracking a buoy on the surface of the sea. For many terrestrial surveys, once sufficient data have been collected to recover the wide-lane ambiguity, no more would be required, except where total elimination of the ionosphere is required, such as for orbit determination and very long baseline recoveries. For these situations, both L_1 and L_2 integers are desired, and geometry changes between satellite and ground receivers are required unless the baseline vectors are already known. The technique of using such short occupation times along with the four-measurement filter to recover wide-lane ambiguities is known as "rapid static surveying." Again, one must be aware that unmodeled multipath can be very detrimental when very short occupation times are utilized.

An example of the use of Eq. (14) in a least-squares algorithm is illustrated in Fig. 6. Here four measurements ρ_1, ρ_2, Φ_1, Φ_2 were collected every 120 s at the Penticton, Canada, tracking station. Although the integer nature of the ambiguity can only be identified after double differencing, the one-way measurements (satellite-to-station) can be smoothed separately, and the biases combined later to yield the double difference ambiguities. Figure 6 shows the difference between the linear combination involving ρ_1, ρ_2, Φ_1, Φ_2 to yield the wide-lane ambiguity on an epoch-by-epoch basis with the estimated values. The reader will notice that individual epoch values deviate little from the mean or least-squares estimate; the rms of these values is 0.06 cycles. The three-measurement combinations are discussed in the next section.

Table 1 shows the estimates of the double-difference ambiguities formed from the combination of one-way bias estimates between Canadian locations Penticton and Yellowknife, which are 1500-km apart. The integer nature of the wide-lane values is clearly seen, while the similar integer values of the L_1 and L_2 bias values

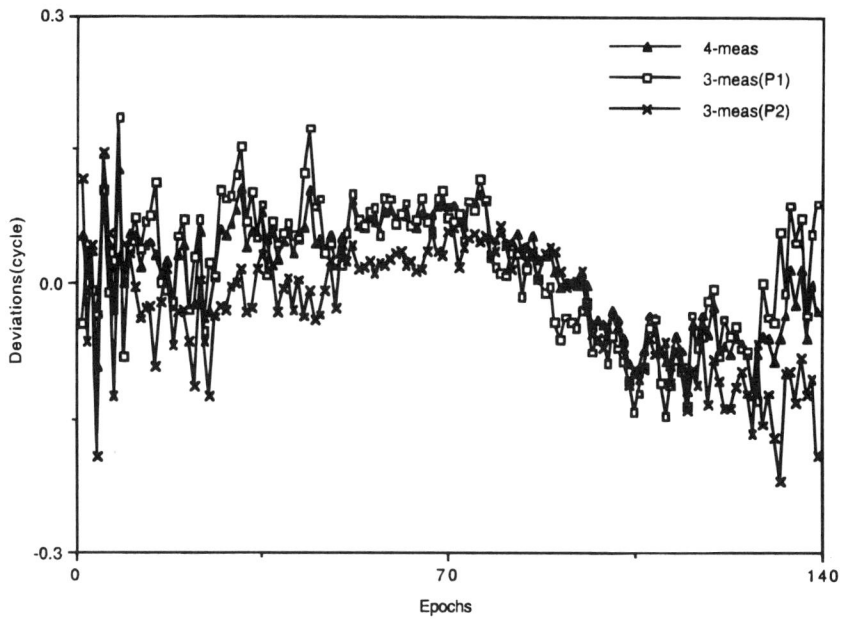

Fig. 6 Deviations from mean values of the four- and three-measurement combinations, Rogue receiver, Penticton, Canada, day 281, 1991, space vehicle (SV) 14.

Table 1 Estimated values of the N_1, N_2, and wide-lane $(N_1 - N_2)$ double-difference ambiguities

Sat	Sat	N_1	N_2	$N_1 - N_2$
2	3	−0.162	−1.177	1.105
2	6	−0.284	−1.250	0.966
2	11	−0.002	−1.044	1.042
2	12	−0.539	−1.497	0.957
2	13	−0.450	−1.396	0.947
2	14	1.544	−0.542	2.086
2	15	−1.492	−2.562	1.070
2	16	0.174	−0.877	1.051
2	17	0.035	−1.015	1.051
2	18	−0.335	−1.382	1.047
2	19	0.905	−0.119	1.024
2	20	−0.214	−1.197	0.983
2	21	0.078	−0.984	1.063
2	23	−0.253	−1.316	1.063
2	24	−0.787	−2.735	1.948

cannot be identified. Clearly, in the processing steps, an integer close to the originally determined bias value has been subtracted from the corresponding phase measurements in an attempt to keep the double-difference ambiguities close to zero. This was not a requirement of the four-measurement technique, however.

B. Three-Measurement Combinations

Here the derivation of the two three-measurement combinations is presented. First, we must use the two phase measurements, Eqs. (13b) and (13d). Next, choose only one of the two pseudorange measurements ρ_1 or ρ_2. Let us choose to examine the selection of either by denoting the chosen measurement as ρ_i, where i denotes either 1 or 2 for the L_1 or L_2 pseudorange, respectively. To simplify the use of the required relations, Eqs. (13a–13d) are rewritten as follows:

$$\rho_i = r^* + I/f_i^2 + \epsilon_{\rho_i} \tag{15a}$$

$$\Phi_1 = r^* - I/f_1^2 + N_1\lambda_1 + \epsilon_{\Phi_1} \tag{15b}$$

$$\Phi_2 = r^* - I/f_2^2 + N_2\lambda_2 + \epsilon_{\Phi_2} \tag{15c}$$

The question to be answered is: What is the final combination of N_1 and N_2 after eliminating the r^* and I terms in Eqs. (15a–15c)? The desired combinations can be expressed as follows:

$$a\rho_i + b\Phi_1 + c\Phi_2 = dN_1 + eN_2 + a\epsilon_{\rho_i} + b\epsilon_{\Phi_1} + c\epsilon_{\Phi_2} \tag{16}$$

where $d = b\lambda_1$, and $e = c\lambda_2$.

To assure the absence of the r^* and I terms, the a, b, and c coefficients must satisfy the following:

$$a + b + c = 0 \tag{17a}$$

$$a/f_i^2 - b/f_1^2 - c/f_2^2 = 0 \tag{17b}$$

One free condition exists. Because it is desirable to compare the resulting linear combinations of N_1 and N_2 to the wide-lane combination, we choose arbitrarily to enforce the following condition:

$$d = d\lambda_1 = 1 \tag{17c}$$

Solving Eqs. (17a–17c) with $i = 1, 2$ yields the two desired three-measurement combinations with noise terms omitted:

$$-1.2844\rho_1 + 5.2550\Phi_1 - 3.9706\Phi_2 = N_1 \\ - 0.9697N_2, \quad \text{for} \quad i = 1 \tag{18a}$$

$$-1.0321\rho_2 + 5.2550\Phi_1 - 4.2229\Phi_2 = N_1 \\ - 1.0313N_2, \quad \text{for} \quad i = 1 \tag{18b}$$

In practice, the coefficients in Eqs. (18a) and (18b) should be evaluated to double precision. The errors in the above combinations are dominated by the pseudorange errors that depend on the receiver characteristics, as discussed earlier. However, when compared to even the most precise GPS pseudoranges, the phase uncertainties are orders of magnitude smaller. Thus, the error in the combination

(18a) in cycles is equal to 1.28 times the uncertainty of ρ_1 (in meters). Similarly, the combination (18b) is equal in cycles to 1.03 times the uncertainty in ρ_2 (in meters). As with the four-measurement combination, averaging can be used to reduce the uncertainty of the estimated combination. Also the two three-measurement combinations possess almost all the desirable characteristics as the four-measurement combination. The same restrictions also apply. For example, simultaneity of code and phase is required; multipath is assumed not to exist; and filtering of the pseudoranges that destroys the validity of Eqs. (15a–15c) is assumed not to be present.

One situation does require some consideration—the magnitudes of N_1 and N_2. That is, in the four-measurement combination, the identification of the wide-lane ambiguity is not hindered by large magnitudes of either N_1 or N_2. However, if either of the two three-measurement combinations differ from the wide-lane integers by 3% of the N_2 value, this difference could be very large if the magnitude of N_2 is large. Thus, some preprocessing is required. For static baseline recovery, this is probably possible by using the estimated biases from the individual wide-lane and ion-free phase solutions. Using these ambiguity estimates, the L_1 and L_2 phase measurements can be modified by adding or subtracting an integer to all the one-way phases so that the new biases are close to zero. With near-zero L_1 and L_2 ambiguities, the magnitude of the 0.03 N_2 deviation from the wide-lane integer should be of no consequence in identifying the integer widelane value.

Furthermore, it appears that the average of the two three-measurement combinations is equal to the four-measurement combination. This is not the case identically; however, again, with small L_1 and L_2 ambiguities, it is true numerically.

To illustrate the power in the three-measurement combinations, the data collected on the Penticton–Yellowknife baseline are used to estimate all three combinations. Table 2 shows the resulting estimates (the last column is discussed later). It is clear that all three combinations round to the same integer values. Also

Table 2 Four-measurement and two three-measurement double-difference ambiguity estimates over the Penticton–Yellowknife baseline

Sat	Sat	$N_1 - N_2$	$N_1 - 1.03N_2$	$N_1 - 0.97N_2$	$N_1 - 1.283N_2$
2	3	1.105	1.052	0.980	1.350
2	6	0.966	1.009	0.933	1.324
2	11	1.042	1.074	1.011	1.337
2	12	0.957	1.003	0.912	1.380
2	13	0.947	0.996	0.909	1.348
2	14	2.086	2.106	2.069	2.261
2	15	1.070	1.149	0.992	1.796
2	16	1.051	1.078	1.025	1.300
2	17	1.051	1.082	1.020	1.338
2	18	1.047	1.090	1.005	1.438
2	19	1.024	1.030	1.015	1.094
2	20	0.983	1.020	0.947	1.321
2	21	1.063	1.086	1.033	1.303
2	23	1.063	1.105	1.023	1.520
2	24	1.948	2.033	1.865	2.723

apparent is that the numerical average of each of the three-measurement estimates equals the four-measurement estimate. Again, this is because of the preprocessing step to ensure that ambiguities are close to zero. Figure 6 shows deviation of the one-way (satellite-station) means from the epoch-by-epoch values. The noise levels seem to be small for all the combinations. Large scatter is noted at lower elevation angles when the satellite rises (low epoch numbers) and sets (large epoch numbers). A cutoff elevation angle of 20 deg was used in the generation of Fig. 6. Moreover, an increase in deviations with the model can be seen at the lower elevation angles.[5] The obvious question is whether this is caused by multipath.

Clearly, if we have all four measurement types, the four-measurement combinations would be used. However, with very little extra effort, all three combinations can be computed, possibly helping to identify potential problems in either the ρ_1 or ρ_2 measurements.

V. Antispoofing?

Under certain assumptions about Y-code structure (AS on), a receiver can compare the two Y codes and obtain an estimate of the difference between the two precise pseudoranges ($\rho_1 - \rho_2$). For this tracking scenario, Eq. (15a) is replaced with the following:

$$\rho_{1-2} = 1/f_1^2 [1 - (f_1/f_2)^2] + \epsilon_{R_1} - \epsilon_{R_2} \qquad (19)$$

Imposing the same restrictions as before on the coefficients a, b, and c, the following is obtained where again the error terms are ignored:

$$(\rho_{1-2}/\lambda_1 + \Phi_1/\lambda_2 - \Phi_2/\lambda_1) = N_1 - 1.2833 N_2 \qquad (20)$$

The recovery of $N_1 - 1.283 N_2$ using differences in pseudoranges from the Penticton–Yellowknife baseline are given in the last column of Table 2. Here, assuming the magnitude of N_2 to be less than or equal to 3, the values of N_1 and N_2 seem to be identifiable in some cases. Using an orbit to recover the ion-free double-difference biases can also be of major importance for those cases where the integer values still remain unknown to within one cycle. In any event, some concern is warranted when we are required to use these measurements. Because $1/\lambda_1 = 5.25$, an amplification of the pseudorange difference uncertainty over the individual pseudorange uncertainty of $\sqrt{2} \times 5.25 = 7.42$ is present, assuming that the pseudorange difference uncertainty is only $\sqrt{2}$ larger than either the L_1 or L_2 individual pseudorange uncertainties. This is far from the expected situation, so clearly, some noisy, but unbiased, C/A-code pseudorange data are highly desirable. The usefulness of these data types when AS is operating is an open question, and no definitive conclusions can be obtained until some actual pseudorange differences and C/A-code pseudoranges are available for testing.

VI. A Look Ahead

It is clear that receivers with precise pseudorange or pseudorange differences can make the job of finding the integer ambiguities far more robust (not depending on the orbit) and easier. Assuming that measurements are always differenced for precise relative positioning, we can envision the day when precise navigation and surveying will merge using the techniques presented here. No longer must precious time be spent prior to motion of the antenna to determine the ambiguities using geometry (change). For example, for a photogrammetric mission, an airplane could begin flight long before the receivers are even powered on. Losses of lock caused by blockages by wings or the tail section create no problem theoretically to restart. Even a temporary loss of power can be accommodated in the data-reduction process. For the first time, buoys arbitrarily placed in the oceans can be tracked to the centimeter level for studies of time variation—a true open ocean tide gauge.

Surveyors with communication gear could transmit data from a master receiver (not moving) to a receiver visiting locations whose coordinates are to be determined and recover the baselines in real time while collecting data. Once the wide-lane biases are determined, the surveyor could be notified that sufficient data have been collected, coordinates could be displayed, and the surveyor could move on to the next site of the survey, turning off the power to the receiver during the motion to extend the life of the batteries. Possibly, the most beneficial use to mankind of the real-time determination of baselines would be the access to real-time monitoring of Earth motions as precursors to earthquakes, if such precursors are, in fact, present.

Many other applications can be identified, such as automatic aircraft landing, automatic steering of a dredge or piloting of a ship, very precise roadway mapping, monitoring of Earth motions for geodynamic purposes, air gravity surveys, satellite orbit determinations, railway leveling measurements, etc. Some of these notions are developed in Chapter 15, this volume.

References

[1]Goad, C. C., and Remondi, B. W., "Initial Relative Positioning Results Using the Global Positioning System," *Bulletin Géodésique*, Vol. 58, 1984, pp. 193–210.

[2]Goad, C. C., and Mueller, A., "An Automated Procedure for Generating an Optimum Set of Independent Double Difference Observables using Global Positioning System Carrier Phase Measurements," *Manuscripta Geodaetica*, Vol. 13, 1988, pp. 365–369.

[3]Counselman, C. C., and Gourevitch, S. A., "Miniature Interferometric Terminals for Earth Surveying: Ambiguity and Multipath with Global Positioning System," *IEEE Transactions on Geosciences and Remote Sensing*, Vol. 19, No. 4, 1981, pp. 244–252.

[4]Blewitt, G., "Carrier Phase Ambiguity Resolution for the Global Positioning System Applied to Geodetic Baselines up to 2000 km," *Journal of Geophysical Research*, Vol. 94 (B8), 1989, pp. 10187–10203.

[5]Euler, H. -J., and Goad, C. C., "On Optimal Filtering of GPS Dual Frequency Observations without Using Orbit Information," *Bulletin Géodésique*, Vol. 65, No. 2, 1991, pp. 130–143.

Chapter 19

Attitude Determination

Clark E. Cohen[*]
Stanford University, Stanford, California 94305

ALTHOUGH originally developed as a means for navigation, GPS has since been shown to be an abundant source of attitude information as well. Using the subcentimeter precision of GPS carrier phase, a receiver can determine the relative positions of multiple antennas mounted to vehicles or platforms so accurately that their orientation may be determined in real time at output rates exceeding 10 Hz. This chapter discusses the fundamentals of attitude determination using GPS. It also describes the mathematics of attitude solution processing, error evaluation, and cycle ambiguity resolution. Finally, it discusses applications and provides a sample of experimental results.

I. Overview

The fundamental principle of attitude determination with GPS and multiple antennas is shown in Fig. 1. The GPS satellite is so distant relative to the antenna separation that arriving wavefronts can be considered as effectively planar. A signal traveling at the speed of light arrives at the antenna closer to the satellite slightly before reaching the other. By measuring the difference in carrier phase between the antennas, a receiver can determine the relative range between the pair of antennas. With the addition of carrier phase measurements from multiple satellites using three or more antennas, the receiver can estimate the full three-axis attitude of an object.

Early experimental work employed TRANSIT satellites for attitude determination.[1] Since then, many GPS receivers have been developed or adapted for carrying out attitude determination. Examples of these include implementations by Magnavox,[2] Trimble,[3] TI,[4] and Ashtech.[5]

In the conventional relative position fix (for example, between two survey receivers), range difference measurements between the two receivers from *four* GPS satellites are required to solve for the three components of Cartesian relative position and receiver clock time bias. For attitude determination, it is possible

---

Copyright © 1995 by the author. Published by the American Institute of Aeronautics and Astronautics, Inc., with permission. Released to AIAA to publish in all forms.

[*]Research Associate, Department of Aeronautics and Astronautics; Manager of GPS Precision Landing.

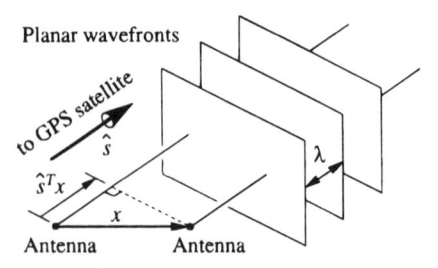

Fig. 1 Attitude geometry.

to configure the receiver so that *only two GPS satellites in view are explicitly required*. There are two reasons for this.

1) *Common Time Reference:* If the receiver is designed as shown in Fig. 2, so that each signal path shares a common time reference, the phase difference measurement precision is maximized. Because any local oscillator variations affect both signal paths identically, these variations cancel out in the final differencing process. Therefore, the measurements are *independent* of the receiver clock bias. Because of the electrical connection between antennas, only signals from *three* GPS satellites are required to find the three Cartesian components of relative antenna position.

2) *Fixed Baseline Configuration:* For attitude determination, the relative mechanical placement of the antennas must be known in advance. Given the additional rigid constraint on relative antenna placement on the vehicle, another satellite measurement can be dropped. Therefore, a minimum of only *two* GPS satellites are required for an attitude fix.

This result provides some very practical benefits. First, overall solution integrity is improved considerably. Because the operational GPS constellation provides at least four satellites in view, attitude solutions are, in general, strongly overdetermined. Occasional cycle slips can be detected and isolated in real time. Second, when the vehicle attitude tips to extremes (such as with an aircraft in a steep angle of bank), attitude solutions are uninterrupted as long as at least two satellites are in view.

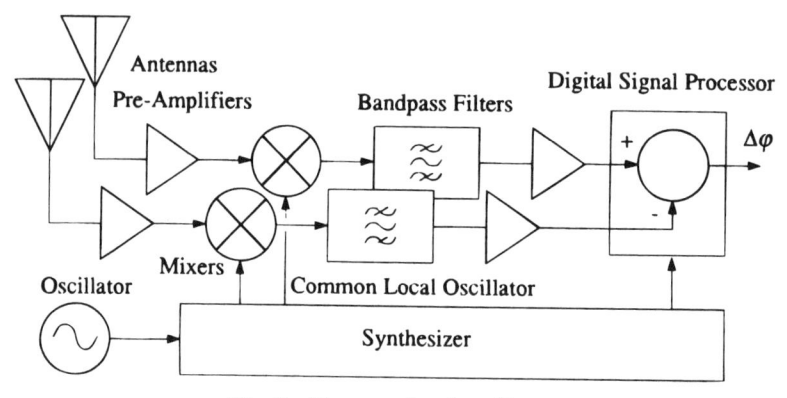

Fig. 2 Common local oscillator.

ATTITUDE DETERMINATION

Referring to Fig. 3, the measured differential phase, $\Delta\varphi$ (measured in wavelengths), is proportional to the projection (vector dot product) of the baseline vector x (3 × 1), measured in wavelengths (cycles), into the line of sight unit vector to the satellite, $\hat{s}$ (3 × 1), for baseline i and satellite j. However, as implied in the figure the GPS receiver initially only measures the fractional part of the differential phase. The integer component k must be resolved through independent means before the differential phase measurement can be interpreted as a differential range measurement. The resulting expression is then $\Delta\varphi_{ij} = \hat{s}_j^T x_i - k_{ij} + v_{ij}$, where v_{ij} is additive, time-correlated measurement noise from the relative ranging error sources discussed in Sec. V. Note that in this chapter as a matter of convention, the integer k_{ij} is treated as a *constant* as long as continuous lock is maintained (i.e., until a cycle slip occurs) on that combination of satellite and baseline. In other words, the initial allocation of integer component between k and $\Delta\varphi$ is *arbitrary*. As the satellite–baseline geometry changes with time, it is assumed that the *receiver tracking loops keep automatic track of integer wrap-arounds* in the $\Delta\varphi$ measurements as they occur (i.e., they track the total change in $\Delta\varphi$, including the integer part). Thus, the only ambiguity is the initial value of the integer k. Also note that the line bias attributable to electrical path length differences is not treated in this expression or those which follow, because it can generally be removed through receiver calibration.

II. Fundamental Conventions for Attitude Determination

Although a full discussion of the mathematical tools generally used for attitude determination is beyond the scope of this book, some introductory material on the fundamental conventions for attitude determination is supplied here as a minimum basis for understanding coordinate transformations and attitude parameterization. For a more in-depth description of these concepts, see Ref. 6.

In attitude determination, we typically are concerned with describing a vehicle system in two separate reference frames: the local horizontal (or, alternatively,

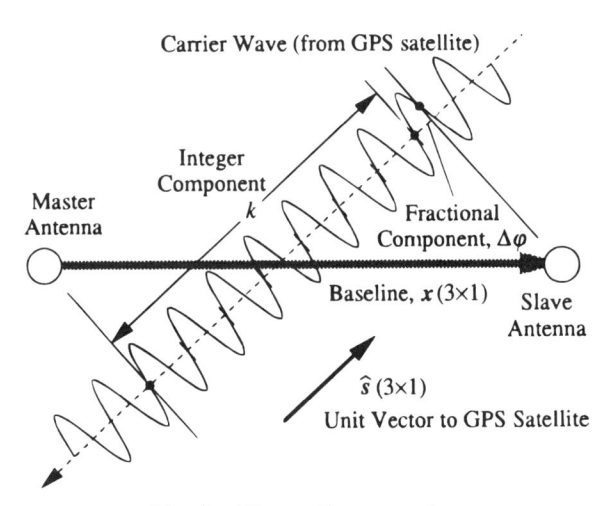

Fig. 3 Observation geometry.

an inertial) coordinate system and the vehicle body coordinate system. As shown in Fig. 4, the vehicle body coordinate axes (dashed; designated as unit vectors x', y', and z') are generally rotated with respect to the local horizontal coordinate frame axes (solid; designated as unit vectors x, y, and z). Each coordinate frame is right-handed (i.e., x crossed into y equals z). A given vector r (3 × 1) is expressed in the local horizontal coordinate frame. The same vector r can also be expressed as r' in the vehicle body (primed) reference frame through a coordinate transformation: $r' = Ar$.

The matrix A (3 × 3) is known as the *attitude matrix* or *direction cosine matrix*. The easiest way to construct the attitude matrix is by assembling the dot products of the orthogonal coordinate frame unit vectors:

$$A = \begin{bmatrix} x' \cdot x & x' \cdot y & x' \cdot z \\ y' \cdot x & y' \cdot y & y' \cdot z \\ z' \cdot x & z' \cdot y & z' \cdot z \end{bmatrix}$$

Although there are nine elements in the matrix, they are not all independent. There are really only three DOF because of the orthonormal constraints ($A^T A = I$) placed on the transformation. The inverse of any attitude matrix is simply its transpose.

The most commonly accepted convention for defining coordinate frames and rotation angles is shown in Fig. 5. Rotations are defined in a specific Euler sequence about the coordinate axes. The axes of the local horizontal frame are such that the x axis points due north, the y axis points due east, and the z axis points directly downward along the local vertical to complete the right-handed set of axes. The body reference frame is fixed to the aircraft so that the x' (roll) axis points out the nose, the y' axis points to the right along the wing, and the z' axis points out the belly to complete the right-handed coordinate axis set. The figure shows the body frame when the pitch and roll angles are zero. For this special case, the pitch axis is aligned with the y' axis, and the heading axis is aligned with the z' axis (local vertical). When the heading, pitch, and roll angles are all zero, the body frame is aligned with the local horizontal frame ($A = I$).

Given this introduction, the three attitude angles (heading, pitch, and roll) may then specify the vehicle attitude. Starting from the reference attitude (where

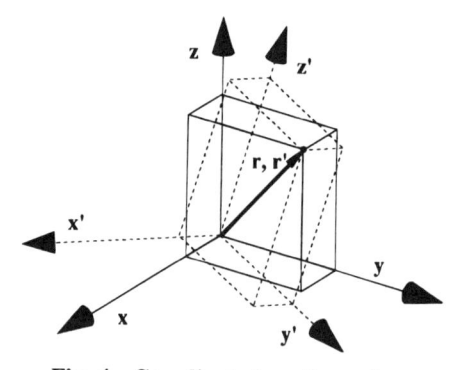

Fig. 4 Coordinate transformation.

ATTITUDE DETERMINATION

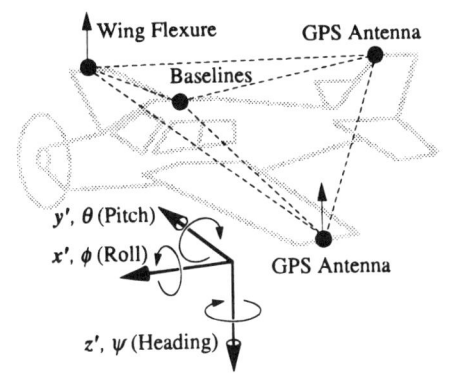

Fig. 5 Aircraft geometry.

the body and local horizontal coordinate frames are aligned), the body frame is rotated (always in a positive, right-handed sense) about the local vertical downward z axis by the heading angle ψ. Then, the body frame is rotated about the new pitch axis by the pitch angle θ. Finally, the body frame is rotated about the roll x' axis by the roll angle ϕ. The resulting attitude matrix A can be shown to be as follows:

$$A = \begin{bmatrix} \cos\theta\cos\psi & \cos\theta\sin\psi & -\sin\theta \\ -\cos\phi\sin\psi + \sin\phi\sin\theta\cos\psi & \cos\phi\cos\psi + \sin\phi\sin\theta\sin\psi & \sin\phi\cos\theta \\ \sin\phi\sin\psi + \cos\phi\sin\theta\cos\psi & -\sin\phi\cos\psi + \cos\phi\sin\theta\sin\psi & \cos\phi\cos\theta \end{bmatrix}$$

III. Solution Processing

This section discusses how differential phase measurements can be converted into attitude solutions. To clarify presentation, it is first assumed that the cycle ambiguities are already known. Discussion of the processes for resolving integers is deferred until Sec. IV. If the integers are known, then the differential *phase* measurements can be treated explicitly as differential *range* measurements through the relationship $\Delta r = \Delta\varphi + k$. Then the process of attitude determination consists of converting these differential range measurements into attitude solutions. An optimal attitude solution for a given set of range measurements Δr_{ij} taken at a single epoch for baseline i and satellite j is obtained by minimizing the quadratic attitude determination cost function:

$$J(A) = \sum_{i=1}^{m} \sum_{j=1}^{n} (\Delta r_{ij} - b_i^T A \hat{s}_j)^2$$

for the m baseline and n satellites, where b (3×1) is the baseline vector defined in the body frame, $\hat{s}$ (3×1) is the line of sight to the GPS satellite given in the local horizontal frame, and A (3×3), the variable to be used in minimization, is the right-handed, orthonormal attitude transformation ($\det A = 1$, $A^T A = I$) from the local horizontal frame to the body frame.

Given a trial attitude matrix A_0, a better estimate may be obtained by linearizing this cost function about the trial solution and solving for a correction matrix δA.

Solving for the best correction matrix during iteration p yields a new and better trial matrix for iteration $p + 1$, so that $A_{p+1} = \delta A_p A_p$. A simple correction matrix can be constructed of small-angle rotations, so that $\delta A(\delta\theta) \cong I + \Theta^\times$, where I (3×3) is the identity matrix, and $\delta\theta$ (3×1) is a vector of small-angle rotations about the following three body frame axes:

$$\delta\theta = \begin{bmatrix} \delta\theta_{x'} \\ \delta\theta_{y'} \\ \delta\theta_{z'} \end{bmatrix}$$

and $\Theta^\times$ (3×3) is the skew-symmetric matrix associated with the vector $\delta\theta$.

$$\Theta^\times = \begin{bmatrix} 0 & -\delta\theta_{z'} & \delta\theta_{y'} \\ \delta\theta_{z'} & 0 & -\delta\theta_{x'} \\ -\delta\theta_{y'} & \delta\theta_{x'} & 0 \end{bmatrix}$$

so that $\Theta^\times b = \delta\theta \times b$. The attitude cost function becomes the following:

$$J(\delta\theta)|_{A_0} \cong \sum_{i=1}^{m}\sum_{j=1}^{n} [\Delta r_{ij} - b_i^T(I + \Theta^\times)A_0\hat{s}_j]^2 = \sum_{i=1}^{m}\sum_{j=1}^{n} (\delta r_{ij} - b_i^T \Theta^\times A_0 \hat{s}_j)^2$$

where $\delta r_{ij} \equiv \Delta r_{ij} - b_i^T A_0 \hat{s}_j$. The measurement geometry is described by the rightmost term, which may be rewritten directly in terms of the three attitude correction angles about each axis $b_i^T \Theta^\times A_0 \hat{s}_j = \hat{s}_j^T A_0^T \Theta^\times b_i = \hat{s}_j^T A_0^T B_i^\times \delta\theta$.

Because the right-hand side of this result can also be written as

$$[(A_0 \hat{s}_j) \times b_i] \cdot \delta\theta,$$

the implication is that the attitude angle sensitivity to a measurement from a given baseline and GPS satellite is simply the *cross-product* of the line-of-sight vector with the baseline vector. The linearized cost function may now be written as follows:

$$\delta J(\delta\theta)|_{A_0} = \|H\delta\theta - \delta r\|_2^2$$

where δr is the vector formed by stacking all measurements, and H is the observation matrix formed by stacking the measurement geometry for each separate measurement:

$$H = \begin{bmatrix} \vdots \\ \hat{s}_j^T A_0^T B_i^\times \\ \vdots \end{bmatrix}$$

The estimate for A is then refined iteratively until the process converges to the numerical precision of the computer.

In cases where the baseline array is non–co-planar, there is an algorithm for carrying out the attitude calculation approximately an order of magnitude faster. This approach is based on solving "Wahba's Problem" of attitude determination using vector observations.[7]

IV. Cycle Ambiguity Resolution

As suggested in Fig. 3, cycle ambiguity resolution is the process of determining the integer number of wavelengths that lie between a given pair of antennas along a particular line of sight. It is the key initialization step that must be performed before attitude determination using GPS can commence.

A. Baseline Length Constraint

Consider a platform with a single baseline constructed from two antennas. The baseline vector originates at the master antenna and ends at the slave antenna. Because differential positioning is employed, no generality is sacrificed by assuming that the tail of the vector stays fixed in space, as shown in Fig. 6. The possible positions of the slave antenna are constrained to lie on the surface of a virtual sphere of radius equal to the baseline length.

B. Integer Searches

The most brute force method of resolving the integers is the search method. In an integer search, all possible combinations of candidate integers (which can number in the hundreds of millions for antenna separations of even just a few meters) are systematically checked against a cost function until (it is hoped) the correct set is found.

Although search techniques work accurately and quickly for smaller baselines (on the order of several carrier wavelengths), they are vulnerable to erroneous solutions with longer baselines or when few satellites are visible. Although many creative techniques have been synthesized for maximizing the execution speed of the search process,[8-10] searches still occasionally suffer from ambiguous results.

The search technique is depicted to scale in Fig. 7 for a single 4λ baseline, three satellites, and 3σ multipath error. The instantaneous satellite line-of-sight vectors are depicted with arrows. Possible integer values are shown as concentric bands about the line-of-sight vectors. The correct integer set is indicated in white. Any place on the sphere where the concentric bands for all three satellites intersect (indicated in black) is a viable baseline orientation candidate. *Note that at any one instant, there is not a unique solution.*

C. Motion-Based Methods

Although lacking the near-instantaneous start-up time of integer search methods, motion-based methods are unmatched for providing the highest level of

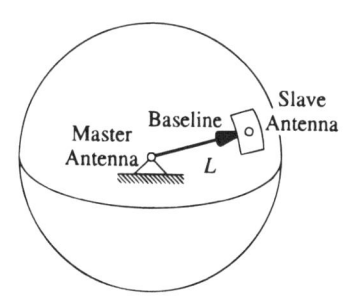

Fig. 6 Length constraint.

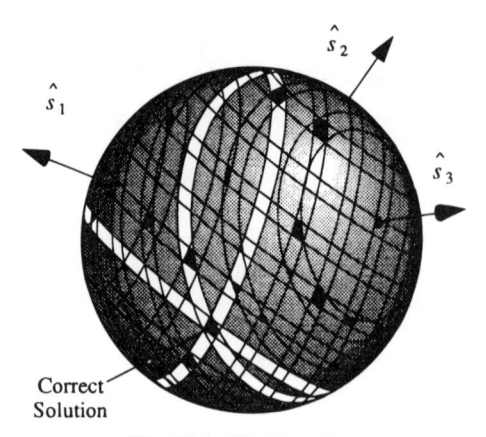

Fig. 7 Integer search.

overall solution integrity.[11] Motion-based integer resolution algorithms make use of the abundance of information provided by platform or GPS satellite motion. This attitude motion modulates the relative carrier phase with a signature that may be used to identify the cycle ambiguities. If the motion occurs rapidly enough, this process is complete within seconds. Rather than constraining cycle ambiguities to lie on integer values, motion-based methods estimate the cycle ambiguities as continuous biases. Checking without imposing the integer constraint that the bias values indeed lie near integer values provides a unique, unambiguous solution with extraordinary integrity—even when there are only a few satellites in view.

For aircraft applications, natural attitude motion consists of banks, turns, or attitude perturbations excited by turbulence. If the baseline vectors are non–coplanar, even a turn on the ground (about a single axis) is sufficient for cycle ambiguity resolution.

Without knowledge of the integers, it is possible to determine the Cartesian position of the slave antenna relative to its unknown starting point. The differential phase measurement equation may be expressed in compact vector and matrix notation for the n satellites in view (neglecting ranging noise):

$$\Delta \varphi = S^T x - k$$

where

$$\Delta\varphi(n \times 1) = \begin{bmatrix} \Delta\varphi_1 \\ \Delta\varphi_2 \\ \vdots \\ \Delta\varphi_n \end{bmatrix}, \quad S(3 \times n) = [\hat{s}_1 \; \hat{s}_2 \; \cdots \; \hat{s}_n], \, k(n \times 1) = \begin{bmatrix} k_1 \\ k_2 \\ \vdots \\ k_n \end{bmatrix}$$

Suppose a baseline moves from Cartesian position $x^{(0)}$ to $x^{(1)}$. The measured change in differential range is given by $\Delta\varphi^{(1)} - \Delta\varphi^{(0)} = S^T[x^{(1)} - x^{(0)}] = S^T \Delta x$. Because the integer ambiguity k cancels out of the above expression, and the satellite line-of-sight vectors are known, we can solve for the displacement vector

Δx (3 × 1), explicitly using a linear least-squares fit. For the integer resolution processing $n = 3$ satellites are used to determine the relative location of the slave antenna undergoing motion. (After cycle resolution is complete, two satellites are required for attitude determination.) It has been assumed that the position displacement is occurring on a much faster time scale than that of the satellite line-of-sight vectors $\hat{s}$.

As baseline motion is occurring, it is possible to accumulate a set of displacement vectors over a short interval of time. If the platform moves by a large angle, the set of displacement vectors can be used to calculate an initial guess to initialize a nonlinear least-squares fit.[4]

A two-dimensional representation of the rigid body antenna mounting constraints is shown in Fig. 8. Suppose that the baseline vector x rotates in space. Here the baseline vector is moved (displaced) by the vector Δx to two different locations at two different times, 1 and 2. If a line is constructed perpendicular to each Δx displacement vector passing through its midpoint, the center of the circle must be included on that line. By simultaneously considering each Δx vector, the center of the circle can be located, along with the initial position x of the baseline.

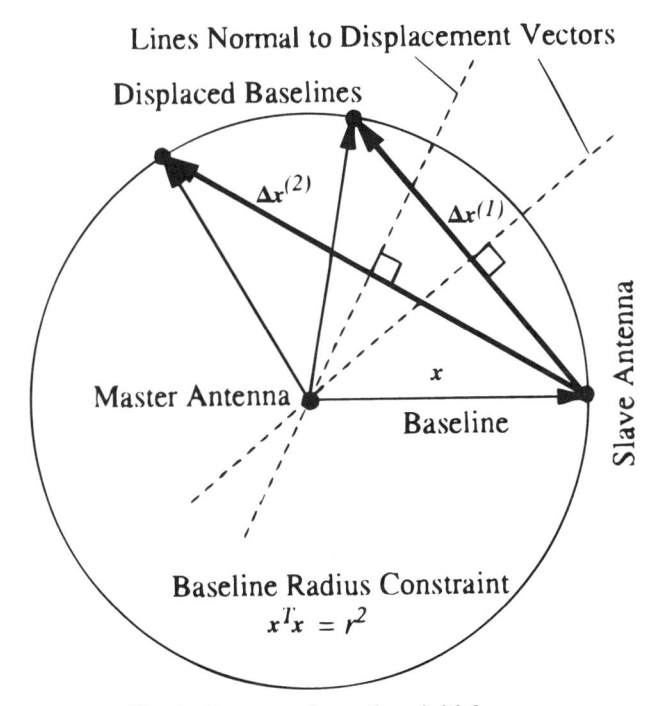

Fig. 8 Large angle motion: initial guess

Mathematically, the solution may be developed by constructing the square of the norm of the rotated baseline vector $x + \Delta x$, given as follows:

$$(x + \Delta x)^T(x + \Delta x) = x^T x + 2\Delta x^T x + \Delta x^T \Delta x$$

Noting that the left side is equal to the square of the baseline length, as is $x^T x$, the two terms may be canceled, leaving $2\Delta x^T x = -\Delta x^T \Delta x$.

Different Δx vectors taken at N different times (indicated by a superscript in parentheses) may be stacked into matrix form as follows:

$$2\begin{bmatrix} \Delta x^{(1)T} \\ \Delta x^{(2)T} \\ \vdots \\ \Delta x^{(N)T} \end{bmatrix} x = -\begin{bmatrix} \Delta x^{(1)T} \Delta x^{(1)} \\ \Delta x^{(2)T} \Delta x^{(2)} \\ \vdots \\ \Delta x^{(N)T} \Delta x^{(N)} \end{bmatrix}$$

Then the baseline solution x may be obtained through a linear least-squares fit. Each Δx vector defines a subspace in which the slave antenna must lie. The baseline is then the point that comes closest to this condition in a least-squares sense. A convenient shorthand notation for this least-squares fit is given by $2\Delta X^T x = -\text{diag}(\Delta X^T \Delta X)$, ΔX $(3 \times N) \equiv [\Delta x^{(1)} \quad \Delta x^{(2)} \quad \cdots \quad \Delta x^{(N)}]$ and $\text{diag}(\cdot)$ defines a vector $(N \times 1)$ comprised of the diagonal elements of the argument matrix $(N \times N)$.

Unfortunately, for the single baseline case, large-angle rotation about the two axes perpendicular to the baseline is always required to resolve completely the integer ambiguities using motion. To avoid the shortcoming of requiring two-axis motion perpendicular to each baseline, information from multiple baselines can be combined into a single simultaneous estimation equation.[12] The constraint equation between different baselines i and j yields $(x_i + \Delta x_i)^T (x_j + \Delta x_j) = x_i^T x_j + x_i^T \Delta x_j + \Delta x_i^T x_j + \Delta x_i^T \Delta x_j$.

Again, the dot product of each baseline pair is constant; hence, the corresponding term may be canceled from both sides of the equation, leaving $\Delta x_j^T x_i + \Delta x_i^T x_j = -\Delta x_i^T \Delta x_j$.

Combining Δx measurements from N different times (typically 10–30 epochs), this form may be expanded as follows:

$$\begin{bmatrix} \Delta x_j^{(1)T} \\ \Delta x_j^{(2)T} \\ \vdots \\ \Delta x_j^{(N)T} \end{bmatrix} x_i + \begin{bmatrix} \Delta x_i^{(1)T} \\ \Delta x_i^{(2)T} \\ \vdots \\ \Delta x_i^{(N)T} \end{bmatrix} x_j = -\begin{bmatrix} \Delta x_i^{(1)T} \Delta x_j^{(1)} \\ \Delta x_i^{(2)T} \Delta x_j^{(2)} \\ \vdots \\ \Delta x_i^{(N)T} \Delta x_j^{(N)} \end{bmatrix}$$

Invoking the same matrix notation as above, the entire initial guess for the case of the three baselines shown in Fig. 9 can be combined into a single, unified least-squares fit equation:

$$\begin{bmatrix} \Delta X_2^T & \Delta X_1^T & 0 \\ \Delta X_3^T & 0 & \Delta X_1^T \\ 0 & \Delta X_3^T & \Delta X_2^T \\ 2\Delta X_1^T & 0 & 0 \\ 0 & 2\Delta X_2^T & 0 \\ 0 & 0 & 2\Delta X_3^T \end{bmatrix} \begin{bmatrix} x_1 \\ x_2 \\ x_3 \end{bmatrix} = - \begin{bmatrix} \text{diag}(\Delta X_1^T \Delta X_2) \\ \text{diag}(\Delta X_1^T \Delta X_3) \\ \text{diag}(\Delta X_3^T \Delta X_2) \\ \text{diag}(\Delta X_1^T \Delta X_1) \\ \text{diag}(\Delta X_2^T \Delta X_2) \\ \text{diag}(\Delta X_3^T \Delta X_3) \end{bmatrix}$$

The left-hand matrix is now $6N \times 9$, the solution vector is 9×1, and the right-hand vector is $6N \times 1$. This same matrix structure applies to any number of baselines. For the case of three or more baselines, two important advantages fall out of this approach.

First, with motion about any two *arbitrary* axes, no a priori information about antenna placement is required to unambiguously solve for all three of the three initial baseline vectors (three components each, nine total dimensions). It is the measurements themselves that are providing all the geometrical information. Therefore, this approach could be adapted to perform *in situ* self-calibration of GPS baselines during normal operation.

Second, by incorporating the known baseline constraints in the case of non–coplanar baseline configurations, *motion about a single axis of rotation is entirely adequate for an unambiguous baseline vector solution.*

1. Measurement Refinement

To refine the initial guess iteratively, a new cost function (modeled after the one defined in Sec. III) is employed:

$$J(A^{(1)}, A^{(2)}, \ldots, A^{(N)}, k) = \sum_{\ell=1}^{N} \sum_{i=1}^{m} \sum_{j=1}^{n} (\Delta\varphi_{ij}^{(\ell)} + k_{ij} - b_i^T A^{(\ell)} \hat{s}_j^{(\ell)})^2$$

where $A^{(\ell)}$ is the attitude matrix at each epoch ℓ, and k is a vector of the cycle ambiguities for all the baseline and antenna combinations. The ambiguities are estimated as continuous variables.

The problem is to find the independent attitude matrices at each epoch and the set of integers (which applies to all epochs) that minimize the stated cost function. For each epoch, it is possible to convert the initial guess of baseline position given above into an initial guess for the attitude A_0. As shown in Sec. III, the cost function can then be linearized about this initial guess. The attitude component of the state variables to be estimated consists of perturbations in the vehicle attitude about all three axes for every epoch l under consideration. The

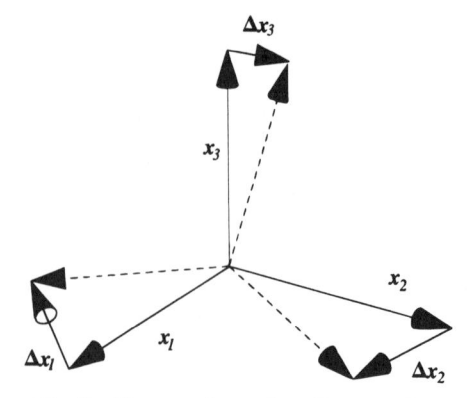

Fig. 9 Non–co-planar baseline rotation.

resulting linearized equations are as follows:

$$\begin{bmatrix} H_1 & 0 & 0 & 0 & | & -I \\ 0 & H_2 & & 0 & | & -I \\ \vdots & & \ddots & \vdots & | & \vdots \\ 0 & 0 & \cdots & H_N & | & -I \end{bmatrix} \begin{bmatrix} \delta\theta^{(1)} \\ \delta\theta^{(2)} \\ \vdots \\ \delta\theta^{(N)} \\ \hline k \end{bmatrix} = \begin{bmatrix} \delta\varphi^{(1)} \\ \delta\varphi^{(2)} \\ \vdots \\ \delta\varphi^{(N)} \end{bmatrix}$$

where each $H_l(mn \times 3)$ is the sensitivity matrix of changes in measured differential phase with respect to rotations about each of the three axes of attitude, and I ($mn \times mn$) is the identity matrix. As was shown in Sec. III, each row of H is given by $(A_0^{(l)} S_j^{(l)}) \times b_i$. The other state variables are as follows:

$$\delta\theta^{(\ell)}(3 \times 1) = \begin{bmatrix} \delta\theta_x^{(\ell)} \\ \delta\theta_y^{(\ell)} \\ \delta\theta_z^{(\ell)} \end{bmatrix} \quad \text{and} \quad k(mn \times 1) = \begin{bmatrix} k_{11} \\ k_{12} \\ \vdots \\ k_{mn} \end{bmatrix}$$

which are vectors of small-angle, three-axis attitude rotation corrections for each time sample l, and a vector of integer biases, respectively, for all combinations of m baselines and n GPS satellites. The right-hand side of the matrix equation is a vector of differential phase residuals, so that the following results:

$$\delta\varphi^{(\ell)}(mn \times 1) = \begin{bmatrix} \delta\varphi_{11}^{(\ell)} \\ \delta\varphi_{12}^{(\ell)} \\ \vdots \\ \delta\varphi_{mn}^{(\ell)} \end{bmatrix}$$

where

$$\delta\varphi_{ij}^{(\ell)} = \Delta\varphi_{ij}^{(\ell)} - b_i^T A_0^{(\ell)} \hat{s}_j^{(\ell)}$$

and $A_0^{(\ell)}$ is the current best estimate of the platform attitude for epoch ℓ.

ATTITUDE DETERMINATION

To this point, it has been assumed that the time scale of platform motion is very much faster than that of the GPS satellites. It is also possible for the time scales of motion for the two to be comparable, such as in space or marine applications. That *quasi static* case can be treated by applying exactly the same nonlinear, least-squares fit equations. The principal difference is that the time interval of measurement collection is increased.

2. Static Integer Resolution

In the static case, the solution may also be refined iteratively by linearizing the observation equation. However, because the static platform attitude is the same for all epochs l, the form of linearized observation equation is as follows:

$$\begin{bmatrix} H_1 & | & -I \\ H_2 & | & -I \\ \vdots & | & \vdots \\ H_N & | & -I \end{bmatrix} \begin{bmatrix} \delta\theta \\ \hline k \end{bmatrix} = \begin{bmatrix} \delta\varphi^{(1)} \\ \delta\varphi^{(2)} \\ \vdots \\ \delta\varphi^{(N)} \end{bmatrix}$$

Again, the right-hand side is a vector of differential phase residuals. The final solution for the vector k yields the integer ambiguities. For non–co-planar baselines, the estimation process usually has enough information to resolve the ambiguities reliably after about 10 min of satellite motion—even with only two satellites in view.

D. Alternative Means for Cycle Ambiguity Resolution

Although motion-based methods for cycle ambiguity resolution are certainly not the only means for system initialization, they undoubtedly have the highest integrity—especially if no external information is available. There are at least two other "instantaneous" approaches that may also be employed with the potential disadvantage that they require more hardware to implement:

1) *Multiple GPS Antennas:* By using small and large baselines together, it is possible to resolve cycle ambiguities in an explicit sequence by starting with the small baselines (where there is little ambiguity) and working one's way out to the larger baselines. Although the entire process is rapid, additional antennas are required.

2) *Multiple GPS Observables:* Another alternative is to use code ranging to establish the dual-frequency carrier, wide-lane ambiguity, allowing the L_1 cycle ambiguities to be resolved. However, on rare occasions the method may still fail to establish the correct wide-lane ambiguity, and a more expensive dual-frequency receiver is required.

V. Performance

This section examines key aspects of the overall performance of attitude determination using GPS and quantifies the most significant sources of error in attitude determination.

A. Geometrical Dilution of Precision for Attitude

The H matrix from Sec. III is the best means for evaluating the attitude fix accuracy. In general, the attitude error is a function of the satellite geometry, baseline geometry, *and* instantaneous vehicle attitude. Given a differential ranging error of σ (typically 5mm), an estimate of the attitude covariance matrix P is given by $P = (H^T H)^{-1} \sigma^2$, where the 1 σ pointing error (in radians) for any given attitude axis is given by the square root of the corresponding diagonal element of this 3 × 3 covariance matrix. The diagonal elements correspond to small rotations about the x', y', and z' body frame coordinate axes, respectively.

For generality, the baseline and satellite line-of-sight vectors can be concatenated into matrices B (3 × m) and S (3 × n), where $B = [b_1 \ b_2 \ \cdots \ b_m]$ and $S = [\hat{s}_1 \ \hat{s}_2 \ \cdots \ \hat{s}_n]$.

For the ideal baseline configuration where $BB^T = L^2 I$ (where L is the effective baseline length), I is the 3 × 3 identity matrix, and each of the n GPS satellites is in view of all the antennas on the vehicle, it can be shown that the attitude covariance simplifies to

$$P = [nI - ASS^T A^T]^{-1} \left(\frac{\sigma}{L}\right)^2$$

This form suggests a convenient means for characterizing the suitability of the constellation geometry for attitude determination. As an analog to GDOP, ADOP, the geometric dilution of precision for attitude, is defined by considering the geometrical component of the covariance matrix. Invoking the invariance of the matrix trace with respect to coordinate rotations, the resulting total angular pointing error σ_θ may be written as follows:

$$\sigma_\theta = (\text{ADOP}) \frac{\sigma}{L}$$

where

$$\text{ADOP} \equiv \sqrt{\text{trace}[(nI - SS^T)^{-1}]}$$

The quantity ADOP is defined even when there are only two satellites in view, the minimum number required for three-axis attitude determination. Its value is generally around unity or smaller, indicating that the GPS constellation consistently provides a favorable geometry for attitude determination.

Thus, a further approximation for the attitude error can be made by simply neglecting the satellite geometry term (ADOP) and considering it to be near unity, so $\sigma_\theta \cong \sigma/L$.

The remaining issue is determining what to use for the value of σ. Table 1 offers typical numbers for relative positioning. In all but the highest regimes of dynamics, the largest error source is multipath.

B. Multipath

Multipath is without question the largest source of error in attitude determination using GPS. Although the actual error it produces is highly deterministic (i.e., a function of the specific environment, materials, antenna gain pattern,

Table 1 Attitude determination ranging error sources (1σ)

Sources	Range error m
Multipath (differential range error for a given pair)	~5 mm
Structural distortion (flexure, thermal expansion)	Application-specific
Troposphere	Modelable
Carrier-to-noise ratio	<1 mm
Receiver-specific errors (crosstalk, line bias, interchannel bias)	<1 mm
Total rss differential ranging error (1σ), excluding distortion	~5 mm

geometry, and other factors), practical experience suggests the approximate rule of thumb that the differential ranging error between a pair of hemispherical microstrip patch antennas is about 5 mm, 1 σ.

In most cases, the most practical and cost-effective approach to systems engineering is simply to use GPS attitude determination in those applications for which this standard multipath error of 5 mm would be acceptable. In cases where more accuracy is required, a number of techniques have been proposed for improving multipath errors. A partial list of techniques includes multipath calibration or antenna pattern shaping,[11] inertial aiding,[13] and mathematical multipath modeling.[14] However, such performance enhancements also carry a penalty of cost or complexity.

C. Structural Distortion

Structural distortion (caused by thermal or flexural bending) can be an issue in certain applications. In most cases, it can either be neglected, modeled, or estimated. (An example of estimating wing flexure on an airplane is given in Fig. 12.)

D. Troposphere

The troposphere can often be a source of error in attitude determination. The simplified model in Fig. 10 shows ray propagation from the vacuum of space down toward the Earth's surface. Refraction of the GPS ranging signal causes the ray to bend downward as atmospheric density (and index of refraction) increases. The simplified slab model depicted in Fig. 10 treats the atmosphere as a block of uniform density and index of refraction n_2. Using Snell's law of refraction

$$\frac{\sin \theta_2}{\sin \theta_1} = \frac{n_1}{n_2}$$

where n_1 is unity, and n_2 (depending upon the atmospheric and water vapor

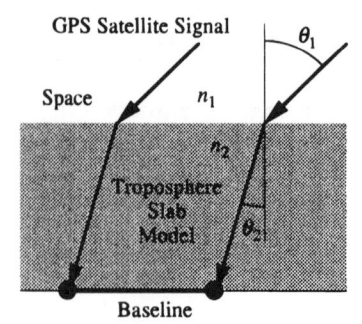

Fig. 10 Atmospheric refraction.

conditions) is somewhere around 1.00026. This simple model can be used to adjust the apparent line-of-sight vectors of the GPS satellites to account for the troposphere in attitude determination applications.

E. Signal-to-Noise Ratio

In applications where it is desirable to track higher dynamics, the tracking loop bandwidth can be opened up (within limitations). The noise on the reconstructed carrier, which dictates the differential range measurement error, is given by the following white noise equation:

$$\sigma = \sqrt{\frac{f_N}{C/N_0}} \frac{\lambda}{2\pi}$$

where f_N is the noise bandwidth of the carrier tracking loop, and C/N_0 is the carrier to noise ratio.[15] Typically, this error is dominated by multipath and is smaller than a millimeter for typical tracking parameters (C/N_0 = 40 dB-Hz, f_N = 10 Hz).

F. Receiver-Specific Errors

Receiver-specific errors, including crosstalk, line bias, and interchannel bias, can be significant sources of error if they are not treated appropriately in a receiver design. Crosstalk between the radio frequency paths for each antenna is an issue, because there is often more than 100 dB of gain along each signal path. Line bias is the nearly constant offset in phase from one antenna to another. A function of both cable length and temperature, line bias is all that remains of the relative clock offset in the design of the common local oscillator. Finally, interchannel bias results from using different hardware channels to measure the carrier phase for each satellite. State-of-the-art receivers employ special techniques to render the errors from all of these effects much smaller than those from multipath.

G. Total Error

Because multipath usually dominates all other error sources (in the absence of significant structural distortion), an approximate and general rule of thumb for attitude determination angular accuracy (in radians) for a representative baseline length of L (in cm) is simply as follows:

$$\sigma_\theta \text{ (in radians)} \cong \frac{0.5 \text{ cm}}{L \text{ (in cm)}}$$

VI. Applications

The capability to use GPS for attitude determination opens up a new realm of applications and opportunities. In the future, it is very likely that the integration of attitude determination into larger systems—including those that use carrier phase for positioning as well—will play a key role in realizing the full potential of GPS. Applications in aviation, spacecraft, and marine areas are summarized below.

A. Aviation

In aviation, heading and attitude sensing using GPS provides a readout that is completely immune to drift and magnetic variation. Many researchers have carried out aircraft experiments to test attitude determination using GPS.[16-19] Figure 11 shows the agreement in roll attitude between GPS and an inertial navigation unit (INU) at a 10-Hz output rate. This flight experiment employed the GPS attitude system that was developed by Stanford University and built

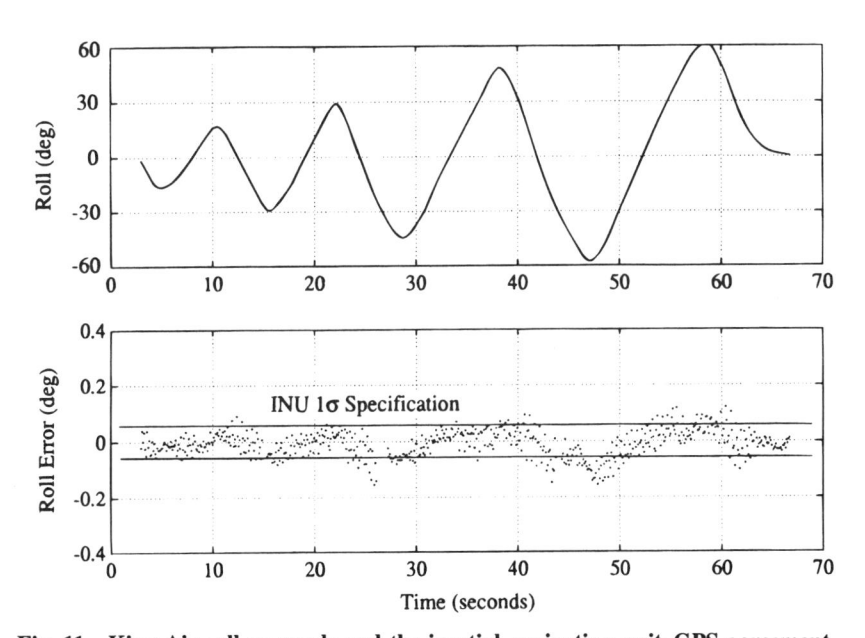

Fig. 11 King Air roll reversals and the inertial navigation unit–GPS agreement.

around a Trimble TANS Quadrex receiver. The system was flown on a NASA Ames King Air, carrying an INU specified to a one sigma accuracy of 0.05 deg. The antennas were mounted on the fuselage, wing tips, and tail, giving the roll component a 16-m baseline. As the aircraft performs roll reversals up to a 60-deg angle of bank, the disparity between roll attitude measured by the two independent sensors seldom exceeds the INU specification. At the steepest angles of bank, the airframe is blocking many of the satellites in view. Sometimes tracking as few as two GPS satellites, the system flawlessly hands off the integers in real time. *In a time span of just a few seconds, the system is tracking a completely new set of satellites with a completely new set of integers.*

A new application of GPS attitude determination is identification of the aircraft dynamic model. The pilot can supply inputs to the controls that excite a dynamic response. The GPS sensor then measures this response to reconstruct an accurate model of the aircraft dynamics. The model then serves as the foundation for optimal autopilot synthesis. Figure 12 shows an example of the characteristic pitch response of an aircraft to a stick input.[12] The "phugoid" response of the aircraft reveals the natural frequency and damping of this mode. It is possible that such model estimation can be carried out continuously in flight, providing a new set of constraints to the position fixes performed by the navigation equipment and, thus, a means for additional integrity checking.

Figure 12 also shows how instantaneous wing flexure is measured to a precision of 1.4-mm rms. The error was evaluated with respect to a best-fit second-order response. Using the same GPS attitude determination system and the antenna arrangement shown in Fig. 5, the structural deformation of the airframe can be used as an indirect means of measuring vertical acceleration.

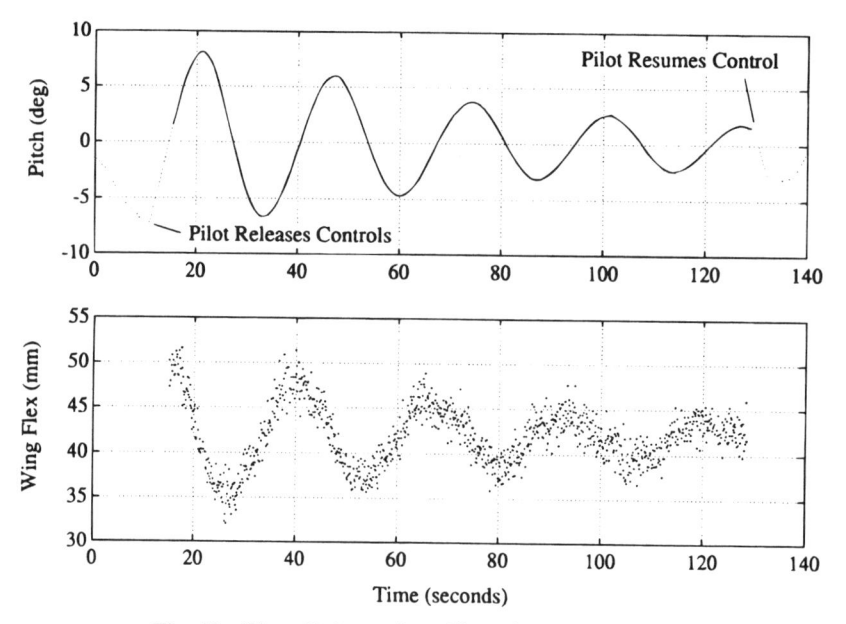

Fig. 12 Piper Dakota phugoid mode state estimates.

By integrating attitude with the enroute navigation, precision landing, collision avoidance, automatic dependent surveillance functions, *a single GPS sensor has the potential to perform the functions of a significant fraction of cockpit instruments currently in use.*

B. Spacecraft

The state-of-the-art in attitude receivers is small (1300 cc), light (~1.5 kg), and low power (~3.5 W), so that one can be carried on just about any spacecraft. Initial experiments with attitude determination on spacecraft[20] indicate that for many types of missions, the GPS may offer significant cost savings. See Chapter 16, this volume, for more information on closed-loop space applications.

C. Marine

In the marine area, the standard for comparison is the gyrocompass. The GPS offers low-cost heading indicator output with rapid start-up times and all-latitude operation. Some of the marine work in attitude determination is described in Refs. 3 and 21.

Attitude determination also provides the potential to point antennas or other directional devices (such as weaponry) on ship-based or other moving platforms. Applying closed-loop control stabilizes the platform against changes in the vehicle orientation.

This partial list of applications hardly begins to address the ultimate potential of attitude determination using the GPS. Given trends in lowering cost, size, weight, and power of GPS technology, it is not inconceivable that backpackers could carry a hand-held portable direction finder that complements the GPS positioning capability. With attitude capability in such a small package, many more applications will undoubtedly arise.

References

[1]Albertine, J. R., "An Azimuth Determination System Utilizing the Navy Navigation Satellites," *Navigation* Vol. 21, No. 1, 1974.

[2]Joseph, K. M., and Deem, P. S., "Precision Orientation: A New GPS Application," International Telemetering Conference, San Diego, CA, Oct., 1983.

[3]Kruczynski, L. R., Li, P. C., Evans, A. G., and Hermann, B. R., "Using GPS to Determine Vehicle Attitude: U.S.S. Yorktown Test Results," *Proceedings of ION GPS-89* (Colorado Springs, CO), Institute of Navigation, Washington, DC, Sept. 1989.

[4]Brown, R., and Ward, P., "A GPS Receiver with Built-in Precision Pointing Capability," *Proceedings, IEEE PLANS, 90, IEEE Position, Location, and Navigation Symposium,* (Las Vegas, NV), Institute of Electrical and Electronics Engineers, New York, March 1990, pp. 83–93.

[5]Kuhl, M., Qin, X., and Cotrell, W., "Design Considerations and Operational Results of an Attitude Determination Unit," *Proceedings of ION GPS-94* (Salt Lake City, UT), Institute of Navigation, Washington, DC, Sept. 20–24, 1994.

[6]Wertz, J. R. (ed.), *Spacecraft Attitude Determination and Control,* Reidel, Boston, MA, 1985.

[7]Cohen, C. E., Cobb, H. S., and Parkinson, B. W., "Two Studies of High Performance Attitude Determination Using GPS: Generalizing Wahba's Problem for High Output Rates and Evaluation of Static Accuracy Using a Theodolite," *Proceedings of ION GPS-92* (Albuquerque, NM), Institute of Navigation, Washington, DC, Sept. 16–18, 1992.

[8]Hatch, R., "Instantaneous Ambiguity Resolution," KIS Symposium 1990, Banff, Canada, Sept. 1990.

[9]Knight, D., "A New Method of Instantaneous Ambiguity Resolution," *Proceedings of ION GPS-94* (Salt Lake City, UT), Institute of Navigation, Washington, DC, Sept. 20–24, 1994.

[10]Brown, R. A., "Instantaneous GPS Attitude Determination," *IEEE Aerospace and Electronics Magazine*, June 1992, p. 3.

[11]Cohen, C. E., "Attitude Determination Using GPS," Ph.D. Dissertation, Stanford Univ., Stanford, CA, Dec. 1992.

[12]Cohen, C. E., and Parkinson, B. W., "Aircraft Applications of GPS-Based Attitude Determination: Test Flights on a Piper Dakota," *Proceedings of ION GPS-92* (Albuquerque, NM), Institute of Navigation, Washington, DC, Sept. 16–18, 1992.

[13]Braasch, M., and van Graas, F., "Guidance Accuracy Considerations for Real-Time Interferometric Attitude Determination," *Proceedings of ION GPS-91* (Albuquerque, NM), Institute of Navigation, Washington, DC, Sept. 1991.

[14]Axelrad, P., Comp, C., and MacDoran, P., "Use of Signal-to-Noise Ratio for Multipath Error Correction in GPS Differential Phase Measurements: Methodology and Experimental Results," *Proceedings of ION GPS-94*, (Salt Lake City, UT), Institute of Navigation, Washington, DC, Sept. 20–24, 1994.

[15]Rath, J., and Ward, P., "Attitude Estimation using GPS," National Technical Meeting ION, San Mateo, CA, Jan. 1989.

[16]Purcell, G. H., Jr., Srinivasan, J. M., Young, L. E., DiNardo, S. J., Hushbeck, E. L., Jr., Meehan, T. K., Munson, T. N., and Yunck, T. P., "Measurement of Aircraft Position, Velocity and Attitude using Rogue GPS Receivers," Fifth International Geodetic Symposium on Satellite Positioning, Las Cruces, NM, March 1989.

[17]van Graas, F., and Braasch, M., "GPS Interferometric Attitude and Heading Determination: Initial Flight Test Results," *Navigation*, Vol. 38, Fall 1991, pp. 297–316.

[18]Cohen, C. E., McNally, B. D., and Parkinson, B. W., "Flight Tests of Attitude Determination Using GPS Compared Against an Inertial Navigation Unit," *Navigation*, Vol. 41, No. 1, 1994.

[19]Cannon, M. E., Sun, H., Owen, T. E., and Meindl, M. A., "Assessment of a Non-Dedicated GPS Receiver System for Precise Airborne Attitude Determination," *Proceedings of ION GPS-94* (Salt Lake City, UT), Institute of Navigation, Washington, DC, Sept. 20–24, 1994.

[20]Cohen, C. E., Lightsey, E. G., Feess, W. A., Parkinson, B. W., "Space Flight Tests of Attitude Determination Using GPS," *International Journal of Satellite Communications*, Vol. 12, Sept.–Oct. 1994, pp. 427–433.

[21]Lu, G., Cannon, M. E., Lachapelle, G., and Kielland, P., "Attitude Determination in a Survey Launch Using Multi-Antenna GPS Technologies," ION National Technical Meeting, San Francisco, CA, Jan. 1993.

Chapter 20

Geodesy

Kristine M. Larson*
University of Colorado, Boulder, Colorado 80309

I. Introduction

GEODESY is the discipline devoted to the measurement of the shape of the Earth and its gravity field in three-dimensional space and time. An ideal geodetic system would provide absolute coordinates of points on the Earth at whatever temporal spacing is required by the geodesist. Until space geodetic techniques were developed, geodetic measurements only indirectly measured changes in the Earth's shape. For example, classical geodetic systems, such as triangulation and trilateration, do not measure position, and thus have significant limitations. In order to resolve changes in the shape of the Earth, the individual angle or length observations must be combined into a network. The final network estimate is formed from an adjustment of the observations, so that the network "closes."[52] Obviously these measurement systems require intervisibility of the observing geodetic stations, making the measurements local in scale (1–50 km). Even at these distances, systematic errors in both triangulation and trilateration grow rapidly with increasing baseline length. Additionally, rotation and translation of the network are indeterminate. In order to interpret the deformation of the network over time, a strain analysis is done. Although useful for interpretation for geodetic data in active tectonic regions, strain analysis is limited by the assumptions of uniformity over the whole network.

Space geodetic techniques such as VLBI (very long baseline interferometry) and SLR (satellite laser ranging) are both more flexible and precise. Visibility requirements for VLBI and SLR are skyward rather than to other observing sites, so that global scale measurements are feasible. Both VLBI and SLR observables can be analyzed to determine three-dimensional station positions. The major drawback of both systems is cost. The VLBI and SLR systems cost millions of dollars to build and millions to maintain and operate. Specially trained personnel

---

Copyright © 1994 by the American Institute of Aeronautics and Astronautics, Inc. All rights reserved.
*Assistant Professor, Department of Aerospace Engineering Sciences.

are required to operate the systems, and mobility is limited by their size (multiple vans) and weight (several tons).

The Global Positioning System (GPS) has revolutionized geodesy through its great accuracy, convenience, and global availability. It is fully three-dimensional, and translation and network rotation can be determined, as long as common sites are observed and a stable reference frame is used. With a full GPS constellation, satellite signals, and thus positioning ability, are globally available 24 h a day. The cost of GPS geodesy is several orders of magnitude less than VLBI and SLR. A high-quality GPS receiver costs ~$25,000 and can be installed at a fixed site with minimal cost. GPS data are then easily and inexpensively collected continuously without requiring human intervention. For specific geodetic experiments, GPS receivers can be easily deployed, with the equipment weighing ~30 lb. With proper modeling of the GPS observables, the shape and surface dynamics of the Earth can be unambiguously determined with great precision and accuracy.

In this chapter we concentrate on reviewing GPS geodetic positioning work that has already been published in the refereed literature. The GPS contribution to the measurement of the gravity field is just now being realized, with encouraging initial results from the TOPEX/Poseidon mission,[47] but is not discussed here. First, we briefly summarize the error sources and discuss reference frames issues of interest for precise geodetic analysis with GPS. Following that, we discuss recent geodetic results, focusing in particular on measuring motions of the Earth's crust and its rotation axis.

II. Modeling of Observables

The GPS observables are corrupted by numerous errors, as is discussed in detail in other sections of this book. For high-quality receivers, the phase measurement error is no worse than 1% of the carrier wavelength, or 2 mm. The P-code pseudorange measurement error is several orders of magnitude worse. Both observables, which are modeled as the distance plus the time offset between the satellite and receiver, are corrupted by satellite and receiver instrumental delays and path errors associated with the atmosphere. An accurate receiver location can be determined only with equivalently accurate satellite positions. Because these error sources are discussed in other sections of this book, we simply summarize techniques that are commonly used in precise geodetic softwares.

Although not technically an "error" source, the treatment of "cycle slips" is an important issue in precise geodetic applications. These breaks in the carrier phase data are caused by the receivers or by obstructions in the path of the signal. A reliable algorithm is required to identify and, if possible, repair such slips. Blewitt[4] developed an algorithm that uses the P-code pseudorange to repair slips for undifferenced data. An alternative algorithm was developed by Freymueller[21] to remove cycle slips for receivers without P code. Most software repairs cycle slips in the double differenced data.[25]

Clock drifts, for both satellites and receivers, can be estimated directly at each epoch,[32] or removed via differencing of the data.[25] As discussed in Chapter 16, this volume, direct estimation of the clocks yields greater flexibility and more independent measurements. One limitation of estimating the clocks is that cycle slips must be repaired in the undifferenced data, which can be difficult with codeless receivers when selective availability is active. For determination of relative positions, clock errors are not a limiting error source. Selective availability (SA), which to the GPS analyst is equivalent to a noisy clock, does not have a significant effect on relative positioning accuracy.

The most significant path errors for precise GPS geodetic applications are associated with the atmosphere, which are discussed in the companion volume. The first-order ionospheric errors are eliminated using an appropriate linear combination of the $L1$ and $L2$ phase data. The troposphere has been a more troublesome error source. The component most affected by the tropospheric error is the vertical component, and to a lesser extent the east component.[51] Although it has been suggested that water vapor radiometers (WVRs) could calibrate the tropospheric error, there has been little success in using WVR data in precise geodetic applications without also estimating a parameter representing the zenith troposphere delay. Recently Ware et al.[54] suggested that a WVR can be used to model the wet zenith delay after proper calibration of the WVR. These lengthy calibration procedures would seem to preclude the use of WVRs in general surveying, but they may prove useful at permanent global tracking sites.

Although it is well known that the wet troposphere delay varies appreciably over the course of typical GPS experiments, many of the original studies of GPS precision estimated only one troposphere zenith delay parameter per site per observing session (e.g., Refs. 2 and 17). Most software has since been modified to allow estimation of a piecewise constant troposphere delay. Lichten and Border[32] showed that the wet troposphere zenith delay is accurately characterized as a time-varying parameter with the statistical properties of a random walk.

The remaining path error, multipathing of the GPS signals off objects near the receiver, seems slight for carrier phase data. Multipathing for the pseudorange data is a more serious problem. Although it has been suggested that site-dependent multipath filters could be developed,[12,22] it would seem that significant reduction of multipath will be attributed to improvements in antenna design, receiver signal processing, and better site selection.

The phase center of a GPS antenna varies as a function of the elevation and azimuth angles to the satellite. The magnitude of this variation ranges from several millimeters to several centimeters. To date, most GPS analysis softwares have used a mean GPS phase center to model the phase observables. This was convenient when all GPS antennas were of the same manufacture. Little error was introduced because the variations canceled to first order.[55] Now that numerous GPS antennas are being used for precision geodetic research, phase center variations corrections are required.[46] If this is not done correctly, an error with an elevation angle dependence becomes apparent. This will be incorrectly modeled as part of the wet troposphere delay, which also exhibits elevation angle dependence. The end result is degraded vertical accuracy.

Ambiguity resolution is important for the most accurate and precise determinations of station coordinates using GPS. Ambiguity resolution converts a precise, yet biased, phase observable into an unbiased range observable of the same precision. Work published by Blewitt[3] and Dong and Bock[17] both find a factor of ~2.5 improvement in baseline precision of the eastern component due to ambiguity resolution for baselines from 50 to 500 km. They also suggest that ambiguity resolution improves agreement with VLBI. Network design is particularly important for successful ambiguity resolution. The confidence limit for resolving an ambiguity is dependent on the baseline length between the two receivers that formed the double difference. Ambiguities are resolved sequentially, meaning the solution covariance matrix is updated with the new information after each ambiguity is resolved. Being able to resolve the first few ambiguities may trigger successful resolution of the entire network. Thus, to ensure successful ambiguity resolution, an experiment coordinator might artificially introduce several short, ~10 km baselines. High-quality dual-frequency pseudorange is helpful for ambiguity resolution.

The final error source that requires attention is the GPS orbit. To first order, the orbit error dr maps onto baseline errors dx as

$$\frac{|dx|}{x} \cong a \frac{|dr|}{r} \qquad (1)$$

where x is the baseline length, r is the altitude of the satellites, and a is a constant. The constant a has been found to have a value of approximately 0.2.[32] Apparently, this is because orbit errors are highly correlated, and this cancels some of the magnitude of their effects. Further reduction of the GPS orbit error depends on the ability of the user to define an accurate reference frame and on the sophistication of the models available.

The principles of orbit determination are simple and have been discussed in many textbooks. The true orbit of the GPS satellite differs from a pure Keplerian ellipse because of several perturbing forces, which include nonsphericity of the Earth's gravity field, attraction of the Sun and Moon, atmospheric drag, solar radiation pressure, and tides. The effects of atmospheric drag are negligible at the altitude of GPS satellites. The accelerations caused by the masses of the Sun and Moon are well known, and models can be used to describe the Earth's gravity field. Models for solar radiation pressure have been developed for both Block I and Block II satellites.[19] For arcs longer than ~12 h, solar radiation pressure bias parameters must be estimated. The limiting errors of GPS orbit determination are solar pressure and thermal radiation effects on the satellites themselves, particularly those satellites that are eclipsing.

III. Reference Frames

The fundamental aim of positional geodesy is to determine the location of a point on the Earth. The position of this point consists of three coordinates in a well-defined and accessible coordinate system. One common definition of a point on the Earth is the latitude, longitude, and altitude. Embedded in any definition of position is the concept of a reference frame. A Cartesian frame is useful for illustrative purposes. We seek to determine the location of a point in space; i.e., we seek to define a vector position r_e. The coordinates of r_e must be referred to an origin. Additionally, we must define the orientation of the three-dimensional orthogonal axes. Finally, the scale, or vector length, must be defined. Thus, a total of seven parameters are required to define a reference frame: three terms for the origin, three terms for the orientation of the coordinate axes, and a scale. For a Cartesian terrestrial reference frame, the coordinate system origin is placed at the Earth's center of mass, or geocenter. The z axis by convention is aligned along the rotation axis. The x and y axes are orthogonal to the z axis, with the x axis traditionally defined at Greenwich.

If we want to measure changes in the position of a point on the Earth over time, we must ensure that what we measure is in fact the motion of the point fixed in the Earth's crust, and not motions associated with the reference frame. Thus, we seek to refer our geodetic measurements to an inertial frame. The motion of the Earth relative to the fixed stars, presumed to be an inertial reference frame, is well understood. Precession is the slow motion of the Earth's pole with respect to inertial space—with a period of approximately 26,000 years. Nutation refers to the oscillations of the pole over shorter periods. In addition, the Earth is rotating at a variable rate, and the pole of rotation moves. We refer to the variation in rotation as $UT1$ or U. The X and Y pole positions are defined by the rotation matrix p. The transformation between the inertial vector r_i and the Earth-fixed vector r_e can then be defined as a combination of these four individual transformations:

$$r_i = PNUp \, r_e \quad (2)$$

where N and P are nutation and precession, respectively. A more detailed discussion of these transformations can be found in Lambeck[27] or Vanicek and Krakiwsky.[52] Because all coordinates in these systems depend on the rotation of the Earth, one of the critical undertakings of geodesy is the study of variations of the Earth's rotation rate and motion of its axis of rotation.

Although only seven parameters are required to define the reference frame, in practice, more information is needed to realize this reference frame. A terrestrial reference frame is, by convention, defined by thousands of observations of station positions over years, such as has been determined by the International Earth Rotation Service in Paris using VLBI and SLR. The number of constraints that will be required to define the reference frame for the GPS analyst will be strongly dependent on the strength of the GPS constellation and the ground-tracking network.

For GPS geodesy, the reference frame is strongly linked to the issue of accurate orbit determination for the GPS satellites. As the GPS constellation and global tracking network have grown, different strategies have been developed. In its initial experimental Block I phase, only seven satellites were visible, over short periods of time. The GPS orbits were phased to favor tracking in the southwest United States. Figure 1 displays the satellite tracks visible in the late 1980s centered over Southern California. For regional (<500 km) experiments in Southern California, it was found that the orbits could be adequately determined by fixing the positions of just three receivers, or nine parameters. Thus, the positions of all other observing GPS receivers were estimated relative to the fixed sites. Because the fixed receivers determined the accuracy of the estimated positions, the receivers were often called a fiducial or "truth" network.[13] Figure 2 shows a typical fiducial network used in the 1980s for measuring crustal motions in a Southern California regional network. The receivers at Westford, Richmond, and Goldstone were operated by CIGNET (Cooperative International GPS Network). The coordinates of each site were determined by collocating the GPS receiver with VLBI or SLR monuments nearby. While the GPS constellation shown in

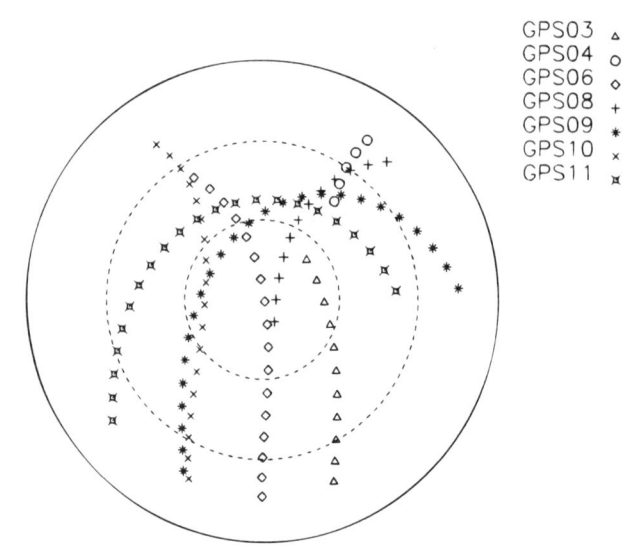

Fig. 1 Block I GPS sky tracks, centered over Southern California. These seven satellites were visible for a period of approximately 7 h. The dashed lines represent elevation angles of 30 and 60 deg, and the observations have been cut off at 15 deg.

Fig. 1 has a pronounced North–South orientation, the fiducial network has no such preference. Lichten et al.[34] and Larson et al.[30] both found that errors in the fiducial network, either through geometry or scale, could produce an appreciable systematic error in the network solution. If the coordinates of all available fiducial sites were known perfectly, of course, more than three sites should be fixed. Because of collocation errors, the coordinates of fiducial sites available to the GPS analyst in the late 1980s were only known with an accuracy of 3–4 cm. Because errors at fiducial sites directly affect the accuracy of the estimated receiver positions, it was desirable to fix the minimum number of sites required.

As GPS experiments became larger and more ambitious, ground tracking networks expanded. The 1988 Central and South American (CASA) experiment attempted to measure plate motions over distances of more than a 1000 km. A fiducial network centered over North America of approximately the same scale as the "regional" network would be insufficient. Thus, an international cooperative effort resulted in a global tracking network which extended to Europe and Australasia. This resulted in improvements in orbit accuracy and baseline precision.[26] Eventually CIGNET was expanded in Europe, Australia, New Zealand, and Japan through international cooperation. Other important sites in the southern hemisphere (Chile and South Africa) were added as part of the TOPEX/Poseidon precise orbit determination network. The global tracking network was expanded for the GPS experiment for the International Earth Rotation Service (IERS), also known as the GIG '91 experiment, as shown in Fig. 3. The GPS constellation consisted of 15 satellites at the time. The experiment lasted three weeks. Blewitt et al.[6] used data from GIG '91 to demonstrate a recent development in defining the GPS reference frame. Instead of fixing the coordinates of three or more sites,

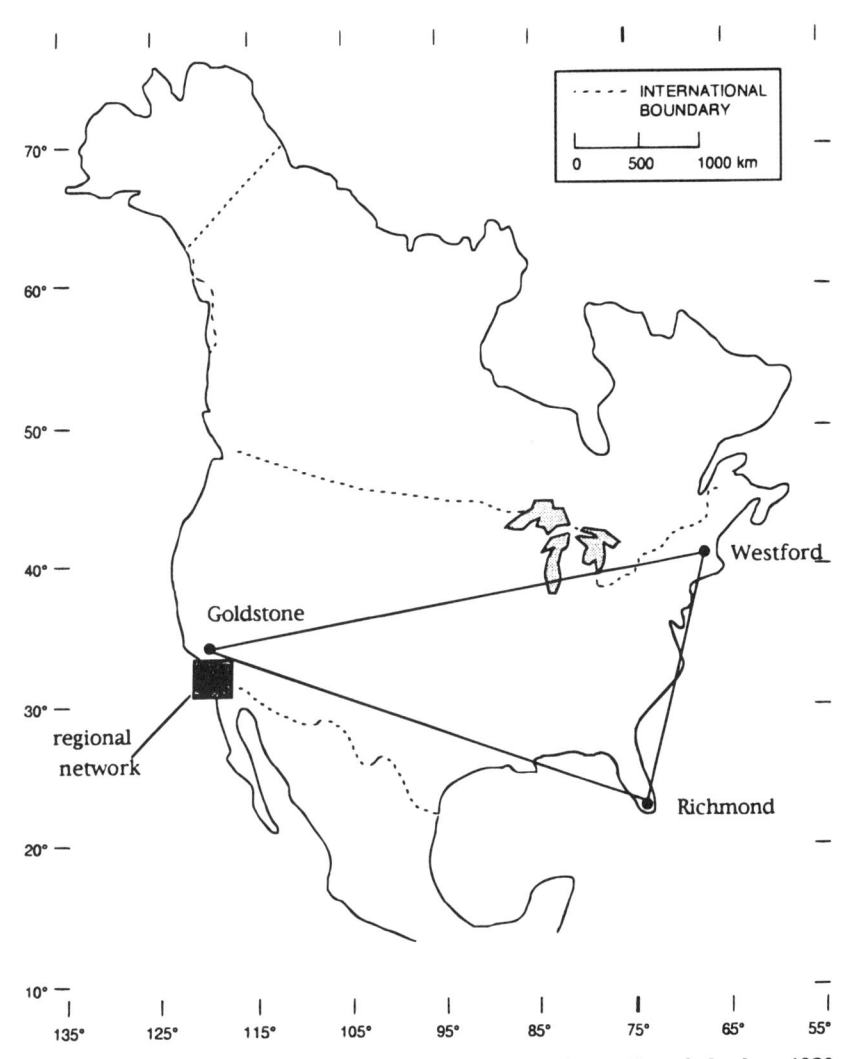

Fig. 2 Fiducial network for North American geodetic studies of the late 1980s. Cooperative International Network provided permanent GPS receivers at Westford, Goldstone, and Richmond. The Canadian Geodetic Survey provided data from Yellowknife.

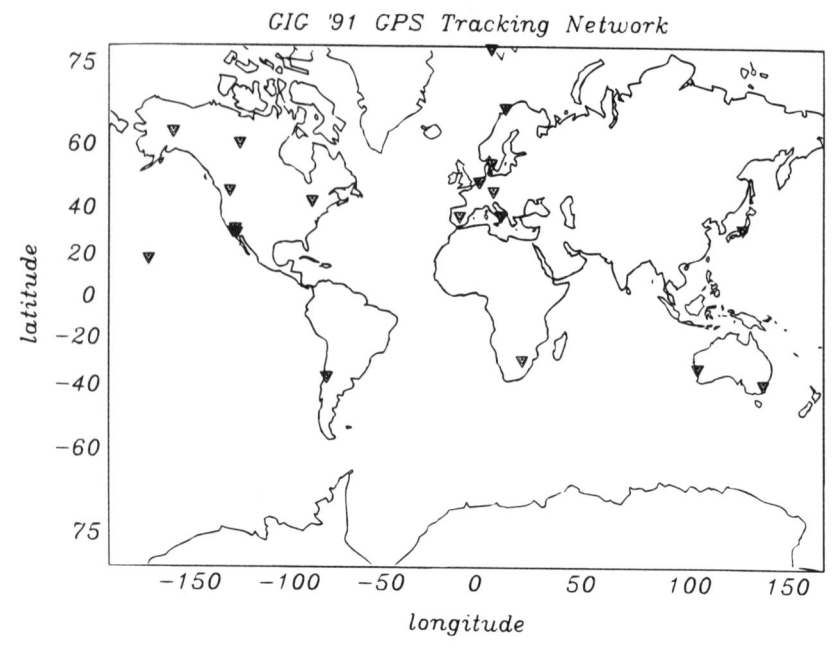

Fig. 3 Global tracking network for the GPS experiment for the International Earth Rotation Service and geodynamics.

they estimated all receiver coordinates. The scale and origin of the reference frame are implied by the satellite force model, propagation model, and the observables.[23] After estimation, Blewitt et al.[6] mapped the resulting station coordinates using a seven-parameter transformation into a frame defined by the ITRF (International Terrestrial Reference Frame),[10] which was itself derived from a joint VLBI-SLR analysis. They achieved accuracy levels commensurate with the errors in the ITRF. Because GIG'91 included 13 precisely determined globally distributed sites, Blewitt et al.[6] could use the information of all 13 sites, rather than only 3 or 4 sites.

The global GPS tracking network continues to expand on a monthly basis. Figure 4 shows some of the tracking stations that provide data to the IGS (International GPS Service for Geodynamics) at the present time. These data are available to all interested users through internet. The positions of these sites are also available from ITRF, derived from combined VLBI, SLR, and GPS estimates. X and Y pole positions and $UT1$ are distributed by IRIS (International Radio Interferometric Surveying) and IERS (International Earth Rotation Service). If desired, an analyst can retrieve GPS data from the IGS archives and estimate GPS orbits for the appropriate time period. Many organizations are also making precise GPS ephemerides available through the IGS. These orbits can be used to determine the relative positions of regional sites of interest. Because the GPS orbits are determined in the ITRF, the regional sites will also be defined in the correct reference frame, but without the need to estimate the satellite state.

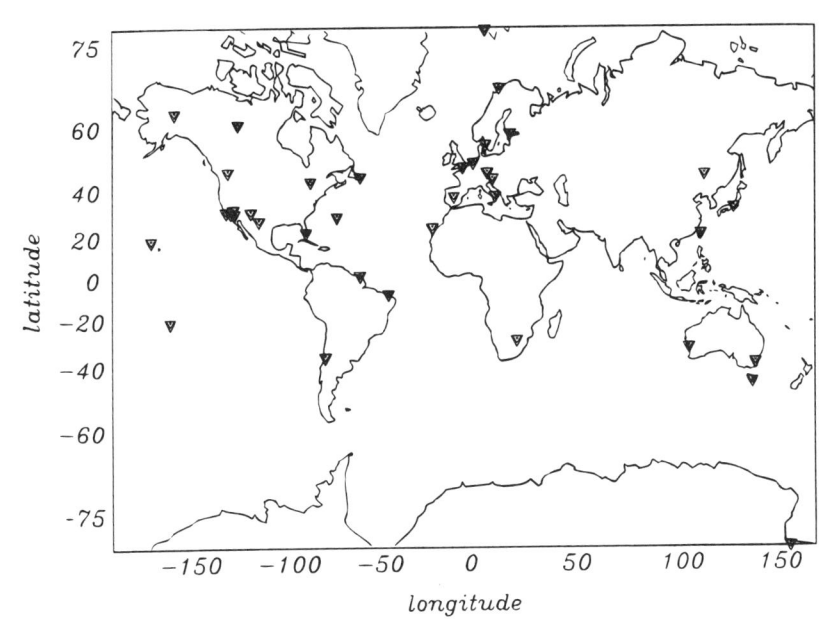

Fig. 4 GPS global tracking network, 1993.

Depending on the accuracy required, the analyst could also use a smaller fiducial network of three or four sites to determine the GPS orbits.

IV. Precision and Accuracy

Because its widescale use for precise geodesy began 10 years ago, GPS has quickly equaled the accuracy of mobile VLBI and SLR over regional and continental scales.[29,43] With the expansion of the constellation and improved modeling, global scale baselines also compare favorably with SLR and VLBI.[1,6]

The limits on baseline precision are determined by the geometry of the GPS constellation, strength of the tracking network, ability to model or correct error sources discussed in the previous section, and measurement noise. Refer again to Fig. 1, which displays the satellite constellation over Southern California in the late 1980s. The most precisely determined component was the North–South component, simply because more GPS sky tracks were North–South, with less variation in the East–West direction. Each site will have its skyplot characteristics, which are determined by the site latitude. Figure 5 shows the tracks for a site at the equator, with an extreme North–South preference, but without a hole in coverage between N30°W and N30°E, as appears over Southern California. Figure 6 shows the GPS sky tracks over the permanent GPS receiver at McMurdo station in Antarctica (77°S, 166°E). In this case, the satellite tracks are not preferentially aligned in either the East–West of North–South directions. The major limitation is that there are no satellite tracks whatsoever above an elevation angle of 60 deg.

Vertical precision is limited by skyview: we can look up but not down. The vertical component is also more sensitive to errors in the atmospheric path delay.

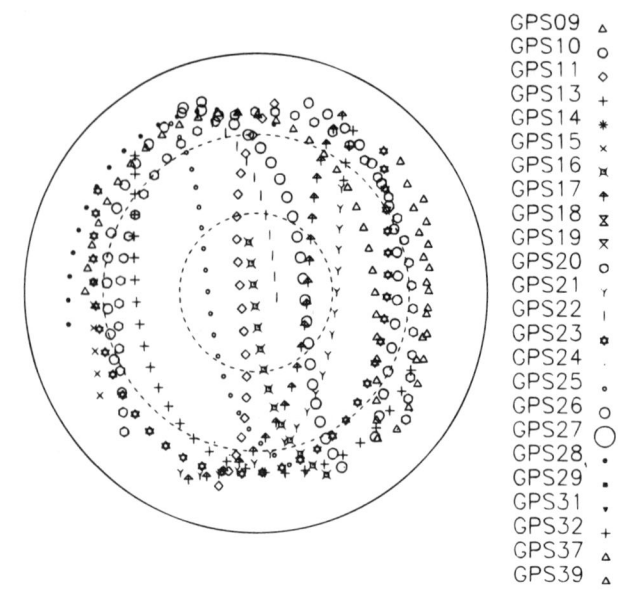

Fig. 5 GPS sky tracks centered at the equator over 12 h, August 1993.

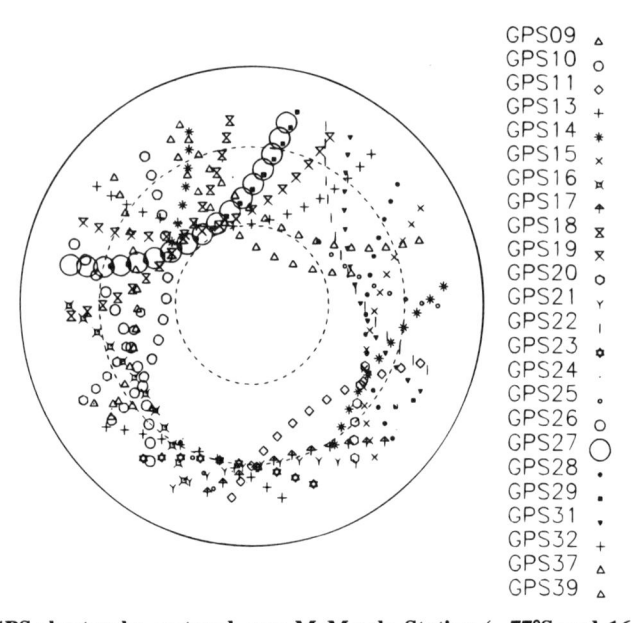

Fig. 6 GPS sky tracks centered over McMurdo Station (−77°S and 166°E), over 12 h, August 1993.

In order to differentiate between the atmospheric error and the vertical station position, a wide range of elevation angles should be observed. (A more detailed discussion of this issue can be found in Yunck.[56])

Precision is generally assessed by our ability to repeat a measurement, and thus it is often referred to as "repeatability." The repeatabilities are simply the weighted rms scatter about the weighted mean value. Because certain systematic errors do not manifest themselves at periods of a few weeks, we might expect that "short-term" repeatability, based on measurements over a few days, would be significantly better than "long-term" repeatability, based on measurements over many seasons and years. Because we are interested in system precision, unless qualified, the precision results we quote are long-term repeatabilities.

Repeatability for a regional network, where baselines spanned 50–350 km, in California was shown to be 2 mm + 6 × 10^{-9} L, for the north component, 2 mm + 13 × 10^{-9} L for the east component and 17 mm for the vertical, where L represents baseline length.[29] Davis et al.[14] studied baselines from 100 m to 225 km with comparable results. Over continental scales, Lichten and Border[32] found precisions of 0.6 and 2.6 cm for the North and East components, respectively, for a 1300-km baseline.

Although important, these early assessments of precision are being revised as the GPS constellation changes. The preliminary analyses all relied on the Block I constellation with four channel receivers and a maximum of 8 h of tracking. With a global network of receivers, a full constellation, and eight channel receivers, all visible satellites can be tracked simultaneously at a large number of sites. Under these new conditions, the greatest improvement in precision has been seen in the vertical component. With four channel receivers, the zenith troposphere delay parameter was highly correlated with the vertical component. This correlation is significantly reduced when the number of observed satellites increases from four to five or six.[5] Recent results from a permanent array of receivers spaced ~200 km apart indicate horizontal precisions of 4 mm and vertical precisions of 10–15 mm.[35] Over longer baselines, Heflin et al.[23] report baseline length repeatabilities of 2 mm + 4 × 10^{-9} L, up to 12,000 km. Anderson et al.[1] find similar precision over global scales.

Accuracy of GPS is determined by comparison with a truth standard. Most accuracy assessments for GPS have consisted of comparisons of baseline components with VLBI. Since the VLBI reference frame has been adopted for the fiducial coordinates used to define the GPS reference frame, these comparisons are self-consistent. If the VLBI and GPS observations had been made at the same ground monument, the comparison would be fairly simple to carry out. Unfortunately, the uncertainty in survey ties between VLBI and GPS monuments often corrupt the comparison. Although blunders, say, 10 cm, are fairly easy to determine, small systematic errors of several centimeters' magnitude can be incorrectly attributed to the new measurement system. Lichten and Bertiger[33] compared continental scale baselines, finding an agreement of 1.5 parts in 10^8 for 2000-km baselines. Suggesting that some of the VLBI–GPS comparisons in California were contaminated with survey errors, Larson and Agnew[29] instead compared four baseline rates and found agreement to better than one standard deviation. Davis et al.[14] compared electronic distance measurements to GPS and found no significant differences between the two systems.

With the permanent, continuously observing, global tracking network, there have been further improvements in precision and accuracy of GPS geodesy. Blewitt et al.[8] report absolute station coordinate accuracy of 15 mm, with weekly repeatabilities of better than 5 mm in latitude and longitude, and 10 mm in height. Furthermore, the independently determined GPS scale agrees with both VLBI and SLR to within 1 part in 10^9, well within the uncertainties of all three systems.[6] These Jet Propulsion Laboratory (JPL) results are summarized in Fig. 7.

V. Results

A. Crustal Deformation

Much of the interest in the geodetic accuracy of GPS is driven by scientists investigating deformation of the Earth's surface. Measurements of the Earth's surface provide direct tests of geophysical models used to describe the dynamics of the Earth. One important hypothesis tested by geodetic measurements is the theory of plate tectonics. Although sea-floor rocks that recorded magnetic reversals occurring over time scales of millions of years provided the first strong evidence of this phenomenon, contemporary measurements by space geodetic systems have shown that those long-term rates are statistically consistent with rates determined over a decade.[11,50] Other fundamental questions that should benefit from geodetic surveys include identification of earthquake hazard zones, mountain-building processes, and estimates of mantle viscosity via accurate measurements of postglacial isostatic rebound.

Nearly all crustal deformation surveys to date have concentrated on determining the horizontal deformation rates of the Earth's crust. This has been particularly successful across transform boundaries, such as the San Andreas Fault (SAF) in

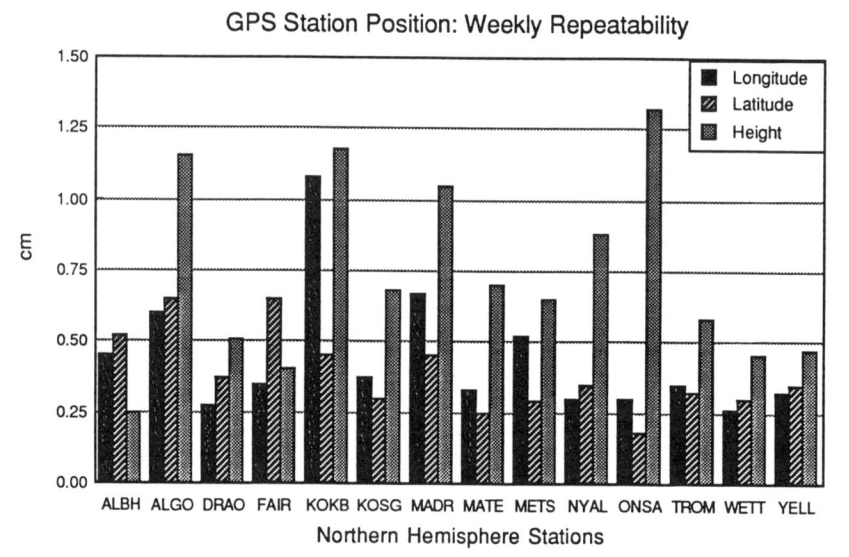

Fig. 7a Weekly repeatability of 14 northern hemisphere sites over 13 weeks during the summer of 1992. The typical daily repeatability is 3–5 mm in the horizontal and 5–10 mm in the vertical.

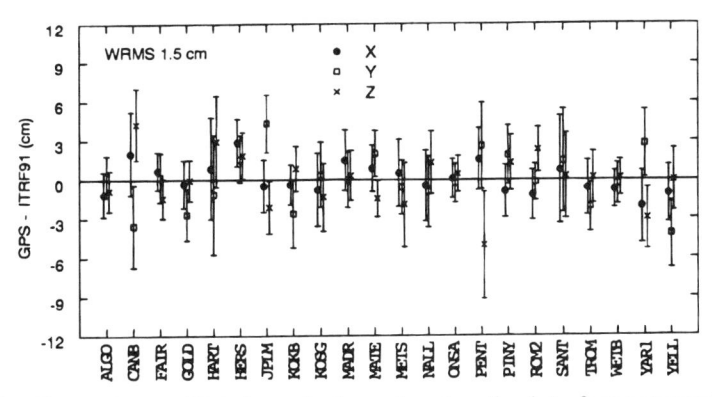

Fig. 7b Comparisons with independent very long baseline interferometry and satellite laser ranging measurements indicate an absolute 1-sigma accuracy of about 1.5 cm.

California, where no vertical deformation is expected. Prescott et al.[44] discussed GPS derived baselines that crossed the SAF in what became the rupture zone of the 1989 Loma Prieta earthquake. The NW–SE relative station velocities, characteristic of the SAF in California, were apparent in the three years of measurements conducted by the U.S. Geological Survey. Deformation rates across the SAF obtained with GPS were consistent with measurements made by both electronic distance measurements (trilateration)[37] and VLBI.[11]

GPS has also been used to address other important tectonic questions. Following the controversy regarding the magnitude of the Pacific-North American plate rate, Dixon et al.[16] began making GPS measurements across the Gulf of California. Over three and one-half years, they estimate a plate rate of 47 ± 7 mm/yr at a direction of −57 ± 6 deg. Their contemporary measurements agree better with the NUVEL1 global plate prediction[15] than with the previous standard plate motion model.[42] Although nearly 75% of the Pacific–North American relative plate motion is taken up on the SAF, it is not understood where the remaining portions of the deformation are distributed. Characterizing the deformation to the west of the San Andreas Fault has been the focus of a recent collaborative effort.[18] Their results are summarized in Fig. 8. Larson and Webb[31] found deformation rates on the order of 5 mm/yr across the eastern Santa Barbara Channel, which is consistent with seismic evidence of deformation.

In Japan, Shimada and Bock[48] were able to monitor the convergence of the Eurasian, Pacific, North American, and Philippine Sea plates, yielding westward motion of 28 mm/yr and significant vertical uplift. Freymueller et al.[21] describe the interactions of the Nazca, Cocos, Caribbean, and South American plates. As shown in Fig. 9, the geodetic results are in good agreement with NUVEL-1, although there are significant differences. Meertens and Smith[41] report on deformation associated with the Yellowstone Caldera.

Many geophysicists are also interested in the potential of GPS for measuring accurate vertical components. To date, there have been no significant estimates of secular vertical rates, such as postglacial isostatic rebound, with GPS although the installation of tracking stations by the Canadian Geodetic Survey at Penticton,

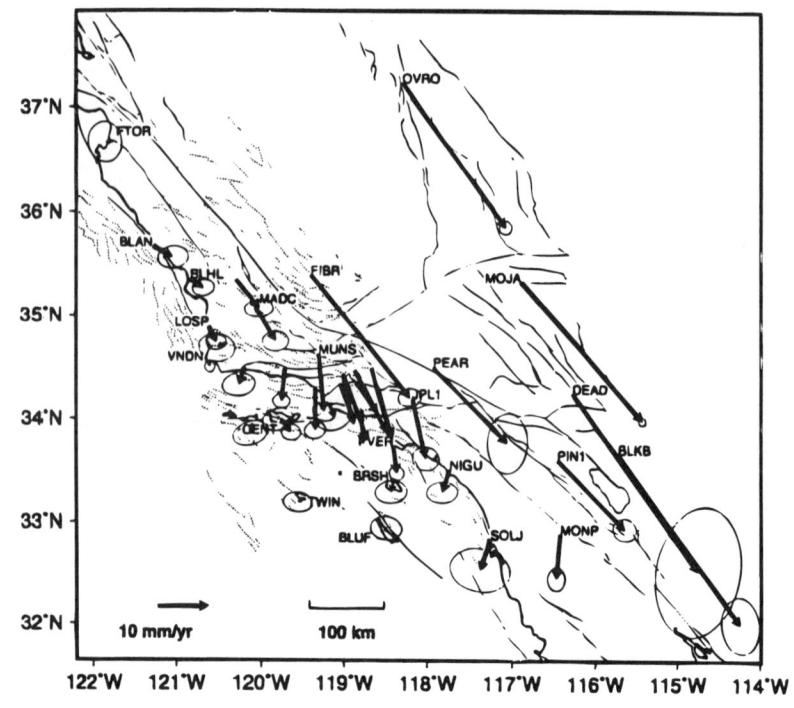

Fig. 8 Observed velocity of stations relative to the Pacific plate estimated from the combined global positioning system and very long baseline interferometry data set. The ellipses denote the region of 95% confidence, after scaling the formal uncertainties by a factor of 2. For clarity, the ellipses are not shown for the sites in the Ventura Basin. Reprinted from Feigl et al.[18] with the permission of the American Geophysical Union.

Algonquin, St. Johns, and Yellowknife should provide important constraints in the coming years.

Abrupt motions, such as those associated with earthquakes and volcanoes, have also been measured with GPS. Larsen et al.[28] analyzed data collected in the Imperial Valley of Southern California before and after the 1987 Superstition Hills earthquake, yielding a detailed map of the surface displacements. They were able to model surface displacements into right lateral slip of 130 ± 8 mm and 30 ± 10 mm of left-lateral slip. The 1991 Costa Rican earthquake (surface wave magnitude of 7.6) caused displacements of up to 2.4 m.[38] Analysis of GPS data was useful in determining a new dislocation model for the earthquake. Displacements associated with the 1992 Landers earthquake were the subject of work by both the JPL[7] and a group of university researchers.[9] These displacements are now available to the geodetic community to improve geodetic control in the Southern California region. Magma chamber deflation associated with the 1991 Hekkla volcanic eruption was discerned by repeated GPS measurements discussed by Sigmundssen.[49]

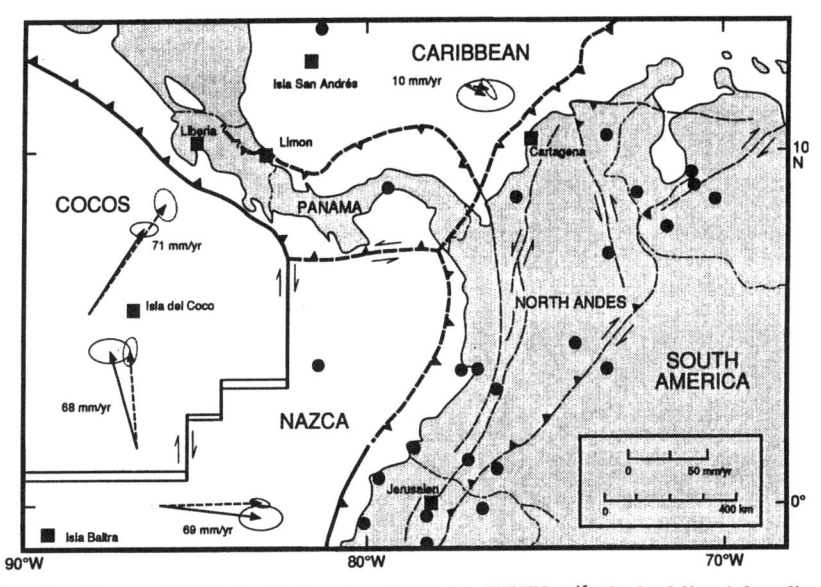

Fig. 9 Observed GPS (solid lines) and model NUVEL-1[15] (dashed lines) baseline rates of change, with their 95% confidence ellipses. Baseline rates of change measure the relative motions of two GPS sites. The Caribbean–North Andes plate motion is based on only two epochs of GPS data (1988 and 1991), and the uncertainty given is conservative. GPS sites used in this study are indicated by squares; other GPS sites are indicated by circles. Reprinted from Freymueller et al.[20] with the permission of the American Geophysical Union.

B. Earth Orientation

The most recent precise geodetic contribution by GPS has come about with the advent of global tracking networks established in the early 1990s. Global tracking networks and a larger GPS constellation now yield sufficient sensitivity to allow estimation of the geocenter and polar motion. One goal of the GIG '91 experiment discussed in Sec. III was to investigate the potential of GPS to determine Earth rotation parameters. A 21-station subset of the GIG '91 network was analyzed by independent groups. The GPS constellation consisted of 15 satellites at the time. Groups at MIT and JPL were able to estimate daily X and Y pole positions to an accuracy of better than 1 milliarcsec.[24,36] Variations in $UT1$ were estimated by Lichten et al.[34] with an accuracy of a few hundredths of a millisecond.

The geocenter was the subject of a companion study by JPL.[53] They found an offset with respect to ITRF of -8.3 ± 2.7, 13.4 ± 2.4, and -7.7 ± 13.7 cm in X, Y, and Z, respectively. With an expanded constellation and global tracking network (particularly in the southern hemisphere) in 1992, Malla et al.[40] reported much improved X and Y components, 0.0 ± 1.4, 1.5 ± 1.3 cm, but the Z estimate was still weak, -8.2 ± 3.0 cm. With the addition of TOPEX GPS data, Malla[39] was able to improve the Z component agreement with ITRF to 0.1 ± 1.5 cm.

VI. Conclusions

Geodetic science has greatly benefited from the GPS. With the appropriate analysis strategy and reference frame, the geodesist can now determine absolute geocentric positions with an accuracy of better than 2 cm, with even greater accuracy for relative positions. Permanent GPS tracking sites can now be used to maintain the terrestrial reference frame, including subdaily resolution of Earth orientation parameters. Within the global network, geophysicists will be able to measure plate motions and test assumptions of plate rigidity. Regional "mobile" GPS experiments will be used to study complicated seismic zones, such as the Himalayan collision zone, uplift in Chile, and subduction in Japan. Within five years there should be significant estimates of postglacial rebound from monitoring programs begun in Canada and Fennoscandia. Finally, GPS will play an important role in monitoring sea level by tying tide gauges into the global terrestrial reference frame.

Acknowledgments

This manuscript was begun while I was a visiting scientist at the Branch of Earthquake Geology and Geophysics of the U.S. Geological Survey and completed at the Laboratory for Terrestrial Physics of the NASA Goddard Space Flight Center. The work is supported by NASA NAG 5-1908 and NSF EAR-9209385. I am grateful to Robert B. Miller, Geoff Blewitt, Jim Davis, Jeff Freymueller, Steve Nerem, George Rosborough, and Mark Tapley for helpful discussions. I thank Tom Yunck and Steve Lichten for reviewing the manuscript and helping me improve it.

References

[1] Anderson, P. H., Hauge, S., and Kristiansen, O., "GPS Relative Positioning at a Precise Level of One Part per Billion," *Bulletin Geodesique,* Vol. 67, 1993, pp. 91–106.

[2] Beutler, G., Bauersima, I., Gurtner, W., Rothacher, M., and Schildknecht, T., "Evaluation of the 1984 Alaska Global Positioning System Campaign with the Bernese GPS Software," *Journal of Geophysical Research,* Vol. 92, 1987, pp. 1295–1303.

[3] Blewitt, G., "Carrier Phase Ambiguity Resolution for the Global Positioning System Applied to Geodetic Baselines up to 2000 km," *Journal of Geophysical Research,* Vol. 94, 1989, pp. 10,187–10,283.

[4] Blewitt, G., "An Automatic Editing Algorithm for GPS Data," *Geophysical Research Letters,* Vol. 17, 1990, pp. 199–202.

[5] Blewitt, G., "Advances in Global Positioning System Technology for Geodynamics Investigators, 1978–1992," *Contributions of Space Geodesy: Technology,* edited by D. Smith and D. Turcotte, Vol. 25, Geophysical Monograph Series, American Geophysical Union, Washington, DC, 1993, pp. 195–213.

[6] Blewitt, G., Heflin, M. B., Webb, F. H., Lindqwister, U. J., and Malla, R. P., "Global Coordinates with Centimeter Accuracy in the International Terrestrial Reference Frame using the Global Positioning System," *Geophysical Research Letters,* Vol. 19, 1992, pp. 853–856.

[7]Blewitt, G., Heflin, M. B., Hurst, K. J., Jefferson, D. C., Webb, F. H., and Zumberge, J. F., "Absolute Far-Field Displacements from the 28 June 1992 Landers Earthquake Sequence," *Nature,* Vol. 361, 1993, pp. 340–342.

[8]Blewitt, G., Heflin, M. B., Hurst, K., Jefferson, D., Vigue, Y., Webb, F., and Zumberge, J., "Viewing the Earth as a Rotating, Deforming Polyhedron, Using the Global Positioning System," *EOS,* Vol. 74, No. 16, 1993, p. 48.

[9]Bock, Y., et al., "Detection of Crustal Deformation from the Landers Earthquake Sequence Using Continuous Geodetic Measurements," *Nature,* Vol. 361, 1993, pp. 338–340.

[10]Boucher, C., and Altamimi, Z., "ITRF 89 and other Realizations of the IERS Terrestrial Reference System for 1989," *IERS Technical Note 6,* Observatoire de Paris, 1991.

[11]Clark, T. A., Gordon, D., Himwich, W. E., Ma, C., Mallana, A., and Ryan, J. W., "Determination of Relative Site Motions in the Western United States Using Mark III Very Long Baseline Interferometry," *Journal of Geophysical Research,* Vol. 92, 1987, pp. 12,741–12,750.

[12]Cohen, C., and Parkinson, B., "Mitigating Multipath Error for GPS-Based Attitude Determination," *Keystone Guidance and Contral Conference, Proceedings of the American Astronomical Society,* (Keystone, CO), Feb. 1991.

[13]Davidson, J. M., Thornton, C. L., Vegos, C. J., Young, L. E., and Yunck, T. P., "The March 1985 Demonstration of the Fiducial Network for GPS Geodesy: A Preliminary Report," *Proceedings of the First International Symposium on Precise Positioning with GPS,* edited by C. Goad, NOAA, Rockville, MD, 1985, pp. 603–612.

[14]Davis, J. L., Prescott, W. H., Svarc, J. L., and Wendt, K. J., "Assessment of Global Positioning System Measurements for Studies of Crustal Deformation," *Journal of Geophysical Research,* Vol. 94, 1989, pp. 13,635–13,650.

[15]DeMets, C., Gordon, R., Argus, D., and Stein, S., "Current plate motions," *Geophysical Journal International,* Vol. 101, 1990, pp. 425–478.

[16]Dixon, T. H., Gonzalez, G., Lichten, S., Tralli, D., Ness, G., and Dauphin, J., "Preliminary Determination of Pacific-North America Relative Motion in the Southern Gulf of California using the Global Positioning System," *Geophysical Research Letters,* Vol. 18, 1991, pp. 861–864.

[17]Dong, D., and Bock, Y., "GPS Network Analysis with Phase Ambiguity Resolution Applied to Crustal Deformation Studies in California," *Journal of Geophysical Research,* Vol. 94, 1989, pp. 3949–3966.

[18]Feigl, K., et al., "Space Geodetic Measurement of Crustal Deformation in Central and Southern California," *Journal of Geophysical Research,* Vol. 98, 1993, pp. 21, 677–621, 712.

[19]Fliegel, H. F., Gallini, T. E., and Swift, E. R., "Global Positioning System Radiation Force Model for Geodetic Applications," *Journal of Geophysical Research,* Vol. 97, 1991, pp. 559–568.

[20]Freymueller, J., Kellogg, J., and Vega, V., "Plate Motions in the North Andean Region," *Journal of Geophysical Research,* Vol. 98, 1993, pp. 21, 853–821, 863.

[21]Freymueller, J., "Phasedit—A GPS Cycle Slip Detector," *IOM,* Stanford University, Stanford, CA, 1993.

[22]Genrich, J. F., and Bock, Y., "Rapid Resolution of Crustal Motion with Short-Range GPS," *Journal of Geophysical Research,* Vol. 97, 1992, pp. 3261–3270.

[23]Heflin, M. B., et al., "Global Geodesy Using GPS without Fiducial Sites," *Geophysical Research Letters,* Vol. 19, 1992, pp. 131–134.

[24]Herring, T. A., Dong, D., and King, R. W., "Sub-Milliarcsecond Determination of Pole Position Using Global Positioning System Data," *Geophysical Research Letters,* Vol. 18, 1991, pp. 1893–1986.

[25]King, R. W., Masters, E. G., Rizos, C., Stolz, A., and Collins, J., *Surveying with GPS,* School of Surveying Monograph 9, University of New South Wales, Kensington, Australia, 1985.

[26]Kornreich-Wolf, S., Dixon, T. H., and Freymueller, J. T., "The Effects of Tracking Network Configuration on GPS Baseline Estimates for the CASA UNO Experiment," *Geophysical Research Letters,* Vol. 17, 1990, pp. 647–650.

[27]Lambeck, K., *Geophysical Geodesy, the Slow Deformations of the Earth,* Clarendon, Oxford, 1988.

[28]Larsen, S., Reilinger, R., Neugebauer, H., and Strange, W., "Superstition Hills Earthquake," *Journal of Geophysical Research,* Vol. 97, 1992, pp. 4885–4902.

[29]Larson, K. M., and Agnew, D., "Application of the Global Positioning System to Crustal Deformation Measurement: 1. Precision and Accuracy," *Journal of Geophysical Research,* Vol. 96, 1991, pp. 16,547–16,565.

[30]Larson, K. M., Webb, F. H., and Agnew, D., "Application of the Global Positioning System to Crustal Deformation Measurement: 2. The Influence of Errors in Orbit Determination Networks," *Journal of Geophysical Research,* Vol. 96, 1991, pp. 16,567–16,584.

[31]Larson, K. M., and Webb, F. H., "Active Deformation in the Santa Barbara Channel Inferred from GPS Measurements," *Geophysical Research Letters,* Vol. 19, 1992, pp. 1491–1494.

[32]Lichten, S. M., and Border, J. S., "Strategies for High-Precision Global Positioning System Orbit Determination," *Journal of Geophysical Research,* Vol. 92, 1987, pp. 12,751–12,762.

[33]Lichten, S. M., and Bertiger, W. I., "Demonstration of Sub-Meter GPS Orbit Determination and 1.5 Parts in 10^8 Three-Dimensional Baseline Accuracy," *Bulletin Geodesique,* Vol. 63, 1989, pp. 167–189.

[34]Lichten, S. M., Bertiger, W. I., and Lindqwister, U. J., "The Effect of Fiducial Network Strategy on High-Accuracy GPS Orbit Determination and Baseline Determination," *Proceedings of the 5th International Symposium on Satellite Positioning,* (Las Cruces, NM), 1989.

[35]Lindqwister, U. J., Zumberge, J. F., Webb, F. H., and Blewitt, G., "Few Millimeter Precision for Baselines in the California Permanent GPS Geodetic Array," *Geophysical Research Letters,* Vol. 18, 1991, pp. 1,135–1,138.

[36]Lindqwister, U. J., Freedman, A. P., and Blewitt, G., "Daily Estimates of the Earth's Pole Position with the Global Positioning System," *Geophysical Research Letters,* Vol. 19, 1992, pp. 845–848.

[37]Lisowski, M., Savage, J. C., and Prescott, W. H., The Velocity Field Along the San Andreas Fault in Central and Southern California, *Journal of Geophysical Research,* Vol. 96, 1991, pp. 8369–8389.

[38]Lundgren, P., Kornreich Wolf, S., Protti, M., and Hurst, K. J., "GPS measurements of crustal deformation associated with the 22 April 1991, Valle de la Estrella, Costa Rica earthquake," *Geophysical Research Letters,* Vol. 20, 1993, 407–410.

[39]Malla, R. P., "Breaking the ΔZ Barrier in Geocenter Estimation," Jet Propulsion Laboratory, JPL Interoffice Memorandum 335.8-93-018, Aug. 18, 1993.

[40]Malla, R. P., Wu, S. C., and Lichten, S. M., "Use of Global Positioning System Measurements to Determine Geocentric Coordinates and Variations in Earth Orientation," NASA TDA Progress Rept. 42-114, 1993.

[41]Meertens, C. M., and Smith, R. B., "Crustal Deformation of the Yellowstone Caldera from first GPS Measurements: 1987–1989," *Geophysical Research Letters*, Vol. 18, 1991, pp. 1763–1766.

[42]Minster, J. B., and Jordan, T. H., "Present-day plate motions," *Journal of Geophysical Research*, Vol. 83, 1978, pp. 5331–5354.

[43]Murray, M. H., "Global Positioning System Measurement of Crustal Deformation in Central California," Ph.D. Thesis, Massachusetts Institute of Technology, Cambridge, MA, 1991.

[44]Prescott, W. H., Davis, J. L., and Svarc, J. L., "Global Positioning System Measurements for Crustal Deformation: Precision and Accuracy," *Science*, 1989, pp. 1337–1340.

[45]Rocken, C., Johnson, J. M., Neilan, R. E., Cerezo, M., Jordan, J. R., Falls, M. J., Nelson, L. D., Ware, R. H., and Hayes, M., "The measurement of atmospheric water vapor, radiometer comparison and spatial variations," *IEEE Transactions on Geoscience and Remote Sensing*, Vol. 29, 1991, pp. 3–8.

[46]Schupler, B. R., and Clark, T. A., "How Different Antennas Affect the GPS Observable," *GPS World*, Vol. 2, No. 10, 1991, pp. 32–36.

[47]Schutz., B. E., Tapley, B. D., Abusali, P. A. M., Rim, H. J., "Dynamic Orbit Determination Using GPS Measurements from TOPEX/POSEIDON," *Geophysical Research Letters*, (submitted for publication).

[48]Shimada, S., and Bock, Y., "Crustal Deformation Measurements in Central Japan Determined by a Global Positioning System Fixed-Point Network," *Journal of Geophysical Research*, Vol. 97, No. B9, 1992, pp. 12,437–12,456.

[49]Sigmundsson, F., Einarsson, P., and Bilham, R., "Magma chamber deflation recorded by the Global Positioning System: the Hekkla 1991 eruption," *Geophysical Research Letters*, Vol. 14, 1992, pp. 1483–1486.

[50]Smith, D. E., et al., "Tectonic Motion and Deformation from Satellite Laser Ranger to LAGEOS," *Journal of Geophysical Research*, Vol. 95, 1990, pp. 22,013–22,042.

[51]Tralli, D. M., Dixon, T. H., and Stephens, S. A., "The Effect of Wet Tropospheric Path Delays on Estimation of Geodetic Baselines in the Gulf of California Using the Global Positioning System," *Journal of Geophysical Research*, Vol. 93, 1988, pp. 6545–6557.

[52]Vanicek, P., and Krakiwsky, E. J., *Geodesy: The Concepts*, North Holland Publishing Company, Amsterdam, 1986.

[53]Vigue, Y., Lichten, S. M., Blewitt, G., Heflin, M. B., and Malla, R. P., Precise determination of the Earth's center of mass using measurements from the Global Positioning system, *Geophysical Research Letters*, Vol. 19, 1992, pp. 1487–1490.

[54]Ware, R., Rocken, C., Solheim, F., Van Hove, T., Alber C., and Johnson, J., "Pointed Water Vapor Radiometer Corrections for Accurate Global Positioning Surveying," *Geophysical Research Letters*, Vol. 20, 1993, pp. 2635–2638.

[55]Wu, J., Wu, S., Hajj, G., Bertiger, W., and Lichten, S. "Effects of Antenna Orientation on GPS Carrier Phase," *Astrodynamics 1991*, edited by B. Kaufman et al., Vol. 76, *Adv. in Astron. Sci.*, 1992, pp. 1647–1660.

[56]Yunck, T. P., "Coping with the Atmosphere and Ionosphere in Precise Satellite and Ground Positioning," *Environmental Effects on Spacecraft Trajectories and Positioning*, edited by A. Vallance-Jones, Geophysical Monograph 73, IUGG Vol. 13, 1993, pp. 1–16.

Chapter 21

Orbit Determination

Thomas P. Yunck*
Jet Propulsion Laboratory, California Institute of Technology, Pasadena, California 91109

I. Introduction

AN Earth satellite collecting GPS data with an onboard receiver can compute its state (position and velocity) in a diversity of ways, the choice depending in part on the type of orbit and mission requirements. Tracking and navigation requirements can include real-time state knowledge and active control during launch and orbit insertion[1] and during re-entry and landing; real-time relative navigation between vehicles during rendezvous[2,3]; autonomous stationkeeping and near-real-time orbit knowledge for operations and orbit maintenance[4]; rapid postmaneuver orbit recovery[5]; and after-the-fact precise orbit determination for scientific analysis.[6,7] Orbit accuracy requirements can range from hundreds of meters or more for routine operations to a few centimeters for precise remote sensing. Among existing tracking systems, only GPS can meet the most stringent of these needs for the most dynamically unpredictable vehicles. An overview of GPS space applications is given in Ref. 8.

The GPS signal beamwidths extend roughly 3000 km beyond the Earth's limb, enabling an Earth orbiter below that altitude to receive continuous three-dimensional coverage. This chapter focuses on orbit estimation for satellites in low circular orbits, below a few thousand kilometers, with emphasis on the high accuracy that GPS so ably provides. Real-time techniques fall under what we call *direct* GPS orbit determination, in which only the GPS data collected by the orbiter are used in the solution. For precise after-the-fact solutions, we turn to a global form of *differential* GPS in which data collected at multiple ground sites are combined with the onboard data to reduce the major errors. We also examine briefly the adaptation of GPS tracking techniques to satellites in highly elliptical and geosynchronous orbits.

Copyright © 1994 by the American Institute of Aeronautics and Astronautics, Inc. The U.S. Government has a royalty-free license to exercise all rights under the copyright claimed herein for Governmental purposes. All other rights are reserved by the copyright owner.
*Deputy Manager, Tracking Systems and Applications Section.

The potential of GPS to provide accurate and autonomous satellite orbit determination was noted early in its development.[9] Early studies of direct GPS-based tracking include those in Ref. 10, which surveyed applications from near Earth to beyond geosynchronous altitudes; Ref. 11, which examined GPS tracking of the Space Shuttle; Ref. 12, which focused on autonomous near Earth navigation; Ref. 13, which described NASA's first planned GPS orbital application to Landsat-4; Ref. 14, which compared the potential of GPS and NASA's Tracking and Data Relay Satellite System (TDRSS) for onboard navigation; Ref. 15, which discussed flight receiver requirements and expected onboard orbit accuracies from near Earth to geosynchronous altitude; and Ref. 16, which explored geosynchronous applications. The first reported results from direct GPS tracking were those of the Landsat-4 experiment,[17,18] which achieved approximately 20 m accuracy during the relatively brief periods of good GPS visibility at that time.

Among the first descriptions of precise orbit determination at the level of several decimeters or better by differential GPS techniques are those in Ref. 19, which proposed a subdecimeter carrier phase-based technique for the TOPEX (later TOPEX/Poseidon) ocean altimetry mission; Ref. 20, which examined differential tracking of a low-altitude orbiter; Ref. 21, which proposed differential techniques for high-altitude satellites; and Ref. 22, which surveyed a variety of differential GPS applications. Since then, several important refinements have been introduced which better exploit the unique signals and the unprecedented observing strength GPS provides.

II. Principles of Orbit Determination

Point positioning with GPS is as accurate in low orbit as on the ground: typically 50–100 m for the GPS standard positioning service (SPS) user (under nominal levels of selective availability) and 10–20 m for the precise positioning service (PPS) user. Corresponding velocity solutions from carrier phase rate may reach 0.5 m/s (SPS) and better than 0.1 m/s (PPS). Although those levels are adequate for many purposes, instantaneous solutions have their limitations. They may be impossible during periods of restricted visibility, for example, and their accuracy may be inadequate for orbit prediction or for some real-time needs. Some scientific instruments require real-time position knowledge of meters to tens of meters for accurate pointing, while after-the-fact requirements can be far more stringent. To reduce the instantaneous position and velocity error, the traditional tools of dynamic orbit estimation can be brought to bear.

A. Dynamic Orbit Determination

Classical dynamic orbit determination exploits orbital mechanics—the physics underlying orbital motion—and filtering theory to yield a stable and accurate orbit solution from generally sparse and noisy measurements. This approach has, in fact, been necessary with conventional tracking systems, which, unlike GPS, seldom if ever provide sufficient information at one time for a geometric solution, and can provide no data at all over much of the globe. (An exception is the use of range and angle data from a single site to determine the instantaneous position of geostationary satellites, although the accuracy of that technique is far worse than that of GPS.) An orbit model must, therefore, be introduced to supply the missing information. In dynamic orbit determination, the orbit model is derived from models of the forces acting on the satellite and the laws of motion.

The process begins with a set of tracking measurements (range or Doppler, for example) along with mathematical models of the forces acting on the satellite and of the satellite physical properties. The major forces include gravity, aerodynamic drag and lift, solar radiation pressure, and active thrusting. Lesser contributions may come from outgassing, satellite thermal radiation, sunlight reflected from the Earth, and electromagnetic effects. The force and satellite models are used to compute a model of satellite acceleration over time, from which, by double integration, a nominal trajectory is formed. In principle, all that is then needed to produce the orbit solution is to determine the two vector constants of integration—position and velocity at one time point—also known as the epoch state. That is done through an estimation procedure that finds the epoch state for which the resulting model trajectory best fits the tracking data according to some optimality criterion, usually minimizing the mean square fitting error. To improve the fit, we can simultaneously estimate various force parameters, such as drag, solar radiation, and gravity coefficients; geometrical parameters, such as tracking station positions and Earth rotation; or empirical parameters, such as nonspecific once- and twice-per-orbit accelerations. The resulting solution is still a trajectory derived from force models, and its accuracy depends on how faithfully those models, fixed or adjusted, describe the real forces acting on the satellite.

More formally, to construct a nominal or a priori satellite trajectory we begin with Newton's second law of motion

$$f = m\,a = m\,\ddot{r} \tag{1}$$

or

$$\ddot{r} = f/m \tag{2}$$

where r is the satellite position vector. This fundamental equation of mechanics provides the dynamical constraint governing the orbit solution. The true acceleration $\ddot{r}$ at any instant depends on the satellite position and velocity at that instant, and on many other parameters that characterize the forces at work. In the orbit solution, those parameters may take the form of spherical harmonic gravity coefficients, drag and lift coefficients, solar flux and reflectivity, a geomagnetic index, and so on. Let $(r_o, \dot{r}_o)$ be the true satellite epoch state to be estimated. We first select a nominal epoch state $(r_{on}, \dot{r}_{on})$, perhaps from an instantaneous GPS state solution, and construct an acceleration model $\ddot{r}_n(t)$ from the force and satellite models. The nominal trajectory $r_n(t)$ is then generated by double integration of the acceleration model,

$$r_n(t) = \iint \ddot{r}_n(t)\,dt + \dot{r}_{on}\,t + r_{on} \tag{3}$$

The least squares solution procedure will then estimate corrections to the nominal epoch state (and possibly to selected model parameters) that bring the model trajectory into better agreement with the tracking data. If only the six-element epoch state and a few other parameters are adjusted, as is commonly the case, then in principle only a relatively few measurements around the orbit are needed to yield a well-determined solution, and a sparse tracking network will suffice. This is the great power and appeal of dynamic orbit determination. Since the first days of space exploration this technique has made practical the accurate tracking of Earth satellites and deep space probes.

Observe, however, that the resulting orbit solution depends intimately on the (possibly adjusted) acceleration model $\ddot{r}_n(t)$. Where high-accuracy orbits are required, high-accuracy models must be found. This can be enormously costly and may be a practical impossibility for low-altitude or maneuvering vehicles. In the mid-1980s it was recognized that the continuous three-dimensional coverage given by GPS offers an escape from this dynamical bind. Before describing these orbit determination techniques, we first review some principles of optimal estimation theory.

B. Batch Least Squares Solution

An enduring technique for estimating celestial orbits is the method of least squares, first employed by Gauss in 1795. Let z be a vector of observations $(z_1, \ldots, z_n)^T$ made over an interval of time, or tracking arc. The objective is to find that trajectory, among all possible trajectories satisfying the dynamical constraint [Eq. (2)], which minimizes the mean square difference between the actual observations z_i and theoretical observations $\tilde{z}_i$ derived from the solution trajectory. That is, we want to find the trajectory $r(t)$ that minimizes the functional

$$J = \sum_{i=1}^{n} \{z_i - \tilde{z}_i [r(t)]\}^2 \qquad (4)$$

As this is a nonlinear problem, we reformulate it as one of computing a linear correction to the nominal trajectory $r_n(t)$ given by Eq. (3). First, we compute theoretical observations $\tilde{z}_i$ from the nominal trajectory, then form the differences $\delta z_i = z_i - \tilde{z}_i$. These *prefit residuals* become the observations to be used in a linear adjustment of the nominal trajectory. (Strictly speaking, this is still not a linear problem; but if the nominal trajectory is sufficiently close to the true trajectory, it will be in the "linear regime," where a linear correction is adequate, if not perfect. If greater accuracy is needed, a linear correction to the new solution can be computed, and so on for multiple iterations, until the solution converges.) The familiar linear equation can be written as follows

$$\delta z = A\,x + n \qquad (5)$$

where x is the vector of parameters to be estimated, n is the vector of random measurement noise on the observations δz, and A is a matrix of partial derivatives of the observations with respect to the elements of x. Here x includes, at a minimum, the adjustments to the six epoch state parameters, and may include adjustments to various dynamic, geometric, and clock parameters as well. Equation (5) is called the regression equation and A is the matrix of regression coefficients.

A detailed discussion of the construction of A is beyond the scope of this chapter, but a simple overview is in order. An element a_{ij} of A is given by the following equation

$$a_{ij} = \frac{\partial z_i}{\partial x_j} \tag{6}$$

where, for simplicity, z_i now represents the differential element δz_i. This partial derivative relates an observation z_i at one time to state parameter x_j at a possibly remote reference time. The A matrix, thus, contains the state transition information from the reference epoch to all times in the data arc and must, therefore, embody the dynamical constraint of Eq. (2). To compute the a_{ij}, we first write

$$\frac{\partial z_i}{\partial x_j} = \frac{\partial z_i}{\partial \boldsymbol{x}_{ci}} \frac{\partial \boldsymbol{x}_{ci}}{\partial x_j} \tag{7}$$

where $\boldsymbol{x}_{ci}$ represents the satellite state at the time of observation z_i. This explicitly introduces the current state $\boldsymbol{x}_{ci}$ and its relation to both the current observation z_i and the state variables x_j. The partial $\partial z_i / \partial \boldsymbol{x}_{ci}$ contains no dynamical information and can be computed directly. The partial $\partial \boldsymbol{x}_{ci} / \partial x_j$ relates the satellite state at the observation time to the epoch state and, thus, embodies the dynamical constraint. To determine that partial, we differentiate the equation of motion (3) with respect to the epoch state parameters, producing a set of linear second-order differential equations in $\partial \boldsymbol{x}_{ci} / \partial x_j$. These *variational equations* are then integrated numerically to obtain the partial derivative and, thus, the final regression coefficients.

The well-known least squares solution to the regression equation (5) is given by

$$\hat{\boldsymbol{x}} = (A^T R_n^{-1} A)^{-1} A^T R_n^{-1} \boldsymbol{z} \tag{8}$$

where

$$R_n = E[\boldsymbol{n}\,\boldsymbol{n}^T] \tag{9}$$

is the covariance matrix associated with the measurement noise vector $\boldsymbol{n}$. This is known as the batch least squares solution because it requires that all observations over a data arc be collected and processed as a batch. In practice, when many parameters are estimated, Eq. (8) will require large matrix inversions, which can cause numerical instability. Most orbit estimators today employ more stable techniques.

C. Kalman Filter Formulation

A spaceborne GPS user may require a continuous real-time state solution more accurate than point positioning can provide. Although filtering is needed to achieve this, a batch solution is generally inappropriate because it may require a long accumulation of measurements and a large amount of computation at once. In such cases, a sequential estimator is called for, a popular example of which is the Kalman filter.

A sequential filter continually updates the current state estimate with each new measurement. The computation needed for each update is small compared with that for a full batch solution (although for a properly formulated filter, the computation required for many hours of updates is comparable to that for the same size batch solution); hence, an onboard processor can maintain the solution in real time. It should be noted that the sequential current state estimate employs only data from the past up to the present, whereas a batch filter may estimate a state with data from both before and after an epoch. In non-real-time uses, the final sequential state estimate can be mapped to all times in the data arc, just as in a batch solution, to achieve an equivalent result.

The conventional Kalman filter is formulated in discrete time recursion relations. Suppose the filter has produced a state estimate $\hat{x}_i$ at time t_i (using data up through time t_i) and that the estimated covariance matrix for $\hat{x}_i$ is $\hat{P}_i$. The state solution $\hat{x}_{i+1}$ at time t_{i+1} is derived in two steps: 1) the *time update*, in which a predicted or a priori solution $\tilde{x}_{i+1}$ and covariance matrix $\tilde{P}_{i+1}$ are generated from their estimated values at time t_i, with no new data yet added; and 2) the *measurement update*, in which the new estimates $\hat{x}_{i+1}$ and $\hat{P}_{i+1}$ are generated from the data at time t_{i+1}, as corrections to the predicted values.

The time update is given by

$$\tilde{x}_{i+1} = \Phi_i \hat{x}_i \tag{10}$$

and

$$\tilde{P}_{i+1} = \Phi_i \hat{P}_i \Phi_i^T \tag{11}$$

where Φ_i is the transition matrix derived from the equation of motion (or other appropriate transition models) relating the state at t_i to the state at t_{i+1}. The measurement update is then

$$\hat{x}_{i+1} = \tilde{x}_{i+1} + G_{i+1} (z_{i+1} - A_{i+1} \tilde{x}_{i+1}) \tag{12}$$

and

$$\hat{P}_{i+1} = \tilde{P}_{i+1} - G_{i+1} A_{i+1} \tilde{P}_{i+1} \tag{13}$$

where z_i is the measurement vector at time t_i, A_i is the matrix of measurement partials with respect to x_i, G_i is the so-called Kalman gain, given by

$$G_i = \tilde{P}_i A_i^T (A_i \tilde{P}_i A_i^T + R_{ni})^{-1} \tag{14}$$

and R_{ni} is the error covariance of the measurement vector z_i. (In some cases, for example when onboard computing is limited, a suboptimal *fixed gain* filter is employed, in which G is predetermined.) Note from Eq. (14) that, like the batch formulation, the conventional current state Kalman filter involves matrix inversion. Various alternative approaches have been devised that employ *pseudo-epoch state* factorized formulations.[23,24] These avoid matrix inversion by factoring P into either upper triangular and diagonal matrices (*U-D* formulation) or its square root matrices (square root information filter formulation). Factorized filters have been incorporated into several of NASA's high-performance orbit determination systems. For more on these techniques see Ref. 41.

Comparison of the batch and sequential formulations reveals that the latter is simply a recursive equivalent of the former. For a given data arc, the final sequential and batch solutions, when mapped to the same epoch, will be identical. As presented here, both are dynamical formulations that depend fundamentally on physical force models to produce the solution trajectory. It is worthwhile at this point to examine the principal errors that arise in the dynamic state solutions.

D. Dynamic Orbit Error

The typical accuracy of instantaneous point positioning is 10–20 m for the PPS user; the major error contributors are GPS orbit and clock error and pseudorange measurement error. Filtering reduces the position error by smoothing measurement error against an orbit model over the fitting arc. Meter-level random errors may readily be reduced to decimeters or below. At the same time, key systematic errors—GPS orbits and clocks, multipath—may be largely uncorrelated with the low orbiter dynamics and, therefore, attenuated in the solution. (Errors that correlate strongly with the orbit, such as once-per-orbit ionospheric effects, may be amplified.) Dynamic filtering can lower real-time orbit error to a few meters for the PPS user and to 20 m or better for the SPS user.

This improvement does not come without a cost. As the filter smooths measurement error, it introduces dynamic model error. Because force models are imperfect, the model trajectory will be imperfect. Model adjustments made during the solution may offer only partial improvement. Any remaining model errors will appear directly in the orbit solution. Gravity and drag model errors are often dominant, and both increase rapidly as the satellite altitude is reduced. Thus, accurate dynamic orbit estimation becomes problematical at low altitudes. For example, the motion of Lageos, a dense, inert sphere at about 6000 km altitude, can be modeled to within a few centimeters over periods of weeks; the motion of TOPEX/Poseidon, a larger vehicle at 1336 km, to about 10 cm over 10 days; the motion of SEASAT, at 800 km, to one or two meters over one day; and the motion of the Shuttle, at 300–400 km, to roughly 10 m over an orbit. At very low altitudes, dynamic filtering may offer little advantage over simple point positioning.

Model errors often reveal themselves in the postfit residuals; that is, they create systematic discrepancies between the actual measurements and theoretical measurements derived from the solution trajectory. Imagine a case in which a force varies randomly from one time step to the next and is, therefore, unpredictable, but can be observed in the postfit residuals. At some level, a number of forces (drag, gravity anomalies) can seem to behave that way. What is needed, then, is a means of extracting information in the residuals to recover unmodeled motion. The Kalman filter provides such a means in the form of process noise modeling.

E. Kalman Filter with Process Noise

To observe unmodeled motion, we model the time-varying satellite force as the sum of a deterministic component (our standard dynamic model) and a stochastic component. The latter is often called a process noise model. Augmenting a Kalman filter with a process noise model is a way of telling the filter that the state transition information in Φ is incomplete—that there is another component that the filter cannot predict, but that it can try to observe in the data and estimate at each time step.

In the context of orbit determination, this means that at each time step, in addition to applying the standard dynamic updates, the filter will examine the discrepancy between the dynamic state estimate and the apparent state as indicated geometrically by the measurements. From that discrepancy, it will estimate a local correction to the dynamic model, valid only over the update interval (t_{i-1},

t_i). When added to the dynamic model, that correction will reduce the disagreement between the observations and the solution trajectory at time t_i. As it proceeds through the data, the filter will generate a sequence of local force model corrections, one at each update time, bringing the solution trajectory into better agreement with the observations. That may be good or bad, depending on the quality of the observations and the accuracy of the models. We must, therefore, take care to hinder the local corrections from chasing after bad measurements.

The process noise model can take many forms, and various constraints may be applied to limit the freedom of each new correction to depart from the dynamic model or from the previous correction. The stochastic correction may be introduced by augmenting the state vector x_i with a vector p_i representing the local force model adjustment to be estimated at time t_i. For this discussion, we let $p_i = (p_{i1}, p_{i2}, p_{i3})^T$ denote a three-dimensional force that is constant over the interval (t_{i-1}, t_i) and zero elsewhere. This force will be estimated to minimize the discrepancy between the dynamic solution update and the observations at time t_i. The augmented state vector X is given by the following equation

$$X = \begin{bmatrix} x \\ p \end{bmatrix} \quad (15)$$

An effective realization of the process noise sequential filter used extensively by NASA in orbit estimation is given below.[23,26]

The time update requires an important modification. We have

$$\tilde{X}_{i+1} = \Phi_i \hat{X}_i \quad (16)$$

and

$$\tilde{P}_{i+1} = \Phi_i \hat{P}_i \Phi_i^T + B\, Q_i B^T \quad (17)$$

where now

$$\Phi_i = \begin{bmatrix} \Phi_x(i+1, i) & \Phi_{xp}(i+1, i) \\ 0 & M_i \end{bmatrix} \quad (18)$$

$$B = \begin{bmatrix} 0 \\ I_p \end{bmatrix} \quad (19)$$

Φ_x is the dynamic transition matrix of Eq. (10); $\Phi_{xp}(i+1, i)$ is the transition matrix relating $\hat{x}_{i+1}$ to the process noise parameters p_i; M_i is a 3×3 diagonal matrix with the jth element

$$m_j = \exp[-(t_{i+1} - t_i)/\tau_j] \quad (20)$$

Q_i is a diagonal covariance matrix associated with a white noise process, with the jth element

$$q_j = (1 - m_j^2)\, \sigma_j^2 \quad (21)$$

and I_p is a 3×3 identity matrix. The measurement update equations are identical to Eqs. (12–14), except that now we use the augmented state vector X and its associated covariance matrix P.

This is a first-order Gauss-Markov process noise model. (For some other possibilities, see Ref. 27.) Note that M_i is the transition matrix for the process noise parameters, and that the transition is in the form of a decaying exponential correlation. The time constant τ_i in Eq. (20) can be chosen to reflect the correlation in the dynamic model error (and thus in the desired correction) over one update interval. If τ_i is much smaller than the update interval, then m_i is small; the model error, therefore, is regarded as uncorrelated from batch to batch, and this becomes a white noise error model. There is one other selectable parameter, the steady state variance σ_i^2. Through Eq. (21), σ_i^2 scales the batch-to-batch variance q_i, which further constrains the correction. In the case of a white noise model, this constrains each independent correction with respect to the dynamic model, with no dependence on the previous correction. If $\sigma_i = 0$, the local force correction is constrained to zero and the conventional dynamic solution is obtained. In summary, the constraint is determined by τ_i (through m_i) and σ_i^2 as they combine through Eq. (21) to form the weighting matrix elements q_i.

Stochastic force models introduce an additional complication for non-real-time applications in which an optimal solution over an entire data arc is desired. It is no longer sufficient simply to map the final state solution to other times by means of the final dynamical models. The local force corrections have been determined with data only up to the times they occur and, thus, have not benefitted from later measurements. To complete the estimates of the local forces it is necessary to filter the data in the reverse direction as well, a process called smoothing, before mapping to all time points. The combined estimator is known as a filter/smoother.

With conventional (sparse) tracking data we must be careful when employing process noise model corrections. The data acquired at any one time are often weak (or nonexistent), and insufficient by themselves to determine position. A relaxed constraint on the process noise estimate may result in a large and erroneous adjustment to the state, or may cause the solution to fail. Care must be taken to constrain the corrections within the observability limits of the data. This has traditionally meant relatively long correlation times and tight sigmas.

III. Orbit Estimation with GPS

We are now in a position to examine the powerful advantages GPS brings to estimating satellite orbits. First we look at a simple geometric technique that can improve point-positioning accuracy without dynamic filtering by combining the continuous carrier phase and pseudorange observables.

A. Carrier-Pseudorange Bias Estimation

When pseudorange and continuous carrier phase are brought together, point position accuracy can be improved by exploiting the absolute pseudorange information to estimate the bias in carrier phase. The bias is estimated simply by averaging the difference between phase and pseudorange for as long as the phase remains continuous (and the bias remains constant). This converts biased phase to a precise pseudorange with a small residual bias, preserving the detailed information on range change in carrier phase. The concept is illustrated in Figs.

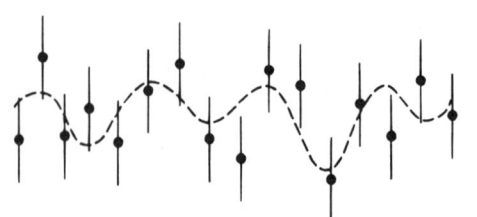

a) Successive pseudorange measurements

b) Successive measurements of continuous but biased carrier phase

c) Absolute phase derived by adjusting mean of b to mean of a

Fig. 1 In carrier-pseudorange bias estimation, the phase bias is estimated by averaging the difference between continuous carrier phase and independent pseudoranges. The result retains the precision and time resolution of carrier phase, while reducing the bias to a fraction of the pseudorange error.

1a–1c. A sequence of N independent pseudorange measurements $\hat{x}_k$ is shown in Fig. 1a. The true time-varying pseudorange is represented by the dashed line. If x_k is the true pseudorange at time t_k, we can write the following:

$$\hat{x}_k = x_k + n_k \tag{22}$$

where, for simplicity, we assume n_k is a white noise process with standard deviation σ_n. Figure 1b shows the record of pseudorange change obtained by tracking carrier phase over the same arc. This can be regarded as a series of pseudorange measurements, $\bar{x}_k$, having a much smaller random error and a common bias. Thus, we can write

$$\bar{x}_k = x_k + b + e_k \tag{23}$$

where b is the bias and e_k is a white noise process with standard deviation σ_e. We estimate the bias b by averaging the difference between the $\bar{x}_k$ and $\hat{x}_k$

$$\hat{b} = \frac{1}{N} \sum_{k=1}^{N} \bar{x}_k - \hat{x}_k \tag{24}$$

or

$$\hat{b} = b + \frac{1}{N} \sum_{k=1}^{N} e_k - n_k \tag{25}$$

Because σ_e is typically 100 times smaller than σ_n, the approximate component error on the bias estimate is the following:

$$\sigma_b \cong \frac{\sigma_n}{\sqrt{N}} \qquad (26)$$

Thus, meter-level random noise on 1-s pseudorange data can give a decimeter-level bias estimate within 2 min. Subtracting Eq. (25) from Eq. (23) eliminates the bias in the phase measurement to give a precise record of absolute pseudorange. As shown in Fig. 1c, the corrected phase measurements sit close to, and have nearly the exact shape of, the true pseudorange sequence. The corrected phase measurements will have an approximate error

$$\sigma_x = (\sigma_b^2 + \sigma_e^2)^{1/2} \qquad (27)$$

where σ_b represents the residual bias common to all phase points and σ_e is the point-to-point random error. A sequence of position solutions derived from the corrected phases will have the precision of a pure carrier solution, with a bias that is a fraction of the typical point position error.

This technique, which was first proposed in Ref. 28, can be readily generalized to provide real-time recursive estimation of the position of an unpredictably moving vehicle. Consider a receiver that produces an instantaneous point position solution $\hat{x}_k$ at time t_k, and a position change solution $\Delta\hat{x}_k$ obtained by continuously tracking carrier phase from t_{k-1} to t_k. An estimate $\hat{X}_{n+1}$ of the position at time t_{n+1} is given by the following

$$\hat{X}_{n+1} = \frac{n}{n+1}(\hat{X}_n + \Delta\hat{x}_{n+1}) + \frac{1}{n+1}\hat{x}_{n+1} \qquad (28)$$

Note that this is a variation on the recursive formula for a simple average

$$\bar{P}_{n+1} = \frac{n}{n+1}\bar{P}_n + \frac{1}{n-1}P_{n+1} \qquad (29)$$

The position change information $\Delta\hat{x}_{n+1}$ maps the current position estimate $\hat{X}_n$ forward to the next time point for averaging with the point position $\hat{x}_{n+1}$ computed at that time. Carrier phase, in effect, inertially aids the sequential averaging of point position solutions to refine the phase bias estimate. The procedure can be tuned by weighting each $\hat{x}_k$ by its inverse covariance.

A principal virtue of this technique is its extreme simplicity. A filter to track unpredictable motion (or the relative positions of multiple vehicles) can be realized in a relatively few lines of code. It is, however, suboptimal. Correlation between the $\Delta\hat{x}_k$ is not properly accounted for, and it does not fully exploit the information in the carrier phase. More refined strategies are presented in Ref. 29. Another drawback is its exclusion of external information about platform dynamics. The solution becomes vulnerable to outages that might easily be bridged with simple dynamic models. These weaknesses are remedied in a more robust technique that employs the Kalman filter formalism.

B. Kinematic Orbit Determination

When a Kalman filter is applied to GPS data from a low orbiter, the full advantage of continuous three-dimensional coverage may not be realized without an aggressive use of process noise corrections. If we assume a full GPS constellation, a flight receiver having six or eight parallel channels, and a relatively wide field of view, strong instantaneous observing geometry is assured. Inclusion of continuous carrier phase data vastly increases the potential precision of the estimates. It then becomes possible to relax or eliminate constraints on the process noise force corrections and track the true motion of the vehicle with great precision.

The concept is illustrated in Fig. 2. The dashed curve represents the irregular path of a low orbiter subject to varying forces. With GPS data collected by the orbiter, we can execute a traditional dynamic orbit solution to produce the smooth orbit estimate shown by the solid line. This leaves a set of (possibly large) postfit residuals. Because GPS provides continuous three-dimensional coverage, the postfit residuals at each time point suffice to reconstruct the observed satellite position (its departure from the dynamic solution) by purely geometric means. The observed trajectory can then be recovered by adding the geometrically determined correction to the dynamic solution at each time point. Force model error, reflected in the initial postfit residuals, is thereby eliminated.

This can be thought of as two distinct steps. First, a conventional dynamic solution produces a reference trajectory and postfit residuals; the residual path is then constructed geometrically, point by point, and added to the dynamic solution. In practice, this can be done in one estimation step in a Kalman filter with process noise. The estimated process noise parameters p in Eq. (15) can provide the geometric corrections to the dynamic solution. In ordinary tracking applications, those parameters would be tightly constrained and geometric information only weakly expressed. But the full observability offered by GPS allows all constraint to be removed. The correlation time τ_i can be set to zero (white noise model) and the steady-state variance σ_i^2 to a large value. The filter will then estimate a three-dimensional force correction for each interval (and a corresponding change in the current state) to account for the geometric discrepancy between the measurements and the dynamic solution. This is called nondynamic or *kinematic* orbit determination, although both terms are somewhat misleading because the technique builds on an underlying dynamic formulation.

As we see in more detail later, the kinematic solution can be carried out with pseudorange data alone, with carrier phase data alone, or with the two in

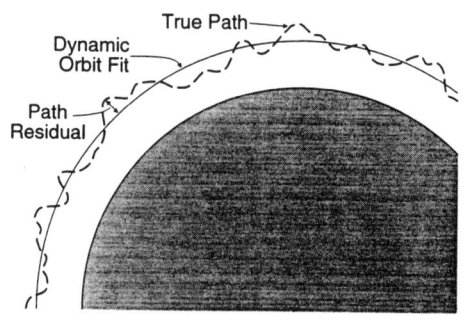

Fig. 2 The kinematic orbit determination technique effectively reconstructs the observed trajectory from the residuals of a dynamic orbit solution.

combination. Observe, however, that as the dynamic constraint is relaxed to allow the geometric correction, the effect of measurement error increases. Instead of being smoothed against the dynamic model, single-point measurement error is fully expressed in the geometric correction. Thus, when pseudorange alone is used, the kinematic solution becomes a series of point positions with full pseudorange noise. For precise applications, continuous carrier phase is essential.

C. Reduced Dynamic Orbit Determination

Because the kinematic correction is geometric, it is vulnerable to weak geometry. Momentary data outages or large position dilutions of precision (PDOPs) will cause the error to grow or the solution to fail. The kinematic solution, moreover, makes little use of dynamic information—it is an empirical result constructed from the measurements. Often, however, dynamic information is at hand, which, if properly treated, can improve the result. When geometry weakens or fails, dynamic information can then carry the solution with little loss of accuracy.

We can achieve a balance of dynamic and geometric information in the orbit solution by imposing a judicious constraint on the process noise parameters. In an optimal solution (under the assumption of a Gauss-Markov process noise model), the time constant τ_i will reflect the actual correlation time of dynamic model errors, and the steady-state variance σ_i^2 the actual error in the dynamic model. The geometric corrections will not be free to follow the measurements wherever they lead, but will be bound by the constraint to the dynamic model. Relative weight will, in fact, shift back and forth between dynamic and geometric information as observing strength varies. When geometry is weak, the process noise constraint will hold the correction close to the dynamic solution; if there are no observations at all, no correction can be computed and the dynamic solution is produced. This optimized technique is known as *reduced dynamic orbit determination*.[42]

Another interpretation is given in Fig. 3, which illustrates the relative significance of random and systematic error in the solution trajectory. In the dynamic solution, random error is minimized (because the fewest parameters are adjusted), while dynamic error is fully expressed. This is reversed in the kinematic solution as many parameters are adjusted, amplifying the effect of data noise while absorbing dynamic error. The reduced dynamic solution seeks the optimal balance to minimize overall error.

This raises the question of how we choose the process noise weighting. Often there is some prior knowledge of the quality of the force models in use and the consequent position error expected. Computer simulations or covariance analysis can then suggest a reasonable a priori weighting. When real data become available, a variety of strategies for tuning the reduced dynamic constraints become possible. One approach is to observe the magnitude of the process noise corrections; if they approach the constraints, the constraints should be relaxed; if they fall well short, the constraints can be tightened. Another technique is to compare orbit solutions on short overlapping segments and tune the constraints to minimize the discrepancy.

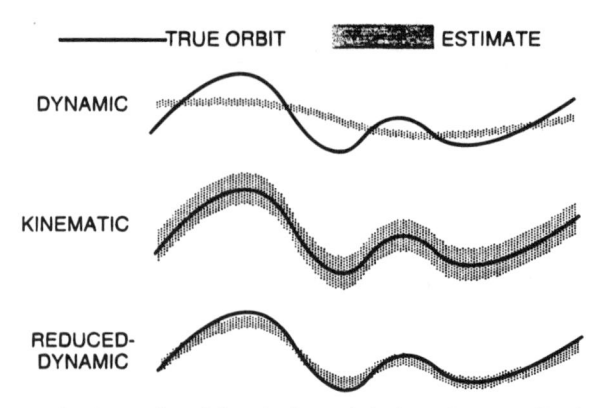

Fig. 3 The purely dynamic orbit solution minimizes the contribution of random error, while dynamic model error is fully expressed; this is reversed in the kinematic solution. The reduced dynamic solution yields an intermediate level of each error and can minimize overall error.

The reduced dynamic technique is one realization of the concept depicted in Fig. 2; others are possible. For example, we might compute position rather than force corrections. An approach along those lines proposed in Ref. 30 has certain advantages for gravity recovery. Force corrections, however, directly augment the dynamic model and have the virtue that, although discontinuous (piecewise constant) themselves, they yield a continuous trajectory when integrated.

D. Orbit Improvement by Physical Model Adjustment

The reduced dynamic solution employs local corrections to dynamic models. Often it is more efficient to adjust physical model parameters; fewer adjustments may be needed and data strength preserved. Adjustment of drag and radiation pressure terms, for example, is common. Particularly attractive with GPS tracking data is *gravity tuning,* or adjustment of gravity field coefficients. The geopotential is commonly represented as a spherical harmonic expansion containing anywhere from a few terms to a few thousand terms, depending on the fidelity required. Each gravity harmonic is a global function representing a permanent component of the geopotential.

Many geopotential models are derived from satellite tracking data, which are often sparse in some regions. GPS, however, leaves no coverage gaps. Because a polar orbiter overflies the entire globe, GPS tracking of such a satellite can enable improvement of the full global model. That will in turn reduce the dynamic model error and permit tighter constraints on the process noise models in subsequent orbit solutions. Gravity tuning has elements in common with reduced dynamic orbit estimation. Both techniques adjust a large number of force parameters to bring the solution trajectory into closer agreement with the data. Where gravity is the dominant model error, gravity tuning is a desirable first step, because it yields a permanent model improvement.

IV. Direct Orbit Determination with GPS

Sophisticated estimation strategies may be of little value in direct GPS-based orbit determination, where only the onboard observables and broadcast data are used in the solution. Although measurement noise can be reduced to centimeters by filtering, final user orbit error will be dominated by GPS ephemeris and clock error (with possibly large contributions from the ionosphere and selective availability) at a level of meters to tens of meters. Evaluating the expected accuracy of the filtered orbit solution is not always straightforward.

Consider the batch least squares dynamic solution of Eqs. (8) and (9). It is easily shown that the error covariance P_x on the estimate $\hat{x}$ is given by the following

$$P_x = (A^T R_n^{-1} A)^{-1} \qquad (30)$$

This is the formal error attributable to the random measurement noise vector n (Eq. 5), sometimes called the *commission* error. It does not take into account other errors present in the solution, such as those attributable to GPS orbit and clock errors, sometimes called *omission* errors. To examine the effect of such errors, we can include the relevant parameters and their relation to the observations explicitly in the regression Eq. (5) by writing the following equation

$$\delta z = Ax + By + n \qquad (31)$$

where y is the vector of omission error parameters and B is a matrix of partial derivatives of the observations δz with respect to y. When the solution given by Eq. (8) is applied to Eq. (31) we have the following

$$\hat{x} = x + (A^T R_n^{-1} A)^{-1} A^T R_n^{-1} B y + \tilde{n} \qquad (32)$$

where $\tilde{n}$ is the transformed random measurement noise. The long expression multiplying y in Eq. (32) describes the response of the estimate $\hat{x}$ to the error parameters y, and is called the sensitivity matrix S

$$S = (A^T R_n^{-1} A)^{-1} A^T R_n^{-1} B \qquad (33)$$

The total error covariance P_{tot} of the estimated vector $\hat{x}$ is given by the following

$$P_{tot} = P_x + S P_{om} S^T \qquad (34)$$

where P_{om}, the a priori covariance matrix for the omission errors, must be derived through careful analysis of those errors.

Because many omission errors are physically unrelated and can be regarded as uncorrelated, P_{om} can often be (and almost invariably is) set up as a diagonal matrix. The errors on the elements of a dynamic satellite state solution, however, are strongly correlated. (Note, for example, that there is a direct relationship between satellite altitude and in-track velocity; in a dynamic solution, an error in one will appear as a compensating error in the other.) A diagonal covariance matrix is therefore inadequate to assess the effect of GPS ephemeris error on a dynamic user orbit solution and, in fact, can be shown to give a highly pessimistic estimate of the error that actually arises. To evaluate the effect of GPS orbit error on a dynamic user solution, a full covariance matrix is needed for the GPS state parameters. One way to obtain such a matrix is to simulate the GPS orbit determination process as it is carried out with ground data to produce the GPS orbits available to the user.

One such study[31] showed that errors in the GPS orbits were attenuated by roughly a factor of two in the dynamic solution for an orbiter at 1300 km. That is, GPS orbit errors of 1 m resulted in errors of about 0.5 m in the user orbit. (When a diagonal GPS covariance matrix was used, this error was overestimated by a factor of about 20.) Because the satellite was at an altitude of 1300 km, model error was small and the full benefit of dynamic filtering could be gained. With a typical GPS broadcast ephemeris error of 5 m, we could expect to achieve a real-time orbit accuracy of 2–3 m for such a satellite, limited by the GPS orbit error.

At lower altitudes, dynamic model error grows. At 500 or 600 km dynamic error may equal GPS ephemeris error. Below 500 km, dynamic error will dominate, and the optimal filter will deweight dynamics. For the Space Shuttle at 300 km, the optimal solution will be nearly kinematic, and position error, dominated by GPS orbit error, will be little better than with simple point positioning. Because GPS orbit errors change slowly, the direct kinematic error will be highly correlated from one second to the next.

The most accurate direct orbit solutions are therefore obtained by dynamic filtering for satellites above about 800 km (and below 3000 km), with the accuracy limited by GPS orbit error. The best GPS orbits produced today (available typically several days after the fact) are accurate to better than 1 m. In principle, such accuracies can be achieved nearly in real time. Moreover, accuracies of 1–2 m can be reached for GPS orbits and clocks predicted several hours into the future[25] and, thus, available for true real time use. If dynamic filtering reduces the resulting user error by a factor of two, real-time dynamic tracking could be made accurate to about 1 m above 800 km. At the lowest altitudes, where the kinematic solution is optimum, accuracy could reach a few meters. For further improvement at all altitudes we must reduce GPS orbit error.

V. Precise Orbit Determination with Global Positioning Systems

A few classes of mission require orbit accuracies ranging from 1 m (land altimetry, precise imaging) down to a few centimeters (ocean altimetry, gravity field modeling). For that we must turn to the techniques of differential GPS. As it has been developed for precise orbit determination, differential GPS is intended for non-real-time applications and differs considerably from the real-time differential techniques used for regional navigation.

A. Global Differential Tracking

The fundamental concept is illustrated in Fig. 4. In addition to the flight receiver, a network of reference receivers around the world continuously tracks all GPS satellites in view. The flight receiver and at least one ground receiver must share common visibility of several GPS satellites at all times. Only about six well-distributed ground sites are needed to ensure this, although in recent experiments with TOPEX/Poseidon a dozen or more have been used. Several ground receivers may be at *fiducial* sites—sites with accurately known absolute

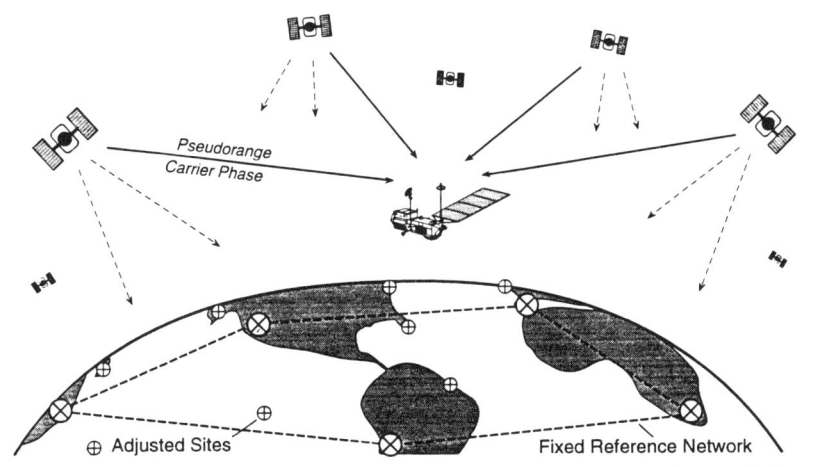

Fig. 4 In precise orbit determination with differential GPS, user and ground observations of GPS are combined to determine user, GPS, and some ground positions with respect to a subset of ground reference or "fiducial" sites.

positions that will be held fixed during the solution. The best current fiducial sites [those of the International Terrestrial Reference Frame (ITRF) maintained by the International Earth Rotation Service (IERS) in Paris] are known relatively to 1–2 cm, and absolutely (with respect to the geocenter) to better than 3 cm.

Pseudorange and carrier phase data from the flight and ground receivers are processed together to produce a single grand solution. The solution strategy can vary greatly in detail, but typically includes estimation of all GPS satellite orbits, the user orbit, transmitter and receiver clock offsets, all carrier phase biases, nonfiducial ground site positions, atmospheric propagation delays at all ground sites, and such satellite force parameters as atmospheric drag and solar radiation pressure. Data arc lengths may range from a few hours to many days. Because only the fiducial sites are held fixed, they establish the reference frame in which all other positions are determined.

One variation permits *all* ground sites to be adjusted within a moderate constraint, typically 10–1000 m on each site. This severs the tie to a predetermined reference frame and allows the entire solution to rotate within the limits of the overall constraint. The solution is then mapped into a chosen reference frame (such as ITRF) through a seven-parameter transformation (translation, rotation, and scale), which minimizes the three-dimensional rms difference between all ground site solutions and their values in the chosen frame. This removes dependence on a particular subset of sites to define the reference frame and reduces reference station error in the total error budget. A less powerful variation processes the ground and user data separately. The ground data first determine accurate GPS orbits and clock offsets, which are then applied in a direct user solution. This is less powerful than a true simultaneous solution, but may offer greater flexibility and convenience.

Although global differential tracking constitutes a major logistical departure from direct tracking, the basic filter equations needed to carry it out, Eqs. (16–21), remain unchanged. What changes is the definition of the estimated state vector X. To the user state and other adjusted parameters we now append state elements for all GPS satellites, clock offsets for all transmitters and receivers, ground site positions, atmospheric delays, and so on. The matrices of measurement partials and a priori covariance are correspondingly augmented, and the solution becomes more computationally demanding. It is worth examining in more detail how some of the key parameters are treated.

B. Fine Points of the Global Solution

When carrier phase data are used in a grand solution, either alone or together with pseudorange, the effective data noise (random measurement error) is typically below 1 cm. This can be seen in the postfit residuals of global geodetic solutions, which for the combined dual-frequency phase observable are typically 3–6 mm. As revealed in numerous covariance studies [Eq. (34)], random measurement error will contribute on the order of 2 cm to the user position error—somewhat higher for purely kinematic solutions and lower for purely dynamic solutions. In the grand solution, the major systematic model errors that plague the direct solution (GPS orbits and clocks) are reduced. Note, however, that if GPS satellite dynamics and clocks are poorly modeled, the GPS orbit and clock estimates will degrade and systematic errors will still arise in the user state solution. Fortunately, the high-altitude GPS satellite dynamics can be well modeled over 24 h, and standard dynamic GPS solutions generally suffice. For longer arcs, a weak stochastic adjustment of the GPS solar pressure parameter may be advantageous.

For clock solutions, we have several options. If high-quality atomic clocks are used in all receivers and transmitters, simple quadratic models might suffice over many hours. Because real clock behavior can be unpredictable, common practice is to allow for the worst by solving for all clock offsets independently at each time-step under a loose constraint. This is equivalent to modeling clock behavior as a white noise process with large variance, in analogy with our treatment of the process noise force parameters in the kinematic orbit solution. It is also similar to the popular practice of double differencing to eliminate clock parameters; however, when global data sets are used, as they must be for precise orbit determination, the white noise clock model is more powerful, as it retains more data.[32] Just as purely kinematic orbit determination fails to exploit known dynamics of the satellite, white noise clock models fail to exploit known (and perhaps smooth) clock behavior and, thus, must be regarded as a conservative strategy.

Computer simulations, covariance studies, and results with TOPEX/Poseidon have shown that the grand solution strategy can reduce user satellite position errors caused by GPS orbit and clock errors to a few centimeters. What, then, becomes the dominant error in the user solution? One candidate is the error in modeling atmospheric propagation delay at the ground sites, or, rather, the variable wet component of that delay. When standard seasonal models (supported by surface weather data) are used to calibrate the atmospheric delay, the error is typically 3–5 cm at zenith, which may translate into 2–10 cm of user state error, depending on the solution technique. This can be reduced by periodically solving for a zenith delay at each site. The most effective strategy yet developed is to model the zenith delay as a stochastic process (a random walk, for example) and adjust it at each time-step under a constraint derived from the observed power spectrum of atmospheric delay variation. Typical zenith delay accuracy with this technique is about 1 cm.

Finally, we note that each carrier phase observable contains an arbitrary bias corresponding to integer cycle ambiguities at each frequency. Those biases must be estimated (or eliminated by time differencing) when the phase observable is used. In precise ground-based geodesy, an effort is often made to determine the exact integer cycle ambiguities in the differential observables and then fix the biases at those values. Resolving ambiguities between an orbiter and ground sites is demanding and, when many hours of data are used, can be shown to contribute little to solution strength, because at that point data noise is not a dominant error. The differential strategies described here attempt no cycle ambiguity resolution, and instead treat each bias as a continuous variable.

C. Precise Orbit Determination Performance Analysis

It may now be evident that the general strategy for achieving high accuracy with GPS is to exploit the great strength of GPS data to observe and correct systematic errors that threaten to dominate. Just how the data will stand up to this demand depends on many details of system configuration and solution strategy. To illustrate those dependencies, we present the results of covariance studies for several real or proposed missions. All studies include both commission and omission errors in an attempt to arrive at realistic final error estimates.

The first example is taken from error studies conducted for TOPEX/Poseidon years in advance of its launch in August of 1992. TOPEX/Poseidon is a U.S.-French ocean altimetry mission flying at an altitude of 1336 km, where dynamic model errors are now well below 10 cm. The GPS configuration for TOPEX/Poseidon includes a six-channel (dual-frequency) flight receiver with a hemispherical field of view, and a six-site ground network. The assumptions, error model, and estimation strategy are summarized in Table 1. Note that a reference frame error of 5 cm per component for each of three fiducial sites was assumed, far greater than that error today.

Figure 5 shows the predicted rms altitude error for three solution strategies—dynamic, kinematic, and optimized reduced dynamic—as a function of the gravity model error. Because the kinematic solution eliminates dynamic error, it is independent of the gravity model. Its total error is divided almost equally between data noise and reference site error. The dynamic solution error depends strongly on gravity error and becomes limited by data noise and reference frame errors only when gravity and other dynamic errors are small. The reduced dynamic strategy surpasses both kinematic and dynamic—the latter even when the gravity error is zero, because other dynamic errors are still reduced.

Also shown in Fig. 5 are three circles representing actual results from TOPEX/Poseidon obtained during the first year of the mission. The circles give the rms altitude agreement over 30 days between purely dynamic solutions made with ground-based laser ranging and Doppler data, and the GPS reduced dynamic solutions. An rms agreement of about 6 cm was obtained with the final prelaunch

Table 1 Error model for TOPEX/Poseidon orbit determination analysis

System characteristics	
Orbit (circular):	1334 km, 66-deg inclination
Number of ground sites:	6 (including 3 fiducial sites)
Number of GPS satellites:	18
Flight antenna field of view:	Hemispherical
Flight receiver tracking capacity:	6 channels ($L1$ and $L2$)
Data types:	$L1$ and $L2$ pseudorange
	$L1$ and $L2$ carrier phase
Data interval:	5 min
Smoothed data noise:	5-cm pseudorange
	1-cm carrier phase
Adjusted parameters and a priori errors	
TOPEX/Poseidon epoch state:	1 km; 1 m/s, each component
GPS satellite states:	2 m; 0.2 mm/s, each component
Carrier phase biases:	10 km
GPS and receiver clock biases:	3 ms (modeled as white noise)
Non-fiducial ground locations:	20 cm each component
Fixed errors evaluated	
Fiducial site positions:	5 cm each component
GM of Earth uncertainty:	1 part in 10^8
Earth gravity error model:	0–100% GEM10–GEML2 (20 × 20)
Zenith atmospheric delay error:	1 cm (modeled as random walk)
Atmospheric drag error:	10% of total
Solar radiation pressure error:	10% of total

gravity model, known as JGM-1, which has a quality roughly in the center of the range shown. This improved to about 3.5 cm with the JGM-2 model, which had been tuned with laser and Doppler data by the Goddard Space Flight Center several months after launch.[33] The agreement improved further, to about 2.5 cm, when the JGM-2 model was tuned with GPS data by investigators at the University of Texas.[34] At this point, the rms altitude error resulting from gravity mismodeling is believed to be no more than 2 cm.

Past ocean altimetry missions have been plagued by what are known as geographically correlated orbit errors, that is, orbit solutions that are consistently biased in different geographic regions. Such errors can confound the construction of global circulation models from the altimetry data. Geographically correlated orbit errors are often a consequence of geographic biases in the gravity model, although other factors may also play a role. Studies reported in Ref. 35 showed that kinematic and reduced dynamic orbits, by reducing dependence on force models in general, can virtually eliminate any geographic correlation in orbit errors resulting from the gravity model. That result was confirmed with TOPEX/Poseidon. Laser/Doppler dynamic orbit solutions with JGM-1 and JGM-2 showed consistent geographic discrepancies from the GPS reduced dynamic solutions.

ORBIT DETERMINATION

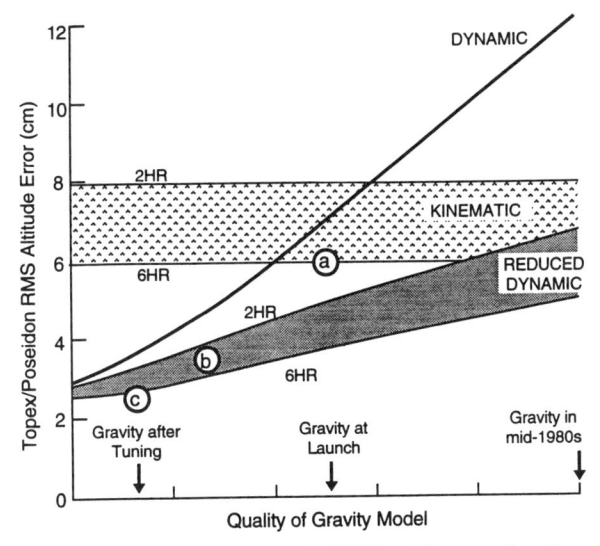

Fig. 5 Predicted rms altitude error for TOPEX/Poseidon as a function of the quality of the gravity model, for three different solution strategies (see Table 1). Circles show actual rms altitude agreement between GPS reduced dynamic and laser/Doppler dynamic solutions made with the prelaunch gravity model ⓐ, a laser/Doppler-tuned gravity model ⓑ, and a GPS-tuned gravity model ⓒ.

In later dynamic solutions made with the GPS-tuned gravity model, geographic discrepancies had all but vanished.

A second example is taken from the Earth Observing System, a suite of scientific Earth probes planned to fly at about 700 km beginning in the late 1990s. Because dynamic errors may grow large at that altitude, a purely kinematic analysis is presented. This time the reference site error is reduced to 3 cm per component. Other assumptions that differ from the TOPEX/Poseidon analysis are shown in Table 2. Figure 6 shows the resulting predicted altitude error as a function of data arc length for several different GPS data combinations. The data type called "carrier-quality pseudorange" is a fictitious pseudorange measurement having the precision of carrier phase, and serves to establish a performance bound.

Table 2 Changes from Table 1 for Earth Observing System kinematic orbit determination analysis

Orbit (circular):	705 km, 98-deg inclination
Number of GPS satellites:	24
Flight receiver tracking capacity:	All in view (within hemisphere)
Zenith atmospheric delay error:	Adjusted as random walk
Fiducial location error:	3 cm each component
Earth gravity error model:	100% GEM10–GEML2 (20 × 20)

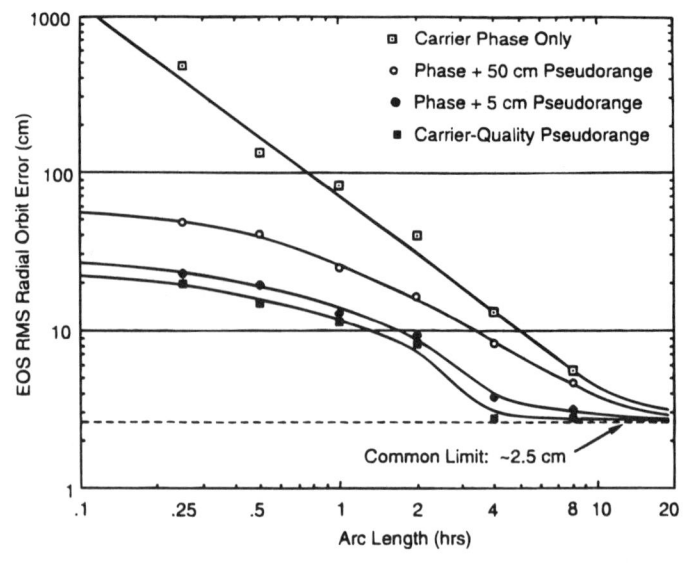

Fig. 6 Predicted rms altitude error for the Earth Observing System as a function of data arc length, for purely kinematic orbit determination and four data combinations. Key assumptions are shown in Table 2.

Figure 6 indicates that few-centimeter accuracy is possible for dynamically complex platforms, and that even the biased carrier phase observable used by itself can approach the performance of the strongest possible data type. It may seem surprising that the kinematic solution can succeed with carrier phase alone, considering that the grand solution must estimate phase and clock biases, which are nicely constrained by pseudorange data. But the dynamic core of the kinematic solution allows the biases to be reliably estimated, just as they are in an integrated Doppler dynamic solution. This illustrates a fundamental difference between the process noise Kalman filter formulation and the simple carrier-pseudorange bias estimation of Sec. III-A. The latter depends entirely on pseudorange to estimate the phase bias, whereas the former can recover the bias dynamically (while correcting the model kinematically) when range information is absent.

A third study explores the limits of kinematic performance with a stringent tracking challenge: the Space Shuttle at 300 km. For a given phase noise, kinematic tracking accuracy is limited largely by observing geometry, which we strengthen by assuming a full sky field of view (each Shuttle is equipped with GPS antennas top and bottom to permit this), a flight receiver able to track all satellites in view (typically 13–15), and 11 ground sites, with reference site error of 1.5 cm per component. Other assumptions are given in Table 3. As shown in Fig. 7, the limiting error in all components now approaches 1 cm, although in reality, dynamic errors in the GPS satellite orbit solutions might degrade this somewhat. This opens up new possibilities for near-Earth ocean altimetry and other precise Earth observations on platforms of opportunity, and for short-duration testing of precise instruments on the Space Shuttle.

Table 3 Changes from Table 1 for Shuttle kinematic orbit determination analysis

Orbit (circular):	300 km, 28-deg inclination
Number of GPS satellites:	24
Number of ground sites:	11 (including 3 fiducial sites)
Flight antenna field of view:	Full sky
Flight receiver tracking capacity:	All in view
Smoothed data noise:	5-cm pseudorange
	5-mm carrier phase
Zenith atmospheric delay error:	Adjusted as random walk
Fiducial location error:	1.5 cm each component
Earth gravity error model:	50% GEM10–GEML2 (20 × 20)

D. Single-Frequency Precise Orbit Determination

The carrier-only kinematic solution is more than a curiosity. It enables accurate orbit determination with simple codeless receivers, bypassing the effects of anti-spoofing, and can be used to achieve fair orbit accuracy with single-frequency data as well.[44] In the examples thus far we have assumed dual-frequency elimination of ionospheric delay; but the ionosphere can also be removed by averaging the $L1$ phase and pseudorange observables. Consider these simplified expressions for the phase delay and group delay (pseudorange) observables

$$\tau_\phi \cong \tau - \frac{k \cdot \text{TEC}}{f^2} + \text{bias} + \epsilon_\phi \tag{35}$$

and

$$\tau_{\text{grp}} \equiv \tau + \frac{k \cdot \text{TEC}}{f^2} + \epsilon_{\text{grp}} \tag{36}$$

where TEC is the total electron content along the raypath, f is the observing frequency, k is a constant, ϵ is the random measurement error; and τ is the common delay caused by geometry and factors other than the ionosphere. Note

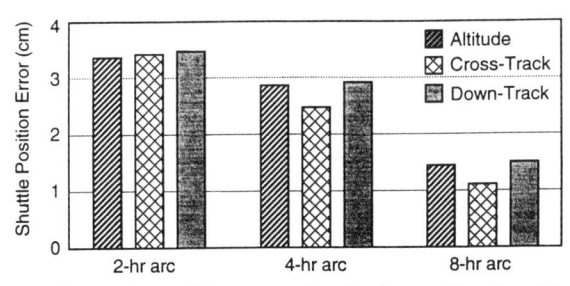

Fig. 7 Predicted kinematic tracking error for the Space Shuttle with a robust GPS observing system. Key assumptions are shown in Table 3.

that the ionosphere term is identical in both equations but appears with opposite sign. Forming the simple average of Eqs. (35) and (36) we obtain

$$\frac{\tau_\phi + \tau_{grp}}{2} \cong \tau + \text{bias}' + \frac{\epsilon_\phi + \epsilon_{grp}}{2} \qquad (37)$$

The ionosphere term is canceled, and the resulting observable has the form of the biased carrier phase delay [Eq. (35)]. This is sometimes called the GRAPHIC (Group And Phase Ionospheric Correction) observable. Because ϵ_{grp} is much greater than ϵ_ϕ the effective measurement error on Eq. (37) is half that of pseudorange. Note that the conventional dual-frequency correction *increases* raw data noise by a factor of three; thus, if single-frequency phase is 100 times more precise than pseudorange, dual-frequency phase will be only 17 times more precise than GRAPHIC data.

Modern receivers that employ 20 MHz C/A-code processing can recover C/A pseudorange with a precision of better than 50 cm in 1 s. The GRAPHIC observable reduces this by half. Smoothing over 60 s can bring the error below 10 cm. Figure 8 shows the predicted three-dimensional rms position error for the Shuttle at 300 km with three solution strategies: dual-frequency dynamic, GRAPHIC kinematic, and dual-frequency kinematic. Key assumptions are shown in Table 4. Note that drag and gravity errors make the dynamic solution worse than simple point positioning. The kinematic solutions improve orbit accuracy by two to three orders of magnitude, reaching about 2 cm per component with dual-frequency phase. The order-of-magnitude difference between the two kinematic cases is explained by the higher data noise on the GRAPHIC observable.

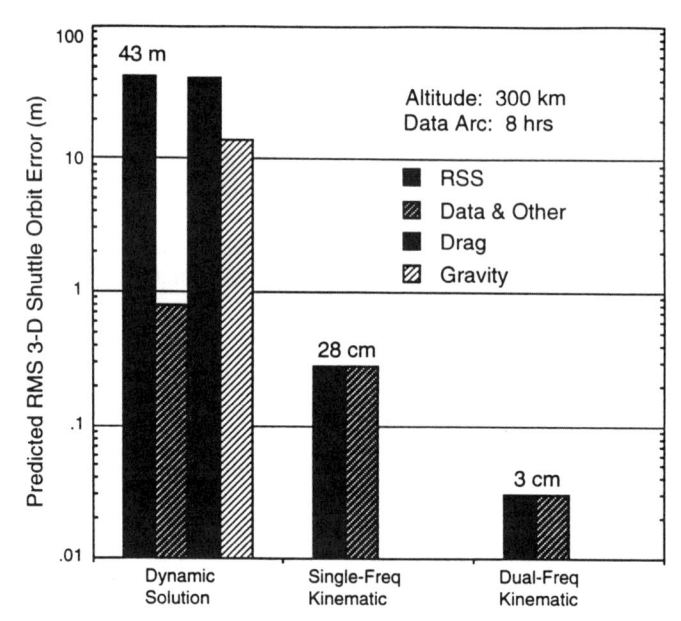

Fig. 8 Predicted three-dimensional error for the Space Shuttle at 300-km altitude, with three different solution strategies.

ORBIT DETERMINATION

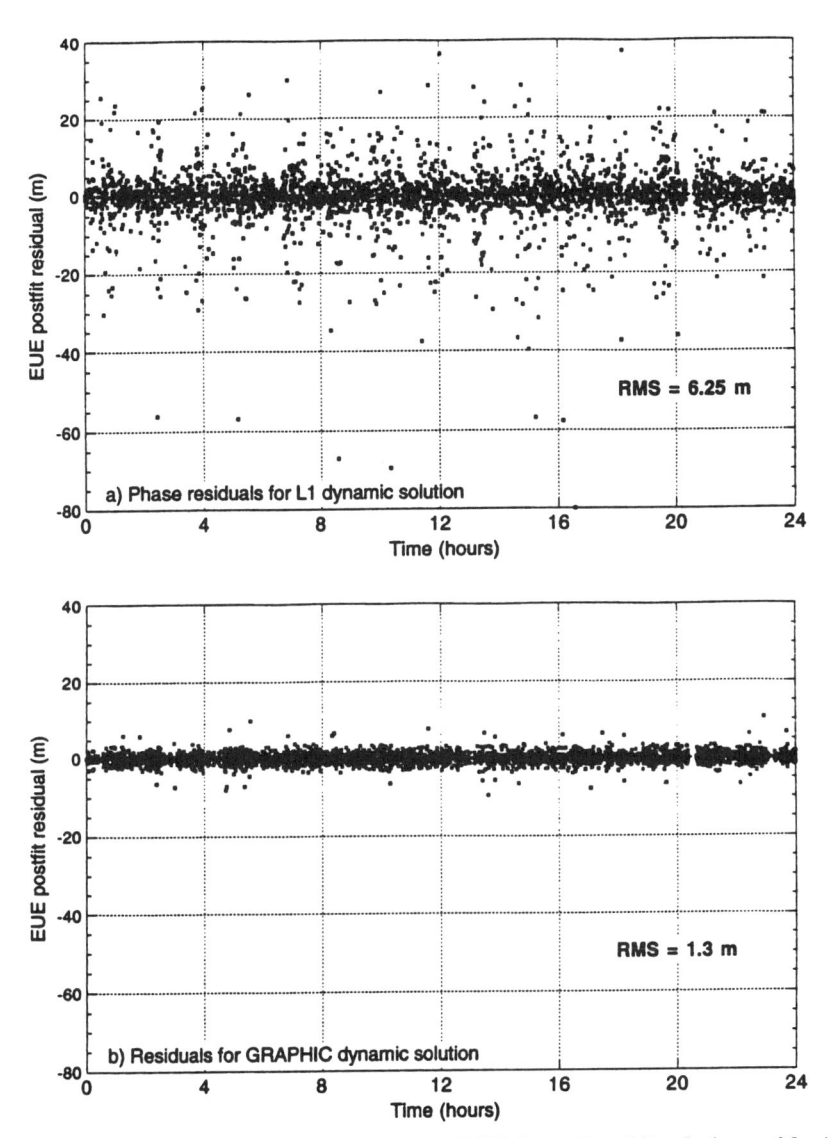

Fig. 9 Postfit residual plots for GPS-based EUE dynamic orbit solutions with a) single-frequency carrier phase data and b) the single-frequency ionosphere calibrated GRAPHIC observable.

Table 4 Key assumptions for Shuttle single-frequency kinematic orbit determination analysis

Orbit (circular):	300 km, 98-deg inclination
Number of GPS satellites:	24
Number of ground sites:	6 (including 3 fiducial sites)
Flight receiver tracking capacity:	All in view (within hemisphere)
Smoothed data noise:	10-cm pseudorange (single-frequency)
	1-cm carrier phase
Zenith atmospheric delay error:	Adjusted as random walk
Fiducial location error:	3-cm each component
Earth gravity error model:	30% GEM10–GEML2 (20 × 20)
Atmospheric drag error:	10% of total

A variant of this technique was first described in Ref. 45. More recently, single-frequency ionosphere calibration was demonstrated on an Earth satellite in Ref. 36. The Extreme Ultraviolet Explorer (EUE), flying at about 500 km, is equipped with a 12-channel $L1$-only receiver and two oppositely directed antennas, providing a full sky field of view. Many GPS tracks acquired by EUE look down through the ionosphere, where the added delay can exceed 50 m. Figure 9 presents typical postfit residual plots for EUE GPS orbit solutions. In Fig. 9b, large ionospheric excursions are entirely absent. Direct comparison of orbit overlaps indicates an rms altitude error of less than 1 m in a differential reduced dynamic EUE solution with the single-frequency GRAPHIC observable.

E. Extension to Higher Altitude Satellites

Above about 3000 km, an orbiter begins to lose coverage from GPS. However, because dynamic model error can be small at high altitudes, the dynamic orbit solution can remain strong. By looking downward to catch the signal spillover from satellites on the other side of the Earth, an orbiter can exploit GPS from well above the GPS satellites themselves, out to geosynchronous altitude and beyond. Alternatively, high satellites can carry GPS-like beacons to be tracked from the ground, with the GPS satellites serving as reference points, a technique known as *inverted* GPS. Figure 10 (from Ref. 30) plots the average number of GPS satellites that can be tracked by a circular orbiter as a function of altitude for both upward- and downward-looking vehicles, where each is assumed to have a hemispherical field of view. Note that above about 2000 km, the down-looking user can track more. The figure also plots the average number of ground sites that can track a beacon on a circular orbiter, assuming a ten-site global network.

Studies of direct orbit determination with down-looking GPS were carried out for NASA's geosynchronous Tracking and Data Relay Satellites.[37,43] With GPS orbit error assumed to be 7 m and clock error 2 m (SA off), a predicted three-dimensional TDRS position accuracy of 12 m is reached dynamically after 8 h, improving to 7–9 m after 24 h. With SA assumed on, the error jumps to over 100 m at 8 h, declining to 60–75 m at 24 h.

A ground tracking network of 12 GPS receivers can provide submeter GPS orbit accuracy. If those receivers were to track an additional high-altitude beacon,

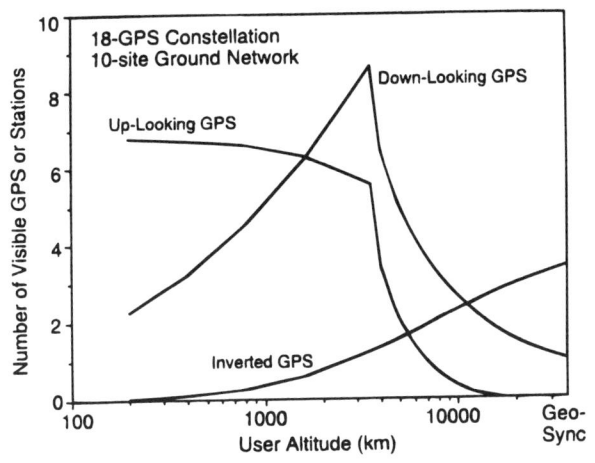

Fig. 10 Average numbers of GPS satellites visible with upward and downward looking hemispherical antennas and average number of sites from a 10-site ground network that can track an orbiting beacon, as a function of altitude.

all orbits could be estimated in one solution with comparable relative accuracy, scaled for distance. The same studies examined this inverted GPS tracking for TDRS. With assumed data errors of 25 cm for pseudorange and 1 cm for carrier phase at 30-min intervals, and a six-site ground network, the predicted three-dimensional rms orbit error over a 24-h dynamic solution arc was about 3 m for the TDRS satellites.

F. Highly Elliptical Orbiters

Because the preferred tracking modes and solution techniques differ for high and low orbiters, the application to highly elliptical orbiters, which may descend to a few hundred kilometers and rise to tens of thousands, presents a special challenge. Up- and down-looking GPS combined with ground-based Doppler during the high-altitude phase can provide particularly strong coverage. The proposed MUSES-B spacecraft, part of the Japanese very long baseline interferometry (VLBI) Space Observatory Program, was studied by Ref. 43. The MUSES-B would move from a perigee of 1000 km to an apogee of 20,000 km. The investigators applied a reduced dynamic strategy while the satellite was below 2000 km, and a purely dynamic strategy elsewhere. Combined omnidirectional differential GPS and ground-based Doppler gave a predicted orbit error of 50 cm for all position components at apogee, falling to less than 10 cm at perigee.

A similar mission, the proposed International VLBI Satellite, would have a perigee of 5000 km, enabling a purely dynamic solution around the orbit. Reference 43 found that with omnidirectional differential GPS and ground-based Doppler, as the apogee increases from 40,000 to 150,000 km, position error at apogee increases from about 15 cm to over 2 m. Such accuracy is needed for only select missions. Direct GPS orbit determination with omnidirectional reception could provide 10 m or better accuracy for nearly all highly elliptical orbiters.

VI. Dealing with Selective Availability and Antispoofing

The two GPS security features, antispoofing (AS) and selective availability (SA), can pose problems for SPS users. Over the years, various strategies have been devised to address them.

A. Antispoofing

Antispoofing is the encryption of the P code to prevent mimicking of the signal by others. In the presence of AS, a conventional SPS receiver would be able to track only the C/A code, recovering pseudorange and carrier phase on $L1$ only. This would prohibit computation of the standard dual-frequency ionospheric correction. The ionospheric effect, of course, depends on the altitude and field of view of the user. Above about 1000 km (assuming an upward-directed hemispherical field of view), the rare ionosphere permits submeter single-frequency orbit accuracy, even with no correction. At lower altitudes, as shown on EUE, the GRAPHIC calibration used in a reduced dynamic differential solution can give an orbit accuracy of about 1 m. That approach holds promise in both real-time and postprocessing uses for all but the most demanding requirements.

Low orbiters seeking subdecimeter performance must turn to dual-frequency calibration. There are several GPS receiver designs that operate in a codeless or quasicodeless mode; that is, they produce carrier phase and pseudorange at both frequencies without knowledge of the precise codes. Although codeless data are of degraded precision, the typical codeless phase error of about 1 cm in 1 s is consistent with the assumptions used in the studies presented here. Phase measurement noise, moreover, is generally not the dominant error in an orbit solution. Therefore, tracking performance will be largely unaffected by a switch to codeless operation.

B. Selective Availability

Selective availability consists of two measures to degrade positioning accuracy to the unauthorized user: the insertion of errors into the broadcast ephemeris and clock parameters, and "dithering" of the fundamental oscillator. Neither of these measures poses a problem for differential GPS tracking as we have defined it here. Receiver sampling times can be synchronized so that dither effects are common to all measurements and drop out of the differential solution. When sampling is not synchronized, quadratic interpolation to a common epoch can still achieve a high degree of dither cancellation, provided the sampling interval is no longer than about 30 s.[38] Ephemeris and clock errors do not come into play because those quantities are solved for with the ground data, either together with the user orbit or in advance.

Real-time direct GPS users encounter more difficulty. Those without access to accurate GPS orbits computed elsewhere will have to rely primarily on dynamic smoothing to reduce the effects of the ephemeris and clock errors and dithering. At nominal SA levels, the broadcast orbit error is 50 m or less on each satellite. (In recent years, the broadcast ephemeris has remained uncorrupted even when dither is active.) From the vantage of the user, the corrupted orbits will appear to some degree inconsistent with the GPS measurements and the user's dynamics—and the dynamic solution will then attenuate GPS orbit error. Early studies suggest that the error reduction will be about a factor of two. A 50-m GPS error may yield a 25-m user error. The actual reduction will depend on the solution strategy, receiver capacity, field of view, and other factors.

Smoothing of dither error by dynamic orbit estimation has been more thoroughly analyzed. Reference 39 simulated the dither process to examine the dynamic error attenuation as a function of the arc length. Figure 11 shows the net three-dimensional error (caused by dither only) for smoothing periods ranging from zero (point positioning) to 6 h. The receiver is assumed to track all satellites within a hemisphere, and dither is set at its nominal level. We see that with no smoothing, the rms dither error is about 30 m. After 2 h, this falls to 5 m, and after 6 h to less than 3 m. In a real-time application, in which the current state estimate cannot be smoothed with future data, the filtered dither error could be two or three times larger than shown in Fig. 11. Several investigators have attempted to mitigate dither effects by estimating a time-varying (stochastic) range bias to each GPS satellite using stochastic models tuned with actual SA data. But rapid dither variations are all but unobservable when the user state and clock are also estimated; such strategies tend, therefore, to sap data strength and degrade the orbit solution. Simple dynamic orbit estimation with a constant phase bias adjustment is generally more effective.[40]

A satellite such as TOPEX/Poseidon, which has well-modeled dynamics, can realize the full benefit of dynamic SA smoothing. At altitudes below about 600 km, dynamic model error will begin to offset the gain from smoothing, and at typical Shuttle altitudes the optimal direct solution may be little better than the point position solution under SA. To improve real-time accuracy at low altitudes, some form of near-real-time correction must be applied. This could be carried out, as is now commonly done for air and surface navigation, with

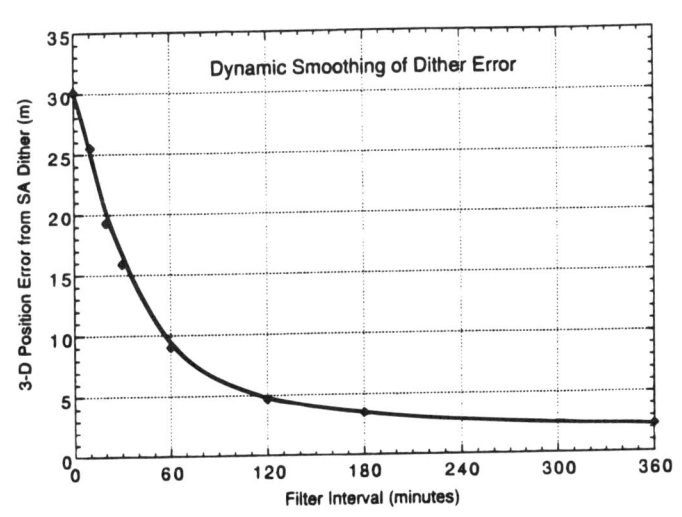

Fig. 11 Simulation results showing the three-dimensional position error that results from SA dithering, set at its nominal level, as a function of dynamic smoothing interval. No other errors are shown.

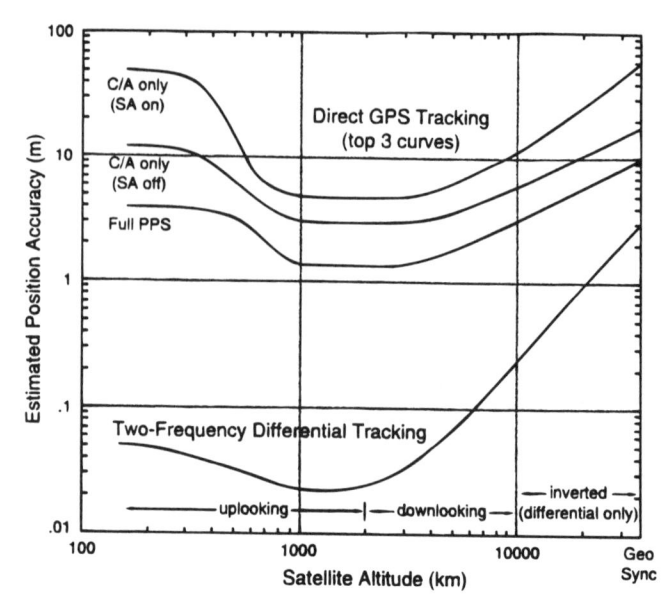

Fig. 12 Summary of the estimated orbit accuracies currently achievable with both differential and real-time direct GPS techniques. The direct solutions assume the use of precomputed GPS orbits of 2–3 m accuracy.

pseudorange corrections broadcast directly to users. Systems may soon be in place to send such corrections over wide areas through geosynchronous satellites (see Chapter 1 of this volume). A low orbiter equipped to receive those corrections could then achieve real-time position accuracy of about 1 m under SA.[29]

VII. Summary

The positioning strength provided by GPS is transforming Earth satellite orbit determination. With even the simplest receiving equipment it is now possible to determine the position of a low orbiter instantaneously to tens of meters, sufficient to meet the needs of most missions. The classical framework of dynamic orbit estimation can be adapted for GPS-equipped satellites in virtually any orbit to deliver orbit accuracies beyond the previous state of the art. Many low Earth satellites now in the planning stages will carry GPS for basic navigation and timing, and in some cases for direct scientific uses. Before GPS, orbits below 700 or 800 km could not be considered for satellites seeking accuracies of a decimeter or better. GPS promises to deliver few-centimeter accuracy at the lowest altitudes and for the most dynamically ill-behaved platforms. This creates the opportunity for low-power, low-mass, low-cost altimetry at an altitude of a few hundred kilometers, and for demonstrating precise sensing instruments on the Space Shuttle.

Figure 12 summarizes the performance that can be achieved as a function of altitude for both real time direct and after-the-fact differential GPS-based orbit determination. The curves reflect the optimal estimation strategy for each case.

For satellites above 10,000 km the standard differential technique is replaced by inverted GPS, where the orbiter carries a beacon tracked from the ground. The differential curve is consistent with the assumption of a dual-frequency codeless (SPS) receiver and is, therefore, unaltered by the presence of SA (eliminated by differencing) or AS. All curves for direct estimation assume the use of high-quality (2–3 m) GPS orbits and clocks distributed by civilian services, rather than the broadcast ephemeris. Thus, only dither error is included in the SA-on case. These curves are necessarily approximate; actual performance will depend on specifics of the GPS tracking configuration and satellite dynamics. But they offer a glimpse of the new standard GPS brings to orbit determination for missions of every description.

Acknowledgments

I am grateful to Catherine Thornton, Sien-Chong Wu, Willy Bertiger, Stephen Lichten, Ron Muellerschoen, Kenneth Gold, and Penina Axelrad for valuable discussions and other contributions. Portions of the work described here were carried out by the Jet Propulsion Laboratory, California Institute of Technology, under contract with NASA.

References

[1]Axelrad, P., and Parkinson, B. W., "Closed Loop Navigation and Guidance for Gravity Probe B Orbit Insertion," *Navigation,* Vol. 36, 1989, pp. 45–61.

[2]Hesper, E. T., Ambrosius, B. A. C., Snijders, R. J., and Wakker, K. F., "Application of GPS for Hermes Rendezvous Navigation," *Spacecraft Guidance, Navigation, and Control Systems,* European Space Agency, 1992, pp. 359–368.

[3]Axelrad, P., and Kelley, J., "Near-Earth Orbit Determination and Rendezvous Navigation Using GPS," *Proceedings, IEEE PLANS '86,* IEEE Position Location and Navigation Symposium (Las Vegas, NV), Institute of Electrical and Electronic Engineers, New York, Nov. 4–7, 1986, pp. 184–191.

[4]Chao, C. C., Bernstein, H., Boyce, W. H., and Perkins, R. J., "Autonomous Stationkeeping of Geosynchronous Satellites Using a GPS Receiver," *Proceedings of the American Astronautical Society/AIAA Astrodynamics Conference* (Hilton Head, SC), AIAA, Washington, DC, Aug. 10–12, 1992, AIAA CP-92-4655, pp. 521–529.

[5]Lichten, S. M., Edwards, C. D., Young, L. E., Nandi, S., Dunn, C., and Haines, B. J., "A Demonstration of TDRS Orbit Determination Using Differential Tracking Observables from GPS Ground Receivers," *Proceedings of the Third American Astronautical Society/AIAA Spaceflight Mechanics Meeting* (Pasadena, CA), American Astronautical Society, San Diego, CA, Feb. 22–24, 1993, AAS Paper 93–160.

[6]Yunck, T. P., Wu, S. C., Wu, J. T., and Thornton, C. L., "Precise Tracking of Remote Sensing Satellites with the Global Positioning System," *IEEE Transactions on Geoscience and Remote Sensing,* Vol. 28, 1990, pp. 108–116.

[7]Schreiner, W. S., Born, G. H., Larson, K. M., and MacDoran, P. F., "Error Analysis of Post-Processed Orbit Determination for the Geosat Follow-On Altimetric Satellite Using GPS Tracking," *Proceedings of the AIAA/American Astronautical Society Astrodynamics Conference* (Hilton Head, SC), AIAA, Washington, DC, Aug. 10–12, 1992, AIAA CP-92-4435, pp. 124–130.

⁸Munjal, P., Feess, W., and Ananda, M. P. V., "A Review of Spaceborne Applications of GPS," *Proceedings of the Institute of Navigation GPS '92,* Institute of Navigation, Washington, DC, 16–18 Sept., 1992, pp. 813–823.

⁹Parkinson B. W., "Navstar Global Positioning (GPS)," *Proceedings of the National Telecommunications Conference,* Nov. 1976.

¹⁰Farr, J. E., "Space Navigation Using the Navstar Global Positioning System," *Rocky Mountain Guidance and Control Conference* (Keystone, CO), American Astronautical Society, San Diego, CA, Feb. 24–28, 1979, AAS Paper 79–001.

¹¹Van Leeuween A., Rosen, E., and Carrier, L. "The Global Positioning System and Its Applications in Spacecraft Navigation," *Navigation,* Vol. 26, 1979, pp. 204–221.

¹²Tapley, B. D., "A Study of Autonomous Satellite Navigation Methods Using the Global Positioning System," NASA-CR-162635, Dept. of Aerospace Engineering, Univ. of Texas at Austin, Austin TX, 1980.

¹³Wooden, W. H., and Teles, J., "The Landsat-D Global Positioning System Experiment," *Proceedings of the AIAA/American Astronautical Society Conference* (Danvers, MA), AIAA, New York, Aug. 11–13, 1980, AIAA CP-80-1678.

¹⁴Kurshals, P. S., and Fuchs, A. J., "Onboard Navigation—The Near-Earth Options," *Proceedings of the Rocky Mountain Guidance and Control Conference* (Keystone, CO), American Astronautical Society, San Diego, CA, Feb. 17–21, 1981, pp. 67–89.

¹⁵Masson, B. L., Ananda, M. P. V., and Young, J., "Functional Requirements of the next Generation Spaceborne Global Positioning System (GPS) Receivers," *Proceedings of the Institute of Electrical and Electronics Engineers National Telesystems Conference* (Galveston, TX), Institute of Electrical and Electronics Engineers, New York, 1982.

¹⁶Jorgensen, P., "Autonomous Navigation of Geosynchronous Satellites Using the Navstar Global Positioning System," *Proceedings of the National Telesystems Conference, NTC '82,* Galveston, TX, Institute of Electrical and Electronics Engineers, New York, Nov. 7–10, 1982, pp. D2.3.1–D2.3.6.

¹⁷Heuberger, J., and Church, L., "Landsat-4 Global Positioning System Navigation Results," *Proceedings of the American Astronautical Society/AIAA Astrodynamics Conference, Part I* (Lake Placid, NY), American Astronautical Society, San Diego, CA, Aug. 22–25, 1983 (AAS Paper 83–363), pp. 589–602.

¹⁸Fang, B. T., and Seifert, E., "An Evaluation of Global Positioning System Data for Landsat-4 Orbit Determination," *Proceedings of the AIAA Aerospace Sciences Meeting* (Reno, NV), AIAA, New York (AIAA CP-85-0286), Jan. 14–17, 1985.

¹⁹Ondrasik, V. J., and Wu, S. C. "A Simple and Economical Tracking System with Sub-Decimeter Earth Satellite and Ground Receiver Position Determination Capabilities," *Proceedings of the Third International Symposium on the Use of Artificial Satellites for Geodesy and Geodynamics,* Ermioni, Greece, Sept. 1982.

²⁰Ananda, M. P., and Chernick, M. R., "High-Accuracy Orbit Determination of Near-Earth Satellites Using Global Positioning System (GPS)," *Proceedings, IEEE PLANS '82,* IEEE Position Location and Navigation Symposium (Atlantic City, NJ), Institute of Electrical and Electronics Engineers, New York, Dec. 6–9, 1982, pp. 92–98.

²¹Wu, S. C. "Orbit Determination of High-Altitude Earth Satellites: Differential GPS Approaches," *Proceedings of the First International Symposium on Precise Positioning with the Global Positioning System,* Rockville, MD, April, 1985.

²²Yunck, T. P., Melbourne, W. G., and Thornton, C. L., "GPS-Based Satellite Tracking System for Precise Positioning," *IEEE Transactions on Geoscience and Remote Sensing,* Vol. 23, July, 1985, pp. 450–457.

[23]Bierman, G. J., *Factorization Methods for Discrete Sequential Estimation*, Academic Press, Orlando, FL, 1977.

[24]Thornton, C. L., "Triangular Covariance Factorizations for Kalman Filtering," Jet Propulsion Laboratory Internal Document TM 33-798, Jet Propulsion Laboratory, Pasadena, CA.

[25]Lichten, S. M., and Bertiger, W. I., "Demonstration of Sub-Meter GPS Orbit Determination and 1.5 Parts in 10^8 Three-Dimensional Baseline Accuracy," *Bulletin Géodésique*, Vol. 63, 1989, pp. 167–189.

[26]Wu, S. C., Bertiger, W. I., Border, J. S., Lichten, S. M., Sunseri, R. F., Williams, B. G., Wolff, P. J., and Wu, J. T., "OASIS Mathematical Description," Vol. 1.0, Jet Propulsion Laboratory Internal Document D-3139, Jet Propulsion Laboratory, Pasadena, CA, 1986.

[27]Lear, W. M., "Range Bias Models for GPS Navigation Filters," Rept. JSC-25857, NASA Johnson Space Center, Houston, TX, 1993.

[28]Hatch, R. R., "The Synergism of GPS Code and Carrier Measurements," *Proceedings of the Third International Geodetic Symposium on Satellite Doppler Positioning*, Las Cruces, NM, 1982, pp. 1213–1231.

[29]Wu, S. C., and Yunck, T. P., "Precise Kinematic Positioning with Simultaneous GPS Pseudorange and Carrier Phase Measurements," *Proceedings, Institute of Navigation National Technical Meeting* (Anaheim, CA), Jan. 18–20, 1995.

[30]Wu, J. T., "Orbit Determination by Solving for Gravity Parameters with Multiple Arc Data," *Journal of Guidance, Control, and Dymanics*, Vol. 15, 1992, pp. 304–313.

[31]Bertiger, W. I., and Yunck, T. P., "The Limits of Direct Satellite Tracking with GPS," *Navigation*, Vol. 37, 1990, pp. 65–79.

[32]Wu, J. T., "Elimination of Clock Errors in a GPS-Based Tracking System," *Proceedings of the AIAA/American Astronautical Society Astrodynamics Conference* (Seattle, WA), AIAA, New York, 1984 (AIAA CP-84-2052).

[33]Yunck, T. P., Bertiger, W. I., Wu, S. C., Bar-Sever, Y., Christensen, E. J., Haines, B. J., Lichten, S. M., Muellerschoen, R. J., Vigue, Y., and Willis, P., "First Assessment of GPS-Based Reduced Dynamic Orbit Determination on Topex/Poseidon," *Geophysical Research Letters*, Vol. 21, 1993, pp. 541–544.

[34]Schutz, B. E., Tapley, B. D., Abusali, P. A. M., and Rim, H. J., "Dynamic Orbit Determination Using GPS Measurements from Topex/Poseidon," *Geophysics Research Letters*, Vol. 21, 1993, pp. 2179–2182.

[35]Rosborough, G., and Mitchell, S. "Geographically Correlated Orbit Error for the Topex Satellite Using GPS Tracking," *Proceedings of the AIAA/American Astronautical Society Astrodynamics Conference, Part 2* (Portland OR), AIAA, Washington, DC, Aug. 20–22, 1990 (AIAA CP-90-2956), pp. 655–663.

[36]Gold, K., Bertiger, W. I., Wu, S. C., Yunck, T. P., Muellerschoen, R. J., and Born, G. H., "GPS Orbit Determination for the Extreme Ultraviolet Explorer," *Proceedings of the Institute of Navigation GPS '93*, Salt Lake City, UT, 1993, pp. 257–268.

[37]Haines, B. J., Lichten, S. M., Malla, R. P., and Wu, S. C., "Application of GPS Tracking Techniques to Orbit Determination of TDRS," *Flight Mechanics/Estimation Theory Symposium*, Goddard Space Flight Center, Greenbelt, MD, 1992, (NASA CP-3186), pp. 117–128.

[38]Wu, S. C., Bertiger, W. I., and Wu, J. T., "Minimizing Selective Availability Error on Topex GPS Measurements," AIAA/American Astronautical Society Conference, Portland, OR, 1992, AIAA CP-90-2942.

[39]Bar-Sever, Y., Yunck, T. P., and Wu, S. C., "GPS Orbit Determination and Point Positioning Under Selective Availability," *Proceedings of the Institute of Navigation, GPS '90*, Colorado Springs, CO, Institute of Navigation, Washington, DC, 1990, pp. 255–265.

[40]Lear W. M., Montez, N. M., Rater, L. M., and Zyla, L. V., "The Effect of Selective Availability on Orbit Space Vehicles Equipped with SPS GPS Receivers," *Proceedings of the Institute of Navigation Conference, GPS '92*, Institute of Navigation, Washington, DC, 1992, pp. 825–840.

[41]Lichten, S. M., "Estimation and Filtering for High-Precision GPS Positioning Applications," *Manuscripta Geodaetica*, Vol. 15, 1990, pp. 159–176.

[42]Wu, S. C., Yunck, T. P., and Thornton, C. L., "Reduced-Dynamic Technique for Precise Orbit Determination of Low Earth Satellites," *Journal of Guidance, Control, and Dynamics*, Vol. 14, No. 1, 1991, pp. 24–30.

[43]Wu, S. C., Yunck, T. P., Lichten, S. M., Haines, B. J., and Malla, R. P., "GPS-Based Precise Tracking of Earth Satellites from Very Low to Geosynchronous Orbits, *Proceedings of the National Telesystems Conference* (Ashburn, VA), 1992.

[44]Yunck, T. P., "Coping with the Atmosphere and Ionosphere in Precise Satellite and Ground Positioning," *Environmental Effects on Spacecraft Positioning and Trajectories*, edited by A. V. Jones, Geophysical Monograph 73, IUGG Vol. 13, American Geophysical Union, Washington, DC, 1993, pp. 1–16.

[45]MacDoran, P. F., "A First-Principles Derivation of the Differenced Range Versus Integrated Doppler (DRVID) Charged Particle Calibration Method," *JPL Space Programs Summary 37-62*, Vol. II, March 1970.

Chapter 22

Test Range Instrumentation

Darwin G. Abby*
Intermetrics, Inc., Holloman Air Force Base, New Mexico 88330

I. Background

HISTORICALLY, land-based test facilities have used a combination of radar, distance-measuring equipment (DME), optical trackers such as cinetheodolites, and other miscellaneous instrumentation to provide time–space position information (TSPI) to satisfy test platform positioning requirements. In the early 1970s, laser trackers became available to support test activities. Each of these systems had their strong and weak points, and systems were used depending on the accuracy, area of coverage requirements, and cost considerations. Radar systems could cover fairly large line-of-sight areas, but accuracys were low (25–50 ft), and cost was high. DME and laser systems had limited areas of coverage, were fairly accurate (2–5 m), and had medium cost. Cinetheodolites provided very high accuracy (0.5–1 m) over very limited areas; the cost was very high, and delays of 2–6 weeks for data processing was the norm.

To obtain overland coverage of larger areas, instrumented test ranges were selected at both ends of a flight trajectory and special radar sites were built or FAA air traffic radars were used to cover the enroute areas. Test support for these types of overland flights was very expensive. Examples of these types of support activities were the Edwards AFB, California to Utah Test Range located near Salt Lake City, Utah and a route from Edwards Air Force Base to White Sands Missile Range in New Mexico.

A combination of radar and DME [i.e., the General Dynamics, Inc. Position Location System (PLS) and the Cubic Corporation Air Combat Maneuvering Instrumentation (ACMI)] systems were used to satisfy combat training requirements. Most of these systems were designed to work in real time, and locations of ground transponders and communication links limited the area of coverage. These systems were very expensive to operate and maintain.

Copyright © 1994 by the American Institute of Aeronautics and Astronautics, Inc. All rights reserved.
*Senior Staff Consultant, Navigation Systems Department, Systems and Software Applications Division.

The broad ocean test ranges had a much more difficult problem for obvious reasons. Radars or other types of tracking instrumentation could be located only on coasts, islands, or instrumented ships and aircraft. Highly instrumented terminal areas for missile testing were built using radar and sonobuoys for positioning impacts. These systems met the requirements at the time, but again costs to maintain these systems were very high.

The use of GPS for test and training applications began to surface in studies conducted in the 1979–1981 time frame. A system that could provide highly accurate TSPI over unlimited areas at low cost had the potential of solving a wide range of test and training requirements. In many cases, there were large geographic areas and even global test and training requirements that either could not be met by other candidate systems, or could only be met at very high cost when compared to the cost of a GPS-based system. At the time these studies were conducted, differential GPS was in its infancy and was not as highly developed as it is today. Therefore, GPS accuracies of 10–15 m Spherical Error Probable (SEP) were assumed for evaluating GPS utility for test and training.

In 1979, SRI International[1] completed a study that concluded that exploitation of GPS in the test community for test article TSPI measurement offered significant economic and operating advantages for test and evaluation (T&E) operations. The 10–15 m accuracy met many of the T&E requirements but did not meet positioning requirements for precision weapon system testing. The main advantage was cost reduction and increased flexibility. The potential to achieve the accuracys any where in the world by installing a GPS receiver brought a cost-effective capability to the test community.

Army Captain William Reinhart completed a thesis[2] in 1981 that investigated the use of GPS for replacing positioning systems used in training systems at Fort Ord, California. Operational battlefield training systems use instrumentation to determine who is firing at whom, simulate each engagement mathematically, and provide opposing players with realistic engagement results. A very accurate positioning system is required to support real-time casualty assessment calculations and data for evaluating the success of opposing armies. Tests were conducted, and 10–15 m rms GPS accuracy were demonstrated. Horizontal accuracy requirements were not met; however, the vertical accuracy was better. Advantages of GPS over existing systems were: 1) larger player capacity; 2) ability to perform concurrent experiments; and 3) exportability or ability for worldwide use.

A 1980 MITRE Corporation report[3] evaluated the use of Navstar GPS receiver equipment as Navy range instrumentation for R&D test and air and ground combat training. Five GPS receiver configurations were considered for satisfying test and training requirements ranging from high-velocity–high-acceleration systems, such as air-to-air missiles, to low-dynamic applications, such as soldiers and trucks. They concluded that the five GPS equipment configurations defined could provide the required accuracy over the range of dynamic conditions expected in test and training environments.

The overall conclusion from these studies was that GPS offered cost-effective solutions to most test and training (T&T) applications. With the advent of differential GPS (DGPS), even many "positioning requirements for precision weapon system testing"[3] could be achieved, and the horizontal requirements for training applications could be met. The use of GPS for T&T applications was only a matter of developing and supplying the systems.

The Undersecretary of Defense for Research and Engineering (USDR&E) established a tri-service steering committee (TSSC) in 1981 to evaluate the possible application of GPS as a TSPI source for test and training ranges.[4] The TSSC was also to evaluate test and training requirements and issues and to recommend interim and long-term test range applications of GPS. The TSSC concluded that GPS could satisfy test and training requirements and could significantly improve position and velocity measurement accuracy in most applications.

The TSSC recommended and the USDR&E approved the establishment of a joint program to develop a family of GPS range equipment specifically tailored to the needs of the test and training communities. The Air Force was chosen as executive service and, via AFSC, delegated to the Range Directorate (AD/YI), Missile Systems Division, Eglin Air Force Base, Florida, program management responsibility. In July 1983, AD/YI established the Range Applications Joint Program Office (RAJPO), to continue to analyze technical issues, manage a preliminary test program, evaluate range integration issues, and develop triservice equipment using GPS as the TSPI source for DOD test and training ranges. The RAJPO equipment is described later in the chapter.

A significant conclusion of the TSSC study was that GPS translators would be useful in a variety of range applications, particularly those involving destructive testing (e.g., missiles) and when the number of test articles active simultaneously is limited.[5] GPS translator technology was being developed at the time by the Navy Fleet Ballistic Missile agency for support of Trident I testing. Translators are now used extensively by the strategic missile community, and the technology has advanced significantly.

II. Requirements

TSPI requirements for test and training applications are very similar, and positioning systems developed for one or the other can potentially be used by both. The differences in the applications are primarily in data management and utilization.

A. Test Requirements

Testing of new weapon systems requires very accurate reference systems for position and velocity. In addition, requirements exist for acceleration and attitude truth references on many weapon systems. Department of Defense test ranges use a wide variety of systems including tracking radars, laser trackers, cinetheodolites, and range measurement systems to provide TSPI. Systems under test include such weapons as bombs and missiles, targeting systems, and navigation systems. In the past, most testing has been conducted on instrumented test ranges within a controlled area. Modern weapon systems using new technologies have created new and unique support requirements. The need for cost-effective, disposable TSPI instrumentation for one-time missile shots is one example. Missiles can vary from the small tactical high-speed aircraft missiles to such larger strategic ones as the Minuteman. Many modern systems require testing over extended areas of land and sea, where it can be very difficult to impossible to provide tracking instrumentation. An example would be evaluating active target sensing and recognition sensors in different terrain and climates. Most test programs do not require the high-accuracy TSPI in real time at the test vehicle or at a range control facility; however, in almost all cases, continuous real-time information on the test platform position must be available to satisfy safety requirements. Real-time requirements for position and/or velocity create significant technological and cost impacts.

Reference accuracy requirements vary depending on the weapon system. The rule of thumb is to have a truth reference of approximately a factor of ten more accurate than the system under test. However, the accuracy often is limited by the availability of a cost-effective system.

The TSSC, as a part of its charter, also conducted a survey of twenty-two DOD test ranges. One objective of this survey was to summarize the test and training accuracy requirements using six different parameters and to determine if GPS could meet those requirements. The results of the survey and what GPS can provide are shown in Table 1. At that time, GPS could meet most of the requirements except in the scoring area, where accuracies of 1–10 ft are required. DGPS accuracies of 0.1 m are now available.

The use of GPS for missile tracking presents a unique set of requirements. First, missiles are not recoverable, and the equipment is destroyed on every test which makes cost a major factor. In addition, size and weight are critical for missile applications, which has led to another GPS translator technology area. The two primary requirements for GPS missile-tracking support are for range safety and precise trajectory determination. For range safety, the missile position must be determined and made available in real time so that corrective action can be taken immediately in case of emergency. In general, precise trajectory determination can be performed postflight unless real-time scoring is required.

B. Training Requirements

Training requirements include instrumenting battlefield or air combat participants, (soldiers, tanks, trucks, helicopters, aircraft, etc.) in such a way that mock battles can be conducted. The question of who shot whom can be sorted out for real-time casualty assessment and made available to commanders and troops in real-time and/or postmission for analysis and debriefing. New tactics are evaluated during training exercises to determine effectiveness against new weapon systems. The major instrumentation components are a position- and velocity-determining system, data links and/or recorders, central processors, and displays. Position and velocity accuracy and data rate requirements are a function of participant dynamics (soldier vs F-16) and training objectives (tactics vs troop training).

Frequently, accuracy requirements are driven by the availability of cost-effective instrumentation. Developments continually are in process to try to improve accuracy and minimize cost. The U.S. Army Training and Doctrine (TRADOC), test and experimentation command (TEXCOM), formerly combat development experimentation command (CDEC), Fort Ord, California has been developing instrumentation for training applications for many years.

TEST RANGE INSTRUMENTATION 597

Table 1 TSPI requirements summary (22 ranges)

TSPI performance parameters	Training and OT&E[a] ranges				DT&E[b] and OT&E ranges			
	Air	Land	Sea (fixed)	Sea (moving)	Long-range	Extended range	Short-range (land)	Short range (water)
Real-time accuracy								
Position (ft)	⑤–200	15–30	200	1000	20–100	⑤–20	⑤–100	⑤–100
Velocity (fps)	0.1–15	3–9	100	—	0.5–5	5	1–20	1–20
Data rate (#/s)	1–20	1–10	5	1–5	20	20	1–100	1–20
Post-test accuracy								
Position (ft)	①–200	6–30	10	TBD	10–20	30	②–15	②–15
Velocity (fps)	0.1–15	3–9	0.1–5	—	0.01–0.1	0.01–0.02	0.1–10	0.1–10
Scoring (ft)	①–10	①–6	—	—	50	①	①–5	①–5
Number of test articles	1–90	2–2000	50	60–125	3–10	1	12–20	12–20
Coverage								
Altitude (kft)	0.1–100	0–10	0.1–58	0–60	300	0.1–30	0–100	0–100
Distance (nm)	30–60	30 × 30	75 × 75	350 × 500	150–3000	100 × 600	50 × 150	125 × 200

○ GPS marginal. □ GPS can not satisfy.

TEXCOM's most valuable contributions in weapon system analysis have historically been to detect strengths or weaknesses that were not intuitively obvious. Often these system traits become apparent only when the system was fielded in a combined arms environment, during which the soldiers and/or the weapons systems are task loaded and carry the threat of being killed. This requirement results in a need for high-accuracy, real-time position data. The most important requirement for TEXCOM's development efforts is to provide 1-m position accuracy for non–line-of-sight weapon systems that simulate conditions actually existing in the field. The immediate weapon candidate is the fiber optics guided missile (FOG-M).

III. Range Instrumentation Components

The three major components for a DGPS test range system are a GPS reference station, a data link, and test vehicle instrumentation. The functions of these systems are described in the sections on DGPS. The intent here is to address unique characteristics relative to test applications. The components and function of a translator system are also described.

A. GPS Reference Station

Most DOD test ranges have access to selective availability/antispoof (SA/AS) authorization, and as a result, they can operate in the GPS precise positioning service (PPS) mode. However, this requires that security procedures be implemented to protect classified data.

GPS standard positioning service (SPS) reference stations are also being used by test ranges because they are available off-the-shelf at low cost, and they eliminate the need for security procedures. Differential GPS accuracies of 5–8 m have been achieved using these systems, thus meeting many test requirements.[6]

The optimum GPS reference receiver should track all visible satellites simultaneously and provide corrections for any potential satellite the test platform receiver might track. In the case of a high dynamic aircraft, the tracked satellite subset could be almost any combination of those visible. The generic components of a DGPS reference station include the reference receiver with antenna, data processor and controller for real-time or postprocessing, data-recording system, keyboard and display, printer/plotter, and optionally an external atomic frequency standard. The reference station interfaces with a data link system for real-time operations. The data broadcast on the link are also recorded for any postmission processing requirements. The operator interface is required for initializing the system, controlling the reference receiver, and monitoring data quality. An optional meteorological data interface may be included.

The Radio Technical Commission Maritime (RTCM) has been the leader in attempting to standardize the format for differential corrections produced by a GPS reference station. The RTCM established Special Committee-104 to prepare a standardized format resulting in RTCM SC-104 of January 1, 1990.[7] Most commercial DGPS vendors have adopted the standard and use it for differential GPS systems. On the other hand, NATO, RAJPO and the Range Commanders Council are developing other standards for exchange of GPS data.

B. Data Links

Detailed descriptions of the different data links being used by the ranges are discussed later. In general, the commercial vendors are adapting off-the-shelf communications radios for data links. To accommodate multiple players they use a combination of time division multiple access techniques and multiple frequencies. RS-422 or RS-232 data ports on the receiver are connected to modems and then to the communication radio. The same interface to the remote user receiver is used in reverse. In almost all cases two-way communication is available.

The RAJPO is developing a complete customized system that interfaces with the suite of RAJPO equipment. Details are outlined in Sec. V.A.1.

Other potential links include cellular telephones, L- and S-band telemetry systems, existing weapon system data links, satellite communications, etc. Virtually any rf link could be used to transmit the differential GPS data. Data rates are very low, on the order of 50 baud, but the data link implementation and vehicle dynamics affect the final rate.

C. Test Vehicle Instrumentation

The GPS receiver and associated instrumentation on the test platform can vary significantly depending on the vehicle dynamics, accuracy requirements, data-processing mode (real-time or postmission), and use of PPS or SPS operations. The minimum package for a low dynamic platform would be the GPS receiver and a data recorder for postprocessing and TSPI generation. For a high dynamic platform, the GPS receiver would need to be aided with an inertial navigation system (INS). A receiver dataport would interface with either a recorder or a real-time data link.

D. Translator Systems

Historically, GPS single- or multi-channel receivers have been used on test vehicles. These receivers receive and process the GPS signals and output position and velocity estimates for use onboard the test vehicle or downlinked to additional equipment at the ground facility for display and data processing. Translators should be considered when the test mission has a requirement for time–space–position information (TSPI) at a ground facility and one of the following is true:

1) The program is concerned with the high cost of GPS receivers because the test vehicles will be attrited.

2) There is insufficient volume to support an onboard GPS receiver.

3) Detailed postmission analysis of the GPS data is required.

Various programs using GPS translators have been supported including the Strategic Defense Initiative Office (SDIO) ground-based interceptor (GBI) and the Peacekeeper.

Figure 1 shows the system concept for the GPS translator equipment. The system consists of a user-equipped translator and a ground-based translator processor system (TPS), of which the GPS translator receiver is an element.

The GPS L-band signals are received at the missile, translated to S-band, and then retransmitted to the ground station.[5] The composite of all GPS satellites

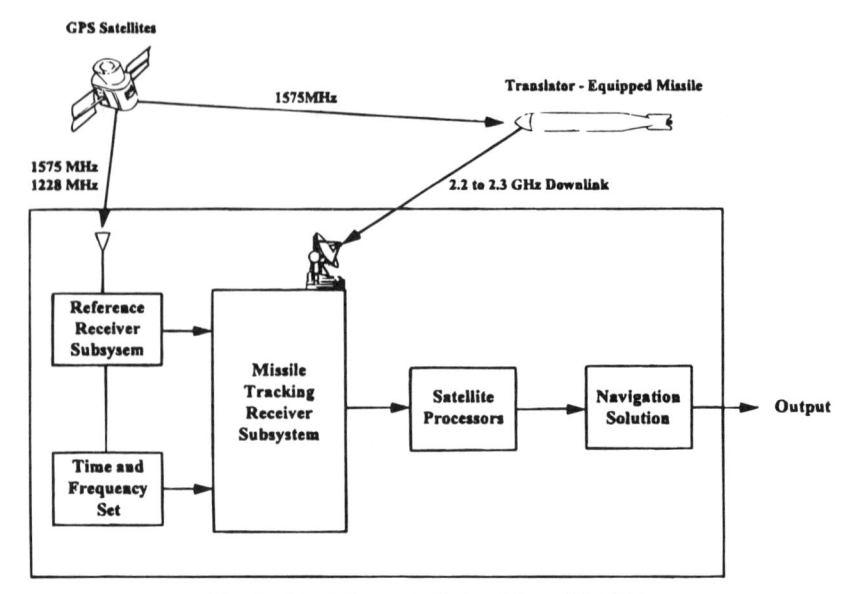

Fig. 1 Real-time missile tracking with GPS.

visible to the missile are received at the ground station for processing. A GPS reference receiver, collocated at the ground station, provides ground station location in GPS coordinates, GPS time, and satellite ephemerides. Because both an S-band target receiver and an L-band reference receiver are used at the ground station, the system becomes a differential navigation system with associated accuracy advantages. The system only uses the satellite L-1 link and C/A code through the missile translator. This configuration was selected for the following reasons:

1) The C/A code has 3 dB more power than the P code, providing a higher quality Doppler (velocity) measurement.

2) The C/A code provides sufficient position accuracy for most missile applications.

3) Translator power output required for the C/A code would have to be increased by a factor of 10 to use P code to retain the same downlink margin.

4) Use of the L-2 link through the translator would add a receiving antenna to the missile, a second receiver channel to the translator, and the required translator output would double.

The composite GPS satellite signals are filtered, amplified, and translated to one of several possible output frequencies in the 2200–2300 MHz range. This selectable output frequency feature allows several missiles to be tracked simultaneously by using frequency division multiple access (FDMA) on the missile to ground links. In addition, a pilot carrier is synthesized in the translator and introduced in close frequency proximity to the translated L-1 spectrum. The combined pilot carrier/translated L-1 is then transmitted to the ground station.

The signal from the missile is received at the ground station by a telemetry receiving antenna. This antenna is typically a high-gain parabolic reflector antenna, which is necessary to maintain a viable telemetry link at long ranges using minimal onboard telemetry transmitter power.

There are both advantages and disadvantages to using GPS translators. Translators are less expensive than equivalent GPS receiver (i.e., space-qualified, etc.). Translators also consume less power, occupy less space and—with a properly designed TPS—can acquire signals and provide a TSPI solution faster than GPS receivers. Translators can (again with a properly designed TPS) track GPS signals to a lower signal level than onboard GPS receivers (because of reference receiver aiding). In addition, a TPS is configured to support postmission processing, thus allowing GPS signals to be tracked through high accelerations (>20 gs), a real challenge for most GPS receivers. The key disadvantages are the large bandwidth downlink requirements (~2 MHz, C/A code; ~20 MHz, P code) and the need for high-gain telemetry antennas to track the user vehicle. A specific system being developed by the RAJPO is described in Sec. V.A.1.e.

E. Digital Translators

In the past four years, NAVSYS Corp. has taken advantage of digital technology to develop a smaller and less expensive system originally called the *Advanced Translator Processing System* and more recently called the *Digital Translator Processing System*.[8–10] The flight unit is 30 in.^{3}, weighs 3 lb, and uses only 28 W of power, vs the older analog units that range from 180 in.^{3} to the more recent analog design that is 40 in.^{3}; weighs 5 lb and requires 56 W of power.

Digital translators take advantage of recent developments in digital microwave radios. A block diagram of the vehicle translator and ATPS system are shown in Fig. 2. As in a conventional analog translator, the L-band GPS signal first passes through an L-band preamplifier. The signal is then filtered to select either the P-code or the C/A code bandwidth. The filtered L-band signal is next sampled and quantized by an A/D converter. The A/D outputs are used to modulate an S-band carrier. A pilot carrier (PC) is then added to the modulated signal. The combined signal is the digital translator output. The digital translator architecture eliminates the need for an IF frequency an significantly reduces the filtering requirements which results in significant savings in size, weight and power dissipation.

A key element of the advanced translator processing systems (ATPS) is the preamplifier/downconverter (P/DC) module developed by NAVSYS to condition the received translator signal so that it can be tracked by a conventional C/A code receiver. The P/DC tracks the pilot carrier from the GPS translator and uses this signal to down-convert the received S-band signal back to the L-band signal received at the translator. This signal can then be processed using a conventional off-the-shelf digital GPS receiver. There are very key functional similarities between the digital translator systems and the newer digital GPS receivers, as shown in Fig. 3.

There are other significant advantages of digital translators. The digital translator approach directly facilitates data encryption. The sampled and quantized L-band signal is a digital stream that can be encrypted by any conventional data encryption technique prior to modulation on the S-band carrier. In addition, the pilot carrier can be optionally modulated with telemetry data that could include

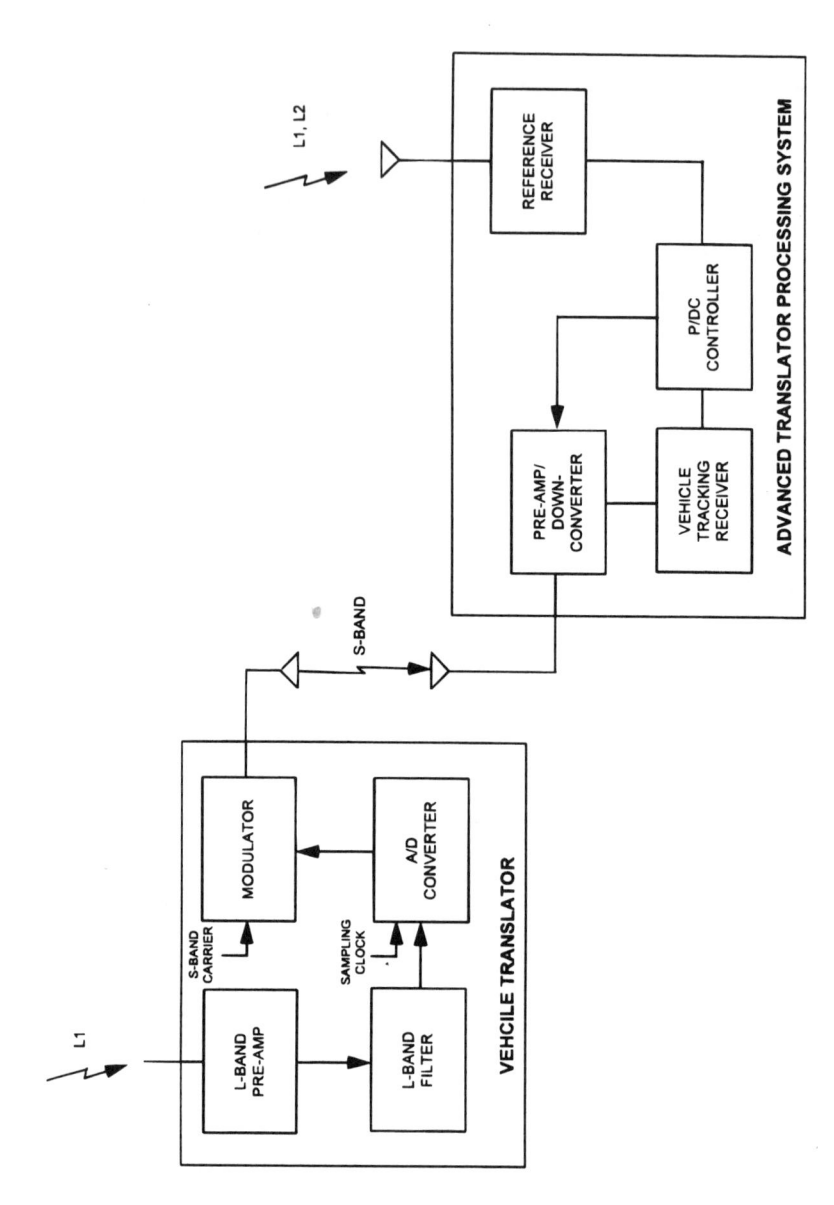

Fig. 2 Digital vehicle translator and translator processing system.

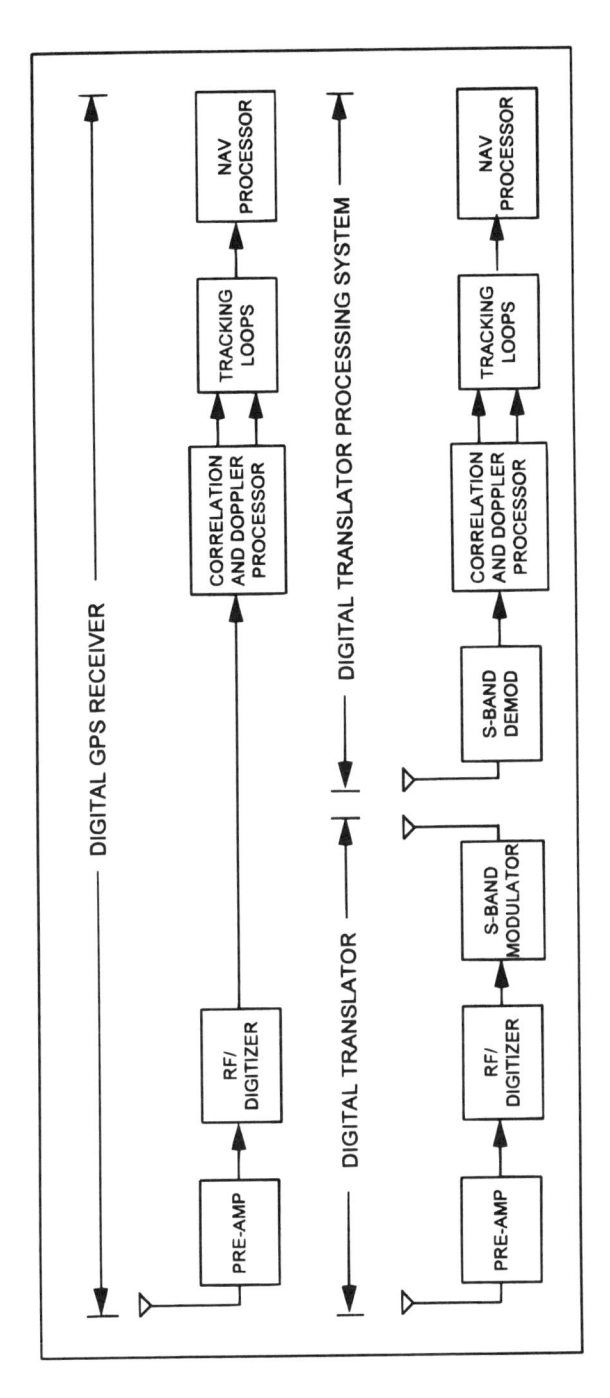

Fig. 3 Digital GPS reciever/digital translator comparison.

inertial reference unit (IRU) data. Intertial reference unit data can be used in post-processing navigation filters to aid the navigation during times of GPS outages.

IV. Differential Global Positioning Systems Implementations

The primary DGPS implementation method used by test range activities is one wherein raw data from the GPS receiver in the test vehicle are recorded or downlinked to the reference station, as shown in Fig. 4. The reference station computes the pseudorange corrections and rate of change of the corrections, applies the corrections to the test vehicle receiver's measurements, and computes the TSPI. This method takes advantage of a ground computer's larger capacity to perform more sophisticated processing and potentially achieve better accuracy. This method also offers the option of sending all data to the range host computer where the data can be combined with other range sensors, processed in a large Kalman filter, and even smoothed to achieve improved results. The data to be downlinked include: pseudoranges and pseudorange rates, satellite ID numbers age of data, ephemeris (AODE), receiver identification, user time, measurement quality estimates (if available) and INS data (attitude, velocities and/or accelerations). In this mode, the link is only one way and the test vehicle does not have the final solution, which in most test programs is not required. If TSPI is not required in real time, l the data can be recorded and postmission processed. This option 1) eliminates the need for a data link, 2) simplifies the instrumentation significantly, 3) eliminates the line-of-sight restriction required for RF links, and 4) allows the test vehicle to cover much larger geographic areas to satisfy many modern weapon system requirements.

V. Existing Systems

Development of DGPS and the applications for test ranges has been ongoing for over 9 years. There are several systems of both PPS and SPS that currently are operational or undergoing acceptance testing. These systems are described in this section.

A. Department of Defense Systems

Department of Defense test facilities have access to the PPS. The GPS receivers used are keyable and remove the effects of SA/AS. The two U.S. Air Force agencies that develop this type of equipment are the RAJPO and the GPS Joint Program Office.

1. Global Positiong Systems Range Applications Joint Program Office

The RAJPO is developing a family of equipment to use the Navstar GPS on DOD test and training ranges to provide TSPI. The individual components are designed so that range systems can be assembled to meet a variety of requirements. The family of equipment includes the high dynamic instrumentation set (HDIS), a reference receiver/processor (RR/P), C/A code receiver (CACR) and a data link system (DLS). Other equipment being developed for range applications

TEST RANGE INSTRUMENTATION

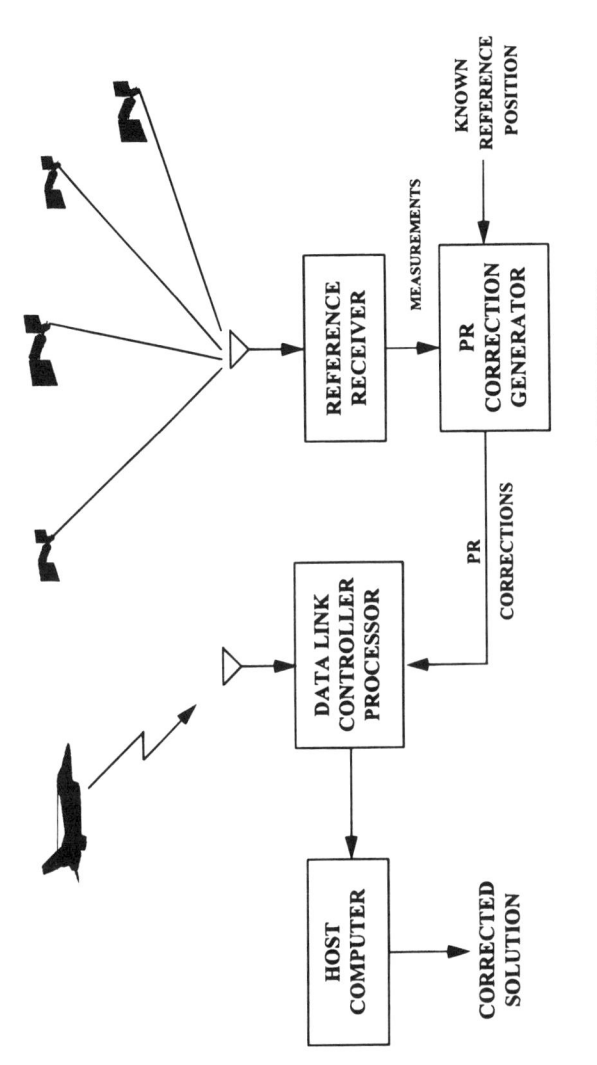

- VEHICLE GPS MEASUREMENTS DOWNLINKED & SENT TO HOST
- THE PRCG GENERATES CORRECTIONS FOR ALL SVs IN VIEW & SEND TO HOST
- HOST PROCESSES DATA AND GENERATES CORRECTED SOLUTION

Fig. 4 Differential GPS method 3.

include translators and translator processing systems and ground transmitters (GTs). A brief description of each component follows.

a. High dynamic instrumentation set (HDIS). The HDIS is five-channel, fully authorized for PPS, GPS receiver designed to fit in a 5-in. air incercept missile (AIM)-9 pod. It includes the antenna system, receiver/processor, range flexible modular interface (RFMI), power conditioning, and all interconnecting cables and connectors. The HDIS has provisions for integration with an IRU, a control display unit (CDU) for operator interface, and support equipment for maintenance. In addition to being able to receive and process GPS satellite navigation signals, the unit can also receive and process GT signals. The HDIS is capable of operating in a differential GPS mode by accepting differential corrections linked from the RR/P.

The integration of an HDIS, an IRU, data link translator (DLT) and solid-state recorder into a 5-in. AIM-9 missile pod provides a small autonomous instrumentation package that can be mounted on aircraft wing weapon stations. The same set of equipment can be configured for a small pallet for installation inside an aircraft. The combination of a pod or pallet system, RR/P and a DLS is called the advanced range data system (ARDS).

b. Reference receiver/processor. The RR/P includes an HDIS receiver, a pseudorange correction generator (PRCG), a navigation correction processor (NCP), an RFMI, a meteorological sensor subsystem, and a control display subsystem. The RR/P provides data outputs to the DLS for transmission to the test or training vehicles and to a host range computer system for processing. The RR/P tracks up to eight satellites and provides pseudorange corrections, rate of change of pseudorange corrections, raw pseudorange, meterological and satellite message data on all satellites. The RR/P operates in either the authorized or unauthorized mode.

c. C/A code receiver. The CACR is a small commercial C/A code receiver that has been modified to interface with all the RAJPO range systems. It will be used primarily for ground vehicle and manpack test and training applications.

d. Range data link (RDL). The RDL operates in segments of the 1350–1530 MHz (L-band) telemetry band; each RDL net uses two frequency channels within that band as shown in Fig. 5. Within a given area, multiple nets can be operated if sufficient channel pairs are available. Multiple nets can be operated independently, if applications are separate and disjoint, or their operation can be coordinated to expand the capacity of the system in a single large application.

An individual RDL net can contain up to 2000 RDL transceivers, ground and airborne. Operation is segmented in time into 330 time slots per second, during each of which a single transceiver transmits one message to one or more of the other transceivers. A transceiver can be assigned as many time slots as necessary to fulfill its communications needs. The assignment of transceivers to time slots is performed automatically by the system in accordance with needs, and it changes dynamically as those needs change. Needs can include relaying if the originator of the message is not within communications range of the recipient, and a small fraction of total net capacity (generally 10% or less) is taken up by internal messages used to control such assignment and reconfiguration.

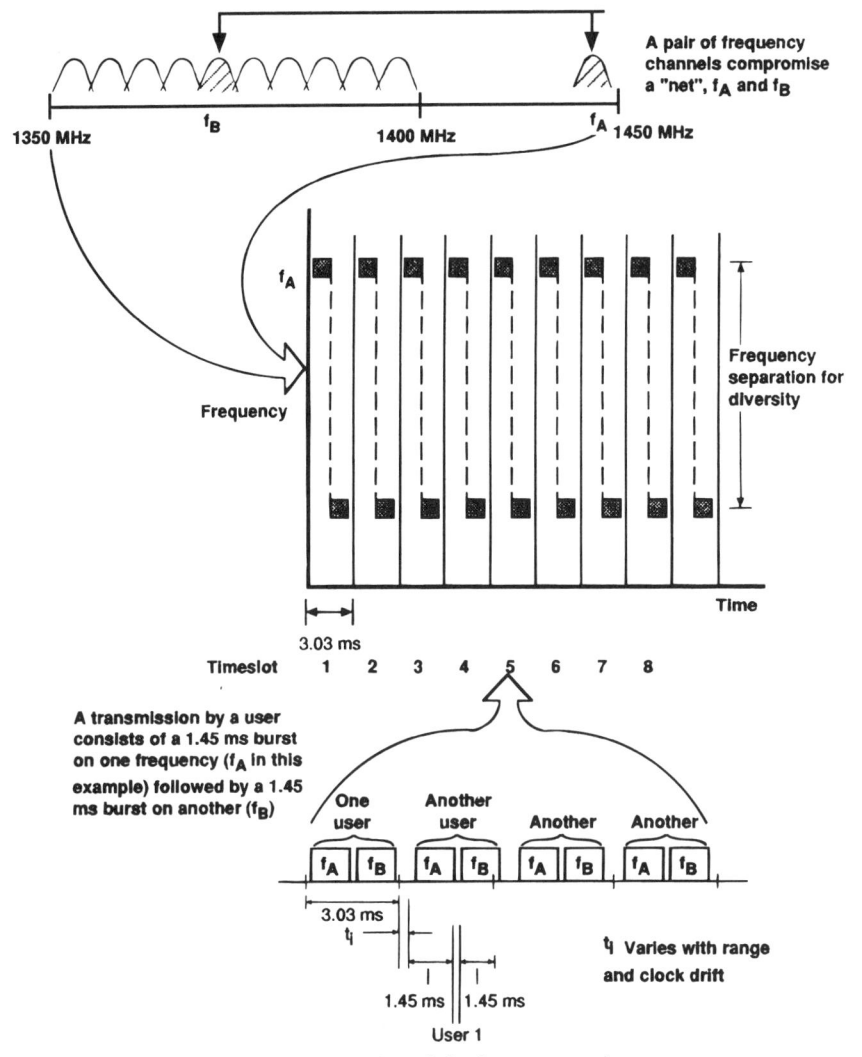

Fig. 5 Illustration of single-net operation.

In multinet coordinated operation, specific common time slots are assigned to the control function on a common frequency channel pair. During the remainder of the time slots, data transmissions from different participants would occur simultaneously on the various assigned channel pairs.

The configuration of the system can be tailored to individual ranges and further to particular test scenarios within those ranges. For example, a small-scale developmental test of a new vehicle with custom telemetry-gathering systems on board that produce, say, 28 kb/s, could be accommodated by interfacing those systems to the data link, defining 700-bit messages to carry the data, and assigning 40 slots per second to this function. An additional 10 slots per second might be

40 slots per second to this function. An additional 10 slots per second might be assigned to downlinking TSPI data from the platform, and three other platforms might also be tracked, at a 10 slot per second rate. Uplink control might require an additional 10 slots per second for the test platform, and for each platform. System control might require a total of 20 slots per second. These figures total 140 slots per second, hence, the system is operating at 42% of the full capacity of one net. Some of that excess capacity might be used for relaying; if the test vehicle were not within line-of-sight of any ground station, an additional 60 slots would be required for that function.

Hardware—Data link units are designed for mounting in an AIM-9 (5-in. diameter) pod as shown in Fig. 6. The 20-W transceiver occupies 14.5 in. An optional 60-W high-power transceiver occupies 18 in. of rail space within the pod, and the associated power supply, an additional 12 in. Antennas for pod-mount and internal aircraft installation are nominally omnidirectional. Stacked dipole antennas are planned for ground applications, omnidirectional in azimuth, but with elevation-plane beamwidths of about 13 deg. Most links within the system will connect airborne units and ground stations.

System control—Up to 17 ground stations are connected to a central facility, called the Data Link Controller (DLC). A variety of connecting links may be employed, using existing range communications systems such as microwave and wireline. The DLC and ground stations perform an accounting/error detection/retransmission protocol in their communication over these connecting links to protect against errors generated in them.

e. GPS translator equipment. The Interstate Electonics Corporation under contract to the RAJPO is one of the major developers of translator equipment. Fig. 7 shows the block diagram for the RAJPO analog translator.

Translator—1) *Operation*: Global positioning system satellites output two primary frequencies, denoted L_1 (1575 MHz) and L_2 (1228 MHz). The L_1 frequency contains two orthogonal pseudorandom codes, the (C/A) code and the precise (P) code. The L_2 frequency contains only the P code. The translator front-end

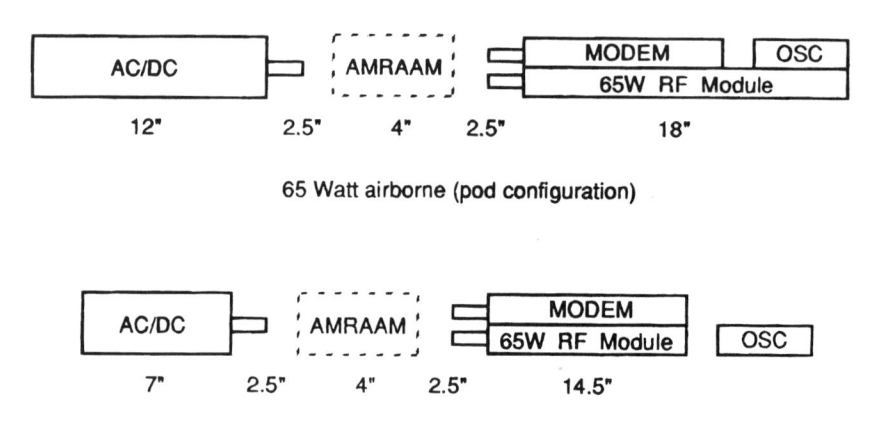

65 Watt airborne (pod configuration)

20 Watt airborne (pod configuration)

Fig. 6 **Data link pod configuration.**

TEST RANGE INSTRUMENTATION 609

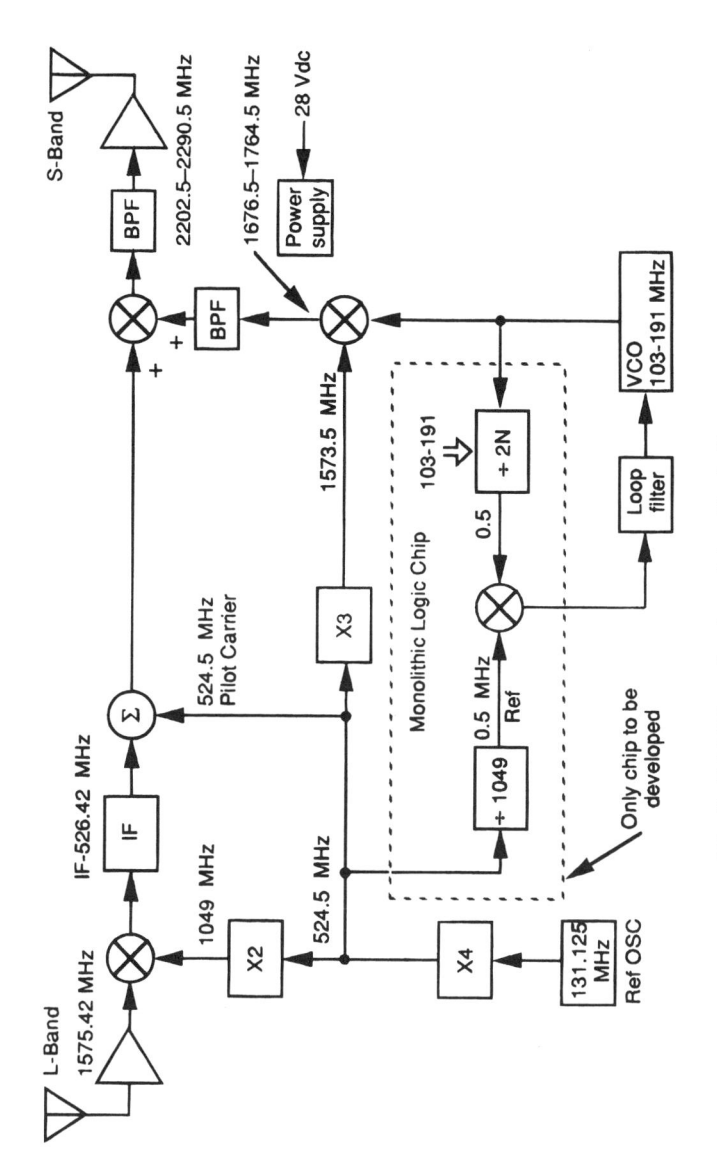

Fig. 7 RAJPO analog translator block diagram.

receives the C/A-code bandwidth portion of the rf spectrum, applies filtering and upconverts the captured spectrum to a fixed, (user-specified) S-band frequency; no signal detection or decoding is performed. The upconversion is coherently tied to a reference oscillator, also part of the translator. A pilot carrier, coherently related to the translator's reference oscillator, is added after the upconversion for TPS removal of the translator reference oscillator effects. The combined signal (S-band C/A code and pilot carrier) is then transmitted to the TPS.

2) *Current translators*: Two translator types have been built and demonstrated; i.e., the analog translator called the ballistic missile translator (BMT), and the dual frequency translator (DFT) developed for the Peacekeeper program. The dual frequency translator is a standard analog translator (L_1, C/A code) with an add-on module to translate a C/A-code bandwidth of the rf spectrum centered about the L_2 signal. Although the L_2 signal only contains the P code, it is possible to correlate a partially captured bandwidth, with only some loss in signal strength, by tracking the L_1 signal and processing the translated signals.

3) *Translator development*: Digital translators are similar to analog translators except that the translated signal is sampled at a high rate (4 million samples/s). This stream of sampled data is then digitized by an encryption device. However, the sampling losses are on the order of 5 dB over that of the analog translator development effort, and various studies have been done to investigate volume reduction. Current analog translators are about 40 in.[3] Proposals are being considered for 30, 20, and as small as 9 cubic inches.

Translator processing system—1) *Operation*: The TPS receives the translated (analog) GPS signals from one or more translators through range-owned telemetry antennas. The telemetry antennas must provide high gain to ensure a sufficient positive downlink signal-to-noise ratio (SNR) to preserve the translated GPS signal level. The translated GPS signals are input to a diversity receiver where the pilot carrier tone is tracked and removed, and the translated signals are downconverted to GPS L-band signals. The received translated signal is also predetect recorded. A reference receiver, part of the TPS, tracks all GPS satellite vehicles in view and provides data to the code-carrier tracker to aid in the translated signal processing. Figure 8 shows the block diagram of the translator processing system/vehicle tracker.

2) *Translator processing system development*: Substantial reductions in size and cost are planned for the next generation of TPS units. A multichannel TPS is planned to be housed in one of two racks. In addition, the analog tape recorders, no longer logistically supported by Ampex Corp., are planned to be replaced with solid state recording, using "flash" memory.

f. Ground transmitter (GT). A ground transmitter (GT) is a ground-based GPS satellite vehicle (SV) (sometimes called a "pseudolite") used on test and training ranges to augment/supplement the GPS constellation. Each GT transmits a GPS satellite-like signal that can be received by RAJPO-developed receivers. The actual transmitted signal consists of two carriers in phase quadrature centered at 1575.42 MHz (i.e., L_1), with each bi-phase modulated by the C/A code (one of 36 codes) and the P code (one of 37 codes). In this respect, the GT signal is identical to a satellite L_1 signal, although GTs do not broadcast L_2 (1227.6 MHz). The key difference between a GT and SV signal is the data content of the message

TEST RANGE INSTRUMENTATION

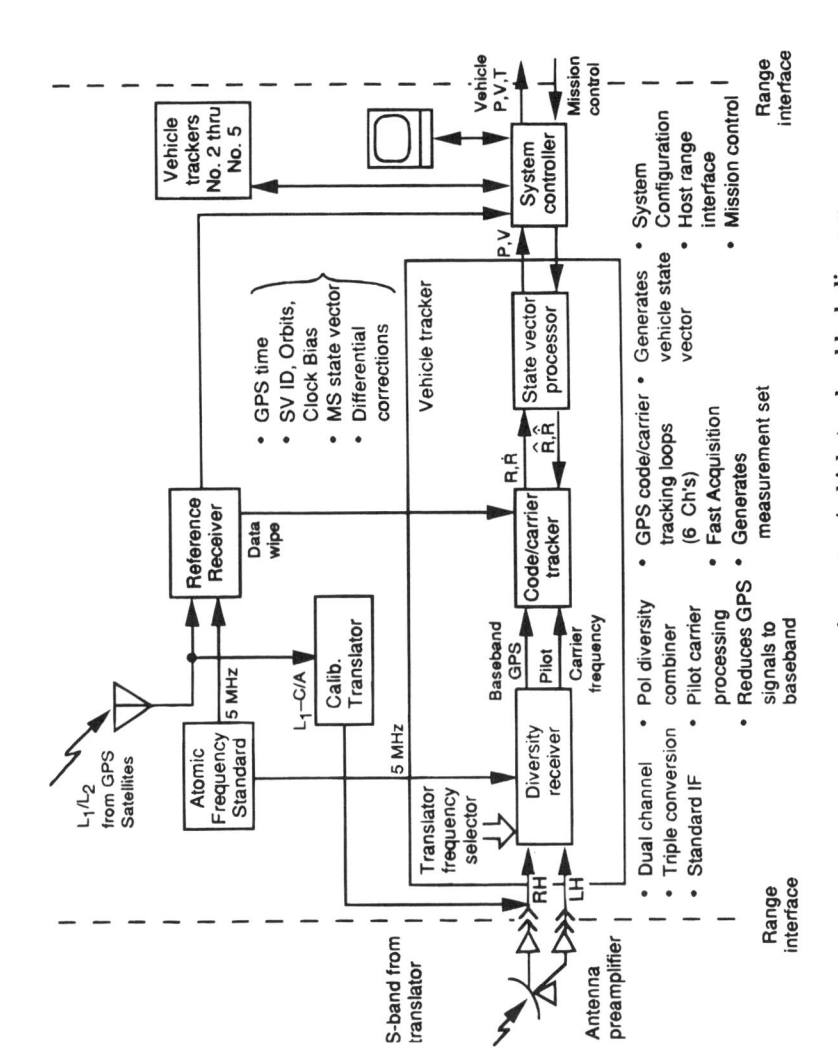

Fig. 8 Translator processing system/vehicle tracker block diagram.

datastream that modulo-two multiplies the codes. In particular, the ephemeris data must be different for a GT vs an SV.

RAJPO contracted Stanford Telecommunications, Inc. to develop GTs primarily for use with SDIO test programs. Ground transmitters are presently used in the Pacific to supplement the SV constellation. Their signals can be processed by the RAJPO receivers and the RAJPO TPS—the ground-based portion of the RAJPO GPS translator system. Figure 9 is an artist's rendition of a RAJPO model 5502 GT. Each GT simulates one GPS SV, and typically several GTs are employed over a test range to augment the GPS constellation or to account for line-of-sight blockage to an SV. GTs perform four basic functions: 1) synchronization to the GPS L_1 C/A- and P-code SV time signals—P code is primary; 2) synchronization to the GPS L_2 P-code time signal; 3) generation of a master timing reference compensated for first-order ionospheric time delay—using the L_1 and L_2 signals; and 4) simultaneous transmission of a simulated L_1 C/A- and P-code signal.

A typical GT scenario is provided in Fig. 10. The figure depicts the stand-alone operation of each GT for initial GPS system time synchronization with visible SVs and the subsequent transmission of GPS signals.

2. Global Positioning Systems Joint Program Office

The first differential GPS test was conducted by the JPO at the U.S. Army Yuma Proving Ground (YPG) in December 1979.[11] The Inverted Range Control Center (IRCC) was modified to operate as a differential GPS reference station, to compute the pseudorange corrections, and to transmit them to a test vehicle via the navigation message from a ground transmitter. The IRCC continued to operate as a reference station until 1987 and was used to monitor the space and control segments and to continue the development of differential GPS techniques.

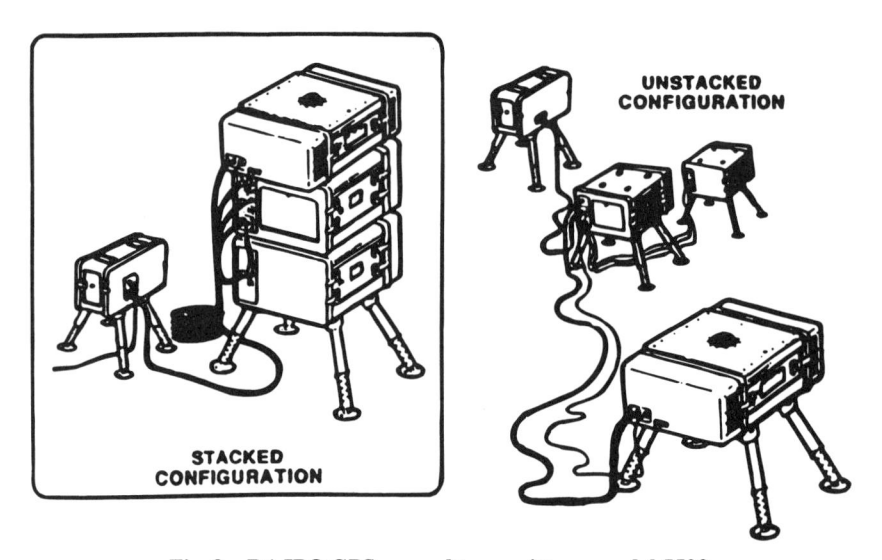

Fig. 9 RAJPO GPS ground transmitter—model 5502.

TEST RANGE INSTRUMENTATION

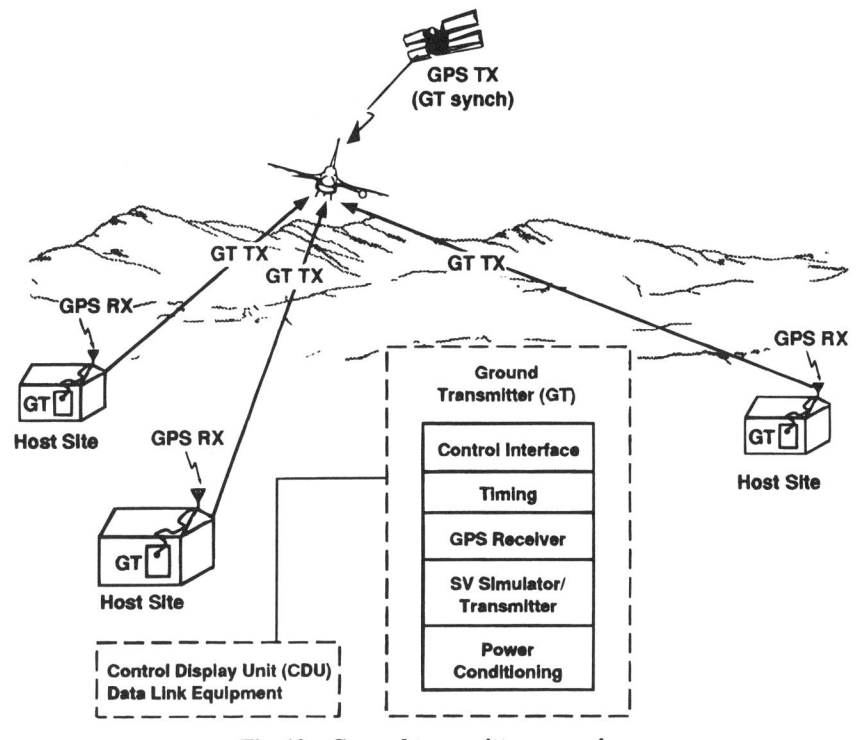

Fig. 10 Ground transmitter scenario.

During this same period, tests were conducted to validate the YPG range and to evaluate JPO Phase II GPS receivers as a reference system for range applications.

In 1985 a study was conducted to replace the IRCC with a dedicated GPS reference station. A system was built using TI-4100 GPS receivers and delivered to the JPO in 1987. An identical system was built and delivered to the B-2 Aircraft Combined Test Force (CTF) at Edwards Air Force Base and was used as the reference station to generate TSPI for test support in November 1987. A TI-4100 was used as the aircraft GPS receiver and TSPI was generated for flight test missions.

The TI-4100 reference receiver was replaced with a Collins 3A receiver in 1990. At that time, General Dynamics Services Company designed[12] and delivered one new reference station to the B-2 CTF and three systems to the JPO for test applications. Collins 3A receivers also replaced the TI-4100s in the flight test aircraft.

In 1991, responsibility for managing the four JPO reference stations called data analysis stations (DAS) was transferred to the 6585th Test Group's guidance test division, also known as the Central Inertial Guidance Test Division (CIGTF) at Holloman, Air Force Base, New Mexico. Three of the DGPS systems have been installed at Holloman Air Force Base, New Mexico, Edwards Air Force Base, California, and Melbourne, Florida to support DOD test programs. The fourth system is installed in a trailer and supports test programs on a mobile basis.

a. Equipment description.

Test platform—The GPS receivers used in the test vehicle are the Collins 3A or the Collins miniaturized airbone GPS receiver (MAGR). Raw measurement data from the RS-422 instrumentation port (IP) is recorded for postprocessing. Data are typically recorded on a PC buffer box (PCBB), which is either a 286/386 PC with a large hard disk or a digital tape recorder. If an analog tape recorder is available, the RS-422 digital data can be recorded on one channel and then downloaded after the mission to a PCBB. After each mission, the data are transported to the GPS reference station for processing and generation of TSPI data.

Ground station—The ground station configuration is shown in Fig. 11. It consists of four principle components: GPS receiver, GPS antenna, computer, and assorted input/output devices.

The GPS receiver is a Collins 3A, five-channel, two-frequency, P-code receiver modified to allow external control of tracking channels and for an external clock input. The raw pseudorange and delta range measurements and other required data from up to 12 satellites are transmitted to the computer for processing and recording. The antenna is a Dorne-Margolin.

The computer is a 80386-based system. Its real-time functions include control of the receiver, selection of satellites to be tracked, correction of measurements for propagation effects, and computation of the pseudorange corrections. In addition, the computer is used in a postprocessing mode to generate the final TSPI product.

The input/output devices are shown in Fig. 11. The primary data recording system is the Bernoulli removable 5 1/4 disk unit. The TSPI output for use by other agencies can be provided on either nine-track tape or Bernoulli disks. The printer is used to generate data products, plots, etc. for analysis.

B. Commercial Systems

The commercial industry has combined the use of differential GPS with low-cost C/A and P-code GPS receiver technology to develop small, lightweight, cost-effective, turn-key systems for range applications. The commercial vendors can provide either 1) complete turn-key systems that can be placed into operation immediately; or 2) hardware and software components that enable users to design a system to meet their requirements. Because the test and training applications have similar requirements, the training agencies are also taking advantage of the commercial equipment.

The generic GPS commercial range system block diagram is shown in Fig. 12. The GPS reference station tracks all visible satellites and pseudorange (PR) corrections are computed for each visible satellite and transmitted via the radio communications link to the mobile units. The mobile unit applies the appropriate PR corrections and performs a real-time computation to derive position and velocity. This solution is available for display in the test vehicle if required and is also transmitted back to the master control station for display and recording. The availability of the very accurate GPS time and the use of TDMA provides the capability to transmit data from 10–100 players (depending on the amount

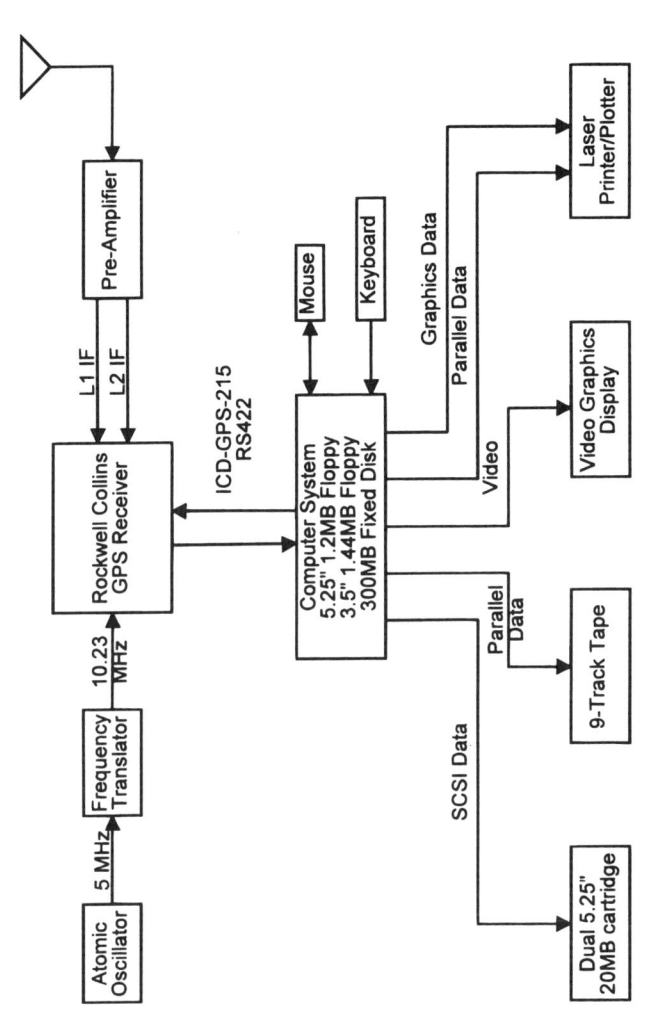

Fig. 11 Hardware configuration.

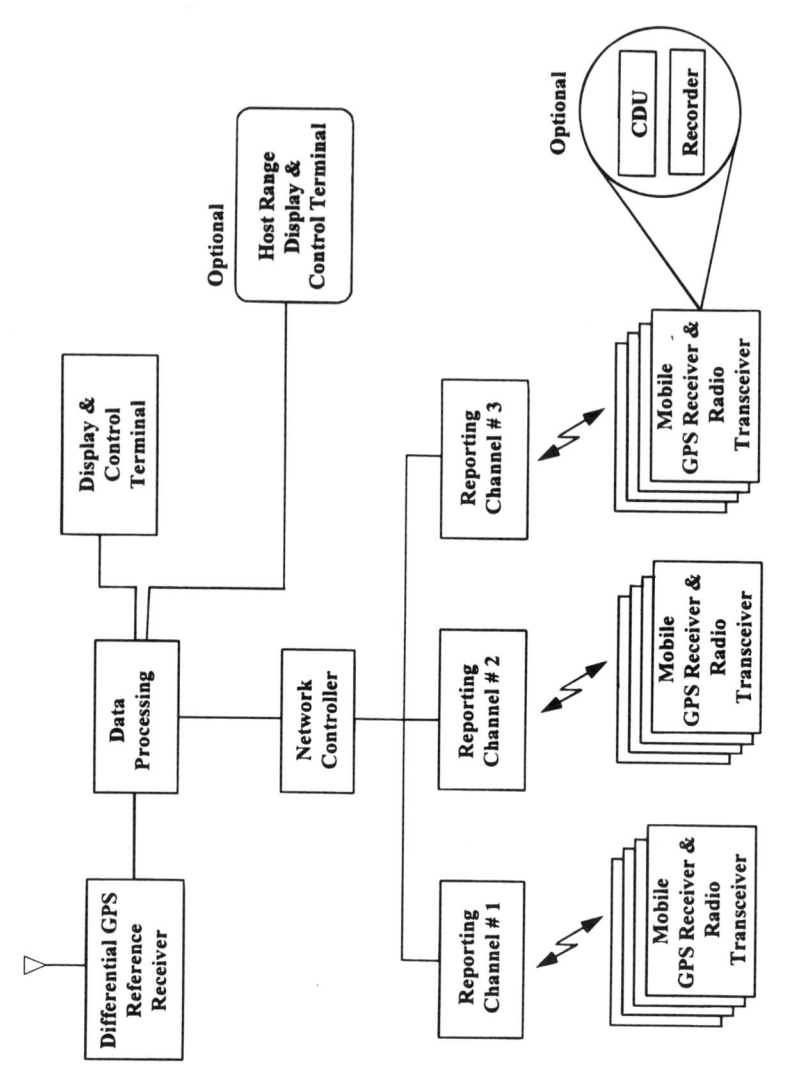

Fig. 12 Generic commercial GPS range system.

of data) on one frequency. The use of multiple frequencies can increase the number of players by the number of channels available.

The U.S. Army YPG is operating a system with a capacity for 24 players to support positioning of aircraft, helicopters, and ground targets to evaluate airborne targeting sensors. The critical problem for the design of a DGPS system was to be able to collect data from ground targets in the rough desert terrain. Yuma Proving Ground is using a system developed by Trimble Navigation to support these requirements. The system uses a Trimble 4000RL differential reference station and six-channel C/A-code receivers for the mobile units. In order to meet the requirement to link data from ground vehicles in rough terrain, Trimble used off-the-shelf low-band vlf communications radios. GPS corrections are broadcast about every 10 s with mobile unit position reports scheduled or polled during the intervening period. In range operations, a base station collects player ID, position, and velocity of each participant and displays this information on a high-resolution color display on a digitized map background to support situation status in real time.

White Sands Missile Range has procured a 10-player GPS range system to support testing of a forward area Air-Defense Command, Control, and Intelligence System. The system was developed by SRI International using off-the-shelf GPS and communication radio equipment. SRI used a NavStar PLM/XR3 for the GPS reference station, Magnavox 4200 GPS receivers for the mobile GPS receivers, and Motorola VHF rf-Modems for the data link.

The positioning systems used for training applications are very similar. However, the total system is more complex because of the requirements for information on war gaming such as RTCA, probability of kill calculations, weapon system data, etc. A generic training system block diagram is shown in Fig. 13. Examples of systems currently deployed or being developed are briefly described, and references are provided.

Training systems currently in development, test, and deployment include the following. Simulated Area Weapons Effects–Radio Frequency (SAWE-RF) is a program that addresses indirect fire weapons, training of mounted and dismounted troops using computer simulated weapons, as well as the multiple integrated laser engagement system (MILES). The Phantom Run Instrumented MILES Enhancement (PRIME) system is being developed to enhance training for armored vehicles. The Army is working on a system that combines features of PRIME and SAWE-RF called Combat Maneuvering Training Center–Instrumentation System (CMTC–IS), primarily for armored vehicle training at the Hohenfels Training area in Germany. In addition, the Army is planning to develop a transportable system that combines all aspects of modern army warfare, including close air support and defense. This system, called Mobile Automated Instrumentation Suite (MAIS) will be designed to be deployed at any location worldwide and to be operational within 5 days. Magnavox MX 7100 and MX 4200 6-channel C/A-code receivers are used by most of these training systems as the differential GPS equipment. Details on these programs can be found in Refs. 13–15.

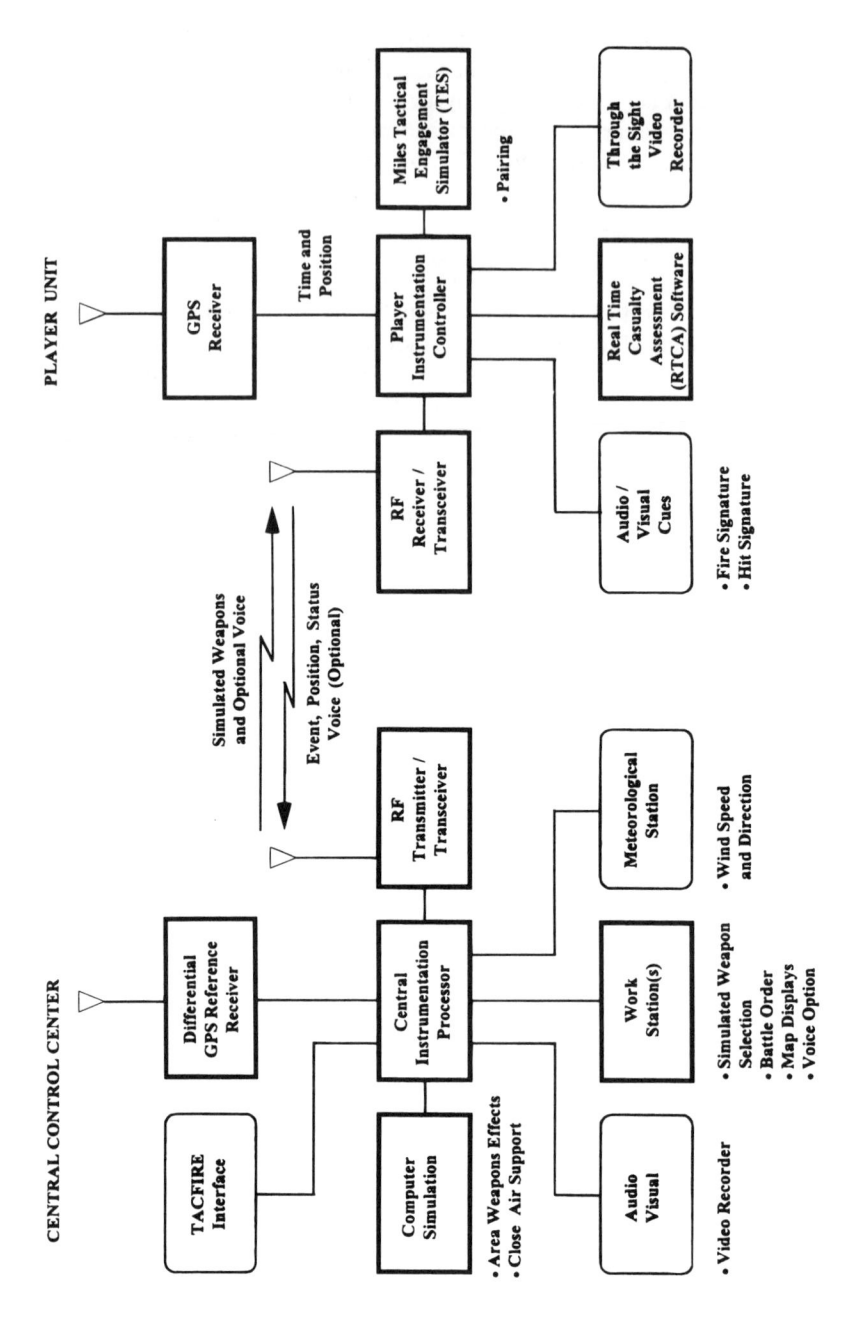

Fig. 13 Generalized battlefield training instrumentation.

C. Data Links

The area that most limits the use of differential GPS in range applications is linking of corrections and/or TSPI to where it is required. The factors must be considered are the following: 1) data rate; 2) test vehicle dynamics; 3) size of area to be covered; 4) cost; 5) number of participants; and 6) DGPS method.

Data link requirements will be addressed by area size progressively. Diameters of areas considered will be 25–50 miles, 50–200 miles, 200–1000 miles, and greater than 1000 miles or what is termed wide-area differential. For the first case, the design issues are minimal, and as previously discussed, off-the-shelf communication radio equipment along with TDMA and use of multiple frequencies can accommodate hundreds of participants at a fairly reasonable cost.

For the 50–200 mile case, the rf line-of-sight limitations become a problem. The RAJPO is using ground relays to transmit data bidirectional from the ground station. The RAJPO data link system is a custom design to handle up to 200 players over these distances. The key word is custom, which results in a high-cost solution to the problem. For limited numbers of low dynamic players, a potential solution would be cellular telephones where coverage is available. Satellite communications is a solution described in the following paragraphs.

The cases of 200–1000 miles and over 1000 miles have the same data link problem but the potential need for additional reference stations comes into play. The most effective data link solution is satellite communications.[16–19] Cost, however, at this point is a limiting factor.

The radius of coverage for one differential station depends on several factors. A P-code reference station can provide coverage over a larger area than C/A-code systems because of the dual-frequency code-tracking capability. The 6585th Test Group at Holloman has verified differential GPS accuracies less than 5 m on test aircraft at distances of up to 600 miles. The coverage for C/A-code reference stations, however, is limited to approximately 50–100 miles. [A concept useful for large or nationwide test beds is wide area DGPS (WADGPS) as described in detail in Chapter 3 of this volume.]

A "network" concept for linking reference stations and generating differential corrections over relatively broad areas is also being studied. Pseudorange corrections (PRC) are measured at each reference station and then processed at a central location to generate corrections as a function of user location. The resultant is an "iso-PRC" contour map for each satellite. Because of the slowly changing error sources and change of the line-of sight vector to each satellite, the contour maps would have to be updated frequently.[20]

A series of tests to evaluate the use of a network of GPS reference stations as a source of differential corrections was conducted for the Burlington Northern Railroad. The tests were conducted over networks of 100, 200, and 300 miles. The results showed that the network concept can be used to cover large areas and achieve accuracy requirements required by the test range community.[21]

VI. Accuracy Performance

A. Position Accuracy

Test results from evaluation of C/A-code range systems against an accepted truth reference are very limited. The Joint Program Office has conducted limited

Table 2 Joint Program Office C/A code differential
GPS test results

					Meters			
Date	No. Missions	Vehicle	Receiver	Code	VLEP	CEP	SEP	3 drms
8/87	5	Static	Ti 420	C/A	6.1	3.1	7.3	5.7
8/87	5	Truck	Ti 420	C/A	4.9	2.8	6.2	6.0
11/87	1	U-21	Tans 2Ch	C/A	3.2	4.7	6.5	6.2
1/89	4	Bac 1-11	Ti 420	C/A	3.7	5.0	7.5	8.1

VLEP, Vertical Linear Error Probable; CEP, Circular Error Probable; SEP, Spherical Error Probable; 3 drms, 3-dimensional root mean square.

testing of C/A-code receivers at YPG, and differential processing and analysis was performed on these receivers. A summary of those results are given in Table 2.

The JPO P-code differential GPS test support capability has undergone extensive testing at YPG under a variety of conditions. The YPG laser system was the truth reference for all tests. The results are summarized in Table 3.

The difference in the position accuracies between P code and C/A code seem to be approximately 3 m. P-code accuracies range from 2–4 m 3drms and C/A-code accuracies range from 6–8 m 3drms. If there is a conclusion to be made, it is that high-accuracy, high-dynamic test and training requirements shown in Table 1 will most likely require P-code systems, which requires more investment and complexity. On the other hand, the low-cost C/A-code systems can meet many of the land and low-dynamic requirements very cost effectively.

B. Velocity Accuracy

Validation of GPS velocity accuracies is even more of a problem because of the lack of accurate truth reference systems. The laser tracker velocity accuracy

Table 3 Joint Program Office P-code differential
GPS test results

					Meters			
Date	No. Missions	Vehicle	Receiver	Code	VLEP	CEP	SEP	3 drms
11/85	1	Static	R/C 3A	P	1.3	1.3	2.1	2.5
10/86	4	Conv 440	TI INAV	P	1.1	2.3	2.8	2.7
11/88	1	B-52	R/C 3A	P	1.8	1.9	3.0	3.5
12/88	1	Truck	R/C 3A	P	1.3	1.1	1.9	2.4
1/89	4	Bac 1-11	R/C 3A	P	2.3	2.5	4.0	5.0
2/89	1	Rc-135	R/C 3A	P	1.8	2.0	3.1	3.6
5/89	1	Bus	R/C 3A	P	1.3	0.8	1.7	2.3
5/89	1	F-16	R/C 3A	P	1.4	1.1	2.0	3.0
6/89	1	Rc-135	R/C 3A	P	1.6	1.2	2.3	3.8
10/89	6	C-141	R/C 3A	P	1.1	1.5	2.1	3.1
9/90	8	T-39	R/C 3A	P	1.4	1.6	2.5	2.9

is only around 0.2–0.3 m/s, which is inadequate to evaluate the GPS specification accuracy of 0.1 m/s. Spot checks have been performed with specialized systems that have verified GPS velocity accuracies of <0.1 m/s.

In 1986, General Dynamics Services Company, the JPO support contractor at YPG was able to acquire Collins 3A data from an Army velocity accuracy test conducted by Draper Laboratories. The Aerial Profiling of Terrain System (APTS) was developed by Draper laboratory for the United States Geological Survey for unique mapping applications.[22] The APTS incorporates an inertial navigation system that produces position and velocity data. Laser ranging to surveyed retroreflectors on the ground provides a companion navigation system that removes the long-term increase of position errors attributable to drift, misalignment, and gravitational anomalies in the inertial solution, and ties that solution to a local geodetic coordinate system. Recorded data is postprocessed to yield very accurate position and velocity. One-sigma errors of postprocessed data are typically 1.0 cm and 0.3 mm/s during lock-on to a retroreflector; errors increase to 50 cm and 5 mm/s 150 s after lock (assuming one retroreflector is acquired every 5 min).

During one of the aircraft missions, a Collins 3A receiver was aboard, and data were recorded. General Dynamics produced postmission differential GPS solutions for position and velocity using a reference station located at YPG. The differential solution was compared with the APTS truth trajectory processed by the Draper Laboratory. The standard deviation of the differences in three axis were the following:

East	0.03 m/s
North	0.03 m/s
Vertical	0.05 m/s

A second velocity verification test was conducted by the 6585th Test Group, Guidance Test Division, also know as the Central Inertial Guidance Test Facility at Holloman Air Force Base, New Mexico using the instrumented test track.[23] Low dynamic, constant velocity tests at from 20–35 mph were conducted using an unaided Collins 3A receiver. The standard deviation of the differences in three axis were the following:

East	0.03 m/s
North	0.03 m/s
Vertical	0.05 m/s

High velocity, 2–3g rocket tests were conducted also, but as expected, the unaided receiver velocity accuracies were much worse. Although the receiver maintained lock, the velocity errors were as much as 2 m/s at maximum acceleration.

VII. Future Developments

A. National Range

The concept of a national range using WADGPS concepts discussed earlier in this chapter has the potential for very cost effectively satisfying many DOD as well as civilian test and training requirements. The DOD SA/AS requirement placed on GPS complicates the implementation of an authorized P-code national range, but it could be done. The next step would be to determine how DOD and civilian applications could both use such a range and yet not violate any security aspects. The national range concept should be strongly considered as the next step in supporting test and training applications.

B. Kinematic Techniques

The other development required to meet the high-accuracy test and training requirements is carrier phase tracking technology. The civilian community is pursuing this area, and progress is good, but the host vehicle dynamics for most civilian applications are very low. Development is required to ensure continuous trajectories in an automated procedure for high-dynamic platforms.

References

[1] Blackwell, E. G., Cline, J. F., and Erb, E. A., "Technical and Economic Feasibility of Airborne and Satellite Instrumentation Systems to Augment National Test and Evaluation Resources," Final Rept., Contr. MDA903-78-C-0405, SRI International, Menlo Park, CA, July 1979.

[2] Reinhart, W. L., "Application of the Navstar Global Positioning System on Instrumented Ranges," Masters Thesis, Naval Postgraduate School, Monterey, CA, March 1981.

[3] Fredericksen, J. N., "Applicability of Navstar GPS to Test and Training," MITRE Corp. Rept., Contr. F19628-80-0001, May 1980.

[4] Sieg, W. D., "Applying GPS to Test Ranges," *ITEA Journal of Test and Evaluation*, Vol. 10, No. 2, 1989, p. 24.

[5] Wells, L. L., "Real-Time Missile Tracking with GPS," *Navigation*, Vol. 28, No. 3, 1984, p. 224.

[6] Mai, R. W., "Air and Ground Vehicle Tracking System," *Proceedings of the Fifth International Technical Meeting of the Satellite Division of the ION*, Institute of Navigation, Washington, DC, Sept. 16–18, 1992, p. 101.

[7] RTCM Paper 134-89/SC 104-68, "RTCM Recommended Standards for Differential Navstar GPS Service, Version 2.0," Jan. 1, 1990.

[8] Brown, A., Sward, W., Pickett, R., Greenberg, R., and Wildhagen, P., "Test Results of the Advanced Translator Processing System," *Proceedings of the Fourth International Technical Meeting of the Satellite Division of the ION*, Institute of Navigation, Washington, DC, Sept. 11–13, 1991, p. 573.

[9] McConnel, J. B., Greenberg, R. H., Pickett, R. B., Wildhagen, P. C., and Brown, A., "Advances in GPS Translator Technology," *Proceedings of the International Technical Meeting of the Satellite Division of the ION*, Institute of Navigation, Washington, DC, Sept. 27–29, 1989, p. 115.

[10] Sturza, M. A., and Brown, A. K., "Digital Translator Design Trades," *Proceedings of the Fifth International Technical Meeting of the Satellite Division of the ION*, Institute of Navigation, Washington, DC, Sept. 16–18, 1992, p. 687.

[11] Teasley, S. P., Hoover, W. M., and Johnson, C. R., "Differential GPS Navigation," Texas Instruments, Inc., Plans Symposium, Dec. 1980.

[12] Robbins, J. E., "Reference Trajectories From GPS Measurements," *Navigation*, Vol. 35, No. 1, 1988, p. 89.

[13]Eastwood, R. A., and Sharpe, R. T. "The Use of GPS to Enhance the Military Training Environment," Paper presented at the NATO GPS Symposium, Brussels, Belgium, Nov. 20–22, 1990.

[14]Peters, R. L., and Lewis, K. M., "Use of GPS as the Position Location Subsystem for the Army's Prime Training Range System," *Proceedings of the Fourth International Technical Meeting of the Satellite Division of the ION,* Institute of Navigation, Washington, DC, Sept. 11–13, 1991, p. 593.

[15]Truog, B., and Ravenis, J., "Combat Manuever Training Center (CMTC) High Dynamic Player Integration," *Proceedings of the Fourth International Technical Meeting of the Satellite Division of the ION,* Institute of Navigation, Washington, DC, Sept. 11–13, 1991, p. 649.

[16]Blanchard, W. F., "Differential GPS Using a Dedicated INMARSAT Satellite Data Link," *Proceedings of the Third International Technical Meeting of the Satellite Division of the ION,* Institute of Navigation, Washington, DC, Sept. 19–21, 1990, p. 237.

[17]Nagle, J. R., "Wide Area Differential Corrections (WADC) from Global Beam Satellites," *Proceedings of the IEEE 0-7803-0468-3/92,* Institute of Electrical and Electronics Engineers, New York, 1992, p. 383.

[18]Slack, E. R., "Towards a Global Differential Service," *Proceedings of the Third International Technical Meeting of the Satellite Division of the ION,* Institute of Navigation, Washington, DC, Sept. 19–21, 1990, p. 323.

[19]Zachmann, G. W., "Differential GPS Transmissions By Geostationary L-Band Satellites," *Sea Technology,* Vol. 31, No. 5, 1990, p. 57.

[20]Loomis, P., Sheynblatt, L., and Mueller, T., "Differential GPS Network Design," *Proceedings of the Fourth International Technical Meeting of the Satellite Division of the ION,* Institute of Navigation, Washington, DC, Sept. 11–13, 1991, p. 511.

[21]Robbins, J., "Evaluation of Differential GPS Capabilities for Railroad Positioning Requirements," Paper presented at the DGPS 91 Symposium, Braunschweig, Germany, 1991.

[22]Greenspan, R. L., "APTS/GPS Measurement Task," Draper Laboratory Final TR, Cambridge, MA, Feb. 1986.

[23]MSD-DP-90-04, "Navstar GPS User Equipment Sled Test," Munitions Systems Division, Data Package, March 1990.

Author Index

Abby, D. G. ...593
Ahn, I-S. ...303
Braff, R. ...327
Brown, R. G. ...143
Cobb, H. S. ...427
Cohen, C. E. ...427, 519
Daly, P. ...243
Dorfler, J. ...327
Elrod, D. B. ...51
Enge, P. K. ...3, 117, 169
Eschenbach, R. ...375
Fitzgibbon, K. T. ...397
French, R. L. ...275
Goad, C. ...501
Greenspan, R. L. ...187
Kee, C. ...81
Klepczynski, W. J. ...483
Larson, K. M. ...539
Lawrence, D. G. ...427
Lee, Y. C. ...221
Lightsey, E. G. ...461
Misra, P. N. ...243
O'Connor, M. L. ...397
Parkinson, B. W. ...3, 397, 427
Pervan, B. S. ...427
Pietrazewski, D. ...303
Powell, J. D. ...427
Sennott, J. ...303
Van Dierendonck, A. J. ...51, 117
van Graas, F. ...169
Yunck, T. P. ...559

Subject Index

Accumulated Doppler, 510
Accuracy summary, 331, 544, 549
Additional secondary factor, 187
Advanced Public Transportation Systems (APTS), 280
Advanced, Rural Transportation Systems (ARTS), 280
Advanced Traffic Management Systems (ATMS), 279
Advanced Traveler Information Systems (ATIS), 280
Advanced Vehicle Control Systems (AVCS), 280
Age, 111
Air traffic control (ATC), 327–329
Aircraft conflicts, 358–361
Aircraft model, 398–399, 401
Airport surface applications, 361
Alarm limit, 117
Alignment
  gyrocompassing, 206
  inflight, 206–213
  leveling, 206, 208
  north-seeking, 208–209
Almanac, GLONASS, 252, 254
Ambiguity
  resolution, 577
  search, 509–510
Antispoofing, 516
Applied range accuracy evaluation, 117–141
Approximate Radial-Error Protected (ARP), 152–153
Area navigation (RNAV), 328
Attitude control, spacecraft
  bandwidth, 461
  dynamics, 468
  results, 468
Attitude and Heading Reference System (AHRS), 393
Attitude determination
  accuracy, 467
  dynamic filtering, 468
  sensor calibration, 466

system considerations, 461
Augmentations, 340
Augmented state vector, 566
Automatic dependent surveillance (ADS)
  availability, 330, 332
Automatic landing
  accuracy, 399, 412
  air speed sensor, 400, 412
  aircraft model, 398–399, 401
  carrier-phase differential GPS (CDGPS), 398, 407–410, 411
  differential GPS (DGPS), 398, 407, 411
  FAA landing requirements, 397, 398
  flight phases, 397, 398, 399–400
  Flight Technical Error (FTE), 409, 410
  glide slope deviation, 400, 404–405
  GPS measurement error, 407–408
  Inertial Measurement Unit (IMU), 397
  integrity, 398, 401
  integrity beacons, 401
  Microwave Landing System (MLS), 397, 399
  Navigation System Error (NSE), 409, 410
  optimal controller, 405–406
  optimal estimator, 398, 421–422
  radar altimeter, 398, 410
  Total System Error (TSE), 410
  tunnel concept, 399
  wind disturbances, 398, 403–404
Automatic Route Control System (ARCS), 283
Automobile navigation, 275–301
Availability, 178–179
Aviation landing
  ILS aids, 377
  MLS aids, 384
  precision types, 381
  SCAT 1 aids, 391
Aviation navigational aids
  DME, 379

SUBJECT INDEX

Aviation navigational aids (*continued*)
 LORAN, 379
 NDB, 381
 Omega, 381
 VOR, 378

Barometric altimeter, 223, 227
Barometric altimeter aiding, 222
Baseline (vector), 501, 506
Batch least squares (BLS), 89, 94, 562–563
Bent-pipe transponder, 141
Bosch Travelpilot, 294–296

Calibration of Loran, 174–176
Carrier
 aiding, 544–554
 frequency, 250
 phase, 26
 pseudorange bias estimation, 567–568
 tracking, 193, 199
Category 1 Precision, 128
Channel, 250
Chayka, 169
Cicada, 247
CIGNET, 544
Clock
 coasting, 222, 228
 errors, 544
Code Division Multiple Access (CDMA), 121, 249
Code phase control, 139
Collision avoidance using GPS, 393
Commercial Vehicle Operations (CVO), 280
Commission error, 573
Common view time transfer, 490
Continuity of service, 332
Coordinate system, 284
Co-seismic deformation, 551
Cross-chain synchronization, 171
Crustal deformation, 550
Current state, 563, 564
Curved precision approaches, 356
Cycle slip, 171

Data, 248
Data links, 598, 599, 619
Dead reckoning, 277, 284
Decision height, 128
Detection level, 151
Detection of failure, 151
DHmax, 153
Differential, 81
Differential GPS (DGPS), 81, 335, 350, 361
Differential odometer, 281, 285–286
Differential pseudorange solution, 510

Digital road maps, 278, 286–287
Dilution of precision (DOP), 508–509
 fixed, 508–509
 float, 508–509
Dispersion, 507–508
Distance/speed sensors, 284
Double difference, 505, 576
Down-looking GPS, 584
Dual-frequency receiver, 85
Dynamic model error, 565
Dynamic orbit determination, 560, 571
Dynamical constraint, 561, 563
Dynamical response, 193, 196

Earth
 deformation, 550
 observing system, 578–580
 orientation, 561
 rotation, 551
Electronic Route Guidance System (ERGS), 283
Embedded GPS, 197–199
Ephemeris, 253
Ephemeris errors, 88, 92
Epoch state, 561
Error states, 193–195, 197, 204
Etak Navigator, 294
Exclusion of failure, 163
Extreme ultraviolet explorer, 582

FAA landing requirements, 397, 398
Failure
 detection and isolation (FDI), 158
 exclusion (FDE), 163
Fast corrections, 132
Frequency division multiplex (FDM), 249
Federal Radionavigation Plan (FRP), 334
Fiducial networks, 544
Filtering
 fixed-gain, 196
First-order Gauss-Markov process, 567
Flight phases, 397, 398, 399–400
Flight Technical Error (FTE), 409, 410
Flight test, 347, 350
Forward error correction, 121
Four measurement filter/smoother, 510–515
Fundamental equation of mechanics, 561

GADACS spacecraft experiment, 462
Geocenter, 553
Geodesy (geodetic), 550
Geometric dilution of precision (GDOP), 509

SUBJECT INDEX

Geostationary navigation message, 133
Geostationary satellite ephemeris estimation, 125
Geostationary satellites, 125
Global, 574
Global differential tracking, 574–577
Global Navigation Satellite System (GLONASS)
  almanac, 252, 254
  clocks, 252
  coverage, 252
  launches, 247
  navigation satellites, 243
  orbits, 243, 244
  performance, 261
  signal design, 249
  signal spectrum, 249
  time, 254
  user range error, 260
GPS augmentation, 340
GPS avionics
  AHRS, 393
  DGPS, 390
  Dzus mount types, 390
  features, 391
  handheld types, 389
  installation, 388
  integration with LORAN, 392
  integration with Omega, 392
  number of channels, 389
  panel mount types, 389
GPS Global Tracking Network, 81
GPS Integrity Channel (GIC), 336
GPS Requirements for Aviation
  Avionics Interface, 384
  certification, 384
  Pilot Interface, 387
GPS-squitter, 367
Gravity tuning, 588
Ground monitoring, 117
Ground network, 117
Ground repetition interval, 169
Ground track repeat, 243
Ground track system, 243
Ground transmitters, 614
Groundwave, 169

Hatch filter, 81–114
Heading/heading-change sensors, 285
Highly elliptical orbiters, 565–589
Horizontal dilution of precision (HDOP), 182
Hyperbolic line of position, 169–185

Ideal oscillator, 501
Ideal pseudorange, 510

Inadmissible geometries, 152
Independent observables, 501
Inertial Measurement Unit (IMU), 397
INMARSAT, 117
Instrument flight rules (IFR), 327
Instrument Landing System (ILS), 328, 377
Integer ambiguity, 501
Integrated Doppler, 243
Integration
  algorithms, 243
  architectures (uncoupled, loosely-coupled, tightly-coupled), 243
Integrity, 130, 143, 331, 398, 401
Integrity beacons, 401
Integrity definition, 143
Integrity monitoring, 259, 267
Intelligent transportation systems, 275
Intelligent vehicle highway systems, 275, 279
International Civil Aviation Organization (ICAO), 224, 334
International GPS Service for Geodynamics (IGS), 546
International terrestrial reference frame, 91
  coordinate frame, 105
Inverted GPS, 584
Ionosphere
  ionosphere errors, 401
  ionospheric correction, 401
  ionospheric refraction, 507
  ionospheric time delay, 89

Joint Program Office (JPO), 604, 618

Kalman filter, 291
  Kalman gain, 564
Kinematic orbit determination, 569–571
Klobuchar's model, 89

Lageos, 565
Landsat-4, 560
Latency, 111
Launch, 247
Law of height vs barometric pressure, 239, 223–227
Least-squares residuals method, 147
Likelihood function, 187
Linearly dependent data, 507
Linearly independent data, 506–507
Local area differential GPS (LADGPS), 3
Long range navigation system, 169–185
Loop threshold, 124

Loran chains, 171, 172
Loran-C, 169–185, 277

Map matching, 283, 288
Marine
  channel clearance, 308
  comparative footprint clearance, 315
  control, 313
  coordinate systems, 306
  DGPS, 322, 304
  filter and controller design, 313
  hazard warning, 305, 323
  requirements, 303
  risk assessment, 306, 321
  sensor model, 307, 310
  sensor/ship bandwidth ratio, 314
  stochastic regulator, 313
  vessel dynamics model, 308
  waypoint steering, 308, 312
Markov process, 98
Masks, 132
Master Control Station, 85
Master station, 85, 89, 94, 98
Matrix of regression coefficients, 564
Maximum separation of solutions method, 150
Measurement update, 564
Message, 252
Message format, 131
Method of least squares, 561, 562
Microwave Landing System (MLS), 328
Midcontinent gap, 169
Minimum baseline, 81
Minimum norm solution, 81
Mobile data communications, 276, 292
Model trajectory, 561, 565
Monitor stations, 83–114, 276
MOPS, 144
Multipath, 467
Multiple access interference, 121

Narrow-lane ambiguity
Navigation
  equations, 179–182
  payload, 117
NAVSTAR, 243
NavTrax fleet management system, 298–299
Networks, 544
Newton's second law of motion, 561

No-ion combination, 508
Noise mitigation, 191, 206
Noncoherent delay lock loop, 29
Nondirectional beacons (NDB), 377–378
Noninterference with GPS, 121
Nonlinear static estimation (NSE), 89
Nonprecision approach, 126, 335
Nuisance variable, 503
NUVEL1, 551

Omission error, 573
Optimal controller, 405
Optimal estimator, 407
Orbit, 243, 244
Orbit error, 546
Outages, 189, 199
Overdetermined, 94

Parallel runways, 355
Parity, 148
Parity method, 148–150
PE-90 geocentric coordinate frame, 260
Phase center variations, 544–554
Photogrammetric mission, 517
Plate tectonics, 550
Polar motion (pole postions), 553
Polynomial, 246
Position, 246, 253
Position dilution of precision (PDOP), 509
Precise orbit determination, 574–576
Precise positioning users, 128
Precision, 547
Precision approach, 128, 344
Prefit residuals, 562
Primary radar, 362
Process noise, 566
Processing transponder, 140
Protection radius, 151
Pseudoepoch state, 564
Pseudolite, 51
Pseudorandom noise (PRN), 250
(P, V, T) Solution, 188

Qualcomm Automatic Satellite Position Reporting System (QASPR), 276

Radar altimeter, 398, 410
Radio determination satellite services (RDSS), 276
Radio-frequency, 244
Radio Technical Commission for Aeronautics (RTCA), 332

National Airspace System (NAS), 327, 361

SUBJECT INDEX

Range applications, 595, 604
Range Applications Joint Program Office, 595
Range decorrelation, 92
Range-comparison method, 146
Rapid static surveying, 512
Receiver autonomous integrity monitoring (RAIM), 143, 164, 221, 336
  aided, 143–144
  availability, 145, 222, 229
  availability of detection function, 221
  availability of identification function, 221
  definition, 143–145
  stand-alone GPS, 143–144
Reduced dynamic orbit determination, 571
Reference correction, 10
Reference frame, 546
Reference station, 613, 619
Reference trajectories, 619
Refraction, 506
Regression equation, 562
Reliability, 117
Remote area operations, 357
Required navigation performance, 329
Right Ascension of the Ascending Node (RAAN), 245
Rotation, 553
Route guidance, 277, 278, 281, 282, 294, 296

Satellite
  clock error, 92
  ephemeris, 129
  laser ranging, 544–554
Satellite Operational Implementation Team, 333
Screening out poor geometries, 152–155
SEASAT, 565
Secondary factor, 174
Secondary surveillance radar, 362
Selective availability, 81, 88
Sensitivity matrix, 564
Sensor error, 335
SGS-85 geocentric coordinate frame, 260
Sidereal, 243
Signature sequences, 122
Single difference, 504
Single-frequency ionosphere calibration, 582
Single-frequency receiver, 85
Skywave, 169
SLOPEmax, 153–154
Solar pressure, 243–271
Spherical harmonic expansion, 572
Spread spectrum signaling, 244, 249

Spread-spectrum multiple access, 121
Square-root information filter, 564
State transition, 563
Steady state variance, 567
Stochastic force model, 587
Strain, 544–554
Surveillance, 362–370

Technical Standard Order C-129 TSO C-129, 222
Terminal, 328
Test and evaluation (T&E), 594
Test instrumentation, 596
Test range, 593
Test range requirements, 594
Three measurement filter/smoother, 514–516
Time
  availability, 125
  difference, 172, 179–180
  interval, 487
  transfer, 490
  of transmission, 176
  update, 564, 566
Time Space Position Information (TSPI), 593
TOPEX, 544
TOPEX/Poseidon, 587
Total System Error (TSE), 335
Toyota Electro-Multivision, 296, 297
Tracking and Data Relay Satellites (TDRS), 584
Traffic Alert and Collision Avoidance System (TCAS), 365
Training requirements, 594, 596
Transit, 519
Transition matrix, 564, 566
Translators, 599, 601
Traveling ionospheric disturbance (TID), 507
TravTek, 297
Troposphere
  refraction, 507, 510
Tunnel concept, 399

Undetermined case, 94
United States Air Force, 462
Universal Time, 484
Unmodeled contributions, 507
User differential range errors, 118

Variational equations, 563
Vector corrections, 118
Vehicle tracking, 275
Velocity, 248, 253
Very High Frequency Omnirange (VOR), 328

Very long baseline interferometry, 549
Visual flight rules (VFR), 362
Volume search technique, 509–510

Weighted HDOP, 182
WGS-84
  geocentric coordinate frame, 260
  SGS-85 transformation, 260
White noise error model, 567

Wide Area Augmentation System (WAAS)
  master stations, 353
  reference stations, 353
Wide Area Differential GPS (WADGPS), 81–114, 350
  WADGPS correction message, 92
Wide-lane ambiguity, 512
Wind disturbances, 398, 403–404
Working Group, 119

PROGRESS IN ASTRONAUTICS AND AERONAUTICS
SERIES VOLUMES

*1. Solid Propellant Rocket Research (1960)
Martin Summerfield
Princeton University

*2. Liquid Rockets and Propellants (1960)
Loren E. Bollinger
Ohio State University
Martin Goldsmith
The Rand Corp.
Alexis W. Lemmon Jr.
Battelle Memorial Institute

*3. Energy Conversion for Space Power (1961)
Nathan W. Snyder
Institute for Defense Analyses

*4. Space Power Systems (1961)
Nathan W. Snyder
Institute for Defense Analyses

*5. Electrostatic Propulsion (1961)
David B. Langmuir
Space Technology Laboratories, Inc.
Ernst Stuhlinger
NASA George C. Marshall Space Flight Center
J.M. Sellen Jr.
Space Technology Laboratories, Inc.

*6. Detonation and Two-Phase Flow (1962)
S.S. Penner
California Institute of Technology
F.A. Williams
Harvard University

*7. Hypersonic Flow Research (1962)
Frederick R. Riddell
AVCO Corp.

*8. Guidance and Control (1962)
Robert E. Roberson
Consultant
James S. Farrior
Lockheed Missiles and Space Co.

*9. Electric Propulsion Development (1963)
Ernst Stuhlinger
NASA George C. Marshall Space Flight Center

*10. Technology of Lunar Exploration (1963)
Clifford I. Cumming
Harold R. Lawrence
Jet Propulsion Laboratory

*11. Power Systems for Space Flight (1963)
Morris A. Zipkin
Russell N. Edwards
General Electric Co.

*12. Ionization in High-Temperature Gases (1963)
Kurt E. Shuler, Editor
National Bureau of Standards
John B. Fenn
Associate Editor
Princeton University

*13. Guidance and Control–II (1964)
Robert C. Langford
General Precision Inc.
Charles J. Mundo
Institute of Naval Studies

*14. Celestial Mechanics and Astrodynamics (1964)
Victor G. Szebehely
Yale University Observatory

*15. Heterogeneous Combustion (1964)
Hans G. Wolfhard
Institute for Defense Analyses
Irvin Glassman
Princeton University
Leon Green Jr.
Air Force Systems Command

*16. Space Power Systems Engineering (1966)
George C. Szego
Institute for Defense Analyses
J. Edward Taylor
TRW Inc.

*17. Methods in Astrodynamics and Celestial Mechanics (1966)
Raynor L. Duncombe
U.S. Naval Observatory
Victor G. Szebehely
Yale University Observatory

*18. Thermophysics and Temperature Control of Spacecraft and Entry Vehicles (1966)
Gerhard B. Heller
NASA George C. Marshall Space Flight Center

*19. Communication Satellite Systems Technology (1966)
Richard B. Marsten
Radio Corporation of America

*20. Thermophysics of Spacecraft and Planetary Bodies: Radiation Properties of Solids and the Electromagnetic Radiation Environment in Space (1967)
Gerhard B. Heller
NASA George C. Marshall Space Flight Center

*Out of print.

*21. Thermal Design
Principles of Spacecraft
and Entry Bodies (1969)
Jerry T. Bevans
TRW Systems

*22. Stratospheric
Circulation (1969)
Willis L. Webb
*Atmospheric Sciences
Laboratory, White Sands,
and University of Texas at
El Paso*

*23. Thermophysics:
Applications to Thermal
Design of Spacecraft
(1970)
Jerry T. Bevans
TRW Systems

24. Heat Transfer and
Spacecraft Thermal
Control (1971)
John W. Lucas
Jet Propulsion Laboratory

25. Communication
Satellites for the 70's:
Technology (1971)
Nathaniel E. Feldman
The Rand Corp.
Charles M. Kelly
The Aerospace Corp.

26. Communication
Satellites for the 70's:
Systems (1971)
Nathaniel E. Feldman
The Rand Corp.
Charles M. Kelly
The Aerospace Corp.

27. Thermospheric
Circulation (1972)
Willis L. Webb
*Atmospheric Sciences
Laboratory, White Sands,
and University of Texas at
El Paso*

28. Thermal
Characteristics of the
Moon (1972)
John W. Lucas
Jet Propulsion Laboratory

*29. Fundamentals of
Spacecraft Thermal
Design (1972)
John W. Lucas
Jet Propulsion Laboratory

30. Solar Activity
Observations and
Predictions (1972)
Patrick S. McIntosh
Murray Dryer
*Environmental Research
Laboratories, National
Oceanic and Atmospheric
Administration*

31. Thermal Control and
Radiation (1973)
Chang-Lin Tien
*University of California at
Berkeley*

32. Communications
Satellite Systems (1974)
P.L. Bargellini
COMSAT Laboratories

33. Communications
Satellite Technology (1974)
P.L. Bargellini
COMSAT Laboratories

*34. Instrumentation for
Airbreathing Propulsion
(1974)
Allen E. Fuhs
Naval Postgraduate School
Marshall Kingery
*Arnold Engineering
Development Center*

35. Thermophysics and
Spacecraft Thermal
Control (1974)
Robert G. Hering
University of Iowa

36. Thermal Pollution
Analysis (1975)
Joseph A. Schetz
*Virginia Polytechnic
Institute*
ISBN 0-915928-00-0

*37. Aeroacoustics: Jet
and Combustion Noise;
Duct Acoustics (1975)
Henry T. Nagamatsu, Editor
*General Electric Research
and Development Center*
Jack V. O'Keefe, Associate
Editor
The Boeing Co.
Ira R. Schwartz, Associate
Editor
*NASA Ames Research
Center*
ISBN 0-915928-01-9

*38. Aeroacoustics: Fan,
STOL, and Boundary
Layer Noise; Sonic Boom;
Aeroacoustics
Instrumentation (1975)
Henry T. Nagamatsu, Editor
*General Electric Research
and Development Center*
Jack V. O'Keefe, Associate
Editor
The Boeing Co.
Ira R. Schwartz, Associate
Editor
*NASA Ames Research
Center*
ISBN 0-915928-02-7

39. Heat Transfer with
Thermal Control
Applications (1975)
M. Michael Yovanovich
University of Waterloo
ISBN 0-915928-03-5

*40. Aerodynamics of
Base Combustion (1976)
S.N.B. Murthy, Editor
J.R. Osborn, Associate
Editor
Purdue University
A.W. Barrows
J.R. Ward,
Associate Editors
*Ballistics Research
Laboratories*
ISBN 0-915928-04-3

---

*Out of print.

41. **Communications Satellite Developments: Systems (1976)**
Gilbert E. LaVean
Defense Communications Agency
William G. Schmidt
CML Satellite Corp.
ISBN 0-915928-05-1

42. **Communications Satellite Developments: Technology (1976)**
William G. Schmidt
CML Satellite Corp.
Gilbert E. LaVean
Defense Communications Agency
ISBN 0-915928-06-X

*43. **Aeroacoustics: Jet Noise, Combustion and Core Engine Noise (1976)**
Ira R. Schwartz, Editor
NASA Ames Research Center
Henry T. Nagamatsu, Associate Editor
General Electric Research and Development Center
Warren C. Strahle, Associate Editor
Georgia Institute of Technology
ISBN 0-915928-07-8

*44. **Aeroacoustics: Fan Noise and Control; Duct Acoustics; Rotor Noise (1976)**
Ira R. Schwartz, Editor
NASA Ames Research Center
Henry T. Nagamatsu, Associate Editor
General Electric Research and Development Center
Warren C. Strahle, Associate Editor
Georgia Institute of Technology
ISBN 0-915928-08-6

*45. **Aeroacoustics: STOL Noise; Airframe and Airfoil Noise (1976)**
Ira R. Schwartz, Editor
NASA Ames Research Center
Henry T. Nagamatsu, Associate Editor
General Electric Research and Development Center
Warren C. Strahle, Associate Editor
Georgia Institute of Technology
ISBN 0-915928-09-4

*46. **Aeroacoustics: Acoustic Wave Propagation; Aircraft Noise Prediction; Aeroacoustic Instrumentation (1976)**
Ira R. Schwartz, Editor
NASA Ames Research Center
Henry T. Nagamatsu, Associate Editor
General Electric Research and Development Center
Warren C. Strahle, Associate Editor
Georgia Institute of Technology
ISBN 0-915928-10-8

*47. **Spacecraft Charging by Magnetospheric Plasmas (1976)**
Alan Rosen
TRW Inc.
ISBN 0-915928-11-6

48. **Scientific Investigations on the Skylab Satellite (1976)**
Marion I. Kent
Ernst Stuhlinger
NASA George C. Marshall Space Flight Center
Shi-Tsan Wu
University of Alabama
ISBN 0-915928-12-4

49. **Radiative Transfer and Thermal Control (1976)**
Allie M. Smith
ARO Inc.
ISBN 0-915928-13-2

*50. **Exploration of the Outer Solar System (1976)**
Eugene W. Greenstadt
TRW Inc.
Murray Dryer
National Oceanic and Atmospheric Administration
Devrie S. Intriligator
University of Southern California
ISBN 0-915928-14-0

51. **Rarefied Gas Dynamics, Parts I and II(two volumes) (1977)**
J. Leith Potter
ARO Inc.
ISBN 0-915928-15-9

52. **Materials Sciences in Space with Application to Space Processing (1977)**
Leo Steg
General Electric Co.
ISBN 0-915928-16-7

53. **Experimental Diagnostics in Gas Phase Combustion Systems (1977)**
Ben T. Zinn, Editor
Georgia Institute of Technology
Craig T. Bowman, Associate Editor
Stanford University
Daniel L. Hartley, Associate Editor
Sandia Laboratories
Edward W. Price, Associate Editor
Georgia Institute of Technology
James G. Skifstad, Associate Editor
Purdue University
ISBN 0-915928-18-3

---

*Out of print.

54. Satellite Communication: Future Systems (1977)
David Jarett
TRW Inc.
ISBN 0-915928-18-3

55. Satellite Communications: Advanced Technologies (1977)
David Jarett
TRW Inc.
ISBN 0-915928-19-1

56. Thermophysics of Spacecraft and Outer Planer Entry Probes (1977)
Allied M. Smith
ARO Inc.
ISBN 0-915928-20-5

57. Space-Based Manufacturing from Nonterrestrial Materials (1977)
Gerald K. O'Neill, Editor
Brian O'Leary, Assistant Editor
Princeton University
ISBN 0-915928-21-3

*58. Turbulent Combustion (1978)
Lawrence A. Kennedy
State University of New York at Buffalo
ISBN 0-915928-22-1

*59. Aerodynamic Heating and Thermal Protection Systems (1978)
Leroy S. Fletcher
University of Virginia
ISBN 0-915928-23-X

60. Heat Transfer and Thermal Control Systems (1978)
Leroy S. Fletcher
University of Virginia
ISBN 0-915928-24-8

61. Radiation Energy Conversion in Space (1978)
Kenneth W. Billman
NASA Ames Research Center
ISBN 0-915928-26-4

62. Alternative Hydrocarbon Fuels: Combustion and Chemical Kinetics (1978)
Craig T. Bowman
Stanford University
Jorgen Birkeland
Department of Energy
ISBN 0-915928-25-6

*63. Experimental Diagnostics in Combustion of Solids (1978)
Thomas L. Boggs
Naval Weapons Center
Ben T. Zinn
Georgia Institute of Technology
ISBN 0-915928-28-0

64. Outer Planet Entry Heating and Thermal Protection (1979)
Raymond Viskanta
Purdue University
ISBN 0-915928-29-9

65. Thermophysics and Thermal Control (1979)
Raymond Viskanta
Purdue University
ISBN 0-915928-30-2

66. Interior Ballistics of Guns (1979)
Herman Krier
University of Illinois at Urbana-Champaign
Martin Summerfield
New York University
ISBN 0-915928-32-9

*67. Remote Sensing of Earth from Space: Role of "Smart Sensors" (1979)
Roger A. Breckenridge
NASA Langley Research Center
ISBN 0-915928-33-7

68. Injection and Mixing in Turbulent Flow (1980)
Joseph A. Schetz
Virginia Polytechnic Institute and State University
ISBN 0-915928-35-3

*69. Entry Heating and Thermal Protection (1980)
Walter B. Olstad
NASA Headquarters
ISBN 0-915928-38-8

*70. Heat Transfer, Thermal Control, and Heat Pipes (1980)
Walter B. Olstad
NASA Headquarters
ISBN 0-915928-39-6

*71. Space Systems and Their Interactions with Earth's Space Environment (1980)
Henry B. Garrett
Charles P. Pike
Hanscom Air Force Base
ISBN 0-915928-41-8

*72. Viscous Flow Drag Reduction (1980)
Gary R. Hough
Vought Advanced Technology Center
ISBN 0-915928-44-2

*73. Combustion Experiments in a Zero-Gravity Laboratory (1981)
Thomas H. Cochran
NASA Lewis Research Center
ISBN 0-915928-48-5

74. Rarefied Gas Dynamics, Parts I and II (two volumes) (1981)
Sam S. Fisher
University of Virginia
ISBN 0-915928-51-5

_____
*Out of print.

75. Gasdynamics of
Detonations and
Explosions (1981)
J.R. Bowen
University of Wisconsin at Madison
N. Manson
Universite de Poitiers
A.K. Oppenheim
University of California at Berkeley
R. I. Soloukhin
Institute of Heat and Mass Transfer, BSSR Academy of Sciences
ISBN 0-915928-46-9

76. Combustion in
Reactive Systems (1981)
J.R. Bowen
University of Wisconsin at Madison
N. Manson
Universite de Poitiers
A.K. Oppenheim
University of California at Berkeley
R.I. Soloukhin
Institute of Heat and Mass Transfer, BSSR Academy of Sciences
ISBN 0-915928-47-7

*77.
Aerothermodynamics and
Planetary Entry (1981)
A.L. Crosbie
University of Missouri-Rolla
ISBN 0-915928-52-3

78. Heat Transfer and
Thermal Control (1981)
A.L. Crosbie
University of Missouri-Rolla
ISBN 0-915928-53-1

*79. Electric Propulsion
and Its Applications to
Space Missions
(1981)
Robert C. Finke
NASA Lewis Research Center
ISBN 0-915928-55-8

*80. Aero-Optical
Phenomena (1982)
Keith G. Gilbert
Leonard J. Otten
Air Force Weapons Laboratory
ISBN 0-915928-60-4

81. Transonic
Aerodynamics (1982)
David Nixon
Nielsen Engineering & Research, Inc.
ISBN 0-915928-65-5

82. Thermophysics of
Atmospheric Entry (1982)
T.E. Horton
University of Mississippi
ISBN 0-915928-66-3

83. Spacecraft Radiative
Transfer and Temperature
Control (1982)
T.E. Horton
University of Mississippi
ISBN 0-915928-67-1

84. Liquid-Metal Flows
and Magneto-
hydrodynamics (1983)
H. Branover
Ben-Gurion University of the Negev
P.S. Lykoudis
Purdue University
A. Yakhot
Ben-Gurion University of the Negev
ISBN 0-915928-70-1

85. Entry Vehicle
Heating and Thermal
Protection Systems: Space
Shuttle, Solar Starprobe,
Jupiter Galileo Probe
(1983)
Paul E. Bauer
McDonnell Douglas Astronautics Co.
Howard E. Collicott
The Boeing Co.
ISBN 0-915928-74-4

*86. Spacecraft Thermal
Control, Design, and
Operation (1983)
Howard E. Collicott
The Boeing Co.
Paul E. Bauer
McDonnell Douglas Astronautics Co.
ISBN 0-915928-75-2

87. Shock Waves,
Explosions, and
Detonations (1983)
J.R. Bowen
University of Washington
N. Manson
Universite de Poitiers
A.K. Oppenheim
University of California at Berkeley
R.I. Soloukhin
Institute of Heat and Mass Transfer, BSSR Academy of Sciences
ISBN 0-915928-76-0

88. Flames, Lasers, and
Reactive Systems (1983)
J.R. Bowen
University of Washington
N. Manson
Universite de Poitiers
A.K. Oppenheim
University of California at Berkeley
R.I. Soloukhin
Institute of Heat and Mass Transfer, BSSR Academy of Sciences
ISBN 0-915928-77-9

*89. Orbit-Raising and
Maneuvering Propulsion:
Research Status and
Needs (1984)
Leonard H. Caveny
Air Force Office of Scientific Research
ISBN 0-915928-82-5

---

*Out of print.

90. **Fundamental of Solid-Propellant Combustion (1984)**
Kenneth K. Kuo
Pennsylvania State University
Martin Summerfield
Princeton Combustion Research Laboratories, Inc.
ISBN 0-915928-84-1

91. **Spacecraft Contamination: Sources and Prevention (1984)**
J.A. Roux
University of Mississippi
T.D. McCay
NASA Marshall Space Flight Center
ISBN 0-915928-85-X

92. **Combustion Diagnostics by Nonintrusive Methods (1984)**
T.D. McCay
NASA Marshall Space Flight Center
J.A. Roux
University of Mississippi
ISBN 0-915928-86-8

93. **The INTELSAT Global Satellite System (1984)**
Joel Alper
COMSAT Corp.
Joseph Pelton
INTELSAT
ISBN 0-915928-90-6

94. **Dynamics of Shock Waves, Explosions, and Detonations (1984)**
J.R. Bowen
University of Washington
N. Manson
Universite de Poitiers
A. K. Oppenheim
University of California at Berkeley
R.I. Soloukhin
Institute of Heat and Mass Transfer, BSSR Academy of Sciences
ISBN 0-915928-91-4

95. **Dynamics of Flames and Reactive Systems (1984)**
J.R. Bowen
University of Washington
N. Manson
Universite de Poitiers
A. K. Oppenheim
University of California at Berkeley
R.I. Soloukhin
Institute of Heat and Mass Transfer, BSSR Academy of Sciences
ISBN 0-915928-92-2

96. **Thermal Design of Aeroassisted Orbital Transfer Vehicles (1985)**
H.F. Nelson
University of Missouri-Rolla
ISBN 0-915928-94-9

97. **Monitoring Earth's Ocean, Land, and Atmosphere from Space– Sensors, Systems, and Applications (1985)**
Abraham Schnapf
Aerospace Systems Engineering
ISBN 0-915928-98-1

98. **Thrust and Drag: Its Prediction and Verification (1985)**
Eugene E. Covert
Massachusetts Institute of Technology
C.R. James
Vought Corp.
William F. Kimzey
Sverdrup Technology AEDC Group
George K. Richey
U.S. Air Force
Eugene C. Rooney
U.S. Navy Department of Defense
ISBN 0-930403-00-2

99. **Space Stations and Space Platforms – Concepts, Design, Infrastructure, and Uses (1985)**
Ivan Bekey
Daniel Herman
NASA Headquarters
ISBN 0-930403-01-0

100. **Single- and Multi-Phase Flows in an Electromagnetic Field: Energy, Metallurgical, and Solar Applications (1985)**
Herman Branover
Ben-Gurion University of the Negev
Paul S. Lykoudis
Purdue University
Michael Mond
Ben-Gurion University of the Negev
ISBN 0-930403-04-5

101. **MHD Energy Conversion: Physiotechnical Problems (1986)**
V.A. Kirillin
A.E. Sheyndlin
Soviet Academy of Sciences
ISBN 0-930403-05-3

102. **Numerical Methods for Engine-Airframe Integration (1986)**
S.N.B. Murthy
Purdue University
Gerald C. Paynter
Boeing Airplane Co.
ISBN 0-930403-09-6

103. **Thermophysical Aspects of Re-Entry Flows (1986)**
James N. Moss
NASA Langley Research Center
Carl D. Scott
NASA Johnson Space Center
ISBN 0-930430-10-X

---

*Out of print.

*104. Tactical Missile Aerodynamics (1986)
M.J. Hemsch
PRC Kentron, Inc.
J.N. Nielson
NASA Ames Research Center
ISBN 0-930403-13-4

105. Dynamics of Reactive Systems Part I: Flames and Configurations; Part II: Modeling and Heterogeneous Combustion (1986)
J.R. Bowen
University of Washington
J.-C. Leyer
Universite de Poitiers
R.I. Soloukhin
Institute of Heat and Mass Transfer, BSSR Academy of Sciences
ISBN 0-930403-14-2

106. Dynamics of Explosions (1986)
J.R. Bowen
University of Washington
J.-C. Leyer
Universite de Poitiers
R.I. Soloukhin
Institute of Heat and Mass Transfer, BSSR Academy of Sciences
ISBN 0-930403-15-0

107. Spacecraft Dielectric Material Properties and Spacecraft Charging (1986)
A.R. Frederickson
U.S. Air Force Rome Air Development Center
D.B. Cotts
SRI International
J.A. Wall
U.S. Air Force Rome Air Development Center
F.L. Bouquet
Jet Propulsion Laboratory, California Institute of Technology
ISBN 0-930403-17-7

108. Opportunities for Academic Research in a Low-Gravity Environment (1986)
George A. Hazelrigg
National Science Foundation
Joseph M. Reynolds
Louisiana State University
ISBN 0-930403-18-5

109. Gun Propulsion Technology (1988)
Ludwig Stiefel
U.S. Army Armament Research, Development and Engineering Center
ISBN 0-930403-20-7

110. Commercial Opportunities in Space (1988)
F. Shahrokhi
K.E. Harwell
University of Tennessee Space Institute
C.C. Chao
National Cheng Kung University
ISBN 0-930403-39-8

111. Liquid-Metal Flows: Magnetohydrodynamics and Application (1988)
Herman Branover,
Michael Mond, and
Yeshajahu Unger
Ben-Gurion University of the Negev
ISBN 0-930403-43-6

112. Current Trends in Turbulence Research (1988)
Herman Branover,
Micheal Mond, and
Yeshajahu Unger
Ben-Gurion University of the Negev
ISBN 0-930403-44-4

113. Dynamics of Reactive Systems Part I: Flames; Part II: Heterogeneous Combustion and Applications (1988)
A.L. Kuhl
R & D Associates
J.R. Bowen
University of Washington
J.-C. Leyer
Universite de Poitiers
A. Borisov
USSR Academy of Sciences
ISBN 0-930403-46-0

114. Dynamics of Explosions (1988)
A.L. Kuhl
R & D Associates
J.R. Bowen
University of Washington
J.-C. Leyer
Universite de Poitiers
A. Borisov
USSR Academy of Sciences
ISBN 0-930403-47-9

115. Machine Intelligence and Autonomy for Aerospace (1988)
E. Heer
Heer Associates, Inc.
H. Lum
NASA Ames Research Center
ISBN 0-930403-48-7

116. Rarefied Gas Dynamics: Space Related Studies (1989)
E.P. Muntz
University of Southern California
D.P. Weaver
U.S. Air Force Astronautics Laboratory (AFSC)
D.H. Campbell
University of Dayton Research Institute
ISBN 0-930403-53-3

---

*Out of print.

117. Rarefied Gas Dynamics: Physical Phenomena (1989)
E.P. Muntz
University of Southernn California
D.P. Weaver
U.S. Air Force Astronautics Laboratory (AFSC)
D.H. Campbell
University of Dayton Research Institute
ISBN 0-930403-54-1

118. Rarefied Gas Dynamics: Theoretical and Computational Techniques (1989)
E.P. Muntz
University of Southernn California
D.P. Weaver
U.S. Air Force Astronautics Laboratory (AFSC)
D.H. Campbell
University of Dayton Research Institute
ISBN 0-930403-55-X

119. Test and Evaluation of the Tactical Missile (1989)
Emil J. Eichblatt Jr.
Pacific Missile Test Center
ISBN 0-930403-56-8

120. Unsteady transonic Aerodynamics (1989)
David Nixon
Nielsen Engineering & Research, Inc.
ISBN 0-930403-52-5

121. Orbital Debris from Upper-Stage Breakup (1989)
Joseph P. Loftus Jr.
NASA Johnson Space Center
ISBN 0-930403-58-4

122. Thermal-Hydraulics for Space Power, Propulsion and Thermal Management System Design (1989)
William J. Krotiuk
General Electric Co.
ISBN 0-930403-64-9

123. Viscous Drag Reduction in Boundary Layers (1990)
Dennis M. Bushnell
Jerry N. Hefner
NASA Langley Research Center
ISBN 0-930403-66-5

*124. Tactical and Strategic Missile Guidance (1990)
Paul Zarchan
Charles Stark Draper Laboratory, Inc.
ISBN 0-930403-68-1

125. Applied Computational Aerodynamics (1990)
P.A. Henne
Douglas Aircraft Company
ISBN 0-930403-69-X

126. Space Commercialization: Launch Vehicles and Programs (1990)
F. Shahrokhi
University of Tennessee Space Institute
J.S. Greenberg
Princeton Synergetics Inc.
T. Al-Saud
Ministry of Defense and Aviation Kingdom of Saudi Arabia
ISBN 0-930403-75-4

127. Space Commercialization: Platforms and Processing (1990)
F. Shahrokhi
University of Tennessee Space Institute
G. Hazelrigg
National Science Foundation
R. Bayuzick
Vanderbilt University
ISBN 0-930403-76-2

128. Space Commercialization: Satellite Technology (1990)
F. Shahrokhi
University of Tennessee Space Institute
N. Jasentuliyana
United Nations
N. Tarabzouni
King Abulaziz City for Science and Technology
ISBN 0-930403-77-0

129. Mechanics and Control of Large Flexible Structures (1990)
John L. Junkins
Texas A&M University
ISBN 0-930403-73-8

130. Low-Gravity Fluid Dynamics and Transport Phenomena (1990)
Jean N. Koster
Robert L. Sani
University of Colorado at Boulder
ISBN 0-930403-74-6

131. Dynamics of Deflagrations and Reactive Systems: Flames (1991)
A.L. Kuhl
Lawrence Livermore National Laboratory
J.-C. Leyer
Universite de Poitiers
A. A. Borisov
USSR Academy of Sciences
W.A. Sirignano
University of California
ISBN 0-930403-95-9

---

*Out of print.

132. **Dynamics of Deflagrations and Reactive Systems: Heterogeneous Combustion (1991)**
A.L. Kuhl
Lawrence Livermore National Laboratory
J.-C. Leyer
Universite de Poitiers
A. A. Borisov
USSR Academy of Sciences
W.A. Sirignano
University of California
ISBN 0-930403-96-7

133. **Dynamics of Detonations and Explosions: Detonations (1991)**
A.L. Kuhl
Lawrence Livermore National Laboratory
J.-C. Leyer
Universite de Poitiers
A. A. Borisov
USSR Academy of Sciences
W.A. Sirignano
University of California
ISBN 0-930403-97-5

134. **Dynamics of Detonations and Explosions: Explosion Phenomena (1991)**
A.L. Kuhl
Lawrence Livermore National Laboratory
J.-C. Leyer
Universite de Poitiers
A. A. Borisov
USSR Academy of Sciences
W.A. Sirignano
University of California
ISBN 0-930403-98-3

135. **Numerical Approaches to Combustion Modeling (1991)**
Elaine S. Oran
Jay P. Boris
Naval Research Laboratory
ISBN 1-56347-004-7

136. **Aerospace Software Engineering (1991)**
Christine Anderson
U.S. Air Force Wright Laboratory
Merlin Dorfman
Lockheed Missiles & Space Company, Inc.
ISBN 1-56347-005-0

137. **High-Speed Flight Propulsion Systems (1991)**
S.N.B. Murthy
Purdue University
E.T. Curran
Wright Laboratory
ISBN 1-56347-011-X

138. **Propagation of Intensive Laser Radiation in Clouds (1992)**
O. A. Volkovitsky
Yu. S. Sedenov
L. P. Semenov
Institute of Experimental Meteorology
ISBN 1-56347-020-9

139. **Gun Muzzle Blast and Flash (1992)**
Gunter Klingenburg
Fraunhofer-Institut fur Kurzzeitdynamik, Ernst-Mach-Institut (EMI)
Joseph M. Heimerl
U.S. Army Ballistic Research Laboratory (BRL)
ISBN 1-56347-012-8

140. **Thermal Structures and Materials for High-Speed Flight (1992)**
Earl. A. Thornton
University of Virginia
ISBN 1-56347-017-9

141. **Tactical Missile Aerodynamics: General Topics (1992)**
Michael J. Hemsch
Lockheed Engineering & Sciences Company
ISBN 1-56347-015-2

142. **Tactical Missile Aerodynamics: Prediction Methodology (1992)**
Michael R. Mendenhall
Nielsen Engineering & Research, Inc.
ISBN 1-56347-016-0

143. **Nonsteady Burning and Combustion Stability of Solid Propellants (1992)**
Luigi De Luca
Politecnico di Milano
Edward W. Price
Georgia Institute of Technology
Martin Summerfield
Princeton Combustion Research Laboratories, Inc.
ISBN 1-56347-014-4

144. **Space Economics (1992)**
Joel S. Greenberg
Princeton Synergetics, Inc.
Henry R. Hertzfeld
HRH Associates
ISBN 1-56347-042-X

145. **Mars: Past, Present, and Future (1992)**
E. Brian Pritchard
NASA Langley Research Center
ISBN 1-56347-043-8

146. **Computational Nonlinear Mechanics in Aerospace Engineering (1992)**
Satya N. Atluri
Georgia Institute of Technology
ISBN 1-56347-044-6

147. **Modern Engineering for Design of Liquid-Propellant Rocket Engines (1992)**
Dieter K. Huzel
David H. Huang
ISBN 1-56347-013-6

---

*Out of print.

148. Metallurgical Technologies, Energy Conversion, and Magnetohydrodynamic Flows (1993)
Herman Branover
Yeshajahu Unger
Ben-Gurion University of the Negev
ISBN 1-56347-019-5

149. Advances in Turbulence Studies (1993)
Herman Branover
Yeshajahu Unger
Ben-Gurion University of the Negev
ISBN 1-56347-018-7

150. Structural Optimization: Status and Promise (1993)
Manohar P. Kamat
Georgia Institute of Technology
ISBN 1-56347-56-X

151. Dynamics of Gaseous Combustion (1993)
A.L. Kuhl
Lawrence Livermore National Laboratory
J.-C. Leyer
Universite de Poitiers
A. A. Borisov
USSR Academy of Sciences
W. A. Sirignano

152. Dynamics of Heterogeneous Gaseous Combustion and Reacting Systems (1993)
A.L. Kuhl
Lawrence Livermore National Laboratory
J.-C. Leyer
Universite de Poitiers
A. A. Borisov
USSR Academy of Sciences
W.A. Sirignano
University of California
ISBN 1-56347-058-6

153. Dynamic Aspects of Detonations (1993)
A.L. Kuhl
Lawrence Livermore National Laboratory
J.-C. Leyer
Universite de Poitiers
A. A. Borisov
USSR Academy of Sciences
W.A. Sirignano
University of California
ISBN 1-56347-057-8

154. Dynamic Aspects of Explosion Phenomena (1993)
A.L. Kuhl
Lawrence Livermore National Laboratory
J.-C. Leyer
Universite de Poitiers
A. A. Borisov
USSR Academy of Sciences
W.A. Sirignano
University of California
ISBN 1-56347-059-4

155. Tactical Missile Warheads (1993)
Joseph Carleone
Aerojet General Corporation
ISBN 1-56347-067-5

156. Toward a Science of Command, Control, and Communications (1993)
Carl R. Jones
Naval Postgraduate School
ISBN 1-56347-068-3

157. Tactical and Strategic Missile Guidance Second Edition (1994)
Paul Zarchan
Charles Stark Draper Laboratory, Inc.
ISBN 1-56347-077-2

158. Rarefied Gas Dynamics: Experimental Techniques and Physical Systems (1994)
Bernie D. Shizgal
University of British Columbia
David P. Weaver
Phillips Laboratory
ISBN 1-56347-079-9

159. Rarefied Gas Dynamics: Theory and Simulations (1994)
Bernie D. Shizgal
University of British Columbia
David P. Weaver
Phillips Laboratory
ISBN 1-56347-080-2

160. Rarefied Gas Dynamics: Space Sciences and Engineering (1994)
Bernie D. Shizgal
University of British Columbia
David P. Weaver
Phillips Laboratory
ISBN 1-56347-081-0

161. Teleoperation and Robotics in Space (1994)
Steven B. Skaar
University of Notre Dame
Carl F. Ruoff
Jet Propulsion Laboratory, California Institute of Technology
ISBN 1-56347-095-0

162. Progress in Turbulence Research (1994)
Herman Branover
Yeshajahu Unger
Ben-Gurion University of the Negev
ISBN 1-56347-099-3

---

*Out of print.

163. Global Positioning System: Theory and Applications Volume I (1995)
Bradford W. Parkinson
Stanford University
James J. Spilker Jr.
Stanford Telecom
Penina Axelrad,
Associate Editor
University of Colorado
Per Enge,
Associate Editor
Stanford University
ISBN 1-56347-107-8

164. Global Positioning System: Theory and Applications Volume II (1995)
Bradford W. Parkinson
Stanford University
James J. Spilker Jr.
Stanford Telecom
Penina Axelrad,
Associate Editor
University of Colorado
Per Enge,
Associate Editor
Stanford University
ISBN 1-56347-106-X

---
*Out of print.